Barron's Review Course Series

Let's Review:

Earth Science— The Physical Setting

Second Edition

Edward J. Denecke, Jr.
Staff Development Specialist
Science Technical Assistance Center
Whitestone, New York

© Copyright 2003, 2002, 2001, 2000, 1995 by Barron's Educational Series, Inc.

All rights reserved.
No part of this book may be reproduced in any form,
by photostat, microfilm, xerography, or any other
information retrieval system, electronic or mechanical,
without the written permission of the copyright owner.

All inquiries should be addressed to:
Barron's Educational Series, Inc.
250 Wireless Boulevard
Hauppauge, NY 11788
http://www.barronseduc.com

Library of Congress Catalog Card No. 00-069775

International Standard Book No. 0-7641-1391-7

Library of Congress Cataloging-in-Publication Data
Denecke, Edward J.
    Let's review : Earth science / Edward J. Denecke, Jr. — 2nd ed.
      p.   cm. — (Barron's review course series)
    Includes index.
    ISBN 0-7641-1391-7
    1. Earth sciences—Outlines, syllabi, etc. 2. Earth sciences—
Study and teaching (Secondary)—New York (State) [1. Earth sciences.]
I. Title: Earth science. II. Title. III. Series.
QE41.D46   2001
550—dc21   00-069775
                                              CIP

PRINTED IN THE UNITED STATES OF AMERICA

987654321

TABLE OF CONTENTS

PREFACE v

CAUTION! DON'T USE THIS BOOK UNTIL YOU READ THIS vii

TOPIC ONE   **Astronomy**   **1**

UNIT ONE   **From a Geocentric to a Heliocentric Universe**   **1**

Chapter 1:	Early Astronomy and the Geocentric Model	1
Chapter 2:	The Development of the Heliocentric Model	18
Chapter 3:	Heliocentric Earth Motions and Their Effects	41
Chapter 4:	Earth's Coordinate System and Mapping	66
Chapter 5:	Our System of Time	98

UNIT TWO   **Modern Astronomy**   **121**

Chapter 6:	Tools of the Modern Astronomer	121
Chapter 7:	Stars, Their Origin and Evolution	135
Chapter 8:	The Solar System	157
Chapter 9:	Theories of the Origin of the Universe	182

UNIT THREE   **Earth's History**   **195**

Chapter 10:	The Origin of Earth and Its Moon	195
Chapter 11:	The Origin and Structure of Earth's Atmosphere	206
Chapter 12:	The Origin and Nature of Earth's Hydrosphere	219
Chapter 13:	The Origin and History of Life on Earth	243

TOPIC TWO   **Geology**   **284**

UNIT FOUR   **Earth Materials**   **284**

Chapter 14:	Minerals	284
Chapter 15:	Rocks	308

UNIT FIVE   **The Dynamic Earth**   **341**

Chapter 16:	Earthquakes and Earth's Interior	341
Chapter 17:	Volcanoes and Earth's Internal Heat	373
Chapter 18:	Plate Tectonics	388

Table of Contents

UNIT SIX — Weathering, Erosion, and Deposition — 420

Chapter 19:	Weathering and Soil Formation	420
Chapter 20:	Erosion	443
Chapter 21:	Deposition	488

TOPIC THREE — Meteorology — 530

UNIT SEVEN — The Atmosphere, Weather, and Climate — 530

Chapter 22:	The Atmosphere	530
Chapter 23:	Weather	565
Chapter 24:	Climate	618

Appendix A: Answers to Review Questions — 646

Appendix B: Earth Science Reference Tables — 685

Glossary of Earth Science Terms — 701

Regents Examination — 719

Answers to Regents Examination — 756

Index — 757

Preface

To the Teacher:

Let's Review: Earth Science is a concise text and review aid for courses based on the New York State Physical Setting/Earth Science Core Curriculum, a comprehensive course of study in earth science on the secondary level. However, the material in this book provides such comprehensive coverage of topics in earth science that it can be used as a review text to supplement virtually any secondary course in earth science taught in the United States, using any major textbook.

This edition has been completely rewritten to reflect the content of the New York State Physical Setting/Earth Science Core Curriculum. It has been reorganized into three major topics: astronomy, geology, and meteorology (with extensive new material added to the astronomy topic). Each topic addresses a key idea of Standard 4: The Physical Setting Earth Science, and is divided into units based upon the performance indicators for that key idea. Each unit is subdivided into chapters that deal with groups of related major understandings underlying the performance indicator. Specific skills identified in Standards 1, 2, 6, and 7 are introduced with the appropriate major understanding. Figures and text have been revised to reflect changes made in the new 2001 edition of the *Earth Science Reference Tables*.

The review questions in this edition have been chosen from previous Regents Examinations to reflect content consistent with the Physical Setting: Earth Science Core Curriculum and suitability for use with the new reference tables. Since the June 2001 examination will include constructed response and extended constructed response items, practice questions of this type have been included at the end of each chapter.

To the Student:

"On the surface, islands may seem separate, but underneath they are all connected."

Earth science is not the study of isolated facts, but rather the development of a deep understanding, and appreciation of, the interconnectedness of Earth phenomena, processes, and systems. In earth science, the key to success is thorough understanding and the ability to demonstrate what you understand. This book is a concise text and review aid in which the author has tried to make earth science as understandable as possible to you, the student, by incorporating the following features:

Preface

- Explanations of concepts and understandings are detailed, yet simply and clearly stated. They are designed to help you grasp the "how" and "why" of an idea, rather than just stating the idea.
- Important terms are printed in *italic type* where they are defined in the text.
- Terms printed in **boldface type** are testable on the Regents examination.
- The illustrations are designed to make difficult ideas easier to understand. Many of the illustrations are similar to those used in Regents Examination questions in order to familiarize you with the types of diagrams you will be asked to analyze and interpret.
- Each unit ends with a wide range of review questions, including constructed response and extended constructed response questions.
- A glossary and a complete index make it easy for you to find the definition of a specific term and the pages in the book where a topic is covered.
- Full-length Regents Examinations provide you with the opportunity to assess your understanding and to test your earth science knowledge and skills before taking the Regents Examination.

Earth science offers the challenge and excitement of new theories, new discoveries, and new problems to be solved. It is a science in which sweeping new theories are being tested and applied to puzzling new observations. It is a science in which revolutionary advances in human knowledge of Earth and the other planets of our solar system are being made almost daily. I hope that studying earth science will fill you with wonder and delight at the complexities of our planet Earth.

I wish to thank my wife, Gerry, for her infinite patience and my children, Meredith, Abigail, and Benjamin, for their loving support during the preparation of this manuscript.

CAUTION!
Don't Use This Book Until You Read This

You are taking earth science and want to get good grades on your exams. Well, before you even look at the earth science subject matter in this book, let's talk about how to use this book. Many students don't really know how to read a textbook or a review book. Suppose you get a homework assignment such as "Read pages 73–82 in the text, and answer questions 1–5 on page 83." What will you do? Typical students turn to page 73 and start reading, sentence by sentence, paragraph by paragraph. Once in a while they may stop to see how much more they have to read. When they finally reach the last sentence on page 82, they consider that they have "studied" the material and then turn to page 83 and try to answer the questions. This simple procedure doesn't require a lot of time or effort. But just reading the text once is a really weak approach to studying. So what is a better way to study?

Understand How the Text Is Organized

First, it is important to realize that textbooks are not written like novels. Textbooks are not meant to be read straight through; instead, they are structured to guide in-depth study. If you understand how a textbook is arranged, you can organize your reading into small, useful blocks.

Let's Review is based upon the New York State Physical Setting Earth Science Core Curriculum and is organized as follows:

ORGANIZATION

Physical Setting/Earth Science Core Curriculum	*Let's Review:* *Earth Science*
STANDARDS tell what you are expected to know and be able to do.	
KEY IDEAS tell you the really important ideas relating to the standard.	**TOPICS** are based upon the major disciplines in earth science—astronomy, geology, and meteorology—that are addressed in the key ideas of the standards.
PERFORMANCE INDICATORS describe what you should be able to do in order to show that you understand the key ideas.	**UNITS** are based upon the elements of the performance indicators for each key idea. For example, Unit 2: Modern Astronomy addresses Performance Indicator 1.2 – Describe current theories about the origin of the universe and solar system.

Caution!

| MAJOR UNDERSTANDINGS list what you need to know in order to do the things described in the performance indicators. | CHAPTERS address groups of related major understandings underlying each performance indicator. For example, Chapter 13: The Origin and History of Life on Earth addresses Major Understandings 1.2h–j.

SECTIONS break chapters into manageable reading blocks to help you better understand what you read. For example, Chapter 19: Weathering has such sections as Physical Weathering, Chemical Weathering, and The Products of Weathering. |

Know How to Spell Success: S-Q-3R

Before you read, you need to realize that studying means not just reading, but also *thinking deeply* about what you have read. One good approach is to take notes while reading and then study your notes before a test. A better approach is to read each chapter subsection once, read it again and highlight the important points, and then study those points after you finish reading each chapter.

One of the best approaches to studying is called **S-Q-3R**, which stands for **S**urvey, **Q**uestion, **R**ead, **R**ecite, and **R**eview. Here's how it works:

1. You **Survey** the chapter by reading through the section and subsection headings to get a mental map of the material.
2. You ask **Questions** about the material by turning each heading into a question.
3. Then you **Read** the text, a subsection at a time, with the purpose of answering these questions.
4. Next, you **Recite** by jotting down brief notes about what you have read, making an outline or, a graphic organizer, or writing a summary. (Note: In this case the word *recite* doesn't mean to "speak publicly"; it means "to re-cite, or cite again." To *cite* is to quote, or mention. Here, *recite* means to list or itemize important ideas you have read.)
5. Finally, you **Review** by rereading your notes and answering questions about the material. They may be questions that you pose to yourself or that have appeared on prior tests.

This five-step approach will take more time than just reading through, but the reward for the extra time you spend will be better grades. Therefore, plan to have enough time to study before you sit down for a session with this book.

Know What's on the Test

The key to success in preparing for any test is to know what will be expected of you so that you can review and practice beforehand. The New York State Physical Setting/Earth Science Regents Examination has four parts:

Part A—Multiple Choice. In a multiple-choice question you are given several choices from which to select the one that best answers the questions or completes the statement. Many practice questions of this type from previous Regents Examinations are included at the end of each chapter of this book. Part A of the exam focuses on earth science content from Standard 4.

Part B—Multiple Choice and Constructed Response. In a constructed response question there is no list of choices from which to select an answer; rather, you are required to provide the answer. Constructed response questions can test skills ranging from constructing graphs or topographic maps to formulating hypotheses, evaluating experimental designs, and drawing conclusions based upon data. Practice questions of this type are also included at the end of each chapter. In Part B you will be asked to demonstrate skills identified in Standards 1, 2, 6, and 7 in the context of earth science.

Part C—Extended Constructed Response. The constructed response questions require more time (15–20 minutes per item) and effort on your part to answer. Questions in Part C require you to apply your earth science knowledge and skills to real-world problems and applications. You may be asked to produce short essays, design controlled experiments, predict outcomes, or analyze the risks and benefits of various solutions to a problem.

Part D—Laboratory Performance Tasks. These tasks test your laboratory skills. You will take this part of the exam sometime during the 2 weeks before the written Regents. Laboratory performance tasks involve skills such as using instruments (e.g., rulers, external protractors, triple beam balances, graduated cylinders, stopwatches), observing properties of Earth materials, performing calculations, and collecting and analyzing data.

Note: The following description represents that information the State Education Department has stated may be shared with students before taking the performance part of the examination. You should be familiar with the skills being assessed because you have used them in laboratory activities throughout the year. However, you will not be allowed to practice the entire test or any of the individual stations before this performance component is administered.

Caution!

Station 1..._Identification_
Using a mineral identification kit and key, the student will determine the characteristics of two mineral samples and identify each sample by name.

Station 2..._Classification_
Using rock identification charts, the student will classify two rock samples as igneous, sedimentary, or metamorphic and state the reason for each classification, in one or more complete sentences.

Station 3..._Angular Measurement_
Using a plastic hemisphere that models the apparent path of the sun, an external protractor, a ruler, and masking tape, the student will locate the position of the sun at a given time and measure the distance between the position and a fixed point.

Station 4..._Mass–Density_
Using a single-pan, triple-beam decigram balance, a mineral density chart, and a calculator, the student will find the density, determine the mass, and calculate the volume of a given mineral sample.

Station 5..._Settling Time_
Using a column of fluid, three sizes of plastic particles of the same density, a stopwatch, and a calculator, the student will determine the average settling time for each of the three sizes of particles.

Station 6..._Graphing_
Using data obtained from Station 5, the student will construct a line graph of average settling time vs. particle diameter and will determine the settling time for another given particle diameter.

Extensive analyses of questions of all types can be found in the companion volume to this book, **Earth Science—*Questions and Answers Explained***. Together, these review books will help you prepare for the New York State Physical Setting/Earth Science Regents Examination by clearly explaining what you should know and be able to do in order to perform well on the exam and by providing you with practice questions from prior exams that are thoroughly explained.

Unit One: FROM A GEOCENTRIC TO A HELIOCENTRIC UNIVERSE

Chapter 1

EARLY ASTRONOMY AND THE GEOCENTRIC MODEL

KEY IDEAS People have observed the stars for thousands of years, using them to find direction, note the passage of time, and express human values and traditions. To an observer on Earth, it appears that Earth stands still and everything else moves around it. Thus, in trying to make sense of how the universe works, it was logical for early astronomers to start with those apparent truths. To comprehend our modern view of the universe, it is helpful to begin by understanding these first attempts to explain the universe in terms of what can be seen from our vantage point on Earth. As technology has progressed, so has our understanding of celestial objects and events.

KEY OBJECTIVES
Upon completion of this chapter, you will be able to:

- Explain the meaning of the term *celestial object*.
- Compare and contrast "apparent" and "real" motion.
- Explain how the celestial sphere model of the sky accounts for the motions of celestial objects.
- Explain how Earth's rotation makes it appear that the Sun, the Moon, and the stars are moving around Earth once a day.
- Locate Polaris in the night sky.

OBSERVING THE SKY

If you kept a list of things observed in the sky, it might include birds, smoke, clouds, rainbows, halos, lightning, stars, the Moon, the Sun, and comets. One of the first ideas that might occur to you is that the sky has depth. Some things in it appear closer, and some appear farther away. Why? Perspective! From everyday experiences you know that closer objects block your view of more distant objects. For example, if you hold your hand in front of your

Early Astronomy and the Geocentric Model

Figure 1.1 Motion in the Sky. (a) Photograph showing the crescent Moon and Venus setting. Exposures were made every 8 minutes, showing the changes in position of these celestial objects over time. Note the motion of the Moon relative to Venus.
Source: Horizons: *Exploring the Universe*, Michael A. Seeds, Wadsworth, 1987.

(b) A time exposure taken with a camera aimed at Polaris over Mauna Kea Observatory, latitude 20°. Note the circular star trails. Source: *Astronomy: The Cosmic Journey*, William K. Hartman, Wadsworth, 1987.

eyes, you cannot see a more distant tree. Therefore, if a bird flying by blocks your view of a cloud, you logically conclude that the bird is closer to you than the cloud. Then you see a cloud move "in front of" the Sun, and you conclude that the cloud is closer to you than the Sun. Or perhaps you see a solar eclipse, and you conclude that the Moon is closer to you than the Sun. In this way, all of the objects on your list could be put in order of distance from an observer.

Careful observation also leads you to realize that many of these closer objects or phenomena are associated with the atmosphere. You feel a wind and see it moving the clouds. You see a rainbow in the spray of a waterfall or in a distant rain shower and realize that it is caused by the interplay of sunlight and tiny droplets of water in the air. You see lightning flash between a cloud and the ground. In this way, you can classify the things seen in the sky into two groups: those that are part of, or occur in, the atmosphere, and those that are beyond the atmosphere.

Celestial Objects

Celestial objects are objects that can be seen in the sky that are not associated with Earth's atmosphere. The most numerous of the celestial objects are

Early Astronomy and the Geocentric Model

the stars. To an observer on Earth, stars are simply points of light that vary in size, brightness, and color. The Sun, the Moon, the planets, and comets are also examples of celestial objects. Clouds, rainbows, halos, and other phenomena seen in the sky that are part of, or occur in, Earth's atmosphere are *not* considered celestial objects.

Celestial Motion

If you observe celestial objects for even a short while, it is clear that they change position in the sky over time. You have probably noticed the Sun in different places in the sky at different times of the day. Thus, it seems that the Sun is "moving." Similarly, if you observe the Moon and stars carefully, you find that they, too, are seen in different places in the sky at different times of the night. Try going out on a moonlit night and noting the Moon's position at 7:00 P.M. If you go out again at 10:00 P.M., you'll notice that the Moon has changed position in the sky. The same is true of stars. When you observe the sky, you find that every celestial object changes position over time, or is in motion. See Figure 1.1 on pages 2 and 3.

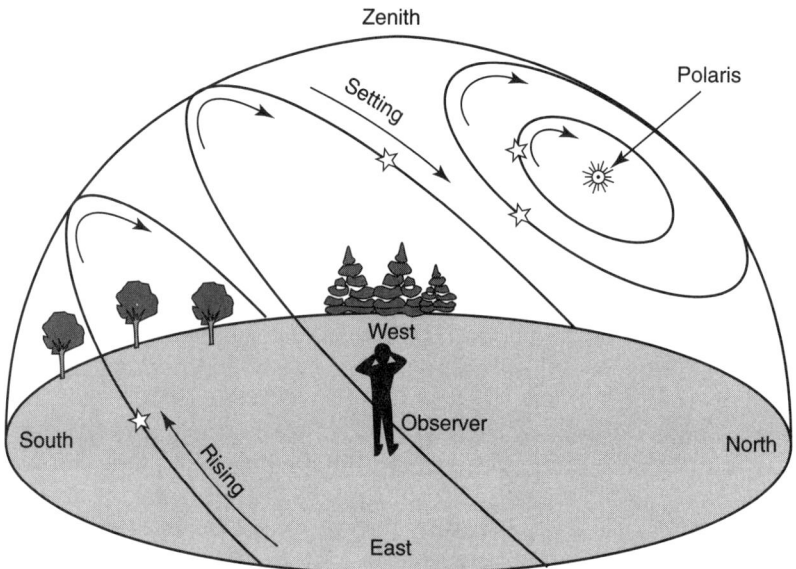

Figure 1.2 The Apparent Motion of Celestial Objects to an Observer in New York State.

If you keep track of this motion, you discover something curious. The motion of celestial objects is not random. They don't all move in different directions at different speeds. Instead, with very few exceptions, every single one of these thousands of objects appears to move in the same general direction—*from east to west*. And if you measure the rates at which all of these celestial objects are moving, you discover something even more curious—with few exceptions (such as the Moon), *they appear to move at the same rate!*

Careful records of this motion reveal that all celestial objects appear to move across the sky from east to west along a path that is an arc, or part of a circle. Since celestial objects appear to follow a circular path at a constant rate of 15 degrees per hour, or one complete circle every day (24 hr/day × 15°/hr = 360°/day), this motion is called *apparent daily motion.*

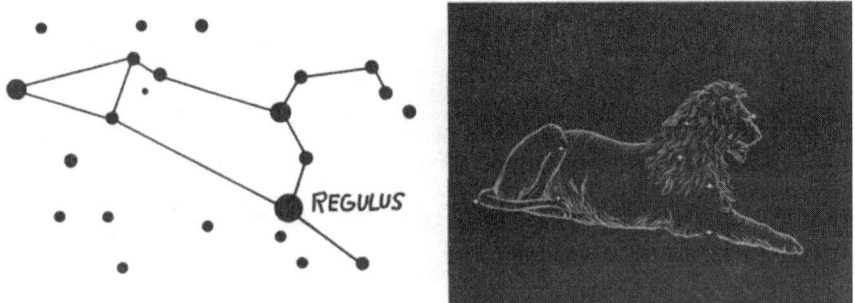

Figure 1.3 Constellations Are Imaginary Patterns of Stars. Source: *Astronomy: A Self-Teaching Guide*, 4th Ed., Dinah L. Moché, John Wiley, 1998.

In the Northern Hemisphere, all the circles formed by completing the arcs along which celestial objects move are centered very near the star Polaris. The apparent circular motion of celestial objects causes them to come into view from below the eastern horizon and to sink from view beneath the western horizon (that is, to rise in the east and set in the west). See Figure 1.2.

Early observers noted that the positions of celestial objects change in a daily and yearly cyclic pattern. They discovered that understanding these patterns of motion was very useful. Since the positions of celestial objects change with time and location, such changes can be used to *determine time* and to *find one's position on Earth.* Since the distribution of stars is random, these observers devised *constellations*, imaginary patterns of stars, to help them keep track of the changing positions of celestial objects. See Figure 1.3.

"Apparent" Versus "Real" Motion

So far, we have used the word *apparent* when referring to celestial motions because the motion of an object is always judged with respect to some other object or point. The idea of absolute motion or rest is misleading because there are several possible reasons why an object may *appear* to an observer to be moving. One possibility is that the observer is standing still and the *object* is moving. Another possibility is that the object is standing still and *the observer* is moving. A third possibility is that *both the observer and the object* are moving, but one is moving faster, or in a different direction, than the other. This is the case when you are in a car speeding down a highway;

as you look out of the car window, trees along the side of the road seem to whiz by. Of course, your brain tells you that the trees are rooted to the ground and that they only seem to whiz by because you are riding in a car. But to your eyes alone, you are not moving; the image of the trees is moving from one side of your window to the other. Now think about driving past a person walking along the sidewalk. The person is moving, but also seems to whiz by your window. Now think of a person sitting next to you in your car. To you, would that person look as though he or she was moving?

By now you should realize that the problem of determining which of the two is moving, the object or the observer, is not always easy to solve. If the signs that tell the body it is moving are removed, an observer may not realize that he or she is in motion. (Do you really feel as if you are moving at 400 miles per hour when watching a movie in an airplane cruising in level flight at that speed?) Without signs telling the observer's body that he or she is moving, any object seen changing position will be interpreted as a moving object by the observer.

THE CELESTIAL SPHERE

Early observers reasoned that when they looked at the sky they were standing still because their senses gave them no signs that they were moving. They *felt* as if they were standing still. Therefore, they interpreted the changing positions of celestial objects to mean that the *celestial objects* were moving. They visualized all celestial objects as revolving around a motionless Earth.

One effect of apparent daily motion is that the sky appears to move as if it were a single object. Here's a simple analogy. If you watch a truck with "Moving Van" painted on its side roll past you, the word "Moving" doesn't get closer to the word "Van" just because the truck is moving in that direction. All of the letters in the two words stay in a fixed pattern even though they are all moving because they are part of a single object—the sign. In much the same way, the stars in the sky stay in a fixed pattern even as you observe them moving through the sky.

It is not surprising, then, that early observers imagined that the sky *was* a single object—a huge dome. Since the "dome" of the sky was in motion, and new parts would come into view as others dropped out of sight, these observers imagined that the dome extended beyond the horizon. As they followed through on this model, they realized that, if the dome were extended far enough, it would form a hollow ball, or sphere, surrounding Earth. They imagined a huge "sky ball," or ***celestial sphere***, slowly spinning around a motionless Earth. See Figure 1.4. To these observers the Sun, Moon, and stars were either holes in the celestial sphere or objects attached to it.

The "celestial sphere" was a nice model because it accounted for many observations. It explained why objects appeared, arced across the sky, disappeared, and then reappeared the next day. Imagine it as a ball tied to a rope and swung in a circle around your head. First the ball arcs across your line of sight as you swing it in front of you, next it disappears as it swings around behind you, and then it reappears as it swings around in front of you again. This model explained why all of the celestial objects moved in the same direction at the same speed. It also explained why the stars remained in fixed positions relative to one another. This Earth-centered, or *geocentric*, model of the universe was used successfully for thousands of years to explain most observations of celestial objects.

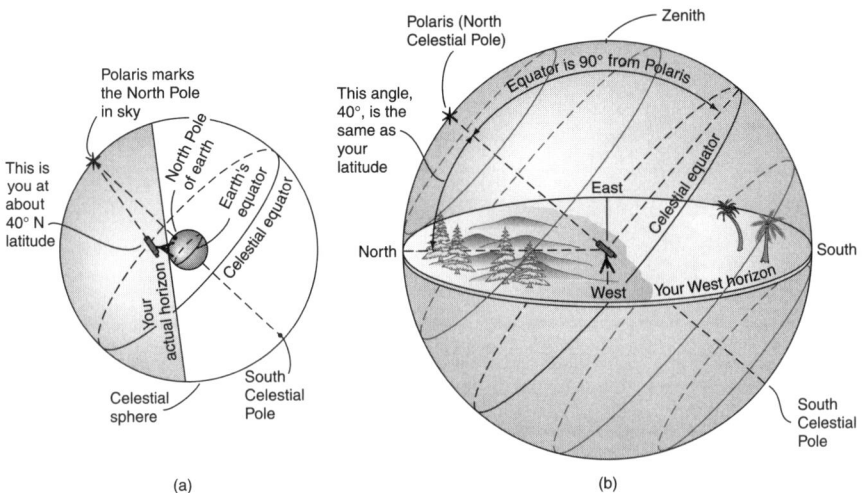

Figure 1.4 The Celestial Sphere, an Imaginary Sphere Surrounding Earth. The most you see at any one time is half of this sphere. Certain reference points on the celestial sphere are defined in relation to reference points on Earth. The celestial poles lie directly over Earth's poles; the celestial equator lies over Earth's equator midway between the celestial poles. Other points are defined by their positions in relation to the observer: the zenith is a point directly above the observer, the celestial meridian is the circle that runs through the celestial poles and the zenith. As Earth rotates from west to east, all objects in the sky appear to move from east to west, revolving around the north celestial pole. (a) View from a spot outside the celestial sphere. (b) **Observer's view.** Sources: *The Practical Astronomer*, Brian Jones, Simon and Schuster, 1990, and *Astronomy: A Self-Teaching Guide*, 4th Ed., Dinah L. Moche, John Wiley, 1998.

Even though we now know that the motion of celestial objects is due to Earth's rotation, it is still sometimes useful, when discussing objects in the sky, to think of them as part of a sphere surrounding Earth. The most that an observer would see at any one time would be half of this sphere; but we still refer to this imaginary half-sphere, or dome, visible over our heads as the celestial sphere. The circle formed by the intersection of the celestial sphere and the ground is called the ***horizon***. The point on the celestial sphere that is right over an observer's head at any given time is the ***zenith***. The imaginary circle that passes through the north and south points on the horizon and through the zenith is the ***celestial meridian***.

Early Astronomy and the Geocentric Model

A Simple Celestial Coordinate System

A useful coordinate system for locating objects on the celestial sphere can be set up by projecting Earth's Equator and poles onto the sky. As shown in Figure 1.5, Earth's Equator, North Pole, and South Pole correspond to a "celestial equator" and "north and south celestial poles" on the celestial sphere. Celestial objects can be located in the sky by their positions in relation to these celestial reference points.

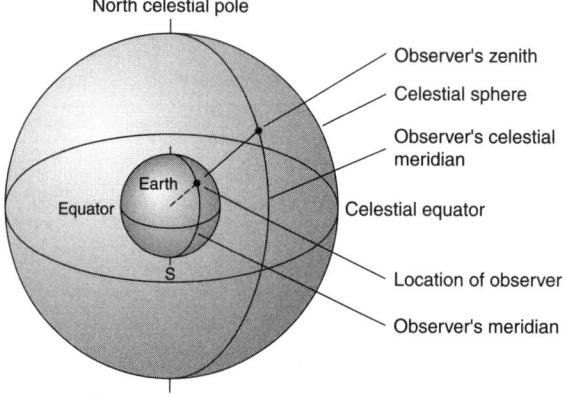

Figure 1.5 Projection of Earth's Latitude-Longitude System onto the Celestial Sphere. Source: *The Practical Astronomer*, Brian Jones, Simon and Schuster, 1990. (page 66).

The star Polaris is located very close to the north celestial pole, making it a convenient reference point for determining the north-south positions of celestial objects in the Northern Hemisphere. Polaris can be located by following the "pointer stars," Dubhe and Merak, in the bowl of the Big Dipper in the constellation Ursa Major. See Figure 1.6.

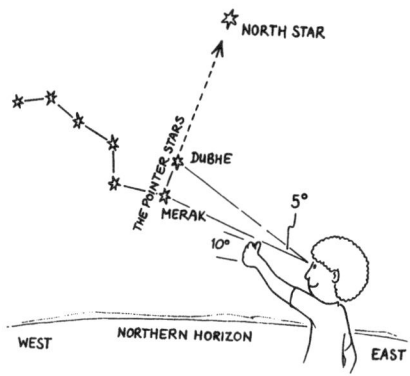

Figure 1.6 The "Pointer Stars," Dubhe and Merak, in the Bowl of the Big Dipper. Use these two stars to find the North Star, Polaris, and also to judge angular distances; they are about 5° apart. Another way to estimate angular distances is to hold a fist out at arm's length; the fist marks about 10°. Source: *Astronomy: A Self-Teaching Guide*, 4th Ed., Dinah L. Moché, John Wiley, 1998.

A convenient reference point for determining the east-west positions of objects on the celestial sphere is the Sun. Objects to the west of the Sun on the celestial sphere will "rise" before the Sun and "set" before it. Likewise, objects to the east of the Sun trail behind it and will "rise" after the Sun and "set" after it. See Figure 1.7.

8

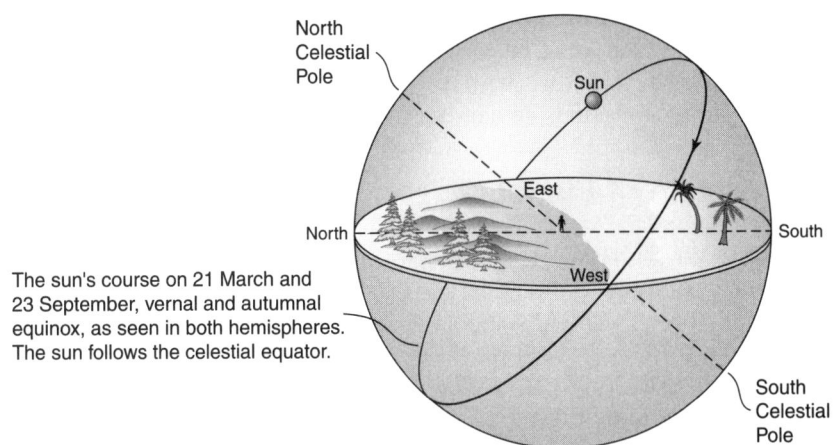

The sun's course on 21 March and 23 September, vernal and autumnal equinox, as seen in both hemispheres. The sun follows the celestial equator.

Figure 1.7 The Sun's Path on March 21 and September 23, the Vernal and Autumnal Equinox. The Sun follows the celestial equator. Source: *The Practical Astronomer*, Brian Jones, Simon and Schuster, 1990.

The Sun's Path

Each day, because of Earth's rotation, the Sun moves along an imaginary path on the celestial sphere. Over the course of a year, however, it also follows an imaginary path on the celestial sphere. As you can see in Figure 1.8, the apparent position of the Sun with respect to the background stars changes continuously as Earth orbits the Sun. When Earth has made one complete revolution in its orbit, the Sun will return to its starting point against the background stars. In other words, the Sun traces out a closed path on the celestial sphere once a year. The apparent path of the Sun through the stars on the celestial sphere over the course of the year is called the ***ecliptic***. Since Earth's axis of rotation is tilted 23½° to the plane of its orbit, the ecliptic is tilted

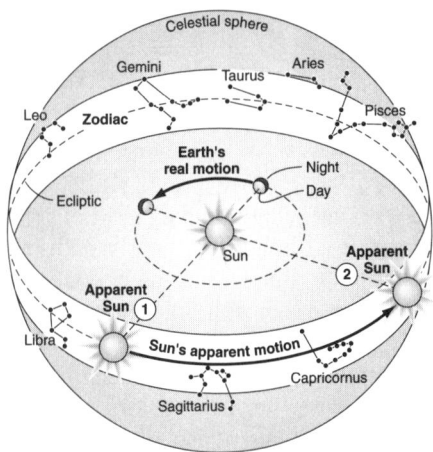

Figure 1.8 During Earth's Annual Journey Around the Sun, We View Stars from a Slightly Different Position from Day to Day. Thus, the Sun appears to travel around the celestial sphere during the course of a year along a path called the *ecliptic*. The part of the sky through which the Sun passes is known as the *zodiac*, and the Sun crosses the celestial equator at the vernal and autumnal equinoxes.

23½° with respect to the celestial equator.

The ecliptic is important because the Sun, the Moon, and the planets are always found near it. As we will see later, this occurs because all of these objects in our solar system lie nearly in the same plane.

The Problem of Planets

There were, however, some problems with the geocentric model. Early astronomers also observed that certain points of light changed position with respect to the background of stars in the sky. They called these points of light *planets*, from the Greek word for "wanderer."

Astronomers working before the invention of the telescope and before anyone understood the present structure of the solar system counted seven such wanderers or planets: Mercury, Venus, Mars, Jupiter, Saturn, the Moon, and the Sun. This list differs from our modern list of planets in several ways:

- Earth is missing, because no one realized that the points of light wandering in the sky and the Earth on which these observers stood were in any way alike.
- The Sun and the Moon were classified as planets because they wandered on the celestial sphere, just like Mars and Jupiter and the other planets.
- Uranus, Neptune, and Pluto are missing because they were not discovered until the telescope made them easily visible. Uranus, which is barely visible to the naked eye, was discovered in 1781. Neptune and Pluto, which can't be seen at all without a telescope, were discovered in 1846 and 1930, respectively.

Planets differ from stars in a number of ways. As already mentioned, the relative positions of stars on the celestial sphere are fixed, while planets move relative to the stars. Stars can be seen anywhere on the celestial sphere; planets are always found near the ecliptic (that imaginary yearly path of the Sun on the celestial sphere). Stars appear to "twinkle," but the brighter planets do not. Even through a telescope, stars appear as points of light, while the larger and nearer planets appear as disks.

These observed differences between planets and stars, particularly the "wandering" of planets on the celestial sphere, attracted a lot of attention from early astronomers. Their attempts to explain these differences ultimately led to the development of a new model of the universe.

MULTIPLE-CHOICE QUESTIONS

In each case, write the number of the word or expression that best answers the question or completes the statement.

1. Which of the following is *not* a celestial object?
   (1) the Sun   (3) a rainbow
   (2) the Moon   (4) a star

2. Why do stars appear to move through the night sky at the rate of 15 degrees per hour?
   (1) Earth actually moves around the Sun at a rate of 15°/hr.
   (2) The stars actually move around the center of the galaxy at a rate of 15°/hr.
   (3) Earth actually rotates at a rate of 15°/hr.
   (4) The stars actually revolve around Earth at a rate of 15°/hr.

3. Which real motion causes the Sun to appear to rise in the east and set in the west?
   (1) the Sun's revolution   (3) Earth's revolution
   (2) the Sun's rotation   (4) Earth's rotation

4. An observer in New York State took a time-exposure photograph from 10 P.M. until midnight of the stars over the *northern* horizon. Which diagram best represents this photograph?

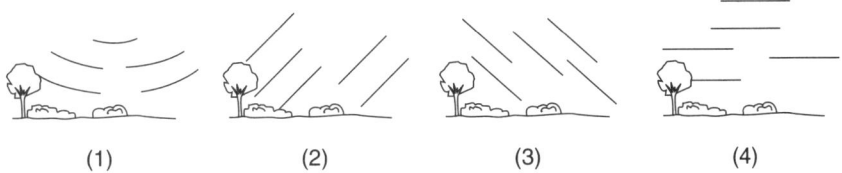

   (1)   (2)   (3)   (4)

5. An observer took a time-exposure photograph of Polaris and five nearby stars. How many hours were required to form the star paths shown to the right?
   (1) 6
   (2) 2
   (3) 8
   (4) 4

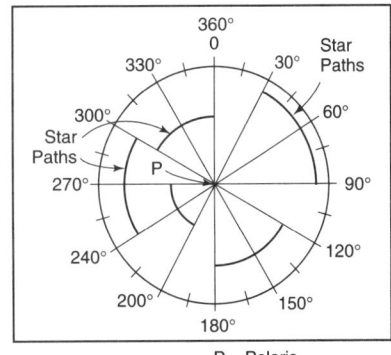

P = Polaris

Early Astronomy and the Geocentric Model

6. Which diagram represents the appearance of a 3-hour time-exposure photograph of stars in the northern sky if Earth did *not* rotate?

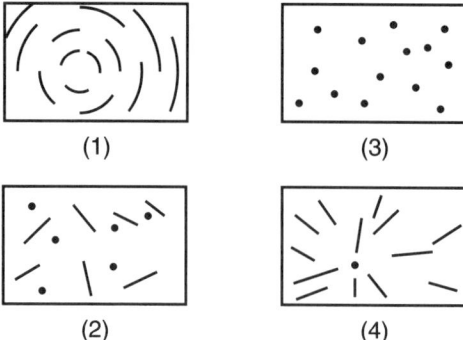

7. What is the total number of degrees that Earth rotates on its axis during a 12-hour period?
   (1) 1°
   (2) 15°
   (3) 180°
   (4) 360°

Base your answers to questions 8 through 12 on your knowledge of earth science and on the diagram, which represents observations of the apparent paths of the Sun in New York State on the dates indicated.

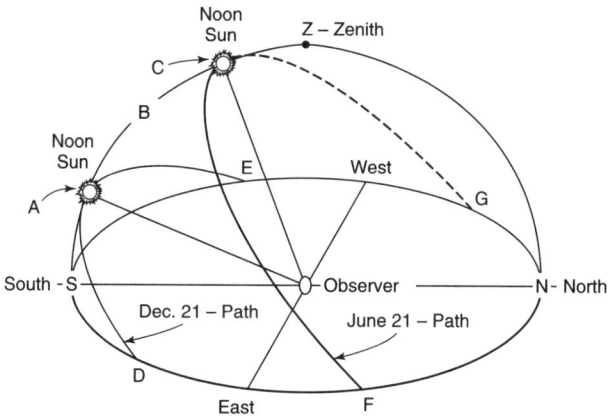

8. On the basis of the diagram, which statement is true?
   (1) The Sun passes through the zenith on December 21.
   (2) The Sun rises due east and sets due west on December 21.
   (3) The Sun passes through the zenith on June 21.
   (4) The Sun rises north of east and sets north of west on June 21.

9. Which statement about the Sun's path is true?
   (1) The Sun's path varies with the seasons.
   (2) The midpoint of the Sun's path is the zenith.
   (3) The angle of the Sun's path to the horizon is greatest on December 21.
   (4) The Sun's path on certain days of the year is shown by line *SZN*.

10. Which arc represents a part of the observer's horizon?
    (1) *DAE*
    (2) *FCG*
    (3) *SBZN*
    (4) *DSEG*

11. On which date will the noon sun be nearest to position *B*?
    (1) September 21
    (2) November 21
    (3) December 21
    (4) January 21

12. Which arc represents part of the observer's celestial meridian?
    (1) *SBZ*
    (2) *DFN*
    (3) *SDF*
    (4) *GCF*

13. The star located almost directly above Earth's North Pole is
    (1) Alpha Centauri
    (2) Betelgeuse
    (3) Polaris
    (4) Sirius

14. An artificial Earth satellite is revolving around Earth's Equator every 2 hours in the same direction as that in which Earth rotates. As viewed from Earth, this satellite rises in the
    (1) north
    (2) south
    (3) east
    (4) west

15. A constellation appears to move from east to west in the sky during the night because
    (1) the constellation rotates
    (2) the constellation revolves in an orbit around the Sun
    (3) Earth revolves around the Sun
    (4) Earth rotates on its axis

Early Astronomy and the Geocentric Model

Base your answers to questions 16 through 20 on the diagrams below and your knowledge of earth science. The diagrams represent four locations on Earth's surface at the present time on March 21. Lines have been drawn to represent the apparent path of the Sun across the sky. The present position of the Sun, the Sun's apparent path, the apparent position of Polaris, and the zenith (Z) are shown for an observer at each location.

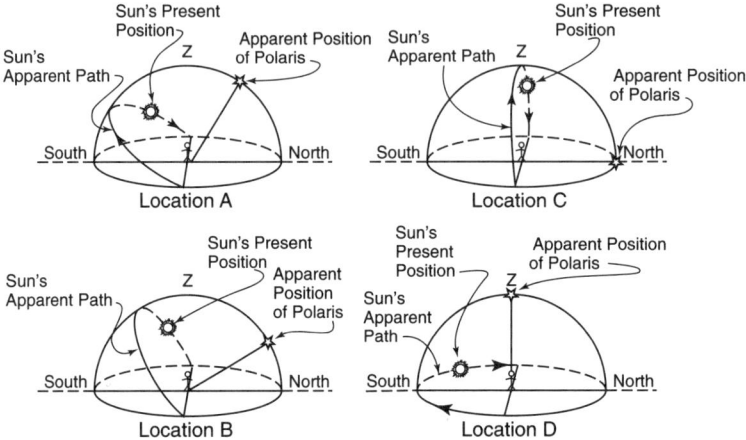

16. To an observer at location A, the Sun will appear to move from
    (1) east to west at 15°/hr
    (2) west to east at 15°/hr
    (3) east to west at 1°/hr
    (4) west to east at 1°/hr

17. The Sun's apparent path through the sky on this day is a direct result of
    (1) the Sun's rotation
    (2) Earth's rotation
    (3) the Sun's revolution around Earth
    (4) Earth's revolution around the Sun

18. At which location is the sun seen highest above the horizon at solar noon?
    (1) A
    (2) B
    (3) C
    (4) D

19. What time of day is shown by the Sun's present position at location A?
    (1) morning
    (2) noon
    (3) afternoon
    (4) midnight

20. Where on Earth's surface is the observer at location *D*?
  (1) The North Pole
  (2) The South Pole
  (3) The Equator
  (4) New York State

CONSTRUCTED RESPONSE QUESTIONS

21. The group of stars known as the Big Dipper can be used to locate the North Star (Polaris) in the night sky. On the diagram of the Big Dipper provided below, draw a straight arrow passing through two stars to indicate the direction to Polaris. [1]

Base your answer to question 22 on the diagram below, which shows the Sun's apparent path as viewed by an observer in New York State on March 21.

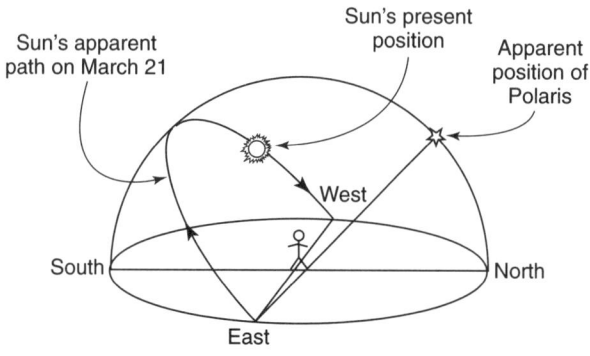

(Not drawn to scale)

22. At approximately what hour of the day would the Sun be in the position shown in the diagram? [1]

Early Astronomy and the Geocentric Model

Base your answers to questions 23 and 24 on the following diagram, which represents the Moon being viewed by an observer in New York State.

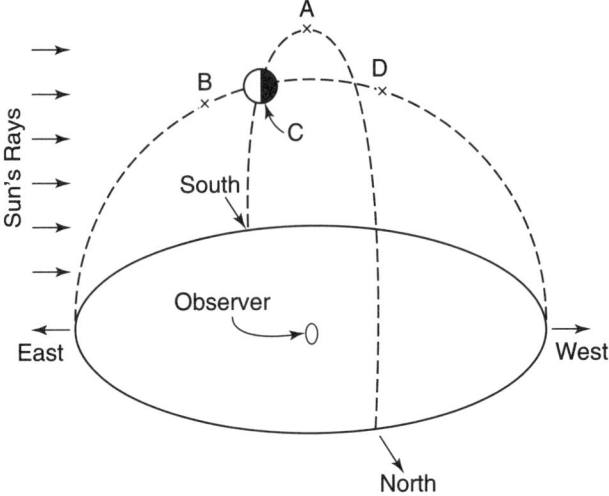

23. Three hours from the time indicated by the diagram, to which position would the Moon be nearest? [1]
24. Twenty-four hours from the time indicated in the diagram, the Moon would be nearest to position *B*. Compare the positions of the Moon in relation to nearby stars on the celestial sphere at the time indicated by the diagram and 24 hours later. [1]

Base your answer to question 25 on the diagram below, which represents an observer's local reference lines on the celestial sphere.[3]

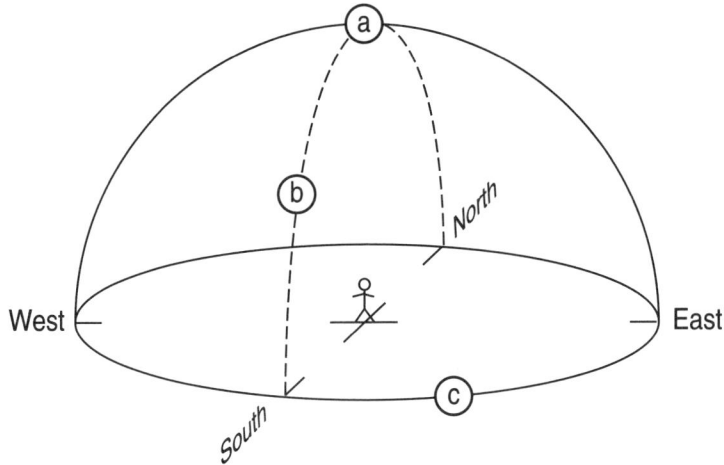

25. Identify the observer's zenith; celestial horizon; and celestial meridian.
    (a) _____ [1]
    (b) _____ [1]
    (c) _____ [1]

Extended Constructed Response Questions

26. Photographs of circumpolar stars, made with a time exposure, showed star trails having arcs of 60 degrees. How long had the film most likely been exposed? (Show all work.) [2]
27. Explain the difference between "real" motion and "apparent" motion. [3]
28. Explain why the stars on the celestial sphere appear to move during the night when you observe them from Earth. [1]
29. List three ways in which planets differ in appearance and/or behavior from stars. [3]
30. Explain why the North Star does *not* appear to change its position during the night. [2]

CHAPTER 2
THE DEVELOPMENT OF THE HELIOCENTRIC MODEL

KEY IDEAS Modern astronomy traces its beginning to the publication in May 1543 by Nicolaus Copernicus of a new *heliocentric*, or Sun-centered, model of the universe. Although Aristarchus of Samos had proposed a Sun-centered model almost 1,800 years earlier, the idea that Earth is moving at great speed had been dismissed as obvious nonsense since no one could feel any motion. Copernicus discarded the idea of a stationary Earth and argued that Earth and the planets circle the Sun. His logical and mathematical arguments paved the way for further investigations. The shift from an Earth-centered to a Sun-centered model was revolutionary and has evolved into our current concept of the universe.

KEY OBJECTIVES
Upon completion of this chapter, you will be able to:

- Compare and contrast the geocentric and heliocentric models of the universe.
- Describe the investigations that led scientists to understand that most of the observed motions of celestial objects are the result of Earth's motion around the Sun.
- Explain how gravity influences the motions of celestial objects.
- Determine the gravitational force between two objects, given their masses and the distance between their centers.
- Analyze the relationships among a planet's distance from the Sun, gravitational force, period of revolution, and speed of revolution.

EARLY MODELS: ARISTOTLE AND PTOLEMY

Ancient Greek thinkers, particularly Aristotle, set a pattern of belief that persisted for 2,000 years—the universe had a large, stationary Earth at its center; and the Sun, the Moon, and the stars were arranged around Earth in a perfect sphere, with all of these bodies orbiting Earth in perfect circles at constant speeds. The Egyptian astronomer Ptolemy refined this concept into an elegant mathematical model of circular motions that enabled astronomers to predict the positions of celestial objects fairly accurately and could account for many of the "problem" observations that plagued Aristotle's model.

Aristotle's Geocentric Universe

Aristotle, a Greek philosopher who lived from 384 B.C. to 322 B.C., wrote about and taught many subjects, including history, philosophy, drama, poetry, and ethics. His wide-ranging knowledge and insight earned him a prominent place among the great thinkers of antiquity.

Aristotle's was a common sense view of the universe. He understood the celestial sphere model and its ability to explain most casual observations, such as the apparent movements of celestial bodies. As records of careful measurements were kept over time, however, some problems arose. The Sun doesn't follow the same path through the sky all year long. The Moon changes position relative to the stars from night to night. Five (actually, nine) "stars," out of the thousands seen in the sky, don't stay in fixed positions relative to the others, but "wander" around in the sky. These moving objects, as explained in Chapter 1, came to be called *planets*, from *planetes*, the Greek word for "wanderer." Aristotle realized that a one-sphere model couldn't explain these "problem" observations, so he revised the model.

Spheres Within Spheres

Aristotle reasoned that, if some objects move differently, they must be on different celestial spheres! Aristotle explained the "problem" observations by proposing a universe consisting of eight crystalline (i.e., transparent) spheres nesting one inside the other like a Chinese box puzzle, with Earth at the very center. The Sun, Moon, stars, and planets were fixed to the surface of separate spheres, which rotated around the unmoving Earth. All motions of the spheres were perfect circles. By having the spheres spinning at slightly different rates and at slightly different angles in relation to one another, most of the "problem" observations could be accounted for. Either the spheres moved because they were self-propelled, or, as was thought more likely, their motion was initiated by a supernatural being. See Figure 2.1.

Common Sense and Parallax

In Aristotle's model, Earth too was a sphere, the perfect shape, as could be seen when its shadow was visible against the Moon during an eclipse. Common sense indicated that Earth wasn't moving because no motion could be felt, but Aristotle believed there was other evidence as well. If Earth moved, objects falling in a straight line should fall to the side of points directly beneath them. Later astronomers argued against a moving Earth by citing a lack of stellar parallax. *Parallax* is the change in apparent position of closer objects in relation to farther objects due to a change in the position

The Development of the Heliocentric Model

Figure 2.1 Aristotle's Model of the Universe. Crystalline spheres were nested one inside the other, with Earth at the center. The spheres and their attached stars and planets rotated around Earth. Source: *Horizons: Exploring the Universe*, Michael A. Seeds, Wadsworth, 1987.

of the observer. If Earth was moving, the apparent positions of the stars should change as Earth (and the observer) moved. See Figure 2.2.

According to Aristotle, the natural state of things on Earth was to be at rest. Natural motion on Earth was towards its center. Unlike the perfect circular motion of the spheres, the motion of objects on Earth was imperfect straight-line motion. The spheres were perfectly clear and were composed of ether, a substance that could not be changed or destroyed. The other four elements were earth, water, air, and fire. All objects were made of mixtures of these four elements and decayed as a result of being forced to move in unnatural directions.

Since Aristotle's universe has Earth at its center, it is called a ***geocentric***, or Earth-centered, model of the universe.

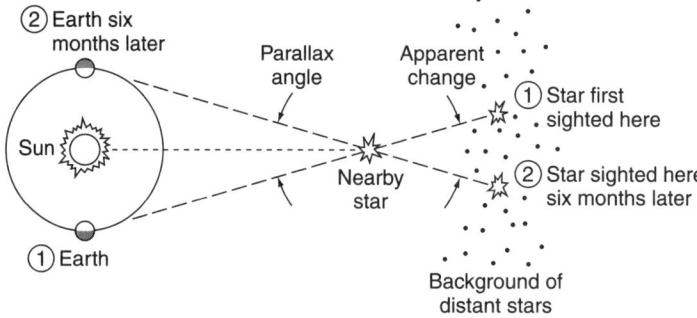

Figure 2.2 Stellar Parallax: the apparent Shift in Position of a Nearby Star Against the Background of More Distant Stars. Stars are so far away that the parallax angle was too small to be detected by early astronomers. Source: *Astronomy: A Self-Teaching Guide*, 4th Ed., Dinah L. Moché, John Wiley, 1998.

Ptolemy's Geocentric Model

There were some obvious problems with Aristotle's view of the universe. The most obvious was visible to the naked eye. There were times when the planets changed course in the sky; for example, at times Mars would stop and then move backwards, a phenomenon called *retrograde motion*. Since the crystal spheres of the Aristotelian universe could not stop or change direction, this observation could not be explained until the second century A.D., when Claudius Ptolemaeus, usually referred to as Ptolemy, proposed an ingenious theory.

Ptolemy, an Egyptian, lived and worked in the Greek settlement at Alexandria in about A.D. 140. There he studied mathematics and astronomy and developed a model of the universe based upon Aristotle's teachings. The details of his model are carefully spelled out in his great book, *Almagest*.

Explaining Retrograde Motion

Ptolemy's view was that each planet was fixed to a small sphere that was in turn fixed to a larger sphere. The smaller sphere and its attached planet turned at the same time that the larger sphere turned. As a result there could be times when, to an observer on Earth, the planet appeared to be moving backward. Ptolemy called the circular motions of the larger spheres **deferents** and the motions of the smaller spheres **epicycles**. He placed Earth's sphere off the center of its deferent. See Figure 2.3.

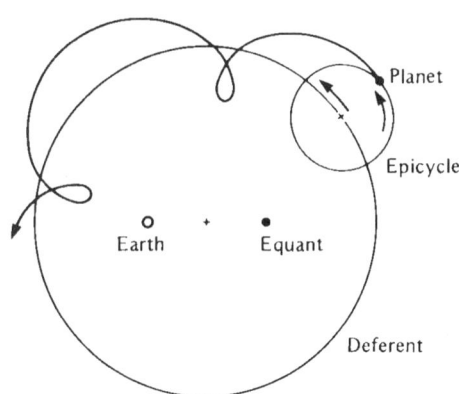

Figure 2.3 Ptolemy's Universe. Ptolemy added epicycles to Aristotle's model to explain retrograde motion and changes in apparent diameter. Source: *Horizons: Exploring the Universe*, Michael A. Seeds, Wadsworth, 1987.

With Ptolemy's ingenious modifications, Aristotle's universe could explain all casual, naked-eye observations of the universe. For 1,000 years astronomers studied and preserved Ptolemy's work, making no changes in his basic theory. It became part of the accepted thinking of the time. See Figure 2.4.

Problems with Predictions

At first, the Ptolemaic system was able to predict the motions of celestial objects with a fair degree of accuracy. However, as the centuries passed, the differences between what the Ptolemaic system predicted and what was actually observed grew so large they could not be ignored. At first, earlier astronomers blamed these discrepancies on poor instruments or inaccurate

The Development of the Heliocentric Model

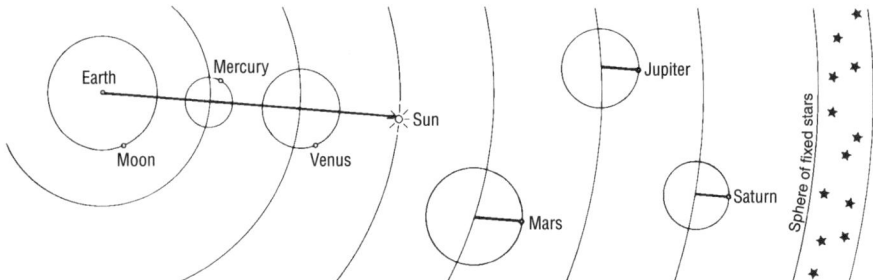

Figure 2.4 Ptolemy's Geocentric Model of the Universe. This model was accepted for well over 1,000 years. Source: *Horizons: Exploring the Universe*, Michael A. Seeds, Wadsworth, 1987.

observations. Arabian and, later, European astronomers corrected the system, recalculated constants, and even added new epicycles. King Alfonso X of Castile paid for the last great correction of the Ptolemaic model. Ten years of observations and calculations were then published as the Alfonsine Tables. By the 1500s, however, the Alfonsine Tables were also inaccurate, often being off by as much as 2°, which is four times the angular diameter of the moon—a significant error.

THE HELIOCENTRIC MODEL

Copernicus

At about the same time that astronomers were struggling with the inaccurate Alfonsine Tables, there was a serious need for calendar reform. By the beginning of the 1500s, the Julian calendar was off by about 11 days. Easter, a major church holiday, was particularly hard to determine. Both the Hebrew calendar, which was based upon the Moon, and the Julian calendar, which was based upon the Sun, had to be used to calculate the phase of the moon, upon which the date of Easter depended. A secretary of Pope Sixtus IV asked Nicolaus Copernicus, a priest-mathematician from Poland (see Figure 2.5), to examine the problem of calendar reform.

Figure 2.5 Nicolaus Copernicus. Source: *Astronomy: The Cosmic Journey*, William K. Hartman, Wadsworth, 1987.

Copernicus recognized that any calendar reform would have to resolve the relationship between the Sun and the Moon. After much study of the problem, Copernicus proposed a mathematically elegant solution in which he suggested a *heliocentric*, or Sun-centered, universe with a moving Earth.

In 1514, he distributed a brief manuscript outlining his ideas, but was discreet because he recognized the potential dangers of questioning church dogma. Not until his death in 1543 was his full argument in favor of a Sun-centered system published. Even then, he avoided heresy charges by crediting classical Greek sources with the idea, thus implying that the concept did not originate with him.

In Copernicus' model of a heliocentric universe, the center of the universe was a point near the Sun. Earth orbited the Sun and spun once a day on its axis. See Figure 2.6.

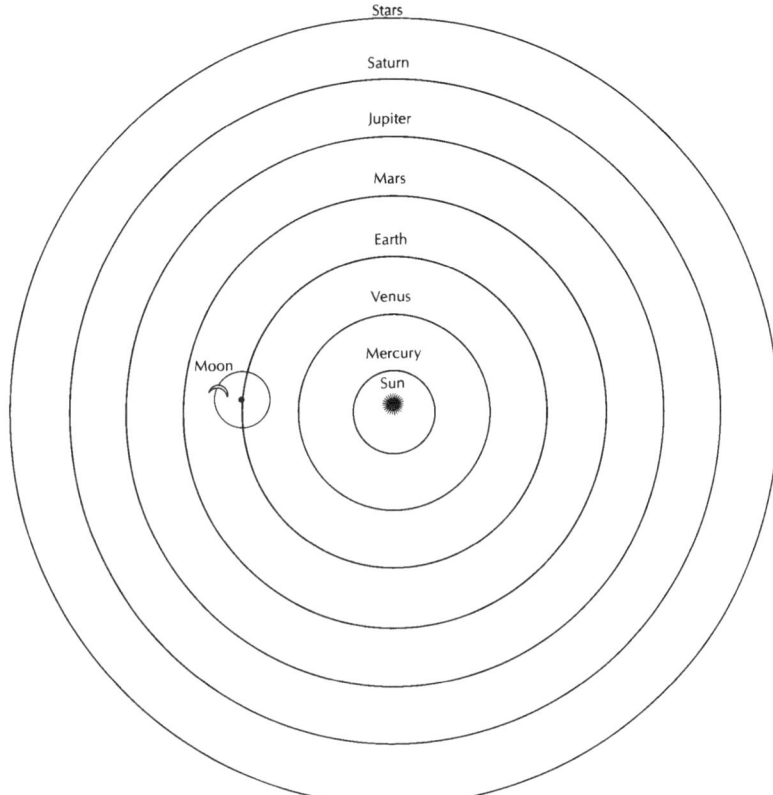

Figure 2.6 The Copernican Heliocentric Universe. Copernicus proposed a Sun-centered model in which all planets and stars moved in perfect circles around the Sun. Source: *Discovering Astronomy*, Robert D. Chapman. W. H. Freeman, 1978. Used with permission.

Copernicus explained the lack of stellar parallax by stating that stars are so distant that parallax is too small to measure. However, he avoided the problem of objects falling to the side of positions directly under them. Retrograde motion occurs because Earth moves faster in its orbit than do planets farther from the Sun. Earth and the other planets all move continuously in their orbits around the Sun, but Earth moves toward an outer planet in one part of its orbit and then passes it and moves away from it. However, planets moving in perfect circles around the Sun could not explain all of the observed details of their motions, and in the end Copernicus, too, resorted to epicycles and did no better at predicting the positions of celestial objects than Ptolemy.

While Copernicus' system was also erroneous, his idea that the universe was heliocentric, or Sun-centered, was correct and gradually gained acceptance.

The Development of the Heliocentric Model

Probably the most important reasons why his theory was eventually accepted were the revolutionary mood of the world in his lifetime and the simple, forthright way in which his model explained retrograde motion. See Figure 2.7.

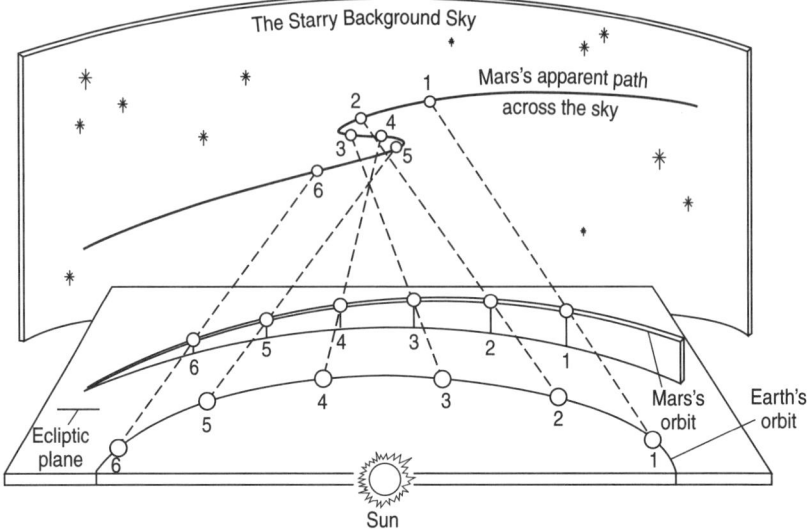

Figure 2.7 Copernicus' Simple, Forthright Explanation of Retrograde Motion. Both Earth and Mars move in a continuous path, but the inner planet (Earth) covers more of its orbit in the same time period, changing its point of view toward the outer planet (Mars). Source: *Astronomy: The Cosmic Journey*, William K. Hartman, Wadsworth, 1987. Used with permission.

Contributions of Tycho Brahe and Johannes Kepler

Tycho Brahe: Precision Observer

Shortly after Copernicus died, a Danish nobleman named Tycho Brahe (see Figure 2.8a) became interested in astronomy. After observing that the Alfonsine Tables were nearly a month off in predicting a conjunction of Jupiter and Saturn, and observing a "new star" produced by a supernova, that is, the explosion of a very large star, Tycho questioned the Ptolemaic system of a perfect, unchanging heaven in a small book he wrote. His book was widely read, and the King of Denmark gave him funds to build a world-class astronomical observatory with no telescopes, but many ingenious devices for precisely measuring celestial motions. When the King of Denmark died, Tycho fell out of favor and accepted a position as court astronomer to the Holy Roman Emperor in Prague, taking with him all of the data from the observatory in Denmark. In Prague, the emperor commissioned him to publish a revision of the Alfonsine Tables. Tycho hired several young mathematicians to help him with his task.

Johannes Kepler: Orbits Are Ellipses, Not Circles

One of Tycho's young assistants was Johannes Kepler (see Figure 2.8b). Shortly after beginning the project commissioned by the emperor, Tycho died unexpectedly. Before he died, however, he recommended Kepler to take over his position. As court

Figure 2.8 (a) Tycho Brahe (1546–1601) (b) Johannes Kepler (1571–1630) Source: *Astronomy: The Cosmic Journey,* William K. Hartman, Wadsworth, 1987. Used with permission.

(a)   (b)

astronomer, Kepler spent six years trying to work out the orbit of the planet Mars, using Ptolemy's system of the planet moving in a small circle that moved in a larger circle around the Sun. But no matter how hard he tried, he could not get the theoretical orbit to match the observed orbit. Finally Kepler realized that the orbit of Mars was elliptical, or oval, and that Mars moved at a speed that varied with its distance from the Sun.

Kepler's Laws of Planetary Motion

After years of studying observations of celestial objects, Kepler made three important discoveries about the motions of planets as they revolve around the Sun.

1. **Each planet revolves around the Sun in an elliptical orbit with the Sun at one focus.**

    An ellipse has a major axis and a minor axis that are lines connecting the two points farthest apart and the two points closest together on the ellipse. It also contains two special points along the major axis, each called a *focus* (plural, foci). The distance from one focus to any point on the ellipse and back to the other focus is always the same.

    As a result it is very easy to draw an ellipse using two tacks and a loop of string. Press the tacks into a board, loop the string around them, and place a pencil in the loop on page 26. Keep the string taut; then, as you move the pencil, it will trace out the shape of an ellipse. See Figure 2.9.

    The closer together the foci, the more nearly circular the ellipse. The farther apart the foci, the flatter the ellipse. The flatness of an ellipse is called its *eccentricity*. Eccentricity is expressed as the ratio between the distance between the foci and the length of the major axis:

    A perfect circle would have an eccentricity of 0; a straight line has an eccentricity of 1.

    $$e = \frac{d}{l}$$

    where e = eccentricity,
    d = distance between foci, and
    l = length of major axis.

The Development of the Heliocentric Model

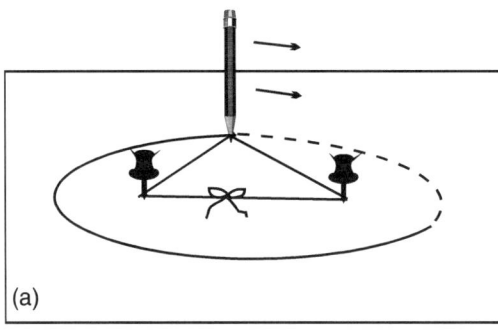

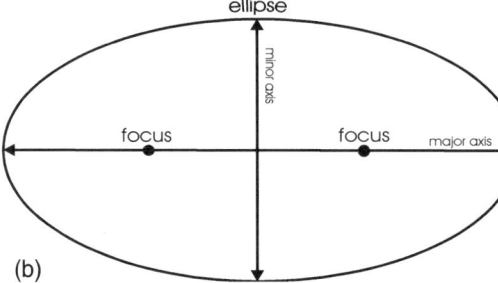

Figure 2.9 (a) The way to draw an ellipse (b) The main parts of an ellipse

Since the orbit of each planet is an ellipse, the distance from each planet to the Sun varies during its orbit. See Figure 2.10. For example, Earth's distance to the Sun varies from 147×10^6 kilometers on January 3 at its closest (perihelion) to 152×10^6 kilometers on July 6 at its farthest (aphelion). The difference between these distances, 5×10^6 kilometers, is the distance between the foci of Earth's elliptical orbit. This measurement is very small compared to the length of the major axis, 299×10^6 kilometers, so the eccentricity of Earth's orbit is very small (0.17), indicating that the orbit is very nearly a circle. Although many illustrations show Earth's orbit around the Sun in a perspective view that exaggerates its eccentricity, if viewed from directly overhead the orbit would appear very nearly circular. (In this connection, it is interesting to note that Ptolemy's system of circular orbits *almost* worked because Earth's orbit is *almost* a circle. However, that slight difference from a perfect circle was enough to throw Ptolemy's system into question over time.)

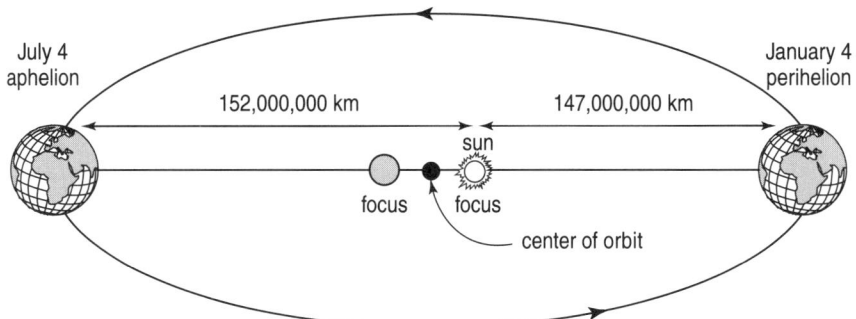

Figure 2.10 View of Earth's Elliptical Orbit with the Sun at One Focus. Earth is closest to the Sun at perihelion and farthest away at aphelion. Distances are approximate.

As the distance between Earth and the Sun changes, the apparent

26

diameter of the Sun changes in a cyclic manner. When Earth is closest to the Sun on January 4, the Sun has its greatest apparent diameter. On July 4, when the Sun is farthest away, it has its smallest apparent diameter.

2. **The planets do not move at a constant velocity.**

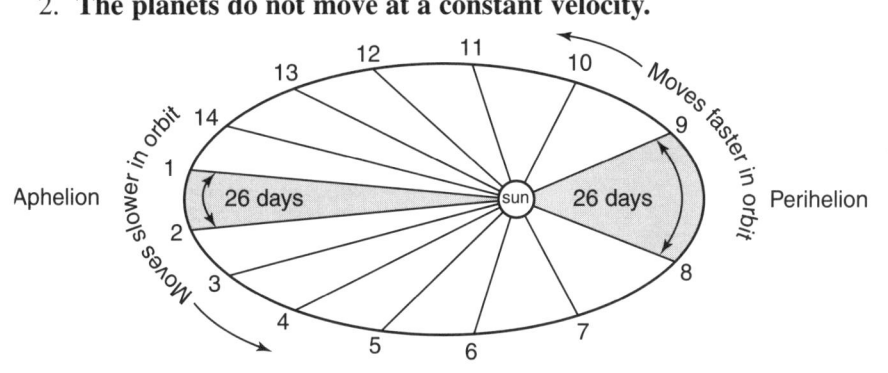

Figure 2.11 Kepler's Law of Equal Areas. A planet sweeps out equal areas of its elliptical orbit in equal periods of time. Since the planet travels less distance in the 26 days it takes to move from 1 to 2 on the ellipse above, it is traveling slower than when it moves from 8 to 9.

In Figure 2.11, the elliptical shape of a planet's orbit is exaggerated to show variation in velocity more clearly. The times the planet takes to move from 1 to 2, from 3 to 4, and from 8 to 9 are all equal. However, if you look at the diagram carefully, you can see that the distance from 1 to 2 is less than the distance from 8 to 9 even though the distances were covered in the same time. This means that the planet is moving fastest when it is closest to the Sun and slowest when it is farthest from the Sun.

Kepler did not know *why* this was so; he only determined that the variation did occur. We now know that this cyclic changing velocity of the planets as they move around the Sun is due to changes in the gravitational force between a planet and the Sun as distance changes. As a planet approaches the Sun (distance decreases), gravitational force increases and, since it is acting in the same direction in which the planet is moving, causes the planet to move faster. As a planet moves away from the Sun (distance increases), gravitational force decreases and, since it is now acting opposite to the direction in which the planet is moving, causes the planet to slow down.

3. **There is a mathematical relationship between the time a planet takes to complete one revolution around the Sun and its average distance from the Sun.**

The time required for a planet to make one revolution is called its ***period of revolution***. It was already known that the farther a planet is from the Sun, the longer its period of revolution. Kepler's careful analysis showed that there is a mathematical relationship between the two factors. See the accompanying table. A planet's period of revolution *squared* is proportional to its distance from the Sun *cubed*:

The Development of the Heliocentric Model

$$T^2 \propto R^3$$

If we measure the period of revolution in units of Earth-years, and let Earth's average distance from the Sun equal 1 unit of distance, called an *astronomical unit* (AU), the relationship simplifies. Then T^2 becomes $(1)^2$, R^3 becomes $(1)^3$, and, since $(1)^2 = 1$ and $(1)^3 = 1$:

$$T^2 = R^3$$

This relationship for the planets in our solar system can be plotted as a curve on a graph. See Figure 2.12.

Planet	T (period of revolution in Earth-years)	R (distance from the Sun in AU)
Mercury	0.24	0.39
Venus	0.62	0.72
Earth	1.0	1.0
Mars	1.88	1.52
Jupiter	11.86	5.23
Saturn	29.46	9.54
Uranus	84.0	19.18
Neptune	164.8	30.05
Pluto	247.7	39.44

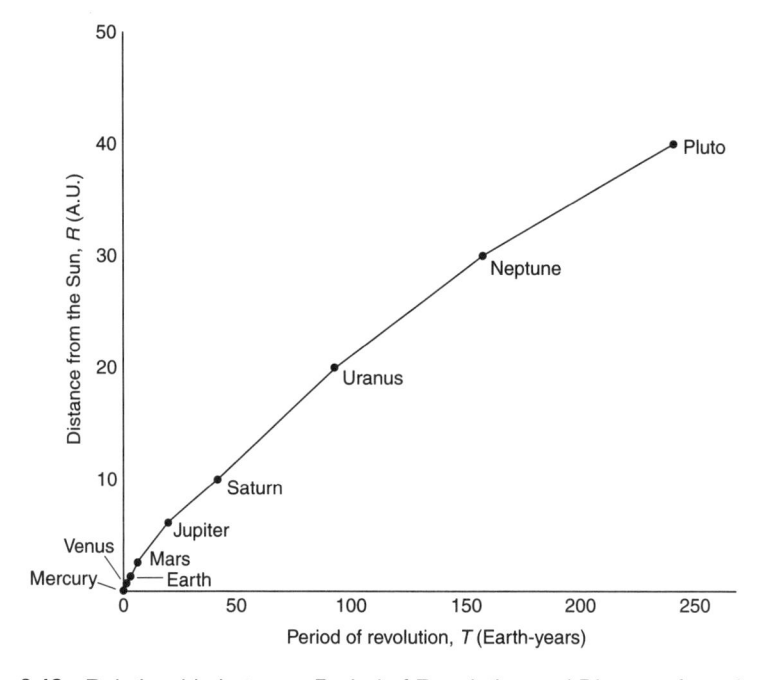

Figure 2.12 Relationship between Period of Revolution and Distance from the Sun

Galileo and Newton: Improving the Heliocentric Model

Galileo: Observations That Challenged Aristotle's Geocentric Universe

Shortly after Kepler's works were published, Galileo Galilei, an Italian astronomer and physicist (see Figure 2.13a), turned his telescope on the heavens and made several discoveries that further undermined the Ptolemaic system. His discovery of imperfections on the Moon's surface, spots on the surface of the Sun, and moons circling Jupiter challenged Aristotle's view of the heavens as perfect and unchanging. He saw Venus go through a full set of phases, which was not possible according to the Ptolemaic system of epicycles. Galileo also studied motion and solved Copernicus' falling-object problem. He argued that, if Earth is in motion, so are all objects on it. Therefore, an object that is dropped moves sideways at the same speed as Earth and falls on a spot directly beneath its point of release. Galileo became an outspoken champion of the heliocentric model.

Figure 2.13 (a) Galileo Galilei (1564–1642) (b) Isaac Newton (1642–1727) Source: *Astronomy: The Cosmic Journey*, William K. Hartman, Wadsworth, 1987. Used with permission.

(a) (b)

Newton: Explaining Motion

Eleven months after Galileo died in 1642, Isaac Newton (see Figure 2.13b) was born in England. Newton brought the discoveries of Copernicus, Galileo, and Kepler together. Using a few key concepts (mass, momentum, acceleration, and force), three laws of motion (inertia, the dependence of acceleration on force and mass, and action and reaction), and the law of universal gravitation, Newton was able to explain both the motions of objects on Earth and the distant motions of celestial objects.

Newton's laws of motion made it possible to predict how an object would move if the forces acting on it were known. Newton thought about the forces that would be needed to keep a satellite moving in orbit around another object. In considering the Moon, Newton realized that the Moon would circle Earth only if some force pulled the Moon towards Earth's center; otherwise it would continue moving in a straight line off into space. Newton's genius was to realize that the force that keeps the Moon in orbit around Earth is the same force that causes objects close to Earth (e.g., apples on a tree) to fall to the ground: gravity. He realized that the force of gravity is universal, that all objects are attracted to one another with a force that depends on the masses of the objects and their distance from each other. Newton expressed this relationship in a simple mathematical formula. See Figure 2.14.

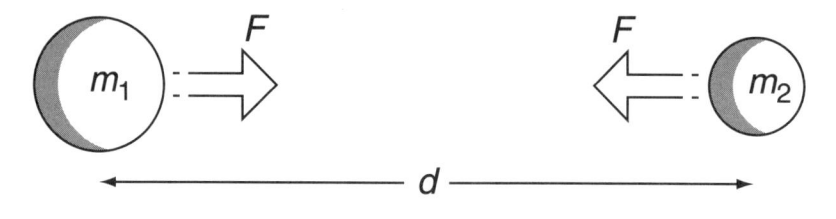

$$F = -G \frac{m_1 m_2}{d^2}$$

Figure 2.14 Newton's Law of Gravity. In this equation, F is the force of gravity acting between two masses, G is the gravitational constant, m_1 and m_2 are the masses, and d is the distance between them. Source: *Astronomy: A Self-Teaching Guide*, 4th Ed., Dinah L. Moché, John Wiley, 1998. Used with permission.

Gravity and Orbital Motion

How does gravity keep satellites moving in a curved orbit? Imagine that a cannonball is shot out of a cannon aimed horizontally. If there were no gravity, inertia would cause the cannonball to fly off horizontally until some force stopped it. However, with gravity pulling the cannonball downward toward Earth's center, as the cannonball is flying horizontally its path curves downward and eventually the cannonball strikes Earth's surface. If a more powerful charge is used in the cannon, the cannonball will travel farther horizontally before it strikes Earth. With a sufficiently powerful charge, the cannonball would travel far enough horizontally that, as its path curved downward because of gravity, Earth's surface would curve away because of its spherical shape, and the cannonball would never strike Earth's surface. Instead, gravity would cause the cannonball to fall downward at the same rate that Earth's surface curves away from it, and it would fall unendingly in a circular path around Earth—it would be in orbit. See Figure 2.15.

Similarly, as Newton explained, it is a combination of two forces that keeps the planets moving in their curved paths around the Sun. The combination of a planet's forward motion and its motion toward the Sun due to gravity results in circular motion—the planet's orbit around the Sun. See Figure 2.16.

Now let's return to our cannonball analogy. For an orbit just above Earth's surface, the cannonball would have to be shot out of the cannon at 7.9×10^3 meters per second, or about 18,000 miles per hour. If the cannonball was propelled at a higher speed, it would travel farther outward before being pulled

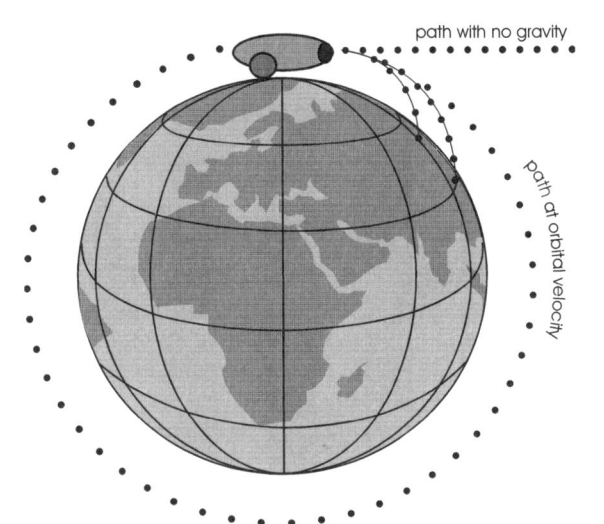

Figure 2.15 Orbital Velocity. As a fired cannonball travels through the air, it is drawn downward by gravity in a curved path until it strikes Earth's surface. If it is fired with more energy, it will travel farther in a curved path before crashing into Earth's surface. If it is fired with enough energy, it's path will curve downward at the same rate that Earth curves away from its path, and it will go into orbit.

back by gravity, and the orbit would be elliptical instead of circular. If the cannonball was shot out at a velocity equal to or greater than 11.2×10^3 meters per second, or about 25,000 miles per hour, it would be able to escape Earth's gravity and fly out of orbit. See Figure 2.17.

With Newton's explanation of the causes of motion, the heliocentric theory began to firmly displace the geocentric theory as the generally accepted model of the universe. Our modern view of planetary motions in the solar system is based upon the heliocentric model.

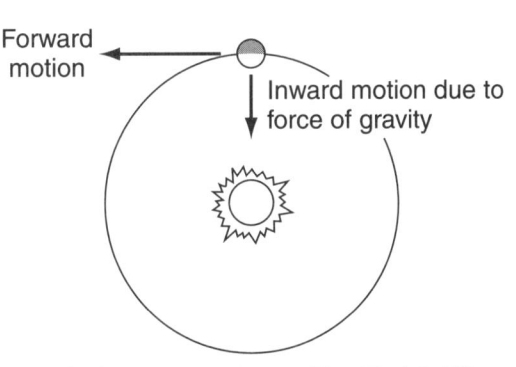

Figure 2.16 The Two Motions That Explain Why a Planet Travels in Orbit. Source: *Astronomy: A Self-Teaching Guide*, 4th Ed., Dinah L. Moché, John Wiley, 1998. Used with permission.

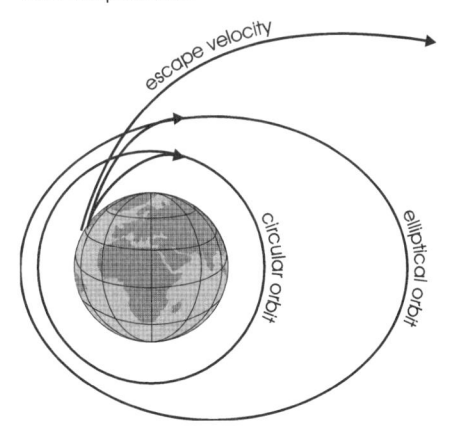

Figure 2.17 Circular Orbit, Elliptical Orbit, and Escape Velocity.

The Development of the Heliocentric Model

MULTIPLE-CHOICE QUESTIONS

For each case, write the number of the word or expression that best answers the question or completes the statement.

The diagram to the right shows one model of a portion of the universe.

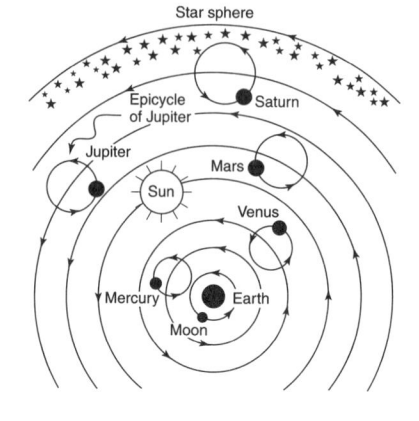

1. What type of model does the diagram best demonstrate?
   (1) a heliocentric model, in which celestial objects orbit Earth
   (2) a heliocentric model, in which celestial objects orbit the Sun
   (3) a geocentric model, in which celestial objects orbit Earth
   (4) a geocentric model, in which celestial objects orbit the Sun

2. Which planetary model allows a scientist to predict the exact positions of the planets in the night sky over many years?
   (1) The planets' orbits are circles in a geocentric model.
   (2) The planets' orbits are ellipses in a geocentric model.
   (3) The planets' orbits are circles in a heliocentric model.
   (4) The planets' orbits are ellipses in a heliocentric model.

3. Which diagram best represents the motions of celestial objects in a heliocentric model?

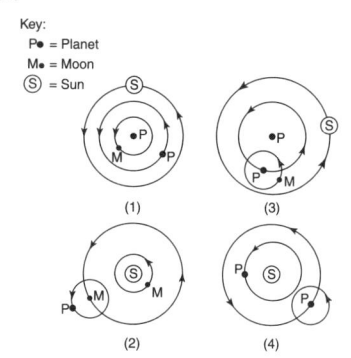

4. What is the exact shape of Earth's orbit around the Sun?
   (1) a slightly eccentric ellipse
   (2) a very eccentric ellipse
   (3) an oblate spheroid
   (4) a perfect circle

Note that question 5 has only three choices.
5. As Earth revolves in orbit from its July position to its January position, Earth's distance from the Sun will
   (1) decrease
   (2) increase
   (3) remain the same

6. If the distance between Earth and the Sun were increased, which change would occur?
   (1) The apparent diameter of the Sun would decrease.
   (2) The amount of insolation received by Earth would increase.
   (3) The time for one Earth rotation (rotation period) would double.
   (4) The time for one Earth revolution (orbital period) would decrease.

7. Which information about a nearby star must be known to determine its distance from an observer?
   (1) size
   (2) color
   (3) temperature
   (4) parallax

8. The force of gravity between two objects will be greatest if their masses are
   (1) small and they are far apart
   (2) small and they are close together
   (3) large and they are far apart
   (4) large and they are close together

Note that question 9 has only three choices.
9. To escape Earth's gravitational field, an object must have a minimum velocity of 12 kilometers per second. What minimum velocity must an object have to escape the gravitational field of the Moon?
   (1) less than 12 km/s
   (2) more than 12 km/s
   (3) 12 km/s

10. The diagram shows Earth (E) in orbit about the Sun. If the gravitational force between Earth and the Sun were suddenly eliminated, toward which position would Earth then move?
    (1) 1
    (2) 2
    (3) 3
    (4) 4

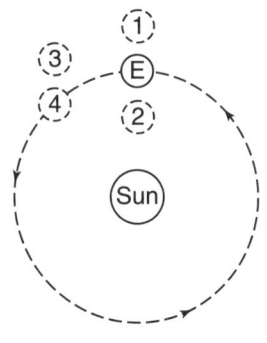

(Not drawn to scale)

The Development of the Heliocentric Model

11. If the distance from Earth to the Sun were doubled, the gravitational attraction between the Sun and Earth would become
    (1) one-fourth as great
    (2) one-third as great
    (3) twice as great
    (4) four times as great

12. Planet A has a greater mean distance from the Sun than planet B. On the basis of this fact, which other comparison can be correctly made between the two planets?
    (1) Planet A is larger.
    (2) Planet A's revolution period is longer.
    (3) Planet A's speed of rotation is greater.
    (4) Planet A's day is longer.

13. According to the *Earth Science Reference Tables*, what is the approximate eccentricity of the ellipse shown to the right?
    (1) 0.50
    (2) 2.0
    (3) 0.25
    (4) 4.0

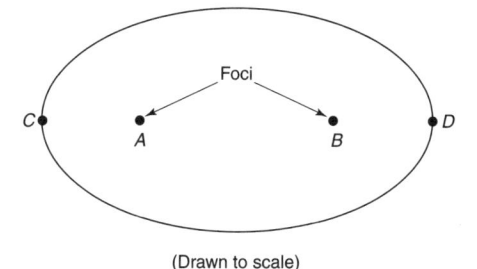

    (Drawn to scale)

14. The diagram below represents the orbit of a spacecraft around the Sun. According to the *Earth Science Reference Tables*, the spacecraft's orbit is
    (1) more eccentric than Earth's orbit but less eccentric than Mars's orbit
    (2) more eccentric than the orbits of planets 300 million km from the sun but less eccentric than the orbits of planets 100 million km from the sun
    (3) more eccentric than the orbit of any planet in the solar system
    (4) less eccentric than the orbits of planets with a density less than 5 g/cm³

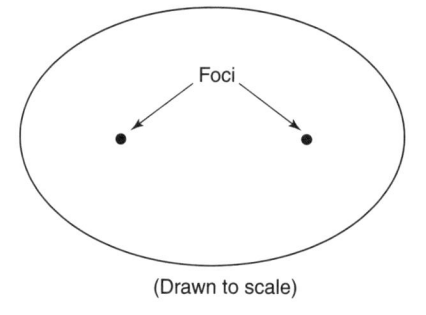

    (Drawn to scale)

Base your answers to questions 15 through 19 on your knowledge of earth science, the *Earth Science Reference Tables*, and the diagrams, tables, and information below. Diagram I represents the orbit of an Earth satellite, and diagram II shows how to construct an elliptical orbit using two pins and a loop of string. The table shows the eccentricities of the orbits of the planets in the solar system.

34

DIAGRAM I

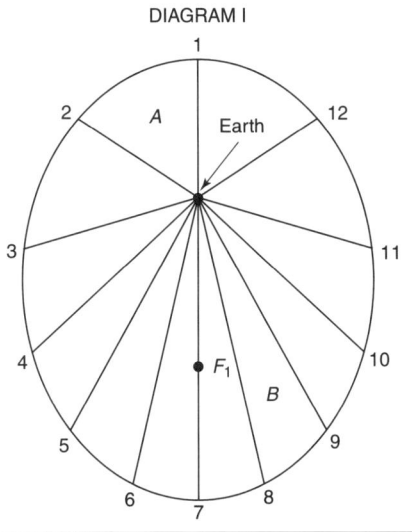

DIAGRAM II

The satellite was at position 1 precisely at midnight on the first day. It arrived at position 2 the next midnight, at 3 the next, and so on.

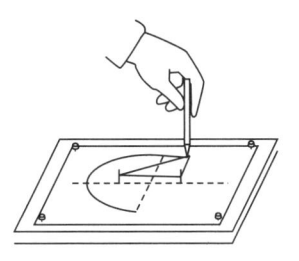

Planet	Eccentricity of Orbit
Mercury	0.206
Venus	0.007
Earth	0.017
Mars	0.093
Jupiter	0.048
Saturn	0.056
Uranus	0.047
Neptune	0.008
Pluto	0.250

15. At which position represented in diagram I would the gravitational attraction between the Earth and the satellite be greatest?
    (1) 1   (2) 7   (3) 3   (4) 11

16. According to the table, the orbit of which planet would most closely resemble a circle?
    (1) Mercury   (2) Venus   (3) Saturn   (4) Pluto

17. What is the approximate eccentricity of the satellite's orbit?
    (1) 0.31   (2) 0.40   (3) 0.70   (4) 2.5

18. The Earth satellite takes 24 hours to move between each numbered position on the orbit. How does the orbital speed of the satellite in section A of its orbit (between positions 1 and 2) compare to its orbital speed in section B (between positions 8 and 9)?
    (1) It is moving faster in section A than in section B.
    (2) It is moving slower in section A than in section B.
    (3) Its speed in section A is equal to its speed in section B.
    (4) It is speeding up in section A and slowing down in section B.

Note that question 19 has only three choices.
19. If the pins in diagram II were placed closer together, the eccentricity of the ellipse being constructed would
    (1) decrease   (2) increase   (3) remain the same

The Development of the Heliocentric Model

Base your answers to questions 20 through 24 on your knowledge of earth science, the *Earth Science Reference Tables*, and the diagram to the right. The diagram represents four planets, A, B, C, and D, traveling in elliptical orbits around a star. The center of the star and the letter *f* represent the foci for the orbit of planet A. Points 1 through 4 are locations on the orbit of planet A.

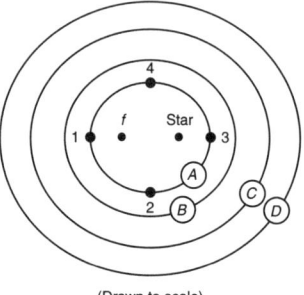

(Drawn to scale)

20. Which is the order of the planets from shortest period of revolution to longest?
    (1) A, B, C, D
    (2) B, A, D, C
    (3) C, D, A, B
    (4) D, C, B, A

21. If planets A, B, C, and D have the same mass and are located at the positions shown in the diagram, the planet that has the greatest gravitational attraction to the star is
    (1) A
    (2) B
    (3) C
    (4) D

22. Planet A travels fastest in its orbit at location
    (1) 1
    (2) 2
    (3) 3
    (4) 4

23. Using the equation and metric ruler in the *Earth Science Reference Tables*, the eccentricity of planet A's orbit is found to be approximately
    (1) 0.10
    (2) 0.20
    (3) 0.50
    (4) 5.0

24. Which graph best illustrates how the apparent diameter of the star varies for an observer on planet A as the planet makes one complete orbit around the star?

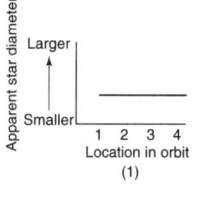

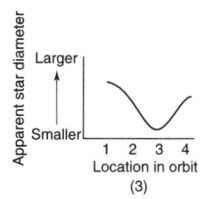

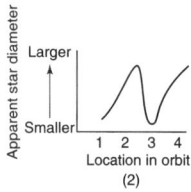

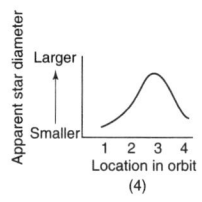

Base your answers to questions 25 and 26 on the diagram to the right, which shows Earth's orbit and the partial orbit of a comet on the same plane around the Sun.

25. Which observation is true for an observer at Earth's Equator at midnight on a clear night for the positions shown in the diagram?
    (1) The comet is directly overhead.
    (2) The comet is rising.
    (3) The comet is setting.
    (4) The comet is not visible.

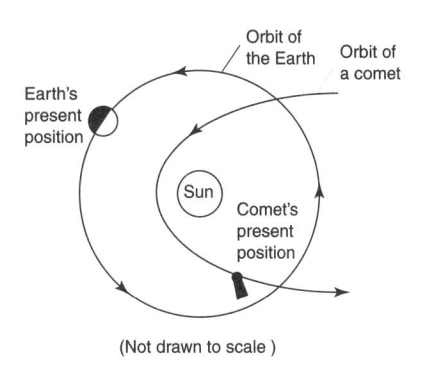

Note that question 26 has only three choices.
26. Compared with Earth's orbit, the comet's orbit has
    (1) less eccentricity
    (2) more eccentricity
    (3) the same eccentricity

Constructed Response Questions

27. Listed below are statements of several theories and observations, some of which may have served as partial bases for Newton's Law of Universal Gravitation. For each statement, (a)–(e), write the number preceding the name of the scientist, chosen from the list below, who first made the observation or proposed the theory.
    (1) Brahe
    (2) Newton
    (3) Copernicus
    (4) Ptolemy
    (5) Galileo
    (6) Aristotle
    (7) Kepler

    (a) The Sun is the center of our solar system. [1]
    (b) Retrograde motions by planets can be explained by the use of epicycles. [1]
    (c) The true orbits of planets are ellipses. [1]
    (d) There is always gravitational force between two masses. [1]
    (e) Falling bodies accelerate at a rate independent of their masses. [1]

The Development of the Heliocentric Model

Base your answers to questions 28 and 29 on the diagram of the ellipse below.

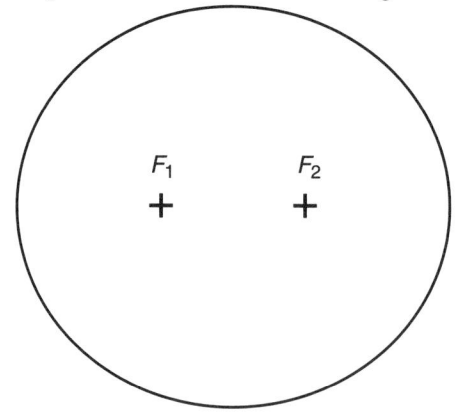

28. Calculate the eccentricity of the ellipse, following the directions below.
    (a) Write the equation used to determine eccentricity. [1]
    (b) Based on measurements of the diagram, substitute values into the equation. [1]
    (c) Calculate the eccentricity of the ellipse. [1]
    (d) State how the eccentricity of the given ellipse compares to the eccentricity of Earth's orbit. [1]

29. Identify the number in the diagram that corresponds to each of the following:

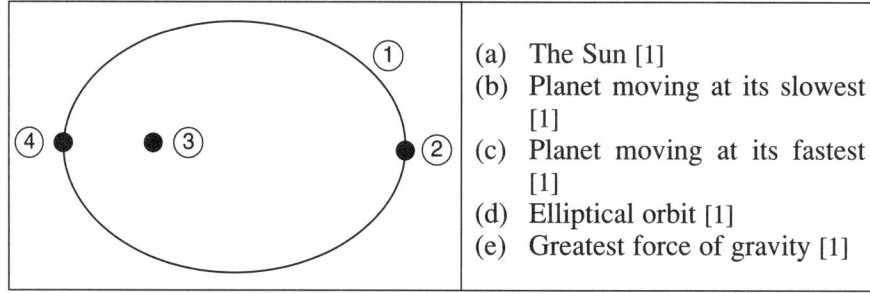

(diagram of ellipse with points 1 (top), 2 (right), 3 (center foci), 4 (left))	(a) The Sun [1] (b) Planet moving at its slowest [1] (c) Planet moving at its fastest [1] (d) Elliptical orbit [1] (e) Greatest force of gravity [1]

EXTENDED CONSTRUCTED RESPONSE QUESTIONS

30. Explain what keeps planets in their orbits around the Sun. [2]

31. Aristotle believed that celestial objects were embedded in solid crystalline shells that rotated one inside the other and carried the celestial objects on their courses. In 1577, Tycho Brahe followed the orbit of a comet across the skies and showed that it moved *through* the solar system, cutting across the orbits of the planets. He later wrote:

"Now it is quite clear to me that there are no solid spheres in the heavens, and those that have been devised by the authors to save the appearances exist only in the imagination, for permitting the mind to conceive the motion which the heavenly bodies trace in their course."

(a) How did Tycho Brahe's observations challenge the geocentric model of the universe? [1]
(b) If the crystalline spheres did not exist, the question arose as to what actually did move the celestial objects and keep them in their orbits. What is the currently accepted answer to this question, and which scientist first proposed it? [2]

32. In his book *Commentariolis*, Nicolaus Copernicus proposed seven "assumptions":

1. All celestial spheres do not have only one common center.
2. The center of Earth is not the center of the universe, but the center of Earth is the center of gravity and of the Moon's sphere.
3. All spheres (except that of the Moon) revolve around the Sun, as though the center were the Sun, so the center of the universe is near the Sun.
4. The sphere of the stars is extremely far away. The distance from Earth to the Sun is insignificant when compared to the distance from Earth to the sphere of stars.
5. What appears to be the motion of the sphere of the stars is not its motion, but that of Earth. Earth, along with its air and water, rotates daily around its poles, while the stars remain motionless.
6. What appear to be the motions of the Sun are not its motions, but the motion of Earth and of the sphere on which it revolves around the Sun in the same manner that the other planets revolve in their spheres.
7. What appears to be retrograde (backward) and forward movements of the planets are not their motions, but the motion of Earth. Earth's motion alone is able to explain many different celestial phenomena.

(a) State one assumption made by Copernicus that does not agree with our current heliocentric model of the universe. [1]
(b) State two commonsense objections people at the time had to the idea that Earth rotates and revolves. [2]
(c) Which part of the geocentric model of the universe did Copernicus keep as part of his new heliocentric model? [1]

The Development of the Heliocentric Model

33. Suppose you landed on a planet with a mass about one-tenth that of Earth and a radius about half that of Earth.
    (a) Compare your weight on that planet with your weight on Earth. [2]
    (b) Refer to the Solar System Data chart in the *Earth Science Reference Tables*, and compare this situation with that of an astronaut on Mars. [2]

Base your answers to questions 34 through 36 on the diagram below, which represents Mars's orbit around the Sun.

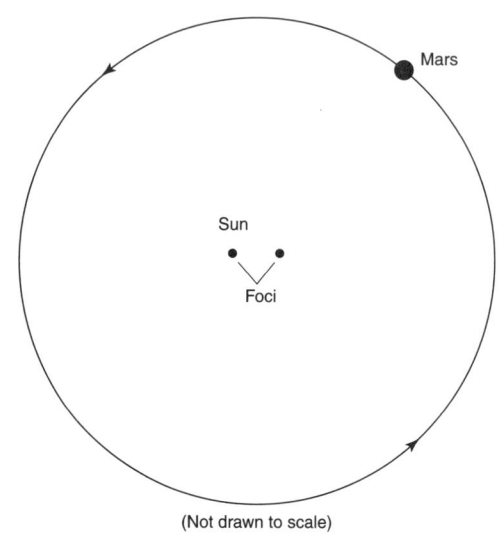

(Not drawn to scale)

34. On the diagram:
    (a) Draw and label the major axis of Mars's orbit. [1]
    (b) Place an X on the orbit to show the location of Mars's greatest orbital velocity. [1]

35. State the difference between the shape (*not* the size) of Earth's orbit and the shape of Mars's orbit. [1]

36. The bar graph to the right shows the equatorial diameter of Earth. Construct the bar that represents the equatorial diameter of Mars. [1]

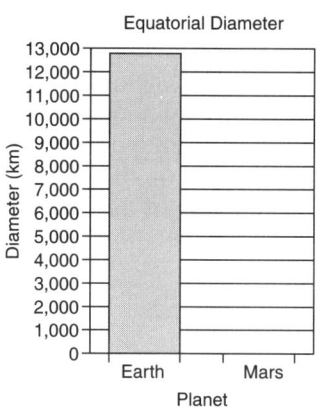

CHAPTER 3

HELIOCENTRIC EARTH MOTIONS AND THEIR EFFECTS

KEY IDEAS The shift from a geocentric to a heliocentric model of the universe changed the way most people saw themselves in relation to the physical universe. The heliocentric model required the apparently immobile Earth to spin completely around on its axis once a day and the universe to be far larger than anyone had imagined; worst of all, Earth lost its position at the center of the universe and became just one of nine planets that orbit the Sun, a typical star in a vast and ancient universe.

However, the heliocentric model has greatly improved human understanding of complex phenomena, such as variations in day length, seasons, ocean tides, eclipses, phases of the Moon, the apparent motion of planets, and the annual traverse of the constellations.

KEY OBJECTIVES
Upon completion of this chapter, you will be able to:

- Explain how the Foucault pendulum and the Coriolis effect provide evidence of Earth's rotation.
- Describe Earth's rate of rotation, the orientation of Earth's axis of rotation with respect to the plane of its orbit, and the effects of Earth's changing position with regard to the Sun.
- Explain how seasonal changes in the apparent positions of constellations provide evidence of Earth's revolution.
- Interpret diagrams of Earth's orbit to determine how the Sun's apparent path through the sky and the relative position of the noon Sun vary with the seasons, and how the length of daylight varies throughout the year at different locations.
- Explain how the changing relative positions of Earth, the Moon, and the Sun account for the phases of the Moon, tides, and eclipses.

EARTH MOTIONS

Rotation

In the modern heliocentric model, Earth rotates at a rate of 15° per hour, or one complete rotation every 24 hours. *Rotation* is a motion in which every part of an object is moving in a circular path around a central line called the *axis of rotation*, like an ice skater spinning around a line through his or her body. Earth's axis of rotation is a line passing through the North Pole, Earth's center, and the South Pole. Earth's axis of rotation is almost directly aligned with the star Polaris and is tilted at an angle of 23½° from a perpendicular to the plane passing through the centers of Earth and the Sun. See Figure 3.1.

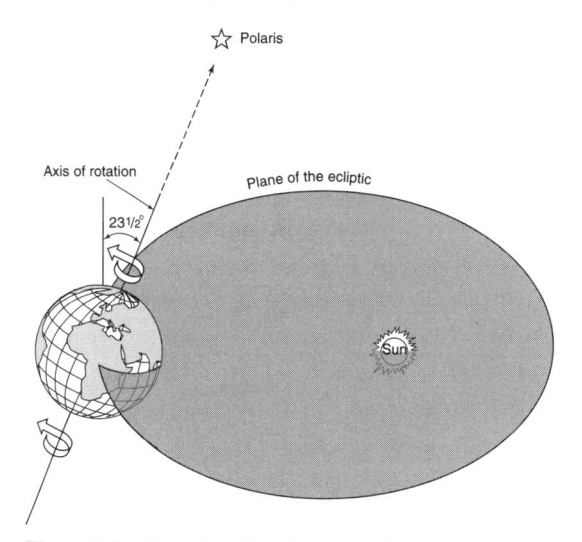

Figure 3.1 Rotation. Earth rotates from west to east once every 24 hours around an axis that runs through the poles.

Evidence of Rotation

The heliocentric theory requires Earth to rotate on its axis once every 24 hours. It is one thing to be able to successfully explain away a problem such as the lack of deflection of a falling object; it is quite another to actually *prove* Earth's rotation. The arguments in favor of rotation may be quite reasonable, but reasonableness is not scientific proof. A French physicist, Jean Foucault, and a French mathematician, Gaspard Coriolis, provided this proof.

The Foucault Pendulum

Jean Foucault used an ingeniously simple method to prove Earth's rotation. Foucault knew that gravity pulls a pendulum only toward Earth's center. It does not act laterally to change the plane of the pendulum's swing. Therefore, if set freely swinging in the absence of any lateral forces, a pendulum should swing in a fixed plane. Foucault further reasoned that the heavier he made the weight of the pendulum, the more force would be required to push it out of its plane of motion, making it unlikely that small breezes would deflect the pendulum. Finally, he realized that, if he made the pendulum really long, it would have a long period and would swing for many hours without needing a push to get it going again (which would introduce a possibility of exerting a lateral force). Foucault argued that, if Earth were motionless, such a pendulum would swing in a plane whose direction would not change. However, if Earth rotated, it would rotate beneath the pendulum and would change position relative to the swinging pendulum. The result would be a pendulum whose plane of swing would appear to rotate relative to the ground in a direction opposite to that of Earth's rotation.

Foucault's idea is easiest to understand if you visualize a pendulum swinging over the North Pole. As the pendulum swings, Earth moves beneath it. In 24 hours the plane of the swinging pendulum will appear to make one rotation. At the Equator, however, the plane of the pendulum will not appear to change. In between, the period of rotation of swinging pendulums will vary from 24 hours at the poles to infinity at the Equator. At 40°N, a pendulum takes 37 hours to make one rotation. See Figure 3.2.

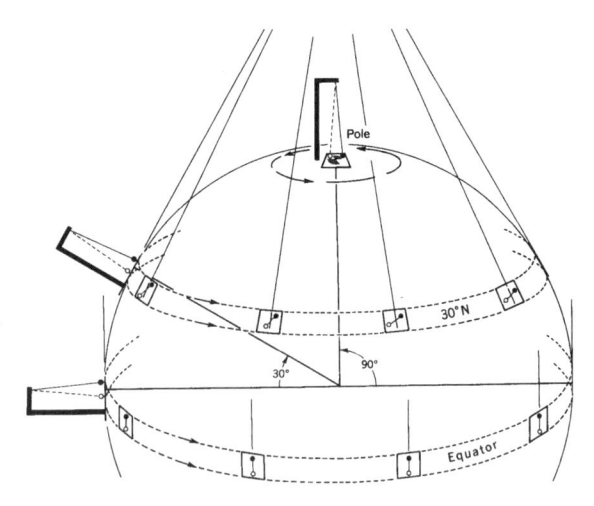

Figure 3.2 The Foucault Pendulum.

In 1851, Foucault suspended a freely swinging pendulum from the inside of the dome of the Pantheon building in Paris. The high dome allowed the use of a very long (60-meter) pendulum. Foucault fastened the pendulum's heavy weight to one side of the room with a thin cord and sealed all entrances to eliminate possible drafts that could exert a lateral force on the pendulum. He then burned the cord to set the pendulum swinging smoothly in a fixed plane. Through windows, observers were able to watch the pendulum rotate slowly in a direction opposite to that of Earth's rotation, thus proving that Earth rotates.

The Coriolis Effect

Gaspard Coriolis first described the behavior of objects moving in a rotating frame of reference. His work successfully predicted the behavior of objects moving near the rotating Earth. In a stationary system, fluids such as the atmosphere or the oceans move directly in a straight line from regions of high pressure to regions of low pressure. However, that behavior does not occur on Earth. Large-scale movements of both the atmosphere and the oceans follow curved paths.

Study the following diagrams in the *Earth Science Reference Tables* (Appendix B): Surface Ocean Currents (page 688) and Planetary Wind and Moisture Belts in the Troposphere (page 698). Note that ocean currents form large circular patterns, called *gyres*, that flow clockwise in the Northern Hemisphere and counterclockwise in the Southern Hemisphere. Note, too, that planetary

Heliocentric Earth Motions and Their Effects

winds follow paths that curve to the right in the Northern Hemisphere and to the left in the Southern Hemisphere. Such motions should not occur if Earth is stationary, but are precisely what would be expected if Earth is rotating. For a full description of why this occurs, refer to the section on the Coriolis effect in Chapter 22. Since the curving paths of ocean currents and weather systems would not occur on a stationary Earth, their existence is considered proof of Earth's rotation.

Revolution

As Earth rotates, it revolves around the Sun once every 365¼ days. *Revolution* is the motion of one body around another body in a path called an *orbit*. The revolving body is termed a *satellite* of the body it orbits. The body that a satellite orbits is its *primary*. Thus, Earth is a satellite of the Sun, its primary. The plane of Earth's orbit is called the *ecliptic* because eclipses occur when Earth, the Moon and the Sun align in this plane.

Evidence of Revolution

Parallelism of Earth's Axis of Rotation
As stated previously, Earth's axis of rotation is tilted 23½° from a perpendicular to the plane of its orbit. Spinning like a top, Earth holds its axis fixed in space as it moves around the Sun. Another way to describe this is to say that Earth's axis of rotation at any point in Earth's orbit is parallel to its axis at any other point. As a result, the Northern Hemisphere is tipped toward the Sun in June and away from the Sun in December.

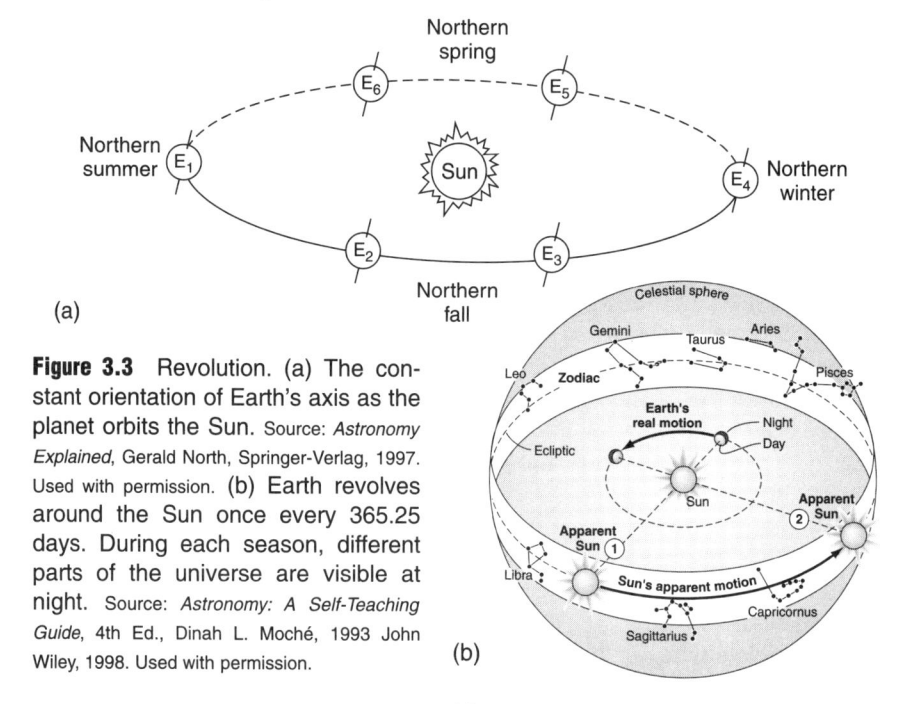

Figure 3.3 Revolution. (a) The constant orientation of Earth's axis as the planet orbits the Sun. Source: *Astronomy Explained*, Gerald North, Springer-Verlag, 1997. Used with permission. (b) Earth revolves around the Sun once every 365.25 days. During each season, different parts of the universe are visible at night. Source: *Astronomy: A Self-Teaching Guide*, 4th Ed., Dinah L. Moché, 1993 John Wiley, 1998. Used with permission.

Annual Traverse of the Constellations

As Earth revolves around the Sun, the side of Earth facing the Sun experiences day and the side facing away experiences night. Since the stars are visible only at night, the portion of the universe whose stars are visible to an observer on Earth varies cyclically as Earth revolves around the Sun. See Figure 3.3.

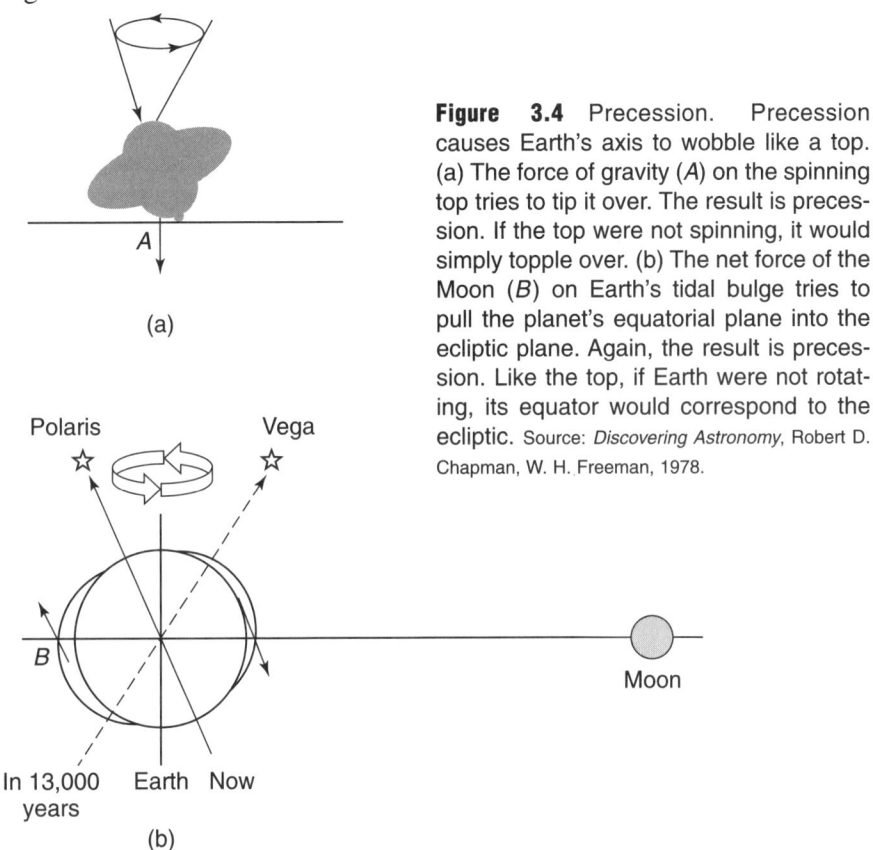

Figure 3.4 Precession. Precession causes Earth's axis to wobble like a top. (a) The force of gravity (*A*) on the spinning top tries to tip it over. The result is precession. If the top were not spinning, it would simply topple over. (b) The net force of the Moon (*B*) on Earth's tidal bulge tries to pull the planet's equatorial plane into the ecliptic plane. Again, the result is precession. Like the top, if Earth were not rotating, its equator would correspond to the ecliptic. Source: *Discovering Astronomy*, Robert D. Chapman, W. H. Freeman, 1978.

Precession

Precession is the very slow change in the direction of Earth's axis of rotation. Earth's axis sweeps around in a cone-shaped path like the wobbling of a spinning top. Almost 26,000 years are required for each sweep, causing Earth's North Pole to move slowly with respect to Polaris. Precession is caused by the gravitational pulls of the Sun and Moon. See Figure 3.4.

Since Earth's precession takes place over such a long period of time, it has little effect during a human lifetime. During your lifetime, Earth's axis will point slightly closer to Polaris. It will point closest to Polaris around the year A.D. 2100 and will then begin to move away from Polaris. In 13,000 years, Earth's axis will point at the star Vega instead of Polaris.

EFFECTS OF EARTH'S MOTIONS

The value of any model lies in its ability to explain past and current observations and to predict future observations. A revolving and rotating planet with its spin axis tilted at 23½° to a line perpendicular to its orbital plane fits all terrestrial and celestial observations.

The Sun's Path

The Sun's apparent path through the sky from sunrise to sunset is an arc, like the paths of all other celestial objects, and is the result of Earth's rotating on its axis. As Earth (and any observer on Earth's surface) rotates from west to east, the Sun appears to move from east to west. See Figure 3.5.

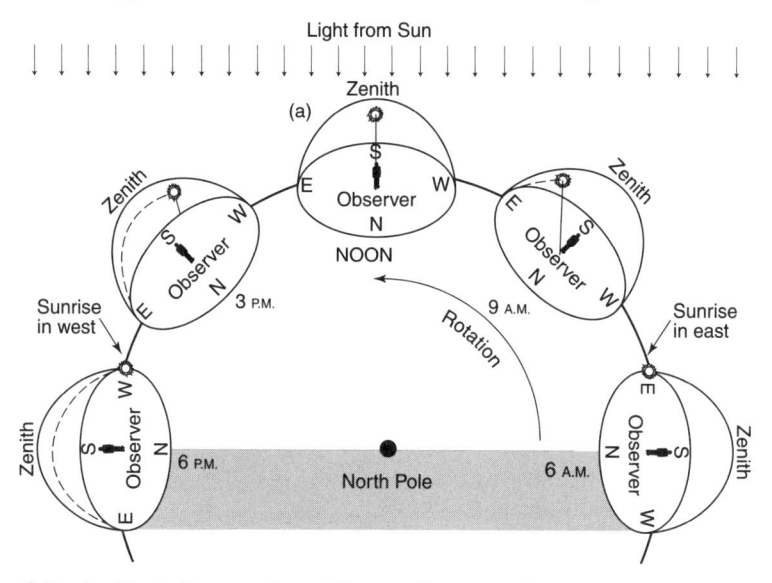

Figure 3.5 As Earth Rotates from West to East, the Sun Appears to Move from East to West.

At 6 A.M., an observer at sunrise would see the Sun to the east and low in the sky, at the horizon. At 9 A.M., Earth has rotated 45° from west to east and an observer would see the Sun to the southeast and higher up, above the horizon. At 12 noon, Earth has rotated another 45° from west to east and an observer would see the Sun to the south and at its highest point above the horizon, its noon position. At 3 P.M., Earth has rotated a further 45° and an observer would see the Sun to the southwest and lower above the horizon than at noon. At 6 P.M., Earth has rotated another 45° and an observer would see the Sun set to the west, low in the sky and at the horizon.

Note that, as Earth rotated through 180° from west to east, an observer saw the Sun move from east to southeast to south to southwest to west. The Sun's path through the sky began at sunrise at the eastern horizon at 6 A.M., arced upward, reaching a high point at noon, and then arced downward, ending at sunset at the western horizon at 6 P.M. What the observer saw as Earth rotated is summarized by path *B* in Figure 3.6.

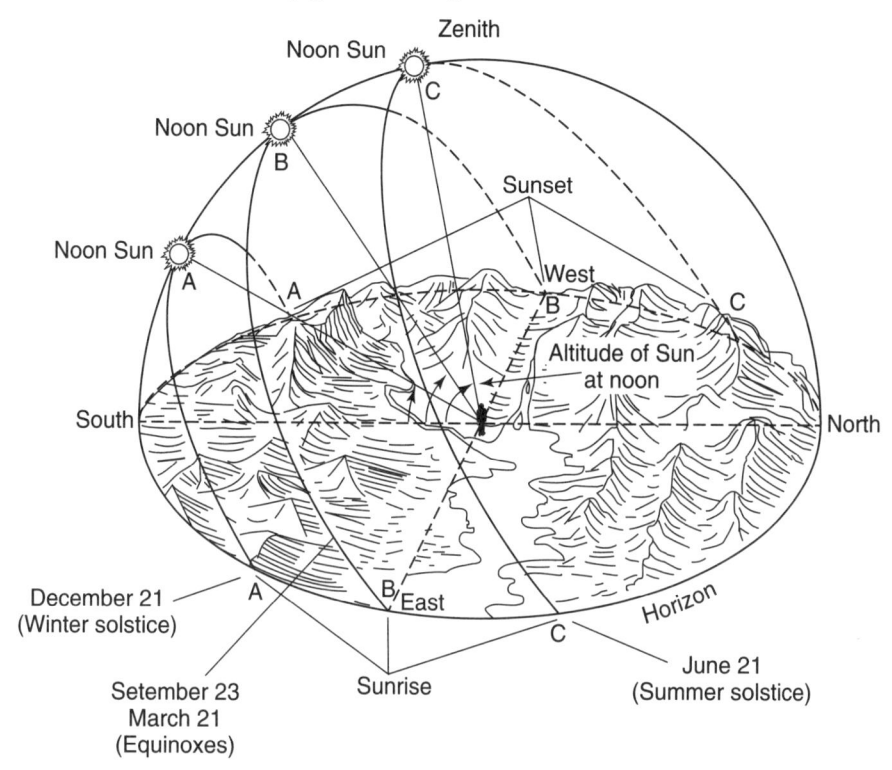

Figure 3.6 The Sun's Apparent Path. An observer at 42° N latitude would see the Sun move along different paths across the sky on different dates. Source: *Earth Science: A Study of a Changing Planet*, Daley, Higham, and Matthias, Prentice-Hall, 1986, and *The Story of Maps*, Lloyd Brown, 1977.

Changes in the Sun's Path

The length and the position of the Sun's path vary with the seasons and latitude. The length of the Sun's path determines the length of the daylight period; the longer the path, the more hours of daylight. The position of the Sun's path determines the angle of insolation and thus the intensity of the insolation. It also determines the altitude of the Sun at noon. See Figure 3.6.

Changes with the Seasons

On different days of the year, an observer will see the Sun follow different paths through the sky. This is so because the Sun's position relative to an observer changes as Earth orbits the Sun. The points of sunrise and sunset vary, as does the altitude of the Sun at noon. See Figure 3.7.

In December, the position of sunrise to an observer in the United States is

Heliocentric Earth Motions and Their Effects

south of east and the position of sunset is south of west. The length of the Sun's path is shortest, so the daylight period is shortest. Because of Earth's tilted axis of rotation, the noon Sun is at its lowest altitude of the year. The combination of a short daylight period and a low angle of insolation causes low temperatures.

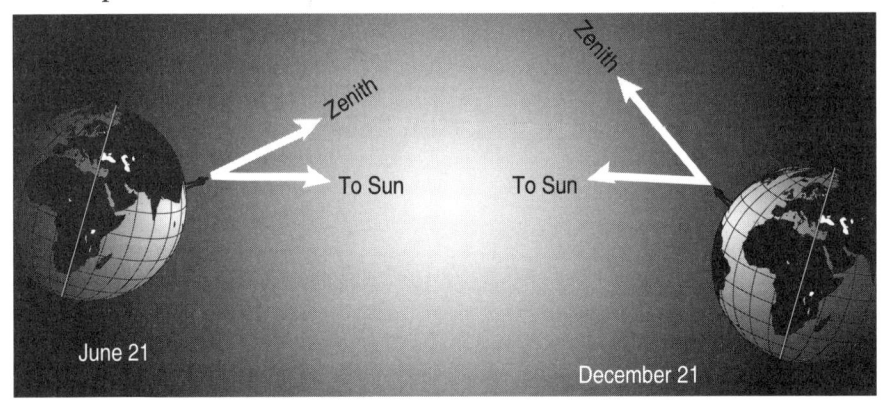

Figure 3.7 Altitude of the Sun at Noon on Two Different Dates. Note that on June 21 the Sun is seen closer to the zenith; that is, it is closer to being directly overhead.

In June, the Sun rises north of east, sets north of west, and rises to a higher altitude. The Sun's path is at its longest, so the daylight period is longest. The high altitude of the Sun results in a more direct angle of insolation. The combination of a long daylight period and a more direct angle of insolation results in high temperatures.

At the equinoxes in March and September, the Sun rises due east and sets due west. The length of the Sun's path results in exactly equal periods of daylight and darkness. The Sun's altitude at noon is halfway between its highest and lowest points. The length of day and the angle of insolation are intermediate between the two extremes; therefore, the temperatures are moderate.

From December to June, the Sun's path gets longer each day and the altitude of the Sun at noon increases. On June 21, the altitude of the Sun at noon stops increasing. Therefore, this date is called the *summer solstice*, meaning summer "Sun stop." From June to December, the Sun's path gets shorter each day and the altitude of the Sun at noon decreases. On December 21, the altitude of the Sun at noon stops decreasing. Therefore, this date is called the *winter solstice*, meaning winter "Sun stop." March 21 and September 21, when the daylight and darkness periods are equal, are called, respectively, the *spring* and the *fall equinox*, meaning spring and fall "equal night." This cyclic pattern of change repeats in an annual cycle.

Changes with Latitude

Now consider what two observers at different latitudes will see on a given day as Earth rotates. As you can see in Figure 3.8, the observer near the Equator will see the Sun at a higher altitude at noon than the observer in New

York State on the same day.

The altitude of the noon Sun at any location can be determined quite easily if you know the latitude of that location and the latitude at which the Sun is directly overhead on that day. To find the altitude of the Sun at noon, find the difference between the latitude of the location and the latitude at which the Sun is directly overhead. Then subtract this number of degrees from 90°.

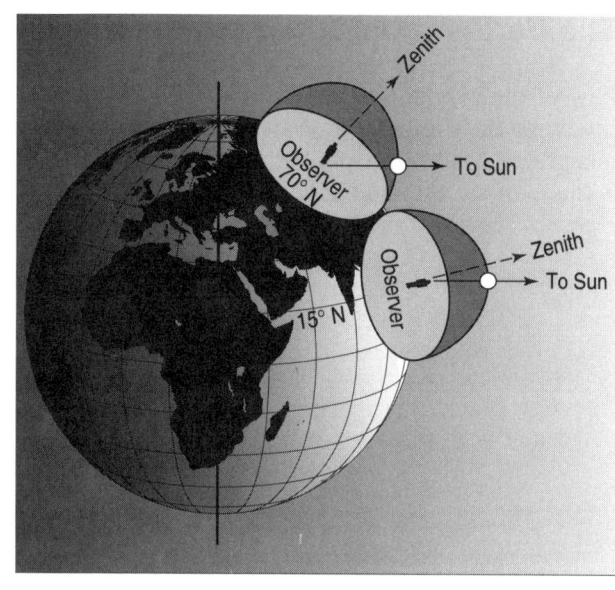

Figure 3.8 Altitude of the Sun at Different Latitudes. Observers at different latitudes will see the noon Sun at different altitudes on the same date.

Example:
On March 21, the Sun is directly overhead at the Equator, 0° latitude. On that day, what is the altitude of the Sun at noon in New York City?

Latitude of NYC − latitude where Sun is directly overhead = difference in latitude

41° N − 0° = 41°

90° − difference in latitude = altitude of Sun at noon in NYC

90° − 41° = 49°

The altitude of the Sun in New York City on March 21 is 49°.

Since the farthest Earth tips toward the Sun is 23½°, the Sun is never directly overhead north of 23½° N (Tropic of Cancer) or south of 23½° S (Tropic of Capricorn). There is always a difference between the latitude of a location in the continental United States and the latitude at which the Sun is directly overhead, and therefore the altitude of the Sun at noon is always less than 90° in the United States.

Because of New York's latitude, an observer in that State sees the noon Sun at its highest altitude on June 21, at about 73° above the horizon, and at its lowest altitude on December 21, at only about 23° above the horizon.

Heliocentric Earth Motions and Their Effects

The Earth-Moon-Sun System

Few of the events that can be observed in the sky are as striking as those involving the Earth-Moon-Sun system. Phases of the Moon are probably the most familiar phenomena, but others involving this system include tides and eclipses of the Sun and Moon.

As Earth revolves around the Sun, the Moon circles Earth in an elliptical orbit with a period of 27.32 days. The Moon's orbit is tilted at an angle of about 5° from the plane of Earth's orbit around the Sun. The Moon moves rapidly in its orbit, covering 13° every day. As a result, each day its position against the backdrop of stars changes by 13°, or about 26 times its apparent diameter. The Moon also rotates on its axis once every 27.32 days. Thus, the same side of the Moon always faces Earth. See Figure 3.9.

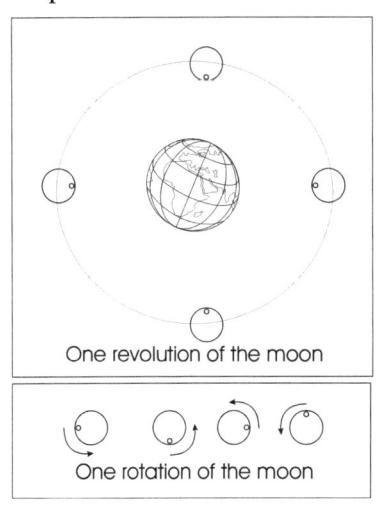

Figure 3.9 Revolution and Rotation of the Moon. The Moon completes one rotation in the same time it makes one revolution. Therefore, the same side of the Moon is always facing Earth.

Phases of the Moon

The Moon does not produce visible light of its own; the Moon is visible only because light from the Sun is reflected from its surface. An observer on Earth is able to see only that portion of the Moon that is illuminated by the Sun. As the Moon moves around Earth, different portions of the side of the Moon facing Earth are illuminated by sunlight, and the Moon passes through a cycle of phases. See Figure 3.10.

Although the Moon makes one revolution in 27.32 days, it takes 29.5 days to go through a complete cycle of phases. Why the extra two days? Let's start at new Moon, when the Moon is between Earth

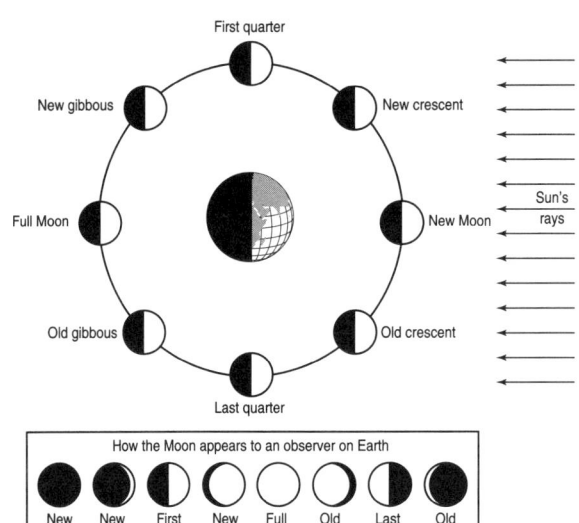

Figure 3.10 The Phases of the Moon.

50

and the Sun. At the same time that the Moon revolves around Earth, Earth is revolving around the Sun at a rate of about 1° per day. Thus, by the time the Moon has completed one revolution, the Earth's position in relation to the Sun is not the same as it was when the Moon started its revolution. In 27 days the Earth has moved 27° in its orbit. Moving at about 13° per day, the Moon takes about 2 days to catch up to Earth and align with it and the Sun in a new Moon phase. The word *month* has its origin in "Moon-th," which referred to this 29.5-day cycle of phases.

Eclipses of the Sun and Moon

A *solar eclipse* occurs when the Moon passes directly between Earth and the Sun, casting a shadow on Earth and blocking our view of the Sun. Both the Moon and the Earth are illuminated by the Sun and cast shadows in space. As seen in Figure 3.11, the shadows cast by Earth and the Moon are extremely long and narrow.

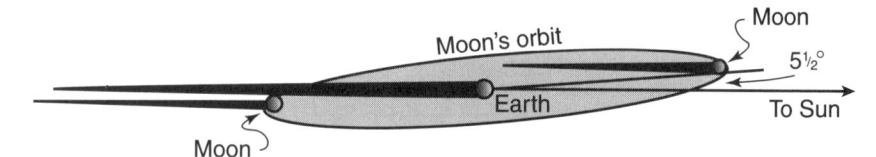

Figure 3.11 The Moon's Shadow. The 5½° tilt of the Moon's orbit and the small size of the Moon's shadow makes it easy for the shadow to miss Earth and not cause an eclipse.

Whether an observer sees a total eclipse or a partial eclipse depends on which part of the Moon's shadow passes over the observer. The Moon's shadow consists of two parts, an umbra and a penumbra. The *umbra* is a region of shadow in which all of the light has been blocked. In the *penumbra* only part of the light is blocked, so the light is dimmed but not totally absent. An observer in the Moon's umbra would see a total solar eclipse. The path of the Moon's umbra over Earth's surface is therefore called the *path of totality*. An observer in the penumbra would see a partial solar eclipse. An observer outside the Moon's shadow would see no eclipse. See Figure 3.12.

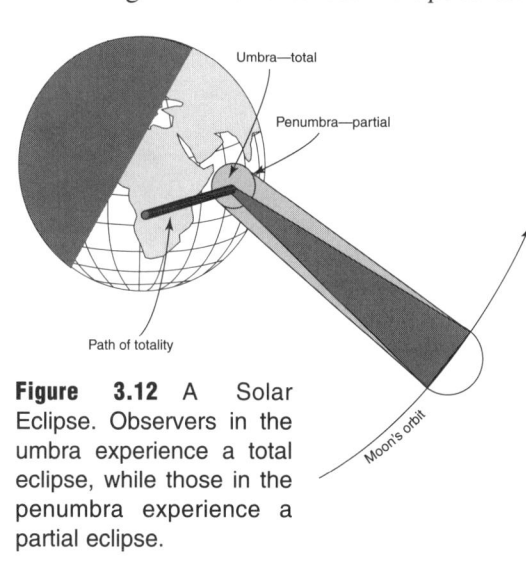

Figure 3.12 A Solar Eclipse. Observers in the umbra experience a total eclipse, while those in the penumbra experience a partial eclipse.

The umbra (i.e., the dark, inner part) of the Moon's shadow barely reaches Earth, and the small, circular shadow that the Moon casts

Heliocentric Earth Motions and Their Effects

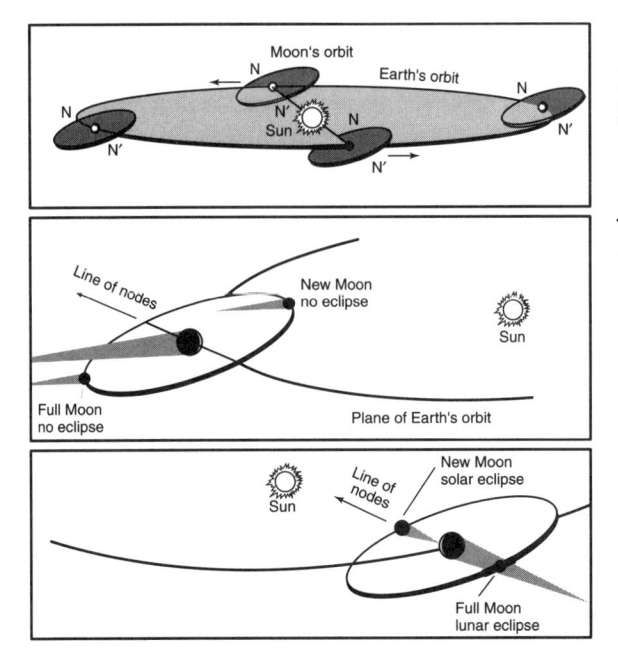

Figure 3.13 Eclipses of the Sun and Moon. Solar and lunar eclipses can occur when the Moon intersects the ecliptic in new-moon or full-moon position. If the Moon is above or below the ecliptic during these phases, no eclipse occurs. Source: *Horizons: Exploring the Universe*, by Michael A. Seeds, Wadsworth Publishing, 1987.

on Earth's surface is never more than 269 kilometers in diameter. The 5° tilt of the Moon's orbit, together with the small size of the shadow that can reach Earth, makes it easy for the shadow to miss Earth at full and new Moon. Thus, total eclipses of the Sun are rare. For an eclipse to occur, the plane of the Moon's orbit must intersect the shadows being cast by both Earth and the Moon. See Figure 3.13.

Since the Moon's orbit is elliptical, the Moon's distance from Earth varies. If the Moon lines up directly with the Sun at a point in its orbit when the Moon is farthest away from Earth, the umbra of the Moon's shadow does not reach Earth's surface. If the umbra does not reach the surface, there is no total solar eclipse. However, a type of partial eclipse called an *annular eclipse* occurs under these conditions. In an annular eclipse the Moon is too far from Earth to completely block our view of the Sun. During the eclipse a narrow ring, or annulus, of light is visible around the edges of the Moon. See Figure 3.14.

The eclipse viewed from space

The eclipse viewed from Earth

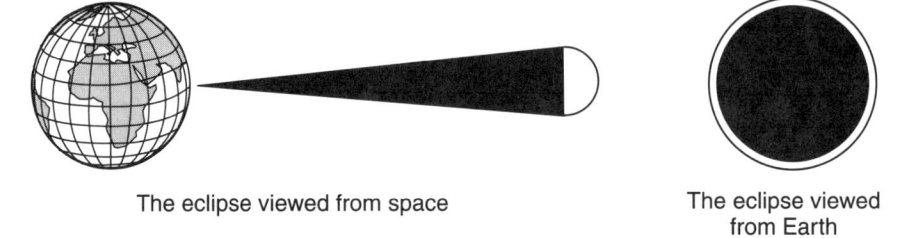

Figure 3.14 An Annular Eclipse and the Way It Appears When Viewed from Space and from Earth.

A *lunar eclipse* occurs when the full Moon moves through Earth's shadow. If the Moon moves into Earth's umbra, a total lunar eclipse is seen. If the

Moon moves into Earth's penumbra, a partial lunar eclipse is seen. When the Moon is totally in Earth's umbra, it does not completely disappear from view. While no direct sunlight reaches the Moon, some light that is bent as it passes through Earth's atmosphere reaches the Moon. Since only the long waves of red light are bent far enough to reach the Moon, the Moon glows with a dull red color during a total lunar eclipse. During a partial lunar eclipse, however, the Moon is only partially dimmed as some of the Sun's light is blocked by Earth. Partial lunar eclipses are not very impressive. See Figure 3.15.

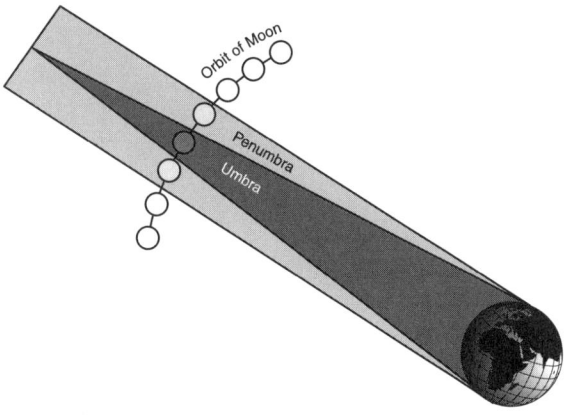

Figure 3.15 A Lunar Eclipse. If the moon passes only through the Earth's penumbra, a partial eclipse of the Moon is observed. If the Moon passes through Earth's umbra, a total eclipse of the Moon is observed.

Tides

Every day you feel the mutual attraction of gravity between Earth and your body pulling you downward with a force called your *weight*. But Earth's gravity is not the only gravity acting on you. While the Moon is farther away from you than Earth's center and has less mass than Earth, it still exerts a measurable force on you and everything else on Earth.

The part of Earth's surface that faces the Moon is about 6,000 kilometers closer to the Moon than is Earth's center. Therefore, the force of gravity the Moon exerts on this surface is stronger than the gravity exerted on Earth's center. Although Earth's surface is solid, it is not absolutely rigid. The Moon's gravity causes Earth's surface to flex outward, forming a bulge several inches high. As the Moon moves around Earth, the bulge moves across the surface as it remains beneath the Moon.

An inches-high bulge in the bedrock spread over half of Earth's surface is barely noticeable. Water is a fluid, however, and can move much more readily than rock in response to the Moon's gravity. Water in the oceans attracted by the Moon's gravity flows into a bulge of water on the side of Earth facing the Moon. A bulge of water also forms on Earth's far side. The Moon pulls on Earth's center more strongly than on Earth's far side, thus attracting Earth away from the oceans on the far side, and the water flows into this space, creating a bulge. The water from the area between these bulges flows into them, creating a deep region and a shallow region in the ocean waters.

As Earth rotates on its axis, the positions of the tidal bulges remain fixed

Heliocentric Earth Motions and Their Effects

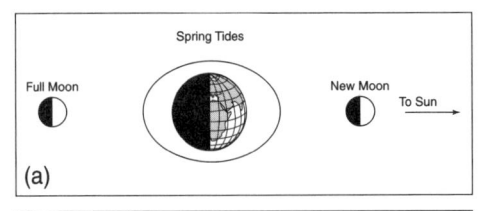

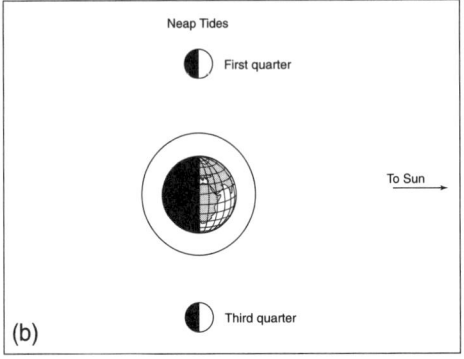

Figure 3.16 Spring and Neap Tides. (a) When the Sun and Moon pull in the same direction, their tidal forces combine and tidal bulges on Earth are larger. Spring tides occur at new Moon or full Moon. (b) When the Sun and Moon pull at right angles, their tidal forces do not combine and tidal bulges are much smaller. Neap tides occur at first- and third-quarter Moon.

in line with the Moon. As the rotating Earth carries a location into a tidal bulge, the water deepens and the tide rises on the beach. As the location rotates out of the tidal bulge, the water becomes shallower and the tide falls. Since there are two bulges on opposite sides of Earth, the tide rises and falls twice a day.

The Sun also produces tidal bulges in Earth's surface and oceans. At new Moon and full Moon, the Sun's tidal bulges and the Moon's tidal bulges align with one another and combine. The result is very high and very low tides, called *spring tides* because they "spring so high," not because they happen during the spring season. See Figure 3.16a. Spring tides occur at every new and full Moon whatever the season. During the first- and third-quarter phases of the Moon, the tidal bulges of the Sun and Moon are at right angles to one another and very nearly cancel each other. The result is *neap tides*, in which there is very little difference between high and low tides. See Figure 3.16b.

MULTIPLE-CHOICE QUESTIONS

In each case, write the number of the word or expression that best answers the question or completes the statement.

1. Some constellations (star patterns) observed in the summer skies in New York State are different from those observed in the winter skies. The best explanation for this observation is that
   (1) Earth revolves around the Sun
   (2) Earth rotates on its axis
   (3) constellations are moving away from Earth
   (4) constellations revolve around Earth

2. Earth's axis of rotation is tilted 23½° from a line perpendicular to the plane of its orbit. What would be the result if the tilt was increased to 33½°?
   (1) an increase in the amount of solar radiation received by Earth
   (2) colder winters and warmer summers in New York State
   (3) less difference between winter and summer temperatures in New York State
   (4) shorter days and longer nights at the Equator

3. On March 21, two observers, one at 45° north latitude and the other at 45° south latitude, watch the "rising" Sun. In which direction(s) must they look?
   (1) Both observers must look westward.
   (2) Both observers must look eastward.
   (3) The observer at 45° N must look westward, while the other must look eastward.
   (4) The observer at 45° S must look westward, while the other must look eastward.

4. On June 21, the altitude of Polaris is observed from New York City and found to be 41°. If the altitude is observed again on December 21, it will be
   (1) 23½°
   (2) 41°
   (3) 49°
   (4) 64½°

5. During a period of 1 year, what is the greatest altitude of the Sun at the North Pole?
   (1) 90°
   (2) 66°
   (3) 23½°
   (4) 0°

6. Which motion causes the Moon to show phases, as seen from Earth?
   (1) the rotation of the Moon on its axis
   (2) the rotation of Earth on its axis
   (3) the revolution of Earth around the Sun
   (4) the revolution of the Moon around Earth

7. The diagram to the right shows four different positions (W, X, Y, and Z) of the Moon in its orbit around Earth. In which position will the full-Moon phase be seen from Earth?
   (1) W
   (2) X
   (3) Y
   (4) Z

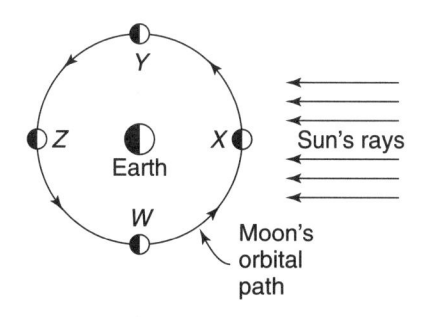

Heliocentric Earth Motions and Their Effects

8. Which diagram best represents the full-Moon phase as seen from Earth?

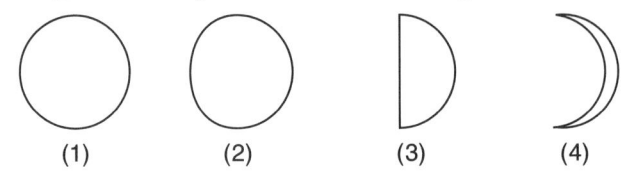

   (1)   (2)   (3)   (4)

9. The diagram below shows the relative positions of Earth, the Moon, and the Sun for a 1-month period.

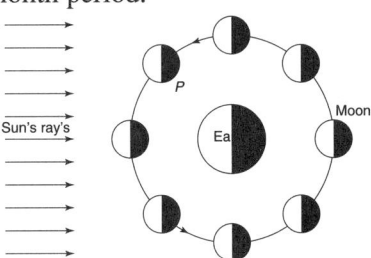

   Which diagram best represents the appearance of the Moon at position *P* when viewed from Earth?

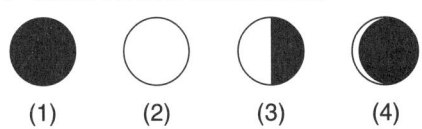

   (1)   (2)   (3)   (4)

10. The diagram to the right shows the relative positions of the Sun, Earth, and the Moon in space. Letters *A*, *B*, *C*, and *D* represent locations on Earth's surface.

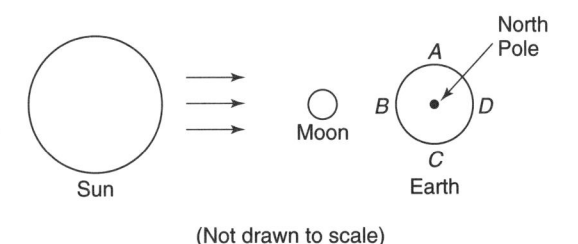

    (Not drawn to scale)

    At which location would an observer on Earth have the best chance of seeing a total solar eclipse?
    (1) *A*   (3) *C*
    (2) *B*   (4) *D*

11. A total lunar eclipse will occur when the Moon moves into the
    (1) umbra of Earth   (3) penumbra of Earth
    (2) umbra of the Moon   (4) penumbra of the Moon

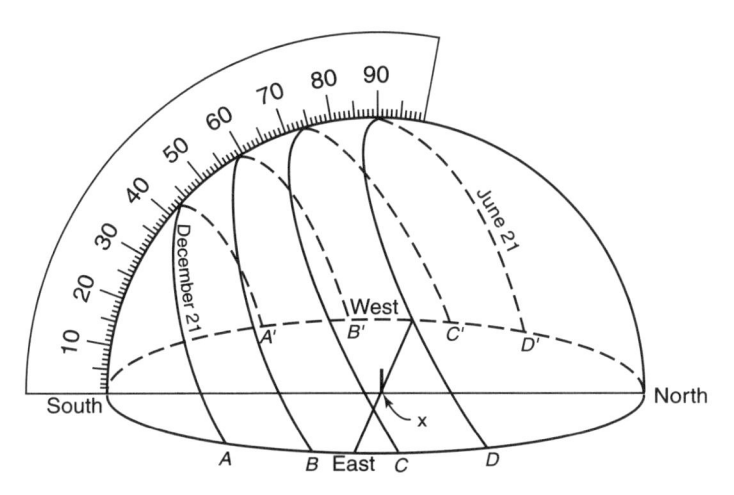

Base your answers to questions 12 through 16 on your knowledge of earth science and the above diagram. The diagram represents a plastic hemisphere upon which lines have been drawn to show the apparent paths of the Sun on 4 days at one location in the Northern Hemisphere. Two of the paths are dated. The protractor is placed over the north-south line. X represents the position of a vertical post.

12. For which path is the altitude of the noon Sun 74°?
    (1) A-A'   (3) C-C'
    (2) B-B'   (4) D-D'

13. How many degrees does the altitude of the Sun change from December 21 to June 21?
    (1) 43°   (3) 66½°
    (2) 47°   (4) 74°

14. Which path of the Sun will result in the longest shadow of the vertical post at solar noon?
    (1) A-A'   (3) C-C'
    (2) B-B'   (4) D-D'

15. Which statement best explains the apparent daily motion of the Sun?
    (1) Earth's orbit is an ellipse.
    (2) Earth's shape is an oblate spheroid.
    (3) Earth is closest to the Sun in winter.
    (4) Earth rotates on its axis.

16. What is the latitude of this location?
    (1) 0°   (3) 66½° N
    (2) 23½° N   (4) 90° N

Heliocentric Earth Motions and Their Effects

Base your answers to questions 17 through 21 on the diagram below and your knowledge of earth science. The diagram represents Earth at a specific position in its orbit. Arrows indicate radiation from the Sun. Points *A* through *D* are locations on Earth's surface.

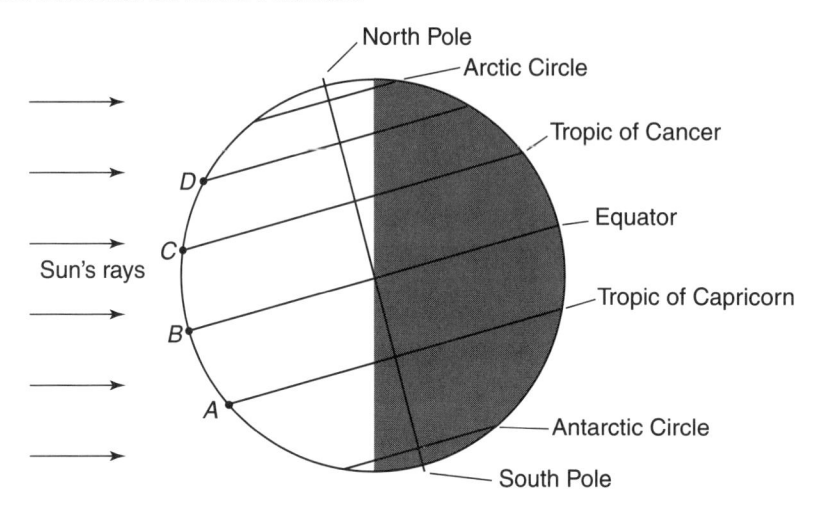

17. During which month could Earth be in the position shown in the diagram?
(1) February
(2) May
(3) October
(4) December

18. Which location will have the greatest number of daylight hours when Earth is in this position?
(1) *A*
(2) *B*
(3) *C*
(4) *D*

19. Locations *A*, *B*, *C*, and *D* all have the same
(1) latitude
(2) longitude
(3) intensity of insolation
(4) noontime altitude of the Sun

20. As an observer travels from position *B* to position *D*, the altitude of Polaris in the nighttime sky will
(1) decrease, only
(2) increase, only
(3) increase, then decrease
(4) remain the same

21. Which diagram below best represents the path of the Sun on this date as seen by an observer at location C?

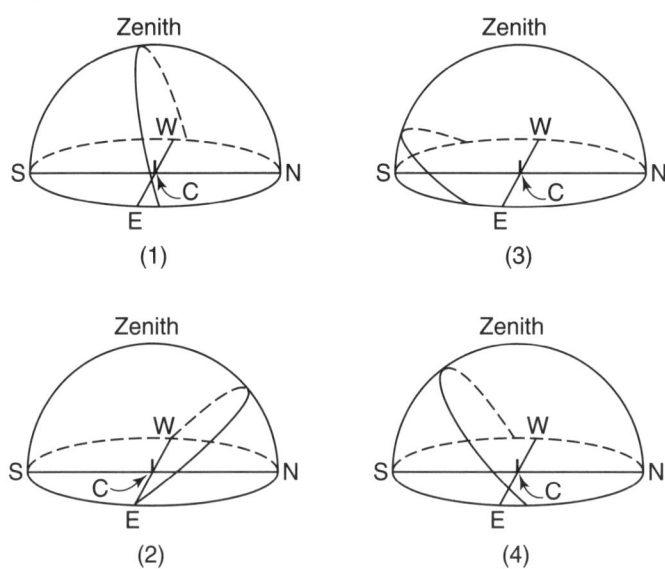

(1) Zenith, W, S, N, E, C

(3) Zenith, W, S, N, E, C

(2) Zenith, W, S, N, E, C

(4) Zenith, W, S, N, E, C

Base your answers to questions 22 through 26 on the *Earth Science Reference Tables*, your knowledge of earth science, and the diagram below, which illustrates an imaginary solar system involving two planets, Planet X and Planet Z.

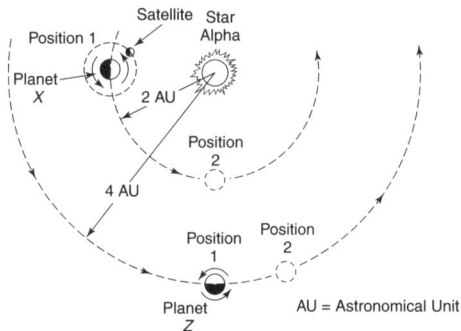

22. Assume that Planet X is being orbited by a small natural satellite that is visible from Planet Z. To an observer on Planet Z, this moon will
    (1) always be fully illuminated
    (2) always be in complete darkness
    (3) be illuminated on one side, only
    (4) be illuminated to varying degrees from nearly full to dark

23. An observer on Planet Z at midnight will
    (1) see only Planet X, not its satellite
    (2) see only Planet X's satellite, not Planet X
    (3) see only Star Alpha
    (4) not be able to see Star Alpha, Planet X, or its satellite

Note that questions 24–25 have only three choices.

24. As the planets move to their second positions, the angular diameter of Planet Z as observed from Planet X will
(1) decrease
(2) increase
(3) remain the same

25. As the planets move to their second positions, the gravitational attraction between Planet X and Planet Z will
(1) decrease
(2) increase
(3) remain the same

26. If the masses of Planet X and Planet Z are equal, the gravitational attraction between Star Alpha and Planet Z, as compared to the gravitational attraction between Star Alpha and Planet X, is
(1) ½ as great
(2) 2 times as great
(3) ¹⁄₁₆ as great
(4) ¼ as great

Base your answers to questions 27 and 28 on the diagram below, which shows four planets orbiting the Sun.

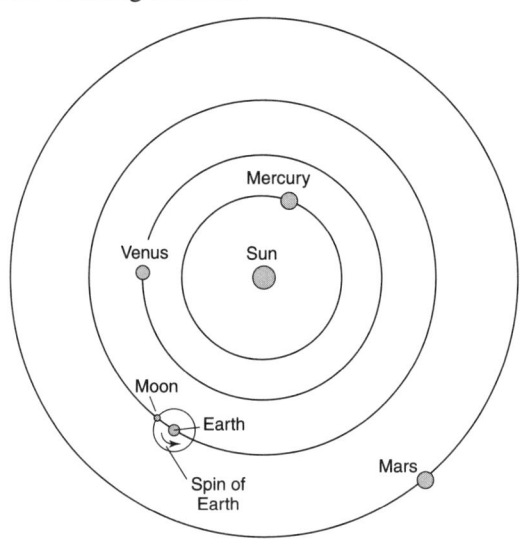

(Not drawn to scale)

27. Which object will be visible from Earth in the eastern sky after midnight?
(1) the Moon   (3) Mercury
(2) Mars   (4) Venus

28. Which diagram best represents how Venus and the Moon would appear in the sky at sunset to an observer in New York State? [Diagrams are not drawn to scale.]

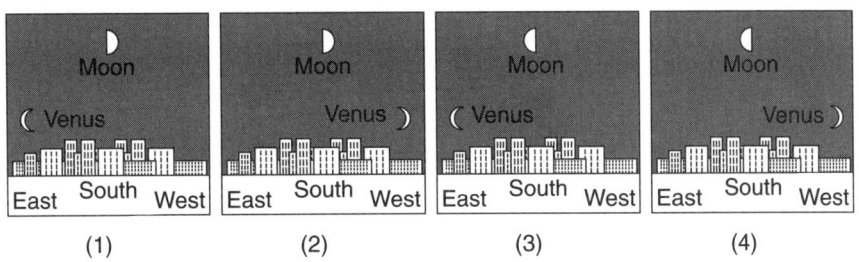

29. The apparent change in direction of a Foucault pendulum is caused by
    (1) star motions
    (2) Earth's rotation
    (3) the Moon's gravitational attraction
    (4) density differences within Earth's mantle

30. Major ocean and air currents appear to curve to the right in the Northern Hemisphere because
    (1) Earth has seasons
    (2) Earth's axis is tilted
    (3) Earth rotates on its axis
    (4) Earth revolves around the Sun

Base your answers to questions 31 and 32 on the diagrams below, which represent two views of a swinging Foucault pendulum with a ring of 12 pegs at its base.

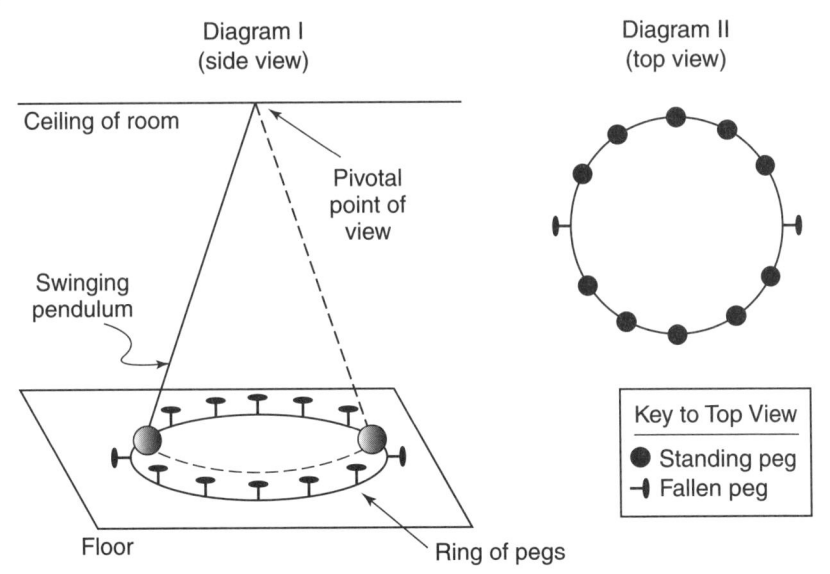

Heliocentric Earth Motions and Their Effects

31. Diagram II shows two pegs tipped over by the swinging pendulum at the beginning of the demonstration. Which diagram below shows the pattern of standing pegs and fallen pegs after several hours?

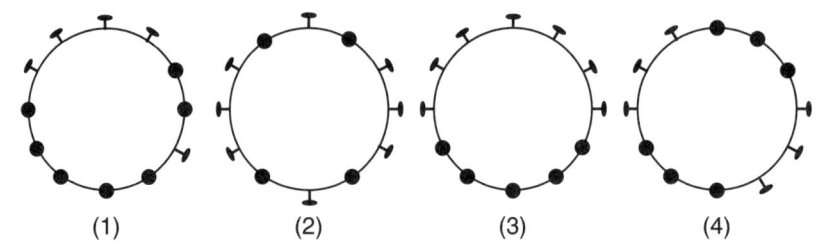

(1)   (2)   (3)   (4)

32. The predictable change in the direction of swing of a Foucault pendulum provides evidence that
    (1) the Sun rotates on its axis
    (2) the Sun revolves around Earth
    (3) Earth rotates on its axis
    (4) Earth revolves around the Sun

Base your answers to questions 33 through 39 on the diagram below, which shows Earth in its relation to the Sun on June 21.

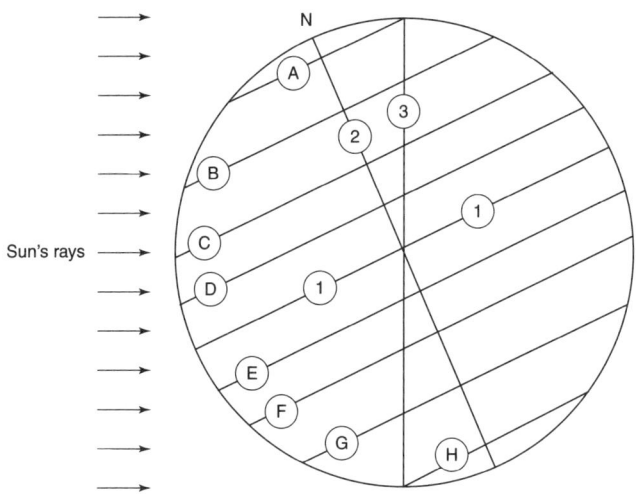

33. The line represented by 1 is the
    (1) Equator
    (2) Tropic of Cancer
    (3) Tropic of Capricorn
    (4) international date line

34. The Arctic Circle is represented line
    (1) H
    (2) E
    (3) B
    (4) A

35. On this date at the North Pole, at how many degrees above the horizon will an observer see the Sun?
(1) 23½ (3) 90
(2) 66 (4) 40

36. On this date, the season beginning in the Southern Hemisphere is
(1) summer (3) spring
(2) winter (4) fall

37. An observer stands on the Equator at noon on this date. His shadow will point
(1) north (3) east
(2) south (4) west

38. The Tropic of Cancer is represented by line
(1) *E* (3) *A*
(2) *G* (4) *C*

39. On this date the vertical rays of the Sun will strike Earth along line
(1) *C* (3) *G*
(2) *E* (4) *H*

CONSTRUCTED RESPONSE QUESTIONS

A total solar eclipse was visible to observers in the southeastern United States on February 26, 1998. The diagram below shows the Sun and Earth as they were viewed from space on that date.

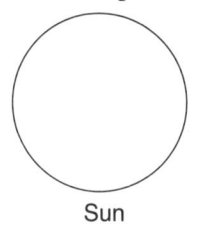

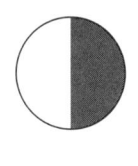

Sun   Earth

(Not drawn to scale)

40. On the diagram, draw the Moon (O), showing its position at the time of the solar eclipse. [1]

Heliocentric Earth Motions and Their Effects

Base your answers to questions 41 through 43 on the diagrams below. Diagram I represents the Moon orbiting Earth as viewed from space above the North Pole. The Moon is shown at eight different positions in its orbit. Diagram II represents phases of the Moon as seen from Earth when the Moon is at position 2 and position 4.

Diagram I

(Not drawn to scale)

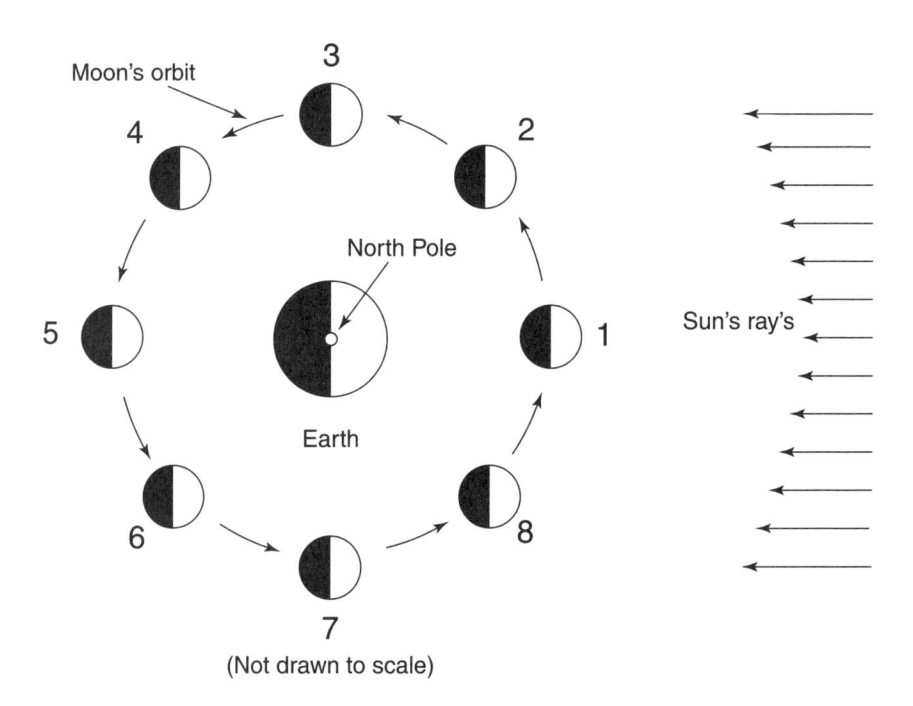

Diagram II

Phase of the Moon as seen from Earth

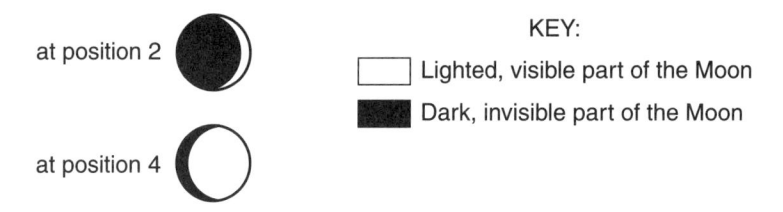

at position 2

at position 4

KEY:
☐ Lighted, visible part of the Moon
■ Dark, invisible part of the Moon

41. Shade the circle provided below to illustrate the Moon's phase as seen from Earth when the Moon is at position 7. [1]

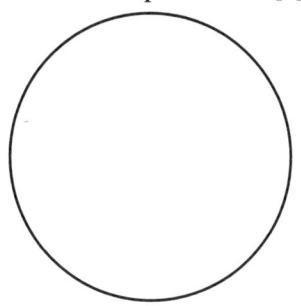

○

42. State the two positions of the Moon at which an eclipse could occur. [1]

43. State the approximate length of time required for one complete revolution of the Moon around Earth. [1]

EXTENDED CONSTRUCTED RESPONSE QUESTIONS

44. Make a diagram showing the positions of the Sun, the Moon, and Earth at a spring tide. Label the Sun, the Moon, and Earth, and show the positions of the high and low tides on Earth's surface. [3]

45. Explain why a lunar eclipse lasts for hours, while a solar eclipse lasts for only minutes. [2]

46. Explain why some different constellations appear in the sky each season. [2]

47. State three noticeable effects of Earth's motions. [3]

48. When an observer on Earth is recording a new Moon, what phase will Earth appear to have to an observer on the Moon? [1]

CHAPTER 4

EARTH'S COORDINATE SYSTEM AND MAPPING

KEY IDEAS Very early on, a variety of evidence led astronomers to conclude that Earth is a sphere. However, if only the shape of Earth is known, celestial observation can be used only to determine relative position—for example, 60° south of the North Pole. But how far is that from other locations: 1,000 kilometers? 10,000 kilometers? To determine actual distances between places on Earth's surface, the planet's size must also be determined. Only then can true-to-scale maps be constructed.

The size and shape of Earth can be readily determined by combining Earth-based measurements with simple observations of the sky. Observations show that Earth is a very slightly oblate sphere with an equatorial diameter of 12,757 kilometers.

The ability to locate and map positions on Earth's surface is essential to a wide variety of human activities ranging from engineering to urban planning to national defense. Topographic maps represent the three-dimensional shape of Earth's surface on a two-dimensional surface. A *field* is a region of space with a measurable quantity at every point. Field maps can be drawn to represent any quantity that varies in a region of space.

KEY OBJECTIVES
Upon completion of this chapter, you will be able to:

- Describe Earth's shape and explain how it can be determined by simple observations.
- Explain how Earth's size can be calculated using Earth-based measurements and simple observations of the sky.
- Use the latitude and longitude coordinate system to locate points on Earth's surface.
- Construct field maps and calculate gradient within a field.
- Read and interpret a topographic map.

THE NATURE OF EARTH'S SHAPE AND SURFACE

In order to map Earth's surface, its size and shape must be known. As stated above, Earth's size and shape can be readily determined by combining Earth-based measurements with simple observations of the sky. Let's begin by considering what we currently know about Earth's shape, and then examine how we arrived at this knowledge.

Earth's Shape

Earth's shape is very nearly a perfect sphere. A perfect sphere has exactly the same diameter when measured in any direction. Actual measurements of Earth's dimensions deviate slightly from this ideal. The polar diameter, or diameter measured from the North Pole through the center of Earth to the South Pole, is 12,714 kilometers. The equatorial diameter, or diameter measured from a point at the Equator through the center of Earth to the Equator, is 12,756 kilometers. *Thus, Earth's spherical shape "bulges" very slightly at the Equator and is very slightly "flattened" at the poles; this shape is called an **oblate (flattened) spheroid**.*

Roundness

Earth's actual shape, however, is so close to being a perfect sphere that the eye cannot detect its oblateness. When viewed from space, Earth appears perfectly round. Any cross section of Earth looks like a perfect circle. To gain an idea of how close Earth is to being perfectly round, let's suppose that we made a scale model of Earth—a globe. If we used a scale of 1 centimeter = 1,000 kilometers, the globe would have a polar diameter of 12.714 centimeters and an equatorial diameter of 12.756 centimeters—a little bigger than a softball. The difference in diameters would be 0.042 centimeter, or less than half a millimeter! We would need a micrometer to measure this difference in diameters.

Look carefully at the circle in Figure 4.1. Its polar diameter is 4.238 centimeters, and its equatorial diameter is 4.252 centimeters. It was drawn exactly one-third the size of the globe described above so that it would fit on this page. Can you tell that it is not perfectly round?

Earth's oblateness is the result of forces produced by Earth's rotation on its axis. Just as a loose skirt will swirl outward if the wearer spins around, planet Earth "swirls" outward when it rotates. However, since Earth is much

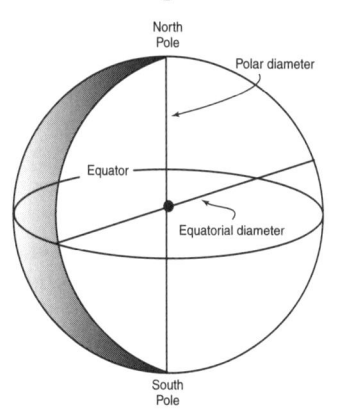

Figure 4.1 Schematic Diagram of Earth Showing Polar and Equatorial Diameters.

stiffer than a skirt, the distance it moves outward is much less.

Smoothness

In addition to being almost perfectly round, Earth is also very smooth. Compared with its diameter, the irregularities on Earth's surface (mountains and ocean basins) are relatively small. Mount Everest rises 8,848 meters, or about 8.8 kilometers, above sea level. That is about 7/10,000 of Earth's diameter (8.8 km/12,756 km). If you were to draw Mount Everest to scale on the circle in Figure 4.1, it would protrude from the surface less than 1/20 of a millimeter. Mark off a millimeter on a blank piece of paper, and then try to make dots small enough so that 20 will fit between the lines. If you put one of those dots on the outside of the circle, it would stick out of the circle as much as Mount Everest protrudes from Earth. As you can see, the dot barely affects the smoothness of the circle.

When you consider that most of the protrusions and indentations on the Earth's surface are much smaller than Mount Everest, you can see why Earth appears smooth. The only reason that Earth's mountains and valleys seem so large to us is that, compared to Earth, we are tiny. If you were to draw humans to scale on the model globe described in the preceding section, you would need a microscope to see them.

Evidence of Earth's Shape

Earth's true shape is a question that has captured the minds and imaginations of humans for thousands of years. Contrary to the popular belief that Columbus was among the first to believe that Earth is a sphere, Aristotle reached this conclusion in the third century B.C.

Ships and Eclipses

Aristotle based his conclusion upon simple observations of the sky and Earth-based measurements. This use of observation to support his ideas, rather than merely stating them as theories, places Aristotle among the earliest scientific thinkers. First, Aristotle observed that Earth's shadow during a lunar eclipse is definitely round. Although this could also happen if Earth was a cone or a cylinder, there was additional evidence supporting a spherical shape. Travelers reported that, as they went farther north or south, some stars appeared lower and lower in the sky. Furthermore, Aristotle noted that ships disappeared bow-first over the horizon, no matter in which direction they sailed. These observations could occur only if the ships were constantly moving along a curved surface that changed the angle at which they were viewed by an observer.

Today, a number of observations provide additional evidence of Earth's shape.

Photographs Taken from Space

The most direct evidence of Earth's shape consists of photographs taken from space. These photographs show that Earth is indeed spherical. When very precise photographs are taken, they can be analyzed and precise measurements of Earth's image in the photographs can be made. These measurements indicate that the Earth's polar and equatorial diameters are indeed slightly different, confirming that its shape is an oblate spheroid.

Observations of the Altitude of Polaris (the North Star)

Observations of Polaris provide us with information that can be interpreted, using simple geometry, to provide evidence of Earth's spherical shape and its oblateness. Polaris is a distant star that is located almost exactly over Earth's North Pole and is almost in line with its axis of rotation. Polaris is very far away from Earth, literally trillions of Earth-diameters distant. If it were observed from two points farthest apart on Earth's surface, the directions in which it was seen would vary by only a minute fraction of a degree. See Figure 4.2.

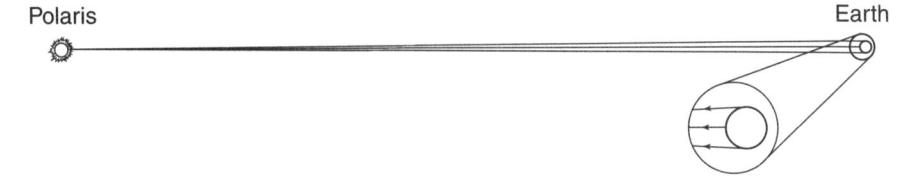

Figure 4.2 Direction to Polaris.

The direction to Polaris from all points on Earth is not measurably different.

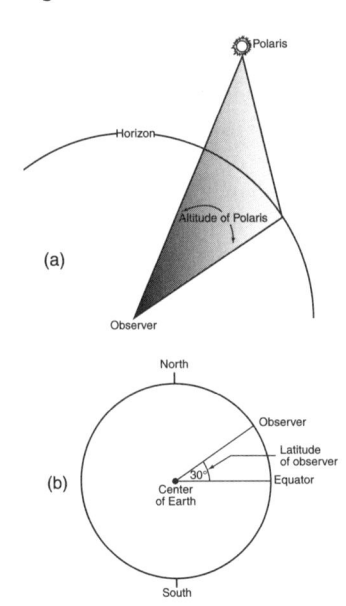

Figure 4.3 (a) Altitude of Polaris (b) Latitude of an observer

The *altitude* of Polaris is the angle between the star and the horizon with the observer at the vertex, as shown in Figure 4.3a. The latitude of an observer is the angle of the observer north or south of the Equator; see Figure 4.3b.

If Earth were a flat disk, the altitude of Polaris would be almost exactly the same at all locations and at all times. See Figure 4.4a.

It is true that, to a fixed observer, as Earth rotates, the altitude of Polaris does not appear to change, as shown in Figure 4.4b. However, if the observer travels north or south of the original location, the altitude does change, a result that would be expected if Earth is spherical. See Figure 4.4c.

If Earth were a perfect sphere, the altitude of Polaris would be the same as an observer's latitude and the distance traveled to change the altitude of Polaris by 1° would always be the same. See Figure 4.5.

Earth's Coordinate System and Mapping

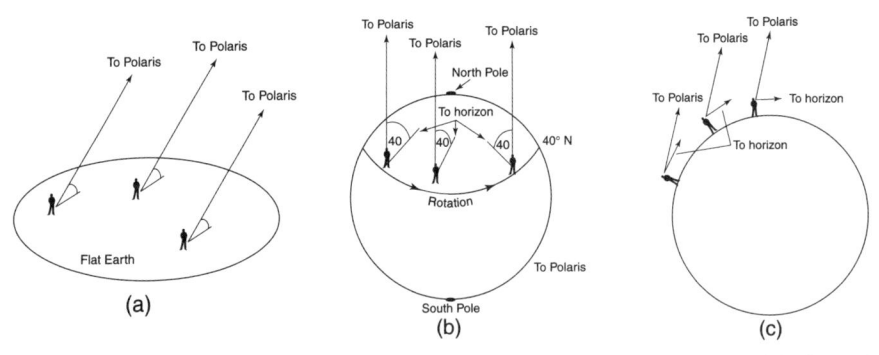

Figure 4.4 The Altitude of Polaris on a Flat and on a Spherical Earth. (a) If Earth were a flat disk, the altitude of Polaris would always be almost exactly the same at any location and at all times. (b) To a fixed observer on a spherical Earth, the altitude of Polaris does not appear to change as Earth rotates. (c) However, if the observer travels north or south, the altitude does change, a result that would be expected if Earth is spherical.

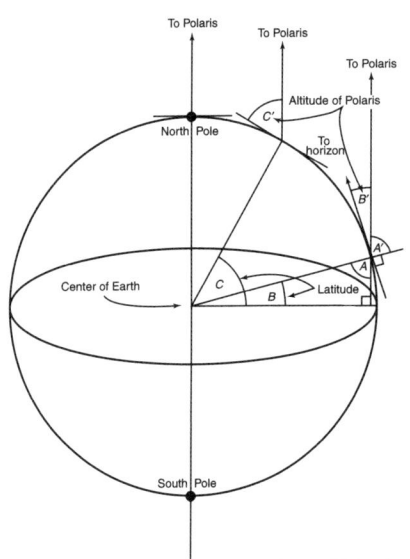

Figure 4.5 Relationship of the Altitude of Polaris to the Latitude of an Observer. By simple geometry it can be shown that the altitude of Polaris is the same as an observer's latitude. The sum of the angles of a triangle is equal to 180°. Similarly, since a straight line has a measure of 180°, <A + <B + 90° = 180°. Also, <A = <A' because they are alternate interior angles. Therefore, <B (latitude) = <B' (altitude of Polaris).

However, what is observed is that the altitude of Polaris is *not* always exactly equal to an observer's latitude, and the distance traveled to change the altitude of Polaris by 1° also varies. This is evidence that Earth is an oblate spheroid. See Figure 4.6.

Measurements of Gravity

Gravity is the force of attraction that exists between any two objects. It is proportional to the mass of the objects and inversely proportional to the square of the distance between their centers. Simply stated, the more massive the objects, and the closer their centers, the stronger the gravitational attraction between them.

The gravitational attraction between Earth and any object can be measured with a spring scale and is called the object's *weight*. If Earth were perfectly spherical, the distance between its center and any point on its surface would always be the same. Therefore, if we measured the weight of the same object at any point on the Earth's surface, the object's weight should always

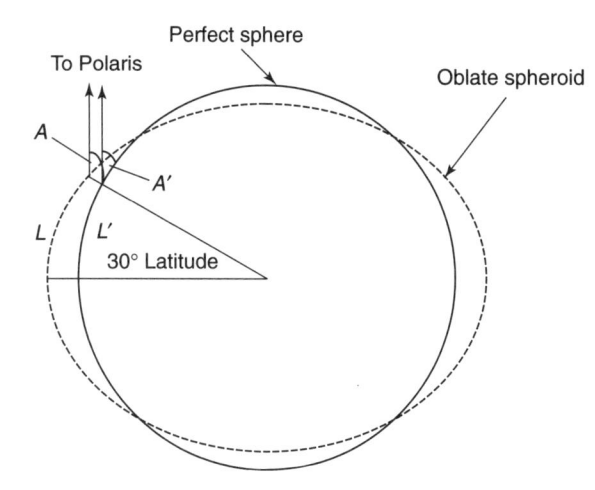

Figure 4.6 Altitude Differences at the Same Latitude on a Perfect Sphere and on an Oblate Spheroid. For the same latitude, the altitude of Polaris on a perfect sphere (A') differs slightly from what it would be on an oblate spheroid (A). Similarly, for an identical change in latitude the distance covered on a perfect sphere (L') would differ from the distance covered on an oblate spheroid (L).

be the same. If, however, Earth is an oblate spheroid, then the distance between the center of an object and the center of Earth will be greater at the Equator than at the Poles. If the distance is greater, the gravitational attraction is weaker and the object will weigh less. Thus, if we measure the weight of the same object at the Equator and at the Poles, it will weigh less at the Equator and more near the Poles. Even taking into account the outward force on an object at the Equator due to Earth's rotation, this is exactly what is observed, as shown in Figure 4.7, and is further evidence of Earth's oblateness.

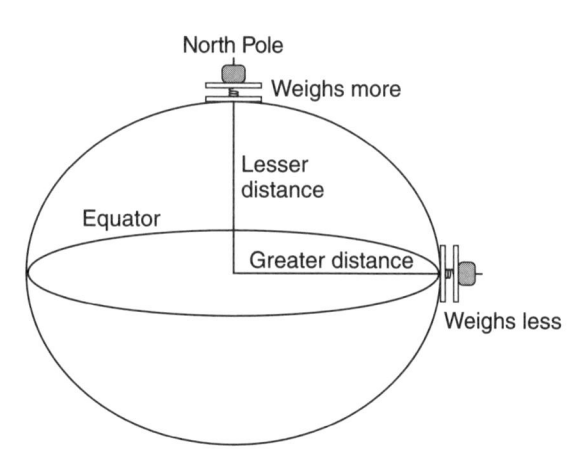

Figure 4.7 Differences in Weight of the Same Object at the Equator and near the Poles.

Determining Earth's Size

Modern measuring instruments based upon lasers allow us to measure Earth with great precision. However, Earth's size was estimated quite accurately more than 2,000 years ago. The principle is simple: if Earth is a sphere, then its circumference is a circle and all of the mathematical relationships between the various parts of a circle hold true for Earth.

Eratosthenes' Method

These mathematical relationships make it possible to determine Earth's

Earth's Coordinate System and Mapping

circumference without having to measure the entire circumference directly. If it is a circle, then it consists of 360° and a 1° angle will intersect 1/360 of Earth's circumference. Therefore, if observers at two locations on a north-south line simultaneously determine the altitude of Polaris to differ by 1°, Earth's circumference must be 360 times the distance between those locations.

In 235 B.C., Eratosthenes used this technique but substituted the Sun for a star. Eratosthenes lived in Alexandria and had reliable reports that on June 21, the summer solstice, the Sun was directly overhead at midday in the city of Syene. Eratosthenes believed that Alexandria was due north of Syene. He also knew that a camel caravan traveling 100 stadia a day typically took 50 days to get to Syene. This meant that Syene was 50 × 100 or 5,000 stadia from Alexandria. The final piece of information he needed was the angle of the Sun at midday on June 21 in Alexandria. Using a gnomon, a vertical column erected perpendicular to the horizon, Eratosthenes determined the angle to be 7 degrees 12 seconds, or about 1/50 of 360°. Hence the distance between Syene and Alexandria was about 1/50 of Earth's circumference: 5,000 stadia × 50 = 250,000 stadia. The problem is that there is some dispute as to exactly how long a stadium, or stadion, a Greek unit of length, was. The likely figure is about 10 stadia to a mile, making Eratosthenes' estimate of Earth's circumference 25,000 miles (40,230 km) compared to the actual figure of 24,862 miles (40,010 km)—a remarkable achievement for his time. See Figure 4.8.

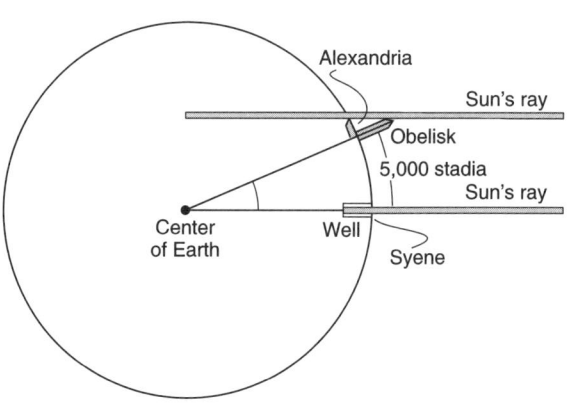

Figure 4.8 Eratosthenes's Method for Determining the Circumference of Earth.

Once Earth's circumference is known, many other dimensions of Earth can be calculated using the formulas given below, where

C = circumference, d = diameter, r = radius, A = surface area, V = volume, and π (pi) = 3.1462:

$C = \pi d$, $d = 2r$, $A = 4\pi r^2$, $V = \frac{2}{3}\pi r^3$

MAPPING EARTH'S SURFACE

Maps

A map is a model of Earth's surface. Although models are different from the real thing, they are tools for learning about the situations or objects they represent. A map is meant to communicate a sense of place, of where one point is in relation to another. A map can be anything from a quick sketch showing a friend how to get to the park from school to an elaborate scale model of Earth complete with mountain ranges and ocean basins. The nature of a map depends on the purpose for which it was created.

Coordinate Systems

A *coordinate system* is a method of locating points by labeling them with numbers called coordinates. *Coordinates* are numbers measured with respect to a system of lines or some other fixed reference. The most commonly used coordinate system is the *Cartesian* system, in which two lines on a flat surface intersect at right angles. Each line is called an *axis*, and the point where they intersect is the *origin*. The horizontal axis is usually referred to as the *x*-axis, and the vertical axis as the *y*-axis. Two coordinates describe any point on the flat surface, each defining one of the intersecting lines. For example, a fixed point can be located on a graph by describing the two lines $x = 4$ and $y = 3$. As shown in Figure 4.9a, these two lines intersect at a single point. By convention, coordinates in this system are written with the *x* value first and the *y* value second—(*x,y*), so the coordinates of the point shown are (4,3). The coordinate axes, in terms of which position is specified, form the *frame of reference* of a coordinate system.

Many coordinate systems can be devised to locate points on a surface. Another common example is the *polar* coordinate system, in which the frame of reference consists of a line and a point. Each point has coordinates (*r*,θ), which refer to a line called the *axis* and a point on that line called the *pole*. The *r* coordinate is the distance from the pole to the point. The θ coordinate is the angle between the axis and the line formed by joining the point and the pole measured counterclockwise. Thus, the coordinates of point P_1 are (4,30) and those of point P_2 are (3,240). See Figure 4.9b.

(a) The Cartesian coordinate system

(b) The polar coordinate system

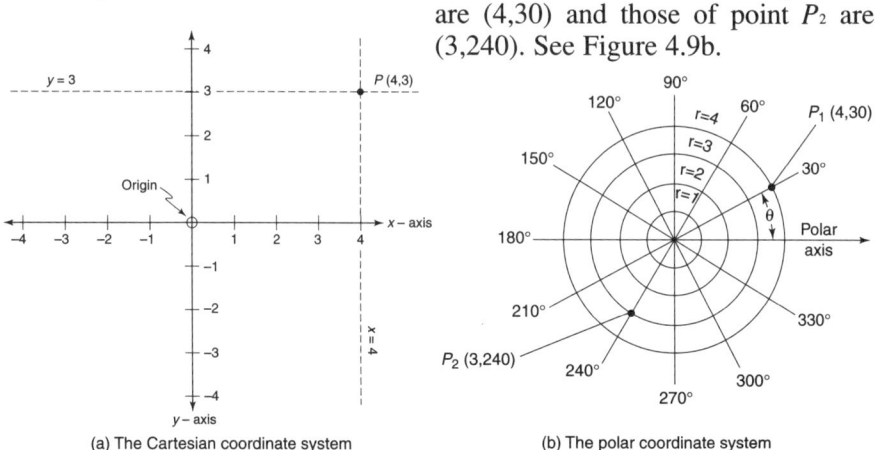

Figure 4.9 Two Common Coordinate Systems. (a) The Cartesian coordinate system (b) The polar coordinate system.

Earth's Coordinate System and Mapping

The Latitude-Longitude Coordinate System

In order to describe the position of any point on Earth's spherical surface, a coordinate system has been set up that uses two coordinates known as latitude and longitude. The latitude-longitude system consists of two sets of lines that cross each other at right angles. *Latitude* lines (called *parallels*) run in an east-west direction, and *longitude* lines (called *meridians*) run in a north-south direction. This type of system and the words *latitude* and *longitude* were already being used to describe such lines at the time of the Egyptian astronomer Claudius Ptolemy in A.D. 150.

Parallels

Since Earth is a sphere, east-west lines on its surface, such as the Equator, actually form circles. The circles formed by lines of latitude are called *parallels* because, if you drew a series of east-west lines, the circles formed would all be parallel to one another. See Figure 4.10a. Parallels lie in planes that are at right angles to Earth's axis of rotation. See Figure 4.10b.

To create a coordinate system on a globe or map, it is necessary to have a frame of reference, or fixed reference lines. The fixed reference line for latitude in this system is the **Equator**, an east-west line midway between the North and South Poles. Parallels are described by their angular distances north or south of the Equator as measured from the center of Earth. See Figure 4.10c.

Latitude is defined as **positive** for locations north of the Equator and **negative** for locations south of the Equator. Of course, latitude may also be given as **North (N)** or **South (S)** latitude. Thus, latitude is 0° at the Equator, and may be written as +90°, or 90°N, at the North Pole and –90°, or 90°S, at the South Pole. The latitude of New York City is about +41°30' or 41°30' N. Notice that the farther a location is from the Equator, the smaller the circle and the shorter its length. The east-west line at 60°N is only half as long as the Equator.

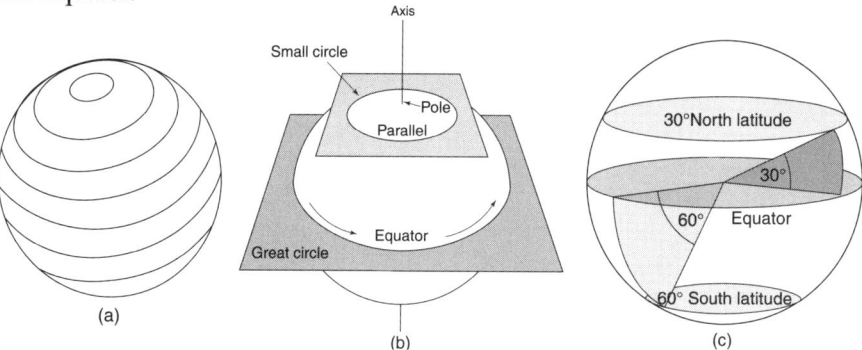

Figure 4.10 (a) Parallels are east-west lines on Earth's surface. (b) The Equator and all parallels lie in planes that are at right angles to the axis of rotation. (c) Latitude is a measure of the angle between the plane of the Equator and a line joining the center of Earth with a point on its surface.

Great Circles and Meridians

Earth's axis and poles also provide the reference points for lines of longitude. To understand longitude, you must first grasp the concept of a great circle. A *great circle* is a line drawn around a sphere to form a circle whose plane passes through the center of the sphere. If the plane of a circle doesn't pass through the center of the sphere, a small circle is formed. See Figure 4.11. An infinite number of great circles can be drawn on a sphere. The Equator, however, is the *only* great circle whose plane is at right angles to Earth's axis.

One interesting property of great circles is that they all cut a sphere exactly in half (e.g., the Equator divides Earth into the Northern and Southern hemispheres). Another is that the arc of a great circle connecting any two points on a sphere's surface is the shortest distance between those points! This is why airliners travel along "great circle routes."

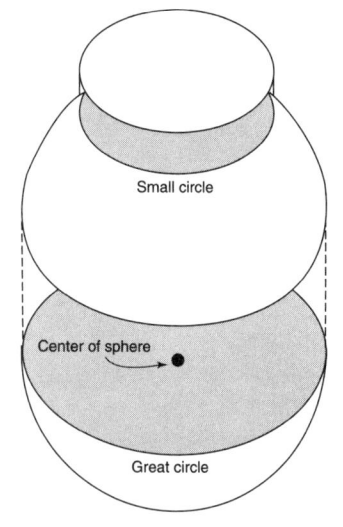

Figure 4.11 A Great Circle and a Small Circle.

On Earth, any great circle that passes through both the North and the South poles is called a *meridian*. Earth's axis of rotation lies in the plane of every meridian. See Figure 4.12. *Meridians of longitude* go only halfway around Earth, from pole to pole. Although a meridian of longitude is actually only half

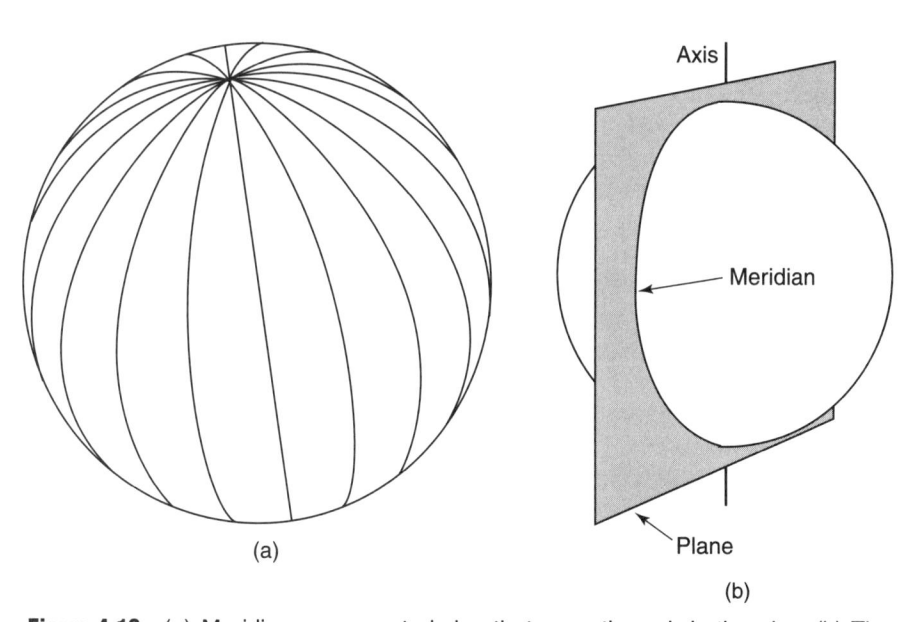

Figure 4.12 (a) Meridians are great circles that pass through both poles. (b) The plane of a meridian passes through Earth's axis.

Earth's Coordinate System and Mapping

of a meridian, it is often referred to simply as a meridian.

The *longitude* of a point is the angular distance between two meridians: a fixed reference meridian and the meridian passing though the point. It is measured from the center of Earth east or west along the Equator between the fixed reference meridian of longitude and the meridian of longitude passing through the point. See Figure 4.13.

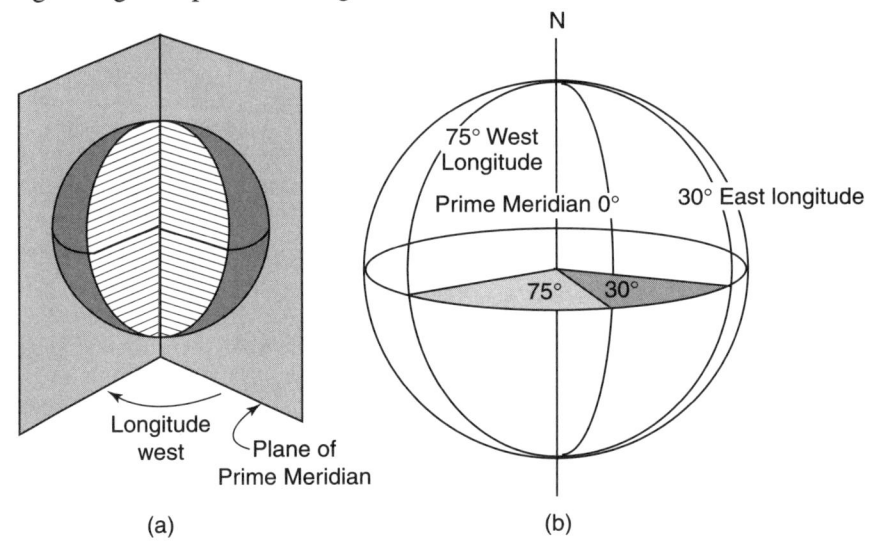

Figure 4.13 (a) Longitude can be thought of as the angle between the planes of two meridians. (b) Longitude is measured from the center of Earth as an angle east or west of the Prime Meridian.

Unlike the Equator, which is the *only* line halfway between the poles, the reference meridian, or **Prime Meridian**, is an arbitrarily chosen line and could be almost anywhere. Over the years, on different maps, it *has* been located in different places. Mapmakers usually began numbering the lines of longitude at whichever meridian passed through the site of their national observatory. As recently as 1881, 14 different prime meridians were being used on topographic survey maps. In 1884 the United States hosted an international conference in Washington whose purpose was to agree upon a single prime meridian to be used by mapmakers worldwide. This conference agreed that the ***Prime Meridian*** would be the meridian of longitude that runs through the Royal Observatory in Greenwich, England. Longitude would be measured in degrees east or west of this reference line up to 180°. East longitude would be positive, and west longitude would be negative. This choice was made mainly because the Prime's twin, the 180° meridian, runs through the Pacific and touches almost no habitable land. Thus, the 180° meridian is a good choice for an international date line, that is, a line marking the transition from one date to the next.

Any meridian will cross the Equator and all other parallels at right angles. Therefore, once reference lines are chosen, a system of meridians and parallels forms a grid of lines that intersect at right angles—a neat coordinate system! See Figure 4.14.

Determining Latitude and Longitude

Both the latitude and the longitude of a location can be determined by making simple observations.

Determining Latitude

As described earlier and shown in Figure 4.5, observers in the Northern Hemisphere can determine their latitudes by measuring the altitude of Polaris. Observers in the Southern Hemisphere use the positions of other stars and make corrections for their deviation from Earth's axis of rotation.

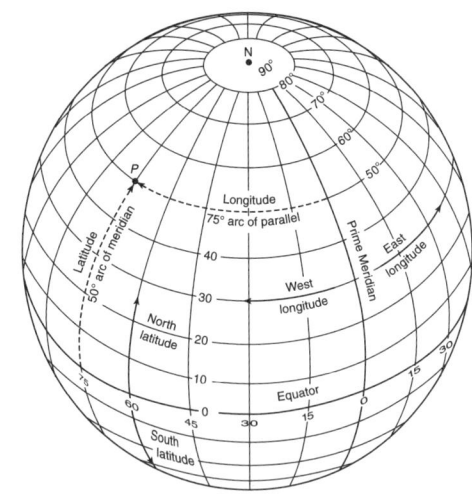

Figure 4.14 The geographic grid of parallels and meridians that allows any point on Earth's surface to be located with a set of coordinates. Point *P* has a latitude of 50°N and a longitude of 75°W.

Determining Longitude

The longitude of a particular location is determined by observing the time and the position of the Sun. Local noon is the time at which the Sun is at its highest altitude of the day. At that moment, a line drawn from the Sun to the center of Earth will pass through that location's meridian. A few moments later, Earth will have rotated from east to west and another meridian will align with the Sun and experience local noon. Thus, different longitudes will experience local noon at different times.

Since Earth makes one complete 360° rotation every 24 hours, the relationship between time and degrees rotated can be calculated. For example, in 1 hour Earth will rotate 360°/24, or 15°. Now, suppose you were at a location and the time was exactly local noon. One hour later, Earth would have rotated 15° and a location 15° of longitude away would be experiencing local noon. Two hours later, Earth would have rotated 30° and a location 30° of longitude away would experience local noon. By comparing the difference in time between local noons at any two locations, you can easily calculate their difference in longitude. Every hour of difference equals 15° of longitude, but this doesn't tell you a location's actual longitude, only its longitude relative to another point.

Suppose, however, that one of the times you knew was the time at which local noon occurred in *Greenwich*! Then you would know the difference in longitude at your location from *zero* longitude, and that would be your actual

Earth's Coordinate System and Mapping

longitude. Therefore, in order to measure longitude at a particular location you need to know two times—local noon and the time of local noon in Greenwich. Each hour of difference between local time and Greenwich time equals 15° of longitude. Since Earth rotates from west to east, locations east of Greenwich will experience local noon earlier than Greenwich, while those west of Greenwich will experience local noon later than Greenwich. For example, if Old Forge, New York, experiences local noon about 5 hours later than Greenwich, England, its longitude is 5 × 15°, or 75°. Since local noon in Old Forge occurs after local noon in Greenwich, it is west longitude.

To measure longitude, then, you need a radio broadcast or a chronometer (a watch set to Greenwich time) to tell you what time it is in Greenwich when you experience local noon. You can then look up the time of local noon in Greenwich on that day in an astronomical table and, using the difference in times, calculate longitude. The need for such clocks drove the development of ever more precise timekeeping devices, culminating with today's atomic clocks.

Field Maps

A *field* is a region of space that has a measurable quantity at every point. Some examples of field quantities are temperature, pressure, magnetism, gravity, and elevation. Field maps can be used to represent any quantity that varies in a region of space.

Isolines

One way to represent field quantities on a two-dimensional field map is to use isolines. *Isolines* connect points of equal field values. For example, a temperature field map shows lines connecting points of equal temperature, or isotherms. See Figure 4.15.

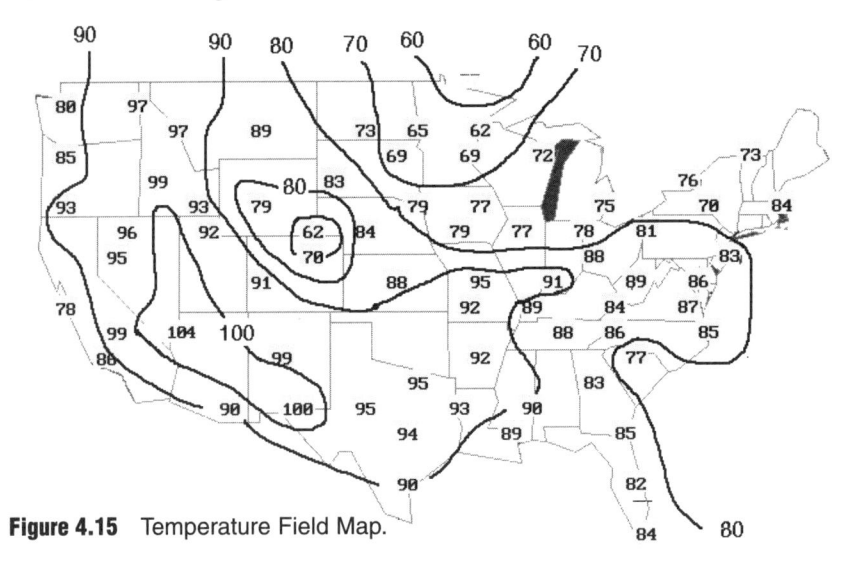

Figure 4.15 Temperature Field Map.

Gradient

Within a field, field values change as you move from place to place. The rate at which a field value changes is called its *gradient*. Gradient can be calculated as follows:

$$\text{Gradient} = \frac{\text{amount of change in field value}}{\text{distance through which change occurs}}$$

Example:

Figure 4.16 shows pollutant levels measured in Lake Meredith. Isolines connect points of equal pollutant concentration, measured in parts per billion (ppb). Abby's Island and Ben's Island are 2 kilometers apart, and the pollutant concentration between them changes from 30 to 50 parts per billion. The rate of change, or gradient, is:

$$\text{Gradient} = \frac{50 \text{ ppb} - 30 \text{ ppb}}{2 \text{ km}}$$

$$= \frac{20 \text{ ppb}}{2 \text{ km}}$$

$$= 10 \text{ ppb/km}$$

The gradient is 10 parts per billion per kilometer.

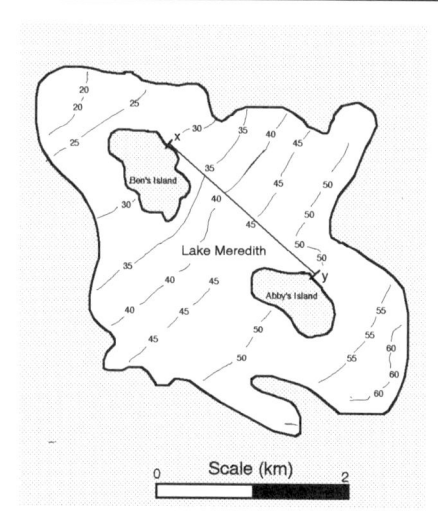

Figure 4.16 Meredith Lake Containing Abby's Island and Ben's Island.

Isosurfaces

Field maps can also be represented in three dimensions using isosurfaces. Isolines show field values in a particular two-dimensional plane (e.g., air pressure at ground level). An *isosurface* is a three-dimensional surface on which every point has the same field value. Isosurfaces provide insights into the nature of a field that it may not be possible to gain by viewing only a two-dimensional map. See Figure 4.17.

Fields seldom remain unchanged over time. A field map shows a field at a particular point in time, the time at which its field values were measured. For example, the temperature field over the United States changes drastically from July to December. Also, Earth's magnetic field has changed position many times since Earth formed. Therefore, field maps need to be updated periodically in order to represent current field conditions. On

Earth's Coordinate System and Mapping

Magnetic Field Isolines

Magnetic Field Isosurfaces

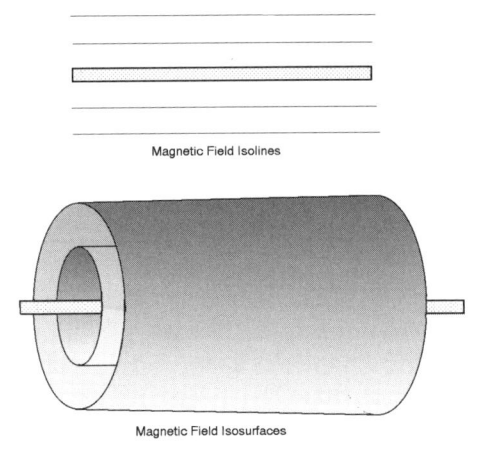

Figure 4.17 Magnetic Field Isolines Versus Isosurfaces.

Contour Lines

Isolines that connect points of equal elevation are called *contour lines* because they represent the shapes, or contours, of Earth's surface. A contour line shows the shape that would be formed if the land surface were sliced by a horizontal plane at a particular elevation above sea level. An unvarying vertical distance called the *contour interval* separates successive contour lines. See Figure 4.18.

Contour lines around an enclosed depression have *hachure marks* pointing into the depression in order to distinguish them from small hills. See Figure 4.19.

Map Symbols

Topographic maps provide a view of the ground as seen from vertically above. Surface features are represented by map symbols. An extensive list of these symbols is shown in Figure 4.20. Symbols are blue for bodies of water, black and red for human-made structures, and brown for contour lines and other relief symbols.

weather maps, which represent rapidly changing fields, values are updated as often as once an hour.

Topographic Maps: Elevation Field Maps

Topographic maps are scale models of a part of Earth's surface that show its three-dimensional shape in two dimensions. A topographic map is actually a type of field in which the field value is elevation above sea level of points on Earth's surface.

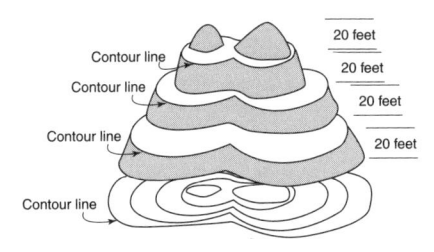

Figure 4.18 Three-Dimensional Mountain Sliced by Horizontal Plane, Showing Contour Line.

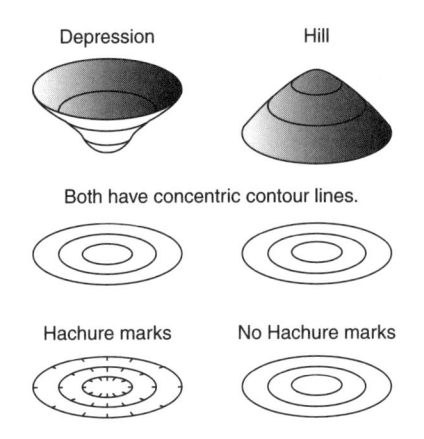

Figure 4.19 Depression Contours Versus Regular Contours.

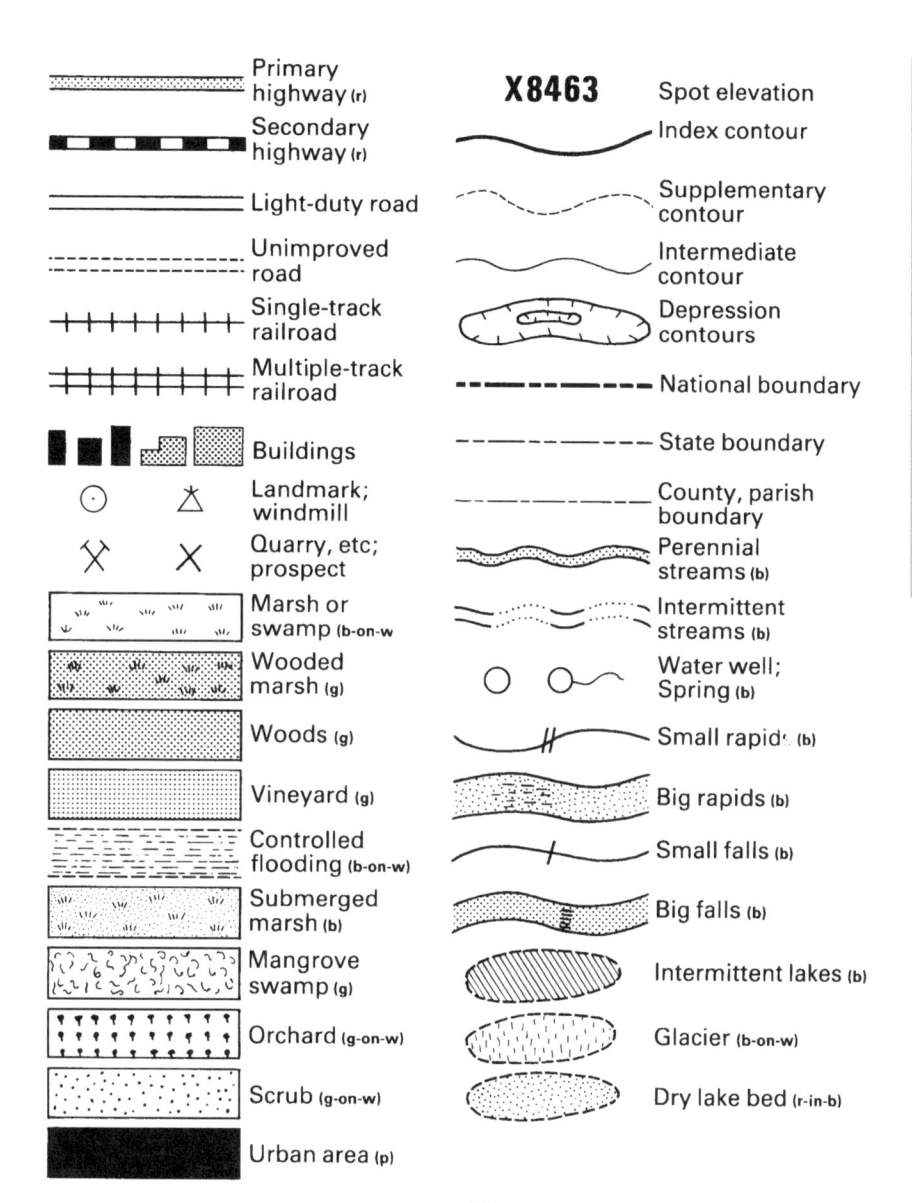

Figure 4.20 Selected Topographic Map Symbols. Source: *Earth Science on File.* Copyright © 1988 by Diagram Visual Information Ltd. Reprinted by permission of Facts on File, Inc.

Map Scale

A map scale is the ratio between the distance shown on a map and the actual distance on the ground. For example, if the map scale is expressed as 1:100,000,

Earth's Coordinate System and Mapping

Figure 4.21 A Graphic Map Scale. The ratio 1:62,500 means that 1 unit of distance on the map equals 62,500 units of distance on the ground. Thus, 1 inch on the map is 62,500 inches on the ground, which is slightly less than 1 mile (1 mil = 63,360 in.). One centimeter on the map equals 62,500 centimeters on the ground (0.625 kilometer). The graphic scale is printed so that each unit is precisely the correct size for the distance indicated. (In this case 1 mile on the scale measures 1 inch in length.) The first unit on the graphic scale is subdivided so that it can be used to make more precise measurements.

then 1 unit of length on the map is equal to 100,000 of the same units on the ground. Both numbers in the ratio have the same units, but those units can be anything. For example, the map scale indicates that 1 centimeter on the map equals 100,000 centimeters on the ground. But it also means that 1 inch on the map equals 100,000 inches on the ground. On most topographic maps, a graphic scale such as the one shown in Figure 4.21 is used to represent the map scale. The graphic scale can be used as a ruler to measure distances on the map.

Map Direction

The convention used in most topographic maps is that the top of the map is north. However, since north is determined with a compass, and geographic north differs from magnetic north, most topographic maps include an arrow showing each. The map in Figure 4.22 shows magnetic north with an arrow labeled "MN" and geographic North with an arrow labeled "GN".

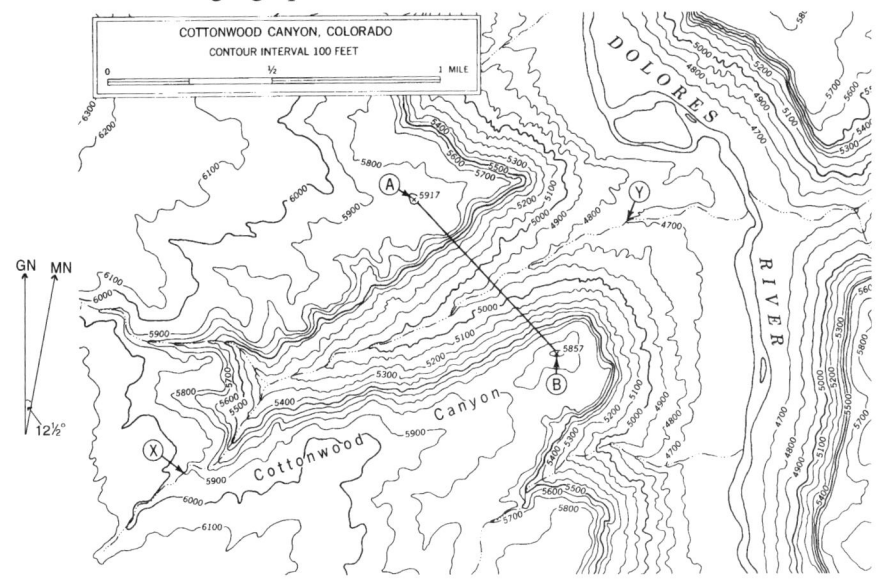

Figure 4.22 Topographic Map. Source: *Exploring Earth Science*, Walter A. Thurber and Robert E. Kilburn, Allyn and Bacon, 1965. Used with permission of Prentice-Hall.

Figure 4.23 Steps in Constructing a Map Profile. Source: *Exploring Earth Science*, Walter A. Thurber and Robert E. Kilburn, Allyn and Bacon, 1965. Used with permission of Prentice-Hall.

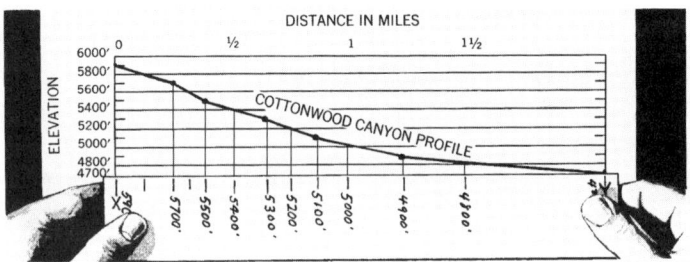

Map Profiles

It is often useful to construct a profile from a topographic map. A map profile shows what a cross section of land looks like between two points. Figure 4.23 shows how a profile can be constructed.

1. The two points between which the profile is to be drawn are chosen, and a line is drawn to connect them.
2. The edge of a piece of paper is placed along the line, and the edge of the paper is marked wherever it intersects a contour line.
3. The paper is moved to a piece of graph paper on which a vertical scale has been marked to match the contour lines intersected. At each point where a contour line crosses the edge of the paper, a line of the appropriate height is drawn on the graph paper.
4. The endpoints of the lines are connected in a smooth line to form the finished profile.

Conventions Used in Topographic Maps

When topographic maps are drawn or read, the following conventions apply:

- All points on a contour line have the same elevation.
- Every fifth line, called an *index line*, is generally indicated in black ink or bold type and labeled with the line's elevation.
- All contour lines are closed, but they may run off the map.
- Two contour lines of different elevations may not cross each other.
- Contour lines may merge at a cliff or waterfall.
- The spacing of contour lines indicates the nature of the slope. The closer

the contour lines are spaced, the steeper the slope. Even spacing indicates a uniform slope.
- Where contour lines cross a stream, they always form a V whose apex points up the valley.
- Where contour lines cross a ridge between valleys, they often form a V whose apex points down the valleys.

MULTIPLE-CHOICE QUESTIONS

In each case, write the number of the word or expression that best answers the question or completes the statement.

1. Which diagram most accurately shows the cross-sectional shape of Earth?

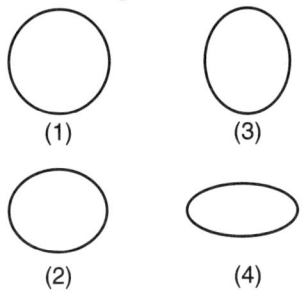

   (1)   (3)

   (2)   (4)

2. The north-south distance between Earth's Equator (0°) and the North Pole (90° N) is 10,002 kilometers. The distance between 0° and 10° N is 1,106 kilometers. Which statement is best supported by this information?
   (1) The shape of Earth is not perfectly spherical.
   (2) The lines of longitude are not parallel.
   (3) The north-south distance for every 10° of latitude is a constant value.
   (4) Earth's equatorial radius and polar radius are equal.

3. The diagrams below represent photographs of a large sailboat taken through telescopes over time as the boat sailed away from shore out to sea. The number above each diagram shows the magnification of the telescope lens.

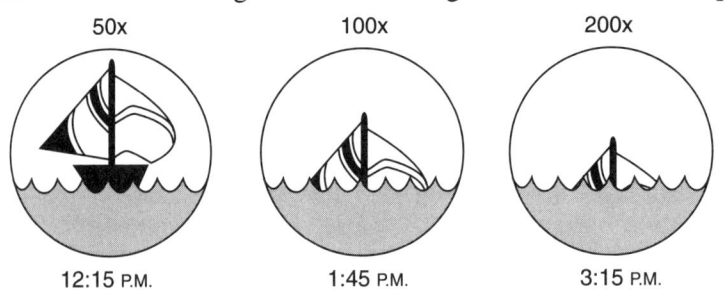

   50x   100x   200x

   12:15 P.M.   1:45 P.M.   3:15 P.M.

Which statement best explains the apparent sinking of this sailboat?
(1) The sailboat is moving around the curved surface of Earth.
(2) The sailboat appears smaller as it moves farther away.
(3) The change in density of the atmosphere is causing refraction of light rays.
(4) The tide is causing an increase in the depth of the ocean.

4. At sea level, which location would be farthest from the center of Earth?
(1) the North Pole   (3) 45° N
(2) 23½° S   (4) the Equator

5. The best evidence of Earth's nearly spherical shape is obtained through
(1) telescopic observations of other planets
(2) photographs of Earth from an orbiting satellite
(3) observations of the Sun's altitude made during the day
(4) observations of the Moon made during lunar eclipses

6. Measurements taken with gravity meters would be most useful in providing information concerning Earth's
(1) magnetic field intensity   (3) tidal range
(2) shape   (4) distance from the Sun

7. According to the data below, what is the exact shape of Earth?
   Actual Dimensions of Earth
   Equatorial radius   12,756 km
   Polar radius   12,714 km
   Equatorial circumference   40,076 km
   Polar circumference   40,008 km
(1) slightly flattened at both the Equator and the poles
(2) slightly bulging at both the Equator and the poles
(3) slightly flattened at the Equator and slightly bulging at the poles
(4) slightly flattened at the poles and slightly bulging at the Equator

8. Two students are located 800 kilometers apart on the same meridian of longitude. They measure the altitude of the Sun at noon on the same day. They can use these measurements to calculate
(1) Earth's circumference   (3) Earth's mass
(2) Earth's density   (4) the Sun's diameter

Earth's Coordinate System and Mapping

9. The diagram below shows stations A and B, which are located on a north-south line. At noon, the altitude of the Sun is 90° at station B and 30° at station A. What is the distance between stations A and B? [Earth's circumference is 40,000 km.]

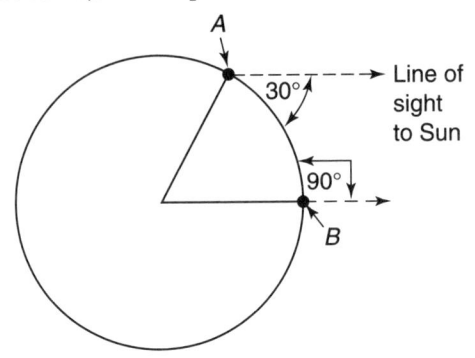

(1) 30 km
(2) 2,222 km
(3) 6,666 km
(4) 10,000 km

10. According to the Generalized Bedrock Geology of New York State map in the *Earth Science Reference Tables*, Triassic bedrock is found in New York State at approximately
(1) 41° 05' N, 74° 00' W
(2) 44° 05' N, 74° 00' W
(3) 74° 00' N, 41° 05' W
(4) 74° 00' N, 44° 05' W

11. Polaris is used as a celestial reference point for Earth's latitude system because Polaris
(1) always rises at sunset and sets at sunrise
(2) is located over Earth's axis of rotation
(3) can be seen from any place on Earth
(4) is a very bright star

12. What is the latitude of the observer shown below?

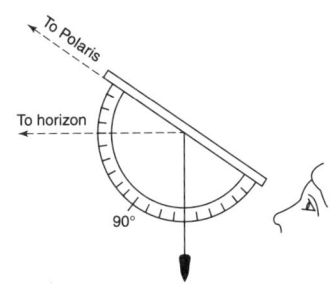

(1) 35° N
(2) 55° N
(3) 90° N
(4) 25° N

The diagrams below illustrate systems that can be used to determine position on a sphere. Refer to them to answer questions 13 and 14.

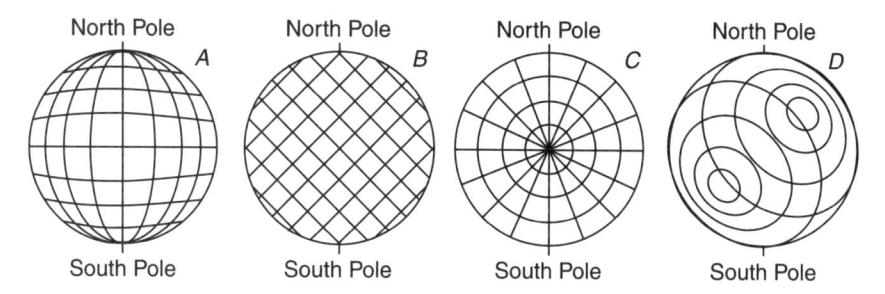

13. Systems of line like those illustrated above are called
    (1) latitude systems
    (2) coordinate systems
    (3) great circle systems
    (4) axis systems

14. Which of the systems illustrated is most like the latitude-longitude system used on Earth?
    (1) A
    (2) B
    (3) C
    (4) D

15. Based on the Generalized Bedrock Geology of New York State map in the *Earth Science Reference Tables*, what could be the approximate location of an observer if he measured the altitude of Polaris to be 41° above the horizon?
    (1) Watertown
    (2) Massena
    (3) Buffalo
    (4) New York City

16. The diagram below represents a portion of Earth's latitude and longitude system.

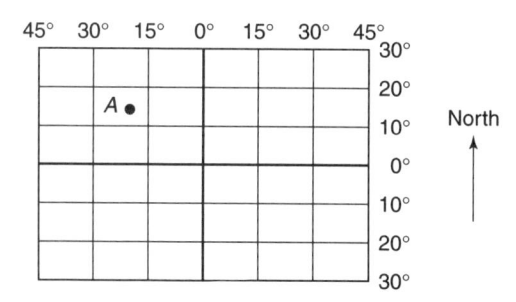

    What are the approximate latitude and longitude of point A?
    (1) 15° S, 20° W
    (2) 15° S, 20° E
    (3) 15° N, 20° W
    (4) 15° N, 20° E

Earth's Coordinate System and Mapping

17. According to the *Earth Science Reference Tables*, what is the straight-line distance, in kilometers, from Elmira, New York, to Buffalo, New York?
    (1) 100
    (2) 120
    (3) 150
    (4) 190

18. On the Generalized Bedrock Geology of New York State map in the *Earth Science Reference Tables*, what similar pattern is found at 43° 00' N by 77° 30' W?

    (1)   (2)   (3)   (4)

19. According to the *Earth Science Reference Tables*, most of the surface bedrock in New York State south of latitude 43° N and west of longitude 75° W was formed during which period?
    (1) Silurian
    (2) Devonian
    (3) Cambrian
    (4) Ordovician

20. According to the *Earth Science Reference Tables*, what type of landscape region is located at 44°30' N and 74°30' W?
    (1) plateau
    (2) plain
    (3) coastal lowland
    (4) mountainous area

21. An observer in New York State measures the altitude of Polaris to be 44°. According to the *Earth Science Reference Tables*, the location of the observer is nearest to
    (1) Watertown
    (2) Elmira
    (3) Buffalo
    (4) Kingston

22. Which measurements would have the same values if taken at the same moment by one observer at 42° N, 74° W and another observer at 44° N, 74° W?
    (1) altitude of Polaris
    (2) angle at which sunlight strikes Earth's surface
    (3) age of the surface bedrock
    (4) time of solar noon

23. Which statement is *not* true about fields?
    (1) They do not change with time.
    (2) They are often illustrated by the use of isolines.
    (3) Any one place in a field has a measurable value at a specific time.
    (4) Gradients indicate the degree of change from place to place in a field.

24. In the diagram below, the thermometer held 2 meters above the floor shows a temperature of 30°C. The thermometer on the floor shows a temperature of 24°C.

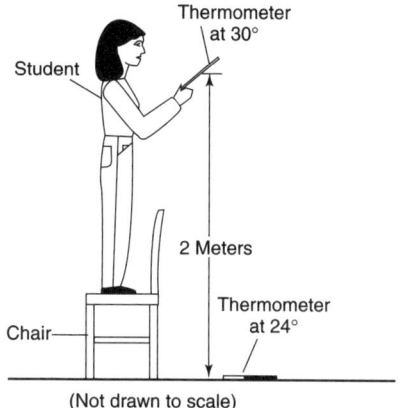

(Not drawn to scale)

What is the temperature gradient between the two thermometers?
(1) 6C°/m
(2) 2C°/m
(3) 3C°/m
(4) 4C°/m

25. The diagram below shows the isothermal pattern obtained at a height of 1 meter above the floor in a classroom during a temperature study in November.

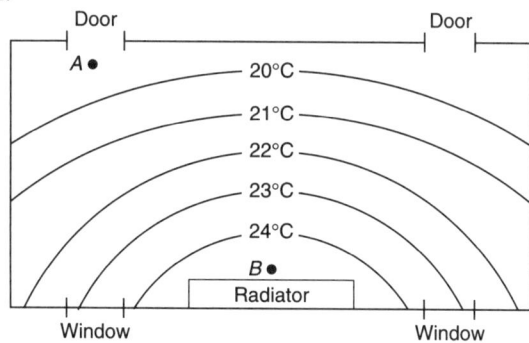

Which conclusion is best supported by this diagram?
(1) A heat source is located at A.
(2) A heat sink is located at B.
(3) Room temperature at the floor and at the ceiling can be determined.
(4) The isotherms indicate the temperatures at only one level in the room.

26. A topographic map is a two-dimensional model that uses contour lines to represent points of equal
(1) barometric pressure
(2) temperature gradient
(3) elevation above sea level
(4) magnetic force

Earth's Coordinate System and Mapping

27. The contour lines on the map below represent a hill.

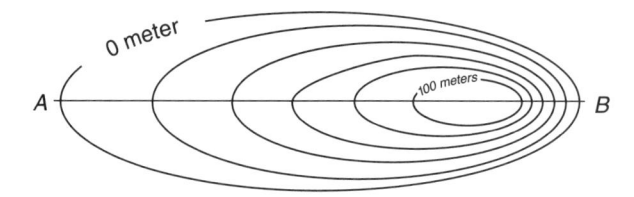

Which hill shape shown below best represents a profile drawn along line *AB* of the contour map?

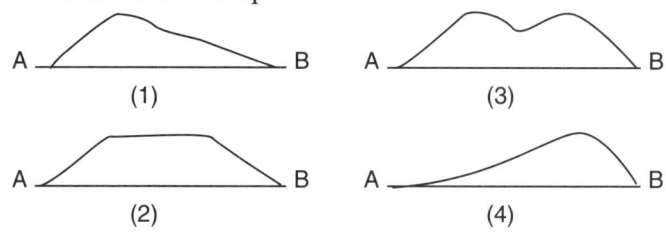

(1)
(2)
(3)
(4)

28. The temperature data (in °C) shown on the map below were taken at the same elevation and time in a room. Letters *A* through *D* represent the corners of the room.

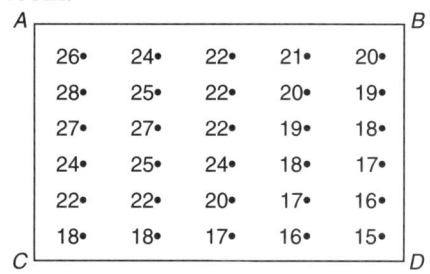

26	24	22	21	20
28	25	22	20	19
27	27	22	19	18
24	25	24	18	17
22	22	20	17	16
18	18	17	16	15

Which diagram best represents the temperature field in the room?

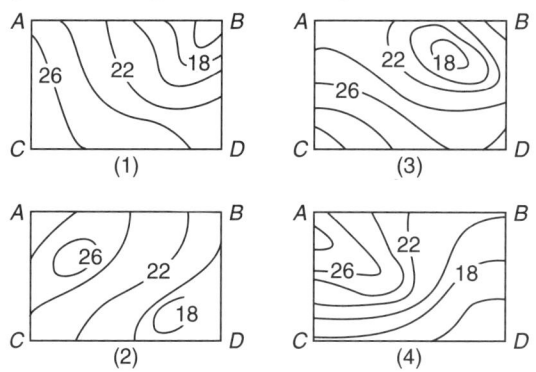

(1)
(2)
(3)
(4)

90

29. A stream has a source at an elevation of 1,000 meters. It ends in a lake that has an elevation of 300 meters. If the lake is 200 kilometers away from the source, what is the average gradient of the stream?
   (1) 1.5 m/km
   (2) 3.5 m/km
   (3) 10 m/km
   (4) 15 m/km

Base your answers to questions 30 and 31 on the diagram below, which represents a contour map of a hill.

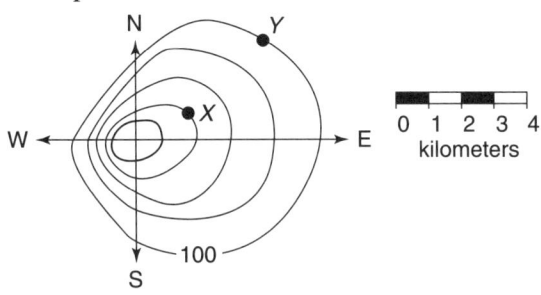

Contour interval = 10 meters

30. On which side of the hill does the land have the steepest slope?
    (1) north
    (2) south
    (3) east
    (4) west

31. What is the approximate gradient of the hill between points X and Y? [Refer to the *Earth Science Reference Tables*.]
    (1) 1 m/km
    (2) 10 m/km
    (3) 3 m/km
    (4) 30 m/km

32. The elevation of a certain area was measured for many years, and the results are recorded in the data table below.

Year	Elevation (m)
1870	102.00
1890	102.25
1910	102.50
1930	102.75
1950	103.00

If the elevation continued to increase at the same rate, what was most likely the elevation of this area in 1990?
   (1) 103.25 m
   (2) 103.50 m
   (3) 103.75 m
   (4) 104.00 m

Earth's Coordinate System and Mapping

Base your answers to questions 33 through 37 on your knowledge of earth science and the information provided by the diagram. The diagram represents a sketch drawn in a notebook by an earth science student. Lines A, B, C, D, E, and F are isolines. The points along the isolines indicate the only locations where actual measurements were made. Points W, X, Y, and Z are reference locations in the field diagram.

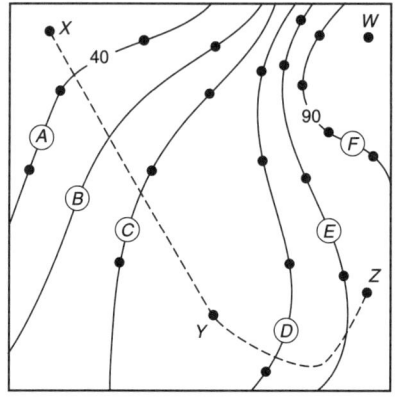

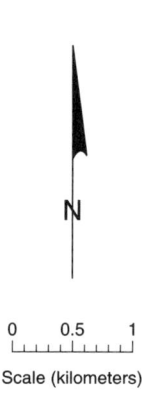

33. The value of point Y might be
    (1) 55   (3) 75
    (2) 65   (4) 95

34. What is the approximate distance from point X to point Z along the dashed line XYZ?
    (1) 1.5 km   (3) 4.0 km
    (2) 3.5 km   (4) 4.5 km

35. Between which two points is the greatest gradient?
    (1) X and W (3) Y and W
    (2) X and Y (4) Z and W

Note that question 36 has only three choices.

36. Which is the best description of the value of the measured quantities along isoline C on the map?
    (1) The values decrease from north to south.
    (2) The values increase from north to south.
    (3) The values remain the same.

37. The isolines of this field diagram are for snow depth in centimeters. In which part of the diagram is the snow deepest?
    (1) southeast corner   (3) southwest corner
    (2) northeast corner   (4) northwest corner

Base your answers to questions 38 through 42 on the topographic map below and your knowledge of earth science.

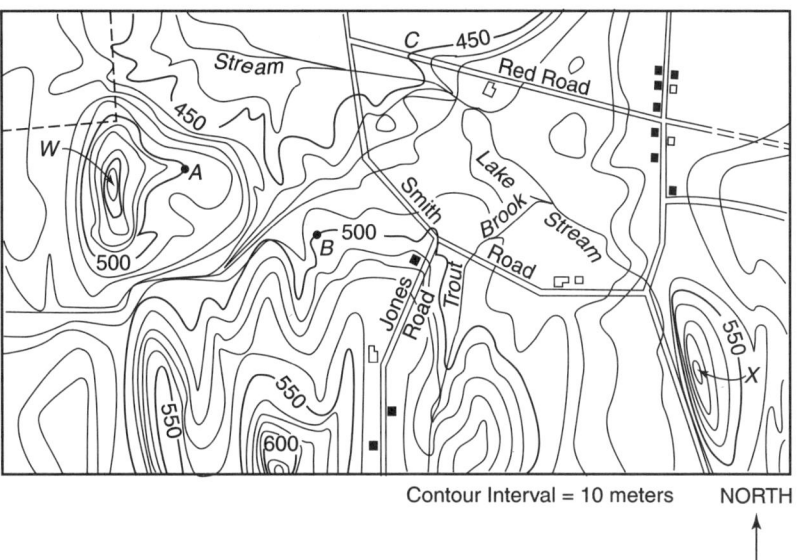

38. What is the elevation at the intersection of Jones Road and Smith Road?
  (1) 450 m   (3) 550 m
  (2) 500 m   (4) 600 m

39. What is the elevation of the highest contour line on hill *W*?
  (1) 440 m   (3) 550 m
  (2) 510 m   (4) 610 m

40. On which side of hill *X* is the steepest slope?
  (1) north   (3) east
  (2) southeast   (4) southwest

41. In which general direction is Trout Brook flowing when it passes under Smith Road?
  (1) northeast   (3) northwest
  (2) southeast   (4) southwest

Earth's Coordinate System and Mapping

42. Which diagram best represents the profile along a straight line between points A and B?

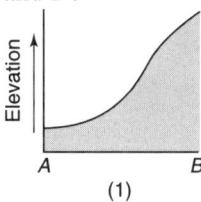

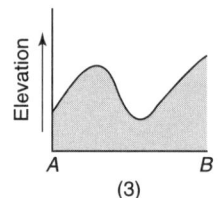

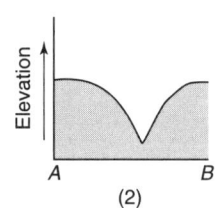

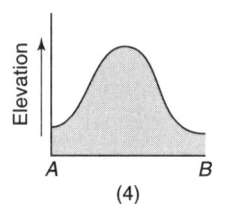

(1) (2) (3) (4)

CONSTRUCTED RESPONSE QUESTIONS

Base your answers to questions 43 through 46 on the topographic map of Cottonwood, Colorado, below. Points A, B, X, and Y are marked for reference.

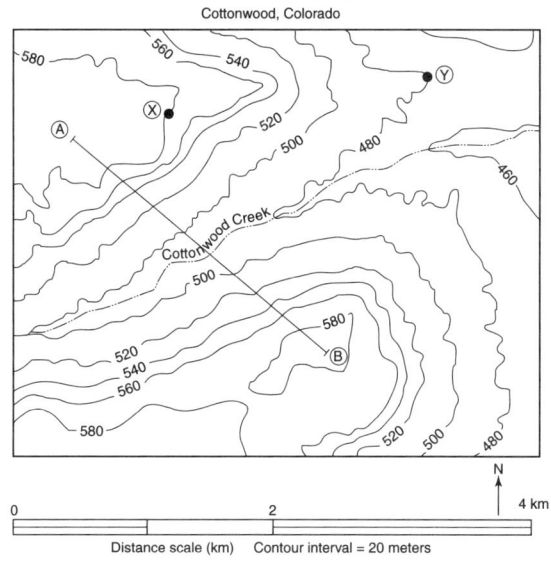

43. State the general direction in which Cottonwood Creek is flowing. [1]

44. State the highest possible elevation, to the *nearest meter*, for point B on the topographic map. [1]

94

45. On the grid provided below, draw a profile of the topography along line *AB* shown on the map. [2]

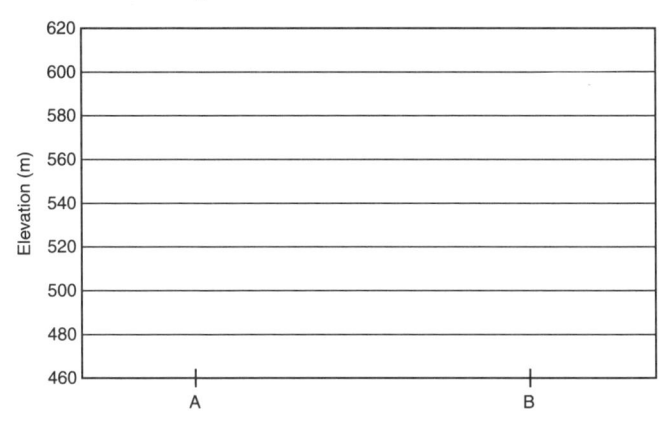

[Grid with Elevation (m) on y-axis from 460 to 620 in increments of 20, with points A and B marked on x-axis]

46. In the space provided below, calculate the gradient of the slope between points *X* and *Y* on the topographic map, following the directions below:
   1. Write the equation for gradient. [1]
   2. Substitute data from the map into the equation. [1]
   3. Calculate the gradient and label it with the proper units. [2]

47. State one observation from which you might *infer* that Earth is a sphere. Explain how this observation could also be interpreted as indicating that Earth has a different shape. [2]

48. Define latitude, and describe a procedure by which latitude can be determined. [2]

49. Define longitude, and explain how it is found. [2]

50. New York City is located at 73° W. Rio DeJaniero, is located at 43° W. When it is noon in New York City, what time is it in Rio DeJaniero? [1]

51. Compare and contrast the information found on a topographic map with the information found on a typical road map. [2]

Earth's Coordinate System and Mapping

EXTENDED CONSTRUCTED RESPONSE QUESTIONS

Base your answers to questions 52 and 53 on the temperature field map below. The map shows 25 measurements (in °C) that were made in a temperature field and recorded as shown. The dots represent the exact locations of the measurements. *A* and *B* are locations within the field.

Temperature Field Map (°C)

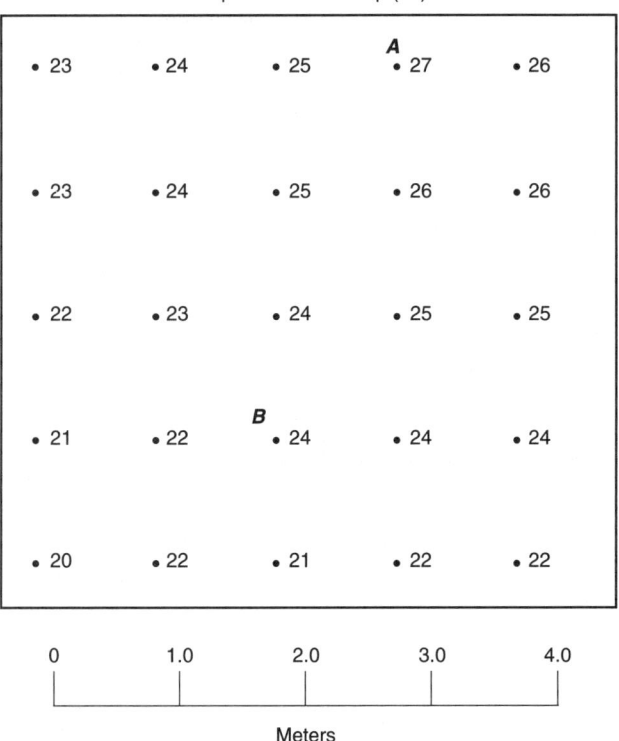

```
                          A
  • 23    • 24    • 25    • 27    • 26


  • 23    • 24    • 25    • 26    • 26


  • 22    • 23    • 24    • 25    • 25

                    B
  • 21    • 22    • 24    • 24    • 24


  • 20    • 22    • 21    • 22    • 22
```

0   1.0   2.0   3.0   4.0

Meters

52. On the temperature field map, draw three isotherms: the 23°C isotherm, the 24°C isotherm, and the 25°C isotherm. [2]

53. Calculate the temperature gradient between locations *A* and *B* on the temperature field map, following the directions below:
1. Write the equation for gradient. [1]
2. Substitute data from the map into the equation. [1]
3. Calculate the gradient and label it with the proper units. [1]

54. Base your answers to parts (a) through (c) on the information and the field map below.

A water-soluble pesticide was applied to the ground surface at location *A* as shown on the map. Precipitation occurred, and some of the pesticide that was carried into the ground dissolved in the water. As the groundwater moved through the ground, it distributed the pesticide. A scientist took samples of the groundwater and determined the concentration of the pesticide in parts per billion (ppb). The field map shows the concentrations of the pesticide.

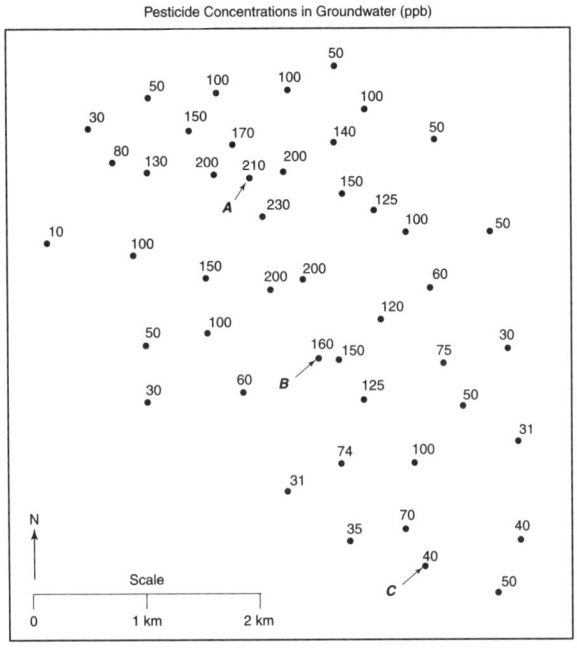

(a) Draw an accurate isoline map on the field map, following the directions below. [4]
 1. Use an interval of 50 ppb.
 2. Begin your drawing with the 200-ppb isoline.
(b) Using one or more complete sentences, briefly describe the general pattern of pesticide concentration as the distance from location *A* increases. [2]
(c) Use the isoline map you have made to determine the gradient of pesticide concentration (in ppb) between point *B* and point *C*, following the directions below. [4]
 1. Write the formula for gradient.
 2. Substitute data into the formula.
 3. Calculate the gradient.
 4. Label your answer with proper units.

Chapter 5
OUR SYSTEM OF TIME

> **KEY IDEAS** For thousands of years, people have used the changing positions of celestial objects to note the passage of time. The frame of reference for our system of time has changed over the years; however, it is still based upon the changing positions of celestial objects due to Earth's motions. Earth's rotation provides a basis for the day and our system of local time; meridians of longitude are the basis of time zones. The revolution of the Moon is the basis for the month, and the revolution of Earth around the Sun provides the basis for the year and our current calendar system.

KEY OBJECTIVES
Upon completion of this chapter, you will be able to:

- Explain how Earth's rotation forms the basis for the day and our system of local time.
- Compare and contrast the solar day and the sidereal day.
- Explain the difference between a synodic month and a sidereal month.
- Describe how meridians of longitude are used to define time zones.
- Analyze the seasonal changes in the Sun's apparent path through the sky.

THE DAY

The seemingly endless cycle of daylight and darkness was probably the earliest timekeeping unit—the day. It is the most basic of all natural cycles. Few things influence our lives as deeply as the change from light to dark, from day to night, from being asleep and being awake. The problem with the day is where to begin and end it and how to divide it up.

The earliest way of expressing "time of day" was simply to point to a place in the sky where the Sun is found at that time. Here is where the Sun rises; over there is where it sets. Morning, noon, and afternoon can be easily determined simply by looking at which side of the sky the Sun is in. Selecting a specific landmark with which the Sun aligns, such as a hut, a tree, or a distant mountain peak, can further refine this system. Using such markers, it is not difficult to divide daytime into 6, 9, or even 12 divisions.

Dividing the Day into Hours

Timekeeping became a bit more precise with the simple invention of a stick stuck vertically into the ground. In sunlight, a vertical stick casts a shadow. As the Sun changes position, the shadow also changes position. The path of the Sun through the sky could be recorded as the line traced on the ground by the tip of the stick's shadow. This line could then be divided into equal units that, in turn, could be used to divide daytime into uniform time periods. The next small step was to realize that stars could also be sighted along the tip of a vertical stick and their motions tracked at night. In turn, these motions could then be used to divide nighttime into uniform time periods.

More than 3,000 years ago, the Egyptian priests of the Sun god Ra decided to divide the day into 12 hours of daytime and 12 hours of night. They chose 12 because they believed this number was mystical and symbolized the gods. The problem they faced was that most days did not have daylight and darkness periods of equal length. In New York State, for example, daylight varies from fewer than 9 hours in the winter to more than 15 hours in the summer. Consequently, dividing daylight into 12 equal time periods would yield a longer "hour" in the summer and a shorter "hour" in the winter. Eventually, the Egyptians discovered that there are only 2 days, the equinoxes, during which daylight and darkness are of equal length, so daylight and darkness were merged into a single 24-hour day with days and nights that vary in length. But the Egyptians still had to measure the length of that 24-hour day.

You might think that mea-

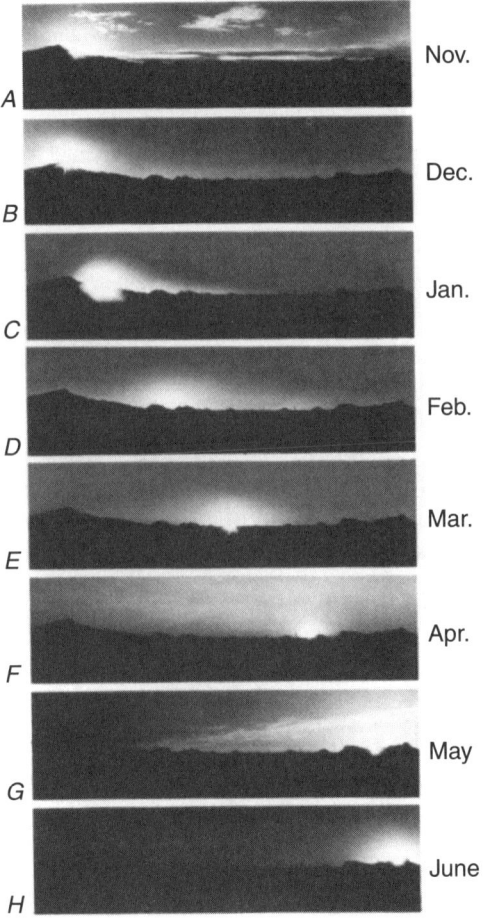

Figure 5.1 Changes in the Sunset Position along the Horizon During an 8-Month Period, Moving South from Position A to the Winter Solstice (B), Then North Again in C through the Spring Equinox (E) to the summer solstice (H).
Source: *Astronomy: The Cosmic Journey*, William K. Hartman, Wadsworth, 1987.

Our System of Time

suring the length of a day is easy—simply pick a fixed point during the day, and then measure the time that elapses until that same point in the day is reached again. Then divide this time into 24 equal "hours." Great idea, but the question now becomes, At what point should you start measuring the day?

Suppose you decide to use sunrise to sunrise, or sunset to sunset, as your fixed time interval. You soon discover that the time of sunrise is not fixed. During part of the year the Sun rises a bit later each day, and during another part of the year it rises a bit earlier each day. See Table 5.1.

TABLE 5.1
TIME OF SUNRISE AND SUNSET DURING A TYPICAL YEAR AT MASSENA, NEW YORK (44°56'N, 74°51'W)

Date	Sunrise (A.M.) (Eastern Standard Time)	Sunset (P.M.) (Eastern Standard Time)
January 21	7:30	4:52
February 21	6:52	5.35
March 21	6:00	6:14
April 21	5:04	6:53
May 21	4:24	7:29
June 21	4:13	7:50
July 21	4:34	7:38
August 21	5:09	6:55
September 21	5:46	5:58
October 21	6:24	5:04
November 21	7:06	4:25
December 21	7:35	4:21

You also find that the positions of sunrise and sunset also change. See Figure 5.1 for changes in the sunset position over an 8-month period. If the times *and* the positions of sunrise and sunset vary, how, then, did people find a fixed point from which to measure the length of an entire day? The answer lies in the apparent path of the Sun. Look at Figure 5.2, which shows the Sun's apparent path through the sky on three different days of the year. Notice that, although the positions of sunrise and sunset and the length of the Sun's path vary, the midpoint of all three paths occurs when the Sun crosses the line running due

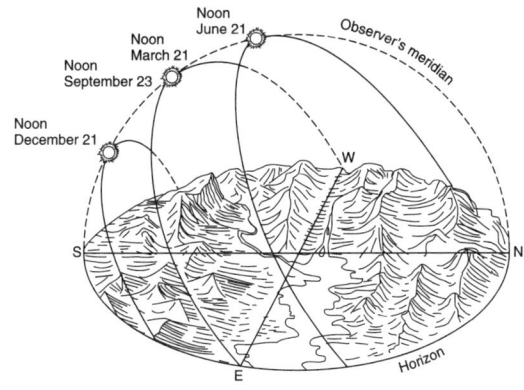

Figure 5.2 The Apparent Path of the Sun as Viewed by an Observer in New York State Varies from Season to Season, but the Sun Always Crosses the Observer's Meridian at Noon.

north-south through the observer's zenith.

You recall from Chapter 4 that this imaginary line is the observer's meridian. Also notice that, when the Sun crosses the observer's meridian, it is at its highest point in the sky for that day and a vertical stick will cast the shortest shadow. The moment when the Sun is at its highest point in the sky is called *solar noon*. No matter how the number of daylight hours varies, solar noon always occurs exactly in the middle of the daylight period. Also, the Sun is always seen in the same direction at solar noon.

The Solar Day Versus the Sidereal Day

The position of solar noon can be marked using two fixed points, one on the horizon and one nearby, such that they line up with the Sun at solar noon. (Since the Sun crosses the observer's meridian at solar noon, a line connecting these markers will run due north-south; a shadow cast by a vertical stick will also lie along a north-south line—a handy way of determining direction without a compass.) Whenever the Sun aligns with these two markers, it is solar noon. The time Earth takes to rotate from one solar noon until the next is called a *solar day*; that is, a solar day is the time from when the Sun crosses an observer's meridian until the Sun next crosses the observer's meridian.

A similar technique can be used to track stars at night. The time from when a star crosses an observer's meridian until the same star next crosses the observer's meridian is called a *sidereal day*. However, a solar day and a sidereal day are not equal in length. How can *that* be?

The answer lies in the fact that stars are very much farther away from Earth than is the Sun. As a result, the direction to a star stays virtually the same throughout Earth's orbit, while the direction to the Sun changes drastically. Look at Figure 5.3. Even in this small diagram, you can see that from one place in Earth's orbit to another the direction to the star changes only slightly. Imagine how little it would change if the star was shown trillions of miles away! There would be almost no detectable difference in the direction to the star from any two places in Earth's orbit.

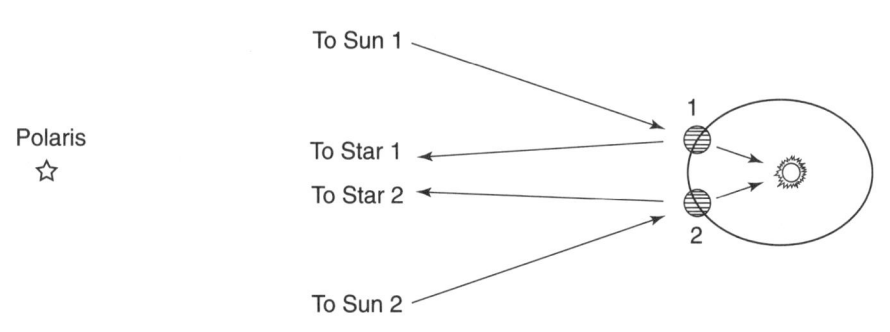

Figure 5.3 As Earth Orbits, the Change in the Direction to a Star Is Tiny Compared to the Change in Direction to the Sun.

Now look at Figure 5.4. Let's begin with Earth at position *A* when the Sun

Our System of Time

and a star are at an observer's meridian, and mark the direction to the Sun and to a star. When Earth has rotated exactly 360°, the star is back on the observer's meridian, marking the end of one sidereal day. But remember that, as Earth rotates on its axis, it also revolves around the Sun. Therefore, by the time an observer reaches position *B*, Earth's position is different too, and so is the direction to the Sun. Therefore, Earth must rotate *more* than 360° for the Sun to reach the observer's meridian and the next solar noon. Thus, a solar day is *more* than one Earth rotation long. As you can see, Earth must rotate an additional 1° per day (which takes about 4 minutes) in order for the Sun to reach the observer's meridian again. Thus, a solar day is about 4 minutes longer than a sidereal day.

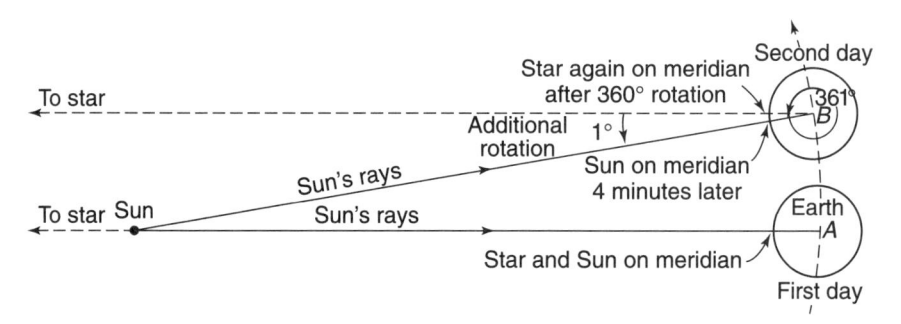

Figure 5.4 The Solar Day Is about 4 Minutes Longer than the Sidereal Day. (Angles are greatly exaggerated in this diagram.)

Which day is better to use as the basis for a timekeeping system? If the sidereal day is used, the problem shown in Figure 5.5 occurs. If we call the time at location *M* midnight when Earth is at position 1, then, by the time Earth has revolved to position 2, midnight would be occurring during the daytime! For that reason, the solar day is used as the basis of our timekeeping system. No matter where Earth is in its orbit, solar noon always occurs where it belongs—in the middle of the daytime.

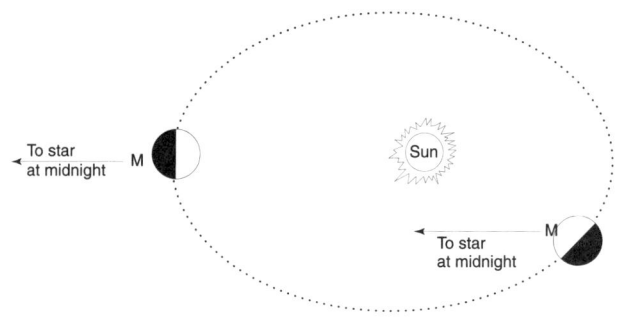

Figure 5.5 Midnight (12:00 A.M.) Sidereal Time at Location *M* at Two Times During the Year. Note that sidereal time gets "out of synch" with our periods of daylight and darkness because of Earth's changing position in relation to the Sun.

The Mean Solar Day

With the development of ever more accurate devices for measuring time, it became clear that the length of a solar day varies throughout the year. If we set a clock to 12:00 at solar noon one day, it will not read 12:00 at solar noon the next day. How can this be, since we do not detect the stars or the Sun speeding up or slowing down as they move through the sky because of Earth's rotation?

The answer lies in the relative motions of Earth and the Sun due to Earth's revolution. Although the rate at which Earth rotates remains steady, the rate at which it orbits the Sun doesn't. Remember that Johannes Kepler determined that Earth travels farther during a fixed time period in one part of its orbit than during another part. This, in turn, means that Earth is revolving faster during one part of its orbit and slower during another part.

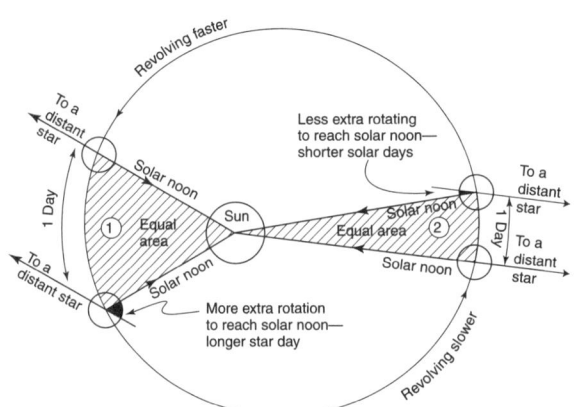

Figure 5.6 Varying Length of the Solar Day. Position 1: The faster rate of revolution causes a greater daily change in Earth's position, requiring more rotation to reach solar noon. The result is a longer solar day. Position 2: A slower rate of revolution results in less of a daily change in Earth's position; thus less rotation is required to reach solar noon and the solar day is shorter. (Angles are greatly exaggerated in this diagram.)

In Figure 5.6, the effect of this difference in orbital speed is shown. At position 1, Earth has to rotate farther between solar noons than it does in position 2. Thus, the solar day is longer at position 1 than at position 2. Making clocks that would speed up or slow down in time with the solar day is very difficult. Clocks are generally constructed to mark off absolutely uniform periods of time. As a result, if you were to set your watch to 12:00 at solar noon on one day, your watch would read earlier than 12:00 at solar noon the next day for part of the year, and later than 12:00 at solar noon for the remainder of the year. The solution has been to split the difference and use the *average* length of a solar day, or the ***mean solar day***, as the basis for timekeeping. Therefore, clocks are constructed to complete a 24-hour cycle in a mean solar day.

Local Time

So far, we have talked about the day in a very general way, with solar noon occurring when the Sun crosses an observer's meridian. However, an observer's meridian depends on his or her longitude, meaning that each

Our System of Time

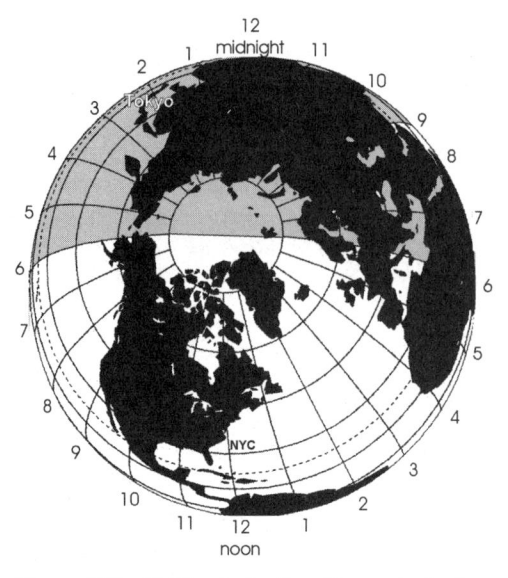

Figure 5.7 At Precisely the Same Instant, a Person in New York City Experiences Solar Noon While in Tokyo, Japan, the Time is 2:00 A.M.

observer has his or her own time scale based upon his or her location. A careful look at Figure 5.7 should make it clear that, while someone in New York City is experiencing solar noon, someone in Tokyo, Japan, is asleep in bed at 2 A.M. Our solar noon in New York (~75°W) occurs some 3 hours earlier than solar noon in California (~120°W)—a 45° difference in longitude—because Earth takes 24 hours to spin through 360°, rotating through 15° every hour.

In the past, every locality set its clocks to its own solar noon. Clocks in Syracuse would be set to a different local time from those in Buffalo. As soon as people began traveling long distances rapidly, by stagecoach, train, or steamboat, however, local timekeeping created severe problems for companies that wanted to set up arrival and departure timetables.

Time Zones

In 1883 and 1884, the countries of the world agreed to set up a series of time zones. In a *time zone* the same time is used throughout the zone, instead of the local time at each place within the zone.

Since Earth rotates 15° per hour, time zones were set up around a series of meridians spaced at 15° intervals starting at the Prime Meridian, which, you have learned, passes through Greenwich, England. Time zones officially cover 7½° on either side of these meridians, so places between longitude 7½° W and

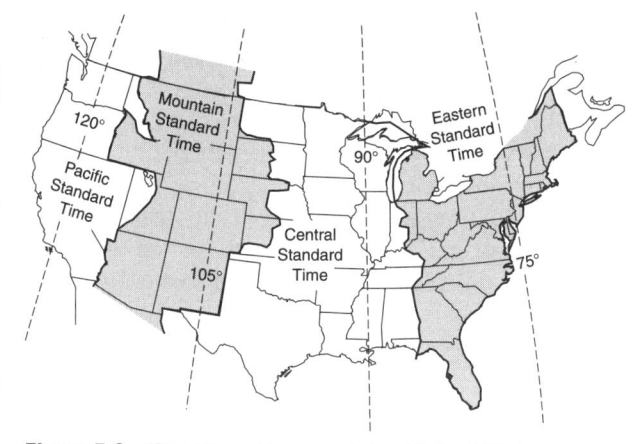

Figure 5.8 The Time Zones of the United States. Source: *ESCP, Investigating the Earth, Teachers Guide*, Part 1, p. 135, American Geological Institute, 1967.

7½° E all set their clocks to Greenwich Mean Time (GMT). The next zone to the west extends from longitude 7½° W to 22½° W, and all clocks in that zone are set 1 hour behind GMT. New York State falls in the time zone surrounding the 75° W meridian, which extends from 67½° W to 82½° W. This zone sets its clocks 5 hours behind GMT.

To avoid inconveniences, the boundaries are shifted where they pass over land so that small countries or whole states fall within the same time zone. Figure 5.8 shows how the time zone boundaries in the United States have been shifted to follow state lines.

The International Date Line

As you travel west of Greenwich, England, you pass through 12 time zones before reaching longitude 180°, ending up 12 hours *behind* GMT. Likewise, as you travel east of Greenwich, England, you also pass through 12 time zones to reach longitude 180°, but now you are 12 hours *ahead* of GMT. How can you be 12 hours behind and 12 hours ahead of GMT at the same location? The answer is that the international date line is located here, at 180° longitude, and marks the boundary between 2 consecutive days; as you cross the date line from east to west, you advance 1 day. The time of day doesn't change, only the date.

To understand how the date line works, consider that the change from one day to the next is made at midnight. Look at Figure 5.9a. At Greenwich, England, it is solar noon on Monday, and the international date line is at midnight. Point *A* is at 1:00 A.M. Monday, and point *B* is at 11:00 P.M. Monday. Now (see Figure 5.9b), Earth rotates for 1 hour, and Greenwich goes from noon

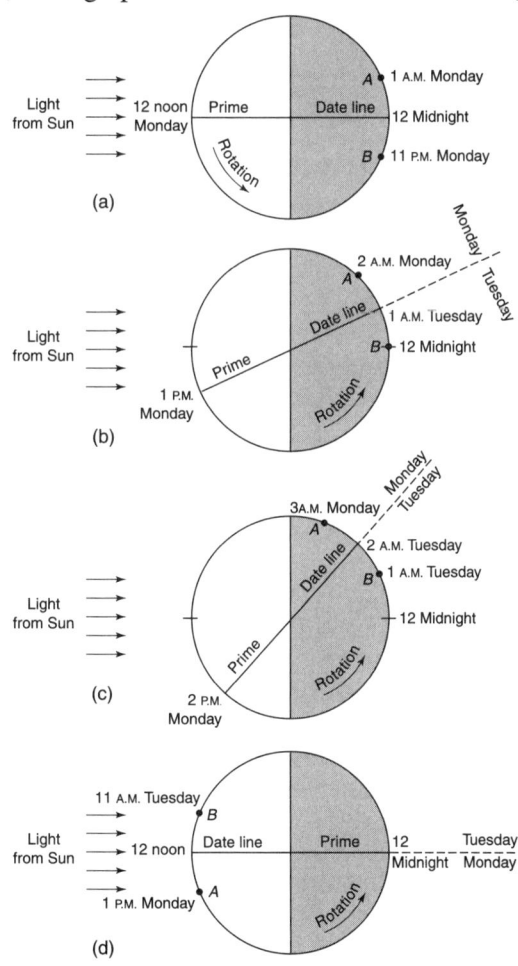

Figure 5.9 Understanding the International Date Line.

Our System of Time

on Monday to 1:00 P.M. on Monday afternoon. Point *A* is now at 2:00 A.M. Monday and point *B* goes from 11:00 Monday to midnight, but the date line goes from midnight on Monday night to 1:00 P.M. on *Tuesday* morning. The date line is still 12 hours behind GMT (1:00 P.M. − 12 hr = 1:00 A.M.) and at the same time is 12 hours ahead of GMT (1:00 P.M. + 12 hr = 1:00 A.M.). But 12 hours behind 1:00 P.M. Monday is 1:00 A.M. Monday, while 12 hours *ahead* of 1:00 P.M. GMT is 1:00 A.M. on *Tuesday*. Thus, the date line marks the boundary between the two different days.

Now (see Figure 5.9c), Earth rotates yet another hour. Greenwich goes to 2:00 P.M. Monday, point *A* is at 3:00 A.M. Monday and the date line is at 2:00 A.M. Tuesday morning, but now point *B* is at 1:00 A.M. *Tuesday*. At this time it is Monday at point *A*, but Tuesday at point *B*. Let's continue this all the way to noon on Tuesday at the international date line (see Figure 5.9d). Now Greenwich is at midnight, finally entering Tuesday, and point *A* is at 1:00 P.M. Monday, but point *B* is at 11:00 A.M. Tuesday. The dividing line between Monday and Tuesday is the international date line. East of the date line it is Tuesday; west of the date line it is Monday.

This sequence continues until the date line once again rotates to the midnight position, when the boundary changes from Monday/Tuesday to Tuesday/Wednesday. With each successive rotation of Earth, the boundary advances another day.

THE MONTH

Just as the concept of a day arose from the natural cycle of light and darkness, the concept of a month originated in the natural cycle of lunar phases. (See Chapter 4.) The Moon, with its changes in shape, brightness, and position, has always fascinated humans. Even in the distant past, observers also perceived a connection between lunar phases and a natural human cycle—the menstrual cycle; the word *menstrual* means, literally, "recurring once a month" (and is derived from the Latin word *mensis* meaning "month"). Indeed, there is evidence that the Moon has been used from the very earliest days of human civilization as a measuring tool for time. A Cro-Magnon eagle bone found in Le Placard, France, with notches cut at regular intervals and marked into groups, suggests that the Moon was the first celestial object used by humans for timekeeping.

When people first began to use arithmetic to count time, it was not difficult to determine that the Moon went through its full cycle of phases in 29 or 30 days. By 430 B.C., the Greek astronomer Meton had determined that the new Moon occurred on the same day each year in a 19-year cycle. A hundred years later, another astronomer, Callippus of Cyzicus, had discovered that the 19-year cycle worked only if a day was dropped once every 76 years.

Synodic and Sidereal Months

We now know that we see the phases of the Moon because the Moon revolves around Earth in an elliptical orbit that is tilted at an angle of 5½° to the plane of the ecliptic. (Refer to Figure 3.11.) We also know that the Moon goes through its cycle of phases, from one new Moon until the next, in precisely 29.530588 mean solar days, or 29 days, 12 hours, 44 minutes, and 3 seconds. This 29½-day month, from one new Moon phase until the next, is called a *synodic month*.

Like a solar day, the synodic month is based upon the Moon's position in relation to the Sun. New Moon occurs when the Moon is aligned between the Sun and Earth. Because Earth moves in its own gigantic orbit around the Sun while the Moon is making its much smaller orbit of Earth, the Moon has to travel more than 360° around Earth to align again with the Sun. See Figure 5.10.

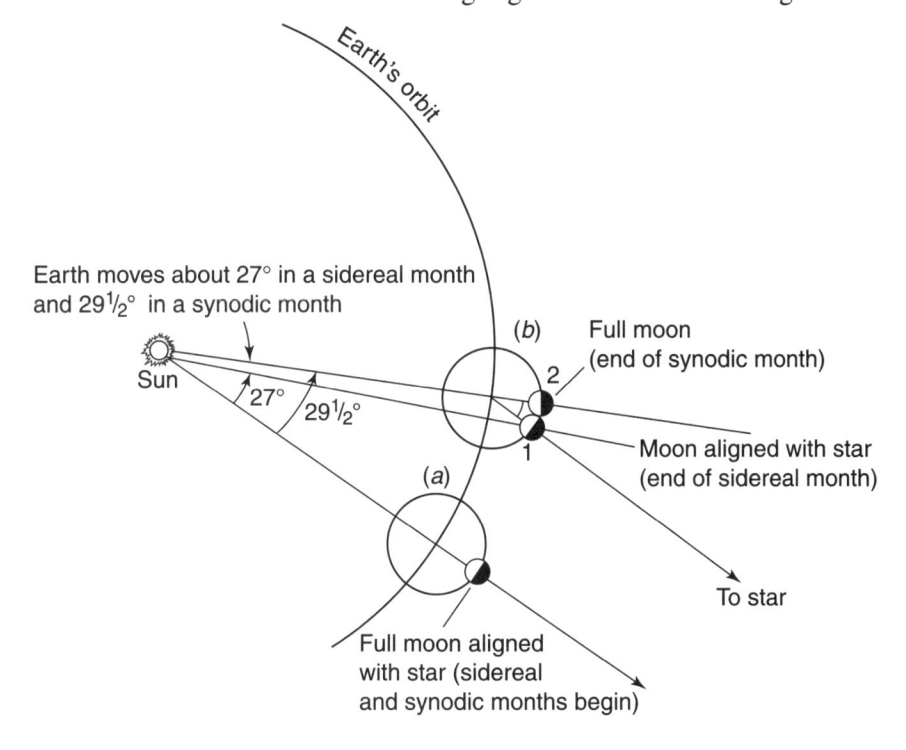

Figure 5.10 Difference in Length of Sidereal and Synodic Months. (a) The full Moon is observed to be near some bright star. (b) Roughly 27⅓ days later, at position 1, the Moon is again near the same bright star, completing a *sidereal month*. But Earth and the Moon have moved together around the Sun during that time. The Moon must revolve for 2⅙ more days before the Moon, Earth, and the Sun are again lined up so that the Moon is full again (position 2), completing a cycle of phases, or a *synodic month*. Thus, the time between full Moons, or the synodic month, is 27⅓ + 2⅙ = 29½ days.

If, instead of the Sun, we use alignment with a distant star, we find that the Moon completes a 360° orbit of Earth every 27.32166 mean solar days, or 27 days, 7 hours, 43 minutes, and 11 seconds. This 27⅓-day month, based upon alignment with a distant star and representing a 360° orbit of Earth, is called a *sidereal month*.

Since the phases of the Moon are so universally visible, the 29½-day synodic month has become the basis for all calendar months. However, lunar months will not fit neatly into a solar year without fractional division, nor will days fit exactly into lunar months. Since the solar year is 365.242199 mean solar days long, you can't get exactly a whole number of days into a solar year either. How, then, can the year be divided into months and weeks? A convenient week will have a number of days that will divide equally into 365. Yet, no matter how you look at it, the week is an artificial unit of time and could have been 5, 6, or even 10 days long. For example, there could have been a 73-week year made up of 5-day weeks for a total of 365 days, because 5 is one of the few practical numbers that divides into 365.

The subdivision of the month into 7-day weeks can be traced back to the Genesis story of creation. It coincides roughly with dividing the month into four equal parts: new Moon to first quarter, first quarter to full Moon, full Moon to last quarter, and last quarter back to full Moon. And 7 days works nicely because $7 \times 52 = 364$, very close to the actual length of a year. But the 7-day week really became firmly entrenched because, in the Genesis story, God created the world in 7 days, which gave this number a sacred significance that survives to this day.

THE YEAR

The concept of a year originated in the natural cycle of seasons. In the Middle East, that cradle of civilization, there were probably two seasons—a hot, dry summer season and a cool, wet winter season. The Egyptians had a built-in time marker in the annual flooding of the Nile River during the wet season. But seasonal changes in temperature and rainfall are highly variable. Sometimes the Nile floods earlier in the season, sometimes later. And who in New York State hasn't experienced occasional stretches of balmy weather in March, only to encounter freezing temperatures and a late spring snowstorm in April? Thus, a more reliable indicator of the time of year is needed. Here again, the observation of celestial objects provided a valuable tool, and the Sun takes center stage.

Seasons and the Sun's Path

Because of Earth's rotation, the Sun appears to move in an arc across the sky from east to west. The daily path that the Sun traces through the sky goes through a cycle of changes with the seasons. Also, although the Sun always rises to the east of an observer and sets to the west, the exact positions of sunrise and sunset on the horizon change cyclically with the seasons too. Three factors combine to produce these cyclic changes: Earth's *revolution* around the Sun, the *tilt* of Earth's axis of rotation, and the *parallelism* of Earth's axis of rotation (the position of the axis at any given time is always parallel to its position at any other time). Figure 5.11 shows how these factors result in important changes in Earth's position relative to the Sun on key dates in the seasonal cycle of changes.

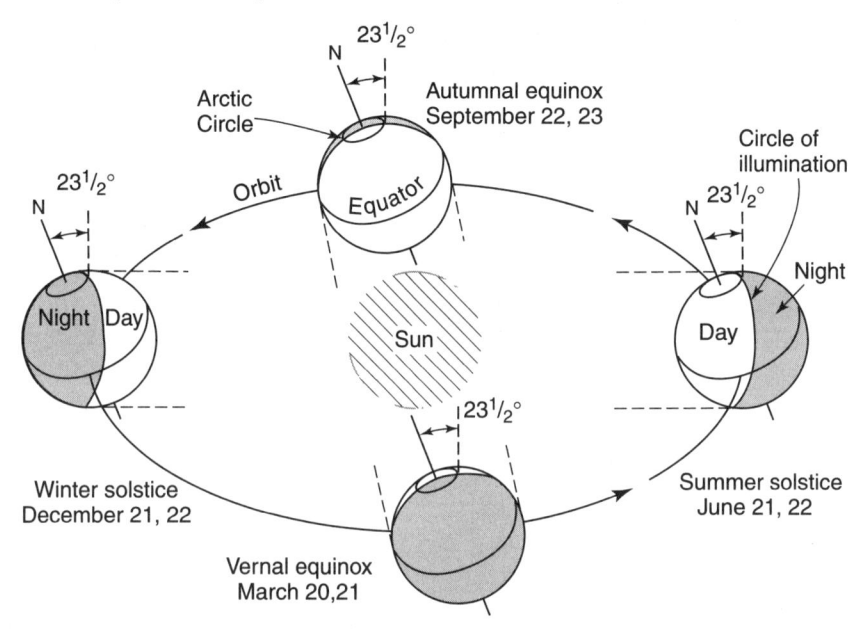

Figure 5.11 Earth's Changing Position in Relation to the Sun.

The Solstices

As Earth orbits the Sun, it also rotates on its axis. At one end of its axis is the North Pole, and at the other end the South Pole. Earth's axis of rotation is tilted at an angle of 23½°, so that during one part of its orbit the North Pole is leaning away from the Sun and the South Pole is leaning toward the Sun. The day on which the North Pole leans farthest *away from* the Sun is December 21, and this day is called the **winter solstice**. As you can see in Figure 5.12a, because of Earth's spherical shape only one parallel, the Tropic of Capricorn receives the Sun's direct rays (the direct rays strike Earth perpendicular to its surface).

On December 21, the direct rays of the Sun reach farthest south of the Equator—23½° S—and this parallel is given a special name, the Tropic of Capricorn. However, when the Sun's rays strike the Tropic of Capricorn at 90°, they are striking New York City, which is 64.5° of latitude away, at 90° − 64.5°, or only 25.5°—the lowest angle of the year. At positions north of New York City, the Sun's rays strike at an even smaller angle. Therefore, the shadow cast

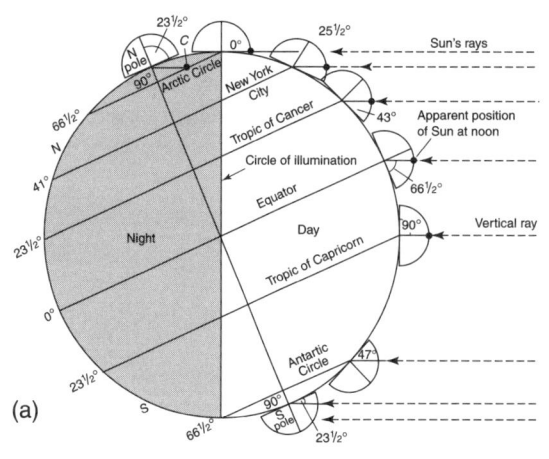

Figure 5.12 (a) Altitude of the noon Sun at the winter solstice (b) Altitude of the noon Sun at the equinoxes

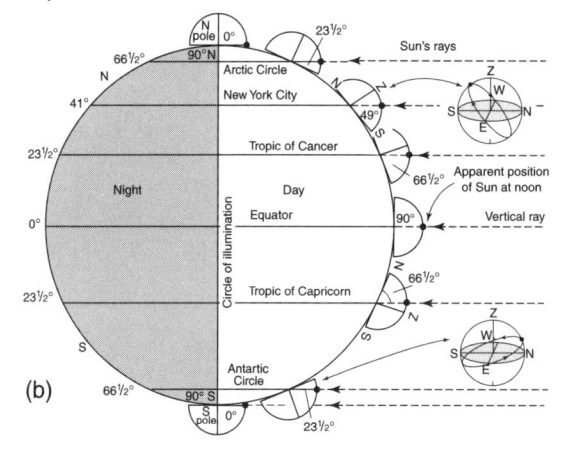

by a stick at noon in New York State is at its longest on December 21.

As Earth orbits the Sun, its axis remains tilted at this angle, so that, when it reaches the *other* side of the Sun, it is parallel to its previous position and the South Pole now leans away from the Sun while the North Pole leans toward it. The day on which the North Pole leans farthest *toward* the Sun is June 21, and this day is called the **summer solstice**. On this day, the direct rays of the Sun reach farthest north of the Equator—23½° N—and this parallel is given a special name, the Tropic of Cancer. Locations north of the Tropic of Cancer never receive the Sun's direct rays, but on June 21 the rays are the most nearly direct of the year. Therefore, the shadow cast by a stick at noon in New York State is at its shortest, and the shadow falls due north.

The Equinoxes

Midway between these ends of Earth's orbit, however, neither the North nor the South Pole leans toward the Sun. To be sure, the poles are still tilted, but to the side, rather than toward or away from the Sun. In fact, at these midpoints, Earth's shadow cuts precisely through the poles. (Refer to the vernal and autumnal equinoxes shown in Figure 5.11.) In other words, the edge of Earth's shadow forms a great circle, cutting Earth precisely in half. And that great circle runs through the poles—it is a meridian! As Earth rotates, every point on its surface spends exactly one-half rotation in the light and one-half rotation in Earth's shadow; day and night are equal in length. These midpoints occur on March 21 and September 23; and these dates are called, respectively, the spring and fall **equinoxes** (*equinox* is the Latin word for "equal night"). On these dates, the direct rays of the Sun strikes the Equator. See Figure 5.12b.

The Sun's Changing Path

At each of the positions shown in Figure 5.11, an observer on Earth's surface would see the Sun follow a different path as it rose, arced through the sky, and set. Figure 5.13 shows the changing path of the Sun to an observer at 42° N as Earth orbits the Sun.

Path *A* is the Sun's path on December 21, the winter solstice. Note that on this date the Sun rises at a point south of east on the horizon and sets south of west, and the arc traced by the Sun through the sky is the shortest. The shorter the Sun's daily path through the sky, the shorter the period of daylight, so December 21 is the shortest day of the year—less than 9 hours in New York State. The altitude of the Sun at noon is also at its lowest for the year, so sunlight strikes Earth's surface indirectly. Indirect sunlight is less intense than direct sunlight and therefore has less of a warming effect. Indirect sunlight, coupled with a short daylight period, is the cause of the winter season's lower temperatures.

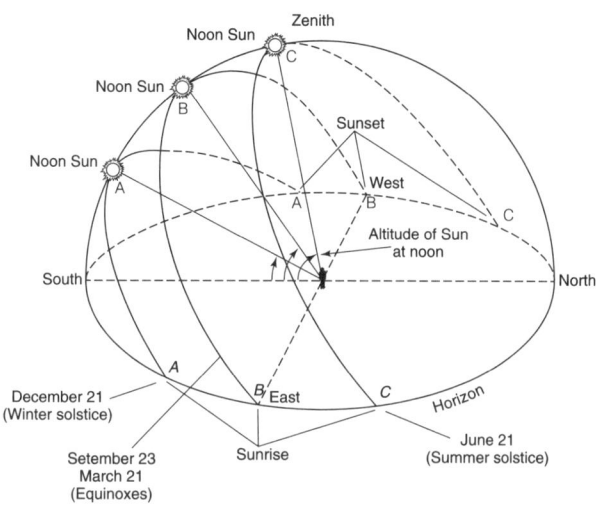

Figure 5.13 The Changing Path of the Sun to an Observer at 42° N.

Path *B* is the Sun's path on March 21 and September 23, the equinoxes. After the winter solstice, the sunrise and sunset points begin to migrate northward a bit each day, and the arc traced by the Sun slowly lengthens. When the Sun reaches path *B* on March 21, sunrise is due east, sunset is due west, and the daylight period is exactly 12 hours long; this is the spring equinox. After the spring equinox, the points of sunrise and sunset keep migrating northward, and the Sun's path keeps lengthening, until June 21, the summer solstice.

Path *C* is the Sun's path on June 21, the summer solstice. On this date, the Sun rises north of east and sets north of west on the horizon. The Sun's path is the longest of the year; therefore, the daylight period is also the longest of the year—more than 15 hours in New York State! The noon Sun also reaches its greatest altitude of the year, and sunlight strikes Earth's surface more directly and intensely than at any other time of the year. More direct, intense sunlight, coupled with longer daylight hours, is the cause of the summer season's higher temperatures.

After June 21, the points of sunrise and sunset begin to migrate southward each day, and the arc traced by the Sun shortens. By September 23, the Sun is back to path *B*. Once again the Sun rises due east and sets due west, and the daylight period is exactly 12 hours in length. After the fall equinox the Sun's path continues its steady southward migration. Each day it rises and sets farther south and traces a shorter arc through the sky until the cycle is complete on December 21, roughly 365 days later. Then the cycle begins anew, moving from path *A* to *B* to *C* and back through *B* to *A* again.

Measuring the Year

Such cyclic seasonal changes in the Sun's path and sunrise-sunset position have long been used by cultures worldwide as fixed points to measure the length of a year and to subdivide it. In principle, the average number of days between successive recurrences of an annual event will yield an estimate for the length of a year.

One method is to use the length of a shadow at noon to find the winter and summer solstices. The Chinese used the number of days between two such solstices to calculate the length of a year.

Between the tropics of Cancer and Capricorn, there is always at least one day each year when the Sun is directly overhead at noon and there are no shadows at noontime. The Incas of Peru used this fact to figure out the length of a year. They noted the days when the Sun could be seen at noon through a skinny vertical tube pointing at zenith, and counted the average number of days between recurrences.

Another method is to note the positions of sunrise and sunset on the horizon. At the solstices, the Sun rises and sets at its northernmost and southernmost positions. Of course, for this method to work, the exact same position must be used for observations and the exact point on the horizon where the sun rises and sets must be carefully marked. It is thought that the circular arrangement of the huge rocks at Stonehenge may have been used for this purpose.

Table 5.2 shows the length of the mean solar year determined by these methods in many different cultures over time—with an amazing degree of accuracy!

TABLE 5.2
ESTIMATED LENGTH OF THE MEAN SOLAR YEAR

Place	Date	Estimated Length of Year (mean solar days)	Error
Babylon	700 B.C.	365.24579	0.00344
Egypt	150 B.C.	365.2466	0.0043
China	8 B.C.	365.25016	0.0078
India	A.D. 476	366.2589	0.0167
Mexico	700	365.2420	−0.0001
Arabia	900	365.24056	−0.00170
Samarkand	1400	365.24223	0.00030
Europe	1500	365.24222	0.00034
China	1620	365.24219	−0.00003
Europe	1600	365.24222	−0.00003
Modern	2000	365.242199	

MULTIPLE-CHOICE QUESTIONS

In each case, write the number of the word or expression that best answers the question or completes the statement.

1. The time required for Earth to complete one revolution around the Sun is
   (1) 1 day
   (2) 1 week
   (3) 1 month
   (4) 1 year

2. In New York State, records will most probably indicate that the length of the shadow cast by a tree at noon each day from January to May
   (1) continuously decreases
   (2) continuously increases
   (3) remains the same
   (4) increases and then decreases

3. In New York State, how do the points of sunrise and sunset change during the course of 1 year?
   (1) They vary with each season in a cyclic manner.
   (2) They move toward the north in the autumn months.
   (3) They move toward the south in the spring months.
   (4) They remain the same during the four seasons.

Our System of Time

4. In New York State at 3 P.M. on September 21, the vertical pole shown in the diagram casts a shadow. Which line best approximates the position of that shadow?
   (1) OA
   (2) OB
   (3) OC
   (4) OD

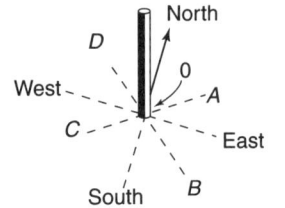

Base your answers to questions 5 and 6 on the diagram below, which represents the apparent daily path of the Sun across the sky in the Northern Hemisphere on the dates indicated.

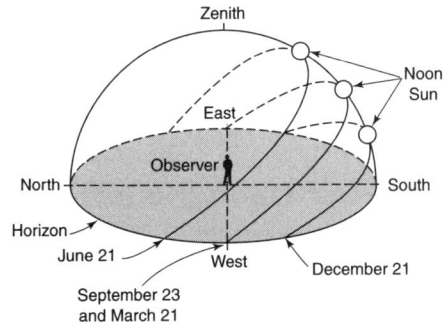

(Not drawn to scale)

5. At noon on which date would the observer cast the longest shadow?
   (1) June 21
   (2) September 23
   (3) March 21
   (4) December 21

6. Which observation about the Sun's apparent path at this location on June 21 does the diagram best support?
   (1) The Sun appears to move across the sky at a rate of 1° per hour.
   (2) The Sun's total daytime path is shortest on this date.
   (3) Sunrise occurs north of east.
   (4) Sunset occurs south of west.

7. Upon which frame of reference is time based?
   (1) the motions of Earth
   (2) the longitude of an observer
   (3) the motions of the Moon
   (4) the real motions of the Sun

8. The time required for one Earth rotation is about
   (1) 1 hour
   (2) 1 day
   (3) 1 month
   (4) 1 year

114

9. A traveler goes from New Orleans, Louisiana (90°04' W) to Miami, Florida (80°11' W). On arrival in Miami, his watch should
    (1) be set ahead 1 hour
    (2) be set ahead 2 hours
    (3) be set back 1 hour
    (4) not be changed

10. It is 10 P.M. Tuesday at 150° W. What is the day and time at 150° E?
    (1) 10 A.M. Monday
    (2) 10 A.M. Tuesday
    (3) 6 P.M. Tuesday
    (4) 6 P.M. Wednesday

11. The time between two successive passages of the Sun across a given meridian is called a
    (1) civil day
    (2) conventional day
    (3) mean solar day
    (4) solar day

12. If a ship's position is 30° N, 60° W at 12 midnight Greenwich mean solar time, what is the mean solar time on shipboard?
    (1) 2 A.M.
    (2) 4 A.M.
    (3) 8 P.M.
    (4) 10 P.M.

13. What is the relationship between apparent solar time and mean solar time?
    (1) Apparent solar time is always ahead of mean solar time by a constant amount.
    (2) Apparent solar time is always behind mean solar time by a constant amount.
    (3) The difference between apparent solar time and mean solar time varies with the seasons.
    (4) There is no difference between apparent solar time and mean solar time.

14. When does local solar noon always occur for an observer in New York State?
    (1) when the clock reads 12 noon
    (2) when the Sun reaches its maximum altitude
    (3) when the Sun is directly overhead
    (4) when the Sun is on the Prime Meridian

15. A sundial measures time based upon the position of the Sun in the sky. This time is called
    (1) apparent solar time
    (2) Greenwich time
    (3) standard time
    (4) mean time

16. Cities located on the same meridian of longitude must have the same
    (1) altitude
    (2) latitude
    (3) length of daylight
    (4) solar time

Our System of Time

Base your answers to questions 17 through 21 on the diagram to the right, which represents Earth in relation to the Moon and Sun during the spring equinox as viewed from a point in space above the North Pole.

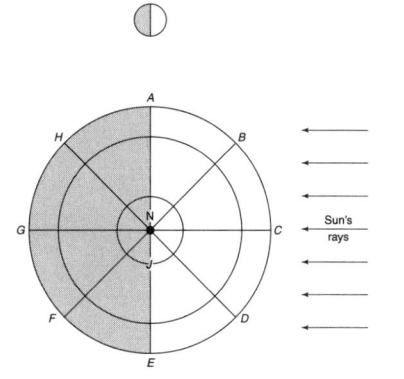

17. The Sun's vertical rays are striking Earth at this time at latitude
    (1) 0°
    (2) 23½° N
    (3) 23½° S
    (4) 66½° N

18. If it is 12 noon at point C, the time at F is
    (1) 3 A.M.   (3) 6 P.M.
    (2) 6 A.M.   (4) 9 P.M.

19. Sunrise on Earth is occurring at the point on the diagram labeled
    (1) A (3) E
    (2) B (4) G

20. If the Prime Meridian is at position B, the longitude at E is
    (1) 90° E   (3) 135° E
    (2) 90° W   (4) 135° W

21. In 7 days the Moon will have moved into a position nearly opposite the meridian indicated by letter
    (1) B (3) G
    (2) E (4) H

Base your answers to questions 22 through 26 on the diagram below and on your knowledge of earth science.

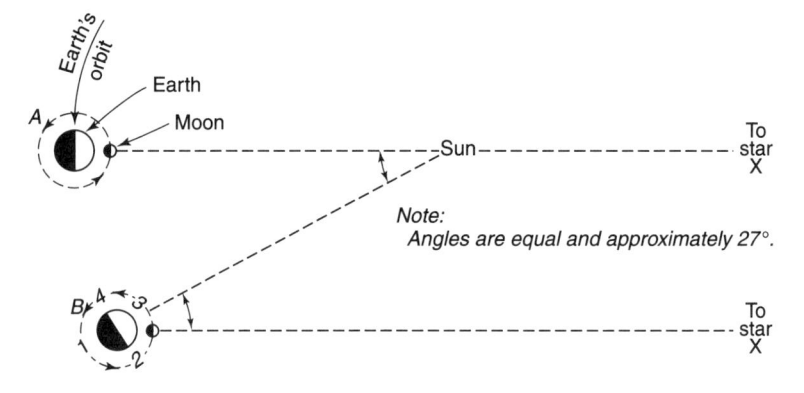

116

22. Earth will travel from *A* to *B* in
(1) 1 day
(2) 2.5 days
(3) 27⅓ days
(4) 29½ days

23. The length of time it takes Earth to travel from position *A* to position *B* is called a
(1) mean solar day
(2) sidereal month
(3) solar year
(4) synodic month

24. How long must the Moon travel beyond where it is shown when Earth is at position *B* in order to be in the same phase as it was at position *A*?
(1) 1 day
(2) 2 days
(3) 6 hours
(4) 7 days

25. In respect to Earth at position *B*, in which phase is the Moon?
(1) full
(2) new
(3) gibbous
(4) crescent

26. At position *B*, the highest tides will occur on Earth when the Moon is at points
(1) 1 and 2
(2) 1 and 3
(3) 2 and 4
(4) 3 and 4

Base your answers to questions 27 through 29 on the diagram below, which shows four lines drawn across a portion of the sky above the southern horizon as viewed from the Northern Hemisphere.

27. Which two lines represent possible daily paths of the Sun?
(1) *A* and *B*
(2) *A* and *D*
(3) *B* and *C*
(4) *C* and *D*

28. If all four lines were possible daily paths of the Sun, which would produce the greatest number of daylight hours?
(1) *A*
(2) *B*
(3) *C*
(4) *D*

Our System of Time

29. If the above diagram were to be adapted to fit observations made at 40° S, one necessary change in the diagram would be to
   (1) change label East to West
   (2) change label East to South
   (3) change label South to North
   (4) change label x to y

Base your answers to questions 30 and 31 on the diagram below, which represents the western horizon as viewed by an observer in New York State. A, B, C, D, and E are possible positions of sunset, and A and E represent the outermost limits beyond which sunset does not occur.

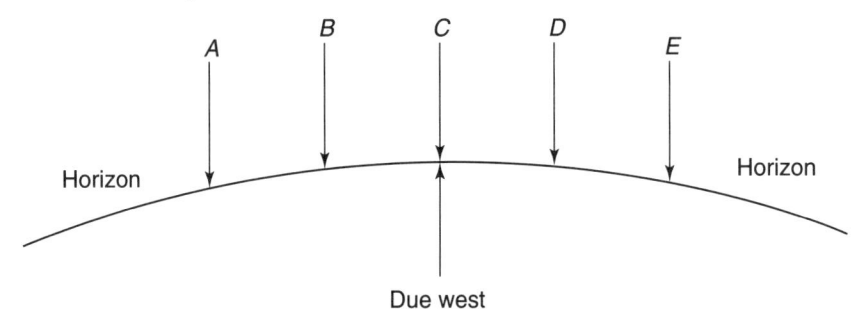

30. Which position most accurately indicates the position of sunset on May 6?
   (1) A   (3) C
   (2) B   (4) D

31. Which sunset position represents the day with the greatest number of daylight hours?
   (1) A   (3) D
   (2) C   (4) E

Constructed Response Questions

32. Base your answers to parts (a) through (c) on the diagram below, which shows Earth's orbit around the Sun.

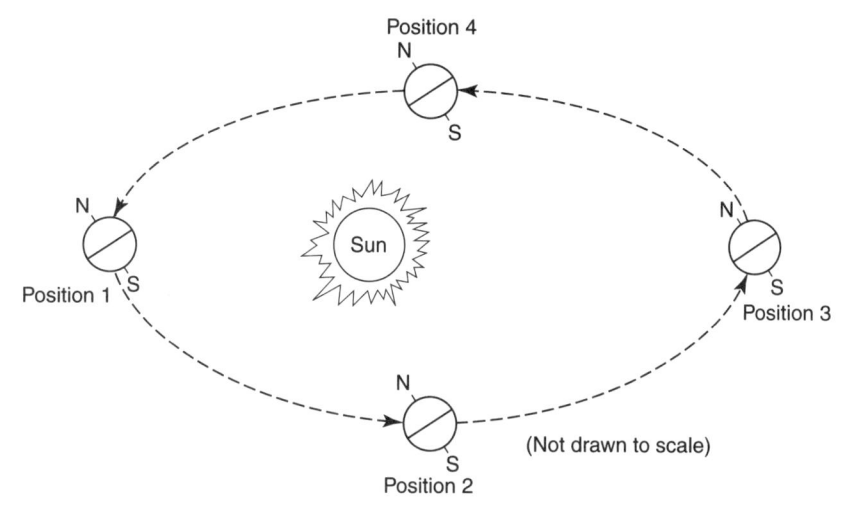

(a) Write the position in the diagram that corresponds to each of the following dates for an observer in the Northern Hemisphere: [4]
   i. Winter solstice
   ii. Summer solstice
   iii. Vernal equinox
   iv. Autumnal equinox

(b) Compare the length of a mean solar day when Earth is at position 1 and at position 3. [2]

(c) Compare the length of daylight hours for an observer in Watertown, New York (44° N) and an observer in Miami, Florida (26° N) when Earth is at position 1. [2]

33. A sundial is located in Philadelphia (75° W). Explain why the sundial indicates time that is different from that shown by a clock. [2]

Extended Constructed Response Questions

34. Why is the solar day used as the basis of timekeeping rather than the sidereal day? [2]

35. Describe a procedure by which an observer's longitude could be determined. [3]

36. Construct a labeled diagram showing Earth's orbit around the Sun. [2]
   (a) On your diagram, indicate the points of greatest and least gravitational attraction and the points of greatest and least orbital velocity. [2]
   (b) Write a short paragraph explaining how the varying length of a solar day relates to the gravitational attraction between Earth and the Sun. [3]

37. A news service dispatch from Tokyo, Japan, is dated December 21, 2000. It appears in a New York newspaper on December 20, 2000. Write a short paragraph accounting for the difference in dates. [2]

Unit Two: MODERN ASTRONOMY

Chapter 6
TOOLS OF THE MODERN ASTRONOMER

> **KEY IDEAS** Humans perceive the universe by the radiation it emits. Through much of recorded history, our understanding of the universe was based upon observations made with the naked eye. This was a great limitation, for our eyes perceive only a small fraction of the radiation emitted by the universe—radiation with wavelengths in the range called *visible light*. Technological advances have greatly extended the scope of human perception and have led to the observations upon which our current theories of the universe are based.

KEY OBJECTIVES
Upon completion of this chapter, you will be able to:

- Give examples of how progress in technology has increased our understanding of celestial objects.
- Explain how refracting and reflecting telescopes work and how they enhance celestial observations made with the unaided eye.
- Describe three types of spectra: continuous, absorption, and bright-line spectra.
- Explain how stellar spectra can be used to classify stars.
- Distinguish among stars, galaxies, and nebulae.

HOW HUMANS PERCEIVE THE UNIVERSE

When an observer "sees" a star or any other object, the lens in his eye has focused light emanating from that object to form an image on his retina. See Figure 6.1. The human retina is lined with specialized nerve cells that respond to light and send information to the brain, where it is interpreted. Nerve cells in the retina respond to two characteristics of light: intensity and wavelength. Intensity is interpreted as brightness; wavelength, as color.

Tools of the Modern Astronomer

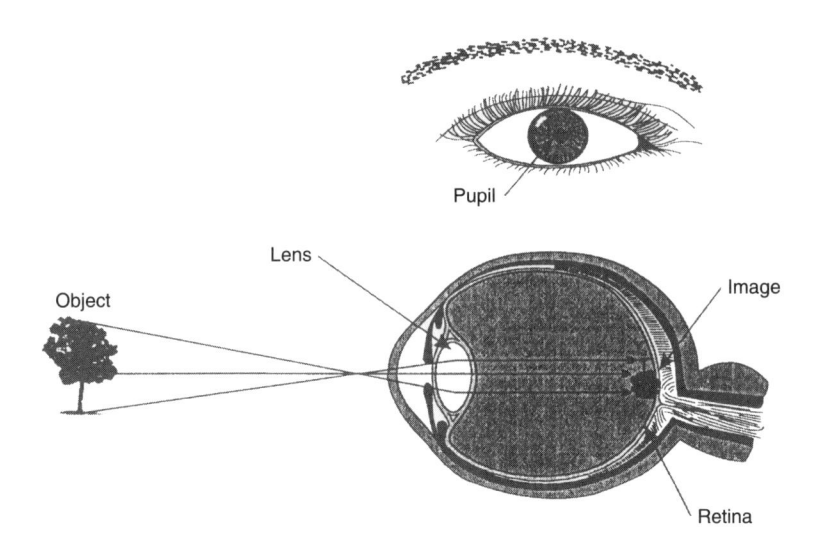

Figure 6.1 The Human Eye Forms an Image. Light from an object enters the human eye through a small opening called the pupil. The light then passes through a lens that focuses it into an image on the retina. The retina is composed of nerve cells that respond to light and send messages to the brain, where the information is interpreted and we "see" the object.

Limits of Human Perception

But the human eye has limitations. The light entering the eye is limited to what passes through the rather small lens. If the light intensity is too low, it doesn't trigger the nerve cells, and we don't "see" the light. If the intensity is too great, the nerves are overwhelmed and signal the pupil to become even smaller.

Also, the nerve cells respond to only a narrow range of wavelengths, commonly called *visible light*. Light of longer or shorter wavelengths doesn't trigger the nerve cells, and again we don't "see" it. Therefore, our eyes limit what we "see" of the universe to objects giving off light in a narrow range of wavelengths and brightness.

Another limitation is resolution. Think of the image sent to the brain as a series of dots, with each dot being the information from a single nerve cell in the retina. Although microscopic, nerve cells are huge compared to light waves. Many billions of light waves, containing lots of detail, can strike a single nerve cell, yet all the light that strikes that cell is perceived as a single dot—the detail is lost. That's the reason why planets look like stars to the naked eye. The image in the eye is too small and the light is too concentrated for us to perceive the planet as a disk. Instead, we see planets as bright dots. Most of the technological advances in observational astronomy have reduced these limitations.

Light and Lenses

The human eye contains a lens that forms images, and many of the instruments used by astronomers are simply improvements on this model. To begin, then, we need to understand how lenses focus images. A beam of light is bent, or *refracted*, whenever it passes at an angle from one substance (such as air) into another (such as glass). A *lens* is a curved piece of glass that bends, or refracts, light. Imagine a lens held vertically. The line through the center of the lens perpendicular to the plane of the lens is called the *optical axis*. A narrow beam of light parallel to the optical axis, shone through the lens, is bent so that it crosses the optical axis on the other side of the lens at a point called the lens's *focal point*. The distance from the center of the lens to the focal point is the lens's *focal length*. See Figure 6.2a.

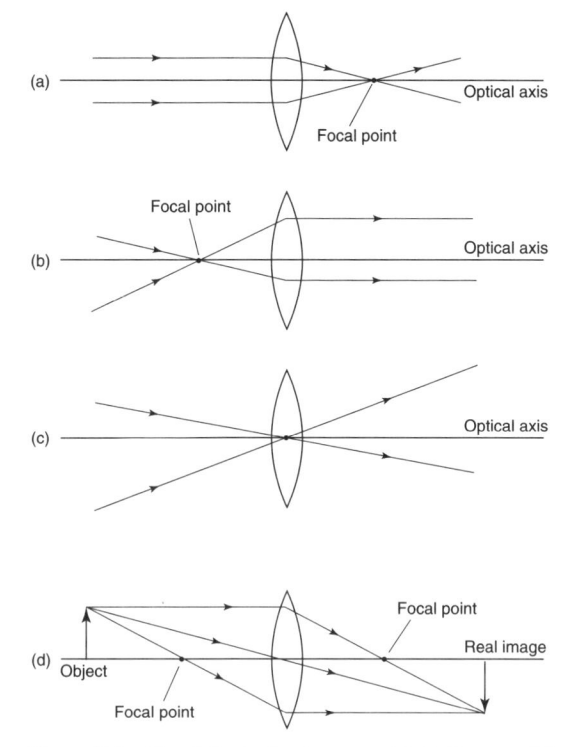

Figure 6.2 The Effect of a Lens on Light Beams. (a) Light beams entering a lens parallel to the optical axis are bent so that they cross at a focal point. (b) Light beams passing through the focal point are bent so that they emerge from the lens parallel to the optical axis. (c) Light beams that pass through the center of a thin lens are not bent. (d) A real image is formed by a lens as the three types of beams shown above meet. Source: *Discovering Astronomy*, Robert Chapman, W. H. Freeman, 1978.

If we shine a light beam through a lens so that the light appears to come from the focal point, it will leave the lens parallel to the optical axis, just the opposite of the result described above. See Figure 6.2b. If we shine a beam directly through the center of the lens, the lens does not bend the light. See Figure 6.2c.

We see an object by light that is emitted or reflected by the object. Whether it gives off its own light, or reflects light from another source, every point on the object sends out light in all directions. Let's place an object, for example, a flagpole, at some distance from a lens. See Figure 6.2d. A point at the tip of the flagpole reflects light in all directions. The light beams include one that is parallel to the optical axis, one that passes through the center of the lens, and one that passes through the focal point. As described above, the lens bends these three light beams so that they all intersect at one point, or converge, on the other side of the lens. The result is the image we see of the point at the tip of the flagpole.

Tools of the Modern Astronomer

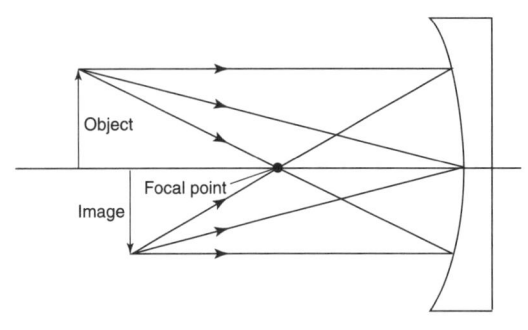

Figure 6.3 Formation of a Real Image by a Curved Mirror. Source: *Discovering Astronomy*, Robert Chapman, W. H. Freeman, 1978.

We can follow the same process with the light coming from the flag, the middle of the flagpole, the bottom of the flagpole, and every point in between. In this way, point by point, an image of the entire object is built up on the other side of the lens. This is a real image; and if we place a piece of paper at the position where the image has formed, we will see the image of the flagpole projected onto the paper.

A curved mirror can also be used to form a real image. Instead of the light converging because it is bent as it passes through a lens, the light is reflected off the surface of a curved mirror in such a way that it converges. See Figure 6.3. However, since the light doesn't pass through the mirror, but is reflected, it converges to form a real image on the same side of the mirror as the object.

TELESCOPES

What if you used a lens to form a real image, and then used a magnifying glass to examine it? Your eye would see an enlarged, close-up image. Such a combination of two lenses, one to form a real image and one to magnify the real image, is called a ***refracting telescope***. See Figure 6.4a. If, on the other hand, a mirror is used to create the real image, which is then magnified, the instrument is a ***reflecting telescope***. See Figure 6.4b. The lens or mirror that forms the real image in a telescope is called the ***objective***. The magnifier used to examine the image is the ***eyepiece***.

The larger the lens or mirror in a telescope, the more light it gathers, and the larger, brighter, and more detailed the image. Light-gathering power is proportional to the area of the collecting surface, or to the square of the aperture (diameter) of the lens or mirror. A human eye has a lens with about a 5-millimeter diameter. The objective of the telescope at the Mount Palomar Observatory, with a diameter of 5 meters, has an aperture that is 1,000 times bigger than the lens in your eye. This means that the telescope on Mount Palomar has a light-gathering power that is $1,000^2$, or 1 million, times greater than that of your eye! Since telescopes can gather more light than the human eye, they allow us to see objects too dim to be visible with the naked eye, and under magnification we can see details in the images not otherwise visible. Some telescopes can even "see" light that is not visible to our eyes. They use sensors that detect electromagnetic radiation invisible to the human eye, and that convert it into images our eyes can see. All of these technological advances have added tremendously to our store of observational evidence about the universe.

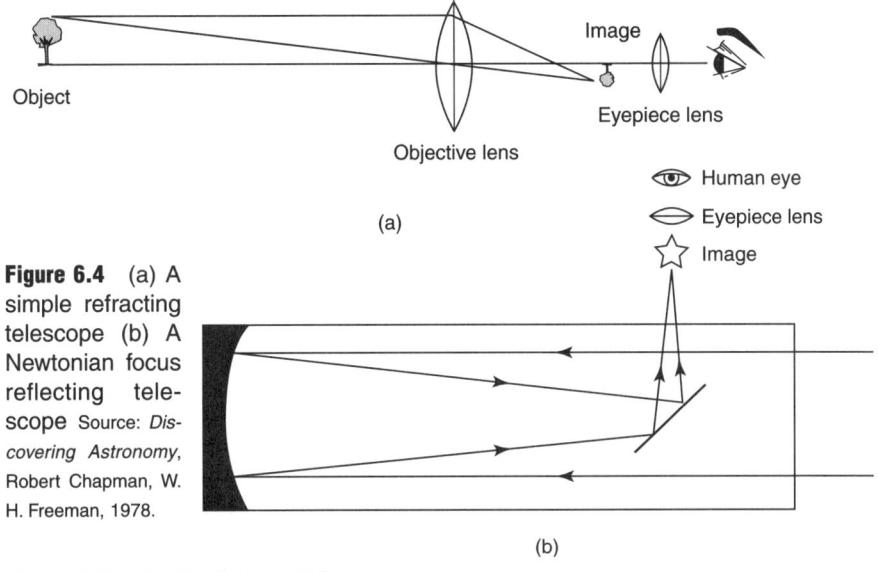

Figure 6.4 (a) A simple refracting telescope (b) A Newtonian focus reflecting telescope Source: *Discovering Astronomy*, Robert Chapman, W. H. Freeman, 1978.

SPECTROSCOPES

Another valuable tool of the modern astronomer is the spectroscope. In 1666, Isaac Newton demonstrated that white light is actually a combination of all colors. He passed light through a triangular piece of glass, or ***prism***, and found that the prism spread the light out into a rainbow of colors, or ***spectrum***. See Figure 6.5. The colors of the spectrum are, in order from longest to shortest wavelength, red, orange, yellow, green, blue, indigo, and violet. (This sequence can be remembered by using the name **Roy G. Biv**; each letter in the name is the first letter of a color, and the letters are in correct order.)

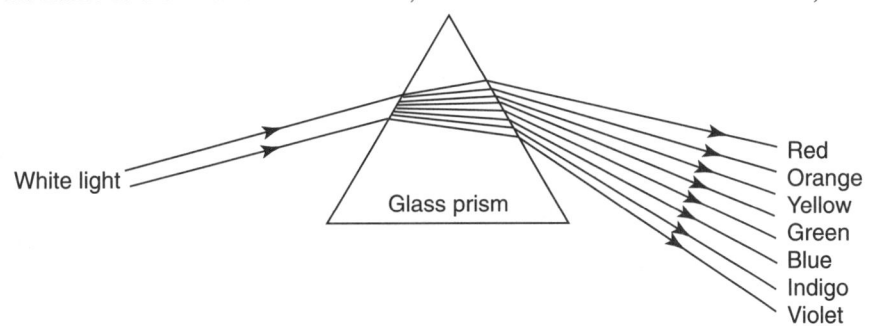

Figure 6.5 A Beam of White Light Entering a Prism Is Dispersed into a Spectrum. Source: *Astronomy Explained*, Gerald North, Springer-Verlag, 1997.

The Prism Spectroscope

A prism separates white light into its separate colors because the degree to which light is refracted depends upon its wavelength. Long-wavelength red

Tools of the Modern Astronomer

light is refracted the least; short-wavelength blue light, the most. The spreading out of light into its component colors is called ***dispersion***.

In a prism spectroscope, light from a source, such as a star, is passed through a slit placed at the focus of a lens. The light passing through the slit acts as was shown in Figure 6.2b. The light beams emerge from the lens parallel to its optical axis and then pass through a prism that disperses the light into a spectrum. Finally, the spectrum enters a telescope, which focuses the light into a real image of the slit, which is examined with a magnifier. See Figure 6.6. If the light source emits light of all wavelengths, the slit images blend together to form a ***continuous spectrum***. If any wavelengths of light are absent, however, these will appear as black lines crossing the continuous spectrum. Where the color should be there is only a dark image of the slit.

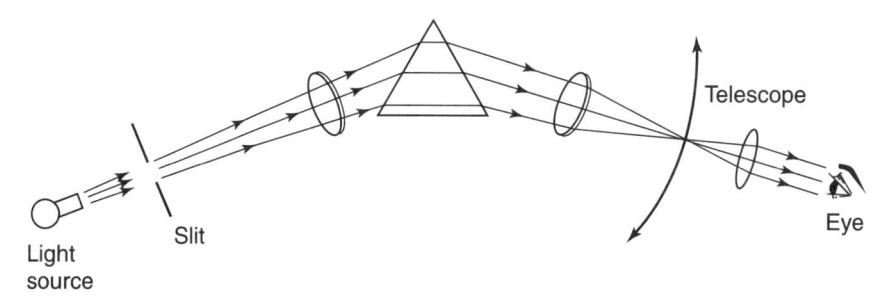

Figure 6.6 A Simple Prism Spectroscope. The telescope pivots around a point centered on the prism so that the viewer can examine all visible wavelengths. Source: *Discovering Astronomy*, Robert Chapman, W. H. Freeman, 1978.

When, in 1814, Joseph von Fraunhofer, a Bavarian optician and physicist, first used such a device to examine light from the Sun, he saw "an almost countless number of strong and weak vertical lines, which are, however, darker than the rest of the color image; some appeared to be almost perfectly black." Puzzled by the black lines, he mapped almost 600 of them (20,000 are recognized today). When he examined the spectra of the Moon, Mars, and Venus he found that they matched the Sun's spectrum exactly. From this discovery, Fraunhofer correctly concluded that these bodies do not produce their own light, but instead reflect the light of the Sun. When, however, he turned the spectroscope on Sirius and other bright stars, the spectra told a different story. The pattern of black lines in each star's spectrum was different, and the spectra of all of the stars

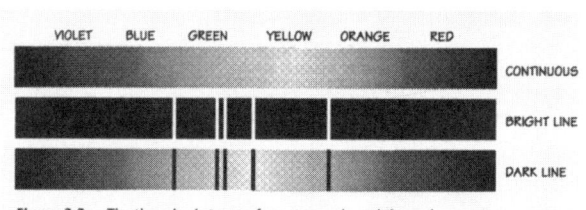

Figure 3.3. The three basic types of spectra as viewed through a spectroscope.

Figure 6.7 Types of Spectra: Continuous, Bright-line, Absorption. Source: *Astronomy: A Self-Teaching Guide*, 4th Ed., Dinah L. Moché, John Wiley, 1998.

differed from the spectrum of the Sun. It appeared that these spectral lines were a kind of "fingerprint," each star having its own unique pattern of lines.

When a spectroscope is used to examine light from chemicals burned in the laboratory, a different kind of spectrum is observed. Only a few colors are present, and each color forms a distinct image of the slit, called a *spectral line*. The brightness of each colored spectral line is due to the amount of light of a particular wavelength that is being emitted by the gas, and the line's position in the spectrum is a function of its wavelength. Over time, it became clear that each element, when burned as a gas, produces a unique pattern of bright spectral lines, or a **bright-line spectrum**, which, like a fingerprint, identifies the element. Note in Figure 6.8 the differences in the spectral lines of sodium and hydrogen. This was a momentous discovery, and scientists soon used spectral analysis to identify the patterns of colored spectral lines for all known elements.

Around 1860, Gustav Kirchhoff, a German physicist, carried out a number of experiments that explained the dark lines discovered by Fraunhofer. He knew that a glowing hot solid, liquid, or gas emits a continuous spectrum. He also knew that the glowing gas of an element forms a spectrum of bright lines. So Kirchhoff decided to view a glowing solid through a cloud of glowing gas to see what type of spectrum

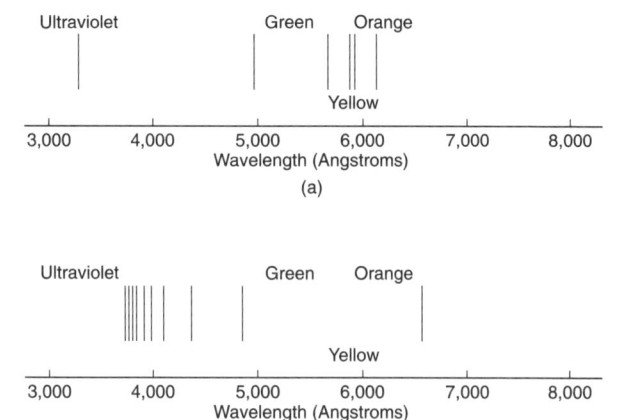

Figure 6.8 Bright-Line Spectra of (a) Sodium and (b) Hydrogen. The pattern of bright lines in every element's bright-line spectrum is unique; this spectral "fingerprint" can be used to identify elements. Source: *Discovering Astronomy*, Robert Chapman, W. H. Freeman, 1978.

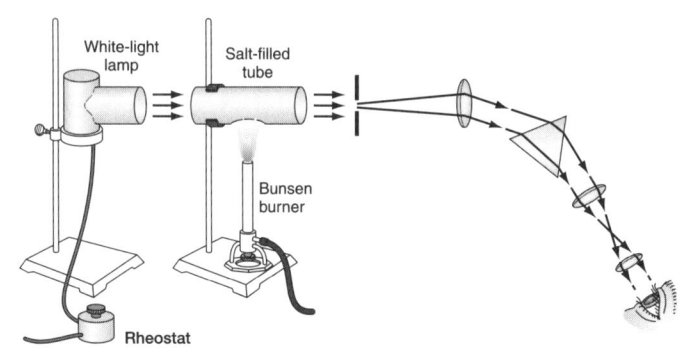

Figure 6.9 Types of Setups Used by Kirchhoff in Experiments that Led to His Discoveries Concerning Absorption Spectra. Source: *Discovering Astronomy*, Robert Chapman, W. H. Freeman, 1978.

Tools of the Modern Astronomer

would result. He set up a very bright lamp whose filament was a glowing hot solid attached to a device like a dimmer switch so that he could vary the lamp's brightness. Then he passed light from the lamp (which produces a continuous spectrum) through a glowing gas (which produces a bright-line spectrum) and made an interesting discovery. When the lamp was turned low, he saw the bright lines of the glowing gas superimposed on the dim background of the lamp's continuous spectrum. When he slowly turned up the lamp so that its continuous spectrum was just as bright as the bright-line spectrum of the gas, the bright lines became virtually invisible.

But when Kirchhoff turned up the lamp to its greatest intensity, he saw dark lines where the bright lines of the glowing gas had been located! He concluded that, when a hot, glowing gas is placed in front of a still hotter source of a continuous spectrum, the gas absorbs light at the same wavelengths that it would emit if viewed alone, producing a spectrum with dark lines, called an ***absorption spectrum***. In other words, the pattern of dark lines is identical to the pattern of colored spectral lines that would be produced by the elements in the gas! See Figure 6.10.

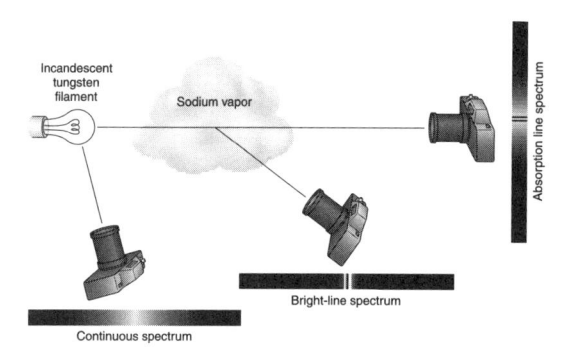

Figure 6.10 The Formation of Continuous, Bright-line, and Absorption Spectra. Source: *Discovering Astronomy*, Robert Chapman, W. H. Freeman, 1978.

Kirchhoff then applied these discoveries to astronomy. He concluded that light from the hot interior of the Sun passes through cooler gases in the Sun's outer atmosphere. Thus, the dark lines in the solar spectrum are the fingerprints of elements in the Sun! Kirchhoff set up an apparatus that allowed him to view, side by side, the dark lines in the Sun's spectrum and the bright lines produced by burning elements found on Earth. By matching lines in the two types of spectra, he identified many elements in the solar atmosphere that also are found on Earth, including hydrogen, sodium, iron, calcium, magnesium, nickel, and chromium. A few years later, during a solar eclipse, a spectrum matching no known element on Earth was found. Scientists called the new element helium, after the Greek word, *helios*, for Sun. Years later, helium was also found on Earth.

Since Kirchhoff's time the spectra of thousands upon thousands of stars have been examined. And although stars vary in composition, they consist of the same elements that are found on Earth and in our Sun.

The Spectrophotometer

In a *spectrophotometer*, a photometer (an electronic device that measures the brightness of light) is attached to a spectroscope and used to view starlight. A spectrophotometer works as follows. After the prism has dispersed the light, a narrow slit lets one color of light at a time pass through to a lens that focuses it onto the photometer. In this way the brightness of each wavelength of light in the spectrum can be very accurately measured. The analysis of the continuous, bright-line, and absorption spectra of light from stars and other celestial objects is a powerful tool, allowing scientists to identify the chemical elements present in them.

The Spectrograph

In 1872, when Henry Draper, a New York physician and amateur astronomer, used photographic plates to record the spectra of stars for later study, the spectroscope became a spectrograph. A *spectrograph* is a spectroscope equipped with a camera, a device that allows a star's spectrum to be photographed. Photographs of spectra can then be saved, analyzed, and categorized.

When Henry Draper died, his widow financed a program to photograph and classify stellar spectra at the Harvard College Observatory. In the 1890s, the Harvard project photographed the spectra of thousands upon thousands of stars. On the basis of similarity of spectral lines, the project director, Edward C. Pickering, devised a scheme for classifying stars with an alphabetical sequence of letters—A, B, C, D, etc.—that was used for a while. Between 1918 and 1924, however, the *Henry Draper Catalogue* of stellar spectra was published, and in it Annie Jump Cannon of the Harvard Observatory classified nearly 400,000 stellar spectra. Based upon her work, some of Pickering's classes were combined or dropped, and the sequence was reordered as O, B, A, F, G, K, M (which can be remembered using the sentence **O**h! **B**e **A** **F**ine **G**irl, **K**iss **M**e.) See Figure 6.11.

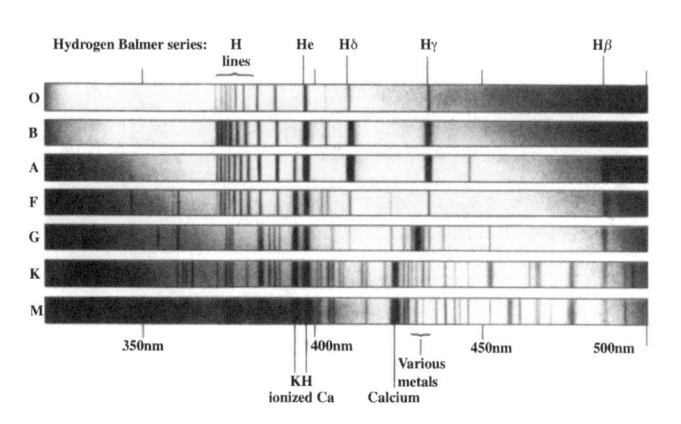

Figure 6.11 Representations of the Spectra of Stars from Each of the Major Spectral Classes. Source: *Astronomy: The Cosmic Journey*, William K. Hartman, Wadsworth, 1987.

Tools of the Modern Astronomer

SOME OBSERVATIONS OF THE UNIVERSE MADE WITH INSTRUMENTS

Using instruments such as telescopes and spectroscopes, astronomers could see more of the universe, and obtain more detailed information about it, than ever before. Let's consider some of the observations made with these instruments that have led to current theories about the universe.

Stars

First, there are a LOT more stars in the universe than the 1,500 or so that can be seen on a really dark night with the naked eye. Since telescopes gather more light than do eyes, stars too dim to be seen with the eye alone *are* visible through a telescope. Patches of sky that look empty to the naked eye are revealed to be filled with stars when viewed through a telescope. As ever-larger telescopes have been built, ever-dimmer stars have been detected, and the number of observable stars has grown from thousands to hundreds of billions.

Galaxies

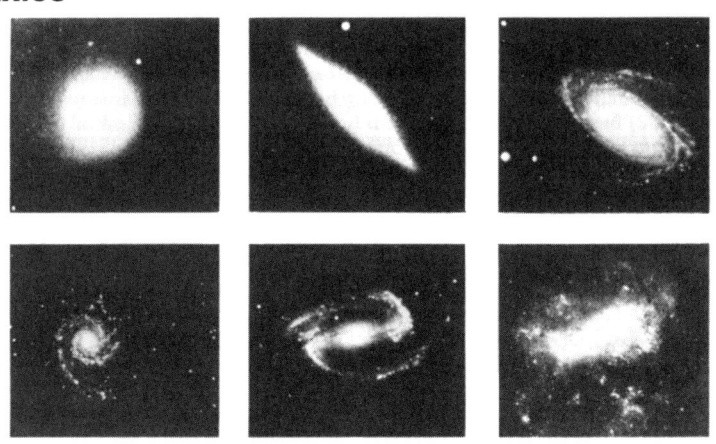

Figure 6.12 Galaxies Are Systems Containing Billions of Stars. The universe is filled with galaxies of all shapes and sizes. Source: *Astronomy: A Self-Teaching Guide*, 4th Ed., Dinah L. Moché, John Wiley, 1998.

The higher resolution of telescopes has revealed much about the structure of the universe. Some objects, which to the unaided eye looked like dim smudges of light, resolve into clusters of billions of stars, called *galaxies*, in systems shaped like spirals, ellipses, and spheres. Astronomers have found a colossal number of galaxies in the observable universe. See Figure 6.12. When astronomers peer at the Milky Way, which looks like a pale band of light to the naked eye, they see that it actually consists of light from countless millions of stars. They realized from the distribution of stars that it is a galaxy, and that our Sun is one of the stars in the Milky Way galaxy.

Nebulae

There are also clouds of gas and dust in the universe. The Orion nebula is a cloud of gas and dust lit by a group of stars within it. See Figure 6.13a. This nebula is dense enough to scatter starlight, and thus we can see it. By spectral analysis, the composition of the dust and gases in the cloud can be determined. When astronomers analyzed invisible radiation emitted by the cloud, they found that it extends out several times the area of the region we can see by visible light.

Sometimes dust is visible in a negative sort of way, as in the Horsehead nebula. There, a dust cloud is so thick that it blocks visible light, forming a black region (shaped like a horse's head) in front of the nebula. See Figure 6.13b. The dust actually absorbs starlight and reradiates it at longer wavelengths. To our eyes, it appears dark; to an infrared detector, it glows brightly. Observations such as these indicate that there is a lot of "dark" matter in the universe, that is, matter that is not visible to the unaided eye.

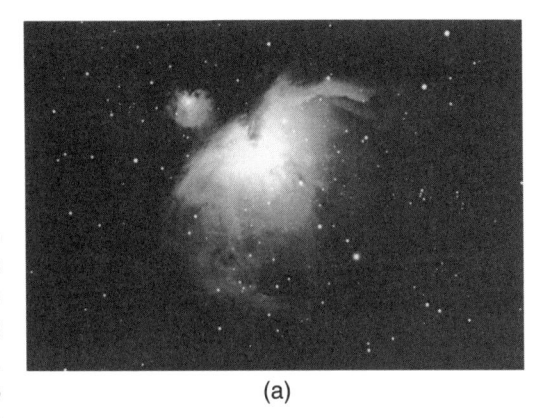

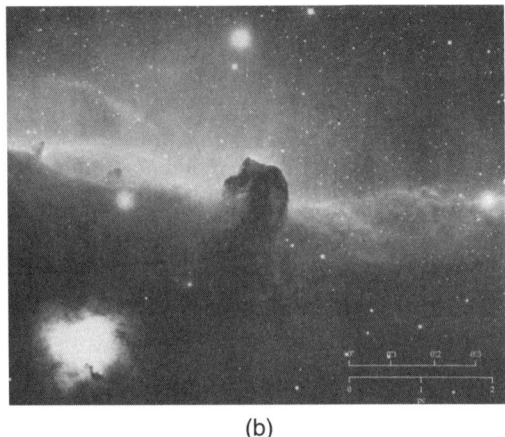

Figure 6.13 Nebulae. (a) Orion nebula. Source: *The Nature Company's Guides: Advanced Skywatching*, Robert D. Burnham et al., Time-Life Books, 1997. (b) Horsehead nebula. Source: *Astronomy: The Cosmic Journey*, William K. Hartman, Wadsworth, 1987.

MULTIPLE-CHOICE QUESTIONS

In each case, write the number of the word or expression that best answers the question or completes the statement.

1. Which instrument is used to study the compositions of stars?
   (1) sextant
   (2) spectroscope
   (3) seismograph
   (4) chronometer

Tools of the Modern Astronomer

2. An observer viewing the sky through a telescope sees a fuzzy, glowing region in the constellation Orion. The region has an irregular shape, and some stars seem to be shining through it. The observer is most likely viewing a
   (1) planet
   (2) comet
   (3) meteor
   (4) nebula

3. In a refracting telescope, the objective lens gathers light to form an image, and the eyepiece
   (1) magnifies the image formed by the objective
   (2) forms a spectrum
   (3) shields the eye from bright light
   (4) is the surface upon which the image forms

4. Compared to stars viewed with the unaided eye, stars viewed with telescopes appear
   (1) larger and brighter
   (2) larger and dimmer
   (3) smaller and brighter
   (4) smaller and dimmer

5. Telescopes reveal many more celestial objects than you can see with the unaided eye because telescopes
   (1) allow the observer to view a larger area of the sky
   (2) collect more light than an observer's eye can
   (3) convert ultraviolet rays into visible light rays
   (4) disperse light into a spectrum

6. The chemical composition of a star can be determined by examining its
   (1) celestial coordinates
   (2) absorption spectrum
   (3) distance from Earth
   (4) apparent brightness

7. When viewed by the unaided eye, the Milky Way looks like a pale band of light. When viewed through a telescope, the Milky Way
   (1) looks like a pale band of light
   (2) appears to be a glowing nebula
   (3) is seen to consist of billions of stars
   (4) appears to be the tail of a large comet

8. The diagram to the right shows light coming from a star's hot interior (A), passing through the star's cooler outer atmosphere (B), and traveling through space to enter a telescope fitted with a prism spectroscope. The star's spectrum is observed to contain many narrow black lines.

   Black lines are seen in the star's spectrum because
   (1) the star's gravity prevents some of the light from escaping
   (2) some wavelengths of light are absorbed by the star's outer atmosphere
   (3) some of the light coming from the star doesn't have enough energy to reach Earth
   (4) Earth's magnetic field blocks some of the star's visible light

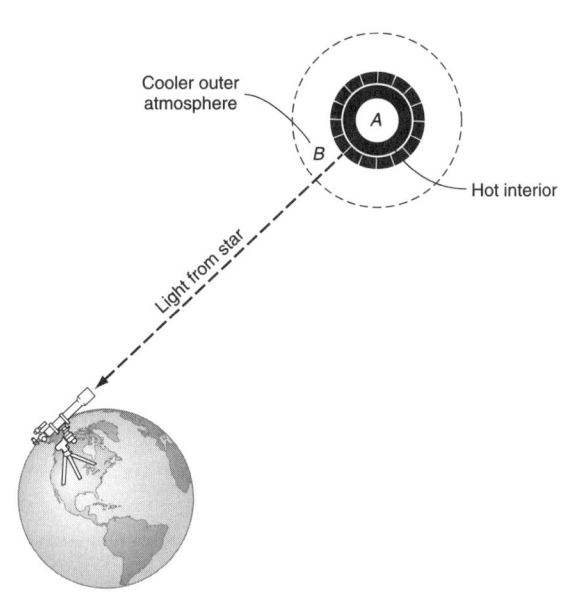

9. Two refracting telescopes, one with a 40-millimeter lens and another with a 160-millimeter lens, are used to observe the red star Betelgeuse in the constellation Orion. Compared to the image seen in the telescope with the 40-millimeter lens, the image seen in the telescope with the 160-millimeter lens will appear
   (1) brighter
   (2) dimmer
   (3) blue in color
   (4) smaller

10. Which of the following is *not* a way in which telescopes can aid human eyes?
    (1) magnification (the ability to make an image bigger)
    (2) light-gathering power (the ability to collect light)
    (3) resolution (the ability to resolve fine detail)
    (4) dispersion (the ability to separate light into its spectrum of colors)

11. In which list are celestial features correctly shown in order of increasing size?
    (1) galaxy, solar system, universe, planet
    (2) solar system, galaxy, planet, universe
    (3) planet, solar system, galaxy, universe
    (4) universe, galaxy, solar system, planet

Tools of the Modern Astronomer

12. Which of the following statements best describes the difference between a galaxy and a nebula?
    (1) A galaxy consists of stars; a nebula consists of dust and gas.
    (2) There are two types of nebula, but only one type of galaxy.
    (3) A galaxy always emits light; a nebula never emits light.
    (4) A galaxy consists of matter; a nebula consists of energy.

Constructed Response Questions

13. Compare the light-gathering powers of a 200-inch telescope and a 20-inch telescope. [2]
14. Describe three advantages of a telescope over the unaided eye. [3]
15. Explain the difference between the way in which a refracting telescope forms an image and the way in which a reflecting telescope forms an image. [2]
16. Construct a diagram showing why the image formed by a refracting telescope is inverted (upside down). [2]
17. Explain why some stars that are invisible to the unaided eye can be observed with a telescope. [1]
18. Compare and contrast continuous, absorption, and bright-line spectra. [3]
19. List the spectral classes of stars, and explain how stars are sorted into these classes. [2]

CHAPTER 7
STARS, THEIR ORIGIN AND EVOLUTION

KEY IDEAS The vast majority of observable objects in the universe are stars. A star is a hot, luminous, gaseous celestial body. Stars form when gravity causes clouds of matter to contract until nuclear fusion of light elements into heavier ones occurs. Fusion releases great amounts of energy over millions of years.

Stars vary in size, temperature, and age. The majority of stars, including the Sun, fall into the main sequence when plotted on the Hertzsprung-Russell diagram according to luminosity and spectral class. Stars undergo a series of changes as they age.

A galaxy is a system of stars, cosmic dust, and gas held together by gravitation. A galaxy typically contains billions of stars and may be thousands of light-years in diameter, and the universe contains billions of such galaxies. Our Sun is a medium-sized star within a spiral galaxy of stars known as the Milky Way.

KEY OBJECTIVES
Upon completion of this chapter, you will be able to:

- Describe the method by which distances to stars are determined using parallax.
- Explain the difference between apparent brightness and luminosity and the relationship of these qualities to distance.
- Describe the Hertzsprung-Russell diagram, and explain the relationship of a star's mass to its luminosity and temperature.
- List the three main steps leading to the birth of a star and the stages in the life cycle of a star such as our Sun.
- Compare and contrast what happens in the later stages of evolution for stars of large and small mass.
- Describe the origins of different chemical elements and the importance of supernovas to new generations of stars.

CHARACTERISTICS OF STARS

A *star* is a hot, luminous, gaseous celestial body. Stars are distant, blazing suns moving through space at different distances from Earth. We see stars by the light they emit. On a clear, dark night you can see about 2,000 stars with the unaided eye. With a telescope you can see billions. Stars differ from each other in size, temperature, and age.

Star Names

In order to keep track of observations of so many stars, each one must be identified in some unique way. Long ago, the brightest stars were given proper names by astronomers; today, modern astronomers use letters and numbers to identify hundreds of thousands of stars.

Many of the proper names of stars come from the Arabic and are usually descriptions of where the star appears in a constellation. For example, Orion, which represents a mighty warrior or hunter, is one of the most eye-catching winter constellations. The left shoulder of Orion is marked by the bright red star Betelgeuse (pronounced "beetle-juice"). This is a form of the Arabic name *Ibt al Jauzah*, meaning "the armpit of the central one." Rigel (pronounced "rye-jel") is from *Rijl Jauzah al Yusra*, meaning "left leg of the central one." But it is not practical to name, and to learn the names of, thousands upon thousands of stars. Clearly some other system had to be devised.

In 1603, Johann Bayer published a star atlas in which each star was named for the constellation in which it appeared and was assigned a Greek letter by brightness (α being the brightest, β the next brightest, and so on). As more and more stars were discovered, though, astronomers soon ran out of Greek letters. Sometime later, the English Astronomer Royal, John Flamsteed, published an atlas in which each star in a constellation was numbered, beginning with 1 for the brightest. Since then, many catalogs have been compiled that list the positions and characteristics of hundreds of thousands of stars.

Today, astronomers use only a handful of proper names—those of the very brightest stars. Then they use Bayer's Greek letter names; and in constellations such as Cygnus, with many bright stars, they use Flamsteed's numbers after Bayer's letters have been exhausted. In one of the more recent catalogs, fainter stars are referred to by their numbers.

The Distance to Stars *(not testable on Regents exam)*

The distances to stars that are relatively close to our solar system can be determined using simple trigonometry and a common optical effect. Hold a finger out at arm's length and alternately close your left and right eyes. The finger appears to shift back and forth against the background of more distant objects. This effect is known as ***parallax***. Now hold your finger about 10 centimeters in front of your nose and repeat the experiment. Notice that the nearer the finger, the greater the parallax.

The same effect is seen when astronomers look at stars that are close to us on one date and then look at them 6 months later, when Earth has traveled halfway around its orbit. (Here the distance between your eyes in the finger experiment is replaced by the 300-million-kilometer diameter of Earth's orbit.) The stars appear to shift slightly in their positions relative to distant stars. Half of the total shift is the star's parallax angle or, simply, its parallax. See Figure 7.1.

The parallax of a star can be used to calculate its distance from Earth. As shown in Figure 7.1, the star and Earth at two positions in its orbit, E_1 and E_2, form an isosceles triangle. The angle at the star equals the total shift in position. Since the Sun is at the center of Earth's orbit, a line from the Sun to the star bisects this angle (which is why parallax is one-half the total shift) and also forms the right triangle star-Sun-E_1. Given the right angle, the parallax angle, and the fact that one side of the triangle is the Earth-Sun distance, simple trigonometry yields the distance to the star.

Stellar parallaxes are very small and are measured in seconds (") of arc. To get a sense of how small an angle that is, look at a protractor and note the size of 1 degree. Now consider that 1 degree = 60 minutes and 1 minute = 60 seconds. Thus, 1 second = 1/3,600 degree! The parallax of the star 61 Cygni is only 3/10 second—the size of a dime viewed from 12 kilometers! Imagine the precision of the instruments needed to measure such a tiny angle!

One **parsec** (from *par*allax *sec*ond) is the distance to an imaginary star that has a parallax of 1 second of arc. One parsec (abbreviated pc) equals about 31 trillion kilometers (19 trillion miles). Another, older unit of distance is the **light-year**, the distance light travels in 1 year. One parsec equals about 3.26 light-years. However, to simplify calculations, most astronomers use parsecs for distances to stars in all their scientific work. To calculate the distance, in parsecs, to a star, they use this formula:

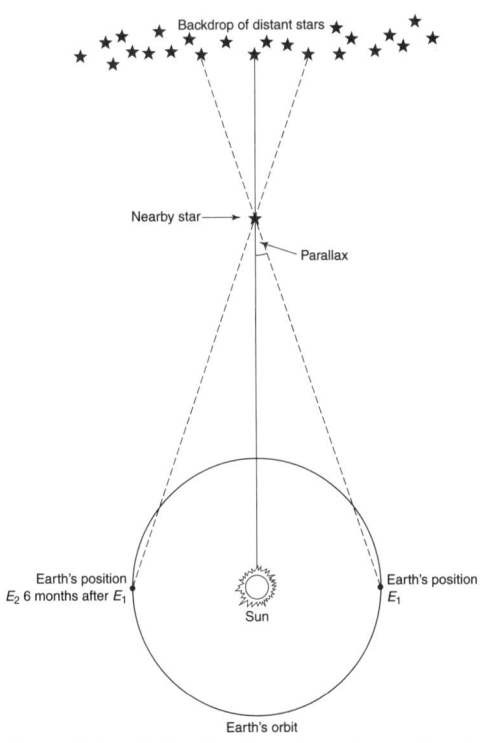

Figure 7.1 Stellar Parallax. A star's parallax is one-half the total angle of its apparent shift in position. Source: *Astronomy Explained*, Gerald North, Springer-Verlag, 1997.

$$\text{Star's distance (pc)} = \frac{1}{\text{parallax (sec)}}$$

Thus, the distance to Proxima Centauri, the star with the largest parallax (0.76 sec), and therefore the closest, is

$$\text{Distance to Proxima Centauri} = \frac{1}{0.76 \text{ sec}} = 1.32 \text{ pc}$$

Stellar parallax decreases with distance. We can measure angles down to about 1/100 second, which corresponds to a distance of about 100 parsecs. For stars more than 100 parsecs away, the parallax angle becomes too small to be measured with any certainty, and more complex methods are needed. Although we will not describe them in detail, it is interesting to note that they range from using the changing position of our solar system over time as the baseline for parallax, to measuring relative velocities of clusters of stars based upon spectral shifts, to calculating distance based upon stellar brightness.

Stellar Brightness

Go outside on any clear night, and you will see that stars vary greatly in *apparent brightness*. Some appear very bright in the sky, while others are so dim you can barely make them out. You might think that the brighter stars appear that way because they are giving off more light, but the explanation is not that simple. Stars may appear bright because they emit more light, *or because they are closer to Earth.* The farther away a source of light is located, the dimmer it appears to an observer. Therefore, astronomers distinguish between a star's apparent brightness and its *luminosity*, or the actual amount of light the star shines into space each second. Since the Sun is the nearest and best known star, the luminosity of other stars is often stated in terms of the Sun's luminosity, which is 3.85×10^{28} watts (the equivalent of 3,850 billion trillion 100-watt light bulbs all shining together). The brightest stars are more than 1 million times as luminous as the Sun, while the dimmest are only 1/10,000 (0.0001) as luminous.

You cannot tell by just looking at stars in the sky which ones have the greatest luminosity. The farther away a source of light is located, the dimmer it appears to an observer. The reason is that the apparent brightness of a light source is measured by the amount of light energy that strikes each unit of area of the detector—be it your eye's retina or a photometer.

As light travels away from a source, it spreads out as shown in Figure 7.2. All of the light that passes through surface 1 also passes through surface 2 and then through surface 3. But note that surface 2 has *four times* the area of surface 1! Therefore, the amount of light that passes through each unit of area in surface 2 is *one-fourth* of the amount that passes through surface 1, and the brightness measured at surface 2 will be one-fourth of the brightness at surface 1. The observed brightness of a source falls off as the square of the distance from the source, a relationship known as the *inverse-square law*. See Figure 7.2. Thus, if two stars have exactly the same luminosity, but one is twice as far away as the other, the distant one will appear (½)^2, or only one-fourth, as bright as the closer one because only one-fourth of the light reaches the eye. And at three times as far away, the distant star will appear only one-ninth as bright!

Over the centuries a scale to describe brightness has evolved. It began in

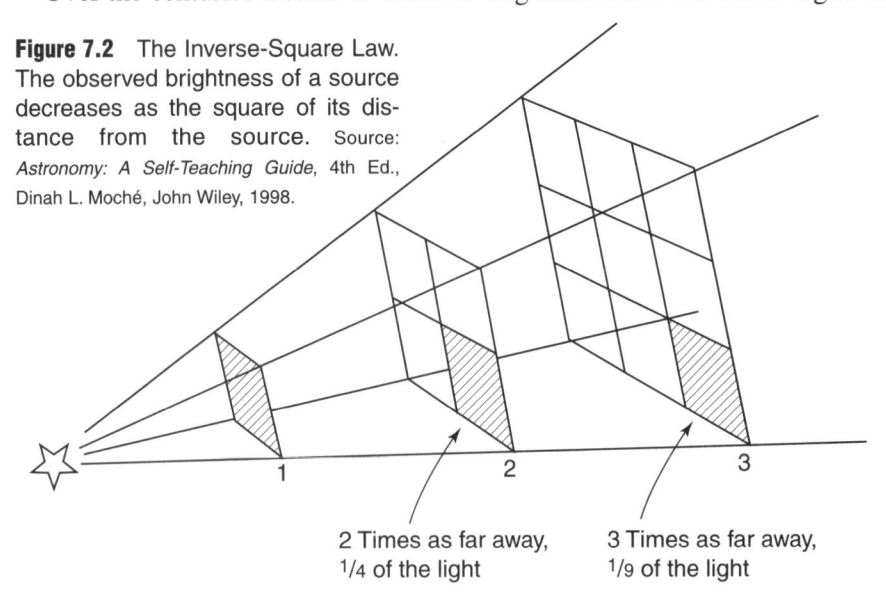

Figure 7.2 The Inverse-Square Law. The observed brightness of a source decreases as the square of its distance from the source. Source: *Astronomy: A Self-Teaching Guide*, 4th Ed., Dinah L. Moché, John Wiley, 1998.

2 Times as far away, 1/4 of the light

3 Times as far away, 1/9 of the light

the second century B.C., when the Greek astronomer Hipparchus classified stars into six brightness classes, or *magnitudes*. Stars of the first magnitude were the very brightest in the sky, and stars of the sixth magnitude were just visible to the human eye. The number assigned to a star's brightness on this scale is called its *apparent magnitude* because it describes how bright a star *appears* to an observer on Earth.

The modern *magnitude scale* defines a first-magnitude star as exactly 100 times brighter than a sixth-magnitude star. Scientific studies show that this is consistent with the way our eyes respond to changes in brightness. What we see as a *linear* increase of one magnitude in brightness is precisely measured as a *geometric* increase in brightness of 2.5 times. If we measure the brightness of stars of different magnitudes with a photometer, we find that, when the magnitude increases by 1, the brightness increases 2.5 times. Thus, a difference of five magnitudes corresponds to a brightness difference of 100 times (2.5 × 2.5 × 2.5 × 2.5 × 2.5 = 100)! Stars that are 2.5 times brighter than first magnitude are zero-magnitude stars. The very brightest stars actually have negative magnitudes. For example, Sirius has magnitude –1.5, and the Sun has magnitude –27!

To overcome this problem, astronomers devised a scale in which stars are

Stars, Their Origin and Evolution

assigned brightnesses based upon how bright they would appear if they were all located at the same distance from Earth. The *absolute magnitude* of a star is the magnitude it would have if it were moved from its actual location to a spot 10 parsecs from Earth. See Figure 7.3.

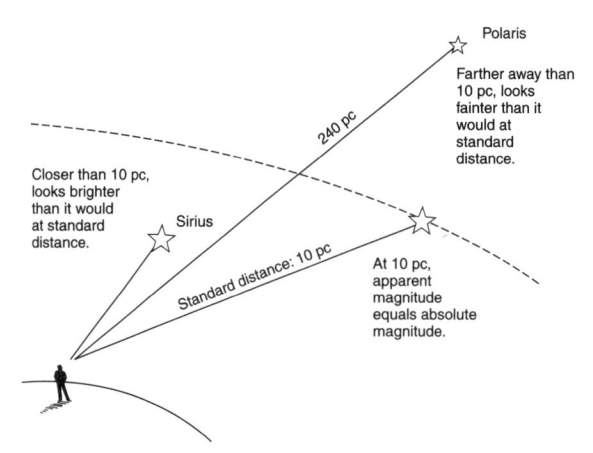

Figure 7.3 Absolute Magnitude and Apparent Magnitude. Source: *Astronomy: A Self-Teaching Guide*, 4th Ed., Dinah L. Moché, John Wiley, 1998.

Star Color and Temperature

If you look carefully at stars, you discover that many show definite shades of color. Betelgeuse is reddish. Another bright star, Arcturus, is a pale orange-red, while Vega, a prominent star in the summer sky, shows a definite blue tint. Why the different colors?

The temperature of a star is related to the average speed of the particles of which it is composed. We say "average" speed because some of the particles are moving faster and some are moving slower. The speed at which matter moves determines the wavelength of the energy it emits. The faster it is moving, the shorter the wavelength of the energy it gives off. Thus, some of the particles in a star emit short-wavelength radiation while others emit long-wavelength radiation. The result is that a star gives off radiation that is a mix of many different wavelengths.

A *radiation curve* shows the amount of radiation a hot object gives off at different wavelengths. The more radiation a star emits at a particular wavelength, the higher the curve at that wavelength. The area under the curve indicates the total amount

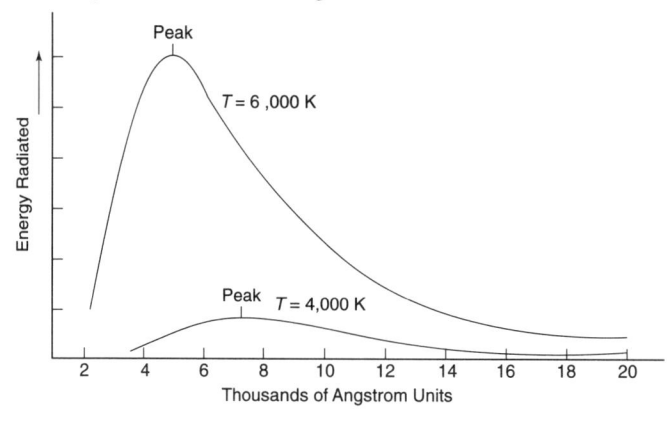

Figure 7.4 The Radiation Curves of a Theoretical Object at Two Different Temperatures. Note that, as temperature increases, the peak radiation shifts. Source: *Discovering Astronomy*, Robert Chapman, W. H. Freeman, 1978. P. 165.

of energy being radiated by the star.

Figure 7.4 shows the radiation curves for a theoretical star at two different temperatures. As you can see, the higher the temperature, the shorter the peak wavelength of the energy emitted. Note, too, that the total amount of energy emitted by the star (the area under the curve) increases with temperature. The relationship between the total energy emitted by a star over all wavelengths and its temperature is directly proportional to its surface area. Thus, if we know the surface temperature of a star, we can calculate the amount of energy being emitted by each square meter of its surface, or its **luminosity**.

Figure 7.5 shows the actual radiation curve of the Sun and the radiation curve of a theoretical star at a temperature of 6,000 K. Note how closely the two curves match. For this reason, scientists believe the Sun has a surface temperature of about 6,000 K. Note, too, that the Sun radiates most intensely in the yellow range, but remember that the *peak* wavelength emitted by a star is not the *only* wavelength it is emitting. Stars emit a mix of wavelengths, so we perceive their light as white tinted the color of the peak wavelength, or as pastel shades of color. Hence, the Sun appears white with a decidedly yellow hue.

The physical conditions in the surface of a star, such as the Sun, are such that theoretical models can be used to derive information about stars. By analyzing the amount of energy emitted at each wavelength in a star's spectrum,

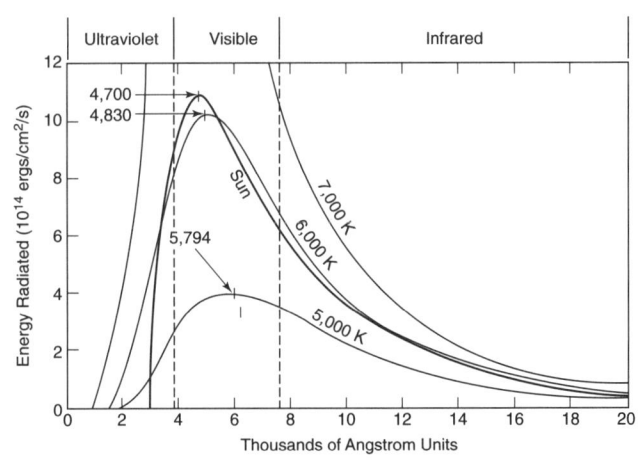

Figure 7.5 The Actual Radiation Curve of the Sun Closely Matches That of a Theoretical Object at 6,000 K. Source: *The Earth Sciences*, Arthur N. Strahler, Harper & Row, 1971.

astronomers can determine the peak wavelength, and thus the temperature, of the star. Also, the pattern of spectral lines reveals which elements are in the star, and measurements of the intensities of various bright lines in the spectrum show how much of each element is present.

Shortly after the publication of the *Henry Draper Catalogue* of stellar spectra, it was recognized that the O, B, A, F, G, K, M sequence was a *temperature* sequence. The surface temperatures of blue O-type stars are high (around 30,000 K), while the red M-type stars are cool (temperatures around 3,000 K). The Sun, with a surface temperature of about 6,000 K, is a yellow G-type star. See Figure 7.6.

Stars, Their Origin and Evolution

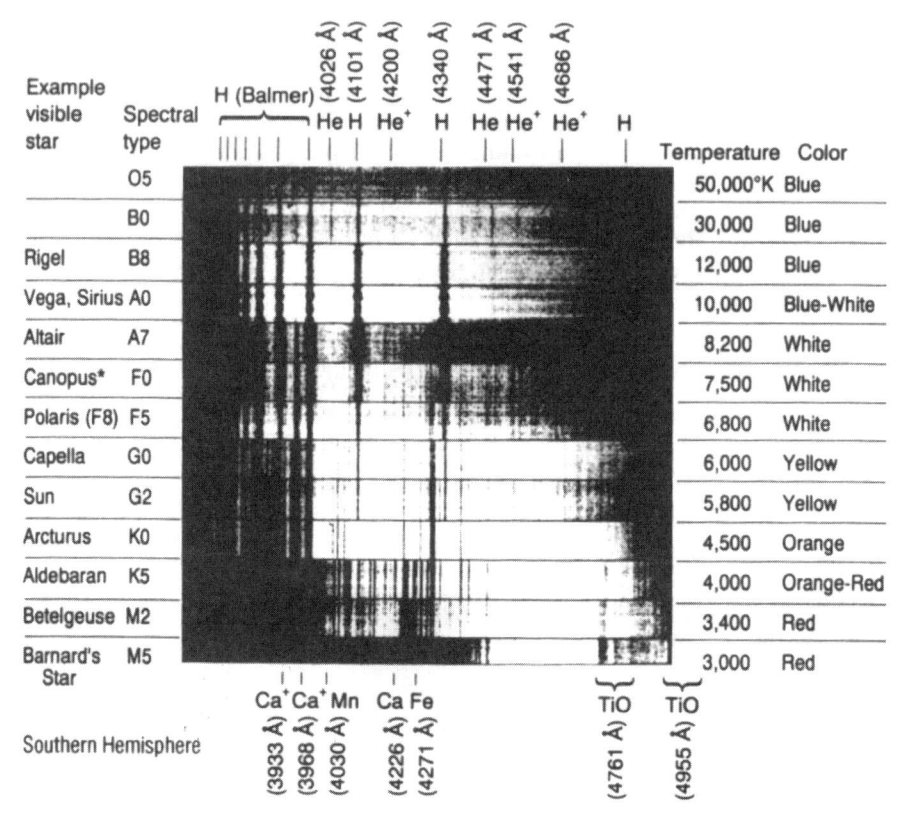

Figure 7.6 Spectra of Stars Representing the Seven Main Spectral Classes Arranged in Order of Decreasing Temperature. Source: *Astronomy: A Self-Teaching Guide*, 4th Ed., Dinah L. Moché, John Wiley, 1998.

The Hertzsprung-Russell Diagram

Early in the twentieth century, the relationship between the luminosities and temperatures of stars was discovered independently by two astronomers: Ejnar Hertzsprung of Denmark and Henry N. Russell of the United States. The ***Hertzsprung-Russell diagram*** (or H-R diagram) is a plot of luminosity versus surface temperature for the stars. See Figure 7.7. Every point on the H-R diagram represents a star. The star's temperature is read along the horizontal axis and its luminosity along the vertical axis. When several thousand stars are chosen at random and plotted on an H-R diagram, the dots do not scatter randomly over the graph. Instead, they fall into definite regions, forming patterns that show a meaningful relationship between a star's luminosity and its temperature.

Roughly 90 percent of all stars fall on a diagonal line across the diagram, called the ***main sequence***, which runs from the upper left (very hot, luminous blue supergiants) to the lower right (cool, dim red dwarfs). Why do stars follow this pattern? As was mentioned earlier, an increase in the temperature of a star results not only in a color shift in the radiation it emits, but also an increase in the total energy it gives off per unit of surface area. This increase in total energy emitted is seen as an increase in the star's brightness.

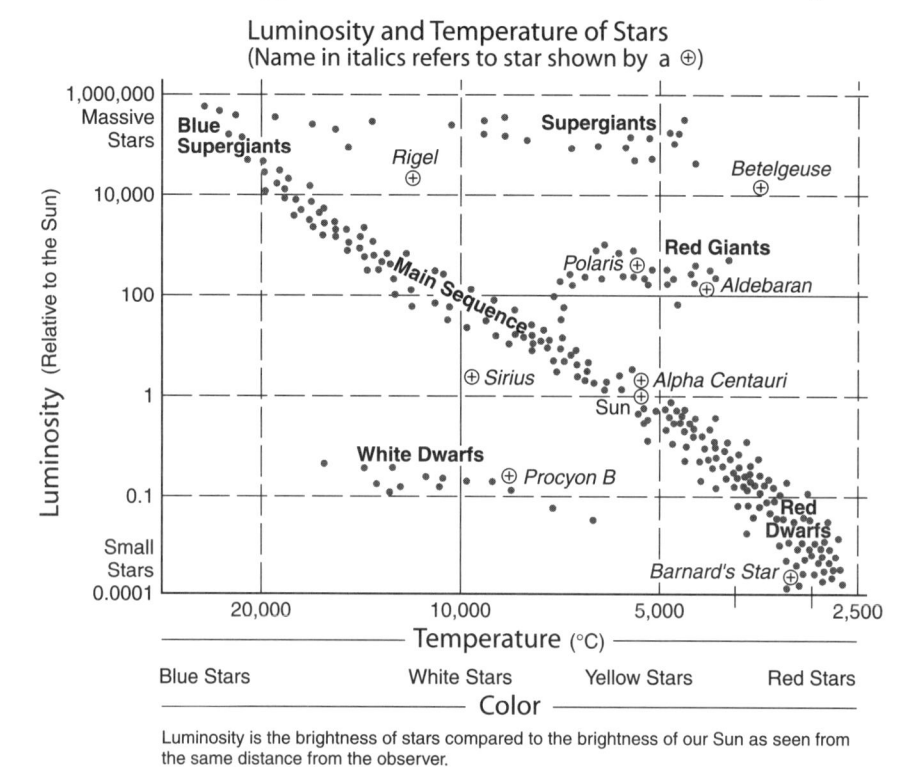

Luminosity and Temperature of Stars
(Name in italics refers to star shown by a ⊕)

Luminosity is the brightness of stars compared to the brightness of our Sun as seen from the same distance from the observer.

Figure 7.7 The Hertzsprung-Russell Diagram. When luminosity is plotted versus temperature, stars fall into distinct regions in patterns that represent the relationship between luminosity and temperature. Source: The State Education Department, *Earth Science Reference Tables*, 2001 Edition (Albany, New York; The University of the State of New York).

Most of the remaining 10 percent of stars fall above and to the right of the main sequence (cool, bright supergiants or red supergiants) or to the lower left (hot, dim white dwarfs). How can a star be cool yet bright, or hot yet dim? Consider, for example, 4,000°C stars. Since they all have the same surface temperature, they emit the same amount of energy per square meter of surface area. Thus, for a star above the main sequence, such as Aldebaran, to be more luminous than a red dwarf at the same temperature on the main sequence, it must have more square meters to radiate; in other words, it must be larger. Similarly, for 10,000°C stars that fall below the main sequence, such as white dwarfs, to be dimmer, they must be smaller than main sequence stars of the same temperature.

This inferred difference in sizes was confirmed after the first successful image of a star's disk was obtained in 1974. Using a technique called *speckle interferometry*, astronomers at Kitt Peak National Observatory measured the angular diameter of Betelgeuse to be 0.06 second of arc, making it 580 times larger than the Sun.

THE ORIGIN AND EVOLUTION OF STARS

Early Ideas

As information about stars accumulated, scientists made the first attempts to explain sunlight (and starlight). The British physicist Lord Kelvin hypothesized that the Sun formed by the collapse of interstellar gas and dust due to gravitational attraction, with the resultant compression causing the particles of matter to heat up. He thought that the Sun was simply radiating away this energy, just as any hot object radiates energy into its cooler surroundings, but his hypothesis ran into a time-scale problem. According to its size, temperature, and the rate at which it was emitting energy, the Sun would cool in tens of millions of years. Evidence indicates, however, that rocks on Earth are at least 10 times older than that.

Nuclear Fusion

By the 1930s, physicists began to understand the workings of the atomic nucleus. They realized that with sufficient force the repulsion between atomic nuclei could be overcome and the nuclei could combine in a process called *fusion*. In the Sun, which is composed mainly of hydrogen, four hydrogen nuclei (each a single proton) combine through a complex series of steps called the *proton-proton reaction*. See Figure 7.8a. During this process, two protons transform to neutrons and the most stable resulting nucleus is that of the isotope helium-4 (^{4}He). However, the resulting ^{4}He nucleus *has 0.7% less mass than the original four hydrogen nuclei*. The missing mass has been converted to energy according to Einstein's formula: $E = mc^2$. This formula says that, if mass, m, is converted to energy, the amount of energy created, E, is equal to m times the speed of light, c, squared. See Figure 7.8b.

If all of the hydrogen in the Sun were converted into helium, enough energy would be created to keep the Sun shining at its present rate for 100 billion years. However, as the ratio of hydrogen to helium in the Sun changes, so will its structure. After about 10 billion years the Sun will begin to undergo profound changes. But assuming that the Sun is about as old as the 4–5 billion-year-old Earth, it is only about halfway through its lifetime—so don't panic.

With star lifetimes measured in billions of years, it is not surprising that the stars we see today are pretty much the same as those observed by early astronomers. However, with all of written human history encompassing little more than 5,000 years, how can astronomers hope to learn anything about the birth, evolution, and death of a star? Luckily, stars did not all form at the same time, nor do they all go through their life cycle at the same rate. Therefore, it

1 The immense heat and temperature of a star's core cause two hydrogen nuclei, each containing one proton, to collide with enough force to overcome the electrical repulsion between them, causing them to fuse into a deuterium nucleus (one proton and one neutron). In the process, one proton is converted into a neutron, and one neutrino and one positron are released.

2 The newly formed deuterium collides with another hydrogen nucleus forming Helium-3 (a nucleus with two protons and one neutron). This reaction releases a gamma ray.

3 Two Helium-3 nuclei collide creating a Helium-4 nucleus (two neutrons and two protons) and releasing two protons.

γ = gamma ray
ν = neutrino
e^+ = positron
N = neutron
P = proton

(a)

Figure 7.8 (a) The proton-proton reaction, one of several nuclear reactions that result in the fusion of hydrogen into helium in stars, releasing energy. (b) One atom of helium has less mass than the four atoms of hydrogen from which it formed. The mass is not lost, but is converted into energy. The amount of energy produced per unit of mass is described by Einstein's famous equation, $E = mc^2$. Source: *Astronomy: A Self-Teaching Guide*, 4th Ed., Dinah L. Moché, John Wiley, 1998.

Heavier mass: Four hydrogen nuclei (before fusion)

Lighter mass: Helium nucleus (after fusion)

Mass difference → Energy → Starlight

(b)

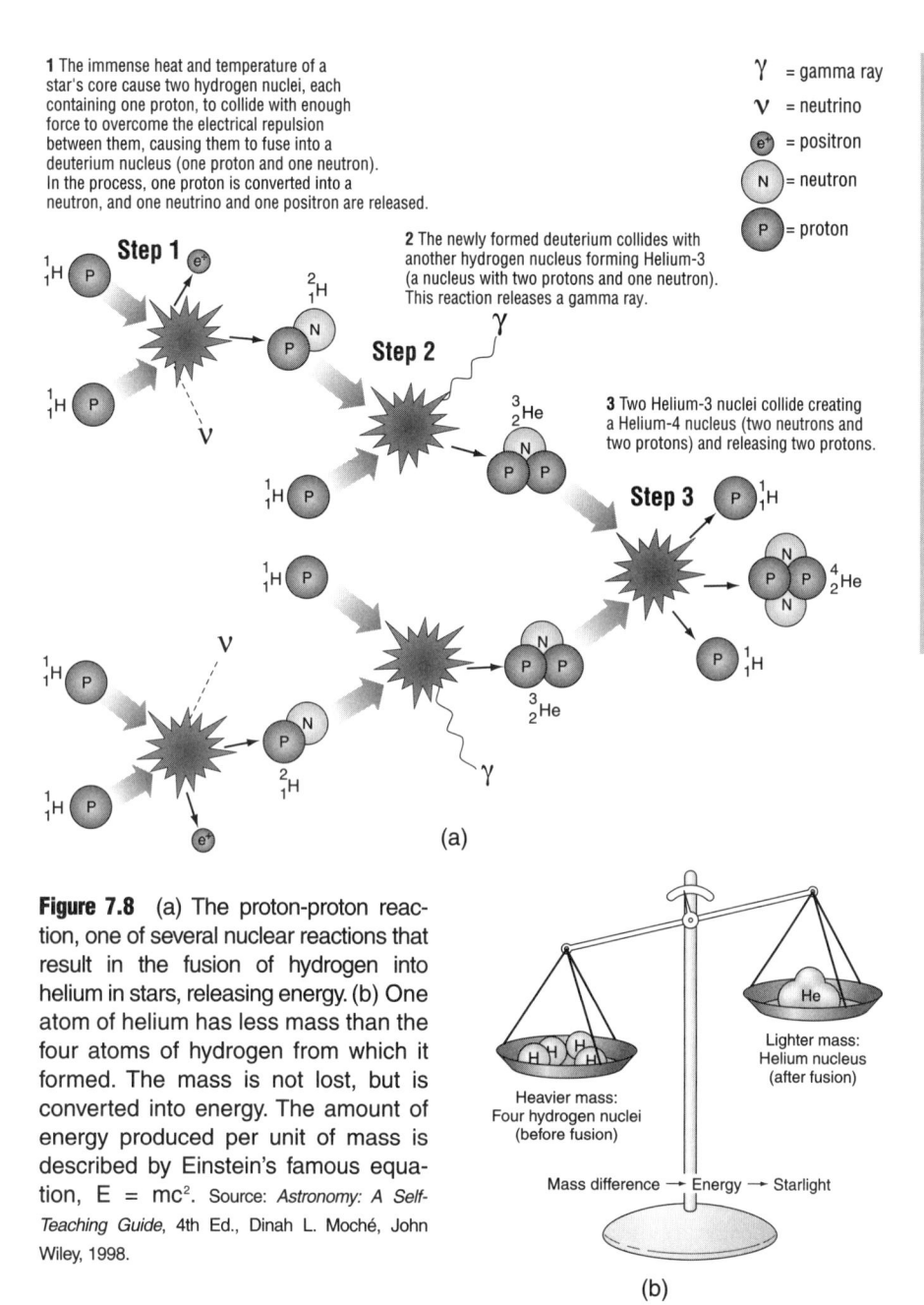

is possible to find stars at all stages of development in the universe and to piece together their life cycles from this evidence.

Stars, Their Origin and Evolution

From Nebula to Protostar

The formation of a star begins in the immense interstellar (between stars) clouds of dust and gas, or ***nebulae***, scattered throughout the universe. Inside these nebulae, which are composed mainly of hydrogen, stars are born through the process of gravitational collapse. However, the temperature of the cloud determines whether or not stars will form. If the temperature is too high, the atoms in the cloud move around too quickly to coalesce under the influence of gravity.

The first step in star formation is fragmentation of the cloud, in which cloudlets of matter form from denser portions of the cloud. The separation of an evenly spread out cloud of matter into cloudlets is probably the result of shock waves from the explosion of nearby stars or the pressure of radiation from hot, young stars nearby. See Figure 7.9.

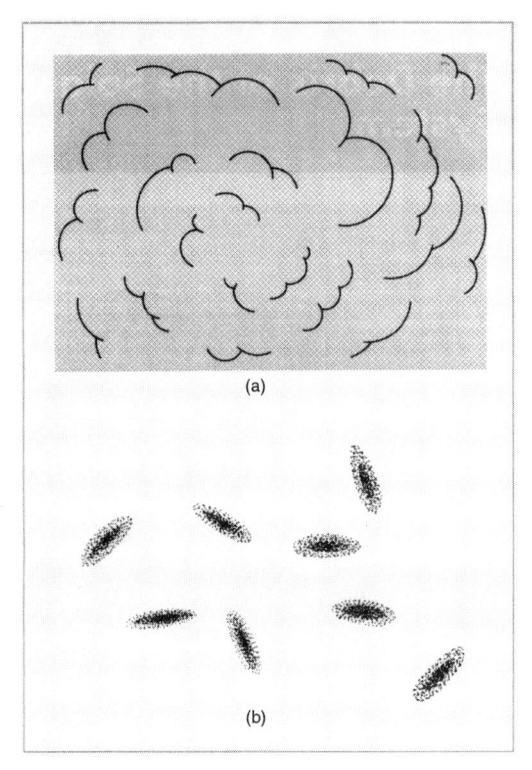

Figure 7.9 A Nebula Condenses into Cloudlets of Matter in the First Stage of Star Formation. Source: *Astronomy Explained*, Gerald North, Springer-Verlag, 1997.

The force of gravity pulls the dust and gas in toward the center of a cloudlet. As the particles of matter come closer together, mutual gravitational attraction strengthens, and the matter contracts and becomes even denser. The cloudlet also develops vortex motions, as matter spirals in toward its center. As this process continues, gravitational attraction strengthens, attracting more and more dust and gas. As more and more matter accretes, gravitational contraction causes temperatures and pressures to rise. Also, the cloudlet further contracts, rotates faster (as an ice skater spins faster as the arms are drawn inward toward the body), and becomes flattened out into a disk.

Heat flows from the hot center of the condensing cloudlet to its cooler surface. At first, the heat generated in the condensing cloudlet is radiated away into the infinite reservoir of the universe around it. As the cloudlet thickens, however, the outer layers begin to absorb radiation. From this point on, the heat generated by its collapse can no longer escape the thickening cloudlet. Now the cloudlet heats up, and its internal pressure increases as the heat energy is shared among its gas and dust particles, causing their velocities to increase. Eventually, the energy of collisions between particles bouncing them apart balances the gravitational energy drawing them together. The rate of compression slows and stops, temperatures reach the point where the whole mass glows, and fragmentation ceases. The cloudlet has now become a hot, relatively dense, disk-shaped region called a *protostar*. The protostar is approximately the size of a solar system, and its surface temperature is about 4,000 K. See Figure 7.10.

The balance between the outward pressure of very hot gases and the inward pull of gravity should prevent the collapse of a protostar. The interplay of magnetic fields and charged particles within the cloudlet, however, permits randomly formed clumps of matter poor in charged particles to collapse into the core, causing it to become increasingly dense and hot.

A Star is Born: Mass Is Destiny

As the core of a protostar collapses, the dust and gas collide more and more vigorously, and in each collision some of the energy of motion is converted into heat. Temperatures at the center of the core reach tens of millions of degrees, and pressures rise to billions of atmospheres. When the temperature in the center of the protostar reaches 10 million K, nuclear fusion reactions are triggered, and a star is born. The energy liberated by the fusion of hydrogen into helium creates a tremendous, new outward pressure in the core. A new balance is then reached between the pressure of matter falling inward toward the core and the pressure of matter being blown outward by nuclear reactions.

If the core expands outward, it cools off, effectively stopping fusion. Too little expansion, on the other hand, causes more infalling of dust and gas, resulting in a denser, hotter core, a faster rate of fusion, and more outward pressure. This newly balanced core of fusing hydrogen is the astronomer's concept of a newly formed star on the main sequence. What happens next depends upon the mass and composition of the protostar.

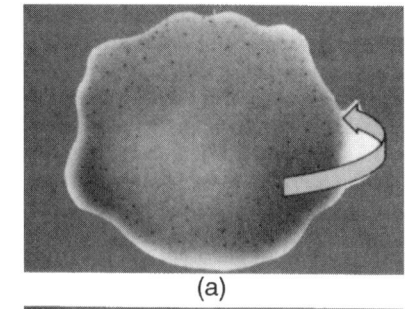

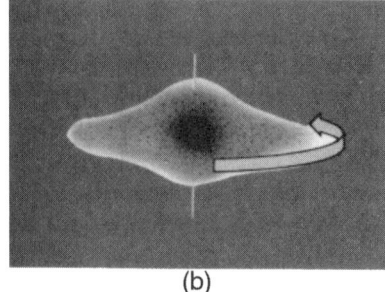

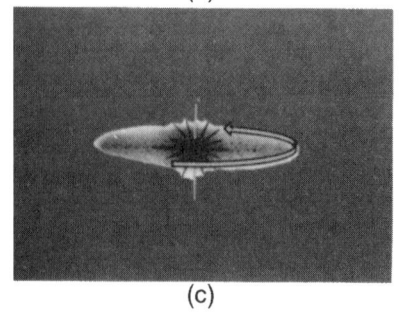

Figure 7.10 Three Stages in the Evolution of a Protostar. (a) An interstellar gas cloud begins to contract because of its own gravity. (b) A central condensation forms, and the cloud rotates faster and flattens. (c) The star forms in the cloud center, surrounded by a rotating disk of gas.
Source: *Astronomy: The Cosmic Journey,* William K. Hartman, Wadsworth, 1987.

Stars, Their Origin and Evolution

Stars that have about the same mass and chemical composition go through the same stages of evolution in about the same time. High-mass stars evolve the fastest, while stars of very low mass take the longest time to evolve.

Red Dwarfs

Low-mass protostars have correspondingly low gravity. Their cores balance at a slow rate of fusion and low temperatures. With less than a third the mass of the Sun, these stars are smaller and cooler and therefore appear red rather than yellow; thus they are called red dwarfs. Red dwarfs fuse hydrogen at such a slow rate that they can remain stable for hundreds of millions of years. The smallest red dwarfs may last for trillions of years before all of their hydrogen is consumed.

Once a red dwarf has converted all of its hydrogen into helium, fusion ends. Without the outward pressure of energy released by fusion, the helium-rich star contracts and heats up; but, because of its small size, it never reaches the temperature needed to trigger the fusion of helium. The star slowly cools, dimming and contracting into an inert ball of gas known as a black dwarf.

Sun-Class Stars

Mid-sized stars such as the Sun have enough mass to escape the dead-end fate of a red dwarf. They are larger, have higher gravity, and, once fusion has begun, stabilize at a higher temperature. Although they contain more hydrogen than dwarfs, their higher temperature causes fusion to occur at a faster rate, exhausting their supply of hydrogen in less than 50 billion years. As the star's hydrogen is converted into helium, the core becomes mostly helium; eventually the star is composed of a helium core surrounded by a shell of hydrogen gas. Without fusion of hydrogen in the core to support it, inward pressure of gravity from the surrounding shell causes the helium core to contract. Heat from the contraction triggers fusion in the hydrogen remaining in the surrounding shell. This fusion, together with heat from the still contracting core, causes the shell to puff out to perhaps 100 times its former size. The surface cools because of the expansion and now appears red. The star has moved off the main sequence and is now a red giant. See Figure 7.11.

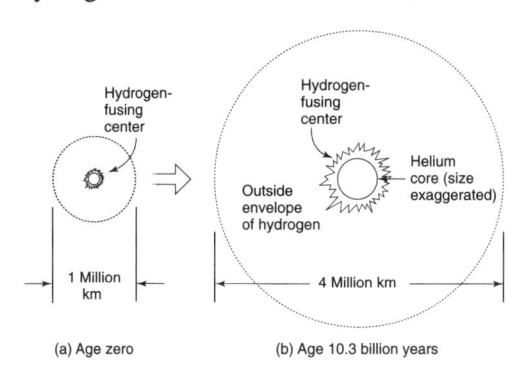

Figure 7.11 (a) At birth, a Sun-class star is mainly hydrogen, which fuses into helium, releasing energy. (b) After several billion years, a star consumes most of its hydrogen and develops a helium core. Heat from contraction of the helium core triggers fusion in the outer shell of hydrogen, causing the star to "puff up" into a red giant. Source: *Astronomy: A Self-Teaching Guide*, 4th Ed., Dinah L. Moché, John Wiley, 1998.

After about 100 million years, odd things begin to happen in the core of a red giant. As the density of the matter in the core increases, the matter changes its behavior and is said to be degenerate. Electrons can no longer move about freely and collide at random. Instead, they exert a damping effect on each other's motion. If heat is added to such a gas, the pressure actually decreases and the gas contracts! In the core of a red giant, degenerate matter contracts and heats up until it reaches the incredible temperature of 100,000,000 K, at which point helium begins to fuse into heavier elements, including carbon. In less than 100 million years this process ends, leaving a hot carbon core. But with insufficient gravity to fuse carbon, the star dies, its envelope of gases drifts off, and its hot, highly compressed core remains as a white dwarf.

Short-Lived Giants

Stars of at least three solar masses shine hot, bright, and blue. At the very upper end of the main sequence are hot, blue supergiants with 30 times the Sun's mass and 100,000 times its brightness. To shine 100,000 times brighter than the Sun, these stars must fuse hydrogen 100,000 times faster, but they have only 30 times the Sun's mass. Thus, a blue supergiant should last for a time period that is 100,000 ÷ 30, or roughly 3,000, times less than the period of time the Sun should exist. Thus, a typical blue supergiant exists for only about 3 million years.

Such massive stars pass quickly through the same stages as a mid-sized star. They are so large, however, that even before all of their hydrogen is consumed, their helium core collapses; and contraction driven by their enormous gravity generates temperatures high enough to trigger the nuclear fusion of helium. Helium is converted into carbon, and eventually the core again collapses until temperatures needed to fuse carbon are reached. This cycle continues with carbon fusing to form neon and magnesium; then neon is converted into oxygen, oxygen into silicon and sulfur, and finally silicon into iron. Such massive stars, at the end of their lives, are layered like an onion. A thick hydrogen sheath

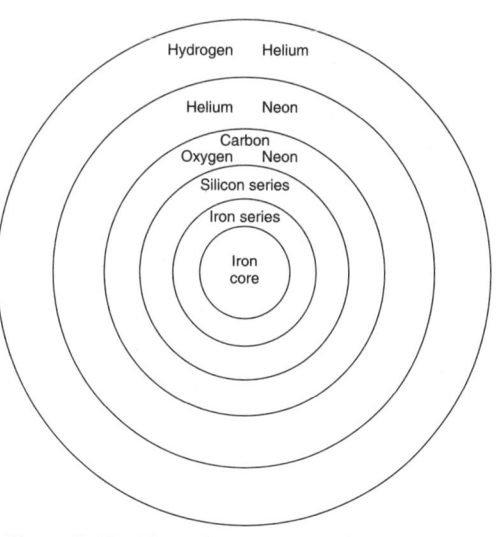

Figure 7.12 The "Onion-Layer" Structure That Develops in a Massive Star Prior to a Supernova. Massive stars undergo a succession of core collapses, each of which triggers fusion of heavier and heavier elements. The result is a star whose core consists of layers of elements produced by fusion. Source: *Astronomy Explained*, Gerald North, Springer-Verlag, 1997.

encases successively thinner layers of helium, carbon, oxygen, and silicon shot through with magnesium, calcium, sulfur, and other elements. See Figure 7.12.

Iron, however, ends the process because the fusion of iron does not release energy, but instead requires energy to occur. When fusion ceases, the massive star's core collapses for the last time. The iron heart of the star crushes in on itself, and temperatures rise to 100 billion degrees while pressures get high enough so that all electrons and protons are squeezed together to form neutrons. Matter in the core then reaches a point at which it can no longer endure further compression. The repulsive force between nuclei overcomes the force of gravity, and the core rebounds like a tightly coiled spring. The decay of protons releases 99 percent of the energy of the explosion in a burst of exotic particles called *neutrinos*, which have little or no mass, have no charge, and can penetrate a solid object as if it were not there. In the wake of this neutrino burst, the recoil of the star's core hurls matter outward in a violent, explosive shock wave, or *supernova*, that blasts through the surrounding layers.

As the explosion speeds through the layers of the core, it sets off fusion of nuclei and new elements are created. When the shock wave erupts from the core, the entire outer envelope of the star is blown away in a violent outburst, spewing energy and huge masses of newly created elements into space. For a time, the exploding star can outshine the rest of the stars in its galaxy. See Figure 7.13. After the outburst, all that remains is a small, superdense sphere composed almost entirely of neutrons—a *neutron star*. A neutron star is so incredibly dense that the entire mass of the Sun may be packed into a sphere only a few kilometers across!

Figure 7.13 A Supernova Captured on Film. These two photographs were taken before and after the supernova, which can be seen as the intensely bright star to the lower right of the galaxy in the photograph to the right. Source: *Discovering Astronomy*, Robert Chapman, W. H. Freeman, 1978.

The elements ejected by a supernova are recycled. Incorporated into interstellar nebulae, they become part of a second and eventually a third generation of stars such as our Sun. In these younger, metal-enriched stars, fusion reactions occur that manufacture still other elements. Thus, all of the elements here on Earth can be traced to supernovas that occurred in the distant past.

If a very massive star (at least 10 solar masses) undergoes supernova, the remaining mass collapses beyond the neutron-star stage. The sphere left behind is so massive that irresistible gravitational contraction causes it to become so dense, with gravity so strong, that not even light can escape; this structure is called a ***black hole***. To an observer, the star simply disappears. See Figure 7.14.

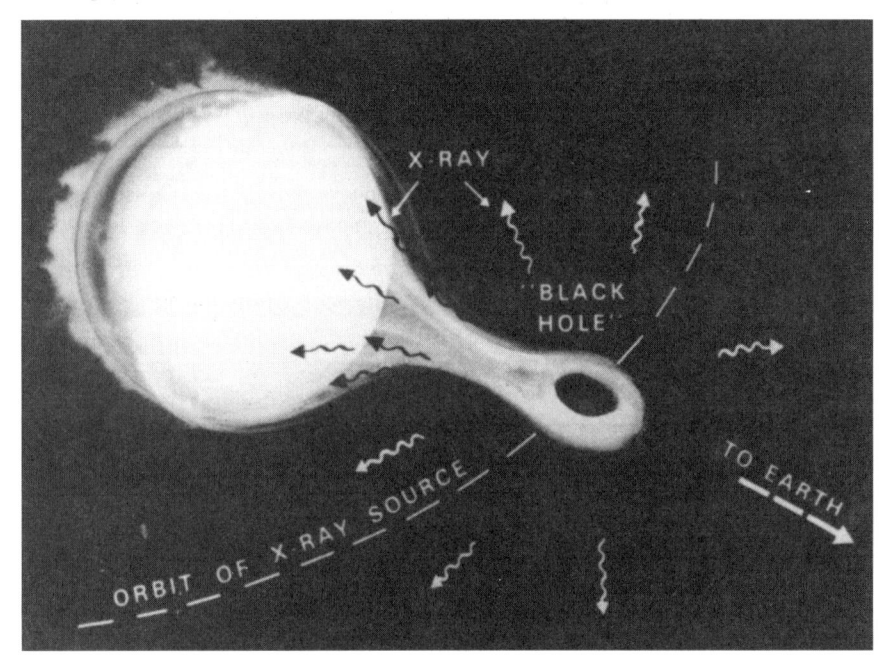

Figure 7.14 An Artist's Conception of a Black Hole. A stream of matter is ripped out of a star by the immense gravity of an orbiting black hole. X rays are emitted as the matter accelerates into the black hole. Source: *Astronomy: A Self-Teaching Guide*, 4th Ed., Dinah L. Moché, John Wiley, 1998.

Multiple-Choice Questions

In each case, write the word or expression that best answers the question or completes the statement.

1. A star differs from a planet in that a star
   (1) has a fixed orbit
   (2) is self-luminous
   (3) revolves about the Sun
   (4) shines by reflected light

Stars, Their Origin and Evolution

2. Which color indicates the lowest star temperature?
   (1) yellow
   (2) white
   (3) blue
   (4) red

3. If three identical 100-watt light bulbs were placed at distances of 1 meter, 10 meters, and 100 meters, respectively, from an observer, each would seem to have a different brightness. Applied to stars, this concept is called
   (1) density
   (2) magnitude
   (3) volume
   (4) twinkling

4. Which factor does *not* directly affect the apparent brightness of Polaris?
   (1) its motion about the north celestial pole
   (2) its distance from Earth
   (3) its mass
   (4) its temperature

Base your answers to questions 5 through 9 on the Luminosity and Temperature of Stars graph in the *Earth Science Reference Tables*. The graph shows the temperatures and relative brightnesses of many stars observed from Earth.

5. According to the graph, the Sun is classified as a
   (1) main sequence star with a temperature of approximately 4,000°C and a luminosity of 100
   (2) main sequence star with a temperature of approximately 6,000°C and a luminosity of 1
   (3) white dwarf star with a temperature of approximately 10,000°C and a luminosity of 0.01
   (4) blue supergiant star with a temperature of approximately 20,000°C and a luminosity of 700,000

6. Stars are believed to undergo evolutionary changes over millions of years. The flowchart to the right shows stages of predicted changes in the Sun.
   According to this flowchart, the Sun will become
   (1) hotter and brighter in stage 2, then cooler and dimmer in stage 3
   (2) cooler and dimmer in stage 2, then hotter and brighter in stage 3
   (3) hotter and dimmer in stage 2, then cooler and brighter in stage 3
   (4) cooler and brighter in stage 2, then hotter and dimmer in stage 3

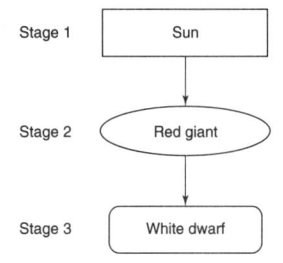

7. According to the graph, the temperature of the Sun is nearest that of
   (1) Sirius
   (2) Polaris
   (3) Barnard's star
   (4) Rigel

8. The Sun is brighter than any star in the group labeled
   (1) supergiants
   (2) red giants
   (3) main sequence
   (4) white dwarfs

9. In which group are the stars most varied in both size and luminosity?
   (1) main sequence
   (2) white dwarfs
   (3) blue supergiants
   (4) red giants

10. Which information about a nearby star must be known to determine the distance of the star from Earth?
    (1) size
    (2) temperature
    (3) color
    (4) parallax

11. A star of high surface temperature and low luminosity is most likely a
    (1) giant star
    (2) main sequence star
    (3) supergiant star
    (4) white dwarf star

Note that question 12 has only three choices.
12. When triangulation is used to measure the distance of stars from our Sun, as the angle of parallax decreases, the distance to the stars
    (1) increases
    (2) decreases
    (3) remains the same

13. As star color changes from blue to red, the temperature of the star
    (1) increases
    (2) decreases
    (3) remains the same
    (4) decreases then increases

Base your answers to questions 14 through 18 on the Luminosity and Temperature of Stars graph in the *Earth Science Reference Tables*.

14. A main sequence star is 10,000 times more luminous than the Sun. The temperature of the star is likely to be most nearly
    (1) 2,500°C
    (2) 5,000°C
    (3) 10,000°C
    (4) 20,000°C

15. A giant star has a luminosity of 300. Its color is most likely to be
    (1) red
    (2) yellow
    (3) white
    (4) blue

153

Stars, Their Origin and Evolution

16. A white dwarf star has a temperature of 13,000°C. What is the probable luminosity of the star?
    (1) 100
    (2) 10
    (3) 1.0
    (4) 0.01

17. A yellow star has a temperature of 6,000°C and is 10,000 times more luminous than the Sun. To which group does it belong?
    (1) giants
    (2) supergiants
    (3) main sequence
    (4) white dwarfs

18. According to the graph, in how many different groups are yellow stars of 5,000°C found?
    (1) 1
    (2) 2
    (3) 3
    (4) 4

19. What does Einstein's equation, $E = mc^2$ (energy = mass times the speed of light squared), indicate?
    (1) Mass and energy are equal.
    (2) Mass is more powerful than energy.
    (3) Energy is the more fundamental quantity.
    (4) Matter and energy are merely different forms of the same thing.

20. A star with a great mass tends to have a short life because it
    (1) becomes a supernova
    (2) consumes its fuel rapidly
    (3) has a core of heavy elements
    (4) collapses to form a white dwarf

21. The most important factors that influence the speed at which a star forms and the type of star that forms when a nebula collapses are
    (1) the volume and mass of the nebula
    (2) the shape and motion of the nebula
    (3) the number and kind of stars in adjacent space
    (4) the location of the nebula in the Milky Way

22. What is the most important factor in determining the life history of a star?
    (1) mass
    (2) volume
    (3) density
    (4) temperature

23. Two nebulae, *A* and *B*, of equal volume are beginning to contract and form stars *A* and *B*. Nebula *A* has 10,000 times the mass of nebula *B*. Which of the following predictions is most accurate?
    (1) Star *A* will use up its fuel faster than star *B*.
    (2) Star *A* will probably be much redder than star *B*.
    (3) Star *B* will be much hotter than star *A*.
    (4) Stars *A* and *B* will be identical in volume.

24. Astronomers study the gas between stars by
    (1) observing supernovas
    (2) observing nuclear fission in the gaseous clouds
    (3) collecting samples with interstellar space probes
    (4) examining the spectra of radiation from distant stars that passes through the gas

25. Jupiter probably approaches the limit of mass of a cold body because
    (1) if it were any larger, it would probably explode
    (2) beyond a certain mass, the heat and pressure of gravitational contraction trigger nuclear fusion
    (3) the solar wind would cause ignition of this cold body
    (4) if it were more massive, centrifugal force would hurl it out of the solar system

Constructed Response Questions

26. State the relationship between luminosity and temperature. [1]

27. State how the Sun generates the energy that it radiates away as its luminosity. [1]

28. State the three main steps in the evolution of a protostar, and construct a diagram showing the shape of a typical protostar. [3]

29. Explain why elements heavier than iron are so much rarer in the universe than elements lighter than iron. [2]

30. How many times brighter is a star of magnitude 1.0 than a star of magnitude 5.0? [1]

Stars, Their Origin and Evolution

EXTENDED CONSTRUCTED RESPONSE QUESTIONS

Base your answers to parts (a) through (c) on the diagram below, which show the theoretical evolutionary paths of contracting protostars of three different masses on the Luminosity and Temperature of Stars graph.

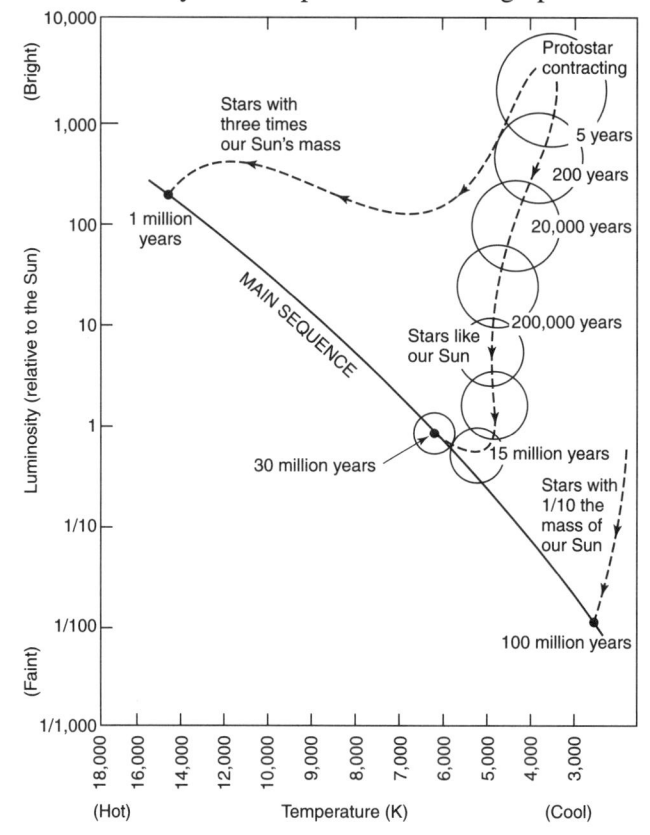

Source: *Astronomy: A Self-Teaching Guide*, 4th Ed., Dinah L. Moché, John Wiley, 1998.

(a) State approximately how long each of the following protostars takes to become a star on the main sequence.
1. a star like the Sun [1]
2. a star with three times the Sun's mass [1]
3. a star with 1/10 the Sun's mass [1]

(b) A protostar at a temperature of 3,000°C emits radiation at a peak wavelength of 10^{-4} centimeter.
1. According to the Electromagnetic Spectrum chart in the *Earth Science Reference Tables*, radiation of this wavelength falls within the range defined as _____. [1]
2. State one reason why light emitted by a protostar when it begins to contract is not visible to a human observer. [1]

(c) State the relationship, according to the diagram, between the mass of a newly formed main sequence star and its temperature. [1]

CHAPTER 8
THE SOLAR SYSTEM

KEY IDEAS Our solar system formed about 5 billion years ago from a giant cloud of gas and debris. The solar system consists of nine large planets, their satellites, and a variety of smaller objects, ranging from asteroids, meteors, and comets to tiny particles of dust and gas, all in orbit around a central star, our Sun.

During the formation of the solar system, gravity caused Earth and the other planets to become layered according to the density of the materials of which they were composed. The distance of each planet from the Sun was a key factor in determining the planet's characteristics. Powerful emissions from the Sun drove off most of the nearby gases, leaving behind the small, dense, rocky terrestrial planets. The more distant Jovian planets had sufficient gravity to retain much of the gas that surrounded them, and thus evolved into the large, low-density, gaseous planets.

Formation of the solar system left numerous smaller objects, such as asteroids, comets, and meteors, orbiting the Sun. From time to time, these objects collide with planets. Impact craters from such collisions have been identified in Earth's crust, and impact events have been correlated with mass extinctions and global climate change on Earth.

KEY OBJECTIVES
Upon completion of this chapter, you will be able to:

- List the members of the solar system.
- Describe the nebular theory of the formation of the solar system.
- Compare and contrast the characteristics of the nine large planets, distinguishing between the terrestrial planets and the Jovian planets.
- Describe the other components of our solar system, such as asteroids, meteors, and comets.
- Explain the evidence linking impact events with mass extinctions and global climate change on Earth.

The Solar System

THE SOLAR SYSTEM'S PLACE IN THE UNIVERSE

We currently know that Earth is one of several planets that orbit a star—the Sun. The Sun is millions of times closer to Earth than is any other star. Light travels at a finite speed of 300,000 kilometers per second. Light from the Sun reaches Earth in less than 10 *minutes*, but light from the next closest star takes several *years* to get to Earth. The light from very distant stars takes several billion years to reach Earth. When we look at distant stars, we see them as they were when the light we now see left each star. Therefore, when we look at distant stars, we look back in time.

The universe is so large that the distance light travels in a year, or a light-year, is used to measure its distances. A light-year is a distance of about 9½ trillion kilometers. Our fastest rockets would take thousands of years to reach the nearest star beyond the Sun. Compared with the vast distances between stars, the distances between the Sun and its planets are small. Most astronomers estimate the universe to be about 15 billion light-years in radius, and it is therefore thought to be about 15 billion years old.

Stars are not scattered evenly throughout the universe; gravity has drawn them together in huge clumps called galaxies. A *galaxy* is a system of hundreds of billions of stars. The universe contains many billions of galaxies, each, in turn, containing billions of stars. Our solar system is located near the edge of a disk-shaped galaxy of stars called the Milky Way galaxy, which gets its name from the faint white band of its stars that can be seen from Earth on a clear, dark night. The Milky Way galaxy contains more than 100 billion stars revolving in huge orbits around the center of the galaxy.

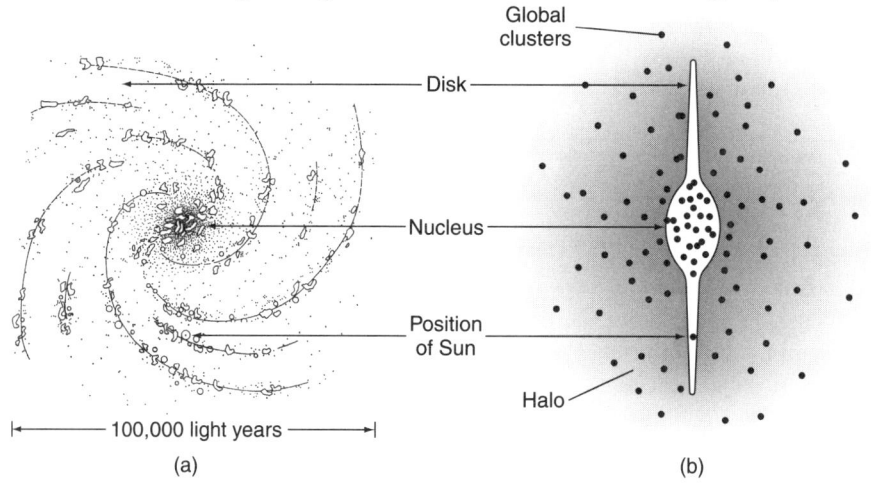

Figure 8.1 The Milky Way Galaxy. Viewed face on (a) and edge on (b), the shapes and locations of the nucleus, disk, and halo can be seen. Note the position of the Sun, about two-thirds of the way out from the nucleus in the Orion arm of the disk.

Sources: *Astronomy: The Cosmic Journey*, William K. Hartman, Wadsworth, 1987. *Horizons: Exploring the Universe*, Michael A. Seeds, Wadsworth, 1987.

THE ORIGIN AND EVOLUTION OF THE SOLAR SYSTEM

Our solar system consists of the Sun, the nine planets, fifty or so moons, and thousands of asteroids, meteors, and countless comets that orbit the Sun. On the whole, though, the solar system *is* the Sun, for the Sun contains 99.9 percent of all the mass of the whole system. It is unlikely that the origin of a family of objects (the planets, moons, etc.) representing such a tiny fraction of the solar system as a whole is not closely linked to the formation of the Sun. Any theory of the formation of the solar system has to account for observational evidence, such as the following:

1. More than 99 percent of the mass of the solar system is contained in the Sun.
2. All the planets (except Pluto) move around the Sun in the same direction and in roughly the same plane.
3. All of the planets (except Venus and Uranus), spin in the same direction, close to the plane of the Sun's equator; and most of the moons also spin in the same direction as their planets, close to the plane of those planets' equators.
4. The planets can be divided by mass and density into the terrestrial and Jovian planets. The terrestrial planets are composed mostly of metal silicates and iron; the Jovian planets, mostly of hydrogen and helium.
5. The planets exhibit a fairly regular spacing of orbits.
6. All the solid bodies of the solar system that have been measured—Earth, the Moon, and meteorites—are roughly 4.6 billion years old, and formed within about 0.1 billion years of each other, that is, at about the same time.
7. Some meteorites, aside from their loss of volatile (easily vaporized) elements such as rare gases, are virtually identical in composition to the Sun.

The question is, Did the solar system form as part of the natural process of star formation or as some later event? As early as the eighteenth century, two theories were proposed to account for the formation of the solar system: the catastrophic theory and the nebular theory.

The *catastrophic theory* suggest that early in the Sun's life there was a collision or near collision between the Sun and a passing celestial body. Such an event would cause a solar upheaval, resulting in streamers and globs of gaseous solar material being thrown off into space. As this material condensed, it would form a great many small bodies, all revolving around the Sun. Some of these would be close enough to each other to aggregate by mutual gravitational attraction to form planets. Others would be left revolving around the Sun as asteroids, planetoids, and comets. However, the catastrophic theory requires an improbable event and does not adequately explain the even spacing of planets and their varying compositions.

In recent years, with the aid of computer modeling, the *nebular theory* has

The Solar System

gained more widespread support. The modern nebular theory proposes that the solar system formed from an interstellar cloud of dust and gas, enriched in heavier elements from earlier supernovae, which condensed into a disk-shaped protostar. The source of planet formation was the disk of dust and gas surrounding the protostar. Turbulence in the dust and gas of the spinning disk caused it to fragment and sort itself out into concentric rings according to the mass and the speed at which the material in it was revolving. See Figure 8.2.

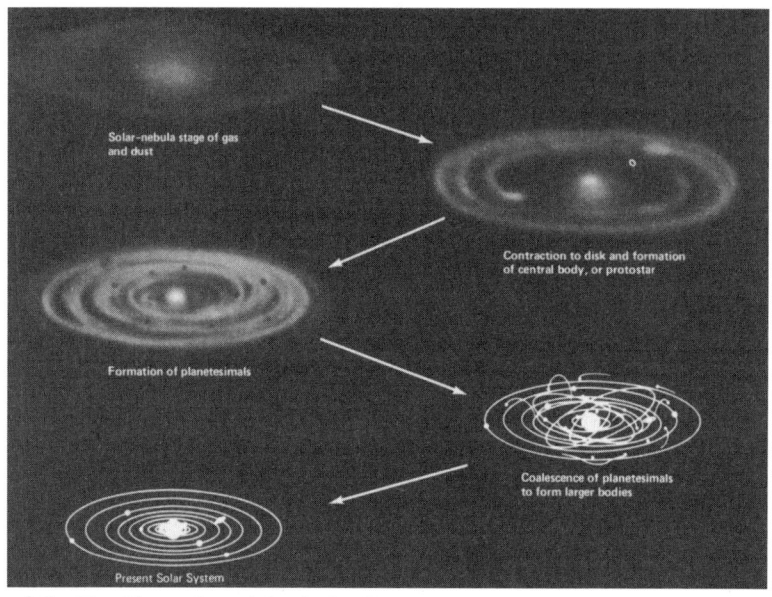

Figure 8.2 The Formation of the Solar System from a Solar Nebula to Our Present Solar System. Source: *Exploring the Cosmos*, Louis Berman and J. C. Evans, Little, Brown, 1986.

Most collisions between celestial bodies tend to cause material to fall inward and end up in the protostar. Some of the dust particles, however, stick together upon collision to form grains. This process of sticking, or accretion, continues, slowly forming larger and larger bodies called ***planetesimals***. Computer simulations indicate that over a period of millions to tens of millions of years, Earth-sized bodies can be formed in this way. See Figure 8.3.

The nature of the dust that accretes depends upon its position in the disk, and it changes over time. The temperature of the disk decreases with movement outward from the core. Where temperatures are lower than 1,500 K, rocky and metallic grains can survive. This region, between 0.5 and 5 astronomical units from the Sun (recall that an AU, or astronomical unit, is the average distance between Earth and the Sun: about 150 million kilometers), is where the terrestrial planets are found today. Beyond about 5 astronomical units, the disk was cold enough so that water ice could condense and survive. Thus, today, in the outer solar system bodies made of rock and ice—the moons of the gas giants— are visible. The Jovian planets probably began forming with the accretion of rock and ice. But, as the body grew larger, it reached a point where it gravita-

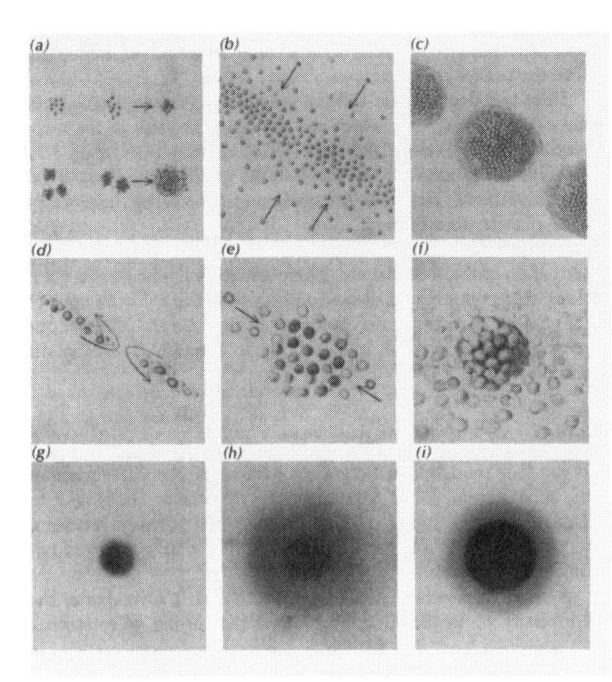

Figure 8.3 Planets Form by the Process of Accretion. Planets form because of the collision and sticking together of small grains (a) in the primordial solar system. The growing particles fall toward the plane of the original cloud (b), forming a loose disk of material. The disk breaks up into asteroid-sized bodies (c), which cluster together (d), collide (e), and coalesce (f) into planet-sized bodies (g). The planet-sized bodies have enough gravity to collect gas from the nebula (h). The result is a primordial planet (i). Source: *The Origin and Evolution of the Solar System*, A. G. W. Cameron, Scientific American, Inc., 1975. All rights reserved.

tionally attracted gas (mostly hydrogen) from the surrounding solar nebula. As the gas built up around the growing planet, the planet's gravity strengthened, attracting yet more gas, rock, and ice. As the giant planets formed, they produced disks of gas and dust of their own, out of which their moons formed. (The formation of Earth's Moon, which is not much smaller than Earth itself, must have occurred in a different way, which will be discussed later.) Computer simulations suggest that Jupiter and Saturn could have formed in this way in just a few million years. Uranus and Neptune, in the colder outer reaches of the disk, would have taken longer, perhaps 10 million years. As the planets drew material from the solar nebula, it dissipated and planet formation ceased as the planets literally ran out of gas and dust to draw upon.

That the terrestrial planets did not acquire huge gaseous envelopes may be due to a number of reasons. The surrounding gas may have been too hot—its molecules moving too vigorously—to form a stable envelope. The solar wind, or stream of charged particles and radiation emanating from the protostar, may have stripped the planets of their hydrogen envelopes as the solar nebula dissipated. Or perhaps the lack of water ice resulted in a slower rate of accretion, so that by the time the terrestrial planets were big enough to attract gas the solar nebula had already dissipated.

Beyond the orbit of Neptune, extending out to about 1,000 astronomical units, lies the ***Kuiper belt***—a region of smaller orbiting objects left over from planet formation. Kuiper belt objects range in size from tiny grains to minor planets hundreds of kilometers in diameter. Pluto and its moon Charon are believed to be Kuiper objects that strayed close enough to the Sun to be drawn into a planetary orbit. See Figure 8.4.

The Solar System

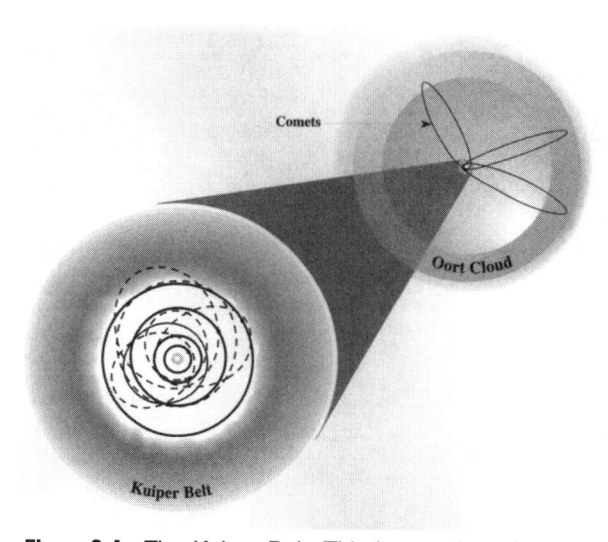

Figure 8.4 The Kuiper Belt. This is a region of objects surrounding the solar system and consisting of particles, ranging in size from dust to planetesimals, that are remnants of solar system formation. It extends from the orbit of Neptune out to about 1,000 AU. Source: *Earth: Evolution of a Habitable World*, Jonathan I. Lunine, Cambridge University Press, 1999.

Even farther out lies a huge, shell-like cloud of icy material called the ***Oort cloud***, after the Dutch astronomer Jan Oort, who proposed its existence in 1950. The Oort cloud is thought to surround the solar system completely and extend to a distance of 100,000 astronomical units (3,000 times the Sun-Pluto distance). See Figure 8.5. The very low temperatures in the Oort cloud allow gases such as methane, nitrogen, and carbon monoxide to form ices that accrete, along with water and dust, and preserve a record of the composition of the original solar nebula. It is believed that the clumps of icy material visible as comets originate within the Oort cloud. The gravitational influence of passing stars causes some of these orbiting ice balls to fall toward the inner reaches of the solar system. Drawn inward by the Sun's gravity, and perturbed by the gravity of major planets, some settle into short-term orbits; one example is Halley's comet, which orbits the Sun once every 76 years. Other, shorter period comets include Tempel 2 (5.3 years) and Encke's comet (3.3 years). Some comets, however, have longer periods, so long that astronomers cannot measure them and may be taken by surprise. One such comet, the Daylight comet of 1910, was probably the brightest seen in the twentieth century. See Figure 8.5.

Although the disk-shaped protostars can be observed in molecular

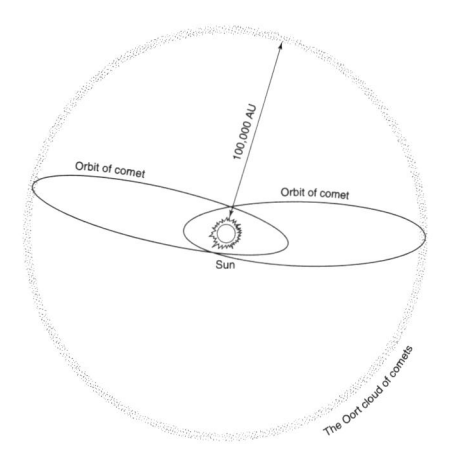

Figure 8.5 The Oort Cloud. A giant cloud of comets is believed to surround the solar system. Stars passing the Oort cloud tear comets from their orbits, sending some off into space and others into long-period orbits around the Sun. Source: *Discovering Astronomy*, Robert Chapman, W. H. Freeman, 1978.

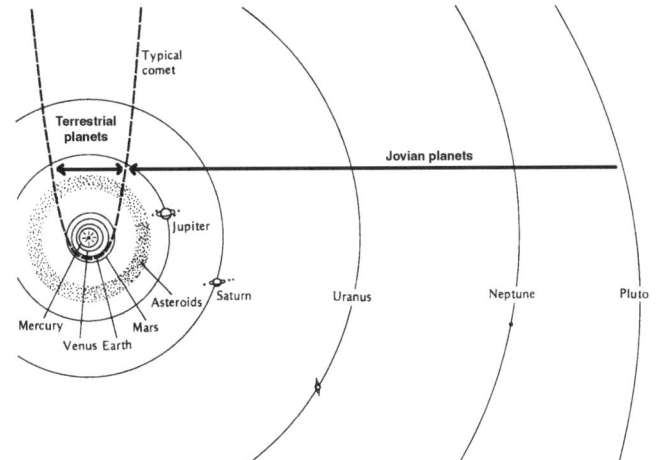

Figure 8.6 The Solar System. Note that in addition to the nine planets and their moons, the solar system includes the asteroid belt, planetoids such as Chiron, and comets that circle the Sun in highly eccentric orbits. *Source: Horizons: Exploring the Universe, Michael A. Seeds, Wadsworth, 1987.*

clouds such as the Orion nebula, planetary formation around stars has never been observed. Of the billions of stars in the universe, only ten or so planets orbiting other stars have been definitively identified. Thus, much of the current thinking about planetary formation is based upon computer modeling and is therefore subject to revision as new observational data are obtained.

THE STRUCTURE OF THE SOLAR SYSTEM

The solar system is defined as the Sun, the nine planets that orbit the Sun, the satellites of those planets, and many small interplanetary bodies such as asteroids and comets. See Figure 8.6. The **Sun** is a star, which is composed of gases and emits electromagnetic radiation produced by nuclear reactions in its interior. **Planets** are bodies that are at least partly solid, orbit the Sun, and emit mainly reflected sunlight. **Satellites** are solid bodies that orbit planets.

Beginning at the center, the main bodies of the solar system are the Sun, Mercury, Venus, Earth, Mars, Jupiter, Saturn, Uranus, Neptune, and Pluto. A common memory aid for this sequence is **M**y **V**ery **E**ducated **M**other **J**ust **S**erved **U**s **N**ine **P**ickles. Some basic information about the bodies in the solar system is summarized in Table 8.1, the Solar System Data chart of the *Earth Science Reference Tables*. A useful way to grasp the range of sizes in the solar system is to remember that Jupiter is about 10 times the size of Earth and that the Sun is about 10 times the size of Jupiter. Conversely, the smallest planet, Pluto, is about one-fourth the size of Earth. When Pluto is omitted, the planets divide into two groups based upon size and composition: the four small, dense, innermost ***terrestrial planets*** and the four large, much less dense outermost ***Jovian planets***. This division provides important clues about the origin and early history of the solar system.

The Solar System

TABLE 8.1 SOLAR SYSTEM DATA

Object	Mean Distance from Sun (millions of km)	Period of Revolution	Period of Rotation	Eccentricity of Orbit	Equatorial Diameter (km)	Mass (Earth = 1)	Density (g/cm³)	Number of Moons
SUN	—	—	27 days	—	1,392,000	333,000.00	1.4	—
MERCURY	57.9	88 days	59 days	0.206	4,880	0.553	5.4	0
VENUS	108.2	224.7 days	243 days	0.007	12,104	0.815	5.2	0
EARTH	149.6	365.26 days	23 hr 56 min 4 sec	0.017	12,756	1.00	5.5	1
MARS	227.9	687 days	24 hr 37 min 23 sec	0.093	6,787	0.1074	3.9	2
JUPITER	778.3	11.86 years	9 hr 50 min 30 sec	0.048	142,800	317.896	1.3	16
SATURN	1,427	29.46 years	10 hr 14 min	0.056	120,000	95.185	0.7	18
URANUS	2,869	84.0 years	17 hr 14 min	0.047	51,800	14.537	1.2	21
NEPTUNE	4,496	164.8 years	16 hr	0.009	49,500	17.151	1.7	8
PLUTO	5,900	247.7 years	6 days 9 hr	0.250	2,300	0.0025	2.0	1
EARTH'S MOON	149.6 (0.386 from Earth)	27.3 days	27 days 8 hr	0.055	3,476	0.0123	3.3	—

Source: The State Education Department, *Earth Science Reference Tables*, 2001 Edition (Albany, New York; The University of the State of New York).

The Terrestrial Planets

The four planets closest to the Sun are Mercury, Venus, Earth, and Mars. These "inner" planets are terrestrial; that is, they resemble Earth in size and rocky composition. They also have about the same density as Earth.

Mercury

Mercury is one of the five planets that can be seen from Earth with the unaided eye. Mercury is very difficult to spot, however, because it is never more than 28° from the Sun, whose glare usually hides it. Through a telescope from Earth, Mercury, like the Moon and Venus, can be seen to exhibit phases, but only dark, fuzzy features can be detected on its surface.

Mercury revolves around the Sun once every 88 days and rotates once every 58.65 days. Because it is the closest planet to the Sun, it receives the most intense radiation. The long rotation period causes the side of Mercury facing the Sun to receive nonstop sunlight for a long period of time. At the point where the Sun is directly overhead of Mercury, the surface temperature reaches 700 K—hot enough to melt lead. At the same time, the side facing away from the Sun is in darkness for long periods and cools to 100 K. Like the Moon, Mercury has no atmosphere because its small gravitational field was unable to hold on to any gases. The lack of an atmosphere allows debris from space to strike the surface of Mercury unhindered. A fly-by of Mercury by the Mariner 10 spacecraft showed a cratered surface remarkably like that of the Moon.

Venus

Venus is one of the brightest objects seen in the sky; only the Sun and the Moon are brighter. Since its orbit is larger than Mercury's, Venus may be seen as far as 47° from the Sun. This feature, together with its brightness, makes it one of the most obvious of all celestial objects.

Venus is almost identical in size to Earth. Like Mercury, Venus can be seen to go through a series of phases. However, Venus rotates in a direction opposite to that of Earth and most other planets.

A telescope can detect little about the surface of Venus because the planet's dense atmosphere, consisting mostly of carbon dioxide, hides the surface from view. This thick, cloudy atmosphere produces a greenhouse effect that traps heat; several Venera probes soft-landed on Venus by the Soviet Union recorded temperatures near 750 K and atmospheric pressures 90 times greater than Earth's. Spectroscopic studies indicate the presence of sulfuric acid, hydrochloric acid, and hydrofluoric acid. In February 1974, the Mariner 10 spacecraft came within 5,800 kilometers of Venus. As it flew by, cameras equipped with special filters and films recorded pictures by ultraviolet light. These photographs revealed details of Venus's atmosphere unseen in visible light, such as cloud motions indicating wind speeds of 100 meters per second (200 mph) in the upper atmosphere.

Earth

Earth is the third planet from the Sun and is the largest of the inner planets. Earth's atmosphere is rich in oxygen and nitrogen. From space, many details of Earth's surface can be seen; more than 70 percent is covered with liquid water. Earth has one natural satellite, the Moon.

Mars

Mars is the fourth planet from the Sun. In the sky, it appears as a reddish star. When Mars is viewed through a telescope, its reddish brown, desertlike surface and polar "ice caps" can be seen. Mars's axis is tilted 24° to its orbit (similar to Earth's 23½ ° tilt); but since its year is nearly twice as long as Earth's, its seasons are longer. The polar "ice caps," consisting mostly of carbon dioxide with some water ice, melt and refreeze as the seasons pass.

Beginning in 1965, a series of Mariner spacecraft passed Mars and photographed its surface, which is pockmarked with craters much like the

Moon's. The photographs also revealed long, sinuous channels that look just like dry riverbeds. Most analysts think these channels were cut by liquid water, but have no idea where the water is now. The Viking probes that landed on Mars found a rocky surface with volcanoes (the largest is four times the size of Mount Everest) and vast, flat plains, but no traces of life. Mars atmosphere is very thin, with less than 1 percent of the surface pressure of Earth. It is composed mainly of carbon dioxide with traces of water vapor, argon, ozone, oxygen, carbon monoxide, and hydrogen.

Mars has two natural satellites, the moons Phobos and Deimos. These moons are tiny; Phobos is only 25 kilometers, and Deimos is just 15 kilometers, in diameter. However, they orbit much closer to the surface than our Moon does to Earth and would therefore appear large—about one-half the size of our Moon—to an observer on Mars. Phobos revolves around Mars in just under 8 hours, faster than the planet rotates, so Phobos's rising and setting are the result of its motion, not the rotation of Mars.

The Jovian Planets

All the more distant Jovian planets—Jupiter, Saturn, Uranus, and Neptune—are gas giants. Although they have more mass than the terrestrial planets, they are less dense. Gas giants have thick atmospheres (hence their name), composed largely of gaseous hydrogen compounds such as water (H_2O), methane (CH_4), and ammonia (NH_3) surrounding a small rocky or liquid core. Despite their large size, the gas giants rotate very rapidly, thereby causing a distinct equatorial bulge. Pluto is a special case, and its origin is still the topic of discussion among astronomers.

Jupiter

Jupiter is the most massive planet in the solar system, containing 70 percent of all the mass outside of the Sun (which is still only one-thousandth of the mass of the Sun). In spite of its large mass, Jupiter is less dense than Earth because its volume is 1,300 times that of Earth.

Through even a small telescope, Jupiter can be seen as a disk crossed by narrow, parallel bright and dark bands. Through more powerful telescopes, Jupiter's great red spot is visible. In 1973 and 1974, Pioneer 10 and 11 flew by Jupiter and took many pictures, which were sent back to Earth. They revealed Jupiter's surface in more detail than had ever been seen.

The visible features of Jupiter's disk are the tops of clouds in the deep atmosphere. A mottled appearance suggests the presence of convective cells. The current thinking is that heated gases from deep within the atmosphere rise and cool, causing clouds to condense and reflect sunlight, thereby forming bright spots. The clouds then spread out to the north and south, sink, and clear up, thus appearing darker. The rapid rotation of Jupiter causes these cloudy and clear regions to swirl together into parallel bands. The great red spot is thought to be the eye of an immense, hurricanelike storm in Jupiter's atmosphere.

Jupiter's interior structure was inferred from gravity measurements made by Pioneer spacecraft. The core is probably a small, rocky ball similar in composition to Earth. The rest of the planet is mostly hydrogen in two distinct layers: an inner layer of liquid hydrogen under such tremendous pressures that it acts like a metal, and an outer layer of liquid hydrogen that acts as hydrogen does on Earth. This outer layer changes gradually from pure hydrogen to hydrogen compounds such as water, ammonia, and ammonium hydrosulfide in the atmosphere.

Jupiter has many natural satellites because of its powerful gravitational field. Four bright moons of Jupiter were first discovered by Galileo. Over the years, ever more powerful telescopes were brought to bear on Jupiter, and by 1975 fourteen moons had been discovered. The four Galilean moons, Io, Europa, Ganymede, and Callisto, are all about the same size as Earth's Moon; the rest are much smaller. Voyager 1 discovered a faint ring around Jupiter, and photographs showed that one moon, Io, has active volcanoes that cover its surface with red and yellow sulfur compounds. Pictures also revealed that Europa has a smooth, ice-covered surface and that Ganymede and Callisto are covered with craters like those on Earth's Moon.

Saturn

Saturn resembles Jupiter in composition and structure, but its cloud markings show less contrast. The surface of Saturn consists mostly of bands of yellowish and tan clouds. The density of Saturn, 0.7 gram per cubic centimeter, is the lowest of any planet, and is, in fact, less than that of water. If a large enough body of water could be found, Saturn would float in it!

Saturn is best known for its ring system. When Galileo first saw Saturn through a telescope in 1610, he drew it as a blurry object with another blurry object on either side and thought it was a triple planet. In 1655, however, Christian Huygens, a Danish physicist and astronomer, discovered that a ring system surrounds the planet. The rings' dimensions are remarkable; they stretch for 274,000 kilometers from tip to tip but are barely 100 meters thick! In fact, the rings are so thin that observers from Earth lose sight of them when they appear edge-on as Earth passes even with their plane.

Saturn's rings are composed of countless particles ranging in size from that of a golf ball to that of a house. Spectroscopic studies have proved that these particles are frozen water or are covered by frozen water. The rings are separated by several large gaps and thousands of finer divisions. The reason that the gaps exist is poorly understood, but one theory is that gravitational effects are responsible. Where did the ring particles come from? Some possibilities are that the particles condensed from gas as Saturn formed, that they are fragments of a satellite that was blown apart by a collision with a comet or an asteroid, and that they are the remains of a comet or asteroid torn apart by tidal forces while passing very close to Saturn.

Like Jupiter, Saturn has an extensive system of natural satellites. It has at least 17 moons, including Titan, which is almost half the size of Earth, and many smaller moons clustered near the rings. Titan is so large it has an atmosphere of its own, consisting mainly of nitrogen with traces of ethane, acetylene, ethylene, and hydrogen cyanide.

Uranus

The English astronomer William Herschel first discovered Uranus in 1781. This planet is so far from Earth that it cannot be seen with the unaided eye and a telescope cannot resolve any markings on it; only a faint greenish color is visible. In 1986 Voyager 2 flew by Uranus and revealed that its atmosphere has an almost featureless blue haze overlying deeper clouds. Voyager 2 recorded a minimum temperature of 51 K (–368°F) and a composition matching that of Jupiter and Saturn. In 1977, evidence of rings around Uranus had been observed. Voyager 2 obtained the first clear pictures of those rings, showing them to be much narrower than those of Jupiter or Saturn.

Uranus's rotation, like that of Venus, is retrograde, and its axis of rotation, pointing toward the Sun, is almost in line with the plane of its orbit. Uranus has five major moons, as well as ten smaller moons only recently discovered by Voyager. The larger moons are composed of ice and black soil. All of the moons are heavily cratered, and cracks and canyons can be seen on several.

Neptune

The discovery of Uranus led astronomers to search for other planets. By 1800, observation of Uranus had revealed that it has irregular motions that run counter to Kepler's laws. Some scientists thought that this phenomenon was caused by a breakdown of gravitation at great distances from the Sun, but others guessed correctly that Uranus was being pulled from its theoretical orbit by an even further planet. Using laborious calculations, astronomers set out to predict where the new planet should be located in order to produce the effects observed on Uranus. In 1846, within 12 hours of beginning a search based upon the calculations of French astronomer Urbain Leverrier, two German astronomers discovered Neptune.

Neptune's atmosphere has been found to contain hydrogen, helium, and some methane, which gives the planet its bluish color. It also has faint cloud patterns resembling the bands of Jupiter and Saturn. Like the other gas giants, Neptune rotates rapidly and has a distinct equatorial bulge. The temperature of its atmosphere is 60 K, warmer than expected for a body so far from the Sun. Its high temperature suggests that Neptune may have an internal source of heat.

Neptune has eight known natural satellites. The largest moon, Triton, revolves around the planet in a direction opposite to that of all other satellites in the solar system. The second largest moon, Nereid, revolves in the normal direction, but its orbit is highly eccentric.

Pluto

Pluto, the farthest planet from the Sun, was not discovered until 1930. Pluto is so far from the Sun that the Sun would not appear as a disk from its surface, but would look like a very bright streetlight observed from across the street. In 1992, it was discovered that this tiny planet is made up of 97 percent nitrogen with small amounts of frozen carbon monoxide and methane. Pluto is about 2,300 kilometers in diameter and has one known satellite, Charon, which is about 1,200 kilometers in diameter. Together, both Pluto and Charon would fit inside the United States.

Pluto's orbit is much more eccentric than the orbits of the other planets and is tilted 17° from the plane of the ecliptic. When Pluto is at perihelion (nearest to the Sun), it is closer to the Sun than is Neptune. These characteristics lead astronomers to believe that Pluto is more likely a comet that was captured into a closer orbit by the Sun's gravity, rather than a planet that formed during the birth of the solar system.

Interplanetary Bodies

In addition to the nine planets, many smaller bodies orbit the Sun, and are therefore also considered part of the solar system. These objects include asteroids, comets, meteors, and meteoroids.

Asteroids

Asteroids are solid bodies having no atmosphere that orbit the Sun. They are tiny planets with well-determined orbits. More than 2,000 asteroids have been found in the solar system; most of them orbit in the gap between Mars and Jupiter. Some asteroids move in very elliptical orbits. The asteroid Icarus comes closer to the Sun than any other asteroid in the solar system.

The composition of asteroids has been studied by examining the wavelength of light that reflects from their surface. If asteroids were perfect mirrors, the light that reflects from them would be identical to the sunlight that strikes their surface. Minerals reflect light of different wavelengths in different ways, however, so that the light that reflects from an asteroid's surface differs from sunlight. By analyzing light reflected from asteroids, astronomers can infer their composition. There are two basic types of asteroids, those with surface reflections characteristic of metals and silicate minerals and those with reflections characteristic of carbonaceous chondrites (stony with a high carbon content).

Meteoroids

In addition to the asteroids between Mars and Jupiter, chunks of matter orbit the Sun in orbits that cross those of the planets. When one of these chunks of matter hits Earth's atmosphere at very high speed, it vaporizes because of friction with the air. As it streaks through the air and vaporizes, it emits light and to an observer on Earth appears as a streak of light. The chunk of matter is a *meteoroid*, the glowing object that streaks through the sky is called a *meteor*, and any of the matter that survives to strike the ground is a *meteorite*.

The Solar System

Meteoroids vary in size from sand-sized particles to chunks weighing many tons. Meteor showers occur when Earth crosses the path of a clump or stream of meteoroids. See Figure 8.7.

Meteorites tell us a great deal about the solar system. There are three main types of meteorites: stony, stony-iron, and iron. Stony meteorites are like Earth's crust, whereas iron meteorites resemble its core. Stony-iron meteorites are like the iron-rich material of Earth's deep mantle. One explanation for this similarity between meteorites and portions of Earth is that a process called ***gravitational differentiation*** took place while the solar system formed. The premise is that the solar system originally had a uniform composition. As gravity drew clumps of the original matter together, however, the material separated into layers because lighter elements do not

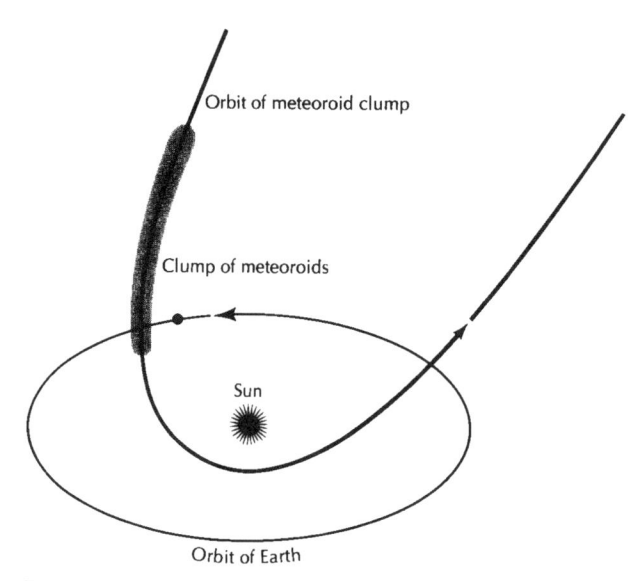

Figure 8.7 Meteor Shower. Source: *Discovering Astronomy*, Robert D. Chapman, W. H. Freeman, 1978. Used with permission.

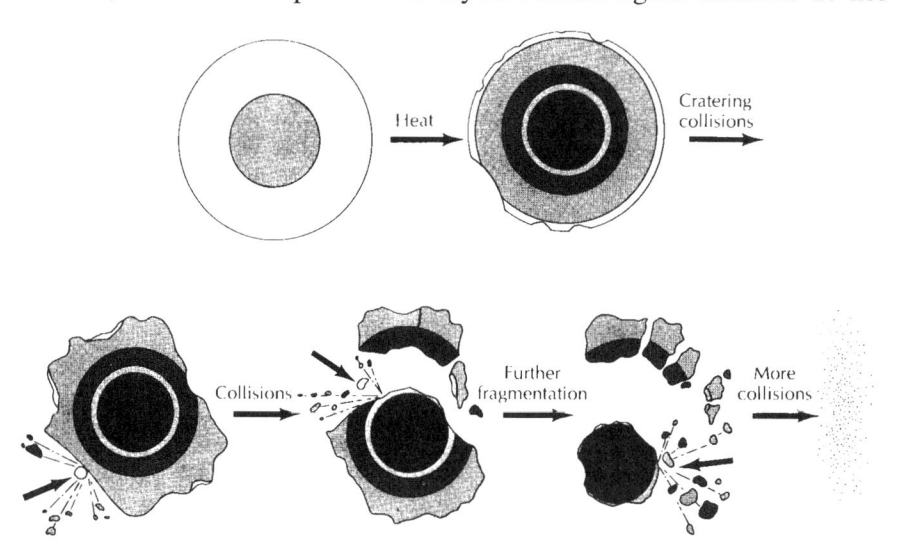

Figure 8.8 Production of Meteoroids. Collisions between large asteroids with layered interiors gave rise to the three types of meteorites found on Earth. Source: *Discovering Astronomy*, Robert D. Chapman, W. H. Freeman and Company, 1978. Used with permission.

experience as strong a gravitational attraction as heavier elements. This separation resulted in Earth's layered interior and also layered interiors for all other bodies in the solar system.

The solar system is thought to have originally contained many large asteroids that collided frequently. With each collision, material was broken off and ejected, some into highly eccentric orbits. If these large asteroids had a layered structure, they would be expected to break into fragments with different compositions reflecting the different layers. See Figure 8.8. For this reason, iron meteorites are considered examples of the composition of Earth's core.

Comets

Comets are another type of object orbiting the sun. Comets, like planets, move in elliptical orbits, but their orbital ellipses are highly elongated. Some comets vary from being 1 astronomical unit from the Sun at perihelion to more than 1,000 astronomical units at aphelion. Kepler's law of equal areas tells us that, although such a comet moves very fast while it is near the Sun, it must move very slowly when it is such a huge distance away. Therefore, some comets take as long as 2 million years to make one orbit of the Sun. Comets are generally divided into long-period and short-period comets. Short-period comets orbit the sun in 200 years or less; long-period comets require more than 200 years to complete an orbit.

Comets are thought to have a solid ***nucleus*** consisting of meteoroid particles embedded in ice. When the nucleus approaches the Sun and heats up, the ice sublimes into a cloud of gas, called a ***coma***, around the nucleus. As the comet gets closer to the Sun, the coma increases in size as more gas sublimes, and particles emitted from the Sun collide with the gas molecules and push them out of the coma, forming a ***tail***. Under the influence of this solar "wind," the tail always streams away from the Sun. It does *not* stream out along the path of the comet like a contrail out of a jet engine. See Figure 8.9.

Impact Events

From time to time, the orbits of asteroids, comets, and meteors are disturbed by collisions or by the gravity of other objects. Their new orbits may then place them on a collision path with a planet such as Earth. When a high-speed object (a comet, asteroid, or meteoroid) collides with a solid surface (a planet or moon), the collision creates a distinctive bowl-shaped hole called an ***impact crater***.

The impact of the speeding object sends out shock waves that spread outward and downward into the ground, and the object's energy of motion is quickly changed to heat. The enormous pressures and heat of the shock wave shatter the ground. Close to the impact, rock is vaporized and melted. Farther away, it is pulverized. The shock waves also raise a rim around the crater and spew material off the sides. When the ground bounces back from the shock wave, it heaves up a central peak in the crater. See Figure 8.10a. These unique characteristics distinguish impact craters from other bowl-shaped

The Solar System

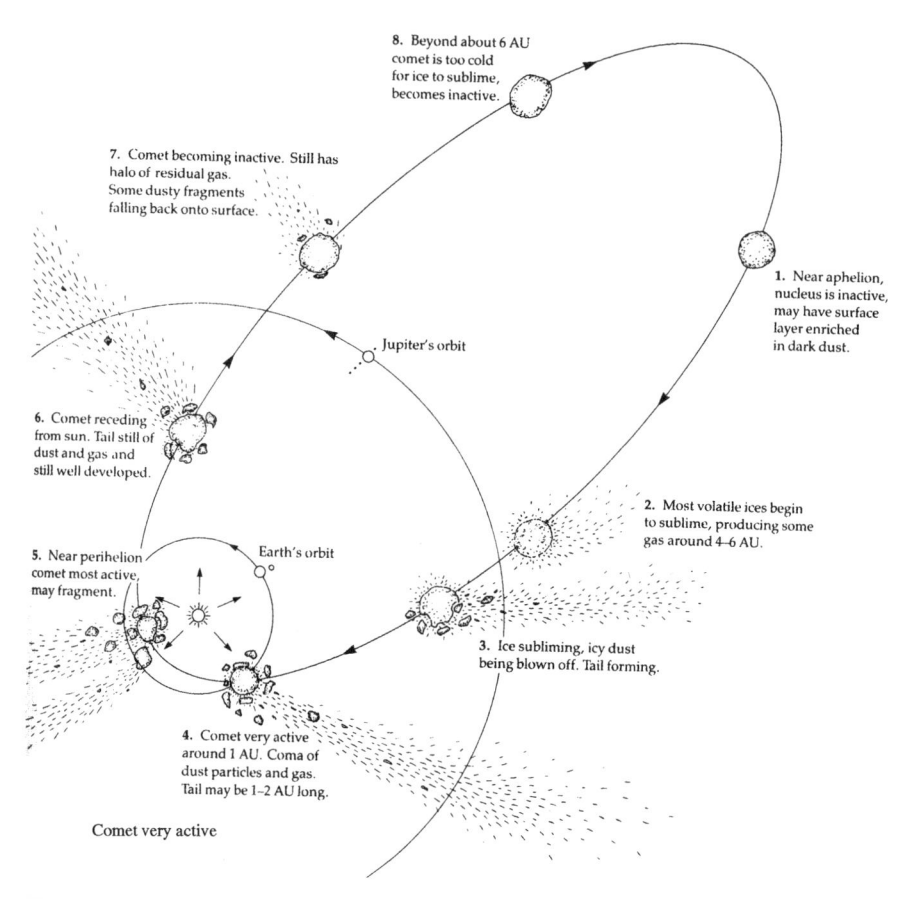

Figure 8.9 Changes That Take Place in a Comet as Its Orbit Carries It Past the Sun.
Source: *Astronomy: The Cosmic Journey*, William K. Hartman, Wadsworth, 1987.

holes, such as volcanic craters and sinkholes.

Photographs of planets and their moons reveal surfaces pock-marked with thousands upon thousands of craters. This evidence shows that impact events are not at all unusual, though they have decreased over time as gravity has swept most debris out of the inner solar system. On Earth, few impact craters have been found because they are quickly erased by water and geologic activity. Of the roughly 200 impact craters discovered on Earth, one of the largest and best known is Meteor Crater, near Flagstaff, Arizona. See Figure 8.10b.

The energy released by the impact of a high-speed object depends upon its mass and the speed of impact. Typical impact speeds on the Moon are about 40 kilometers per second, or about 100,000 miles per hour. Impact speeds would be even higher on Earth because of its stronger gravity. The energy released during a collision between Earth and a 1-kilometer asteroid moving at 100,000 miles per hour would be equivalent to setting off the world's entire nuclear arsenal at once—a dozen times over! Such a huge release of energy has the capability to transform oceans and atmospheres and to destroy life on a planetary scale.

The impact on Earth of an asteroid or comet fragment roughly 10 kilo-

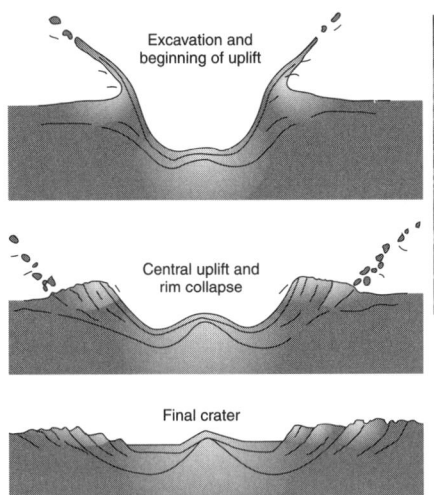

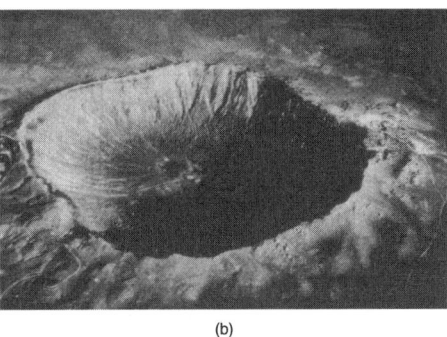

Figure 8.10 (a) Formation of an impact crater (b) Meteor crater near Flagstaff, Arizona Source: *Earth: Evolution of a Habitable World*, Jonathan I. Lunine, Cambridge University Press, 1999. Reprinted with the permission of Cambridge University Press.

meters in diameter would have widespread disastrous effects. Such an impact would gouge a crater more than 100 kilometers in diameter. It would blow a temporary hole the same size in the atmosphere and hurl dust into the upper atmosphere. The dust blown into the stratosphere would darken Earth for months with drastic effects on global climate. When this dust settled, it would fall on land and sea alike, carrying with it the chemical signature of the asteroid. Rock near the site would be "shock-heated," and the shock wave and molten debris thrown out of the crater would knock down trees across thousands of kilometers of land. An ocean impact would create huge waves that would submerge land for hundreds of miles around the impact site.

There is much evidence that an impact event was responsible for the sudden extinction of many life-forms at the end of the Cretaceous period. At that time, about 65 million years ago, 15 percent of shallow-water *families* of organisms became extinct, including 80 percent of all shallow-water invertebrate species. The dinosaurs, too, disappeared around this time. The dividing line between the Cretaceous and the Tertiary, called the K/T boundary (K is the symbol commonly used by geologists for the Cretaceous), is a thin layer of clay that has been identified in sediments worldwide. K/T boundary sediments contain numerous indications of a massive impact event. The clay has an abundance of platinum-group elements—iridium, osmium, gold, platinum, etc.—that is more similar to that found in meteorites than that in Earth's crust. Iridium, in particular, is more abundant than in normal crustal rocks. Other features associated with an impact that are found in this thin boundary of clay include shocked quartz grains, melt spherules (droplets formed from molten rock), graphite and other evidence of burning, and evidence of large waves such as would be caused by an ocean impact.

The Solar System

Solar System Exploration

Exploration of the solar system has yielded information about the origin of Earth and the solar system. Exploration of other planets will help scientists to understand how or whether these planets can benefit humans. In the future, colonization of other planets may be possible and/or necessary. Asteroids may become a valuable source of metals and mineral resources.

MULTIPLE-CHOICE QUESTIONS

In each case, write the number of the word or expression that best answers the question or completes the statement.

1. The diagram below represents a side view of the Milky Way galaxy.

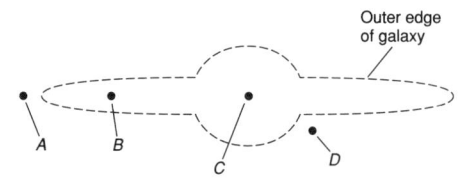

   (Not drawn to scale)

   At approximately which position is Earth's solar system?
   (1) A   (3) C
   (2) B   (4) D

2. According to the *Earth Science Reference Tables*, approximately how many years ago did the solar system originate?
   (1) 570,000,000   (3) 4,600,000,000
   (2) 1,000,000,000   (4) 10,000,000,000

3. Which statement about the planets of our solar system is true?
   (1) All have about the same period of rotation.
   (2) All have about the same period of revolution.
   (3) All have natural satellites.
   (4) All revolve in about the same plane about the Sun.

4. Which of the following best represents the successive stages in the development of the solar system?
   (1) solar nebula → protostar and ringed disk → accretion of planetesimals → present planets
   (2) interstellar nebula → cloudlets → stars → formation of planets by supernova
   (3) protostar → solar nebula → capture of present planets → fragmentation of planets into planetesimals
   (4) protostar → solar nebula → capture of planets by nebula → dissipation of solar nebula

5. In what way are the planets Mars, Mercury, and Earth similar?
   (1) They have the same period of revolution.
   (2) They are perfect spheres.
   (3) They exert the same gravitational force on each other.
   (4) They have elliptical orbits with the Sun at one focus.

Base your answers to questions 6 and 7 on the four graphs below, which represent trends for four characteristics, *A, B, C, D*, of the planets in Earth's solar system. The planets are presented in order of increasing distance from the Sun.

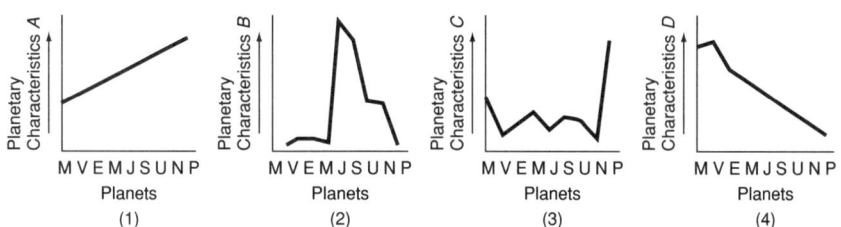

6. Which graph best represents the surface temperatures of the planets?
   (1) 1   (3) 3
   (2) 2   (4) 4

7. Which graph best represents the amount of time each planet takes to orbit the Sun once?
   (1) 1   (3) 3
   (2) 2   (4) 4

8. The Sun's energy is most likely the result of
   (1) the fusion of hydrogen atoms
   (2) transformation of the Sun's gravitational potential energy into heat energy
   (3) burning of hydrocarbons
   (4) radioactive decay of uranium and thorium atoms

9. Earth has fewer impact craters than Mercury because of the
   (1) destruction of meteorites in Earth's upper atmosphere
   (2) more rapid subduction of crustal plates on Mercury
   (3) slower weathering and erosion rates on Earth
   (4) faster rotational speed of Mercury

10. Three of the planets known as gas giants are
    (1) Venus, Neptune, and Jupiter
    (2) Jupiter, Saturn, and Venus
    (3) Jupiter, Saturn, and Uranus
    (4) Venus, Uranus, and Jupiter

The Solar System

11. Which three planets are known as terrestrial planets because of their high densities and rocky compositions?
    (1) Venus, Neptune, and Pluto
    (2) Venus, Saturn and Neptune
    (3) Jupiter, Saturn and Neptune
    (4) Mercury, Mars and Venus

Base your answers to questions 12 and 13 on the four graphs below. The graphs show the compositions of the atmospheres of Venus, Earth, Mars, and Jupiter.

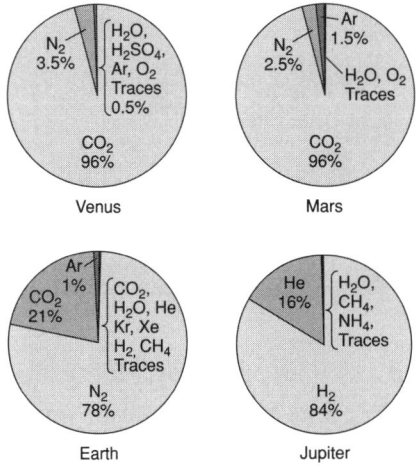

Venus: CO_2 96%, N_2 3.5%, H_2O, H_2SO_4, Ar, O_2 Traces 0.5%

Mars: CO_2 96%, N_2 2.5%, Ar 1.5%, H_2O, O_2 Traces

Earth: N_2 78%, O_2 21%, Ar 1%, CO_2, H_2O, He, Kr, Xe, H_2, CH_4 Traces

Jupiter: H_2 84%, He 16%, H_2O, CH_4, NH_4 Traces

12. Which gas is present in the atmospheres of Venus, Earth, and Mars but *not* in the atmosphere of Jupiter?
    (1) argon (Ar)
    (2) methane (CH_4)
    (3) hydrogen (H_2)
    (4) water vapor (H_2O)

13. Which planet has an atmosphere composed primarily of carbon dioxide (CO_2) and a period of rotation greater than its period of revolution?
    (1) Venus
    (2) Mercury
    (3) Earth
    (4) Mars

14. Why do rock samples brought back from the Moon show no signs of chemical weathering?
    (1) The Moon has no gravity.
    (2) The Moon has no atmosphere.
    (3) Temperatures on the Moon are very cold.
    (4) Temperatures on the Moon are very hot.

15. A person observes that a bright object streaks across the nighttime sky in a few seconds. What is this object most likely to be?
    (1) a comet
    (2) a meteor
    (3) an aurora
    (4) an orbiting satellite

16. Planetary temperatures were recorded by the Voyager I space probe as it traveled from Earth past the outer planets Jupiter, Saturn, and Uranus before leaving the solar system. Which graph best represents the relationship between the temperatures of the planets and their distances from Earth?

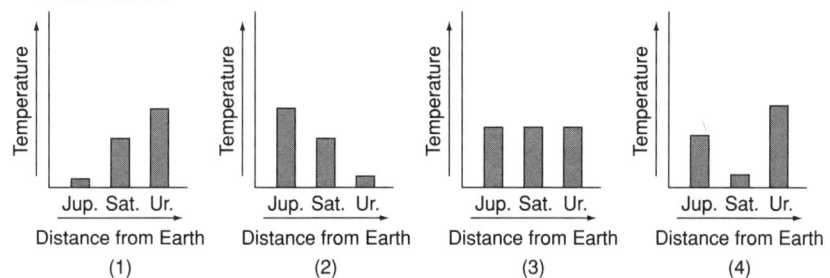

Base your answers to questions 17 through 21 on the table below, which shows data for the various planets of the solar system, and on your knowledge of earth science.

Planet	Diameter (Earth = 1)	Mass (Earth = 1)	Period of Revolution (Earth-years)	Period of Rotation (Earth-days)	Maximum Apparent Magnitude
Mercury	0.38	0.05	0.24	58	−0.2
Venus	0.96	0.81	0.62	247	−4.2
Earth	1.0	1.0	1.0	1.0	...
Mars	0.53	0.11	1.9	1.0	−2
Jupiter	11.2	318.4	11.9	0.41	−2.5
Saturn	9.5	95.3	29.5	0.43	−0.7
Uranus	3.7	14.5	84	0.45	5
Neptune	3.5	17.2	164.8	0.65	7.9
Pluto	1.0	0.9	248	?	14.9

17. Which of the following planets is smallest?
    (1) Jupiter
    (2) Uranus
    (3) Neptune
    (4) Pluto

18. Which planet has traveled around the Sun more than once in your lifetime?
    (1) Mars
    (2) Uranus
    (3) Neptune
    (4) Pluto

19. Which planet would appear as the brightest object in the sky?
    (1) Mercury
    (2) Venus
    (3) Neptune
    (4) Pluto

20. On which of the following planets would the Coriolis effect be greatest?
    (1) Mars
    (2) Mercury
    (3) Jupiter
    (4) Neptune

177

The Solar System

21. If the period of revolution of a planet is directly related to its distance from the Sun, between which two planets is the separation greatest?
    (1) Mercury and Venus
    (2) Venus and Earth
    (3) Earth and Mars
    (4) Mars and Jupiter

22. The surface of Venus is much hotter than would be expected, considering its distance from the Sun. Which statement best explains this fact?
    (1) Venus has many active volcanoes.
    (2) Venus has a slow rate of rotation.
    (3) The clouds of Venus are highly reflective.
    (4) The atmosphere of Venus contains a high percentage of carbon dioxide.

23. The greatest difference in seasons will occur on a planet that has
    (1) a circular orbit
    (2) a slightly elliptical orbit
    (3) its axis of rotation perpendicular to the plane of its orbit around the Sun
    (4) its axis of rotation inclined 4.5° to the plane of its orbit around the Sun

24. On which planet does a natural liquid cover a large portion of the surface?
    (1) Mercury
    (2) Venus
    (3) Earth
    (4) Mars

25. The existence of Pluto and Neptune was accurately predicted through study of the movements of
    (1) comets
    (2) other planets
    (3) stars
    (4) the Sun

26. Impact craters are more obvious on the Moon and Mercury than on Earth because
    (1) meteorites have not struck Earth
    (2) weathering processes on Earth have removed the craters
    (3) Earth is younger than Mercury or the Moon
    (4) all meteorites burn up in Earth's atmosphere

27. Rock samples brought back from the Moon show absolutely no evidence of chemical weathering. This is most likely due to
    (1) the lack of an atmosphere on the Moon
    (2) extremely low surface temperatures on the Moon
    (3) lack of biological activity on the Moon
    (4) large quantities of water in the lunar "seas"

28. The average temperature of a planet
    (1) increases with greater distance from the Sun
    (2) decreases with greater distance from the Sun
    (3) has no relationship to distance from the Sun
    (4) depends only on the planet's atmosphere.

Base your answers to questions 29 and 30 on the diagram below, which shows Earth's orbit and the partial orbit of a comet on the same plane around the Sun.

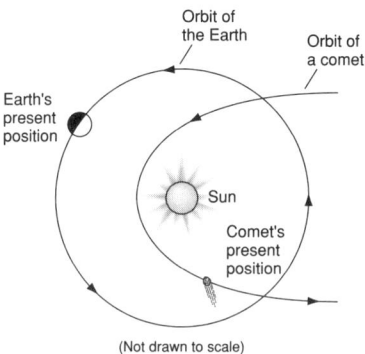

(Not drawn to scale)

29. Which observation is true for an observer at Earth's Equator at midnight on a clear night for the positions shown in the diagram?
    (1) The comet is directly overhead.
    (2) The comet is rising.
    (3) The comet is setting.
    (4) The comet is not visible.

30. Compared with Earth's orbit, the comet's orbit has
    (1) less eccentricity
    (2) more eccentricity
    (3) the same eccentricity

31. Small planetlike bodies revolving in orbit between Mars and Jupiter are called
    (1) satellites   (3) comets
    (2) asteroids   (4) meteoroids

CONSTRUCTED RESPONSE QUESTIONS

32. Classify the following properties of planets into the two groups listed below.
    Properties: far from Sun, small diameter, large mass, low density, short period of revolution, short period of rotation, many moons [1 each]

The Solar System

Groups	
Terrestrial Planets	**Jovian Planets**

33. State two differences between a comet and an asteroid. [2]

34. State two facts that support the nebular theory of the formation of the solar system as opposed to the catastrophic theory. [2]

35. State two fundamental differences between a planet and a star. [2]

EXTENDED CONSTRUCTED RESPONSE QUESTIONS

Base your answers to questions 36 through 38 on the tables below. Table 1 shows the average distances from the Sun in astronomical units (AU) and the average orbital speeds in kilometers per second (km/s) of the nine planets in our solar system. Table 2 lists five large asteroids and their average distances from the Sun.

TABLE 1		
PLANET	**AVERAGE DISTANCE FROM SUN (AU)**	**AVERAGE ORBITAL SPEED (KM/S)**
Mercury	0.4	48.0
Venus	0.7	35.0
Earth	1.0	30.0
Mars	1.5	24.0
Jupiter	5.2	13.0
Saturn	9.6	10.0
Uranus	19.0	7.0
Neptune	30.0	5.1
Pluto	39.0	4.7

TABLE 2	
ASTEROID	**AVERAGE DISTANCE FROM SUN (AU)**
Ceres	2.8
Pallas	2.8
Vesta	2.4
Hygiea	3.2
Juno	2.7

36. On the grid provided below, plot the average distance from the Sun and the average orbital speed for each of the nine planets listed in Table 1. Connect the nine points with a line. [1]

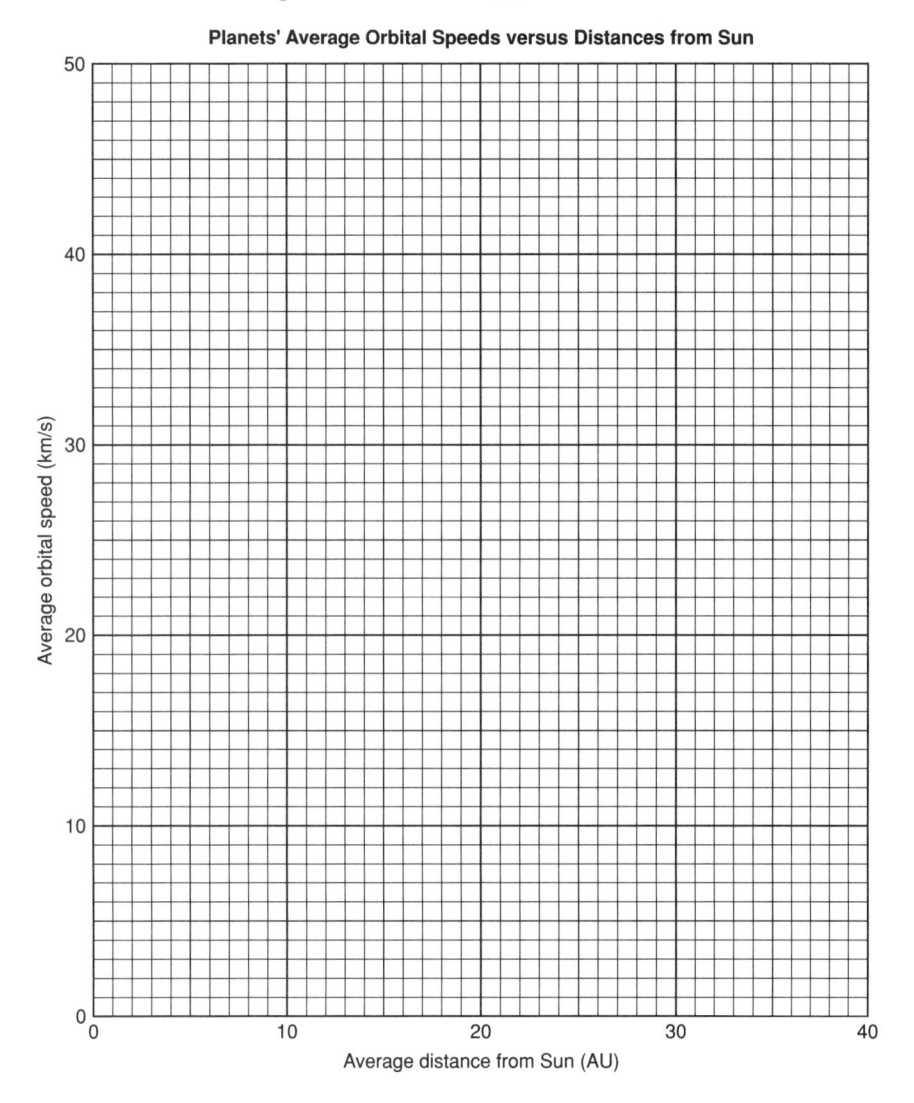

Planets' Average Orbital Speeds versus Distances from Sun

Average orbital speed (km/s) vs *Average distance from Sun (AU)*

37. State the relationship between the planets' average distances from the Sun and the planets' average orbital speeds. [1]

38. The orbits of the asteroids listed in Table 2 are located between two adjacent planetary orbits. State the names of the two planets. [1]

CHAPTER 9
THEORIES OF THE ORIGIN OF THE UNIVERSE

> **KEY IDEAS** The universe is vast and is estimated to be over 10 billion years old. The current theory is that the universe was created by an explosion called the *big bang*. Evidence for this theory includes the red-shifted spectra of distant galaxies and cosmic background radiation. There are several possible models for the future of the universe, depending upon its total mass.

KEY OBJECTIVES
Upon completion of this chapter, you will be able to:

- Identify the underlying assumptions and limitations of cosmology.
- Describe the observed structure of the universe.
- Identify evidence that the universe is expanding.
- Describe the big bang theory and the evidence for it.
- Compare and contrast the future of the universe according to the open and closed models of the universe.

WHAT IS THE UNIVERSE?

The universe, because of its seemingly orderly arrangement, is sometimes called the *cosmos*. **Cosmology** is the study of the nature of the universe as a whole. It is the scientific inquiry into the origin, evolution, and fate of the universe. By making assumptions that are not at odds with the observable universe, cosmologists create models, or theories, that attempt to describe the universe, its origin, and its future. The model is then used to make predictions until an observation is found that contradicts it, whereupon the model is modified or discarded in favor of a new model.

Astronomers define the universe in three different ways. The *observable* universe is all that we can see. It includes all the stars, gas clouds, galaxies, and other objects that we can detect by the radiation they emit. As new and more powerful instruments are developed to detect radiation, more objects can be perceived, and the observable universe grows. The *entire* universe is everything we can see, plus everything else that may be there. We can infer things about the entire universe only on the basis of what we know of the observable universe. We're not even sure that there is an entire universe! The *physical* universe is that part of the entire universe that can be described by the laws of physics. It extends a little beyond the observable universe because it includes objects that we cannot see, but that can be detected by their effects on objects we can see.

The Origin and Evolution of the Universe

Cosmologists assume that the laws of physics are identical throughout the universe and that the universe has the same appearance to all observers no matter where they are located. This assumption means that the universe cannot have an edge; if it did, it would appear different to an observer near the edge than to an observer near the center. The geometry of space must be such that all observers see themselves as being at the center. Cosmologists believe that the only motion that can occur in such geometry is expansion or contraction of the universe.

All this not only sounds mind-boggling; it actually is! The problem with thinking about the universe is that there are too few facts of which we can be certain. Therefore, we must make assumptions about the universe to make progress in cosmology. False assumptions, however, can lead us to nonsensical results! Consider a well-known example, Olber's paradox, shown in Figure 9.1.

The discrepancy between what Olber's paradox theorizes and what we actually observe tells us that one of his assumptions is wrong. We now know that the light from many distant stars moving away from us is red-shifted into the infrared range, which is not visible to human eyes. Also, there are many stars that do not radiate energy in the range of visible light. If we can't see light, it can't make the sky glow. Furthermore, while the universe may be infinitely large, the observable universe is only 16–20 billion light-years in diameter. Light from distances beyond 20 billion light-years has not yet reached us.

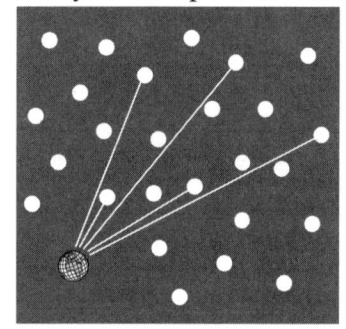

Figure 9.1 Olber's Paradox. If the universe is infinitely large, infinitely old, and uniformly filled with stars, any line of sight from Earth should intersect a star. This assumption predicts that nights should not be dark; instead, the sky should glow with starlight. Source: *Horizons: Exploring the Universe*, Michael A. Seeds, Wadsworth, 1987.

The Structure of the Universe

The distribution of matter, such as stars, dust, and gas, throughout the universe is not uniform; there appears to be a high degree of clustering. On a very large scale, the universe seems to be made up of clusters of objects. Each of these objects is a cluster of smaller objects, which are clusters of yet

Theories of the Origin of the Universe

smaller objects, which are clusters of even smaller ones. See Figure 9.2.

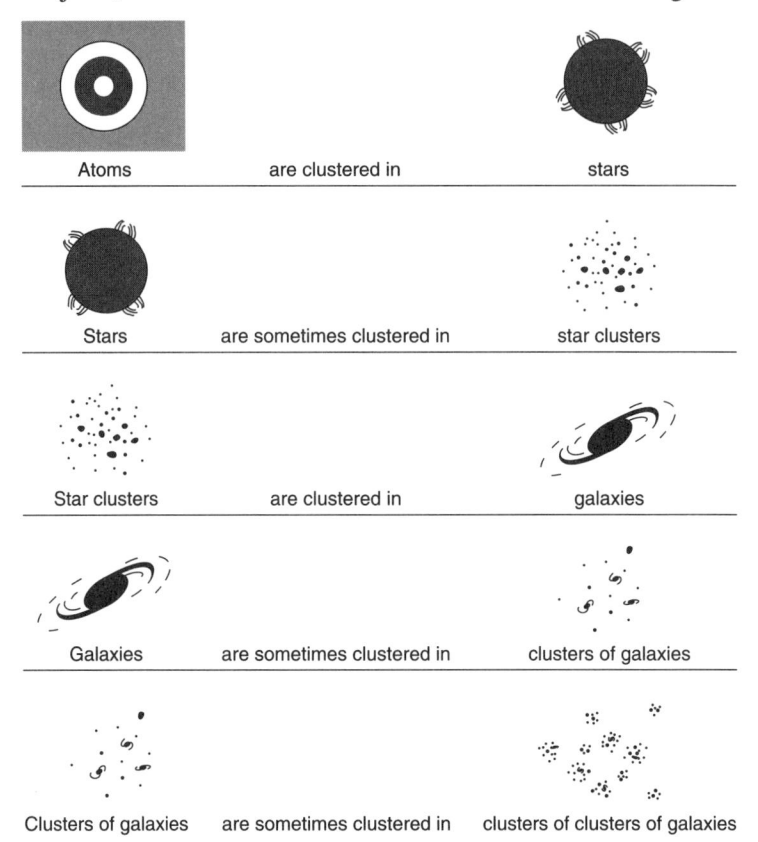

Atoms	are clustered in	stars
Stars	are sometimes clustered in	star clusters
Star clusters	are clustered in	galaxies
Galaxies	are sometimes clustered in	clusters of galaxies
Clusters of galaxies	are sometimes clustered in	clusters of clusters of galaxies

Figure 9.2 Clustering in the Universe. Source: *Concepts of the Universe*, Paul W. Hodge, McGraw-Hill, 1969.

The largest clusters detected so far are clusters of clusters of galaxies, called *superclusters*. Superclusters contain from 5 to 40 clusters of galaxies and are mind-boggling in size. An average supercluster is about 100 million light-years across; a *light-year*, you will recall, is the distance light travels in 1 year at the speed of 300,000 kilometers per second.

Clusters of galaxies are "smaller" objects—only a few million light-years across. Galaxies, which are clusters of stars, are even smaller—only about 100,000 light-years in diameter. But consider that an average galaxy contains about a million million stars, and the average star is 100 or so times the diameter of the Earth and millions of times its volume—not bad for a "small" object! Finally, each star is a cluster of atoms. And what are atoms but clusters of protons, neutrons, and electrons?

Despite the almost unimaginable number of stars, galaxies, and clouds of dust and gas that the universe contains, it is still mostly empty space. Vast distances separate the clusters of matter that can be observed.

Figure 9.3 (a) All galaxies are observed to be moving away from our Milky Way Galaxy (arrows indicate speeds.) Source: *Astronomy: A Self-Teaching Guide*, Dinah L. Moché, 4th Ed., John Wiley, 1998. (b) Expansion of the Universe explains why all galaxies are moving away from us. Source: *Astronomy Explained*, Gerald North, Springer-Verlag, 1997.

The Expanding Universe

Observations of objects in the universe indicate that it is expanding. Other galaxies are moving away from us, and the most distant galaxies are racing away the fastest. See Figure 9.3. The evidence of this motion away from Earth comes from the light reaching us from distant galaxies, which has a lower frequency than would be expected. The explanation for this shift lies in a phenomenon called the *Doppler effect*.

The ***Doppler effect*** is the shift of a spectrum line away from its normal wavelength caused by motion of the source toward or away from the observer. See Figure 9.4. If the source is approaching the observer, there is a blue-shift toward shorter wavelengths of light. If the source is moving away from the observer, there is a red-shift toward longer wavelengths of light.

Why does this occur? Light consists of electromagnetic waves. The human eye distinguishes between one wave and another by wavelength, that is, the distance between one crest of a wave and the next.

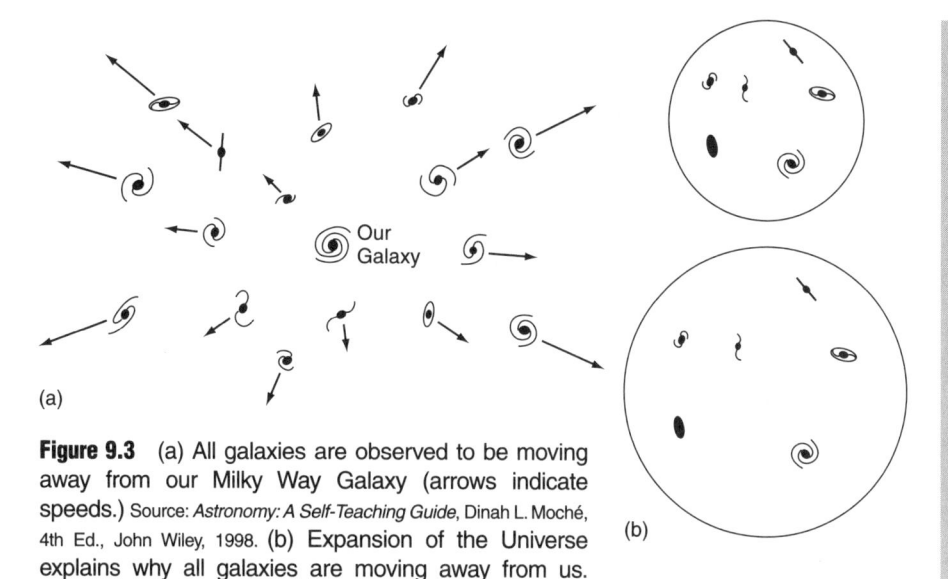

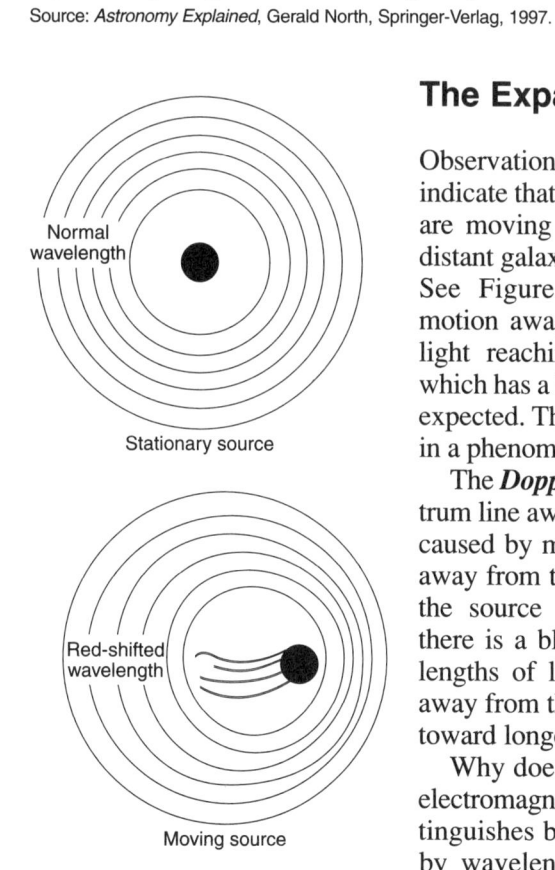

Figure 9.4 The Doppler Effect.

Theories of the Origin of the Universe

Now, suppose a source and an observer are not moving in relation to each other. The source is emitting waves of a particular wavelength; let's say 500 nanometers (billionths of a meter) at the speed of light. The observer will perceive a certain number of crests per second, and the light will be "seen" as a particular color, in this case yellow. Now, if the source moves toward the observer at the same time that it is emitting the waves, more crests will reach the eye of the observer per second. This is the same result that the eye would experience if the light waves had a shorter wavelength. Therefore, the eye will interpret the light as being a different color, let's say blue, even though the source is emitting yellow light. If the source is moving away from the observer, just the opposite happens. Fewer crests per second reach the eye, and the yellow light is interpreted as having a longer wavelength, for example, that of red light.

The amount of shift is proportional to the speed at which the source is approaching or receding from the observer. Therefore:

$$\frac{\text{shift in wavelength}}{\text{normal wavelength}} = \frac{\text{approach or recession speed}}{\text{speed of light}}$$

When the light from distant galaxies is observed, all the spectra are seen to be red-shifted, indicating that the galaxies are moving away from us. See Figure 9.5. Furthermore, the farther away the galaxy, the more it is red-shifted, indicating that distant galaxies are moving away from Earth faster than nearer galaxies. The only explanation for such an observation is that the universe is expanding in all directions.

The Big Bang Theory

Since the universe appears to be expanding, it must have been smaller at some time in the past. If the galaxies could be traced

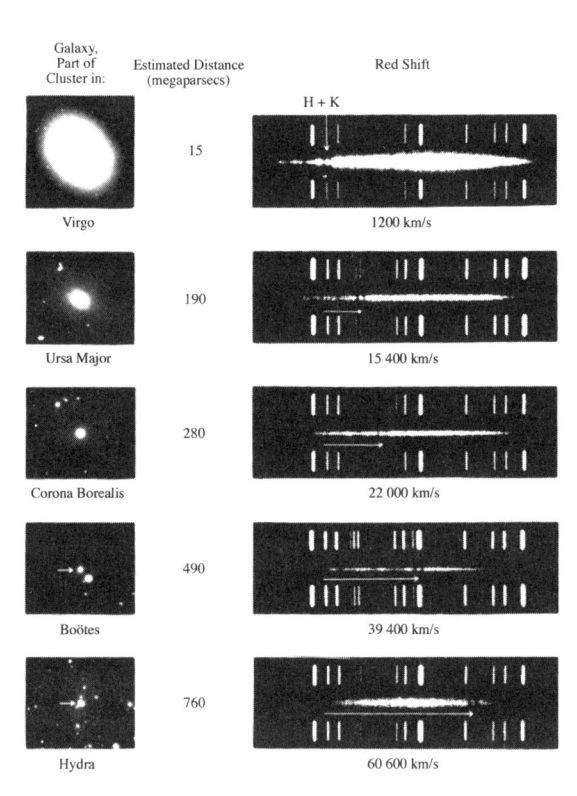

Figure 9.5 Red-Shifted Spectra and Corresponding Velocities of Several Galaxies. Source: *Astronomy: The Cosmic Journey*, William K. Hartman, Wadsworth, 1987.

Galaxy, Part of Cluster in:	Estimated Distance (megaparsecs)	Red Shift
Virgo	15	1200 km/s
Ursa Major	190	15 400 km/s
Corona Borealis	280	22 000 km/s
Boötes	490	39 400 km/s
Hydra	760	60 600 km/s

back in time, there would be a point at which they were all very close to one another. According to the observed expansion rate, this point occurred between 10 and 20 billion years ago. Thus, we have a model in which the universe started out with all of its matter in a small volume and then expanded outward in all directions, much like an explosion. For this reason, this model of the universe is called the ***big bang theory***. See Figure 9.6.

The big bang was, however, quite different from any explosion on Earth. An explosion on Earth can be observed from a distance; it comes from a specific point in space at a particular time. But the big bang did not occur at some pre-existing point in space. Instead, the entire universe, including space and time, originated in the big bang. Before it occurred, there was no place from which an observer could stand and watch the explosion, and there was no time as we know it. With the big bang, both space and time came into existence simultaneously. What we perceive as movement of galaxies away from us is due to the expansion of space, and what we perceive as the elapsing of time is the expansion of time. Thus, galaxies are not hurtling through space; the space itself is expanding and carrying the galaxies apart in the process.

After the big bang, energy and matter spread out as the universe expanded, so temperature and density dropped. The rapidly expanding matter cooled and fragmented into clouds of gas that coalesced into clusters of galaxies 10 to 20 billion years ago. When an observer looks at a distant galaxy today, she is not seeing it as it is now, but as it was long ago when the light now entering her eye first left the galaxy. The redshift in the light seen coming from distant galaxies is due to the expansion of the universe.

The most distant galaxies are so far away that the light reaching us today left them soon after the big bang occurred. Light from beyond these galaxies is so ancient it comes from the original hot clouds of the big bang. The radiation from the original big bang is even more red-shifted than the light from distant galaxies. In fact, it has such a large red-shift that the radiation reaching us from the original big bang arrives at Earth in the range of long-wave infrared or short microwave radio waves. When radiation is red-shifted,

(a) A region of the universe during the big bang

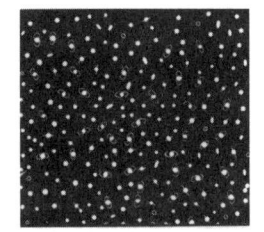

(b) A region of the universe now

(c) The present universe as it appears from our galaxy

Figure 9.6 The Big Bang: An Artist's Conception of the Stages. (a) During the big bang, matter was thrown out uniformly in all directions. (b) Thereafter, gravity caused the matter to coalesce into clumps—galaxies. (c) Near us we see galaxies, farther away we see young galaxies (dots), and at a great distance we see radiation (arrows) coming from the hot clouds of the big bang explosion. Source: *Horizons: Exploring the Universe*, Michael A. Seeds, Wadsworth, 1987.

its wavelength makes it appear that the source is much cooler than it actually is. Thus, the radiation given off by the original big bang, when red-shifted, appears like radiation given off by an object with a temperature of 2.7 K (about 2.7° above absolute zero—very cold!). Since we are looking at the big bang from within the sphere of the expanding shock wave (see Figure 9.6c), that radiation should be coming at us from *all* directions, even from what may appear to be "empty space." Thus, space should be aglow with long-wavelength infrared and microwave radiation. Although predicted in the 1940s, this **cosmic background radiation** glow was not detected until the mid-1960s by Arno Penzias and Robert Wilson at Bell Laboratories in New Jersey. For their discovery of the big bang's primordial background radiation, they received the Nobel Prize in 1978.

The Future of the Universe

The future of the universe depends upon the total amount of matter in the universe. Gravity is a strong and pervasive force. If the density of matter in the universe is greater than a certain critical amount, gravity will be able to slow the expansion of the universe and slowly turn expansion into contraction. The problem is that measurements of the amount of matter that exists in the universe vary so much that none of the possibilities described below can be ruled out. One difficulty is that the matter we can see, that is, the matter that emits light, represents only a tiny fraction of all the matter in the universe. We can perceive matter only by the electromagnetic radiation it emits, and the amount and type of radiation emitted depends upon the temperature of the matter. The "dark" matter, which we cannot see because it doesn't emit electromagnetic radiation, will determine the fate of the universe.

The Open Universe

If the density of matter falls short of the critical amount, there won't be enough matter to exert the gravitational pull needed to stop the expansion. The universe will continue to expand and cool. Although stars will continue to form for a while, the amount of matter available for star formation will dwindle until no new stars form. Eventually, all of the mass in the universe will be contained in cold, dead planets and burned out stars. The universe will end as ever more widely scattered matter in a black void containing no light or heat.

If the universe has exactly the critical density of matter, expansion will continue, but will slow down at an ever-decreasing rate. In this case the fate of the universe will be similar to that of an ever-expanding universe, except that at some point it will stop moving entirely. See Figure 9.7.

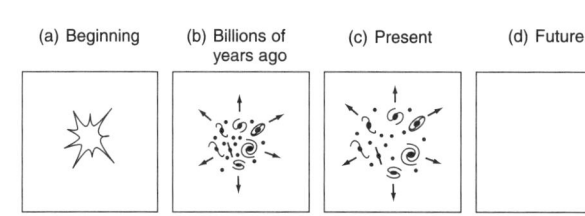

Figure 9.7 Stages of the Open Universe. Source: *Astronomy: A Self-Teaching Guide*, 4th Ed., Dinah L. Moché, John Wiley, 1998.

(a) Beginning (b) Billions of years ago (c) Present (d) Future

The Closed Universe

If the density of matter in the universe even slightly exceeds the critical density, then, as stated previously, the matter will exert enough gravitational attraction to stop the expansion and start a contraction. Ultimately, all of the matter would again collapse back into a small volume—the big crunch. And what then? One theory is that the universe is oscillating and the big crunch would give rise to another big bang. Another theory is that the big crunch would form an enormous black hole that would consume space and time, just as the big bang produced them. See Figure 9.8.

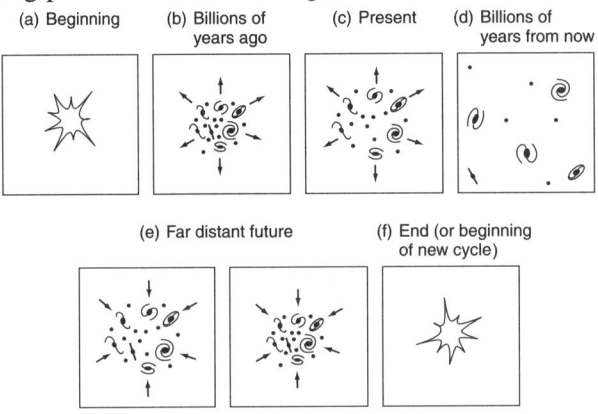

Figure 9.8 Stages of the Closed Universe. Source: *Astronomy: A Self-Teaching Guide*, 4th Ed., Dinah L. Moché, John Wiley, 1998.

Will we ever know which model is correct? Perhaps, but even then we will encounter new questions and continue to search for answers.

Multiple-Choice Questions

In each case, write the number of the word or expression that best answers the question or completes the statement.

1. Which statement best describes how galaxies generally move?
   (1) Galaxies move toward one another.
   (2) Galaxies move away from one another.
   (3) Galaxies move randomly.
   (4) Galaxies do not move.

2. A comparison of the age of Earth obtained from radioactive dating and the age of the universe based on galactic Doppler shifts suggests that
   (1) Earth is about the same age as the universe
   (2) Earth is immeasurably older than the universe
   (3) Earth was formed after the universe began
   (4) the two dating methods contradict each other

Theories of the Origin of the Universe

3. Background radiation detected in space is believed to be evidence that
   (1) the universe began with a primeval explosion
   (2) the universe is contracting
   (3) all matter in the universe is stationary
   (4) galaxies are evenly spaced throughout the universe

4. Earth and our solar system are
   (1) older than the universe
   (2) the same age as the universe
   (3) unable to be age-dated by scientists
   (4) younger than the universe

Base your answer to question 5 on the diagram below and your knowledge of earth science.

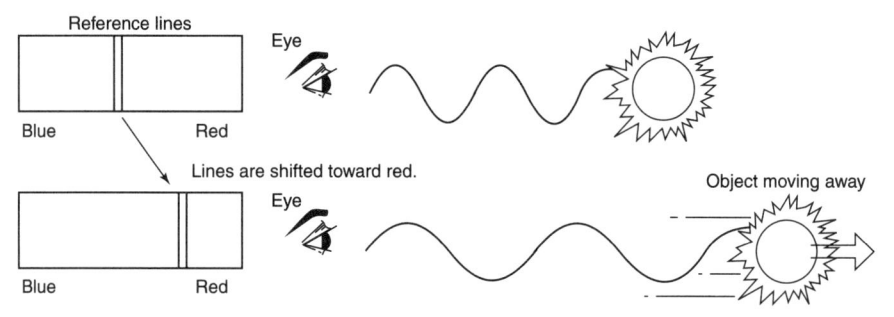

5. The diagram indicates that, if the spectral lines produced by the light from a distant galaxy are
   (1) blue-shifted, the galaxy is moving away from us
   (2) blue-shifted, the galaxy is stationary
   (3) red-shifted, the galaxy is moving away from us
   (4) red-shifted, the galaxy is moving toward us

6. Which of the following characteristics of the universe has been determined with the greatest accuracy?
   (1) age   (3) shape
   (2) size   (4) composition

7. Which of the following characteristics of a star is measured by the Doppler effect?
   (1) color   (3) composition
   (2) temperature   (4) relative velocity

8. In an expanding "open" universe, which of the following would be taking place?
   (1) Materials in the system would remain unchanged.
   (2) Galaxies would recede, leaving great open spaces.
   (3) Galaxies would come together, creating another big bang.
   (4) Galaxies moving away would be replaced by new material.

9. The diagram below represents a standard dark-line spectrum for an element.

   Violet   Red

The spectral lines for this element are observed in light from a distant galaxy. Which diagram best represents those spectral lines?

   Violet   Red
   (1)

   Violet   Red
   (2)

   Violet   Red
   (3)

   Violet   Red
   (4)

10. Observations show that distant galaxies are racing away from us. This means that the Milky Way galaxy is
    (1) a special galaxy at the center of the entire universe
    (2) a galaxy that is located outside the universe
    (3) a typical galaxy, since all galaxies are moving away from one another
    (4) contracting while other galaxies remain stationary

Theories of the Origin of the Universe

11. The symbols below represent the Milky Way galaxy, the solar system, the Sun, and the universe.

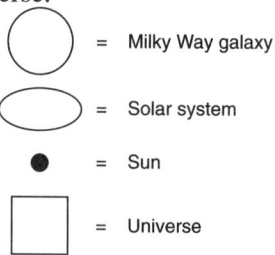

    ○ = Milky Way galaxy

    ⬭ = Solar system

    ● = Sun

    ▢ = Universe

    Which arrangement of symbols is most accurate?

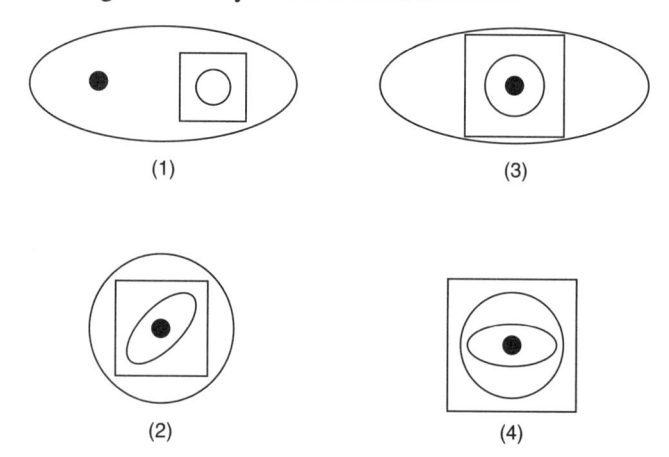

    (1) (2) (3) (4)

12. According to the big bang theory, the universe began as an explosion and is still expanding. This theory is supported by observations that the stellar spectra of distant galaxies show a
    (1) concentration in the yellow portion
    (2) concentration in the green portion
    (3) shift toward the blue end
    (4) shift toward the red end

13. Observations of the universe indicate that the distribution of matter in the universe is
    (1) equally concentrated everywhere
    (2) concentrated in clusters of various sizes
    (3) mostly concentrated in the Milky Way galaxy
    (4) constantly increasing in concentration

14. The observable universe is estimated to be roughly 16–20 billion years old. Which statement best describes why a galaxy located 25 billion light-years from Earth may not be visible to an observer on Earth?
    (1) Galaxies 25 billion light-years away would emit no visible light.
    (2) Light from beyond 20 billion light-years has not yet reached Earth.
    (3) Light from beyond 20 billion light-years passed our galaxy before Earth existed.
    (4) No galaxies are located farther than 5 billion light-years from Earth.

15. If the universe appears to be expanding, which conclusion is most logical concerning the universe?
    (1) The universe must have been smaller at some time in the past.
    (2) The universe must have been larger at some time in the past.
    (3) The density of the universe is increasing.
    (4) The temperature of the universe is increasing.

Constructed Response Questions

16. State two observations successfully explained by the big bang theory. [2]

17. The critical density of the universe is the minimum average density of matter required for the force of gravity to stop the expansion of the universe without reversing it. Infer the future of the universe for the following two conditions:
    (a) density of the universe > critical density. [1]
    (b) density of the universe < critical density. [1]

18. A particular galaxy is located an estimated 15 billion light-years from Earth.
    (a) Decide whether an observer on Earth is seeing the galaxy as it is now, and defend your conclusion. [2]
    (b) Give one reason why an astronomer studying conditions around the time our galaxy formed would study this particular galaxy. [1]

Extended Constructed Response Questions

19. Plot a graph of velocity versus distance based upon the red-shifts of the five galaxies listed in the table. Connect the points in a smooth curve. [2]

Location of Galaxy	Distance (light-years)	Velocity (km/s)
Virgo	78 million	1,200
Ursa Major	1 billion	15,000
Corona Borealis	1.4 billion	22,000
Bootes	2.4 billion	39,000
Hydra	3.8 billion	61,000

(a) State the relationship between distance and velocity shown in the graph. [1]
(b) Compare the rate at which the universe was expanding 8 billion years ago with the rate at which it is expanding today. [1]
(c) Decide whether the graph supports or contradicts the big bang theory model, and support your conclusion. [2]

Unit Three: EARTH'S HISTORY

Chapter 10
THE ORIGINS OF EARTH AND ITS MOON

> **KEY IDEAS** Earth formed as colliding matter accreted. A giant cloud of gas and debris contracted to form our solar system, with Earth reaching its present size about 4.5 billion years ago. The process of accretion heated the young Earth above the melting point of silicate minerals, creating a vast molten region extending downward from the surface. Within this molten region, denser materials sank toward the center and less dense materials floated to the surface, creating Earth's layered internal structure.
>
> Internal convection currents are responsible for Earth's magnetic field and the motion of crustal plates. Earth's Moon probably formed when collision of Earth with another large object caused a chunk of molten Earth material to spin off into a circular orbit.

KEY OBJECTIVES
Upon completion of this chapter, you will be able to:

- Describe the process of accretion by which Earth is thought to have formed.
- Explain how gravity caused Earth to become layered according to the densities of the materials of which it is composed.
- Describe the giant impact theory of the origin of the Moon.

THE ORIGIN OF EARTH

A Fiery Birth

Earth and the other inner, or terrestrial, planets accreted by collisions. When another body impacts a planet, its energy of motion (kinetic energy) is converted to heat. The larger and denser the planet, the stronger the gravitational force exerted on the incoming body, and the greater the energy of the impact. Therefore, of the terrestrial planets, Earth was heated the most, and Mercury the least. The temperature reached depends on the rate of impacts as compared to the rate at which heat is radiated away.

The Origins of Earth and Its Moon

The heat produced by accretion caused Earth, Venus, and perhaps Mars to reach internal temperatures above the melting points of silicate minerals. Thus, the earliest Earth had a vast molten region, extending from the surface partway into the interior. Heat flowed from the hot surface toward the cooler center *and* outward to space. Shortly after accretion brought Earth to its present size, roughly 4.5 billion years ago, interior temperatures became high enough to partially melt the mixed solids of silicates, as well as iron. Once molten, elements that had been locked into crystalline structures were free to recombine with other, compatible elements.

The Layered Earth

Substances in the molten Earth then underwent ***differentiation***, a process by which denser materials (principally iron) sank toward the center, forming the core, while less dense materials (e.g., sodium and potassium ions) floated to the top. As a rock falling off a cliff releases gravitational potential energy as it falls, so the iron that fell toward the core released gravitational energy, creating extra heat. About 32 percent of the mass of our planet is iron, and iron is 50 percent denser than silicates, so the heat generated by formation of the core was substantial and fueled even more melting and differentiation. While this process was going on, about 4.5 billion years ago, an event that resulted in the formation of the Moon occurred, as discussed later in this chapter.

Eventually, Earth's gravity attracted most of the bodies in space surrounding it. With the surrounding space swept clean, the accretion rate fell off. In its molten state, Earth lost heat by convection. With little or no incoming energy from impacts, and removal of heat by convection, Earth cooled, forming a solid crust. The end result of differentiation and cooling was a planet with an internal structure that is layered like an onion, with the least dense materials at the surface and the densest ones at the center.

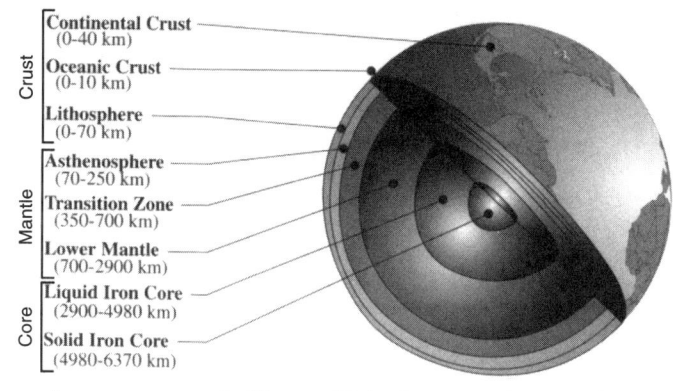

Figure 10.1 Cross Section Showing Earth's Layered Interior. Source: *Earth: Evolution of a Habitable World*, Jonathan I. Lunine, Cambridge University Press, 1999.

Crust:
- Continental Crust (0-40 km)
- Oceanic Crust (0-10 km)
- Lithosphere (0-70 km)

Mantle:
- Asthenosphere (70-250 km)
- Transition Zone (350-700 km)
- Lower Mantle (700-2900 km)

Core:
- Liquid Iron Core (2900-4980 km)
- Solid Iron Core (4980-6370 km)

As you can see in Figure 10.1, there are three main zones within Earth: the crust, the mantle, and the core. The core is composed of iron, nickel, and cobalt and extends from the center of Earth out to a depth of 2,900 kilometers. It consists of two parts: a solid inner core and a fluid outer core. Over the core, out to a depth of about 70 kilometers, lies the mantle. The mantle is rather plastic in nature and is composed mainly of dark, dense silicates. It is subdivided into several layers based upon density, composition, and fluidity. Overlying the mantle is the solid crust, also subdivided by density and composition. Ranging from 0 to 70 kilometers in thickness, it is chemically diverse but contains a lot of lighter, less dense silicates than those in the mantle.

Earth also experiences radioactive heating. Among the elements present in the developing Earth were uranium, thorium, and potassium, each of which has radioactive isotopes. These elements also have large atomic radii and therefore floated to the top and became concentrated in crustal rocks, mainly granite. Decay of the radioactive isotopes in the crust and mantle generates roughly 30 trillion watts of power a year. Today, between radioactive decay and heat transferred to the mantle by the core, there is enough heating to soften mantle rock and allow convective flow to remove heat. The patterns of motion, however, are complex. Simple convection currents in the mantle are interrupted by plumes of hot material driven by heat from the core, which convects separately.

The Earth as a Magnet

The iron core was able to create a magnetic field. Magnetic fields in the solar nebula magnetized some of the rocks and iron grains that later accreted, forming Earth. Thus, some iron in the early Earth was magnetic. But iron is also a conductor of electricity. Convection in the molten core caused the electrically conductive iron to move in the presence of the magnetic field. This induced an electric current, which in turn created a stronger magnetic field. Further convection of the iron in this field induced an even larger current, thereby creating an even stronger magnetic field. Fueled by heat escaping from the core, this self-perpetuating process, called a *magnetic dynamo*, resulted in Earth's magnetic field. See Figure 10.2.

Earth's magnetic field, or **magnetosphere**, extends out into space about four Earth radii. Blasted by the solar wind's steady stream of high-energy charged particles, the magnetosphere sticks

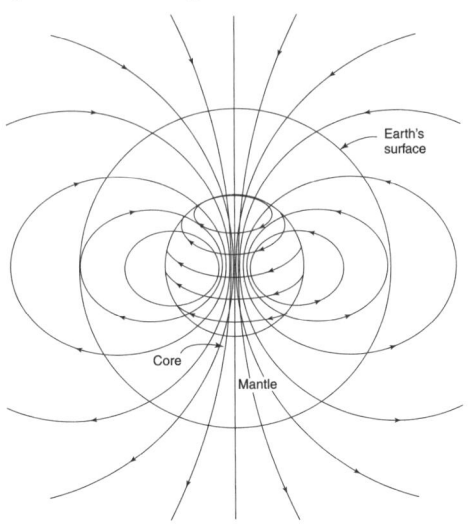

Figure 10.2 Earth's Magnetic Field. Convection motion of the molten iron core induces a magnetic field. Source: *The Earth Sciences,* Arthur N. Strahler, Harper & Row, 1971. Library of Congress Catalog Card Number 78-127335.

The Origins of Earth and Its Moon

out like a tail on the side of Earth facing away from the Sun. See Figure 10.3.

High-energy charged particles can be deadly. Penetrating organisms like a bullet, they can disrupt and destroy living tissue. Fortunately for life on Earth, moving charged particles have a magnetic field of their own, and many of the deadly particles of the solar wind are deflected by the magnetosphere. Also, some charged particles are trapped by Earth's magnetic field and move around rapidly inside two doughnut-shaped regions of the magnetosphere known as the Van Allen belts. See Figure 10.3.

High-energy charged particles moving through the atmosphere stimulate the atoms and ions of the atmosphere to radiate light, producing auroras. The Aurora Borealis, or northern lights, and Aurora Australis, or southern lights, are shimmering bands of light that shine in the night sky. Charged particles tend to be funneled down to Earth near the poles by the magnetosphere, and auroras are seen mainly in Earth's Arctic and Antarctic regions. See Figure 10.4.

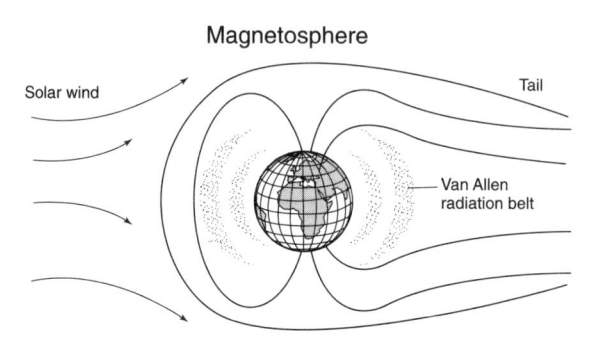

Figure 10.3 Earth's Magnetosphere and Van Allen Belts. The solar wind distorts Earth's magnetosphere. The Van Allen Belts are two doughnut-shaped regions of charged particles trapped by the magnetosphere that surround earth. Source: *The American Heritage Dictionary of Science*, Robert K. Barnhardt, ed., Houghton Mifflin Company, 1986.

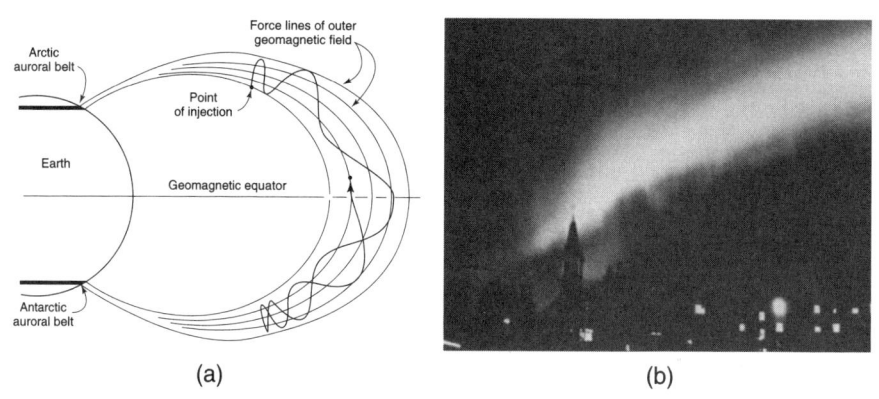

Figure 10.4 (a) Charged particles funneled into the upper atmosphere near the poles by lines of magnetic force surrounding Earth cause shining lights known as auroras. Source: *The Earth Sciences*, Arthur N. Strahler, Harper & Row, 1971. Library of Congress Catalog Card Number 78-127135. (b) An Aurora Borealis photographed during a winter night in Alaska. Source: *The Earth Sciences*, Arthur N. Strahler, Harper & Row, 1971. Library of Congress Catalog Card Number 78-127135.

THE ORIGIN OF THE MOON

Early Ideas

Until the mid-1980s, the origin of the Moon was a difficult issue for astronomers because the Moon is unusually large compared to Earth and moves in a circular orbit. A variety of theories, ranging from capture of a passing body by Earth's gravity, to ejection of a chunk of matter by a rapidly spinning Earth, to formation in place at the same time as Earth, had been suggested. Attempts to model the third theory, formation in place, with computers presents problems. The Moon's density is consistent with little or no iron content. Capture of the Moon, while possible, is unlikely, and capture into a circular orbit is even more unlikely. Finally, there are problems with the physical plausibility of a molten Earth spinning rapidly enough to throw off the Moon.

Apollo's Clues and a New Theory

Rock and dust samples brought back from the Moon by the Apollo missions in the 1980s virtually eliminated all of the early theories. Although the Moon's small size limits possible geologic activity, Moon rocks are more similar to Earth's mantle than to meteorites like those that would have accreted to form the Moon. Moon rocks are also richer in silicates and poorer in metals and volatile elements than Earth rocks. Simplistically put, Moon rocks could be made by heating rocks from Earth's mantle until they vaporize, and then recondensing the less volatile materials.

This geochemical mystery has prompted planetary scientists to consider a new theory—that the Moon is the result of a *giant impact*, a huge collision between Earth and another planet-sized body.

How likely is such an impact? Early in the formation of the solar system, conditions were right for this kind of collision. At first, accreting planetesimals would have been small and would have moved in roughly circular orbits. Collisions would have been gentle, largely resulting in accretion. As planets grew larger, however, close passes of these bodies would have sent them into elliptical orbits, which, in turn, would increase collision speeds. By the time the terrestrial planets had formed, catastrophic collisions with solar-system debris traveling in high-speed elliptical orbits would have been likely. Small bodies hitting big ones would vaporize and melt, adding to the larger body. Big bodies hitting other big bodies would have had disastrous results. It was probably a giant impact that knocked Uranus on its side and spun out a disk from which its moons formed.

The Giant Impact

Computer simulations show that, if a planet as big as Mars, or slightly bigger, struck Earth, a big chunk of Earth's mantle, but very little iron core,

The Origins of Earth and Its Moon

would have been spun off. Also, part of the material ejected would have entered a circular orbit, forming a ring around Earth. See Figure 10.5.

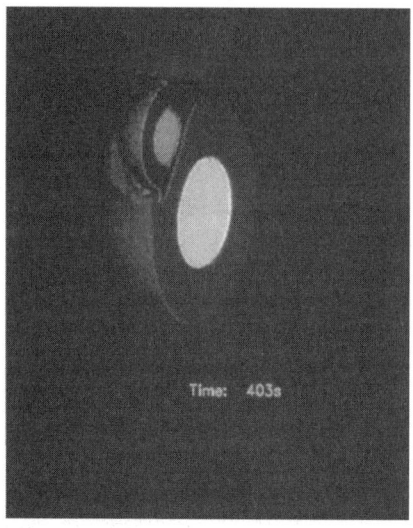

Figure 10.5 Computer Simulation of the Formation of the Moon by a Giant Impact.
Source: *Earth: Evolution of a Habitable World*, Jonathan I. Lunine, Cambridge University Press, 1999.

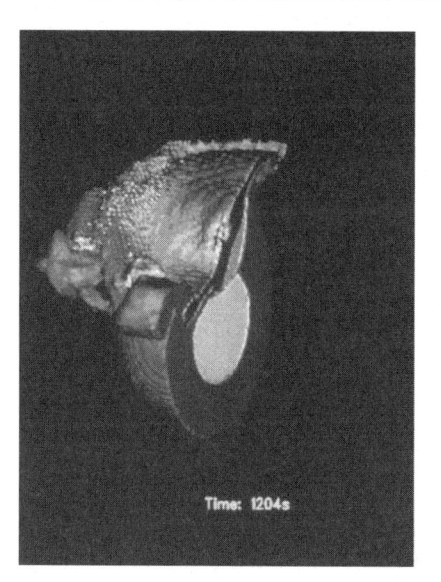

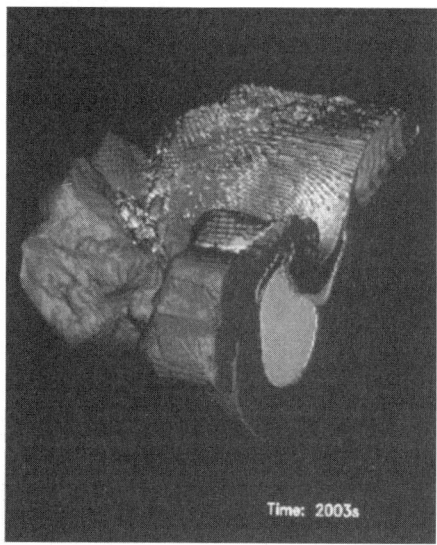

The rest of the ejected material would have reaccreted into Earth (computer simulations of the event indicate that the iron core of the planetesimal would have merged with Earth's in a matter of hours) or been lost into orbit around the Sun. The material ejected would have left a sizable hole, but the spinning of the still malleable Earth would quickly have restored the planet's normal shape.

Most of the material ejected into space would have been vaporized, with only the least volatile materials remaining solid. Much of the volatile material (water and less dense ions) would be lost. This hypothesis is consistent with the lower density minerals with which rocks from the lunar highlands are enriched. Since little of Earth's core was ejected, little iron would be present. Material in the ring then collected to form the Moon.

For the first billion years of its life, the newly formed Moon was bombarded by meteorites. They created great craters and melted the surface that is now the lunar crust. About a billion years into the Moon's life, accretion and radioactive decay produced enough heat to cause melting and differentiation in the upper 500 kilometers or so. Volcanoes erupted, pouring out huge floods of basalt, which now form the maria. By about 3 billion years ago, the Moon had cooled enough that volcanic activity ceased. With the exception of a few major impacts and small lava flows, the Moon has remained virtually unchanged since then.

When did this sequence of events take place? The oldest rocks recovered from the Moon date to 4.4–4.5 billion years ago, so it certainly didn't occur any later. This age also sets a time by which Earth's core must have formed. Formation of the core had to occur before the impact that formed the Moon, because the Moon is poor in iron. Most likely, the Moon formed early in Earth's history, close to or before 4.5 billion years ago.

MULTIPLE-CHOICE QUESTIONS

In each case, write the number of the word or expression that best answers the question or completes the statement.

1. Earth formed by accretion as a giant cloud of gas and debris contracted to form our solar system. During the process of accretion, the energy of motion of bodies that collided with Earth was converted into
   (1) magnetic force
   (2) heat energy
   (3) electrical force
   (4) matter

2. Of the terrestrial planets, Earth was heated the most by collisions during accretion because
   (1) collisions with Earth produced greater impact energy because of Earth's large mass and high density
   (2) Earth was closest to the Sun and received the most intense solar radiation
   (3) bodies that collided with Earth were hotter than bodies that collided with other planets
   (4) bodies that collided with Earth were heated by friction with Earth's magnetic field

The Origins of Earth and Its Moon

3. Within the early Earth's vast molten region, substances underwent a process known as differentiation, during which
   (1) substances of low density rise to Earth's surface, while those of high density sink toward its center
   (2) substances of high density float to Earth's surface, while those of low density sink toward its center
   (3) substances of high and low density chemically combine to form uniformly dense substances
   (4) substances of high density form gases, while those of low density form solids

4. While Earth was in a molten state, gravity caused Earth materials to separate into layers according to their
   (1) colors
   (2) densities
   (3) magnetisms
   (4) ages

5. Earth's magnetic field is likely a result of
   (1) convection currents in Earth's mantle
   (2) convection currents in Earth's core
   (3) a high concentration of iron in Earth's crust
   (4) high-energy particles in the solar wind

6. As depth beneath Earth's surface increases, temperature
   (1) increases
   (2) decreases
   (3) remains the same
   (4) increases, then decreases

7. Much of Earth's internal heat is produced by
   (1) decay of radioactive elements
   (2) fusion of hydrogen into helium
   (3) friction between Earth's surface and atmosphere
   (4) energy absorbed from sunlight

8. Analysis of rock and dust samples brought back from the Moon by the Apollo missions indicates that Moon rocks
   (1) have a composition more similar to that of Earth's mantle than to meteorites
   (2) contain elements never before identified on Earth
   (3) have a composition very different from that of Earth rocks
   (4) have a high concentration of iron, similar to that of Earth's core

9. Which statement about the age of Earth rocks and the age of Moon rocks is true?
   (1) Moon rocks are much younger than Earth rocks.
   (2) Moon rocks are much older than Earth rocks.
   (3) Moon rocks are about the same age as Earth rocks.
   (4) Moon rocks cannot be dated because they contain no radioactive elements.

10. Which of the following models of the Moon's origin is currently considered most likely?
    (1) The Moon was spun off by a rapidly spinning, molten Earth.
    (2) The Moon was a passing body that was captured by Earth's gravity.
    (3) The Moon formed in place at the same time that Earth formed.
    (4) The Moon was ejected from a molten Earth by a giant impact.

11. Measurements of radioactivity in the rocks of Earth's crust currently place the age of Earth closest to
    (1) 100,000 years
    (2) 1.0 million years
    (3) 4.6 billion years
    (4) 16 billion years

12. Displays of auroras (northern lights) are believed to be caused by
    (1) radiation of infrared rays from Earth
    (2) reflection of the midnight Sun on northern snowfields
    (3) gases glowing under the impact of electrically charged particles
    (4) radioactive dust in the upper atmosphere

13. Craters and layers of basalt on the Moon's surface probably formed as a result of
    (1) meteorite impacts and the heating of surface rock due to accretion
    (2) recent volcanic eruptions
    (3) wrinkling of the Moon's molten surface as it cooled and contracted
    (4) bubbles of gas rising through the molten rock just beneath the Moon's surface

14. The statement "The core of the earth is metallic" is
    (1) a fact
    (2) a principle
    (3) an inference
    (4) an observation

15. Considering only places within Earth, which of the following would have its lowest value at Earth's center?
    (1) gravity
    (2) density
    (3) temperature
    (4) pressure

16. What percent of Earth's mass does the crust represent?
    (1) less than 1
    (2) approximately 10
    (3) between 20 and 30
    (4) almost 50

17. Which of the following facts does *not* support the hypothesis that Earth's core consists of iron?
    (1) Iron is very common.
    (2) Iron is a strong metal.
    (3) Many meteorites contain iron.
    (4) Earth's average density is 5.5 g/cm^3, whereas that of the crust is much less.

The Origins of Earth and Its Moon

18. In the layered model of Earth's interior, what happens to layer density from Earth's core outward?
    (1) It increases.
    (2) It decreases.
    (3) It remains the same.
    (4) It decreases to a point and then increases.

19. In which parts of Earth's interior would melted or partially melted material be found?
    (1) stiffer mantle and inner core
    (2) stiffer mantle and outer core
    (3) crust and inner core
    (4) asthenosphere and outer core

Constructed Response Questions

20. Explain the role of gravity in the process of differentiation that created Earth's layered internal structure. [2]

21. State one reason why, in light of current knowledge, it is unlikely that the Moon is a captured passing body. [2]

22. Construct a labeled diagram of Earth's magnetic field. [2]

Extended Constructed Response Questions

23. Base your answers to parts (a) through (c) on the *Earth Science Reference Tables* and the diagram below, which shows Earth's internal structure and its three main layers.

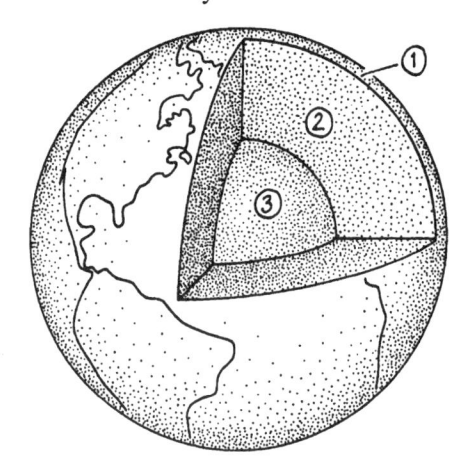

(a) Identify the three main layers of Earth's interior. [3]
(b) State the thickness of each of the three main layers of Earth's interior. [3]
(c) Describe the process by which Earth's layered internal structure is thought to have been formed. [2]

CHAPTER 11

THE ORIGIN AND STRUCTURE OF EARTH'S ATMOSPHERE

> **KEY IDEAS** Earth is surrounded by an atmosphere that extends several hundred kilometers out into space. Earth's early atmosphere formed as a result of the outgassing of water vapor, carbon dioxide, nitrogen, and other gases, in lesser amounts, from its interior. The evolution of life caused dramatic changes in the composition of Earth's atmosphere. Free oxygen did not form in the atmosphere until oxygen-producing organisms evolved. Earth's current atmosphere has a layered structure, with each layer differing in composition and temperature from other layers.

KEY OBJECTIVES
Upon completion of this chapter, you will be able to:

- Describe the process of outgassing that formed Earth's early atmosphere.
- Explain how the evolution of life led to the development of Earth's current oxygen-rich atmosphere.
- Compare and contrast Earth's early reducing atmosphere and current oxidizing atmosphere.
- Describe the layered structure of Earth's current atmosphere.

THE ORIGIN OF EARTH'S ATMOSPHERE

Earth's Lost Atmosphere

In an early stage, while still accreting and forming a core, Earth's atmosphere was a cloud of silicate vapor. Then, as accretion and core formation stopped, Earth cooled and the silicate vapor condensed, forming molten and solid rock. Any hydrogen and other gases left from the solar nebula would quickly have been swept away by the vigorous solar wind, leaving primordial Earth with no atmosphere.

Formation of Earth's Present Atmosphere

Outgassing

Where, then, did the gases of our present atmosphere originate? One likely source was *outgassing*, the release, from the interior of the hot, young Earth, of water and trace gases trapped in rocks. (Recall that the original matter from which Earth accreted was mostly bits of rock and water ice.) Compounds of hydrogen, carbon, oxygen, and nitrogen, such as carbon dioxide, ammonia, methane, and a large amount of water, worked their way to the surface of the molten Earth. Even after Earth had cooled and formed a solid crust, outgassing during volcanic eruptions would have continued. In addition to ash and lava, gases such as hydrogen sulfide, carbon dioxide, and water vapor are still being released during volcanic eruptions today. Table 11.1 compares, in terms of composition, the gases found in Earth's atmosphere and hydrosphere with those given off during volcanic eruptions and emitted by other sources such as hot springs, geysers, and fumaroles.

TABLE 11.1 COMPARISON OF THE COMPOSITIONS OF GASES FROM INTERNAL EARTH SOURCES WITH THOSE OF EARTH'S ATMOSPHERE AND HYDROSPHERE*

Gas	Some Gases of Earth's Hydrosphere and Atmosphere	Gases in Hot Springs, Fumaroles, and Geysers	Volcanic gases from Basaltic Lava of Mauna Loa and Kilauea
Water vapor, H_2O	92.8	99.4	57.8
Total carbon, as CO_2	5.1	0.33	23.5
Sulfur, S_2	0.13	0.03	12.6
Nitrogen, N_2	0.24	0.05	5.7
Argon, A	Trace	Trace	0.3
Chlorine, Cl_2	1.7	0.12	0.1
Fluorine, F_2	Trace	0.03	—
Hydrogen, H_2	0.07	0.05	0.04

*Data from W. W. Rubey (1952), *Geol. Soc. Amer., Bull.*, vol. 62, p. 1137. Figures in table represent percentages by weight.

Cometary Impacts

Another likely source of gases was comets and other bodies from farther out in the forming solar system. The outer solar system is cold enough to condense many volatile materials, and comets are rich in ices of water, carbon dioxide, carbon monoxide, ammonia, and organic compounds. Today, hundreds of Earth-masses of comets orbit the Sun. According to computer models, there were even more comets in the early solar system and they commonly intersected the orbits of Venus, Earth, and Mars. Collisions of comets with planets would have released the comets' icy gases into the atmospheres of the planets they struck. Hundreds of millions of years of cometary collisions could have added large amounts of material to Earth's atmosphere. As stated earlier, in addition to water, comets would have provided carbon dioxide, carbon monoxide, methane, ammonia, nitrogen, and other gases. Carbon dioxide was also released from rocks in Earth's mantle, so this gas probably dominated Earth's early atmosphere after water condensed out.

Virtually no molecular oxygen is found in comets, very little exists on Venus and Mars, and it was absent from Earth's early atmosphere. Evidence of the absence of oxygen comes from ancient rocks containing minerals that would have been unstable in an oxygen atmosphere.

Development of Earth's Atmosphere

There were a number of stages in the development of Earth's present, oxygen-rich atmosphere. Once liquid water had condensed on Earth's surface, the atmosphere began to change.

A Reducing Atmosphere

Carbon dioxide is very soluble in water, and was probably removed from the atmosphere by dissolving in the ocean and then combining with a substance such as calcium to form limestone. Oxygen began to be formed photochemically; water absorbs ultraviolet radiation from the Sun, and the water molecule (H_2O) breaks down into hydrogen gas and oxygen gas. The light hydrogen can escape from the atmosphere, so it does not recombine. The remaining oxygen forms molecular oxygen (O_2) and ozone (O_3). However, the weathering of minerals and reactions with volcanic gases bind the oxygen almost as quickly as it is formed, so that this element remained very scarce, less than one hundred-millionth of present atmospheric levels. Oxygen levels this low formed a *reducing atmosphere*, that is, an atmosphere in which elements and compounds that would combine with oxygen in today's atmosphere remained stable (e.g., iron would not form rust).

An Oxidizing Atmosphere

Once photosynthetic life (such as chlorophyll-containing life-forms) appeared and spread around the planet, however, there was a new source of oxygen. See Figure 11.1. As these organisms multiplied over time, photosynthesis steadily increased. Geological evidence suggests that between 2.2 and 2.0 billion years ago the oxygen level in the atmosphere skyrocketed to roughly one-hundredth of its present level—more than a millionfold increase. Oxygen levels grew high enough to form an *oxidizing atmosphere*, one in which oxidation of elements and compounds occurred but that was still too low to sustain aerobic respiration (the oxidation of food to produce energy). However, at this concentration, enough molecular oxygen, which strays into the upper atmosphere and is chemically changed into ozone by ultraviolet rays, is formed to sustain an ozone shield that blocks incoming ultraviolet rays.

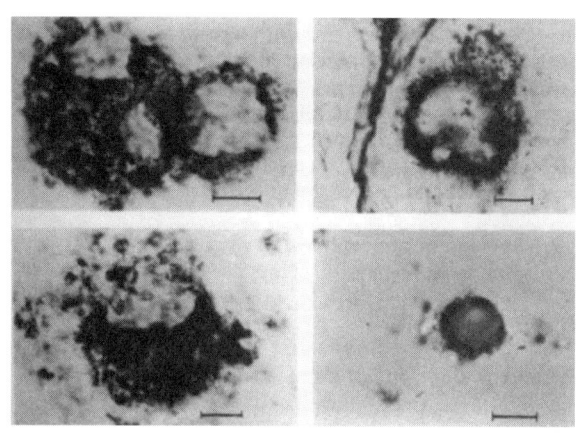

Figure 11.1 Possible Algaelike Fossils Found in Rocks More than 3.1 Billion Years Old. Source: *The Earth Sciences*, Arthur N. Strahler, Harper & Row, 1971.

As the oxygen level of the atmosphere continued to increase, and ozone shielded organisms from damaging ultraviolet rays, photosynthetic organisms proliferated. But these primitive photosynthesizers were anaerobic; their body metabolism did not use oxygen. In fact, anaerobic organisms cannot survive if oxygen levels rise too high. Oxygen is a waste product of anaerobic photosynthesizers, just as urea is a waste product of humans. Most organisms can tolerate waste products in moderate amounts, but high levels are toxic (which is why humans with damaged kidneys undergo dialysis to remove urea from their blood). At first this limited oxygen levels; as a population of anaerobic photosynthesizers would increase, oxygen levels would rise to toxic levels. This, in turn, would cause their population (and oxygen production) to decrease until oxygen levels were once again tolerable. By about 1.7 billion years ago, however, organisms had developed higher and higher oxygen tolerances, so production of this element surged again.

An Aerobic Atmosphere

Eventually, Earth formed an ***aerobic atmosphere***, one with sufficient oxygen to support life-forms that metabolize oxygen. Over time, aerobic organisms (life-forms that use oxygen for body metabolism) evolved to take advantage of this oxygen-rich atmosphere. The number and vigor of photosynthetic life-forms (e.g., algae, green plants) increased. Photosynthesis has increased along with them, but is almost balanced by the consumption of oxygen by these new organisms during respiration. Photochemical oxygen production, as well as loss by weathering and volcanic activity, is now insignificant compared with photosynthesis and respiration. In this new balance, oxygen production slightly exceeds loss, and within the past billion years atmospheric oxygen has steadily risen to its current level. See Figure 11.2.

The Origin and Structure of Earth's Atmosphere

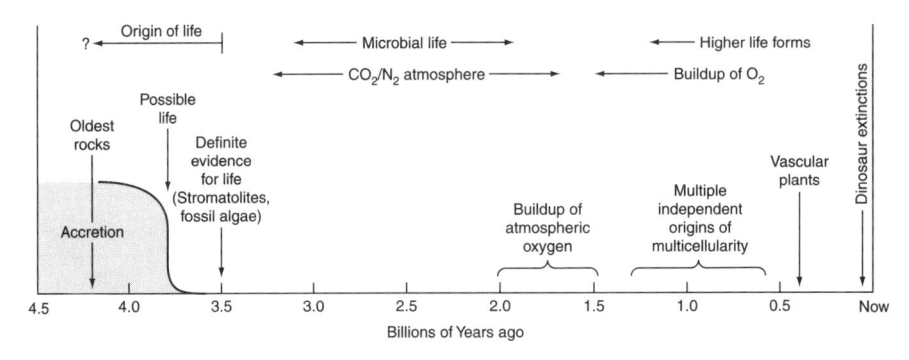

Figure 11.2 The Geologic and Climatological History of the Development of Earth's Atmosphere. Source: McKay, C.P., and Stoker, C.P. "The early environment and its evolution on Mars," in *Review of Geophysics* 27, 189–214, 1989, copyright © 1989 by the American Geophysical Union.

THE STRUCTURE OF EARTH'S ATMOSPHERE

The Composition of Air

The atmosphere, which now has a total mass of about 5,000 trillion tons, is held in place by Earth's gravity and extends several hundred kilometers into space. The atmosphere consists mostly of gases but also contains water, ice, dust, and other particles. This mixture of gases and other substances is called *air*.

Dry air is composed mainly of nitrogen and oxygen, with traces of other gases as well. On average, air today contains about 78 percent nitrogen and 21 percent oxygen; the remaining 1 percent consists mostly of argon, with traces of carbon dioxide and other gases. The water vapor content of air may range from 0 percent over deserts and polar ice caps to as much as 4 percent in a tropical jungle. See Table 11.2. Air also contains varying amounts of water vapor, dust, and chemicals released by industry and microorganisms.

TABLE 11.2 THE COMPOSITION OF AIR IN THE TROPOSPHERE

Gas	Percent by Volume in Dry Air
Nitrogen	78
Oxygen	21
Argon	Almost 1
Neon	Trace
Helium	Trace
Krypton	Trace
Xenon	Trace
Hydrogen	Trace
Ozone	Trace
Carbon dioxide	Trace
Nitrous oxide	Trace
Methane	Trace

Layers of the Atmosphere

With movement upward through the atmosphere, the temperature changes. The atmosphere is divided into four distinct layers according to temperature: the troposphere, stratosphere, mesosphere, and thermosphere. See Figure 11.3.

Figure 11.4 shows the changes in temperature that occur in the atmosphere. Notice that, near the ground, temperature decreases with altitude. Then, at an altitude of about 11 kilometers, the air stops getting colder and starts to become warmer. This change in behavior signals that the boundary from one layer to another—in this case from troposphere to stratosphere—has been crossed. The boundary is called the *tropopause* because, at the top of the troposphere, the temperature stops decreasing, or pauses. You can see from Figure 11.4 that such a change happens twice more, at the stratopause and the mesopause.

These changes in temperature are the result of differences in the compositions of the different atmospheric layers. The upper stratosphere contains the gas ozone. Ozone absorbs ultraviolet light from the Sun, thereby warming the stratosphere. Once past the ozone, the temperature decreases in the mesosphere. In the thermosphere, gases and charged particles absorb ultraviolet light and other wavelengths of energy and turn them into heat energy. Once again, temperatures rise. In fact, the amount of energy absorbed

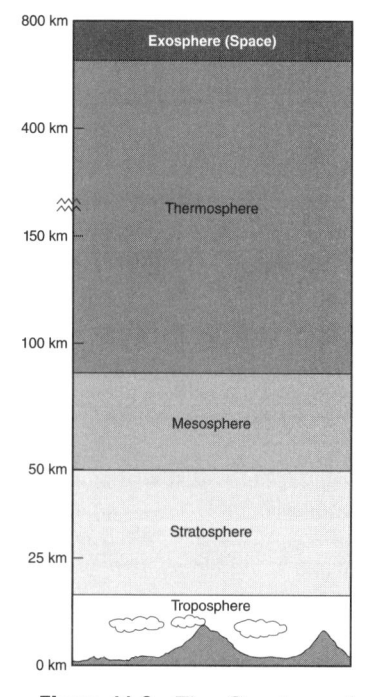

Figure 11.3 The Structure of Earth's Atmosphere. Source: *Astronomy Explained*, Gerald North, Springer-Verlag, 1997. ISBN 3-540-76136-5 (page 30).

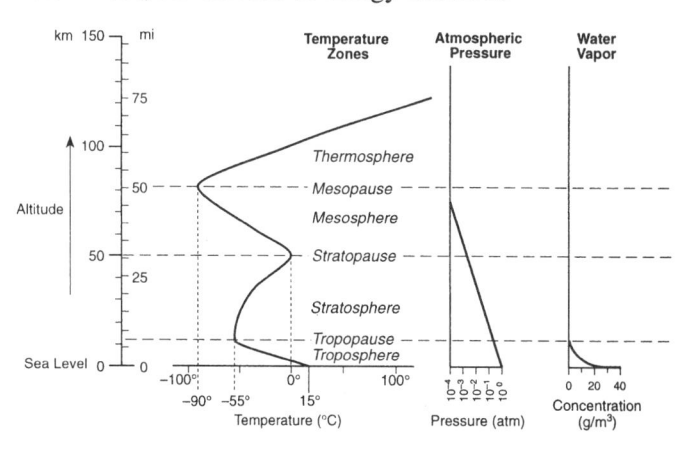

Figure 11.4 Selected Properties of Earth's Atmosphere. Source: The State Education Department, *Earth Science Reference Tables*, 2001 ed. (Albany, New York; The University of the State of New York).

The Origin and Structure of Earth's Atmosphere

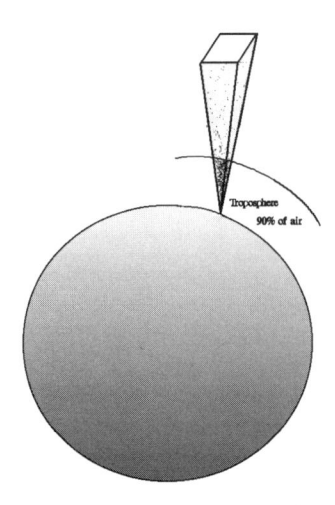

Figure 11.5 A Sectional Slice through Earth's Atmosphere. Most of the gas molecules in the atmosphere are clustered near the Earth's surface forming the troposphere. Only the most energetic molecules can overcome gravity and reach higher levels in the atmosphere.

by this outermost layer is so great that temperatures exceed the boiling point of water. Some gas molecules absorb so much energy they lose or gain electrons and become ions, that is, charged particles. These ions in the thermosphere form a layer, the ionosphere, that can reflect radio waves.

As you can see from the atmospheric pressure and water vapor sections of Figure 11.4, most of the air in the atmosphere and all of the water vapor is confined to the troposphere. Air exerts pressure in proportion to the amount of air that is present in the atmosphere. At the tropopause air exerts only one-tenth of the pressure that it does at sea level. This means that there is only about one-tenth as much air! At the stratopause there is only about one-thousandth of the air that is present at sea level, and by the mesopause less than one ten-thousandth the air. The atmosphere extends for hundreds of kilometers above the lithosphere and hydrosphere, but most of the air is confined to the 8–10 kilometers closest to their surface. See Figure 11.5.

MULTIPLE-CHOICE QUESTIONS

In each case, write the number of the word or expression that best answers the question or completes the statement.

1. The most likely source of Earth's atmosphere is
   (1) gases left over from Earth's formation
   (2) outgassing from Earth's hot, young interior
   (3) gases trapped by Earth's magnetic field
   (4) respiration by living organisms

2. Earth's early atmosphere probably did *not* contain
   (1) carbon dioxide   (3) ammonia
   (2) methane   (4) oxygen

3. The source of most of the oxygen in Earth's current atmosphere is thought to be
   (1) the chemical breakdown of mineral compounds containing oxygen
   (2) the breaking apart of water into hydrogen and oxygen by lightning
   (3) the product of photosynthesis by chlorophyll-containing life-forms
   (4) the breakdown of carbon dioxide by ultraviolet rays from the Sun

4. According to the *Earth Science References Tables*, the element oxygen is found in Earth's
   (1) troposphere, only
   (2) hydrosphere, only
   (3) troposphere and hydrosphere, only
   (4) troposphere, hydrosphere, and crust

5. The greatest mass of Earth's atmosphere is found in the
   (1) ionosphere   (3) troposphere
   (2) exosphere   (4) stratosphere

6. Earth's atmosphere consists
   (1) mostly of oxygen, argon, carbon dioxide, and water vapor
   (2) entirely of ozone, nitrogen, and water vapor
   (3) of gases that cannot be compressed
   (4) of a mixture of gases, liquid droplets, and minute solid particles

7. The atmosphere is bound to Earth by
   (1) magnetic fields   (3) the force of gravity
   (2) atmospheric pressure   (4) chemical bonding at the interface

8. Which statement most accurately describes Earth's atmosphere?
   (1) The atmosphere is layered, with each layer possessing distinct characteristics.
   (2) The atmosphere is a shell of gases surrounding most of Earth.
   (3) The atmosphere's altitude is less than the depth of the ocean.
   (4) The atmosphere is more dense than the hydrosphere but less dense than the lithosphere.

9. According to the *Earth Science Reference Tables*, nearly all the water vapor in the atmosphere is found within the
   (1) mesosphere   (3) troposphere
   (2) thermosphere   (4) stratosphere

10. Which layer of the atmosphere is found immediately above the tropopause?
    (1) ionosphere   (3) hydrosphere
    (2) troposphere   (4) stratosphere

The Origin and Structure of Earth's Atmosphere

11. According to the *Earth Science Reference Tables*, the greatest atmospheric pressure occurs in the
    (1) troposphere
    (2) stratosphere
    (3) mesosphere
    (4) thermosphere

12. According to the *Earth Science Reference Tables*, which circle graph best represents the volume of gases in the troposphere?

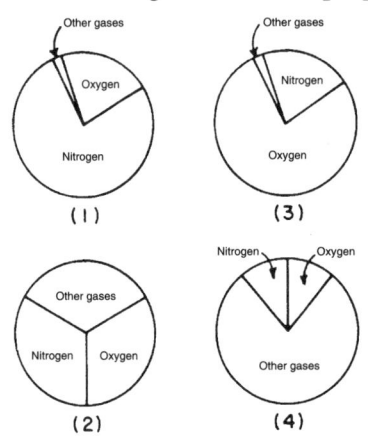

13. At which interface in the diagram below is the moisture content of the air likely to be highest?

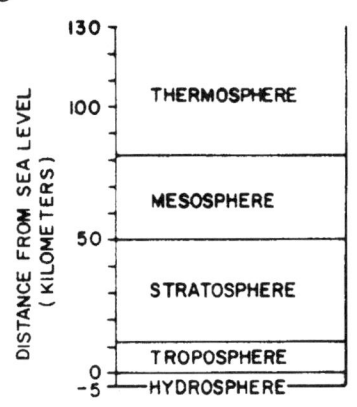

    (1) thermosphere–mesosphere
    (2) mesosphere–stratosphere
    (3) stratosphere–troposphere
    (4) troposphere–hydrosphere

14. According to the *Earth Science Reference Tables*, which graph best represents the relationship between altitude and pressure in Earth's atmosphere?

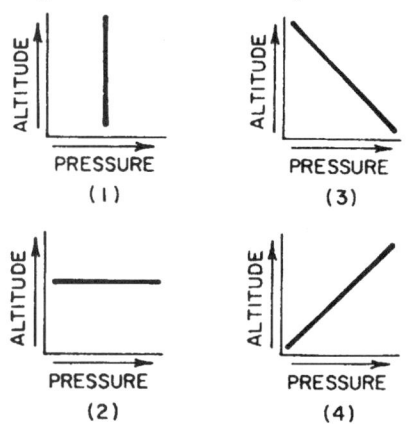

(1) [graph: vertical line — constant pressure with altitude]
(2) [graph: horizontal line]
(3) [graph: line decreasing from upper-left to lower-right]
(4) [graph: line increasing from lower-left to upper-right]

Base your answers to questions 15 through 19 on your knowledge of earth science, the *Earth Science Reference Tables*, and the diagram below. The diagram represents five positions of a balloon after being released from a ship. The drawings of the balloon are not to scale compared to the altitude distances, but are to scale with each other

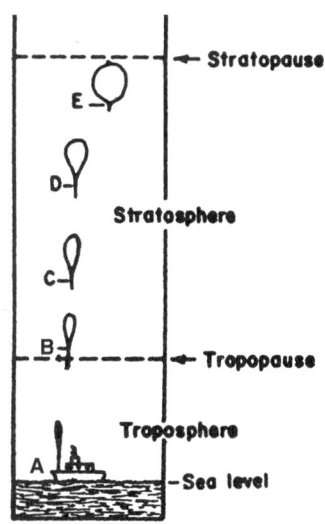

[Diagram: balloon positions A (sea level, on ship), B (at tropopause), C, D (in stratosphere), E (near stratopause)]

15. Which position represents the balloon when it is about 12 kilometers above sea level?
   (1) A
   (2) B
   (3) D
   (4) E

215

The Origin and Structure of Earth's Atmosphere

16. Why is the balloon's appearance at position E different from its appearance at position A?
    (1) There is a partial vacuum inside the balloon at A, but not at E.
    (2) There is more gas inside the balloon at A than at E.
    (3) The outside air temperature is lower at E than at A.
    (4) The outside air pressure is lower at E than at A.

17. At which position would the surrounding air temperature most likely be warmest?
    (1) A (3) C
    (2) B (4) D

18. Which graph best represents the relative amounts of water vapor found in the atmosphere at the different balloon positions?

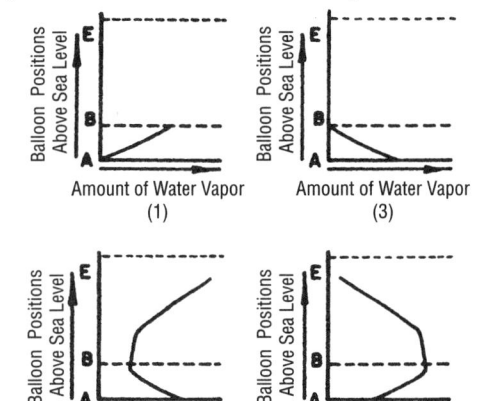

Note that question 19 has only three choices.
19. If the balloon is to rise, the density of the gas inside the balloon must be
    (1) less than the density of the air at sea level
    (2) greater than the density of the air at sea level
    (3) the same as the density of the air at sea level

20. According to the *Earth Science Reference Tables*, the zone of the atmosphere with the smallest vertical extent is the
    (1) troposphere   (3) mesosphere
    (2) stratosphere   (4) thermosphere

21. Ozone is important to life on Earth because ozone
    (1) cools refrigerators and air-conditioners
    (2) absorbs energy that is reradiated by Earth
    (3) absorbs harmful ultraviolet radiation
    (4) destroys excess atmospheric carbon dioxide

Base your answers to questions 22–25 on the graph below, which represents the temperature of a parcel of air as it is lifted through the atmosphere.

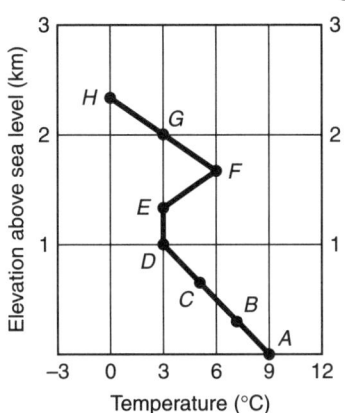

22. The temperature of the air at an elevation of 2 kilometers is most likely to be
(1) –2.5°C (3) 3°C
(2) 2.5°C (4) 0°C

23. As the air rises from the surface to an elevation of 1 kilometer, its temperature
(1) decreases, only
(2) increases, only
(3) increases, and then decreases
(4) remains the same

24. The temperature change is least between points
(1) B and C (3) E and F
(2) D and E (4) F and H

25. Between which two points does the temperature increase with an increase in elevation (a temperature inversion occurs)?
(1) A to D (3) E to F
(2) D to E (4) F to H

Constructed Response Questions

26. State two likely sources of the gases of Earth's current atmosphere. [2]

27. Describe the process by which Earth's early reducing atmosphere evolved into an oxidizing atmosphere. [3]

The Origin and Structure of Earth's Atmosphere

28. How might the development of Earth's atmosphere have been different if water was scarce and no oceans formed on Earth's surface? [2]

EXTENDED CONSTRUCTED RESPONSE QUESTIONS

29. Refer to the Average Chemical Composition of Earth's Crust, Hydrosphere, and Troposphere chart in the *Earth Science Reference Tables* to answer the following questions.
    (a) Complete the pie graph below to show the composition in percent by volume of the troposphere. The percentage of "Other" has already been plotted. [2]

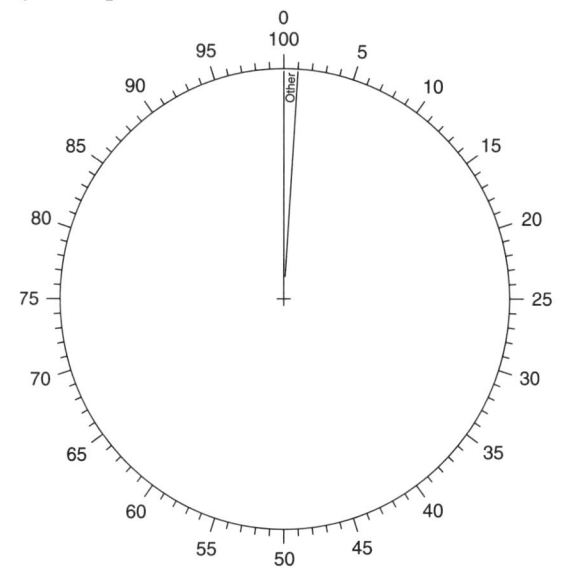

    (b) Compare the percent by volume of oxygen in the crust, hydrosphere (oceans), and troposphere. State one fundamental difference between the oxygen in the atmosphere and the oxygen in the crust and hydrosphere. [2]
    (c) What biological process created most of the oxygen in Earth's current atmosphere? [1]

218

CHAPTER 12

THE ORIGIN AND NATURE OF EARTH'S HYDROSPHERE

> **KEY IDEAS** Approximately 70 percent of Earth's surface is covered by a relatively thin layer of water. Earth's oceans formed as a result of precipitation over millions of years. The presence of early oceans is indicated by sedimentary rocks of marine origin, dating back about 4 billion years.
>
> Earth has continuously been recycling water ever since the outgassing of water early in the planet's history. This constant recirculation of water at and near Earth's surface is described by the water (hydrologic) cycle. Water leaves Earth's surface and enters the atmosphere by evaporation or transpiration from plants. Water is returned from the atmosphere to Earth's surface by precipitation.
>
> A portion of the precipitation becomes runoff over the land or infiltrates into the ground and is stored in the soil or groundwater below the water table. Soil capillarity influences these processes. Porosity, permeability, and water retention also affect runoff and infiltration. The amount of precipitation that seeps into the ground or runs off is influenced by climate, slope of the land, soil, rock type, vegetation, land use, and degree of saturation.

KEY OBJECTIVES
Upon completion of this chapter, you will be able to:

- Describe the origin of Earth's hydrosphere.
- Identify the forms in which water exists on Earth.
- Analyze the water cycle, and describe the transfer of water and energy into and out of the atmosphere at each step of the cycle.
- Explain why fresh water is a limited but renewable resource.
- Give examples of how factors such as porosity, permeability, water retention, and soil capillarity affect infiltration and runoff.
- Describe the formation of the water table and some methods by which groundwater is obtained for use at Earth's surface.

THE ORIGIN OF THE HYDROSPHERE

Earth is often referred to as "the water planet." Most people know that the oceans are full of water, and think water is commonplace because there is so much of it around. On a cosmic scale, however, water is not so common. Earth is the *only* known planet that has liquid water on its surface! This has resulted from a unique combination of Earth's temperature and the unusual properties of water. Water not only covers Earth's surface, but also makes life as we know it possible.

Let us now consider the origin and nature of Earth's oceans and the properties of water that make it such an unusual substance.

The Origin of Earth's Water

The origin of Earth's water is directly linked to the formation of Earth. Meteorites called Type I carbonaceous chondrites provide a clue to the probable origin of Earth's water. They contain 5 percent carbon compounds and 20 percent water tied up in mineral compounds. These meteorites are believed to be debris left over from the formation of the solar system. As such, they represent a type of material thought to have formed while the planets were condensing from a cloud of dust and hydrogen gas. See Table 12.1.

TABLE 12.1 COMPOSITION OF EARTH AND CHONDRITE METEORITES*

Rank	Element	Symbol	Earth Average (% by wt.)	Average of Chondrites (% by wt.)
1	Iron	Fe	34.6 ⎫	27.2 ⎫
2	Oxygen	O	29.5 ⎬ 92.0	33.2 ⎬ 91.8
3	Silicon	Si	15.2 ⎪	17.1 ⎪
4	Magnesium	Mg	12.7 ⎭	14.3 ⎭
5	Nickel	Ni	2.4	1.6
6	Sulfur	S	1.9	1.9
7	Calcium	Ca	1.1	1.3
8	Aluminum	Al	1.1	1.2
9	Sodium	Na	0.57	0.64
10	Chromium	Cr	0.26	0.29
11	Manganese	Mn	0.22	0.25
12	Cobalt	Co	0.13	0.09
13	Phosphorus	P	0.10	0.11
14	Potassium	K	0.07	0.08
15	Titanium	Tl	0.05	0.06

Source: *The Earth Sciences*, Arthur N. Strahler, Harper & Row, 1971.

*Data from *Principles of Geochemistry*, 3rd Ed., Brian Mason, John Wiley, 1966.

As Earth cooled and condensed from a cloud of dust and gas, hydrogen reacted with oxygen to form water, which in turn reacted with crystals of solids that were forming. The products of these reactions were silicate minerals that contain water and oxidized iron. Therefore, the original form in which water existed on Earth was as part of solid compounds—rocks. The common clay minerals found in soil are examples of this type of compound.

As these substances formed, they were drawn inward by gravity and heated. We know that, as rocks are heated above 500°C, the water in their molecules is released in a process called *degassing*. When vented to Earth's surface, the water is released into the atmosphere, a process called *outgassing*, and upon further cooling condenses to form liquid water. Much evidence supports the idea that much of Earth's water was released to the surface soon after it formed. Even today, however, the process continues on a small scale through degassing associated with volcanic eruptions.

From an Ocean of Gas to an Ocean of Water

Outgassing from Earth's interior and collisions of comets with Earth added large amounts of water vapor to the developing atmosphere. Consider that a typical comet contains about 10^{15} (one million billion) kilograms of water. Since all of Earth's oceans hold roughly 10^{21} (one trillion billion) kilograms of water, only a million or so comet collisions could have created Earth's entire water supply ($10^6 \times 10^{15} = 10^{21}$). This is not an unlikely number of collisions, considering the estimated number of existing comets, which are but a fraction of those that crowded the early solar system.

As Earth cooled, water vapor in the atmosphere would have condensed and fallen as torrential rains that ran off high places and filled low-lying basins. At first, striking a still warm Earth, water would vaporize quickly and return to the atmosphere. But as the planet continued to cool, liquid water would remain for longer and longer periods in the basins of Earth's surface. Eventually, the oceans and lakes were permanently filled with water that returned to the atmosphere only because of evaporation by the Sun, as is the case today.

This reservoir of liquid water blanketing Earth is called the ***hydrosphere***. The hydrosphere, however, is very thin. If you dipped a basketball in water and then shook the water off, the thin film of water adhering to its surface would be a good model of the hydrosphere.

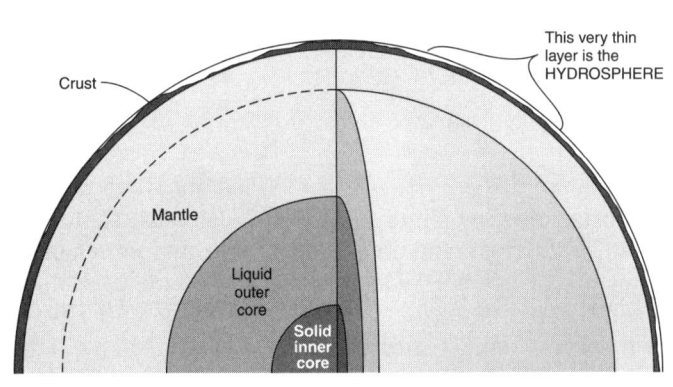

Figure 12.1 The Hydrosphere. The thicknesses of the crust and hydrosphere are greatly exaggerated in this diagram. The hydrosphere is actually much thinner than shown here. Source: *World Geomorphology*, E. M. Bridges, Cambridge University Press, 1990.

Outgassing of water during volcanic eruptions still occurs today, and so do collisions with icy comets. (Consider the 1994 collision of Comet Shoemaker-Levy with Jupiter.) As water continues to work its way to Earth's surface and to be carried in by comets, the oceans may gradually increase in volume until, in a few billion years, the hydrosphere covers the whole planet.

The Origin of Seawater

How did seawater become salty? The general assumption is that the ocean was originally a huge freshwater lake. Over the billions of years that the ocean has existed, soluble ions freed by chemical weathering have been carried into the ocean by rivers, creating the saline waters of today. We know, for example, that the extremely salty Dead Sea and Great Salt Lake have resulted because all types of substances entered them, but only water vapor left them. But at the current rate at which streams add salts to the ocean, it could have reached its present salinity in roughly 100 million years. This time span falls far short of the current estimates, based upon radioactive decay rates and other data, of 3.5–4 billion years for the age of the ocean. Why the discrepancy?

One answer is that the salinity of the ocean does not increase at a constant rate, and probably decreases from time to time. For example, the end of a glacial period would flood the oceans with melting ice, decreasing salinity worldwide. Shallow seas that become landlocked and then evaporate would remove substantial amounts of salt from the ocean. (Salt deposits near Moab, Utah, that formed when a vast inland sea evaporated are as much as 4,000 feet thick!)

Another answer is that the rate at which water has been added by outgassing has not been uniform or continuous. Thus, it appears that salinity, along with sea level, has risen and fallen with the freezing and melting of ice and the removal of huge amounts of salt by evaporation of landlocked seas.

Seawater Today

Seawater is a complex mixture of dissolved inorganic matter, dissolved organic matter, dissolved gases, and particulate matter. *Salinity* is the total amount of dissolved material in seawater. Salinity is measured in parts per thousand (‰) by weight in 1 kilogram of seawater.

Inorganic Matter

By weight, seawater is about 96.5 percent pure water and about 3.5 percent dissolved inorganic substances. See Figure 12.2. Six elements—chlorine, sodium, magnesium, sulfur, calcium, and potassium—make up more than 90 percent of the materials dissolved in seawater. The minor elements include strontium, bromine, and boron. The elements in seawater are almost always present as part of chemical compounds, most of which are salts. Although the salinity of seawater varies from 33 percent to 37 percent, depending upon where in the ocean the sample is taken, the ratio of the major dissolved substances remains about the same.

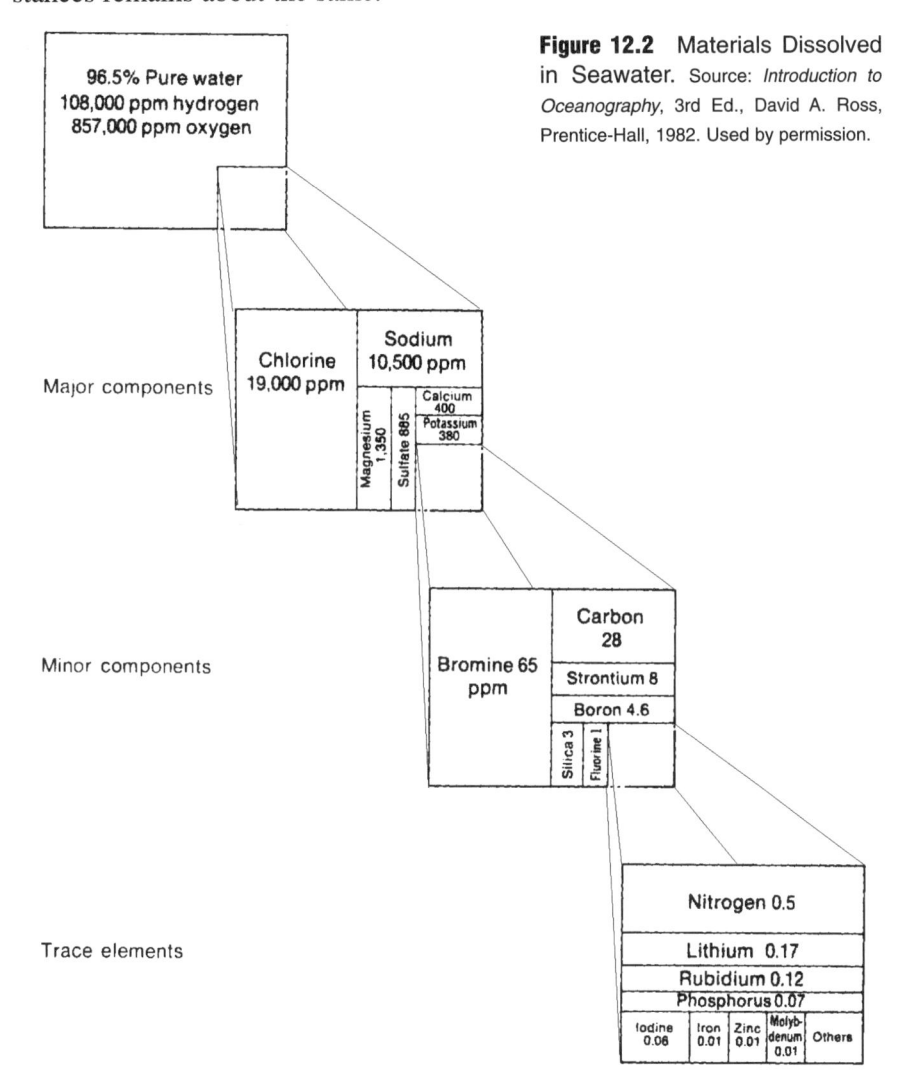

Figure 12.2 Materials Dissolved in Seawater. Source: *Introduction to Oceanography*, 3rd Ed., David A. Ross, Prentice-Hall, 1982. Used by permission.

96.5% Pure water
108,000 ppm hydrogen
857,000 ppm oxygen

Major components:
- Chlorine 19,000 ppm
- Sodium 10,500 ppm
- Magnesium 1,350
- Sulfate 885
- Calcium 400
- Potassium 380

Minor components:
- Bromine 65 ppm
- Carbon 28
- Strontium 8
- Boron 4.6
- Silica 3
- Fluorine 1

Trace elements:
- Nitrogen 0.5
- Lithium 0.17
- Rubidium 0.12
- Phosphorus 0.07
- Iodine 0.06
- Iron 0.01
- Zinc 0.01
- Molybdenum 0.01
- Others

Organic Matter

The dissolved organic matter in seawater comes mainly from wastes excreted by organisms and from the decaying bodies of dead organisms. It is usually present in small amounts (0–6 parts per million) and is highly variable. Dissolved organic matter includes organic carbon, carbohydrates, proteins, amino acids, organic acids, and vitamins. Organic compounds containing nitrogen and phosphorus oxidize to form nitrates and phosphates.

Dissolved Gases

Nitrogen, oxygen, and carbon dioxide are the major gases dissolved in seawater. Dissolved gases should not be confused with gaseous elements that are part of compounds. For example, water is a compound of hydrogen and oxygen, both of which are gases. However, the compound is not a gas. The oxygen that is part of the water molecule is not a dissolved gas; it is bound in the molecule and can't be "breathed in" and used for respiration. Other gases dissolved in seawater include helium, neon, argon, krypton, and xenon. With the exception of oxygen and carbon dioxide, the gases in seawater usually enter it from the atmosphere.

Seawater has two sources of oxygen: the atmosphere and the plants that live in the ocean. Oxygen gas dissolves directly into seawater from the atmosphere. Photosynthesis by plants living in the upper layers of the ocean (mainly phytoplankton) releases oxygen into seawater.

Particulates

Particulates are solid particles of material that is not dissolved in seawater, but is suspended or settling through it. Aside from living organisms, particulate matter consists mainly of fine-grained minerals and decaying organic material. Particulate matter in seawater varies greatly depending on ocean currents, winds, and nearness to rivers and streams.

WATER: A UNIQUE SUBSTANCE

Earth's hydrosphere is the result of a unique combination of Earth's temperature and the properties of water. Water has unusual properties that make it different from any other substance on Earth. These properties enable liquid water to exist on Earth's surface and are responsible for its chemical and physical behavior. If water had properties typical of a substance with its molecular size and weight, Earth would be a very different place indeed.

The Structure of Water

A water molecule consists of one relatively large oxygen atom combined with two smaller hydrogen atoms. The result is a somewhat lopsided molecule that creates an unequal distribution of electrical charges. The hydrogen side of the molecule is slightly positive; the oxygen side, slightly negative. This unusual structure causes water molecules to be attracted to one another and "stick" together. The attractions between molecules of liquid water are called ***hydrogen bonds***. Hydrogen bonds are weak, but they make water behave differently from any other substance on Earth. See Figure 12.3.

Although weak, hydrogen bonds hold water molecules together more strongly than is the case in substances lacking such bonds. Thus, liquid water has a high ***surface tension***; to push liquid water molecules apart and to

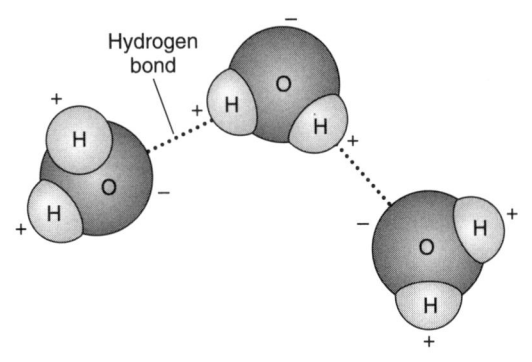

Figure 12.3 The Structure of Water Molecules Causes Hydrogen Bonding. The two ends of a water molecule have opposite electrical charges that attract each other as do opposite poles of a magnet. The weak attractions between the slightly positive end of one water molecule and the slightly negative end of another water molecule are known as hydrogen bonds. Source: *Marine Biology*, 3rd Ed., Peter Castro and Michael Huber, McGraw-Hill, 2000.

penetrate the surface of liquid water, a relatively strong force must be exerted to overcome the hydrogen bonds. Water striders, long-legged bugs that move about on the surface of water, spread their mass over several legs, so that no one leg exerts enough force to break through the surface.

The same "stickiness" that holds water molecules together also makes them adhere to the surfaces of other substances. For example, when water seeps through the ground, some of it "sticks" to the soil particles so that their surface stays "wet" even after most of the water has drained through them. The water that adheres to the surface of soil particles is an important source of water for plant life.

Water's "stickiness" is also responsible for ***capillarity***, the tendency for water's surface to rise in a narrow opening. The water rises because its attraction to the walls of the opening pulls it upward, and surface tension drags the water surface between the walls upward along with it. Capillarity is an important process in the movement of water through porous rocks and soil.

The States of Water

Matter can exist in three different physical forms—liquid, gas, and solid—called the *states*, or *phases*, of matter. Water is the only substance on Earth that occurs naturally in all three phases of matter. Since the majority of the hydrosphere is in the liquid state, we'll begin our discussion with water as a liquid.

Liquid Water

In the liquid phase, molecules of water are in constant motion. Because of hydrogen bonds individual water molecules cluster together in clumps of two to eight molecules. Although weak, hydrogen bonds hold water molecules together more strongly than is the case in substances lacking such bonds.

The average speed of the molecules in a substance is reflected in its temperature; the faster the molecules are moving, the higher the temperature. The addition of heat to water makes its molecules move faster, and thus raises its temperature. However, some of the heat added to the water goes into breaking hydrogen bonds instead of promoting molecular motion. Therefore, a large amount of heat energy is needed to get water molecules to move faster, and thus increase the temperature of water. This heat energy is not lost; it is stored *in* the water. Thus, liquid water holds more heat than do substances without hydrogen bonds; and its ability to hold heat, or **heat capacity**, is high. In fact, water has one of the highest heat capacities of any naturally occurring substance.

Water Vapor

If a water molecule moves fast enough, it can break free of the molecules around it. This process, known as ***evaporation***, marks the change from the liquid phase of water to the gaseous or vapor phase. See Figure 12.4. The molecules in water vapor are so fast moving, and move so much farther apart than in the liquid, that they are not held together by hydrogen bonds. As temperature rises, and liquid molecules move faster and faster, more and more escape, so the rate of evaporation increases. When the temperature reaches 100°C, nearly all of the molecules break free of their neighbors and try to enter the vapor state all at once; in other words, the water boils.

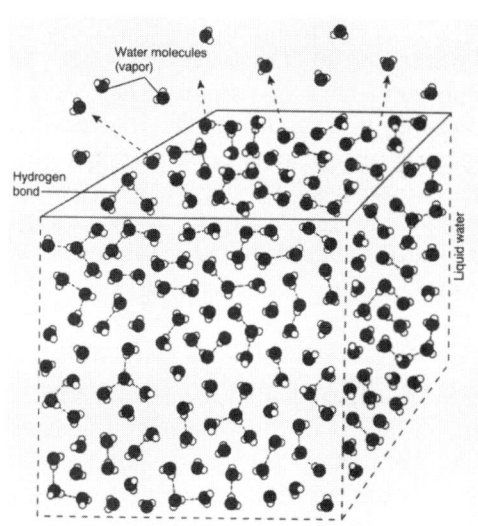

Figure 12.4 Evaporation Takes Place When Molecules Break Free of the Hydrogen Bonds of the Liquid and Enter the Gaseous State.
Source: *Marine Biology*, 3rd Ed., Peter Castro and Michael Huber, McGraw-Hill, 2000.

Once water begins to boil, added heat goes into breaking hydrogen bonds between molecules instead of increasing molecular speed. Therefore, while water is changing phase, it does not change temperature. The amount of heat required to evaporate a substance is called the ***latent heat of vaporization***. The word *latent* means "hidden," and in this term refers to the added heat that is not evident as a temperature change. As long as there is boiling water to be evaporated, the temperature of the liquid remains constant at 100°C. Water absorbs a large amount of heat when it evaporates; in other words, it has a high latent heat of vaporization.

At temperatures lower than the boiling point, only the most energetic, fastest moving molecules can break free of the hydrogen bonds and become a gas. Since the fastest moving molecules leave the liquid, those that are left behind are slower moving. As a result, the average speed of the molecules in the liquid, and thus its temperature, decreases. This effect is known as ***evaporative cooling***. Our bodies use evaporative cooling to regulate body temperature—when sweat evaporates, it cools the skin.

As liquid water cools, the molecules move slower, pack closer together, and take up less space. In other words, the volume of the water decreases. Since the same mass of water is occupying less volume, the water becomes denser. Thus, liquid water gets denser as it gets colder, but only down to a temperature of about 4°C.

Water as a Solid

As water continues to cool, the molecules move more slowly and rebound from one another more weakly. Eventually, the hydrogen bonds between the molecules suppress these motions and begin to hold the molecules in a fixed, three-dimensional pattern. This process, known as *freezing*, marks the change from the liquid phase to the solid phase. Solids in which atoms or molecules are arranged in such a regular, fixed pattern are called *crystals*. The solid, crystalline form of water is, of course, ice.

In ice, the hydrogen bonds of water hold the molecules in a hexagonal, or six-sided, pattern. As shown in Figure 12.5, the molecules of water are spaced

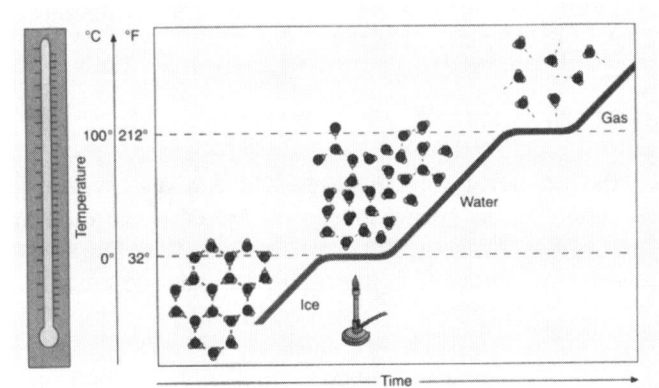

Figure 12.5 The Crystalline Structure of Ice. In ice, hydrogen bonding holds water molecules in a fixed structure in which they are spaced farther apart than in liquid water.
Source: *Marine Biology*, 3rd Ed., Peter Castro and Michael Huber, McGraw-Hill, 2000.

farther apart in ice than in liquid water. What happens might be compared to putting up a tent. When disassembled, the tent poles fit into a small bag; but when they are connected together, they form a large, open framework. Similarly, when water freezes, it expands. Since the same mass of water occupies more volume in ice than in liquid water, ice is less dense than liquid water and therefore floats. It is extraordinarily unusual for the solid phase of a substance to be less dense than the liquid phase. This unique property of water has important consequences for aquatic life. As ice forms in a body of water, it floats and forms an insulating blanket that keeps the subsurface water from cooling rapidly in cold weather. Organisms can live in the liquid water below the surface ice.

If ice were denser than liquid water, it would sink as it formed and build up at the bottom. In colder climates, lakes and even some parts of the oceans would freeze solid. During the summer, only a thin surface layer would thaw, and these bodies of water would be much less hospitable to life.

A large amount of heat is required to melt ice. Water molecules in ice crystals vibrate, but they do not move around freely. For melting to occur, the hydrogen bonds holding ice crystals together must first be broken. As heat energy is added to ice, the molecules vibrate faster and eventually break the hydrogen bonds holding the crystal together; then the ice melts. The molecules then move around as well as vibrate. Since a lot of energy is needed to break the hydrogen bonds holding ice crystals together, ice melts at a much higher temperature than do substances that do not have hydrogen bonds. Similarly, since hydrogen bonds help pull water molecules together into crystals, water does not have to be cooled as much before freezing. Thus, water melts and freezes at a much higher temperature than do substances that lack hydrogen bonds. If water did not have hydrogen bonds, the change from solid to liquid (or liquid to solid) would occur at about –90°C instead of 0°C.

As ice crystals are melting, added heat goes into breaking hydrogen bonds instead of increasing molecular speed. Therefore, while water is changing phase it does not change temperature. The amount of heat required to melt a substance is called the ***latent heat of fusion***. Here, again, the word *latent* means "hidden," and in this term refers to the added heat not evident as a temperature change. As long as there is ice left to be melted, the temperature of the ice-water mixture remains constant at 0°C.

Water: The Universal Solvent

Water is often called the universal solvent because it can dissolve more things than any other naturally occurring substance. Water is particularly good at dissolving molecules held together by the electrical attraction between oppositely charged particles. Common table salt, or sodium chloride, is such a molecule. It is composed of positively charged sodium (Na^+) and negatively charged chlorine (Cl^-). These sodium and chlorine ions have much stronger charges than the slightly charged ends of a water molecule. When a salt crystal is placed in water, the strong charges of the sodium and chlorine ions attract water molecules. Water's positive end is attracted to (or will attract) the negative chlorine; likewise, water's negative end will be attracted to the positive sodium. Thus, water molecules cluster around both ions in a salt molecule. As water molecules cluster around the ions, they cancel some of the attraction between the sodium and the chlorine, and the two move farther apart. Then more water molecules can come between them until the attraction is eliminated. Finally, the sodium and chlorine ions in the salt crystal pull apart, and the crystal ***dissolves***. Once the sodium and chlorine ions separate, water molecules surround them completely, insulating them from each other and preventing them from recombining to form a crystal. See Figure 12.6.

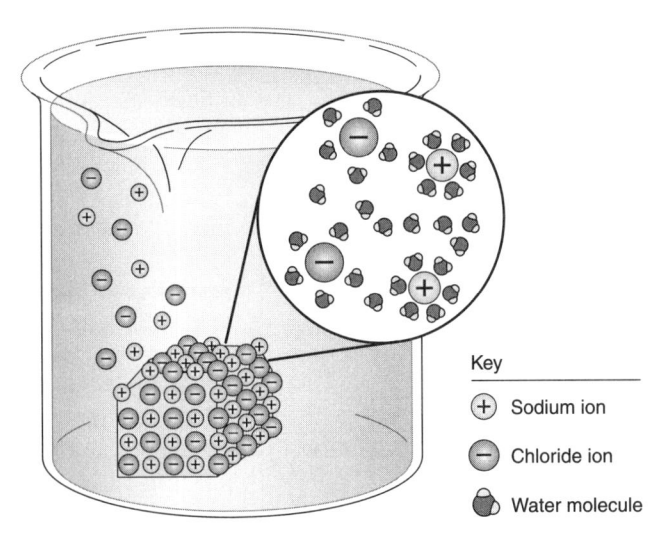

Figure 12.6 Water Molecules Cluster Around the Electrically Charged Sodium and Chlorine ions in a Salt Crystal. This weakens the bonds between the ions, causing them to separate, and the crystal dissolves. Source: *Marine Biology*, 3rd Ed., Peter Castro and Michael Huber, McGraw-Hill, 2000.

Key
(+) Sodium ion
(−) Chloride ion
Water molecule

THE WATER CYCLE

The *water cycle* is the process by which Earth's water moves into and out of the atmosphere. Let's begin by describing this process in the oceans that cover nearly three-quarters of Earth's surface.

1. Each day, insolation (*in*coming *sol*ar radi*ation*) reaching the surface of the oceans provides energy to the water molecules at the surface. As the molecules heat up, trillions of tons of water change from liquid to gas, or *evaporate*. The resulting water vapor enters the atmosphere, where it is circulated by winds and carried upward by rising warm air.
2. When air is cooled to its dewpoint by expansion during rising or contact with cooler surfaces, the water vapor changes from gas back to liquid, or *condenses*, forming tiny water droplets. These tiny water droplets suspended in air form clouds.
3. When the water droplets become too large to remain suspended, they fall back to the surface as *precipitation*. Precipitation may fall on the oceans or the land. Water that falls on the oceans has completed its cycle and may evaporate back into the atmosphere, beginning the entire process anew.
4. A number of things may happen to precipitation that falls on land. It may become *runoff*, water that flows downhill into rivers and streams that eventually carry it back into an ocean or other body of water. Or, it may seep into the ground to become **groundwater**, water that has infiltrated open spaces in the outer part of Earth's crust. Groundwater may move slowly back to the oceans through the ground, but most of it seeps into rivers and streams. It is then carried back to the oceans.
5. Some of the water may evaporate again to be carried farther over land, or be blown back over the oceans.

6. Once the water is back in the oceans, the cycle begins all over again.

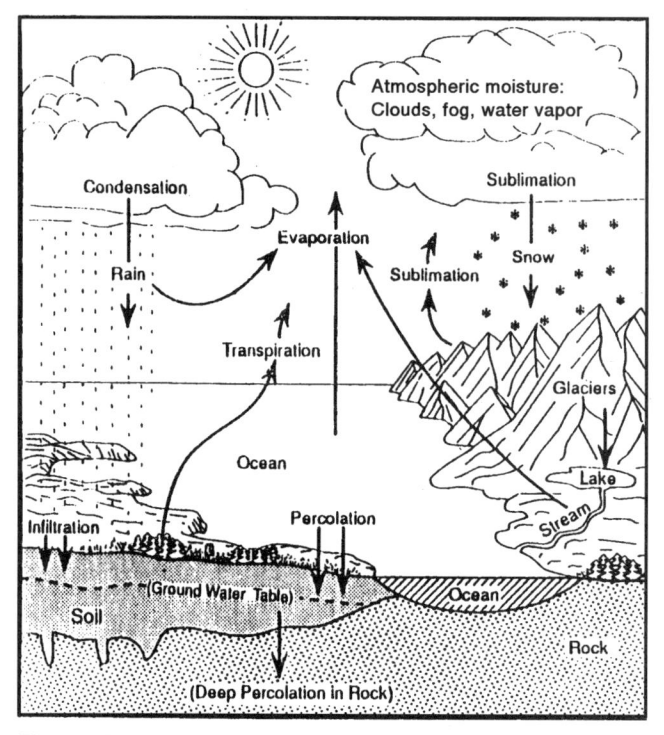

Figure 12.7 The Water Cycle. Source: Permission from Thomas McGuire.

There are many variations in the water cycle. Sometimes the water that evaporates over the oceans condenses there and falls directly back into the oceans as rain. Water falling on land may evaporate almost immediately, and in some cases precipitation evaporates before it ever reaches the ground.

Plants play a role in the water cycle as well. The roots of plants absorb water that has seeped into the soil. Then the water is transported to their leaves and released back into the atmosphere by a process called *transpiration*. In an area of abundant vegetation, such as a forest, transpiration returns more than one-third of precipitation back to the atmosphere as water vapor.

Other variations are possible, but all are part of the endless water cycle that receives its energy from the Sun. The water cycle continually renews our supply of fresh water. Each day an estimated 15 trillion liters of water in the form of rain or snow fall on the United States alone. Earth is a closed system, and water consumed today will eventually be recycled for use by a future generation.

In theory, Earth's water will never be exhausted. If, however, persistent, long-lasting pollutants enter the basins and ground from which Earth's water supply is drawn, water that has been cleansed by evaporation will be polluted as soon as it condenses and falls back into these polluted areas. Consider what will happen if the ground is contaminated with nuclear waste, which remains radioactive for tens or even hundreds of thousands of years. For the entire time the material remains radioactive, any water that infiltrates that ground will immediately become unusable.

GROUNDWATER

Porosity

Earth's surface is not completely solid; rather, it is filled with empty spaces, or *pores*. When precipitation falls on the surface, some of it *infiltrates*, or seeps through openings in the surface and into pore spaces. Pore spaces are usually filled with air unless water, oil, or natural gas has forced the air out. The amount of open space in a material is its ***porosity***. Porosity is expressed as a percentage of the total volume of a given substance. For example, if half of the volume of a sample of rock or sediment is open space, the specimen is said to have a porosity of 50 percent. Earth materials differ greatly in porosity. Factors that affect porosity are shown in Figure 12.8. Loose sediments such as sand, gravel, and clay usually have the highest porosity.

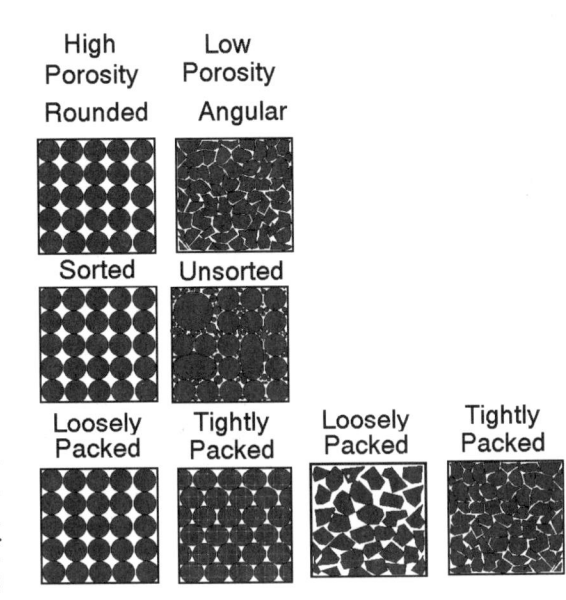

Figure 12.8 Factors That Affect the Porosity of a Material.

Porosity can be calculated by measuring the amount of water required to fill the pores in a substance and then setting up a ratio of the volume of water to the total volume of the sample being tested.

Example:
Water is poured into a 100-cubic-centimeter sample of sand. When 35 milliliters of water have been poured in, the water is just even with the surface of the sample. What is the porosity of the sand? [Note: 1 ml = 1 cm³]

$$\text{Porosity} = \frac{\text{volume of pore space}}{\text{total volume of sample}} \times 100\%$$

$$= \frac{35 \text{ cm}^3}{100 \text{ cm}^3} \times 100\% = 35\%$$

The porosity of the sand is 35%.

Permeability

Permeability is the rate at which water can infiltrate a material. A substance through which water passes rapidly is said to be *permeable*; a substance through which water cannot flow is *impermeable*. The permeability of a substance depends on two factors: the size of its pores and the degree to which they are interconnected. See Figure 12.9. Small pores constrict the flow of water; and if water cannot get from one pore to another, it cannot flow through a substance.

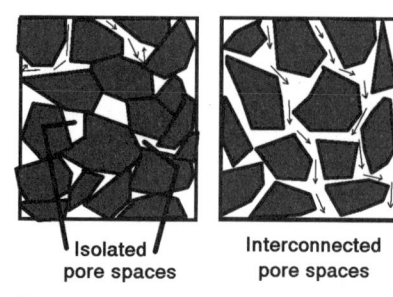

Figure 12.9 Isolated and Interconnected Pore Spaces. Permeable materials have interconnected pore spaces.

Knowing the permeability of a material is very useful. From it, the rate at which rainwater will sink into the ground can be determined, as well as the speed and direction in which sewage in a septic tank will flow. Also, the rate at which a well will refill with water after some has been pumped out can be determined.

The Water Table

When rain falls on land or when snow that has fallen melts, gravity pulls the water down into the ground through interconnected pore spaces. Some of the water clings to particles near the surface, forming a moist layer called the *soil moisture zone*. This is the layer from which plants obtain the water they need to survive. Most of the water that infiltrates continues to trickle downward through the ground until it reaches an impermeable substance. Then the water begins to fill all the pore spaces from the impermeable material upward. The region of permeable ground in which all of the pore spaces are filled with water is the *zone of saturation*.

The top of the zone of saturation, known as the **water table**, is the boundary between pores filled with water and pores filled with air. Just above the water table is a narrow region, called the *capillary fringe*, in which narrow pore spaces are filled with water drawn upward by capillary action. The area

Figure 12.10 Zones of Soil Water and Ground Water. The water table is the boundary between the zone in which pores are filled with air and the zone in which pores are filled with water.

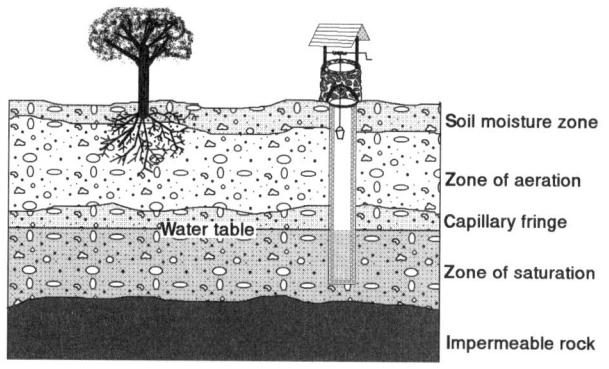

above the capillary fringe, where most of the pore spaces are filled with air, is called the *zone of aeration*. See Figure 12.10.

Groundwater can leave the ground in several ways. It can be drawn up by *capillary action* through tiny interconnected pore spaces and reach the soil moisture zone, where plant roots remove it. Or, instead, it may evaporate and slowly diffuse out of the ground through the zone of aeration, or it may seep through the ground sideways and slowly move out of the area. If the surface of the ground dips beneath the level of the water table, water can seep out and run off, forming streams, or collect in depressions, forming springs, ponds, or lakes. When the water table in a depression is just above the surface, a swamp may form. See Figure 12.11.

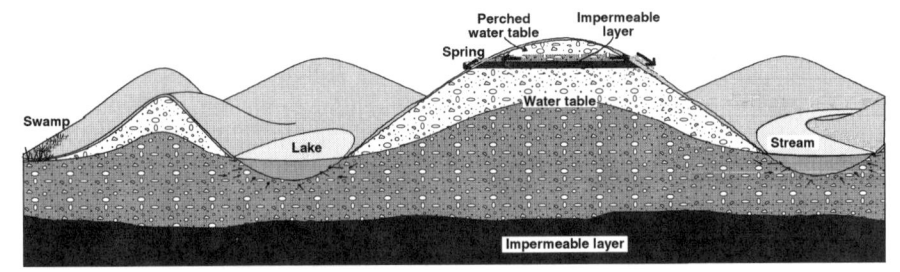

Figure 12.11 Variations in the Subsurface Position of the Water Table. The water table roughly follows the contours of Earth's surface. Where the surface dips beneath the water table, water seeps out of the ground and can collect in depressions, forming a body of water such as a lake or pond. If the water seeping out runs off, it may form a stream. Impermeable layers in hills can create a perched water table, which may seep out of a hillside as a spring.

Wells

Groundwater is an important source of fresh water. There is 37 times as much fresh water beneath the ground as there is on top of it in lakes, streams, glaciers, and other sources. Layers of permeable rock or loose sediments whose pore spaces are filled with water are called *aquifers*. Groundwater in aquifers can be tapped by digging a well, which is a hole sunk into the ground so that it penetrates the water table. Beneath the water table the well acts as a large pore space in which water accumulates. This water can then be pumped up to the surface through the hole.

Artesians

In artesian rock formations, groundwater can be tapped by wells and the water brought to the surface without pumping. In these rock formations an aquifer is sandwiched between two impermeable layers called *aquicludes*. The aquifer slopes downward away from the area where water seeps into it. The aquicludes act like the walls of a pipe, confining the water as it seeps downslope, so that water pressure builds up in the aquifer. If the upper aquiclude is pierced by a well, this pressure may push the water up to the surface or even higher. The amount of pressure in an artesian well depends on the difference in elevation between the well and the water table in the aquifer. See Figure 12.12. Artesian wells get their name from Artois, a former province of France, where the first well of this type was dug.

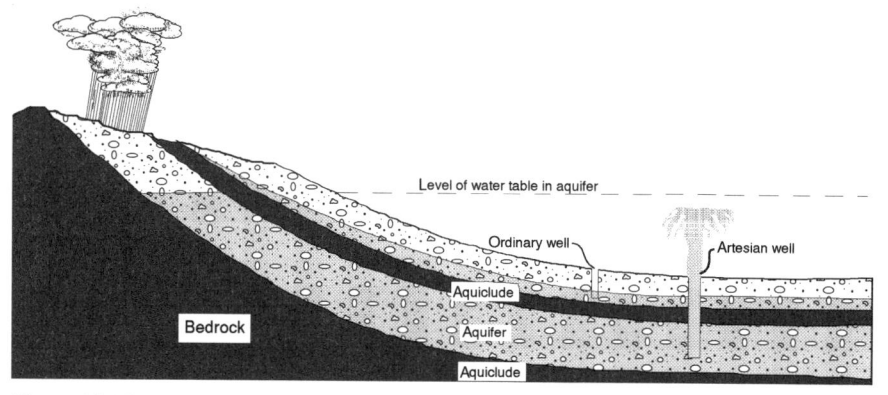

Figure 12.12 An Artesian Formation. The water rises in an artesian well because of hydrostatic pressure created by the weight of the water in the aquifer.

MULTIPLE-CHOICE QUESTIONS

In each case, write the number of the word or expression that best answers the question or completes the statement.

1. Large amounts of water vapor were added to Earth's developing atmosphere by
   (1) outgassing and collisions with comets
   (2) photosynthesis by early life-forms
   (3) dust and hydrogen gas
   (4) electrical discharges in the atmosphere

Note that question 2 has only three choices.
2. The hydrosphere is mostly
   (1) solid rock
   (2) liquid water
   (3) gaseous air

3. The diagrams below represent true-scale models for the solid Earth. Which diagram best shows the hydrosphere as drawn to the same scale?

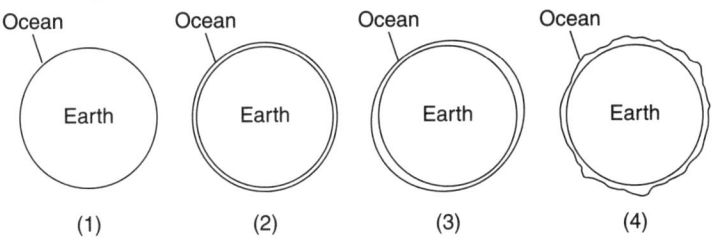

4. Which profile when drawn to true scale most accurately shows the relationship between the ocean width of 1,500 kilometers and a maximum depth of 8 kilometers?

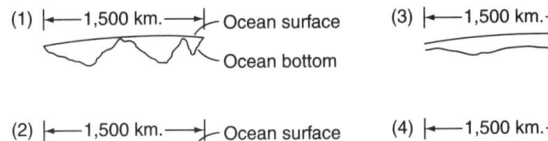

5. Approximately what fraction of Earth's surface do oceans now cover?
 (1) ¼
 (2) ⅓
 (3) ½
 (4) ¾

6. If the deepest parts of the oceans are about 10 kilometers and Earth's radius is about 6,400 kilometers, the depth of the oceans represents what percent of Earth's radius?
 (1) less than 1%
 (2) about 5%
 (3) about 25%
 (4) more than 75%

7. What is the major source of the dissolved minerals that affect the salinity of ocean water?
 (1) ice sheets
 (2) industrial pollutants
 (3) continental erosion
 (4) tropical storms

8. What is the average salinity of ocean water?
 (1) 0.035%
 (2) 0.35%
 (3) 3.5%
 (4) 35%

9. A lake without outlets may increase in salinity if the inflow of fresh water
 (1) is equal to or less than the amount of water lost by evaporation
 (2) is greater than the amount of water lost by evaporation
 (3) carries no dissolved substances into the lake
 (4) is used to power turbines in hydroelectric plants

The Origin and Nature of Earth's Hydrosphere

10. The increase of dissolved materials in the oceans is primarily the result of
    (1) abrasion of the ocean floors
    (2) transporting of materials by rivers
    (3) melting of continental glaciers
    (4) deposition of materials by groundwater

Base your answers to questions 11 and 12 on the graph below, which shows the percentage distribution of Earth's surface elevation above and depth below sea level.

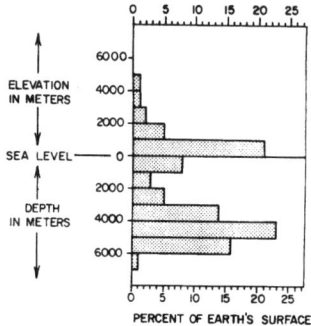

11. Approximately what total percentage of Earth's surface is below sea level?
    (1) 30%   (3) 70%
    (2) 50%   (4) 90%

12. Water is a good solvent because
    (1) a water molecule contains two hydrogen atoms and one oxygen atom
    (2) water is held together by strong molecular bonds
    (3) the two ends of a water molecule have different electrical charges
    (4) water is the most common compound on Earth

13. Which process results in a release of latent heat energy?
    (1) melting of ice
    (2) heating of liquid water
    (3) condensation of water vapor
    (4) evaporation of water

14. A sample of water undergoes phase changes from ice to water vapor and back to ice as shown in the model below. During which phase change does the sample gain the greatest amount of energy?
    (1) A
    (2) B
    (3) C
    (4) D

15. A large amount of latent heat is absorbed by water during
    (1) evaporation (3) condensation
    (2) freezing (4) precipitation

16. If ice were denser than water, lakes would
    (1) freeze no differently than they do now
    (2) freeze from the top down
    (3) freeze from the bottom up
    (4) not freeze at all

Base your answers to questions 17 through 21 on the *Earth Science Reference Tables*, the graph below, and your knowledge of earth science. The graph shows the temperatures recorded when a sample of water was heated from −100°C to 200°C. The water received the same amount of heat each minute.

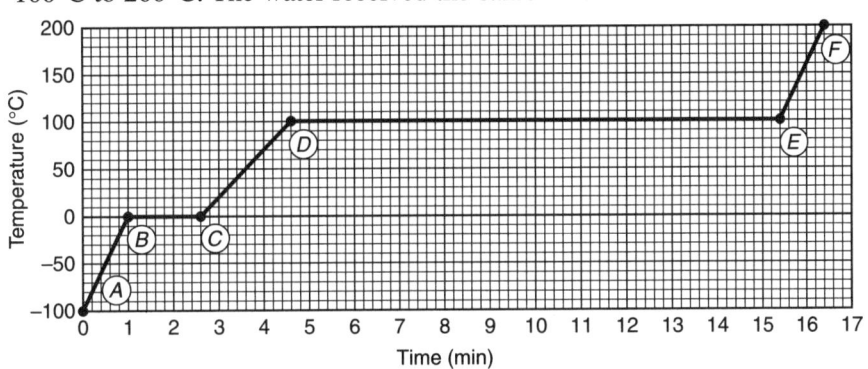

17. During the time on the graph represented by the line from point *B* to point *C*, the water was
    (1) freezing (3) condensing
    (2) melting (4) boiling

18. At which point in time would most of the water be in the liquid phase?
    (1) 1 min (3) 16 min
    (2) 14 min (4) 4 min

19. The greatest amount of energy was absorbed by the water between points
    (1) *A* and *B* (3) *C* and *D*
    (2) *B* and *C* (4) *D* and *E*

20. What is the rate of temperature change between points *C* and *D*?
    (1) 10 C°/min (3) 50 C°/min
    (2) 25 C°/min (4) 150 C°/min

237

The Origin and Nature of Earth's Hydrosphere

Note that question 21 has only three choices.
21. In which phase does water have the greatest density?
    (1) solid
    (2) liquid
    (3) gas

22. A diagram of the water cycle is shown below. Letters *A* through *D* represent the processes taking place.

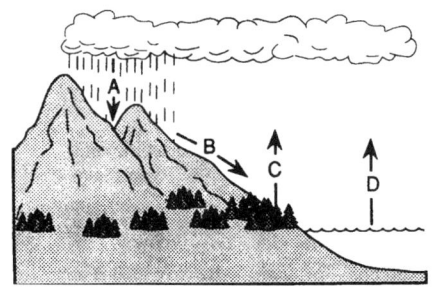

    Which arrow represents the process of transpiration?
    (1) *A* (3) *C*
    (2) *B* (4) *D*

23. By which process does moisture leave green plants?
    (1) convection   (3) transpiration
    (2) condensation   (4) radiation

24. Which is most important in determining the amount of groundwater that can be stored within a rock?
    (1) the rock's geologic age
    (2) the rock's hardness
    (3) the rock's porosity
    (4) the rock's color

25. Water will infiltrate surface material if the material is
    (1) impermeable and unsaturated
    (2) impermeable and saturated
    (3) permeable and unsaturated
    (4) permeable and saturated

26. Which graph best represents the relationship between the particle size of a loose material and the loose material's permeability?

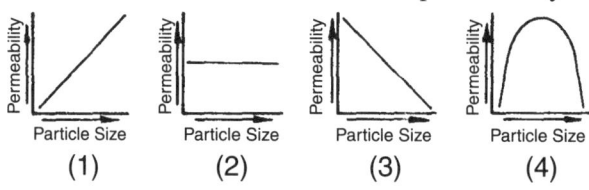

27. Why does water move very slowly downward through clay soil?
    (1) Clay soil is composed of low-density minerals.
    (2) Clay soil is composed of very hard particles.
    (3) Clay soil has large pore spaces.
    (4) Clay soil has very small particles.

28. The diagram below represents two identical containers filled with samples of loosely packed sediments. The sediments are composed of the same material, but differ in particle size. Which property is most nearly the same for the two samples?

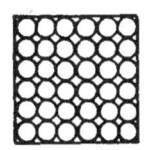

    (1) infiltration rate
    (2) porosity
    (3) capillarity
    (4) water retention

Note that question 29 has only three choices.

29. As the amount of precipitation on land increases, the depth from the surface of Earth to the water table will probably
    (1) decrease
    (2) increase
    (3) remain the same

30. Which diagram best illustrates the condition of soil below the water table?

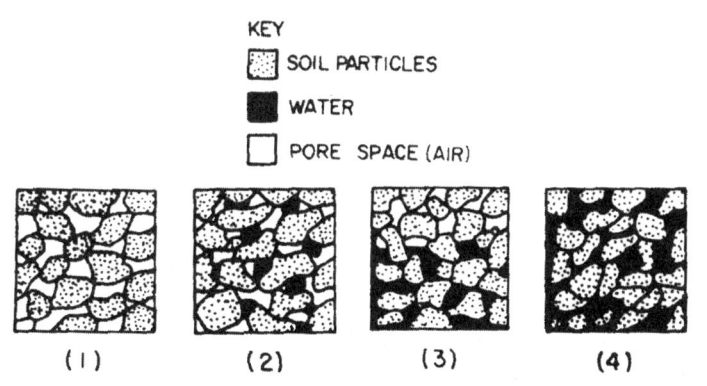

    KEY
    SOIL PARTICLES
    WATER
    PORE SPACE (AIR)

    (1) (2) (3) (4)

The Origin and Nature of Earth's Hydrosphere

Note that questions 31 and 32 have only three choices.

31. As the amount of precipitation on land increases, the level of the water table will probably
    (1) fall
    (2) rise
    (3) remain the same

32. As the amount of vegetation at a certain location increases, the amount of water vapor entering the air at that location will
    (1) increase
    (2) decrease
    (3) remain the same

33. If a 20-foot well is dug at a place where the water table is 15 feet below the surface, how many feet of water will be in the well?
    (1) 0   (3) 15
    (2) 5   (4) 20

34. The ground directly below the water table is always
    (1) below sea level
    (2) drier than the ground above the water table
    (3) saturated with water
    (4) impervious

35. A well in which water may rise above the local water table without pumping is
    (1) an artesian well   (3) a permanent well
    (2) a dug well   (4) a temporary well

CONSTRUCTED RESPONSE QUESTIONS

36. During a recent drought, a town passed an ordinance banning watering of lawns by homeowners whose water was supplied by the local reservoir. However, residents that had their own wells to supply them with water were exempt from the ordinance. One resident of the town decided to drill a well to extract groundwater from beneath his property for watering his lawn.
    (a) State three characteristics of the subsurface materials on his property that would affect the volume of water a well could supply. [3]
    (b) The resident discovered that the soil beneath his property was almost entirely clay. Decide whether drilling a well would be worthwhile, and state at least one reason to support your decision. [1]

37. The diagram below shows the changes in the molecular structure of water that occur as it changes temperature.

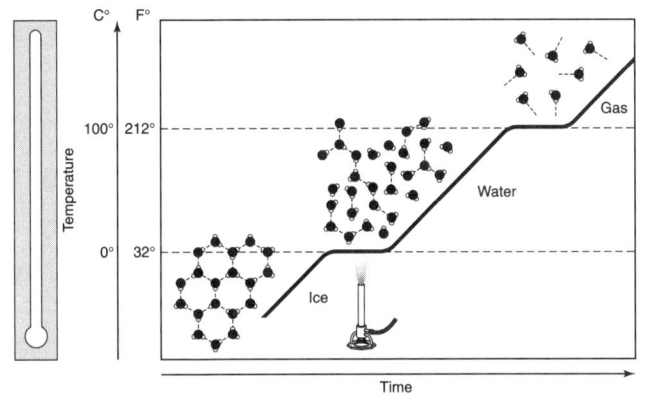

(a) According to the diagram, in which phase would water have the greatest density? Explain your answer. [1]
(b) What would happen to the spacing between molecules of water as a gas if the temperature was increased to 300°C? [1]
(c) Describe how the volume of a particular sample of water changes as the sample changes from water into ice. [1]

EXTENDED CONSTRUCTED RESPONSE QUESTIONS

38. The diagram below shows the water cycle. Letters *A* through *F* indicate natural processes that are taking place.

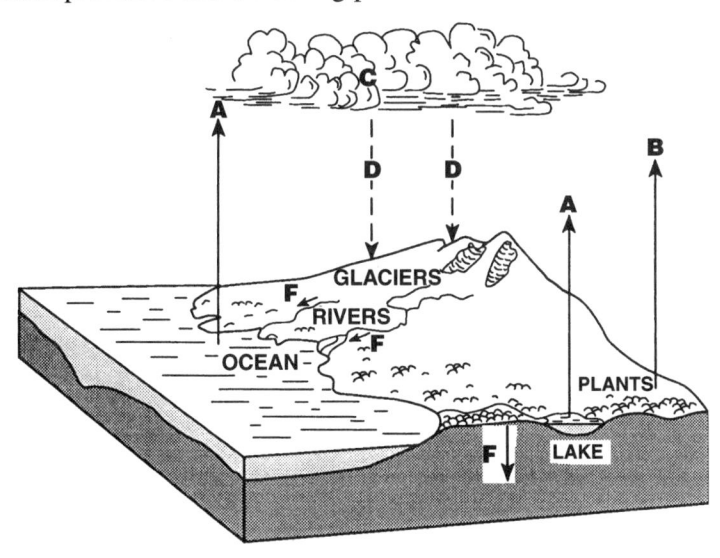

The Origin and Nature of Earth's Hydrosphere

(a) State the natural process of the water cycle that is indicated by each of the following letters:
A. _____[1]
B. _____[1]
C. _____[1]
D. _____[1]
E. _____[1]
F. _____[1]

(b) State one characteristic that the ground surface must have in order for process F to occur. [1]

(c) State two factors that would make water on Earth's surface more likely to undergo process E rather than F. [1]

(d) What is the relationship between the season of the year in New York State and the rate at which process A occurs? [1]

39. Construct a correctly labeled diagram of an ordinary well. Include each of the following in your diagram: zone of saturation, zone of aeration, water table, capillary fringe, soil moisture zone. [6]

40. Construct a diagram of an artesian well. Include each of the following in your diagram: aquifer, aquicludes, height of the water table in the aquifer. [4]

CHAPTER 13
THE ORIGIN AND HISTORY OF LIFE ON EARTH

> **KEY IDEAS** All of Earth's present-day life forms have evolved from common ancestors reaching back about three billion years to the simplest one-celled organisms. Before that time, simple molecules may have formed complex organic molecules that slowly evolved into cells capable of replicating themselves. The history of life on Earth has been pieced together from geological, anatomical, and molecular evidence. The geological evidence has come mainly from the sequence of changes seen in fossils found in successive layers of rock that have formed over more than a billion years. Human existence has been very brief compared to the expanse of geologic time.
>
> Geologic history can be reconstructed by observing sequences of rock types and fossils to correlate bedrock at various locations. The characteristics of rocks indicate the processes by which they formed and the environments in which these processes took place. Fossils preserved in rocks provide information about past environmental conditions.
>
> Geologists have divided Earth history into time units based upon the fossil record. Age relationships among bodies of rocks can be determined using principles of original horizontality, superposition, inclusions, cross-cutting relationships, contact metamorphism, and unconformities. The presence of volcanic ash layers, index fossils, and meteoritic debris can provide additional information. The regular rate of nuclear decay (half-life time period) of radioactive isotopes allows geologists to determine the absolute ages of materials found in some rocks.

KEY OBJECTIVES
Upon completion of this chapter, you will be able to:

- Identify the characteristics of life and two ideas about how it may have originated.
- Explain how a study of the fossil record shows that life-forms have evolved through geologic time.

The Origin and History of Life on Earth

- Describe how fossils provide evidence of past environments.
- Determine the relative ages of a series of rock layers and any igneous intrusions, faults, folds, or fossils they may contain, based upon the principle of superposition.
- Explain the significance of index fossils and volcanic ash deposits in correlating widely separated rock layers.
- Interpret the geologic time scale.
- Determine the absolute age of a rock, given the relative amounts of a radioisotope and decay product in the rock, and the half-life of the radioisotope.

THE ORIGIN OF LIFE *(This section is not testable on the Regents Exam but is included as background information for the reader.)*

What Is Life?

Most people have a pretty good sense of what the word *living* means, but it is hard to come up with an exact definition. About the best we can do is to describe the properties that living things have in common. *All living things grow, metabolize, react to the external environment, and reproduce.* All of these processes require energy. Living things use energy to do work, grow, and maintain themselves. These activities are accomplished through a huge number of chemical reactions, which together are called *metabolism*. Living things also use energy to maintain conditions able to sustain life inside themselves no matter what their external surroundings are like. Energy is also central to the ability to sense and react to the external environment. Finally, living things reproduce to perpetuate their kind, and they pass their characteristics on to their offspring.

Water and Organic Molecules: the Stuff of Life

Life processes involve intricate sequences of interactions among a wide variety of chemicals. The most important substance is water; as the universal solvent, water provides the medium in which all of these other chemicals can dissolve and interact. It is the chemical "soup" in which the chemical reactions of metabolism take place. Most living things are 70–80 percent water, though jellyfish can be up to 95 percent.

Except for water and a handful of other substances, most of the chemicals involved in life processes are organic compounds—molecules consisting of carbon, hydrogen, and usually oxygen. A key characteristic of living things is their ability to use energy to make organic compounds. Organic compounds have a tendency to break down and release the energy that went into forming them. Living things have developed the ability to control this breakdown, usually in order to harness the energy that is released. Four types of organic molecules are the basis of most biological processes and structures: proteins, nucleic acids, carbohydrates, and lipids. See Figure 13.1.

Proteins consist of long chains of smaller molecular units called *amino acids*. Proteins are the primary structural material of living things. Muscles are largely protein, enzymes are proteins that speed up chemical reactions, and hormones are proteins that act as chemical messengers in the body.

Nucleic acids, such as DNA and RNA, are chains of repeating subunits called *nucleotides*. There are five types of nucleotides, and their order in a chain forms a kind of code, like a five-letter alphabet, that contains genetic information. You may not think that much information can be conveyed with a five-letter alphabet, but remember that modern computers use just a two-letter alphabet (binary code) to accomplish everything they do! The DNA molecules in a living thing contain the "recipe" for the organism—all of the detailed instructions needed to build and maintain it. See Figure 13.2.

Carbohydrates are composed of just the three basic elements of organic molecules—carbon, hydrogen, and oxygen. Many living things control the breakdown of simple carbohydrates as a source of energy. Glucose and other simple sugars are among the simplest carbohydrates. More complex carbohydrates, such as starches, consist of long chains of these simple sugars and other components. Still other carbohydrates, such as cellulose and the modified carbohydrate chitin, are important structural molecules used by plants and animals.

Lipids include fats, oils, and waxes. They are often used for energy storage. Also, since waxes and oils repel water, they often serve to protect body parts or to keep water inside an organism.

Figure 13.1 Organic Compounds: The Building Blocks of Life. Source: *Let's Review: Biology*, 3rd Ed., Scott Hunter, Barron's Educational Series, Inc., 2001.

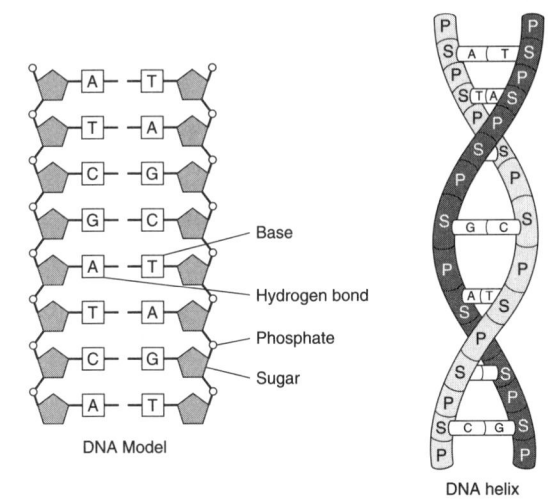

Figure 13.2 The Structure of a DNA Molecule. The sequence of "links" in a chain of DNA is the genetic code that contains all of the information needed to build and maintain the organism. Source: *Let's Review: Biology*, 3rd Ed., Scott Hunter, Barron's Educational Series, Inc., 2001.

245

Thus, many birds and marine mammals use a coating of oil to keep their feathers or fur dry, and cacti have a coating of wax to reduce water loss by evaporation. Some hormones are lipids.

Life's Origin: Two Schools of Thought

It is now generally accepted that early Earth was rich in the building blocks of life, that is, water and organic compounds. After a few hundred million years, a stable liquid ocean had formed. Comets striking Earth brought in a whole range of substances: methane (CH_4), ammonia (NH_3), carbon monoxide (CO), carbon dioxide (CO_2), nitrogen (N_2), and even more complex molecules. All of the raw materials for the formation of complex organic molecules were present. This atmospheric "soup" of gases, when exposed to ultraviolet radiation or the electrical discharges of lightning bolts, was converted into even more complex molecules.

In the 1950s, Stanley Miller and Harold Urey performed a classic experiment in which they filled a flask with some water and a model atmosphere rich in methane and ammonia, with a little carbon dioxide, and then discharged a spark to simulate lightning. When they analyzed the contents of the flask, they found that amino acids (the building blocks of proteins) had formed from the gases and water. However, in order for life to originate, the amino acids had to build in complexity up to that of the genetic coding molecules RNA and DNA. There are two schools of thought about how simple molecules could have produced increasingly complex molecules: autocatalysis/vesicles and natural formation of RNA.

A *catalyst* is a substance that enables or speeds up a reaction without being consumed in the process. *Autocatalysis* is a chemical process in which the product of a reaction acts as a catalyst in its own production. This process can lead to complex chemical reactions. Beginning with a soup containing two substances that react with one another, and supplying enough chemicals to maintain vigorous reactions, progressively more complex molecules can be produced in the soup. Some scientists have argued that autocatalysis turned amino acids and other molecules into more and more complex molecules until primitive proteins and other substances were formed.

In water, certain simple molecules form sheets that can spontaneously wall off tiny bits of the water, forming an "outside" and an "inside." These tiny structures, known as *vesicles*, are isolated environments in which complex molecules could have become more concentrated as autocatalysis continued. This process, in turn, would have favored the production of even more complex molecules, and so on all the way up to primitive proteins. Some scientists think vesicles may have been the precursors of cells.

The second school of thought is that the first step toward life was the formation of RNA by natural chemical processes. RNA is a twisted double chain of small molecules. When the double chain separates, molecules from the surrounding "soup" interact with it in a complex series of steps in which each single chain reforms a double chain, resulting in two RNA molecules identical to the original. Thus, RNA could act like a primitive life-form itself, reproducing and changing as it encountered other complex molecules in the early ocean. The change from RNA to DNA is a relatively minor chemical step. RNA molecules can also serve as catalysts and may have acted in autocatalytic reactions to sustain a primitive form of life.

THE GEOLOGIC RECORD OF LIFE'S HISTORY

Until the nineteenth century, nearly everyone believed that Earth was only a few thousand years old. Scientific evidence, though, indicates that some of the rocks on Earth's surface are several *billion* years old. Observations of patterns in rock layers and the location of various kinds of fossils allow inferences concerning the relative ages of rocks and the events that formed them. The absolute age of a rock can be determined from the relative amounts of a radioisotope and its decay products in the rock. The rock and fossil record reveals that life-forms originated early in Earth's history and that they and the environments in which they lived have changed over time.

How Can the Order in Which Geological Events Occurred Be Determined?

Much of what is known about Earth's history, and that of life, has been learned by studying rocks. Rock layers contain traces of events that occurred in the past and provide clues to the origins of the layers and the environments in which they formed. With careful logic, inferences can be drawn about Earth's past from data obtained from rocks. For example, the presence in a rock layer of sorted sand grains that are rounded implies that the grains were carried by running water. Also, the size of the grains may indicate how fast the water that deposited them was moving. If a rock layer contains shells, the type of shell may show that the rock formed in a lake, not an ocean. Shells can also provide clues to the depth and temperature of the water. Microscopic pollen grains in the rock could identify plants living around the lake when the rock formed. Still other rock layers may contain evidence of glaciers, deserts, or volcanic eruptions.

Such inferences are based upon an assumption proposed by James Hutton in 1795—uniformitarianism. *Uniformitarianism* is the idea that Earth's features, such as mountains, valleys, and rock layers, have formed gradually by processes still underway, not by instantaneous creation. It assumes that Earth has always behaved much as it does now; that water, for example, ran downhill in streams carrying sand and silt 10 million years ago just as it does today. This idea can be summarized in the statement "The present is the key to the past."

According to uniformitarianism, every rock layer contains a record of a short part of Earth's long history. By working out the age of each layer in a series, geologists can determine the order in which the layers formed and can arrange the events they represent in the order in which these events occurred. In this way a history of Earth can be pieced together.

There are two ways of expressing the ages of Earth materials or geologic events, relative age and actual age. *Relative age* is the age of a rock, fossil, geologic feature, or event relative to that of another rock, fossil, geologic feature, or event. Thus, relative age is not an exact age in years; it tells you only that one thing is older or younger than another. If you say that you are older than a first grader but younger than your parents, you are giving your relative age.

Actual age is the age of a rock, fossil, geologic feature, or event given in units of time, usually years. If you state, "I am 15 years old," you are giving your actual age. The actual age of Earth materials and events is usually determined by analyzing radioisotopes in objects.

Determining Relative Age

Two key ideas in determining relative age are the law of original horizontality and the principle of superposition. *The law of original horizontality* states that sediments deposited in water form flat, level layers parallel to Earth's surface.

The principle of superposition is the concept that the bottom layer of a series of sedimentary layers is the oldest, unless it has been overturned or older rock has been thrust over it. Each layer forms atop the one that is already there. (Layers of sediment do not form in midair, nor do they hover halfway between the ocean surface and the ocean bottom!) Therefore, each layer is younger than the one under it and older than the one on top of it. The relative ages of any two layers can be determined based upon which layer is on top and which is on the bottom. See Figure 13.3.

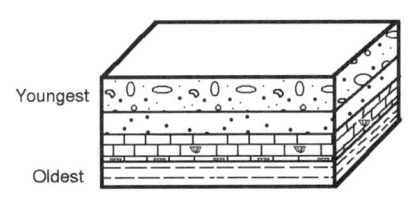

Figure 13.3 Series of Rock Layers Showing Oldest to Youngest.

The principle of superposition must be used with care, however, because there are events that can disturb the positions of rock layers. Forces within Earth may tilt, fold, or fault the layers. Older layers may be pushed on top of younger ones. In such cases, one must work out the original positions of the rock layers before applying the principle of superposition.

In general, a rock layer is older than any joint, fault, or fold that appears in it; the rock had to already exist in order to be folded or faulted. By unfolding or unfaulting the rock layers, one can determine the positions of the layers before they were disturbed, and can then apply the principle of superposition to determine their relative ages.

For example, let's look at Figure 13.4. If we were to judge only by the position of the rock layers in the core sample at (3), we would conclude that layer *C* is the youngest and *A* is the oldest because *C* is on top and *A* is on the bottom. If we look at the entire rock structure, however, we see that the three layers of rock have been folded and overturned. By unfolding the layers, we place the layers in their original position and see that layer *A* is the *youngest* while *C* is the oldest.

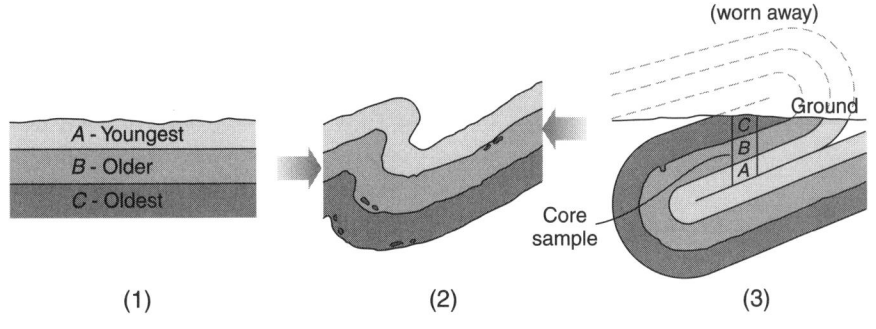

Figure 13.4 The Original Position of Rock Layers Must Be Determined Before Applying the Principle of Superposition. Originally horizontal rock layers (1), may be folded by compression (2), causing older rock layers to appear above younger rock layers (3). Source: *Macmillan Earth Science*, Eric Danielson and Edward J. Denecke Jr., Macmillan Publishing Co., 1989.

The same is true of faults, that is, fractures along which the rocks on either side have moved. If rock layers fracture and then move along that fracture, they are displaced. Any rock displaced by a fault is older than the fault. During faulting, underlying rock layers may be pushed up so that they are found on top of younger rock. See Figure 13.5. Again, to obtain true relative ages, one must work backward to the position the rock layers were in before the fault offset them.

Igneous intrusions or extrusions are often found associated with other types of rock. Igneous intrusions form when molten rock forces its way into preexisting rock, cools, and hardens. Thus, an intrusion is younger than any rock it cuts through. See Figure 13.6. Extrusions form during volcanic eruptions when molten rock flows out onto Earth's surface as lava and hardens, or is blown into the atmosphere and settles on the ground, forming a blanket of volcanic rock particles. Thus, extrusions are younger than the rocks beneath them, but older than any layer that may form over them. Therefore, if an igneous body is found in rock, one must first determine whether it is an intrusion or an extrusion before the relative ages of the layers can be established.

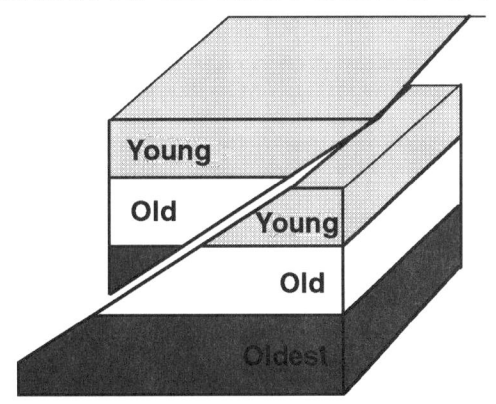

Figure 13.5 Fault, Showing Older Rocks Pushed Atop Younger Ones.

The Origin and History of Life on Earth

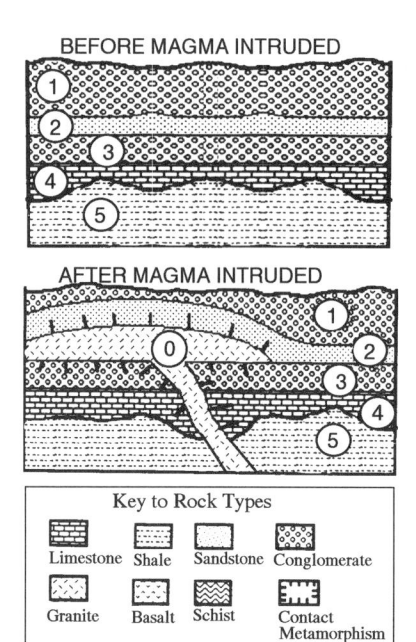

Figure 13.6 Five Rock Layers Before and After Intrusion of Magma. Layer O, which formed from the cooling magma, is younger than the five rock layers into which it was intruded.

Unconformities

Layers of rock are generally deposited in an unbroken sequence. However, if forces within Earth uplift rocks, deposition ceases. Erosion may wear away many layers of rock before the land is low enough for another layer to be deposited. The result is an *unconformity*, a break or gap in the sequence of a series of rock layers. Thus, the rocks above an unconformity are quite a bit younger than those below it.

There are three types of unconformity. See Figure 13.7. *Angular unconformities* form when rock layers are tilted or folded before being eroded. When new layers are deposited, they form horizontally and the layers below the unconformity are at an angle to those above it. *Disconformities* are irregular erosional surfaces between parallel layers of rock. Disconformities occur when deposition stops and layers are eroded, but no tilting or folding occurs. These surfaces are not easy to discern and are often found when fossils of very different ages are discovered in adjacent layers. *Nonconformities* are places where sedimentary layers lie on top of igneous or metamorphic rocks.

How Can Rocks and Geologic Events in One Place Be Matched with Those in Other Locations?

Correlation

With care and logic the relative ages of rock layers in any particular location can be worked out. However, this information provides only a small part of the overall picture of Earth's history, and unconformities may mean that even more details are missing. To broaden the scope of the picture, rocks in one place need to be correlated with rocks in other locations. The process of *correlation* involves determining that rock layers in different areas are the same age. In this way, rocks in one location may fill in gaps in the record in another location. Correlation also allows the relative ages of rocks in widely separated outcrops to be determined.

Walking the Outcrop

Rock layers can sometimes be followed from one location to another by walking the outcrop. See Figure 13.8. *Walking the outcrop* means physically tracing the layers from one place to another. In this way, rock layers in two different places, such as two adjacent mountains or ridges, can sometimes be correlated. Unfortunately, rock layers are rarely continuously visible for any distance. Most rocks are hidden beneath regolith, the loose fragments of rock and soil that blanket Earth's surface. Rock outcrops poke out here and there like islands. As a result, geologists are often faced with the task of correlating rocks in widely separated outcrops.

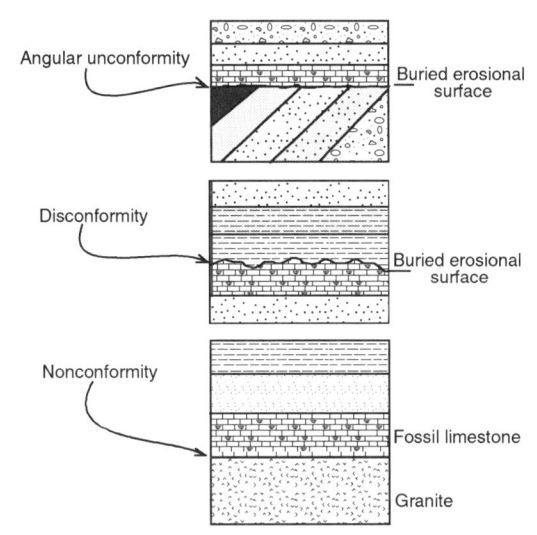

Figure 13.7 Angular Unconformity, Disconformity, and Nonconformity.

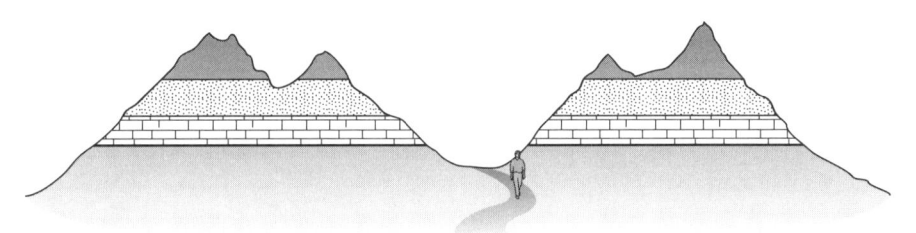

Figure 13.8 Walking the Outcrop.

Matching Physical Characteristics

Rocks can sometimes be correlated on the basis of distinct similarities in physical characteristics such as composition, color, thickness, and fossil remains. Distinctive rocks or sequences of rock types can also be used to correlate rocks.

For example, Alfred Wegener used this type of correlation as part of his evidence for continental drift. He stated: "Igneous rocks in Brazil and Africa [have] no less than five parallels: 1) the older granite, 2) the younger granite, 3) alkali-rich rocks, 4) volcanic Jurassic rock and intrusive dolerite, and 5) kimberlite, alnoite, etc… The last of the rock groups (kimberlite, alnoite) is best known since both in Brazil and South Africa the beds yield the famous diamond finds. In both these regions the peculiar type of stratification known as 'pipes' occurs. There are white diamonds in Brazil in Minas Geraes State and in South Africa north of the Orange River only."

This type of correlation must be done with care, though, because different formations can look almost identical.

Index Fossils

Index fossils are remains of organisms that had distinctive body features, were abundant, and had a broad, even worldwide range, yet existed only for a short period of time. The best index fossils include swimming or floating organisms that evolved rapidly and were distributed widely, such as graptolites, trilobites, and ammonoids. See Figure 13.9. Their distinctive body shapes and broad distribution make index fossils easy to find in widely separated rock layers (see Figure 13.10). Also, their short existence pinpoints the time period during which those rock layers were formed.

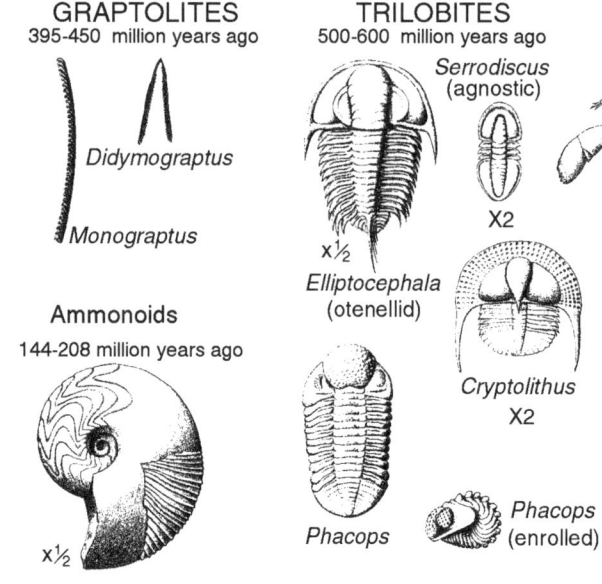

Figure 13.9 Graptolites, Trilobites, and Ammonoids Labeled with Ages. Source: The New York State Museum. Educational Leaflet #28, *Geology of New York, A Simplified Account*, Second Edition. Printed with permission of New York State Museum, Albany, N.Y.

Key Beds

Key beds are well-defined, easily identifiable layers or formations that have distinctive characteristics or fossil content that allows them to be used in correlation. Key beds can be readily identified and are not easily confused with other layers. Materials such as volcanic ash layers that are rapidly deposited over a wide area make excellent key beds.

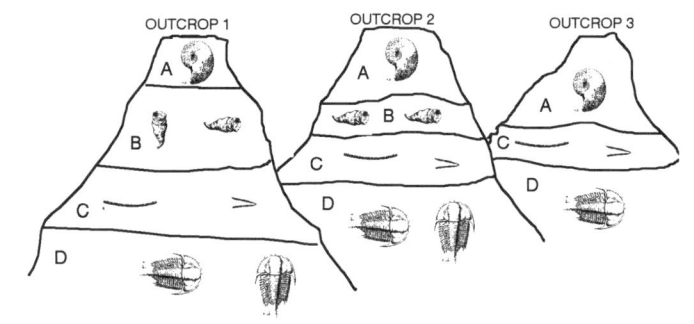

Figure 13.10 Correlation by Fossils of Strata in Several Locations.

One well-known key bed is the iridium-rich layer of rock discovered in Italy, Denmark, and many other places around the world. Iridium is an extremely rare element on Earth, but not in meteorites. The iridium-rich layer is thought to have formed 66 million years ago when a meteorite impact created a dust cloud that encircled Earth and then slowly settled to the ground.

How Can Earth's Geologic History Be Sequenced from the Fossil and Rock Record?

The Geologic Column

Even though the rock record in any one place is incomplete, correlation has made it possible to piece together a fairly complete record of Earth's geologic history by examining rocks in different locations. By the nineteenth century, geologists had correlated rocks worldwide into a single sequence called the *geologic column*. Rocks are still being added to the geologic column as more outcrops are mapped and described.

The Geologic Time Scale

Geologists then divided Earth's history into a sequence of time units called the *geologic time scale*. Geologic time is the entire time that Earth has existed. The geologic time scale divides geologic time into units and subunits

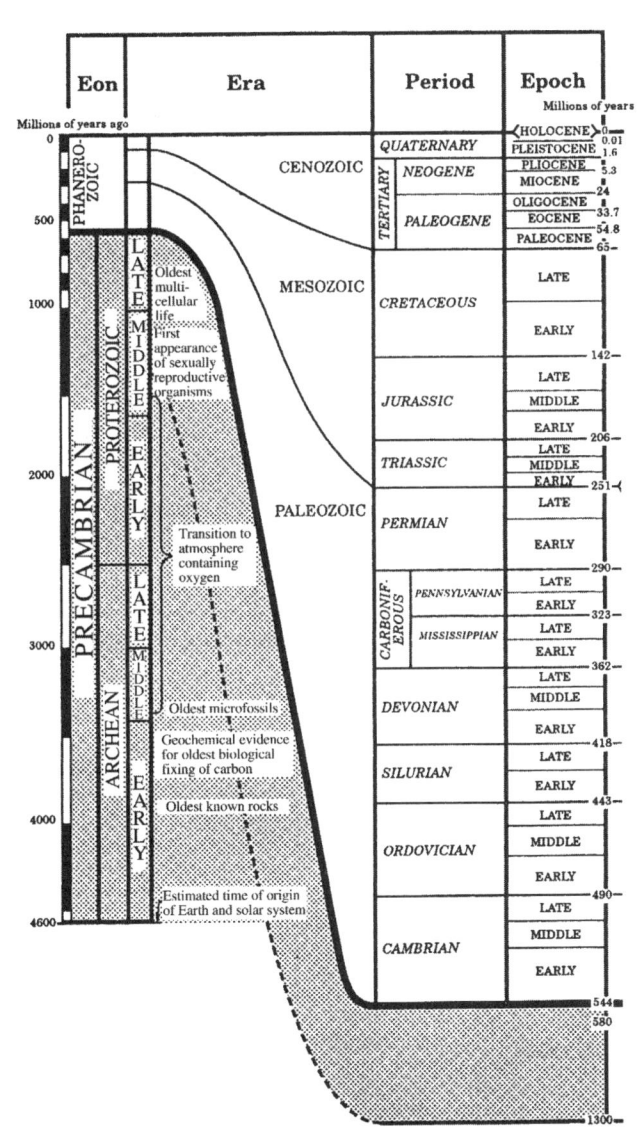

Figure 13.11 The Geologic Time Scale. Source: The State Education Department, *Earth Science Reference Tables*, 2001 ed. (Albany, New York; The University of the State of New York).

253

based upon the fossil record. Geologic time is divided into eons, eons are subdivided into eras, eras are subdivided into periods, and periods are subdivided into epochs. At first, these time units were based only upon the relative ages of fossils in rocks of the geologic column, but radioisotope dating has enabled geologists to assign actual ages to the time units of the geologic time scale as shown in Figure 13.11.

Eons
Eons are the largest time units on the geologic time scale. The oldest is the *Archean* (Greek "ancient"); it covers the time from Earth's formation to the appearance of multicelled organisms. Archean rocks are the oldest known on Earth and contain microscopic fossils of single-celled, bacteria-like organisms. The *Proterozoic* (Greek "earlier life") is the next oldest eon; its rocks contain traces of multicelled organisms that had no preservable hard parts. The most recent eon is the *Phanerozoic* (Greek "visible life"), which provides an abundant fossil record of preserved hard parts.

Before traces of soft-bodied organisms and microfossils were found in Archean and Proterozoic rocks, rocks from these two eons were lumped together under the general term *Precambrian*. The oldest known rocks that contain fossils of hard parts of organisms are called Cambrian because they were found in Wales, which was once called Cambria; therefore, rocks that were older than Cambrian rocks were termed Precambrian.

Eras
Eons are subdivided into eras based upon the fossil record. Since there are so few fossils in Archean and Proterozoic rocks, these eons have not yet been formally divided into eras. During these eons, simple forms of life such as bacteria, microorganisms, and later algae and soft animals (e.g., worms and jellyfish) predominated.

The Phanerozoic eon is subdivided into the *Paleozoic* (old life), *Mesozoic* (middle life), and *Cenozoic* (recent life) eras based upon the types of life-forms that predominated during those time intervals. In general, life-forms developed in complexity over time. During the Paleozoic era, marine invertebrates dominated and the earliest land plants and animals developed. The Mesozoic was dominated by reptiles, such as the dinosaurs, and saw the first mammals develop. The Cenozoic has been dominated by mammals, including humans, and flowering plants have become dominant. Notice, though, that human existence has been very brief in comparison with the expanse of geologic time.

Periods
Eras are subdivided into periods, for which terminology is much less organized. The names for time periods are based upon the names of rock formations in England, Germany, Russia, the United States, and many other countries. Thus, some periods are named for locations, such as the Permian for the city of Perm in Russia, and the Pennsylvanian and Mississippian after

those states in the United States. Other periods are named for characteristics of the rock layers where rocks of this age were first studied, such as the Cretaceous, from the Latin word for chalk.

Epochs
The smallest unit of time on the geologic time scale is the epoch. Epochs are subdivisions of periods, usually into early, middle, and late. The names of epochs of the Tertiary Period are Greek words that denote degrees of recentness; examples are Holocene ("wholly recent"), Miocene ("less recent"), and Eocene ("dawn of the recent").

Figure 13.12 shows the Geologic History of New York State at a Glance. In it, the geologic time scale is shown. Next to, and correlated with, the time scale are a series of columns that contain additional information.

Life on Earth. This column describes the types of life that existed during the different time periods based upon the fossil record.

Rock Record in N.Y. This column shows the time periods for which there is a rock and fossil record in New York State. The thick black line indicates that rock or fossil evidence from that time period exists in New York State. If there is no line, there is no rock or fossil record for that time. These gaps reflect unconformities in the rocks of New York State.

Time Distribution of Fossils (Including Important Fossils of New York). This column contains bold black lines labeled with names and circled letters. The names identify types of fossil organisms. The circled letters are keyed to the illustrations of specific index fossils printed along the top of the chart, and are placed on the line to indicate the approximate time of existence of each specific index fossil. For example, find A at the bottom of the line labeled "Trilobites." Now look to the left, and note that organism A lived at the end of the Early epoch of the Cambrian period.

Tectonic Events affecting Northeast North America. This column uses thick, black lines to indicate the time periods during which major plate interactions occurred. Each line is labeled with the type of plate tectonic event that produced mountain building.

Important Geologic Events in New York. This column describes in more detail how the plate tectonic events shown in the preceding column affected New York State. It describes the origin of well-known features of New York State, such as the Palisades Sill, which was intruded during the late Triassic, and the Adirondacks, which were uplifted during the Pliocene.

Inferred Position of Earth's Landmasses. This column shows the changing positions of landmasses due to plate movements. North America is shown in black to highlight its movements. The latitude and longitude lines show the positions of landmasses in relation to the Equator and poles, as well as to each other. Notice that during the Devonian and Mississippian periods, North America was centered on the Equator, a location that explains how coal beds formed from tropical rain forests.

The Origin and History of Life on Earth

The Geologic History of New York State at a Glance Chart

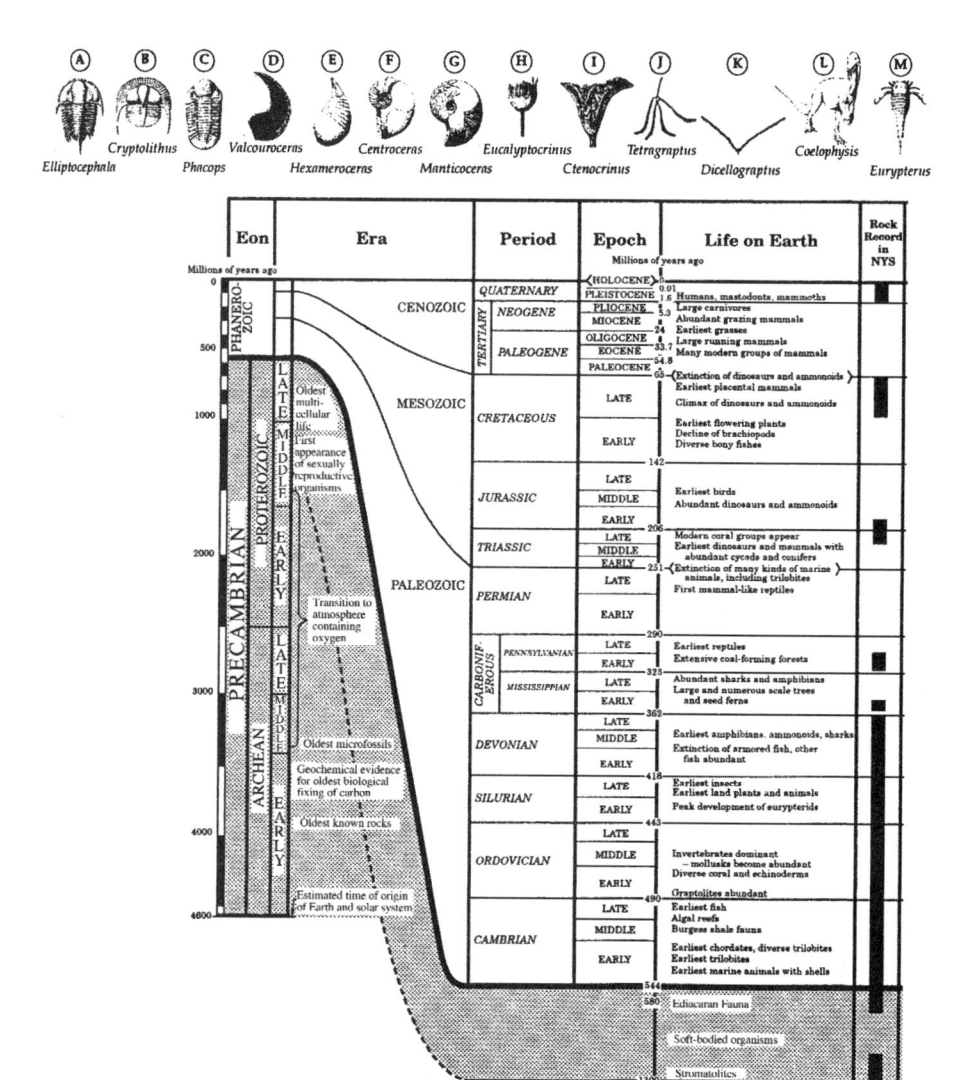

A. Elliptocephala
B. Cryptolithus
C. Phacops
D. Valcourocerus
E. Hexamerocerus
F. Centroceras
G. Manticoceras
H. Eucalyptocrinus
I. Ctenocrinus
J. Tetragraptus
K. Dicellograptus
L. Coelophysis
M. Eurypterus

Eon	Era	Period	Epoch	Life on Earth	Rock Record in NYS
PHANEROZOIC	CENOZOIC	QUATERNARY	HOLOCENE 0.01		
			PLEISTOCENE 1.6	Humans, mastodonts, mammoths	
		NEOGENE	PLIOCENE 5.3	Large carnivores	
			MIOCENE 24	Abundant grazing mammals	
				Earliest grasses	
		PALEOGENE	OLIGOCENE 33.7	Large running mammals	
			EOCENE 54.8	Many modern groups of mammals	
			PALEOCENE 65	Extinction of dinosaurs and ammonoids	
	MESOZOIC	CRETACEOUS	LATE	Earliest placental mammals	
				Climax of dinosaurs and ammonoids	
				Earliest flowering plants	
			EARLY	Decline of brachiopods	
				Diverse bony fishes	
		JURASSIC	LATE 142		
			MIDDLE	Earliest birds	
			EARLY	Abundant dinosaurs and ammonoids	
		TRIASSIC	LATE 206	Modern coral groups appear	
			MIDDLE	Earliest dinosaurs and mammals with abundant cycads and conifers	
			EARLY 251	Extinction of many kinds of marine animals, including trilobites	
	PALEOZOIC	PERMIAN	LATE	First mammal-like reptiles	
			EARLY 290		
		CARBONIFEROUS PENNSYLVANIAN	LATE	Earliest reptiles	
			EARLY 323	Extensive coal-forming forests	
		MISSISSIPPIAN	LATE	Abundant sharks and amphibians	
			EARLY 362	Large and numerous scale trees and seed ferns	
		DEVONIAN	LATE		
			MIDDLE	Earliest amphibians, ammonoids, sharks	
			EARLY 416	Extinction of armored fish, other fish abundant	
		SILURIAN	LATE	Earliest insects	
			EARLY 443	Earliest land plants and animals	
				Peak development of eurypterids	
		ORDOVICIAN	LATE		
			MIDDLE	Invertebrates dominant — mollusks become abundant	
			EARLY 490	Diverse coral and echinoderms	
		CAMBRIAN	LATE	Graptolites abundant	
				Earliest fish	
			MIDDLE	Algal reefs	
				Burgess shale fauna	
			EARLY	Earliest chordates, diverse trilobites	
				Earliest trilobites	
			544	Earliest marine animals with shells	
PRECAMBRIAN (PROTEROZOIC/ARCHEAN)			580	Ediacaran Fauna	
				Soft-bodied organisms	
			1300	Stromatolites	

Precambrian notations:
- Oldest multi-cellular life
- First appearance of sexually reproductive organisms
- Transition to atmosphere containing oxygen
- Oldest microfossils
- Geochemical evidence for oldest biological fixing of carbon
- Oldest known rocks
- Estimated time of origin of Earth and solar system (4600)

Figure 13.12 Geologic History of New York State at a Glance. Source: The State Education Department, *Earth Science Reference Tables*, 2001 ed. (Albany, New York; The University of the State of New York).

Unit Three **EARTH'S HISTORY**

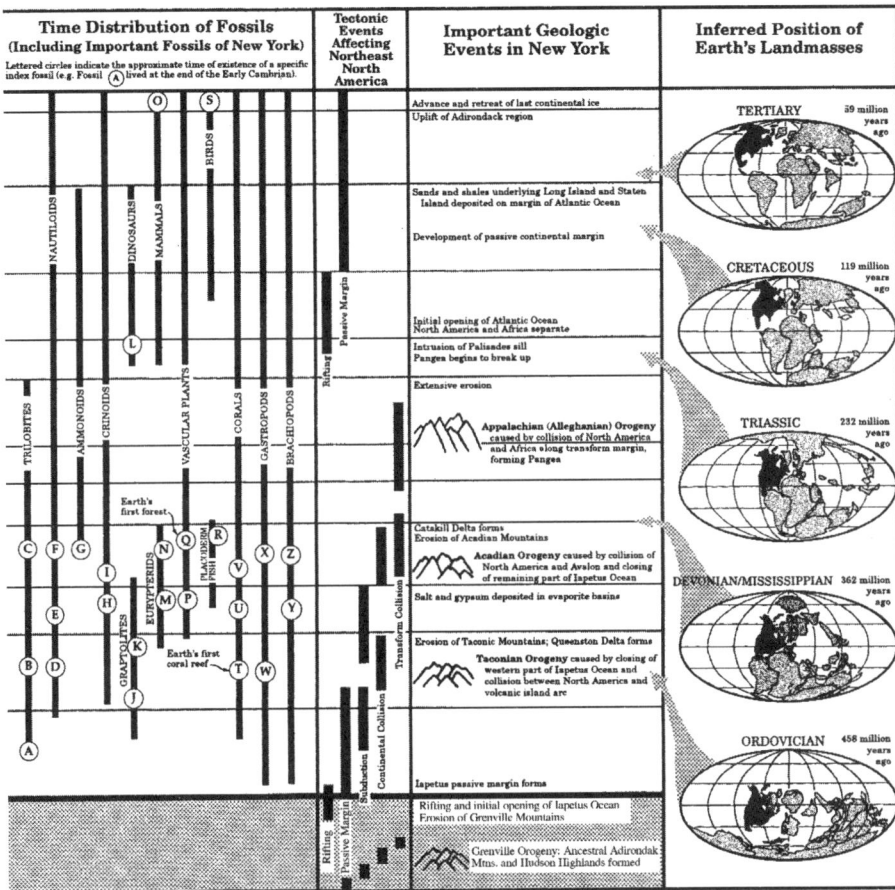

Time Distribution of Fossils
(Including Important Fossils of New York)

Lettered circles indicate the approximate time of existence of a specific index fossil (e.g. Fossil (A) lived at the end of the Early Cambrian).

N	Stylonurus
O	Mastodont / Beluga Whale
P	Cooksonia / Aneurophyton
Q	Naples Tree
R	Bothriolepis
S	Condor
T	Lichenaria
U	Cystiphyllum
V	Pleurodictyum
W	Maclurites
X	Platyceras
Y	Eospirifer
Z	Mucrospirifer

Tectonic Events Affecting Northeast North America

Important Geologic Events in New York

- Advance and retreat of last continental ice
- Uplift of Adirondack region
- Sands and shales underlying Long Island and Staten Island deposited on margin of Atlantic Ocean
- Development of passive continental margin
- Initial opening of Atlantic Ocean; North America and Africa separate
- Intrusion of Palisades sill; Pangea begins to break up
- Extensive erosion
- **Appalachian (Alleghanian) Orogeny** caused by collision of North America and Africa along transform margin, forming Pangea
- Catskill Delta forms; Erosion of Acadian Mountains
- **Acadian Orogeny** caused by collision of North America and Avalon and closing of remaining part of Iapetus Ocean
- Salt and gypsum deposited in evaporite basins
- Erosion of Taconic Mountains; Queenston Delta forms
- **Taconian Orogeny** caused by closing of western part of Iapetus Ocean and collision between North America and volcanic island arc
- Iapetus passive margin forms
- Rifting and initial opening of Iapetus Ocean; Erosion of Grenville Mountains
- **Grenville Orogeny:** Ancestral Adirondack Mtns. and Hudson Highlands formed

Inferred Position of Earth's Landmasses

- TERTIARY — 59 million years ago
- CRETACEOUS — 119 million years ago
- TRIASSIC — 232 million years ago
- DEVONIAN/MISSISSIPPIAN — 362 million years ago
- ORDOVICIAN — 458 million years ago

How Can the Actual Age of a Rock Be Determined?

Early attempts at determining the age of Earth on the basis of uniformitarianism were inaccurate at best. For example, careful measurement of the accumulation of sediments in a shallow sea over objects of known age, such as shipwrecks, shows that about 15,000 to 30,000 years are needed to form a layer of sediment 1 meter thick. Layers of sedimentary rock exist that are more than 2,000 meters thick. If the sediment accumulated at a rate of 1 meter in 20,000 years, then the layers would have taken 2,000 meters × 20,000 years per meter or *40 million years to form*! If this method is applied to the entire geologic column, an estimate for the age of Earth can be obtained.

Another early method used to estimate Earth's age was based upon the salt content of the oceans. Salt is carried into the oceans by rivers, so scientists measured the amounts of salt in all the oceans. Then they measured the amounts of salt all the world's rivers are carrying into the oceans. Finally they calculated how long the rivers would take to carry in all the salt now in the oceans.

Based on such attempts, the age of the Earth was estimated at several hundred million years. We now know that this figure is much too low. Dating by those early methods was not reliable because they depended too heavily on rates that vary widely from place to place and through time. What was needed was a way to measure time by a process that does not vary, a process that runs continuously through time and leaves a record with no gaps in it. The discovery of radioactivity by A. H. Becquerel in 1896 provided the needed process.

All elements are made up of tiny bits of matter called *atoms*. Most elements have a number of isotopes. *Isotopes* are varieties of the same element whose atoms differ slightly in mass. For example, most carbon atoms have a mass of twelve units. This isotope is called carbon-12. Some carbon atoms, however, have a mass of fourteen units. This is isotope is carbon-14.

The most common isotopes of elements are stable, meaning their atoms do not change. However, some isotopes are unstable. In a process called radioactive decay, the atoms of unstable isotopes break apart. During this process the atoms go through a series of changes. They give off energy and some of the small particles that make up their nucleus. Finally, they form a stable isotope of a new element that is not radioactive. This new substance is called the decay product. For example, uranium-238, a radioactive isotope, decays slowly into the stable decay product lead-206.

Radioactive decay takes place at a steady, constant rate. It is not affected by outside factors such as changes in temperature, pressure, or chemical state. The minerals in certain types of rocks contain radioactive isotopes that start to decay when the rock forms. Therefore, the decay process may be used as a clock to determine the actual ages of these rocks.

The decay of a radioactive isotope occurs at a statistically predictable decay rate known as its half-life. The *half-life* of a radioisotope is the time required for one-half of the unstable radioisotope to change into a stable decay product. Table 13.1 lists the half-lives and decay products of four commonly used radioisotopes.

TABLE 13.1 RADIOACTIVE DECAY DATA

RADIOACTIVE ISOTOPE	DISINTEGRATION	HALF-LIFE (years)
Carbon-14	$C^{14} \rightarrow N^{14}$	5.7×10^3
Potassium-40	$K^{40} \rightarrow Ar^{40}$ $\rightarrow Ca^{40}$	1.3×10^9
Uranium-238	$U^{238} \rightarrow Pb^{206}$	4.5×10^9
Rubidium-87	$Rb^{87} \rightarrow Sr^{87}$	4.9×10^{10}

Source: The State Education Department, *Earth Science Reference Tables*, 2001 ed. (Albany, New York; The University of the State of New York).

If the half-life of a radioisotope is known, the age of a rock can be determined from the relative amounts of the radioisotope and its decay product in the rock. Every half-life, the percentage of decay product will increase and the percentage of radioisotope will decrease. Thus, the ratio of radioisotope to decay product indicates how many half-lives have gone by. This information, in turn, reveals how many years the decay process has been going on. In this manner, geologists are able to find the absolute ages of rocks. Figure 13.13 shows how the relative amounts of radioisotope and its decay product change as each half-life elapses.

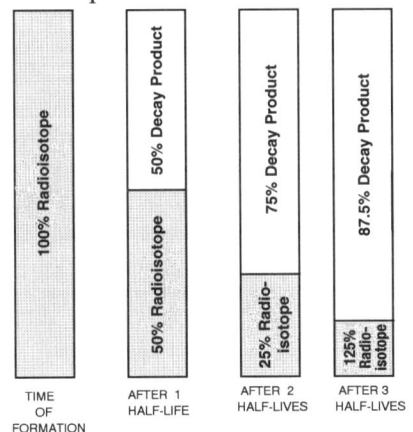

Figure 13.13 Ratios of Isotope and Decay Product after One, Two, and Three Half-Lives.

Example:

A rock contains 50 grams of potassium-40 and 50 grams of its decay product argon-40. How old is the rock?

Since the entire decay product was originally radioactive isotope, the rock originally contained 100 grams of potassium-40 (the 50 g of the radioisotope that still exist plus the 50 g that decayed into argon-40). Thus, exactly one-half of the original radioisotope has decayed; the rock is one half-life old. From Table 13.1 we know that the half-life of potassium-40 is 1.3×10^9 years, or 1.3 billion years.

The rock is 1.3 billion years old.

There are two reasons why a certain isotope may be used to date a rock. It may be used because it is present in the minerals of that rock, or it may be used because its half-life is long or short. Isotopes with long half-lives are used to date very old rocks; those with short half-lives, to date younger objects. For example, potassium-40 can be used to date rocks between 100,000 and 4.6 billion years old. On the other hand, carbon-14 is best for dating objects between 100 and 50,000 years old. Table 13.2 gives data for six radioisotopes used in dating.

The Origin and History of Life on Earth

TABLE 13.2 SIX RADIOISOTOPES USED IN DATING*

Isotopes Parent	Decay Product	Half-life (years)	Dating Range (years)	Minerals or Other Materials That Can Be Dated
Uranium-238	Lead-206	4.5 billion	10 million–4.6 billion	Zircon
Uranium-235	Lead-207	710 billion	10 million–4.6 billion	uraninite
Uranium-232	Lead-208	14 billion	10 million–4.6 billion	
Potassium-40	Argon-40 Calcium-40	1.3 billion	50,000–4.6 billion	Muscovite, biotite, hornblende
Rubidium-87	Strontium-87	47 billion	10 million–4.6 billion	Muscovite, biotite, potassium feldspar
Carbon-14	Nitrogen-14	5,730 ± 30	100–70,000	Wood, peat, grain, charcoal, bone, tissue, cloth, shell, stalacites, glacier ice, ocean water

*From *The Dynamic Earth: An Introduction to Physical Geology*, Brian J. Skinner and Stephen C. Porter, copyright © 1992 by John Wiley & Sons, Inc. Reprinted by permission of John Wiley & Sons, Inc.

Carbon-14 is especially useful because it can be used to date the remains of living things. All living things contain carbon, and some of that carbon is carbon-14, which is present in air, water, and food. As long as an organism is alive, the amount of carbon-14 in its body remains constant. Whatever decays is quickly replaced from its surroundings. However, when the organism dies, the carbon-14 that decays is not replaced. The longer the organism has been dead, the less carbon-14 remains. Carbon-14 can be used to date fossils in which the original material remains unchanged, such as wood, bones, and shells. However, this radioisotope has a short half-life. After about 50,000 years the amount of carbon-14 left in a fossil is too small to be measured. Therefore, other radioisotopes (e.g., potassium-40) must be used to date older rocks.

Example:
A wood branch is found buried beneath layers of sediment in a lake. The wood is found to contain one-fourth of the carbon-14 present in contemporary wood. How old is the layer of sediment in which the wood is found?

After one half-life, a sample would contain one-half the normal amount of carbon-14. After a second half-life, half of that would have decayed, leaving one-quarter the normal amount of carbon-14. Therefore, two half-lives have elapsed since the tree died. Table 13.1 indicates that the half-life of carbon-14 is 5.7×10^3 years, or 5,700 years.

Since two half-lives have elapsed, the wood and therefore the layer of sediment are $5,700 \times 2$ or 11,400 years old.

How Can the Rock Record and Fossil Evidence Reveal Changes in Past Life and Ancient Environments?

Many thousands of layers of sedimentary rock contain evidence of the long history of Earth and the changing life-forms whose fossils are found in the rocks. Fossils show us that a great variety of plants and animals lived on Earth in the past. Fossils can be compared to one another and to living organisms by their similarities and differences. Some fossil organisms are similar to existing organisms, but many are quite different. A study of the fossil record reveals that more recently deposited rock layers are more likely to contain fossils resembling existing life-forms and that most life-forms of the geologic past have become extinct.

Evolution

The history of life on Earth depicted by the fossil record is one of change. The fossil record shows that, since its beginning, many new life-forms have appeared and most old forms have disappeared. It reveals also that life-forms have changed gradually over time, or evolved, so that the current life-forms differ significantly from the earliest ones.

Scientists have traced many sequences (inferred from ages of rock layers) of anatomical forms over time and have observed the accumulation of differences from one generation to the next. These researchers are convinced that *biological evolution*, that is, the development of existing life-forms from earlier, different ones, is what has led to species as different from one another as algae are from whales. See Figure 13.14.

Natural Selection

The theory of natural selection provides a scientific explanation for the evolution of life-forms seen in the fossil record. Natural selection proposes the following mechanism for evolution:

1. Individual organisms of the same species have different characteristics, and sometimes these differences give one organism an advantage in surviving and reproducing.
2. Offspring that inherit the advantage are more likely to survive and reproduce.
3. Over time the proportion of individuals with the advantageous characteristic will increase.

In this way, natural selection leads to organisms that are well suited to their environments. Changes in the environment can affect the survival value of some inherited characteristics. Then the characteristics that are advantageous for survival may change, so changes in the environment can lead to changes in organisms. Small differences between parents and offspring can accumulate over many generations so that descendants may become very

The Origin and History of Life on Earth

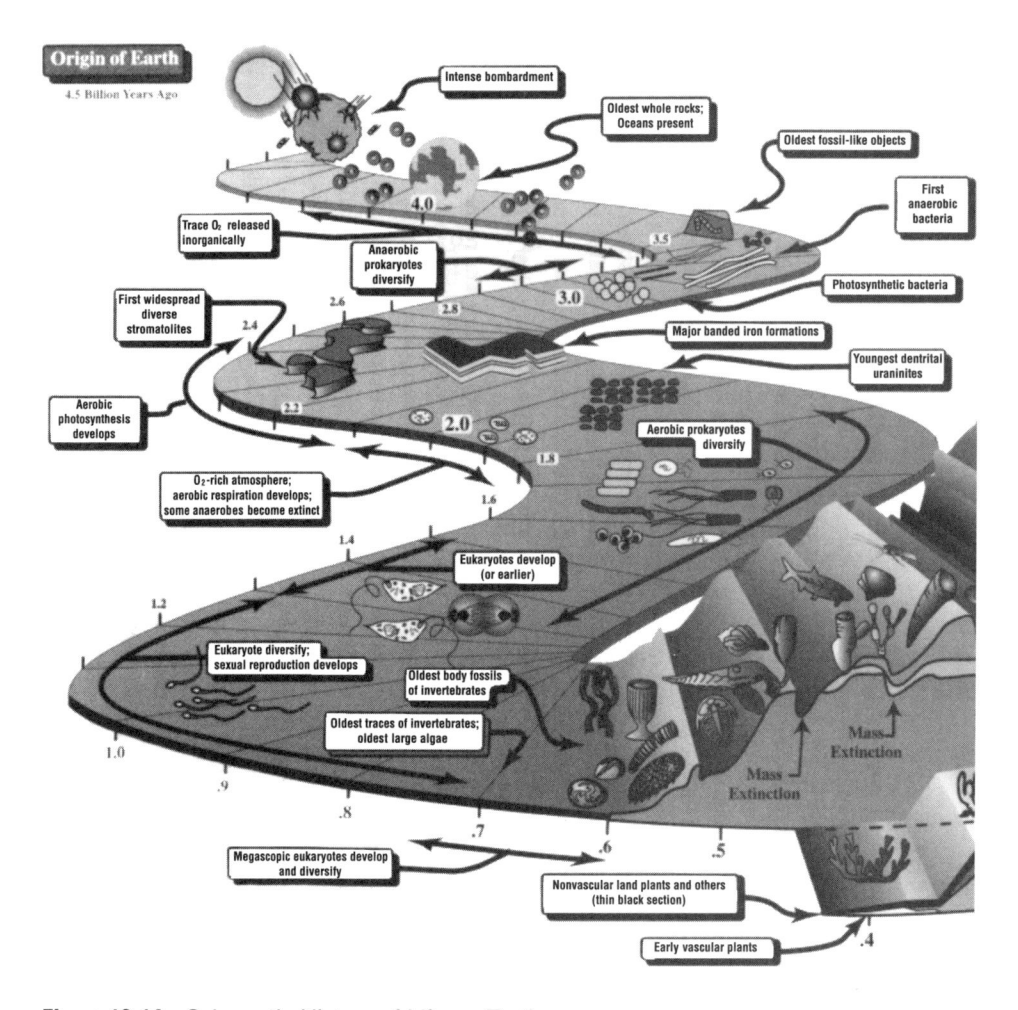

Figure 13.14 Schematic History of Life on Earth.
Source: *Earth: Evolution of a Habitable World*, Jonathan I. Lunine, Cambridge University Press, 1999.

different from their ancestors.

Evolution by natural selection does not imply long-term progress toward a goal, or have a set direction. Nor does evolution always occur gradually. Much recent information indicates that evolution occurs in spurts brought on by sudden, large-scale changes in the environment. A good analogy for evolution is the growth of a hedge. Some branches exist from the beginning of the hedge's life with little or no change. Other branches grow and then die out. Still others grow a little, and then branch apart repeatedly. The end result is a complex network of large and small branches.

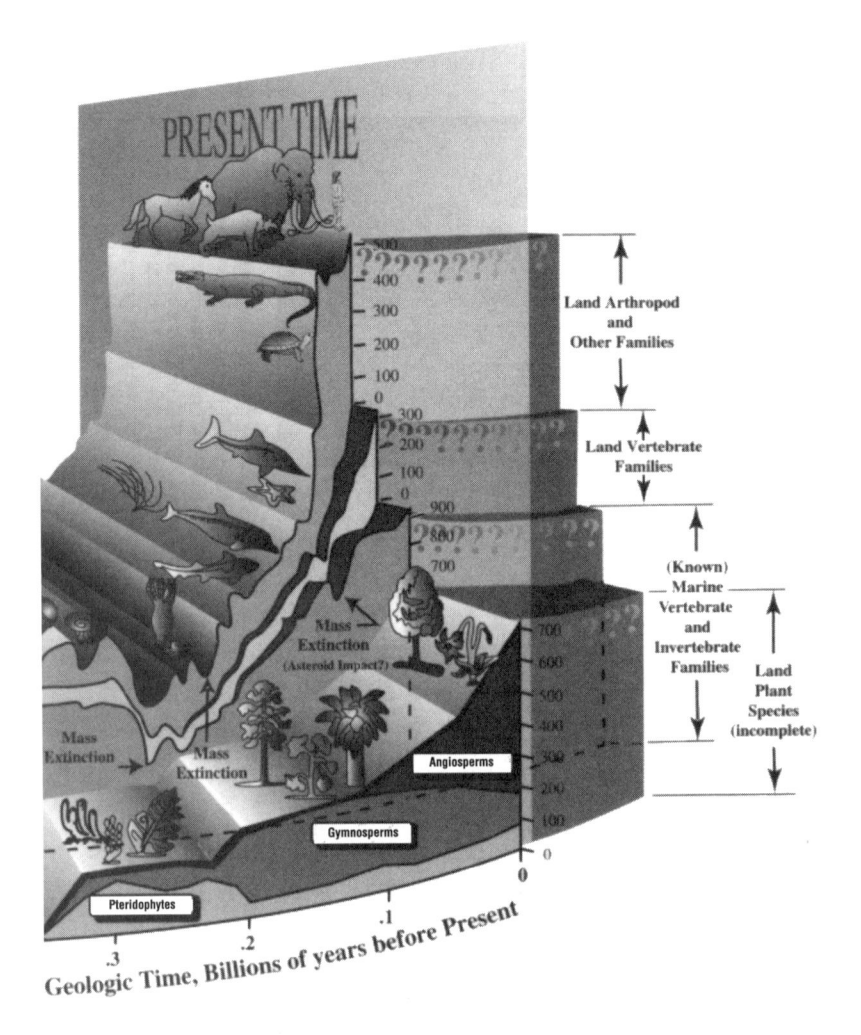

On the basis of the fossil record, life on Earth is thought to have begun about 4 billion years ago as simple, one-celled organisms. During the first 2 billion years, only these one-celled organisms existed. About 1 billion years ago, cells with nuclei began to develop. Since then, increasingly complex multicellular organisms have evolved.

The Origin and History of Life on Earth

One important reason to preserve life on Earth is that evolution builds on what exists. The more variety there is at the present time, the even greater variety can exist in the future. Human behavior that results in the extinction of organisms decreases the variety of life-forms on Earth and may have far-reaching effects in the future.

Ancient Environments

Rocks in which fossils are found provide evidence of ancient environments. Many fossils are clear indicators of particular types of environment. For example, fossil coral that is almost identical to existing corals can be found in some rocks. Corals that exist today can live only in shallow, tropical seas. Therefore, we can infer that rocks containing fossil coral existed in a similar environment. Oak and birch trees inhabit moist, temperate climates. Abundant oak and birch pollen in a sedimentary layer implies that, when the rock layer formed, the climate was moist and temperate. The fossils in coal are the remains of tropical rain forest plants and animals, so layers of coal imply a tropical rain forest origin. Inferences of this type, together with the inferences that can be drawn from the composition of a given sedimentary layer, enable geologists to piece together a model of the environment that existed when that particular rock layer formed. See Figure 13.15.

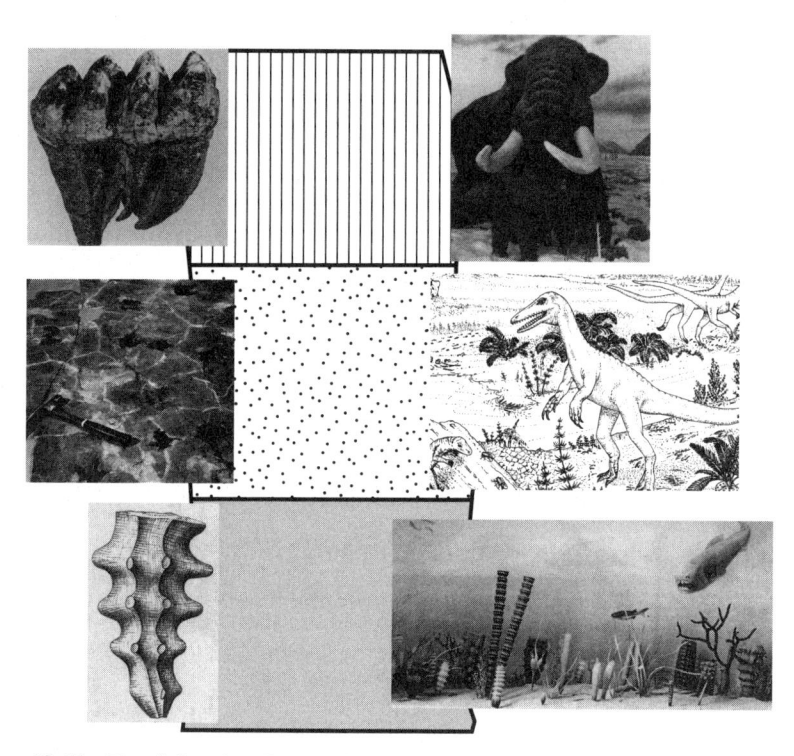

Figure 13.15 Fossil-Bearing Rock Layers with Illustrations of Environments Inferred from Fossil Evidence.

MULTIPLE-CHOICE QUESTIONS

In each case, write the number of the word or expression that best answers the question or completes the statement.

1. It is difficult to state with certainty the way in which life on Earth originated because
   (1) Earth's original life-forms left no fossil record
   (2) no rocks date back to the time when life originated
   (3) the age of rocks formed at the time when life originated cannot be determined
   (4) the process of radioactive dating destroys the fossils found in rocks

2. If life developed spontaneously on Earth, why is it not developing in the same way today?
   (1) There was less salt in the oceans at that time than now.
   (2) The Sun no longer emits the needed type of radiation.
   (3) There are no longer electrical discharges in the atmosphere.
   (4) Once developed, plants changed their environments.

3. Earth's age is most accurately determined by the
   (1) thickness of sedimentary strata
   (2) salinity of the oceans
   (3) study of fossils
   (4) radioactive dating of rock masses

4. Evidence suggests that the geologic processes of the past
   (1) were similar to those of the present
   (2) were different from those of the present
   (3) occurred at a faster rate than those of the present
   (4) occurred at a slower rate than those of the present

5. If the vertical cross section shown in the accompanying diagram represents sedimentary rock layers that have not been overturned, which letter most probably indicates the oldest rock layer?
   (1) *A*   (3) *C*
   (2) *B*   (4) *D*

6. Tilted layers of sandstone found in the Rocky Mountains indicate that
   (1) sandstone was deposited at the tops of the mountains
   (2) the oceans were once higher than they are now
   (3) sandstone layers were formed and then displaced
   (4) the oceans rose, deposited sandstone, and receded

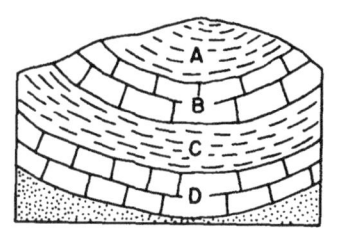

7. Older layers of rock may be found on top of younger layers of rock as a result of
   (1) weathering processes
   (2) igneous extrusions
   (3) joints in the rock layers
   (4) overturning of rock layers

8. Which features in a rock layer are older than the rock layer?
   (1) igneous intrusions
   (2) mineral veins
   (3) rock fragments
   (4) faults

9. The diagram below represents layers of rock.

   Rock layer *A* is inferred to be older than intrusion *B* because
   (1) layer *A* is composed of sedimentary rocks
   (2) parts of layer *A* were altered by intrusion *B*
   (3) layer *B* is located between layer *A* and layer *C*
   (4) parts of layer *C* were altered by intrusion *B*

10. The best method for the correlation of sedimentary rock layers several hundred kilometers apart is to compare the
    (1) index fossils in the layers
    (2) layers by walking the outcrop
    (3) thickness of the rock layers
    (4) color of the rock layers

Base your answers to question 11 on the diagram below, which shows a geologic cross section and landscape profile of a section of Earth's crust.

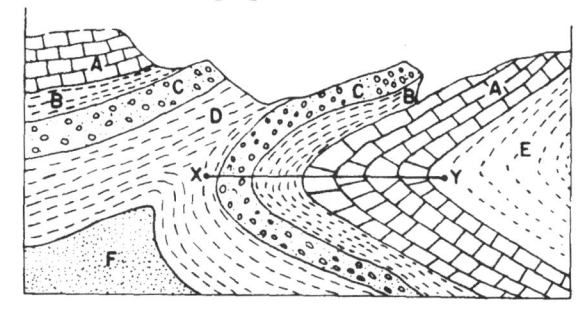

11. Which graph best represents the age of the rocks along line *XY*?

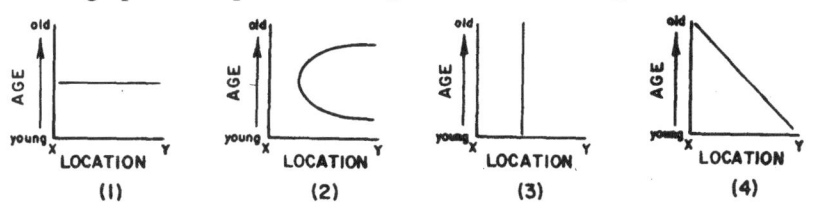

12. Erosion has created gaps in the geologic record preserved in rocks. According to the *Earth Science Reference Tables*, in New York State there are no rocks of the following ages:
    (1) Permian and Tertiary
    (2) Ordovician and Cretaceous
    (3) Ordovician and Cambrian
    (4) Triassic and Jurassic

13. Geologic columns *A*, *B*, and *C* in the diagrams below represent widely spaced outcrops of sedimentary rocks. The symbols indicate fossils found within each rock layer. Each rock layer represents the fossil record of a different geologic time period.

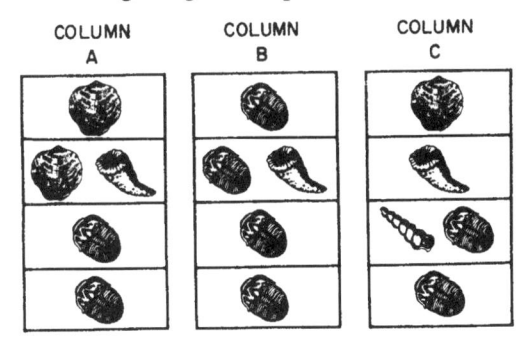

    According to the diagrams for all three columns, which would be the best index fossil?

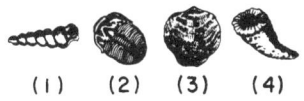

267

The Origin and History of Life on Earth

14. The diagram below represents two geologic rock columns. The color and environment of deposition of each sedimentary rock are indicated beside the rock layers. Which rock layer in the West geologic column is most likely the same as rock layer X in the East column?

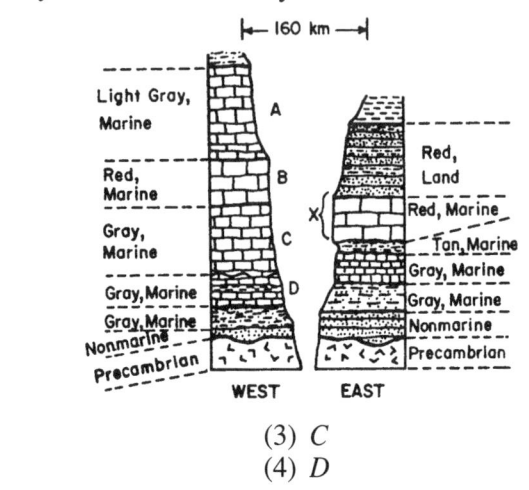

(1) A (3) C
(2) B (4) D

15. Which is the best method of determining the relative ages of a layer of sandstone in western New York State and a layer of sandstone in eastern New York State?
    (1) Compare the thickness of the two layers.
    (2) Compare the colors of the two layers.
    (3) Compare the size of sand particles of the two layers.
    (4) Compare the index fossils in the two layers.

16. Unconformities (buried erosional surfaces) are good evidence that
    (1) many life-forms have become extinct
    (2) the earliest life-forms lived in the sea
    (3) part of the geologic rock record is missing
    (4) metamorphic rocks have formed from sedimentary rocks

17. Geologic time is divided into specific periods and epochs based on
    (1) fossil evidence of Earth organisms
    (2) inferred positions of Earth's landmasses
    (3) rock types found in mountainous areas
    (4) uplift and erosion of New York State bedrock

18. A timeline is made on a strip of paper to illustrate Earth's history. A length of 1.0 centimeter is used to represent 10 million years. According to the *Earth Science Reference Tables*, what distance should be used to represent the length of the Mesozoic Era?
    (1) 0.19 cm   (3) 19 cm
    (2) 1.9 cm   (4) 190 cm

Base your answers to questions 19 and 20 on the cross-sectional diagram of a folded region shown below.

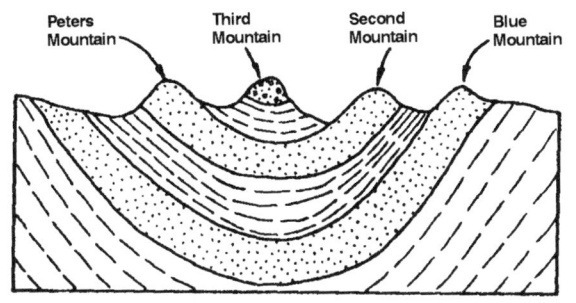

19. Where would a person normally expect to find the same type of fossil remains as are present at the top of Peters Mountain?
    (1) in the sandstone on Blue Mountain
    (2) in the sandstone on Second Mountain
    (3) in the conglomerate on Third Mountain
    (4) in the shale between Second Mountain and Blue Mountain

20. The present surface features known as Peters, Blue, Third, and Second mountains were mainly the result of
    (1) intrusion of igneous material
    (2) faulting
    (3) different erosion rates
    (4) metamorphism

21. Which statement best explains why dinosaur fossils have *not* been found in bedrock in the area around Syracuse, New York? (Refer to the *Earth Science Reference Tables*.)
    (1) Fossils are found only in sedimentary rock.
    (2) In the Syracuse area, dinosaur bones are located deep below Earth's surface.
    (3) The dinosaurs were mobile and left no remains in the Syracuse area.
    (4) No rock record exists in the Syracuse area for the time period when the dinosaurs lived.

22. According to the *Earth Science Reference Tables*, which rock is most likely the oldest?
    (1) conglomerate containing the tusk of a mastodon
    (2) shale containing trilobite fossils
    (3) sandstone containing fossils of flowering plants
    (4) siltstone containing dinosaur footprints

The Origin and History of Life on Earth

23. According to the *Earth Science Reference Tables*, a straight line connecting which two cities would cross surface bedrock of only two geologic periods?
    (1) Watertown and Plattsburgh
    (2) Jamestown and Old Forge
    (3) Kingston and New York City
    (4) Binghamton and Syracuse

24. Why are radioactive substances useful for measuring geologic time?
    (1) The disintegration of radioactive substances occurs at a predictable rate.
    (2) The ratio of decay products to undecayed products remains constant in sedimentary rocks.
    (3) The half-lives of most radioactive substances are shorter than 5 minutes.
    (4) Measurable samples of radioactive substances are easily collected from most rock specimens.

Note that question 25 has only three choices.
25. When the quantity of a radioactive material decreases, the half-life of that substance will
    (1) decrease
    (2) increase
    (3) remain the same

26. According to the *Earth Science Reference Tables*, which radioactive isotopes formed at the time of Earth's origin has just reached about one half-life?
    (1) carbon-14   (3) uranium-238
    (2) potassium-40   (4) rubidium-87

27. Why is carbon-14 not usually used to date objects more than 50,000 years old?
    (1) Carbon-14 has a relatively short half-life, and too little carbon-14 is left after 50,000 years.
    (2) Carbon-14 has a relatively long half-life, and not enough carbon-14 has decayed after 50,000 years.
    (3) Carbon-14 has been introduced as an impurity into most materials older than 50,000 years.
    (4) Carbon-14 has existed on Earth only during the last 50,000 years.

28. An ancient bone was analyzed and found to contain carbon-14 that had decayed for nearly two half-lives. According to the *Earth Science Reference Tables*, approximately how old is the bone?
    (1) 1,400 years   (3) 5,600 years
    (2) 2,800 years   (4) 11,000 years

29. A rock sample contained 8 grams of potassium-40 (K^{40}) when it was formed, but now contains only 4 grams because of radioactive decay. Based on the *Earth Science Reference Tables,* what is the approximate age of this rock?
   (1) 0.7×10^9 years
   (2) 1.4×10^9 years
   (3) 2.8×10^9 years
   (4) 5.6×10^9 years

30. A rock sample containing the radioactive isotope potassium-40 is calculated to be 4.2×10^9 years old. According to the information in the *Earth Science Reference Tables,* how much of the original potassium-40 is left in this rock sample?
   (1) 0
   (2) ⅛
   (3) ¼
   (4) ½

Base your answers to questions 31 through 35 on your knowledge of earth science, the *Earth Science Reference Tables,* and the diagram below, which shows matching geologic columns from three different locations, *A, B,* and *C.* The locations are about 5 kilometers apart, and the layers have not been overturned.

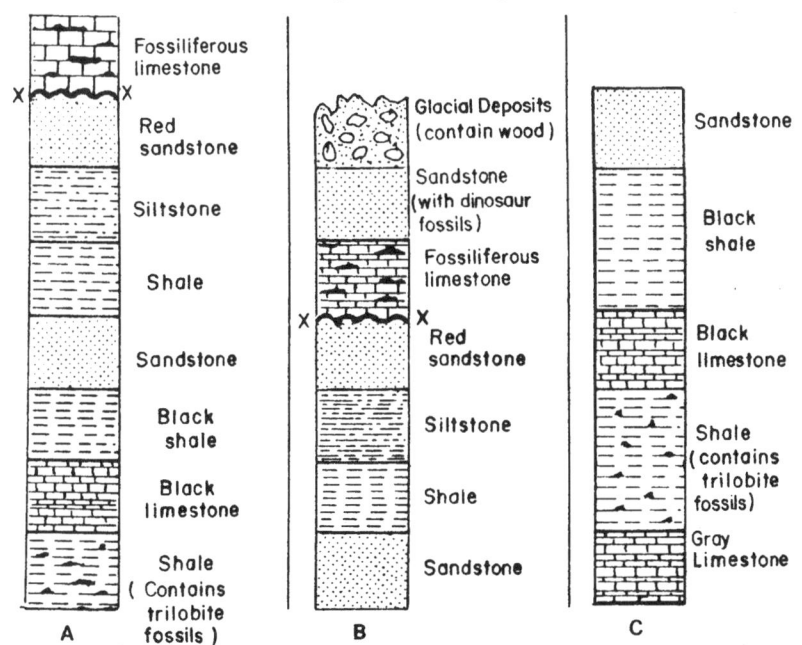

31. Which is the oldest layer shown?
   (1) gray limestone
   (2) sandstone
   (3) glacial till containing wood
   (4) shale containing trilobite fossils

271

32. The shale that contains the trilobite fossils was most likely deposited during which geologic period?
  (1) Cretaceous   (3) Triassic
  (2) Tertiary   (4) Ordovician

33. The formation of the fossiliferous limestone in column *B* was probably due to
  (1) heat and pressure that metamorphosed the limestone and fossils
  (2) deposition of a variety of glacial sediments
  (3) compaction and cementation of skeletons and shells of sea organisms
  (4) cooling and solidification of molten material containing fossils

34. The feature at *X* is a buried erosional surface. Based on this information, what inference can best be supported?
  (1) Faulting has occurred along the boundary between the red sandstone and the fossiliferous limestone.
  (2) No rock layers were ever formed between the red sandstone and the fossiliferous limestone.
  (3) An igneous intrusion has destroyed part of the fossiliferous limestone layer.
  (4) The red sandstone and the fossiliferous limestone do not provide a continuous geologic record.

35. Radioactive carbon-14 would be most useful in determining the age of the
  (1) trilobite fossils in the shale
  (2) wood in the glacial till
  (3) calcite in the black limestone
  (4) iron oxide in the red sandstone

36. For which geologic period are no fossils found in New York State?
  (1) Ordovician   (3) Devonian
  (2) Silurian   (4) Permian

37. According to the *Earth Science Reference Tables*, which area of New York State has the youngest bedrock?
  (1) the area south of the Finger Lakes
  (2) the area around Mt. Marcy
  (3) the area between Syracuse and Rochester
  (4) the area east of Albany

38. According to the *Earth Science Reference Tables*, most of the surface bedrock found in New York State was formed during which era?
  (1) Precambrian   (3) Mesozoic
  (2) Paleozoic   (4) Cenozoic

Base your answers to questions 39 through 43 on your knowledge of earth science, the *Earth Science Reference Tables*, and the diagram below. The diagram represents a cross section on Earth's crust showing several rock layers containing marine fossils. Overturning has not occurred. (Diagram is not drawn to scale.)

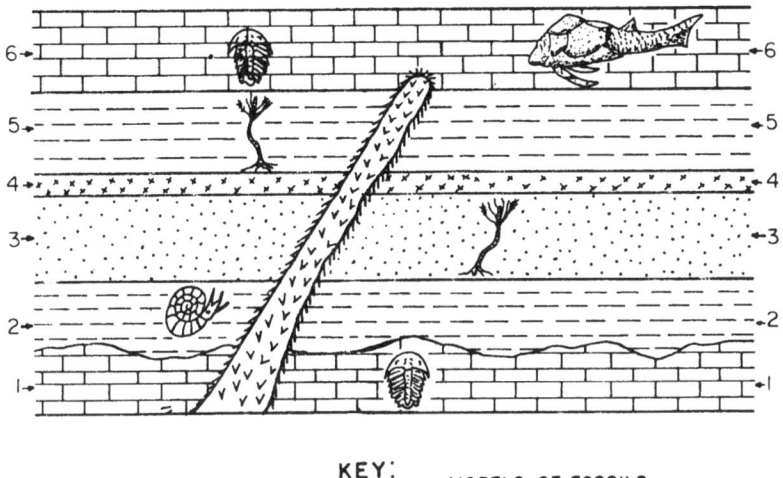

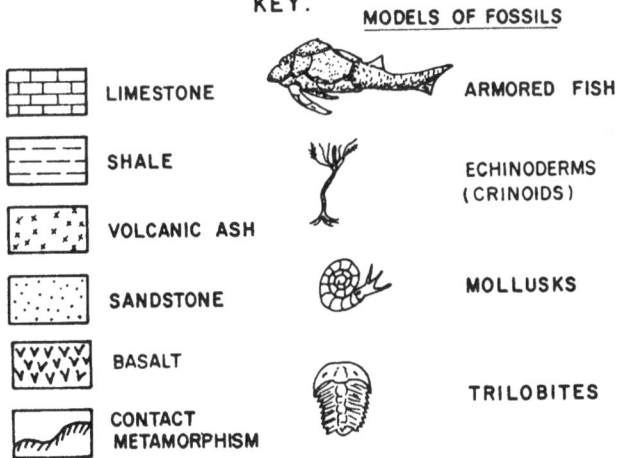

39. Why is layer 4 likely to be a good time marker?
  (1) Volcanic ash is usually a unique gray color.
  (2) Volcanic ash is usually deposited rapidly over a large area.
  (3) Volcanic ash can usually be dated with carbon-14.
  (4) Volcanic ash usually contains index fossils.

40. What could be an approximate age of rock layer 1?
  (1) 110 million years   (3) 510 million years
  (2) 210 million years   (4) 810 million years

273

41. Which best describes the order of events for the formation of this section of Earth's crust?
    (1) intrusion of basalt; deposition of rock layers 1, 2, 3, 4, 5, 6
    (2) deposition of rock layers 1, 2, 3; intrusion of basalt; deposition of rock layers 4, 5, 6
    (3) deposition of rock layers 1, 2, 3, 4, 5; intrusion of basalt; deposition of rock layer 6
    (4) deposition of rock layers 1, 2, 3, 4, 5, 6; intrusion of basalt

42. Which is the best explanation for the irregular surface between layers 1 and 2?
    (1) Layer 1 was folded after 2 was deposited.
    (2) Volcanic actions pushed layer 1 up before 2 was deposited.
    (3) Pressure from the layers above pushed layer 2 into 1.
    (4) Layer 1 was partially eroded before 2 was deposited.

43. Which rock was formed by the compaction and cementation of particles 0.07 centimeter in diameter?
    (1) limestone   (3) shale
    (2) sandstone   (4) basalt

Base your answers to questions 44 through 48 on your knowledge of earth science and on the graph below, which shows the relationship between mass and time for a radioactive isotope during radioactive decay.

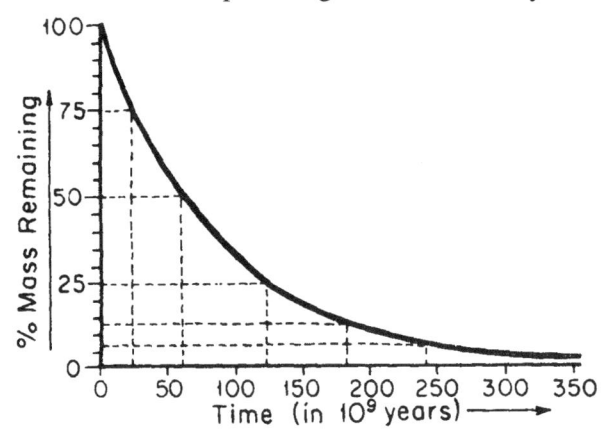

44. What is the half-life of the isotope?
    (1) 20×10^9 years   (3) 125×10^9 years
    (2) 60×10^9 years   (4) 180×10^9 years

45. When 35% of the original radioactive material is left, the sample will be approximately how many years old?
    (1) 5×10^9 years   (3) 100×10^9 years
    (2) 60×10^9 years   (4) 120×10^9 years

46. According to the graph, which statement best describes the decay of the radioactive isotope?
    (1) The actual mass of radioactive material that decays in each half-life decreases with time.
    (2) The rate of decay is greatest between 150×10^9 years and 250×10^9 years.
    (3) There is the greatest loss in mass between 250×10^9 and 300×10^9 years.
    (4) The same number of atoms of radioactive material decay during each half-life.

47. According to the trend indicated in the graph, the radioactive isotope will be completely decayed
    (1) in 60×10^9 years
    (2) in 250×10^9 years
    (3) in 350×10^9 years
    (4) at a time greater than 350×10^9 years

Note that question 48 has only three choices.

48. If the rock containing this radioactive material is buried deeper in Earth's crust and is subjected to an increase in temperature and pressure, the rate of radioactive decay will probably
    (1) decrease
    (2) increase
    (3) remain the same

Base your answers to questions 49 through 53 on your knowledge of earth science and on the block diagram below, which shows rock layers *A*, *B*, *C*, *D*, and *E* with holes I, II, III, and IV drilled through these rock layers.

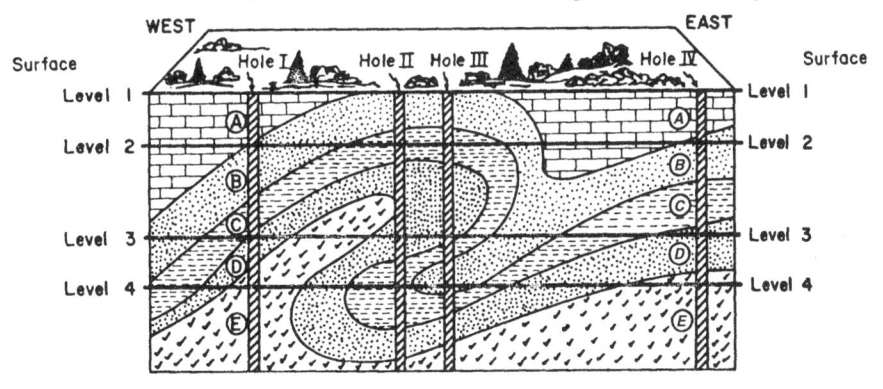

49. Which sequence of rock outcrops would be found on the surface by an observer who walked a straight line from west to east?
    (1) *A, B, A*
    (2) *A, D, A*
    (3) *B, A, B*
    (4) *D, B, A*

The Origin and History of Life on Earth

50. At which level line would all four drill holes show the same rock layer?
    (1) 1   (3) 3
    (2) 2   (4) 4

51. The best evidence for overturned rock layers could be found by examining and correlating rock samples taken from
    (1) hole I, only   (3) holes II and III
    (2) holes I and IV   (4) hole IV, only

52. If rock layer D was formed during the Devonian Period, during which period could rock layer B have been formed? (Use the *Earth Science Reference Tables*.)
    (1) Cambrian   (3) Silurian
    (2) Ordovician   (4) Devonian

53. If rock layer B is more resistant than rock layer A, then, as the surface weathers and erodes away, the probable surface elevation of hole III, as compared to the elevations of hole I and hole IV, will be
    (1) lower than that of Hole I, but higher than that of Hole IV
    (2) higher than that of Hole I, but lower than that of Hole IV
    (3) lower than the elevations of both Hole I and Hole IV
    (4) higher than the elevations of both Hole I and Hole IV

Base your answers to questions 54 through 58 on your knowledge of earth science, the *Earth Science Reference Tables*, and the diagrams below. The diagrams show cross sections of Earth's crust at four widely scattered locations, A through D. Numbers 1 through 10 represent fossils located in the rock layers. (The numbers do not represent the relative ages of the fossils.) The rock layers have not been overturned.

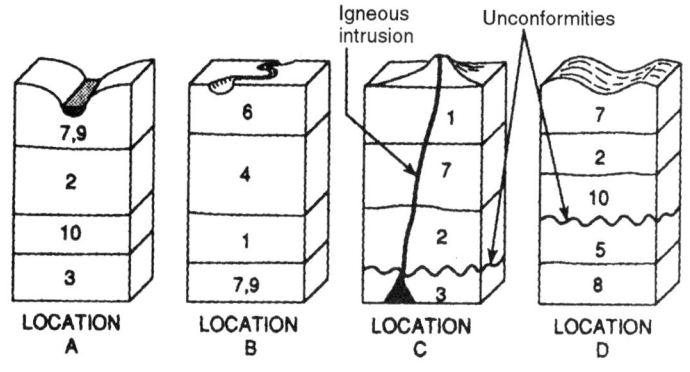

54. What is the most likely cause of the unconformities at locations C and D?
    (1) volcanic activity   (3) faulting
    (2) human activity   (4) uplift and erosion

55. Which location most likely contains the youngest fossil?
    (1) A (3) C
    (2) B (4) D

56. What is the relative age of the igneous intrusion at location *C*?
    (1) younger than the layer containing fossil 10
    (2) older than the layer containing fossil 7
    (3) the same age as the layer containing fossil 1
    (4) the same age as the layer containing fossil 9

57. Index fossils such as 7 are useful for correlating rocks because these fossils
    (1) are found only in sedimentary rocks
    (2) contain radioactive carbon-14, which is used for relative dating
    (3) represent organisms that lived for a relatively short period of geologic time in widespread areas
    (4) represent organisms that lived close to Earth's surface for a relatively long period of time

58. Fossil 8 represents the earliest fish. How many millions of years ago was the rock layer containing this fossil probably formed?
    (1) 560 (3) 300
    (2) 450 (4) 275

59. Theories of evolution suggest that variations between members of the same species give the species greater probability of
    (1) remaining unchanged
    (2) surviving environmental changes
    (3) becoming fossilized
    (4) becoming extinct

60. The changes observed in the fossil record from the Precambrian Era to the Cenozoic Era best provide evidence of
    (1) sublimation   (3) evolution
    (2) radioactive decay (4) planetary motion

61. A skull was discovered that has human characteristics and is about 2.8 million years old. On the basis of this information, during which epoch could early humans have existed? (Refer to the *Earth Science Reference Tables*.)
    (1) Pliocene   (3) Oligocene
    (2) Miocene   (4) Eocene

The Origin and History of Life on Earth

62. Which fact provides the best evidence for the scientific theory of the evolutionary development of life on Earth?
    (1) Fossils are found almost exclusively in sedimentary rocks.
    (2) Characteristics of simpler forms of life can be found in more complex forms of life.
    (3) Only a small percentage of living things have been preserved as fossils.
    (4) Most species of life on Earth have become extinct.

63. Which conclusion can be drawn based on existing fossil evidence?
    (1) Present life-forms have always existed.
    (2) Earth's environment has always been the same.
    (3) Many life-forms have become extinct.
    (4) All life-forms will remain the same in the future.

64. The climate that existed in an area during the early Paleozoic Era can best be determined by studying
    (1) the present climate of the area
    (2) recorded climate data of the area since 1700
    (3) present landscape surface features found in the area
    (4) the sedimentary rocks deposited in the area during the Cambrian and Ordovician periods

65. Trilobite fossils found in shale bedrock in the Albany, New York, area indicate that this area once
    (1) was covered by an ocean
    (2) was covered by a large forest
    (3) had iron-ore deposits
    (4) had many land animals

66. According to the *Earth Science Reference Tables*, at which location is the surface bedrock likely to contain fossils of early fishes, corals, and mollusks?
    (1) Elmira
    (2) Long Island
    (3) Watertown
    (4) Old Forge

67. In Earth's geologic past there were long periods that were much warmer than the present climate. What is the primary evidence that these long warm periods existed?
    (1) United States National Weather Service records
    (2) polar magnetic directions preserved in the rock record
    (3) radioactive decay rates
    (4) plant and animal fossils

68. Earth scientists studied fossils of a certain type of plant. They noted slight differences in the plant fossils throughout geologic time. What inference is best made from this evidence?
   (1) When the environment changed, this type of plant also changed, allowing it to survive.
   (2) When uplifting occurred, the fossils of this type of plant were deformed.
   (3) The processes that form fossils today differ from those of the past.
   (4) The fossils have changed as a result of weathering and erosion.

69. Which best explains why fossil remains of sharks have been found in the sedimentary rocks of Wyoming?
   (1) Sharks used to live in freshwater basins.
   (2) Sharks were once land animals.
   (3) The remains were carried to Wyoming by streams from another area.
   (4) The area was once covered by the sea.

CONSTRUCTED RESPONSE QUESTIONS

Base your answers to questions 86 and 87 on the table below, which shows the results of a student's demonstration modeling radioactive decay. To begin, the student put 50 pennies heads up in a container. Each penny represented one radioactive atom. Then the student placed a top on the box and shook the box. Each penny that had flipped over to the tails-up side was replaced with a bean, which represented the stable decay product. The student continued the process until all of the pennies had been replaced by beans.

Shake Number	Number of Radioactive Atoms (pennies)	Number of Stable Decay Atoms (beans)
0	50	0
1	25	25
2	14	36
3	7	43
4	5	45
5	2	48
6	1	49
7	0	50

The Origin and History of Life on Earth

70. On the grid provided, graph the data shown on the table by following the steps below.
  (a) Mark with a dot each number of radioactive atoms (pennies) after each shake. Surround each dot with a small circle (O). The zero shake has been plotted for you. [1]
  (b) Connect all the dots with a solid line. [1]
  (c) Mark with an X the number of stable decay atoms (beans) after each shake. The zero shake has been plotted for you. [1]
  (d) Connect all the Xs with a dashed line. [1]

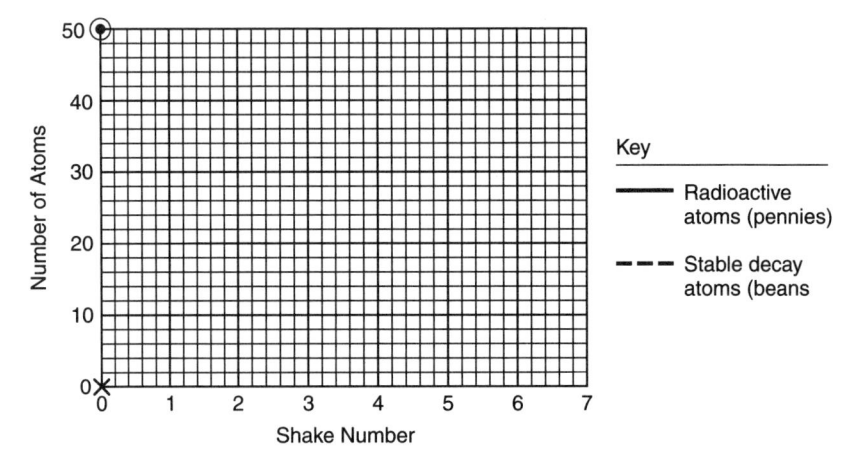

71. Assume that each shake number represents an additional 100 years. State the half-life of the radioactive material in this model. [1]

Base your answers to questions 72 through 75 on the diagrams below. Columns *A* and *B* represent two widely separated outcrops of rocks. The symbols show the rock types and the locations of fossils found in the rock layers. The rock layers have not been overturned.

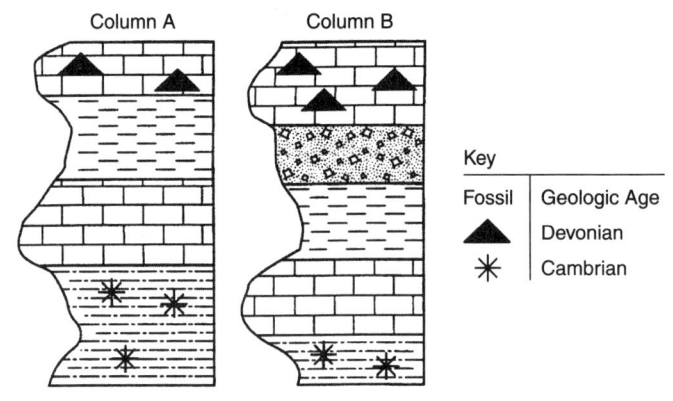

72. State one method used to correlate rock layers found in the outcrop represented by column *A* with rock layers found in the outcrop represented by column *B*. [1]

73. An unconformity (buried erosional surface) exists between two layers in the outcrop represented by column *A*. Identify the location of the unconformity by drawing a thick, wavy line ⌒⌒⌒) at the correct position on column *A*. [1]

74. In one or more sentences, state the evidence that limestone is the most resistant layer in these outcrops. [1]

75. State the oldest possible age, in millions of years, for the fossils in the siltstone layer. [1]

Base your answers to questions 76 and 77 on the diagram to the right and your knowledge of earth science. The diagram represents a profile view of a rock outcrop. The layers are labeled *A* through *H*.

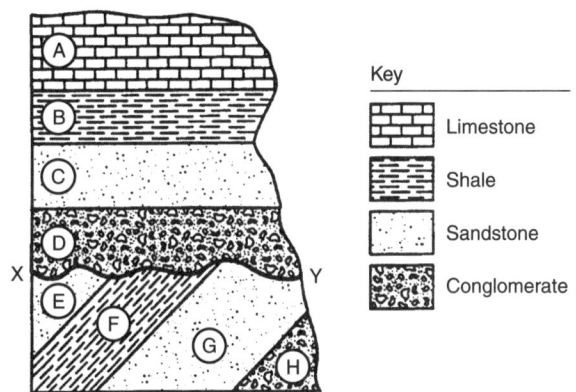

76. Using one or more complete sentences, briefly describe the geologic process that resulted in the boundary represented by line *X–Y*. [2]

77. None of the layers has been overturned. Layer *D* is 505 million years old, and layer *B* is 438 million years old. State the geologic period during which layer *C* could have been formed. [1]

Base your answers to questions 78 through 80 on the geologic cross section below, which shows an outcrop in which the layers have not been overturned. Rock units are labeled *A* through *E*.

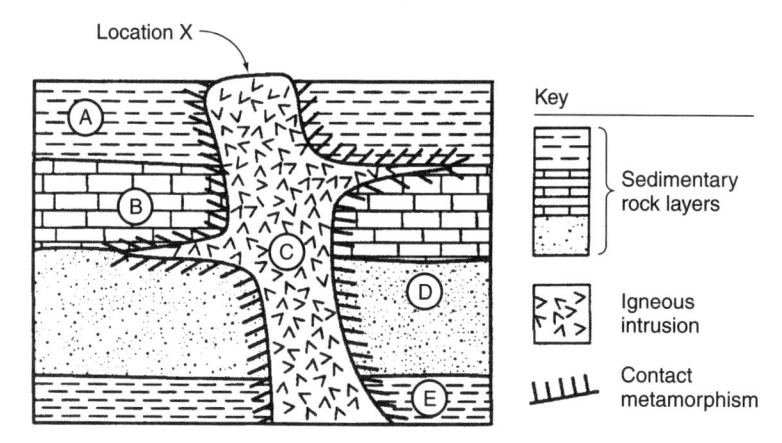

78. Using letters *A* through *E*, list the rock layers in order from oldest to youngest. [3]

79. State the name of the sediment that was compacted to form rock unit *A*. [1]

80. State one observation about the crystals at location *X* that would provide evidence that igneous rock unit *C* was formed by the very slow cooling of magma. [1]

EXTENDED CONSTRUCTED RESPONSE QUESTIONS

Base your answers to questions 81 and 82 on the newspaper article below.

New Fossils Indicate Arctic Climate Used to Be Floridian

The frigid Arctic regions were as warm as present-day Florida some 90 million years ago, according to researchers who found fossils of a crocodile-like animal in northern Canada.

Six hundred miles from the North Pole, researchers from the University of Rochester found the fossilized remains of the champosaur, a toothy, 8-foot-long extinct crocodile.

"We found a whole collection of fossils, from both young and adults," said scientist John H. Tarduno.

"The champosaur is a cold-blooded animal that could not have survived in the current climate of the Canadian Arctic where the fossils were found," Tarduno said.

Temperatures at the fossil site now routinely drop to minus 60 degrees Fahrenheit in the winter. When the champosaur lived there 86 million to 92 million years ago, winter temperatures rarely dropped to freezing and summer readings of 80 degrees were common.

The cold-blooded champosaur depended on the environment for warmth and probably became immobile if the temperature was too cold. Most likely, the champosaur was too small to have migrated seasonally.

A field team from the University of Rochester found the fossils in a layer of sandstone located above a layer of basaltic lava.

81. State the geologic time period in which the champosaur lived. [1]

82. Explain why no champosaur fossils were found within the layer of basaltic lava. [1]

Unit Four: EARTH MATERIALS

CHAPTER 14
MINERALS

KEY IDEAS The solid part of Earth is made up of rocks. Rocks, in turn, are composed of minerals, and most minerals are made of chemical elements. Earth consists of a great variety of minerals, whose properties depend on the history of how they were formed as well as the elements of which they are composed. Most properties of minerals can be explained in terms of the arrangement and properties of the atoms that compose them.

Minerals are made and remade on Earth's surface, in its oceans, and in the hot and high-pressure layers inside Earth. Observing and classifying minerals has helped us understand the great variety and complexity of Earth materials. Through the study of minerals, we have gained insight into Earth's historical development and its dynamics.

Minerals are important to us because many are sources of essential industrial materials, such as iron, copper, aluminum, and magnesium. The abundance of minerals ranges from almost unlimited to extremely rare. Many of the best sources, however, are depleted, making it more difficult and expensive to obtain those minerals. Also, the difficulty of extracting minerals from Earth's crust has important economic and environmental impacts. As limited resources, minerals must be used wisely.

KEY OBJECTIVES
Upon completion of this chapter, you will be able to:

- Explain how the physical properties of minerals are determined by their chemical compositions and crystal structures.
- Describe how minerals can be identified by well-defined physical and chemical properties, such as cleavage, fracture, color, specific gravity, hardness, streak, luster, crystal shape, and reaction with acid.
- Explain that chemical composition and physical properties determine how humans use minerals.
- Describe how minerals are formed inorganically by the process of crystallization as a result of specific environmental conditions. These include:

- Cooling and solidification of magma.
- Precipitation from water caused by such processes as evaporation, chemical reactions, and temperature changes.
- Rearrangement of atoms in existing minerals subjected to conditions of high temperature and pressure.

WHAT IS A MINERAL?

If you walk outside and pick up any earth material—a rock, sand, soil, gravel, or mud—you will hold minerals in your hand. Nearly all rocks are composed of one or more substances called *minerals*. When rocks are broken down into smaller pieces such as pebbles or sand, those smaller pieces are composed of the same minerals that were in the rock.

Nearly every single thing we use is made of, or contains, minerals. Minerals are essential to the life processes of plants and animals. Industry is equally dependent upon an abundant supply of minerals as raw materials. Without minerals, none of the devices and structures that surround us in our daily lives could be made.

Some minerals are rare and highly prized for their characteristics, such as gold for its conductivity, resistance to corrosion, and high luster, and diamonds for their beauty and hardness. Some minerals are raw materials from which industries manufacture the products that are the basis of a nation's wealth. Hematite, an iron ore, is needed to make the steel from which products ranging from cars to skyscrapers are manufactured. As such, minerals are of strategic importance and wars have been fought to gain control over mineral resources. Some minerals, however, are so abundant that they have little commercial value.

The word *mineral* means different things to different people. To a nutritionist, minerals are things to be eaten, along with proteins, carbohydrates, and vitamins. To a jeweler, a mineral is a stone to be cut or polished. To a geologist, a ***mineral*** is a naturally occurring, inorganic compound with a fixed chemical composition and an orderly internal arrangement of atoms.

Defining Characteristics of Minerals

Although there are many different minerals, they all share certain characteristics.

Minerals are naturally occurring. A naturally occurring material is formed as result of natural processes in or on Earth. It is not manufactured in a factory or synthesized in a laboratory. For example, most diamonds are formed naturally in Earth and are minerals. Synthetic diamonds made in the laboratory are not minerals.

Minerals are inorganic matter. Inorganic substances are not alive, never were alive, and do not come from living things. Thus, amber (a tree resin in which insects are often found embedded) and the fossil fuels coal, petroleum, and natural gas are not true minerals. They were formed from organic substances, animal or vegetable material, that once lived on Earth.

Minerals

A chemical symbol or formula can be written for a mineral. Minerals are either elements or compounds. Both elements and compounds have definite chemical and physical properties.

Elements are substances that cannot be broken down into simpler substances by ordinary chemical means. Ninety-two different elements have been found to occur naturally on Earth, each with distinctly different physical and chemical properties. Elements consist of particles called atoms, and all atoms of an element have the same properties. One- or two-letter symbols are used to represent the atoms of elements. For example, the symbol for the element oxygen is O and the symbol for the element silicon is Si.

Table 14.1 shows the relative abundances of different elements in Earth's crust, hydrosphere, and troposphere. Notice that only eight elements make up more than 98 percent of Earth's crust, the two most common elements being oxygen and silicon.

TABLE 14.1 AVERAGE CHEMICAL COMPOSITION OF EARTH'S CRUST, HYDROSPHERE, AND TROPOSPHERE

Element (symbol)	Crust Percent by Mass	Crust Percent by Volume	Hydrosphere Percent by Volume	Troposphere Percent by Volume
Oxygen (O)	46.40	94.04	33.0	21.0
Silicon (Si)	28.15	0.88		
Aluminum (Al)	8.23	0.48		
Iron (Fe)	5.63	0.49		
Calcium (Ca)	4.15	1.18		
Sodium (Na)	2.36	1.11		
Magnesium (Mg)	2.33	0.33		
Potassium (K)	2.09	1.42		
Nitrogen (N)				78.0
Hydrogen (H)			66.0	
Other	0.66	0.07	1.0	1.0

Source: The State Education Department, *Earth Science Reference Tables*, 2001 ed. (Albany, New York; The University of the State of New York)

Although some minerals found in Earth's crust are pure elements, such as native copper and silver, the elements in Earth's crust rarely exist by themselves. Most are chemically combined with other elements as compounds. *Compounds* consist of molecules, that is, groups of atoms joined together in a definite proportion. For example, the mineral calcite is a compound of the elements calcium, carbon, and oxygen (see

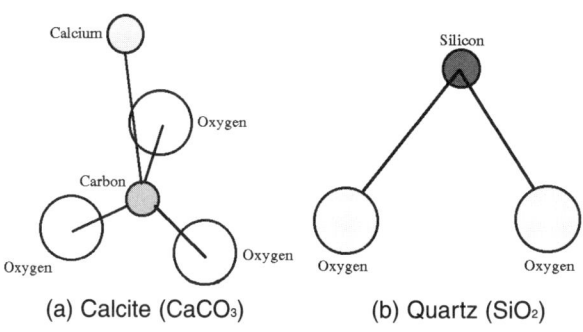

Figure 14.1 Molecular Models of Calcite and Quartz.
(a) Calcite ($CaCO_3$)
(b) Quartz (SiO_2)

Figure 14.1a). Every calcite molecule contains one atom of calcium, one atom of carbon, and three atoms of oxygen. To show the chemical composition of a mineral, a formula can be written. The chemical formula for calcite is $CaCO_3$. The mineral quartz consists of molecules each containing one atom of silicon and two atoms of oxygen (Figure 14.1b). The chemical formula for quartz is SiO_2.

Every compound has distinct properties of its own. Thus, a mineral, which is either an element or a compound, has both a definite composition and distinct physical and chemical properties. Table 14.2 shows the chemical names and formulas of a few common minerals. The last column of the Properties of Common Minerals chart in the *Earth Science Reference Tables* has a more extensive list of the chemical compositions of common minerals.

TABLE 14.2 CHEMICAL NAMES AND FORMULAS OF SOME COMMON MINERALS

Mineral	Chemical Name	Chemical Formula
Calcite	Calcium carbonate	$CaCO_3$
Galena	Lead sulfide	PbS
Gypsum	Calcium sulfate-water	$CaSO_2 \bullet 2H_2O$
Olivine (fosterite)	Magnesium silicate	Mg_2SiO_4
Potassium Feldspar	Potassium aluminum silicate	$KAlSi_3O_8$
Pyrite	Iron sulfide	FeS_2
Quartz	Silicon dioxide	SiO_2

Minerals have a crystalline form. The atoms or molecules of a mineral are the same throughout that mineral. When they are joined in fixed positions as a solid, a definite pattern is formed. A solid having a definite internal structural pattern is said to have a crystalline form. If the pattern is large enough to be seen with the unaided eye, the solid is called a *crystal*. The crystal form of a mineral determines its cleavage, or the way it splits or breaks, as well as many other properties. Mica, for example, splits into thin, flat sheets (see Figure 14.2).

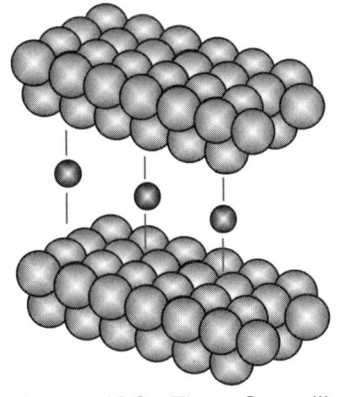

Figure 14.2 The Crystalline Pattern of Mica. Mica consists of sheets of tightly bonded silicon and oxygen atoms held together weakly by metallic ions.

Identifying Minerals

Mineralogists have examined Earth materials and identified more than 2,000 minerals. Minerals can be identified on the basis of well-defined physical and chemical properties. Several important properties commonly used in mineral identification are color, luster, streak, hardness, and cleavage, parting, and fracture. Other properties used to identify minerals include specific gravity, radioactivity, luminescence, and chemical, thermal, electrical, and

Minerals

magnetic properties, as well as elasticity and strength.

Since no single property can be used to identify all minerals, mineral identification is usually a process of elimination. As each property is observed, it becomes evident what a mineral is not, rather than what it is. Step by step, the possibilities are narrowed down until the identity of the mineral is determined.

Color

Color is often the first property noticed about a mineral. When observing color, it is important to use a fresh surface of a mineral since exposed surfaces are often discolored by weathering. Color alone is an unreliable property by which to identify a mineral. Although some minerals are always the same color (e.g., sulfur is yellow), others may occur in a variety of colors. Thus quartz is found in many colors, some of which are rare (these specimens are highly prized as semiprecious gems). For example, amethyst is purple quartz, citrine is yellow quartz, and rose quartz is pink quartz. Therefore, quartz cannot be identified by color alone. Another reason why color is unreliable is that many minerals have almost the same color. For example, calcite, quartz, and halite all occur in white and transparent varieties and can look strikingly alike.

Luster

The *luster* of a mineral is the way light reflects from its surface. Luster can be metallic or nonmetallic. Nonmetallic lusters can be described as glassy, brilliant, greasy or oily, waxy, silky, pearly, or earthy.

Related to the luster of a mineral are its transparency and iridescence, or the play of colors in its interior or exterior.

Streak

Streak is the color of the fine powder left when a mineral is rubbed against a hard, rough surface. A piece of unglazed porcelain, called a *streak plate*, is usually used to create a streak. Although the color of a mineral may vary, the streak of a particular mineral is always the same color. For example, fluorite may range in color from green to blue; yet its streak is always white. This characteristic makes streak a useful property for identifying a mineral. It must be remembered, though, that a mineral's streak is not always the same color as the mineral because powders reflect light in a different way than large crystals. Pyrite crystals have a brassy, yellow color, but their streak is greenish, or brownish black! Most minerals have colorless or white streaks; therefore, a colored streak is very useful in identifying a mineral.

Hardness

Hardness is a mineral's resistance to being scratched. Talc is so soft that it can be scratched by a fingernail. At the other extreme, diamond is so hard that no other mineral can scratch it. The hardness of a mineral is usually stated in terms of Moh's scale (see Table 14.3). On this scale, ten typical minerals are arranged in order from the softest to the hardest.

TABLE 14.3 MOH'S SCALE

Mineral	Hardness	
Talc	1	SOFTEST
Gypsum	2	
Calcite	3	
Fluorite	4	
Apatite	5	
Orthoclase	6	
Quartz	7	
Topaz	8	
Corundum	9	
Diamond	10	HARDEST

To find the hardness of a mineral, you determine what minerals your sample can scratch and what minerals it cannot scratch. For example, tourmaline will scratch quartz or anything softer, but cannot scratch topaz or anything harder. The hardness of tourmaline, therefore, is between 7 and 8 (sometimes expressed as 7.5).

Cleavage, Parting, and Fracture

Some minerals break in ways that help to identify them. *Cleavage* (see Figure 14.3) is the tendency of a mineral to break parallel to atomic planes in its crystalline structure. *Parting* is the tendency to break along surfaces that follow a structural weakness caused by factors such as pressure, or along zones of different crystal types. These cleavage or parting surfaces often occur at very specific angles to one another and can be helpful in identifying a mineral.

Mineral	Cleavage	Appearance
Muscovite mica	Breaks in parallel sheets	
Pyroxene	Breaks along planes at 88° angle to one another in a prismatic shape	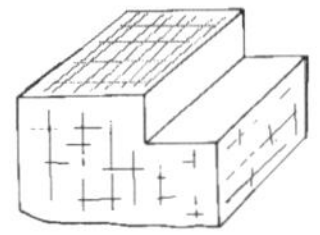
Halite	Breaks along planes at 90° to one another in a cubic shape	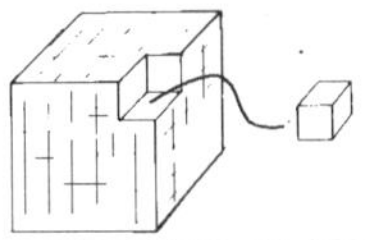
Calcite	Breaks along planes at a 75° angle to one another in a rhombohedral shape	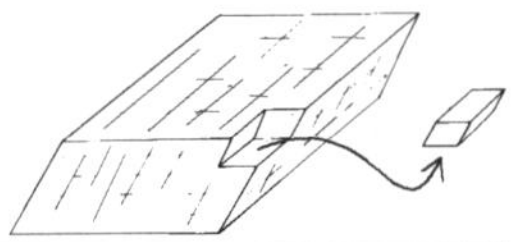

Figure 14.3 Some Common Minerals That Display Cleavage

Minerals

Some minerals have no planes of weakness, or are equally strong in all directions. The *fracture*; that is, they do not follow a particular direction when they break. Although fracture does not occur along smooth surfaces, it can display distinctive patterns that are very helpful in mineral identification (see Figure 14.4).

Patterns of Fracture	Description	Appearance
Conchoidal	Smooth, curved break that looks somewhat like the inside of a shell (e.g., quartz, olivine, gypsum)	
Fibrous or splintery	Fibers, as in asbestos Long, thin splinters (e.g., chrysotile, actinolite, tremolite, satin-spar calcite)	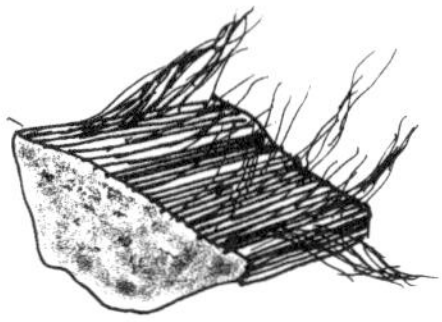
Hackly	Jagged or sharp-edged surfaces (e.g, native metals such as copper, silver, iron, nickel)	
Uneven	Rough, irregular surfaces (e.g., sulfur, anhydrite, barite)	

Figure 14.4 Some Common Minerals That Display Fracture

Specific Gravity

Every mineral has a certain density. If you have samples that are about the same size, you can compare the densities of two different minerals. When you hold one in each hand, the denser mineral will feel heavier than the less dense one. A more exact way to determine the relative densities of different minerals is to compare them all to one standard. *Specific gravity* is the ratio of the density of a mineral to the density of water. It tells us how much heavier than water a mineral is. For example, the density of galena is 7.5 grams per cubic centimeter; the density of water is 1 gram per cubic centimeter. Galena is 7.5 times denser than water and therefore has a specific gravity of 7.5.

Most rocks have a specific gravity of 2.5 to 3.5. Most people have a feeling for how heavy a chunk of mineral of a given size should be. Sometimes, though, when you pick up a sample of a mineral and toss it in your hand, it feels heavier than you expected. This feeling of unexpected weight is termed *heft*. You can usually judge by a mineral's heft that its specific gravity is greater than 4.0.

Chemical Tests

Two chemical tests are commonly used to identify minerals in the field. The acid test uses hydrochloric acid. Some minerals bubble when a drop of hydrochloric acid dropped on the mineral reacts with it. Calcite bubbles vigorously; dolomite, slowly.

The second chemical test is the taste test. Halite tastes the same as table salt, which it is, though not purified. **The taste test should be used with caution and only at your teacher's direction.**

Special Properties

Special properties that are displayed by some minerals and can be used to distinguish them from other minerals are summarized in Table 14.4.

TABLE 14.4 SOME SPECIAL PROPERTIES OF MINERALS

Property	Definition or Description
Magnetism	Mechanical force of attraction associated with moving electricity (magnetite)
Luminescence	Emission of light by a substance that has received energy or electromagnetic radiation of a different wavelength from an external stimulus
includes:	
phosphorescence	Light emitted after stimulus is removed (scheelite autunite)
fluorescence	Light emitted while stimulus is applied (fluorite, willemite)
triboluminescence	Light emitted when pressure is applied (fluorite, lepidolite)
thermoluminescence	Light emitted when heated (chlorophane fluorite)
Piezoelectricity	Electricity emitted when pressure is applied (quartz)
Pyroelectricity	Electricity emitted when heat is applied (tourmaline)
Flame color	Color of the flame when a mineral is intensely heated in a flame
Double refraction	Double image produced when an object is viewed through the mineral (Iceland spar calcite)
Play of colors	Series of colors produced when the angle of the light shining on the mineral is changed (labradorite, opal)
Chatoyancy and asterism	Appearance of being made of tiny parallel fibers (satin spar gypsum, cat's-eye chrysoberyl)
	Light scattered in a pattern that looks like a three- or six-rayed star (star rubies and sapphires)

Minerals

Using the Properties of Common Minerals Chart to Identify Minerals

As mentioned earlier, mineral identification is a process of elimination. As each property of a mineral is observed, possibilities are narrowed down until the identity of the mineral is determined. The Properties of Common Minerals chart in the *Earth Science Reference Tables* is arranged to assist you in this process. From left to right the columns list properties in the order of their usefulness in narrowing down the possible identity of a mineral.

Example:
Which mineral contains iron, has a metallic luster, is hard, and has the same color and streak?
(1) biotite mica   (2) galena   (3) potassium feldspar   (4) magnetite

First, rearrange the properties given in the question in the order in which they appear in the Properties of Common Minerals chart in the *Earth Science Reference Tables*: metallic luster, hard, same color and streak, contains iron. Now, referring to the chart, follow these steps:

- Since the mineral has a metallic luster, all of the minerals with nonmetallic luster are eliminated. This leaves graphite, galena, magnetite, pyrite, and the earthy-red form of hematite (earthy is a nonmetallic luster).
- Since the mineral is hard, graphite and galena are eliminated. This leaves magnetite, pyrite, and the metallic silver form of hematite.
- Since the mineral has the same color and streak, pyrite (brassy yellow color, green-black streak) and hematite (metallic silver color, red-brown streak) are eliminated. This leaves magnetite.
- Since the mineral contains iron, check to see that magnetite contains iron. The last column lists the composition of magnetite as Fe_3O_4; and since the chemical symbol for iron is Fe, magnetite does, indeed, contain iron.

Thus, the mineral in the question is magnetite, choice 4.

Minerals as the Building Blocks of Rocks

Of the thousands of known minerals, about 100 are so common that they make up more than 95 percent of Earth's crust. Not surprisingly, these minerals are generally compounds of the most abundant elements in the crust. Nearly every rock contains one or more of these so-called *rock-forming minerals* (see Table 14.5). Of the rock-forming minerals, fewer than 20 comprise most of the rocks you are likely to find.

TABLE 14.5 SOME MAJOR ROCK-FORMING MINERALS

Mineral or Mineral Group	Composed of These Elements
Olivines	Iron, magnesium, silicon, oxygen
Pyroxenes	Calcium, iron, magnesium, silicon, oxygen
Amphiboles	Sodium, calcium, aluminum, silicon, oxygen
Micas	Potassium, aluminum, iron, magnesium, silicon, oxygen, hydrogen
Chlorite	Magnesium, iron, manganese, aluminum, silicon, oxygen, hydrogen
Kaolinite	Aluminum, silicon, oxygen, hydrogen
Feldspars	Potassium, sodium, calcium, aluminum, silicon, oxygen
Quartz	Silicon, oxygen
Nepheline	Sodium, potassium, aluminum, silicon, oxygen
Hematite	Iron, oxygen
Rutile	Titanium, oxygen
Magnetite	Iron, oxygen
Spinels	Magnesium, aluminum, oxygen
Pyrites	Iron, sulfur
Gypsum	Calcium, sulfur, oxygen
Calcite	Calcium, carbon, oxygen
Dolomite	Calcium, magnesium, carbon, oxygen
Apatite	Calcium, phosphorus, oxygen, hydrogen, fluorine, chlorine
Fluorite	Calcium, fluorine

HOW DO MINERALS DIFFER FROM EACH OTHER?

Minerals are grouped according to their chemical compositions. Major mineral groups include the silicates, sulfides, oxides, carbonates, and sulfates. Minor mineral groups include the nitrates, borates, chromates, phosphates, halides, hydroxides, and native elements.

Major Mineral Groups

Silicates

If we look at the chemical composition of Earth's crust (refer to Table 14.1), we see that oxygen is the most abundant element, by both mass and volume, and silicon is second, by mass. It is not surprising, then, that minerals composed mainly of silicon and oxygen are abundant and widespread. In all silicates, one ion (charged atom) of silicon is always joined to four ions of oxygen in the shape of a tiny tetrahedron (plural, tetrahedra), as shown in Figure 14.5. The bonds in this structure, which we will call the silicon terahedron, are very strong.

Silicon tetrahedra can join with positive ions of common elements in a wide variety of structures. Consider the next most abundant elements in Earth's crust after silicon and oxygen; aluminum, iron, calcium, sodium,

Minerals

magnesium, and potassium are all metals that form positive ions. When the positive ions of these elements join with negative silicon tetrahedra, a tremendous number of different silicates can be formed. Rings, chains, sheets, and frameworks are some of the different structures that silicon tetrahedra can form by joining with positive ions and sharing oxygen atoms.

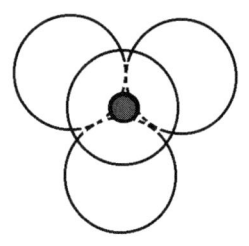

| An overhead view, as though looking through the large oxygen ions at the smaller silicon ion between them. | The natural position of the silicon tetrahedon. Notice that the silicon ion is nestled among the oxygen ions and is hidden from view. | An exploded view, showing the positions of the silicon ion and oxygen ions. |

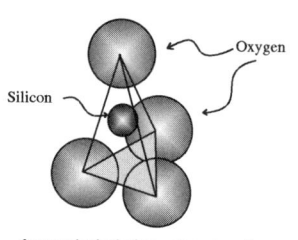

Figure 14.5 Three Views of a Silicon Tetrahedron. The silicon ion has a +4 charge (Si^{4+}), and each oxygen ion has a –2 charge (O^{2-}), so the net charge of the silicon tetrahedron is –4 $(SiO4)^{4-}$.

In the complex structures shown in Figure 14.6, many other elements and groups of elements, such as extra oxygen, water, aluminum and other metal ions, and hydroxide ions, promote electrical and mechanical balance so that these structures are stable.

The physical properties of minerals depend on the arrangement and bonding of their atoms, as is very clearly seen in the structures formed by silicon tetrahedra. Look at a typical sheet silicate, mica, in Figure 14.6. The tetrahedra all share oxygens and are tightly bound together. The metal ions between the sheets, however, are isolated from each other and are less tightly bound. They form a plane of weakness in the structure—a weakness that manifests itself in the tendency of mica to break into thin sheets. Quartz, on the other hand, is a framework silicate. All of the tetrahedra are bound tightly to each other, and no place in the structure is weaker than any other place. Therefore, quartz does not break apart easily; its structure resists outside forces, so it is hard and strong. When it does break, there are no planes of weakness, and the break is random and uneven.

This relationship between the arrangement and bonding of atoms and the physical properties of minerals extends to all mineral compounds. Knowing the structure of a mineral helps us to understand why that mineral displays certain physical properties.

Silicate minerals are so abundant and widespread that almost every rock contains one or more. Although there are more groups of nonsilicate minerals than there are of silicates, nonsilicates make up only a very small part of Earth's crust, and very few are important rock-forming minerals. A brief overview of four nonsilicate minerals follows.

Independent Tetrahedra
e.g., olivine, zircon, garnet

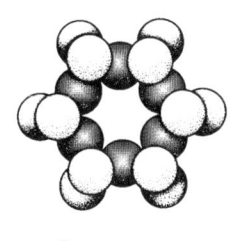

Ring Silicates
e.g., tourmaline, beryl

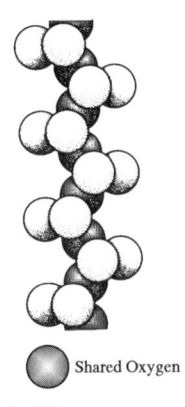

Shared Oxygen

Single-Chain Silicates
pyroxenes: e.g., enstatite, ferrosilite, augite, diopside, jadeite, spodumene

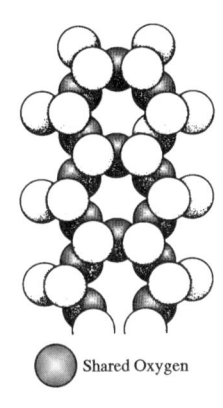

Shared Oxygen

Double-Chain Silicates
amphiboles: e.g., hornblendes, anthophyllite, tremolite, glaucophane

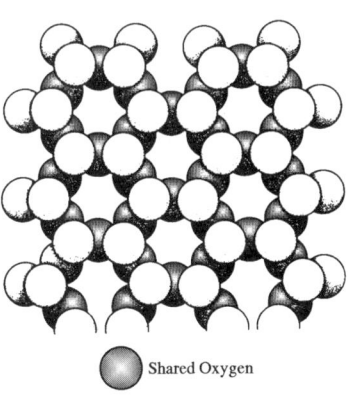

Shared Oxygen

Sheet Silicates
micas: e.g., muscovite, biotite, lepidolite
also, talc, chlorite, serpentine

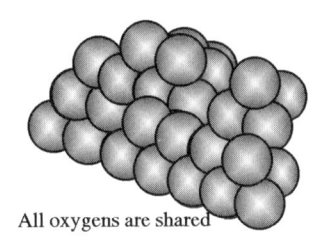

Shared Oxygen

All oxygens are shared

Framework Silicates
e.g., feldspars, quartz, nephaline, sodalites

Figure 14.6 Some Structures That Form with Silicon Tetrahedra

Unit Four **EARTH MATERIALS**

295

Minerals

Sulfides

Sulfides are compounds in which one or more ions of sulfur are combined with a metallic ion. Sulfide minerals usually display a metallic luster, and many are important ores of metals. Some examples of sulfides are galena (PbS), pyrite (FeS), marcasite (FeS_2), cinnabar (HgS), sphalerite (ZnS), and stibnite (Sb_2S_3).

Oxides

Oxides are compounds in which oxygen is joined with ions of other elements, usually metals. Common oxides include hematite (Fe_2O_3), magnetite ($FeO\ Fe_3O_4$), corundum (Al_2O_3), ilmenite ($FeTiO_5$), chromite ($FeCr_2O_4$), and spinel ($MgAl_2O_4$). Magnetite is the only mineral attracted to a magnet. Iron oxides such as magnetite and hematite are mined to make iron and steel. Corundum is extremely hard and is widely used as an abrasive. When it forms with impurities, corundum produces red rubies and blue sapphires.

Carbonates and Sulfates

In oxides, oxygen forms a single ion and acts alone. In other minerals that contain oxygen, the oxygen ions join with other ions to form a group called a polyatomic ion. Two very common polyatomic ions are the carbonate ion, $(CO_3)^{2-}$, and the sulfate ion, $(SO_4)^{2-}$. Polyatomic ions are electrically charged and can combine with ions of the opposite charge to form mineral compounds.

When carbon dioxide dissolves in water, it often forms carbonate ions. These react with other ions in the water to form carbonate compounds that can precipitate out of the water or be left behind when the water evaporates. Some organisms take carbon dioxide from the environment and form, in their bodies as skeletons, compounds containing carbonate ions. In this way, carbonate minerals have accumulated into huge sedimentary rock formations and in some cases have been metamorphosed. Calcite (calcium carbonate) is the chief mineral in limestone and marble. Dolomite (calcium magnesium carbonate) is the chief mineral in dolostone.

Gypsum (hydrated calcium sulfate) is probably the best known sulfate mineral. It is mined for use in making wallboard, plaster, cement, fertilizer, and paint and as a filler in paper. Barite ($BaSO_4$) and celestite ($SrSO_4$) are important sources of barium and strontium, respectively, which are used in medicines and drilling.

Minor Mineral Groups

There are also other, less abundant ions that form minerals: nitrates (nitrogen and oxygen), borates (boron and oxygen), chromates (chromium and oxygen), phosphates (phosphorus and oxygen), and hydroxides (hydrogen and oxygen).

Halides are compounds of metallic ions with halogen elements such as chlorine, fluorine, and bromine. The only major rock-forming mineral in this group is halite (NaCl), a sedimentary mineral that forms when seawater evaporates.

Native elements are found as pure elements in Earth's crust. They include gold, silver, copper, and sulfur. These were among the first minerals to be mined by humans.

CONSERVATION OF EARTH'S RESOURCES

Minerals

The minerals in Earth are a treasure of almost unimaginable value. They help to fill many human needs, which, through time, have become greater. Minerals are used in making products ranging from steel to electric light bulbs. Unfortunately, minerals are a nonrenewable resource; that is, they can be used only once and then they are gone.

Imagine that every week you put a few cents of your allowance into a bank. For three years you do not take any money from the bank. Then you see a video game that you want, you open the bank, and you use the money to buy the game. In a few minutes, you have spent the money it took you three years to save. In the same way, it is easy to use up in a few years minerals that took millions of years to form. But while you will probably save more money in the future, mineral resources may not form again within your lifetime or even a thousand lifetimes. Every time a mineral is used, that much less remains. Thus, Earth's mineral resources will not last forever.

For this reason, people must conserve mineral resources, not just by saving these resources, but also by making sure that there is no waste in using them. Recycling and modernizing factories are parts of a good conservation policy. Another important part is research into better ways of production.

If conservation is practiced by everyone, Earth's mineral resources can be used to best advantage. Today there are recycling centers throughout the United States, and many communities have instituted mandatory recycling. Recycling centers process used aluminum cans, old bottles, scrap metals, and newspapers. These materials are then reused in the manufacture of aluminum, glass, steel, and paper products. You may have noticed on some cans a statement that they were made from recycled aluminum.

Fossil Fuels

At certain times in Earth's history, environmental conditions have enabled plant and marine organisms to grow at a rapid rate. Over millions of years, their bodies stored energy from sunlight that was converted into chemical bonds during photosynthesis. When these organisms completed their life cycles, their remains accumulated faster than decomposers could break them down and recycle them back into the environment. Over time, when these remains became buried by deposition, the pressure of overlying sediments

Minerals

and the increased temperatures at depth converted the layers of organic remains into coal beds and pools of petroleum. Since coal and petroleum are derived from organic remains and are used chiefly as fuels, they are called *fossil fuels*.

Fossil fuels are a vital source of energy and petrochemicals for the global economy. Transportation, urban development, industry, commerce, agriculture, and many other human activities depend on the amount and type of energy available. Most of the energy used today is obtained by burning fossil fuels. While coal was widely used in the past, petroleum and natural gas are now the fuels of choice because they are easier to obtain than coal, have many uses in industry, and are concentrated, easily portable sources of energy for cars, trucks, airplanes, and trains.

Unfortunately, the burning of fossil fuels releases into the atmosphere waste products that threaten the health of living things and have environmental risks associated with them. Current environmental conditions are not resulting in the large-scale accumulation of remains that can be converted into coal or petroleum. Therefore, for the foreseeable future, organic fossil fuels are a nonrenewable resource.

Global Implications of Resource Use and Distribution

The use and the distribution of Earth's resources have global political, financial, and social implications. Highly industrialized nations use more energy and mineral resources than less developed countries. The energy and natural resources we consume contribute to our high standard of living. This consumption, however, has led to a rapid depletion of Earth's natural resources and to mounting environmental crises, ranging from the devastation of marine life by oil-tanker spills to the breakdown of the ozone layer of the atmosphere by the chloroflourocarbons (CFCs) used in refrigerators and air conditioners.

As developing nations industrialize and their urban centers grow, they demand more energy and natural resources, thereby entering into competition with all of the other industrialized nations for these resources. Often this competition has resulted in friction between nations and even in war. If we are to live in peace and maintain our environment, we must develop ways to slow the depletion of natural resources and to use them more efficiently.

The depletion of energy and other natural resources can be slowed by decisions to use efficient technologies, to conserve, and to recycle. These decisions can be implemented at many levels ranging from the national down to the personal. Government can restrict low-priority uses of materials, such as the use of petrochemicals to manufacture purely ornamental packaging that adds to our solid-waste disposal problems. Government can also set standards of technical efficiency, for example, mileage and emission requirements for automobiles or the use of insulation in home construction. Cars that are built with more power than their function warrants waste energy. (If the speed limit is 55 mph, why build cars with the power to do 100 mph?) However, such decisions often involve trade-offs of cost and social values. For example, when other, less damaging compounds are substituted for CFCs, new compressors are required, so older units must be replaced before they have worn out. Reduced emissions depend on catalytic converters filled with toxic heavy metals. The development of wise policies for conserving and managing natural resources is one of the great challenges now facing all nations.

MULTIPLE-CHOICE QUESTIONS

In each case, write the number of the word or expression that best answers the question or completes the statement.

1. Which two elements are most abundant by weight in Earth's crust?
   (1) oxygen and aluminum
   (2) oxygen and silicon
   (3) iron and aluminum
   (4) iron and silicon

2. Which element comprises most of Earth's crust both by weight and by volume?
   (1) nitrogen
   (2) oxygen
   (3) hydrogen
   (4) silicon

3. The physical properties of a mineral are due chiefly to the
   (1) hardness and cleavage of the mineral
   (2) number of atoms present
   (3) number of oxygen-silicon tetrahedra
   (4) internal arrangement of its atoms

4. The hardness of a mineral such as quartz is due to
   (1) the internal arrangement of its atoms
   (2) large amounts of impurities
   (3) its characteristic luster and color
   (4) its formation in certain rock types

5. Which property is most useful in mineral identification?
   (1) hardness
   (2) size
   (3) color
   (4) texture

6. Certain minerals usually break along flat surfaces, while other minerals break unevenly. This characteristic is due to the
   (1) luster of the mineral
   (2) age of the mineral
   (3) internal arrangement of the mineral's atoms
   (4) force with which the mineral is broken

Minerals

7. Which property of a mineral most probably determines the extent to which the mineral will resist being mechanically eroded?
   (1) density
   (2) luster
   (3) streak
   (4) hardness

8. Which element combines with silicon to form the tetrahedral unit of structure of the silicate minerals?
   (1) oxygen
   (2) nitrogen
   (3) potassium
   (4) hydrogen

9. The data table below shows the composition of six common rock-forming minerals.

Mineral	Composition
Mica	$KAl_3Si_3O_{10}$
Olivine	$(FeMg)_2SiO_4$
Orthoclase	$KAlSi_3O_8$
Plagioclase	$NaAlSi_3O_8$
Pyroxene	$CaMgSi_2O_6$
Quartz	SiO_2

   The data table provides evidence that
   (1) the same elements are found in all minerals
   (2) a few elements are found in many minerals
   (3) all elements are found in only a few minerals
   (4) all elements are found in all minerals

10. The diagram below represents a single silicon-oxygen tetrahedron unit. Two different minerals have these same units arranged in different patterns. How will the minerals differ?

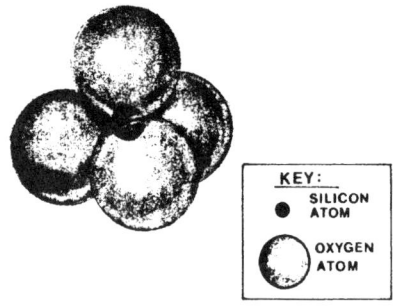

    (1) One mineral will have some physical properties different from those of the other mineral.
    (2) One mineral will be more radioactive than the other.
    (3) One mineral will have the silicon atom outside the tetrahedron, while the other will have it inside the tetrahedron.
    (4) One mineral will have larger silicon atoms than the other.

11. Which object is the best model of the shape of a silicon-oxygen structural unit?

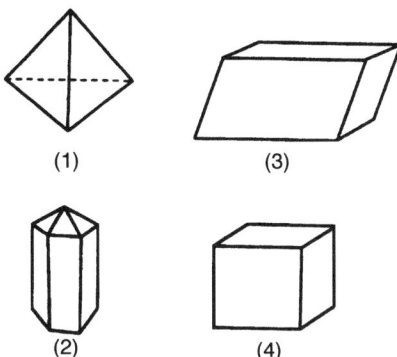

(1) (3) (2) (4)

12. The most abundant minerals found in the rocks of Earth's crust are
(1) phosphates   (3) silicates
(2) carbonates   (4) sulfates

13. The crystal characteristics of quartz shown in the accompanying diagram are the result of the

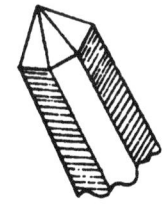

(1) internal arrangement of the elements from which quartz is formed
(2) shape of the other rock crystals in the area where the quartz was formed
(3) amount of weathering that the quartz has been exposed to
(4) age of the quartz crystal

14. The diagrams below represent samples of five different minerals found in rocks of Earth's crust.

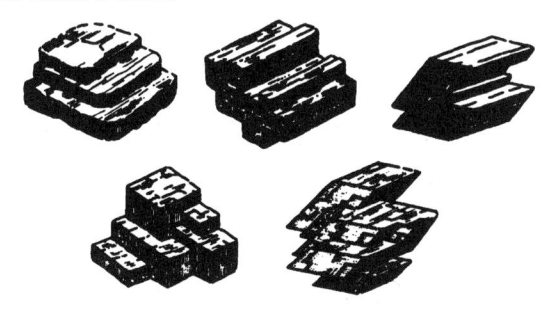

Minerals

Which physical property of minerals is represented by the flat surfaces in the diagrams?
(1) magnetism
(2) hardness
(3) cleavage
(4) crystal size

15. Which mineral is an ore of iron and has a characteristic reddish brown streak?
(1) magnetite
(2) pyrite
(3) hematite
(4) olivine

16. The data table below gives information on mineral hardness.

MINERAL HARDNESS

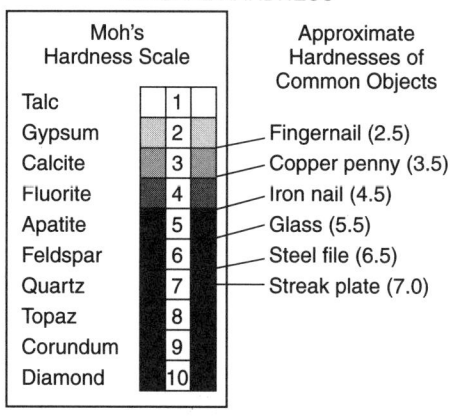

Moh's Hardness Scale		Approximate Hardnesses of Common Objects
Talc	1	
Gypsum	2	Fingernail (2.5)
Calcite	3	Copper penny (3.5)
Fluorite	4	Iron nail (4.5)
Apatite	5	Glass (5.5)
Feldspar	6	Steel file (6.5)
Quartz	7	Streak plate (7.0)
Topaz	8	
Corundum	9	
Diamond	10	

Moh's scale would be most useful for
(1) identifying a mineral sample
(2) finding the mass of a mineral sample
(3) finding the specific gravity of a mineral sample
(4) counting the number of cleavage surfaces of a mineral sample

17. What causes the characteristic crystal shape and cleavage (breaking along flat surfaces) of the mineral halite as shown in the diagram below?

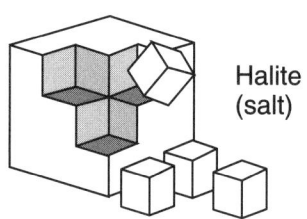

Halite (salt)

(1) metamorphism of the halite
(2) the internal arrangement of the atoms in halite
(3) the amount of erosion the halite has undergone
(4) the shape of other minerals located where the halite formed

18. Two mineral samples have different physical properties, but each contains silicate tetrahedra as its basic structural unit. Which statement about the two mineral samples must be true?
    (1) They have the same density.
    (2) They are similar in appearance.
    (3) They contain silicon and oxygen.
    (4) They are the same mineral.

19. When dilute hydrochloric acid is placed on the sedimentary rock limestone and the nonsedimentary rock marble, a bubbling reaction occurs with both. What does this indicate?
    (1) The minerals of these two rocks have similar chemical compositions.
    (2) The molecular structures of these two rocks have been changed by heat and pressure.
    (3) The physical properties of these two rocks are identical.
    (4) The two rocks originated at the same location.

20. Which common mineral fizzes when dilute hydrochloric acid (HCl) is placed on it?
    (1) calcite   (3) quartz
    (2) feldspar   (4) talc

Base your answers to questions 21 and 22 on the diagram below, the *Earth Science Reference Tables*, and your knowledge of earth science. The diagram shows the elements found in four minerals.

	O	Si	Al	Fe	Ca	Na	C
Quartz	■	■					
Feldspar	■	■	■		■		
Olivine	■	■		■			
Diamond							■

■ = element present

21. Which mineral contains the greatest variety of elements?
    (1) quartz   (3) olivine
    (2) feldspar   (4) diamond

22. Which of the four minerals shown is *not* a silicate mineral?
    (1) quartz   (3) olivine
    (2) feldspar   (4) diamond

Minerals

Base your answers to questions 23 through 27 on the table of minerals below, which shows the physical properties of nine minerals.

Mineral	Color	Luster	Streak	Hardness	Specific Gravity	Chemical Composition
Biotite mica	Black	Glassy	White	Soft	2.8	$K(Mg,Fe)_3(AlSi_3O_{10})(OH_2)$
Diamond	Varies	Glassy	Colorless	Hard	3.5	C
Galena	Gray	Metallic	Gray-black	Soft	7.5	PbS
Graphite	Black	Dull	Black	Soft	2.3	C
Kaolinite	White	Earthy	White	Soft	2.6	$Al_2(Si_2O_5)(OH)_4$
Magnetite	Black	Metallic	Black	Hard	5.2	Fe_3O_4
Olivine	Green	Glassy	White	Hard	3.4	$(Fe,Mg)_2SiO_4$
Pyrite	Brass yellow	Metallic	Greenish black	Hard	5.0	FeS_2
Quartz	Varies	Glassy	Colorless	Hard	2.7	SiO_2

Definitions
  Luster: the way a mineral's surface reflects light
  Streak: color of a powdered form of the mineral
  Hardness: resistance of a mineral to being scratched (soft—easily scratched; hard—not easily scratched)

Chemical Symbols
Al—aluminum, Si—silicon, H—hydrogen, O—oxygen, Pb—lead, Fe—iron, S—sulfur, C—carbon, K—potassium, Mg—magnesium

23. Which mineral has a different color in its powdered form than in its original form?
    (1) pyrite
    (2) graphite
    (3) kaolinite
    (4) magnetite

24. Which mineral contains iron, has a metallic luster, is hard, and has the same color and streak?
    (1) biotite mica
    (2) galena
    (3) kaolinite
    (4) magnetite

25. Why do diamond and graphite have different physical properties, even though both are composed entirely of the element carbon?
    (1) Only diamond contains radioactive carbon.
    (2) Only graphite consists of organic material.
    (3) The minerals have different arrangements of carbon atoms.
    (4) The minerals have undergone different amounts of weathering.

26. Which mineral is commonly found in granite?
    (1) quartz
    (2) olivine
    (3) magnetite
    (4) galena

27. Which mineral would most likely be weathered most after being placed in a container and shaken for 10 minutes?
    (1) pyrite
    (2) quartz
    (3) magnetite
    (4) kaolinite

CONSTRUCTED RESPONSE QUESTIONS

Base your answers to questions 28 through 30 on the data table below, which shows the relationship between the amount of aluminum in a type of rock and the energy needed to extract aluminum from that rock.

Sample	Aluminum Content of Ore (percent)	Energy Needed to Extract 1 Ton of Aluminum (thousands of kilowatt-hours)
1	3	220
2	5	140
3	10	90
4	20	65
5	30	54
6	40	49
7	50	48

28. (a) On the grid below, label the axes and select an appropriate scale for the data. [2]
    (b) Plot the energy needed to extract 1 ton of aluminum versus the aluminum content of the ore from which it is extracted for each of the seven samples. Connect the seven points with a line. [2]

Minerals

29. According to the graph you constructed, how much energy is needed to produce aluminum from rock that is 15 percent aluminum? Express your answer in thousands of kilowatt-hours. [1]

30. Using one or more complete sentences, describe how the amount of aluminum in a rock is related to the amount of energy needed to extract the aluminum from the rock. [1]

31. A student is using a set of minerals for a laboratory exercise. The minerals have numbered labels that correspond to a key on the inside cover of the box in which they came. Somehow, during the exercise, the tags on two of the minerals fell off and were lost. Both unlabeled minerals have a metallic silver color and at first glance look very similar. After matching up the labeled minerals with the key on the box cover, the student determines that the two silver minerals are galena and hematite. Outline a procedure by which the student can determine which mineral is galena and which is hematite.
   (a) Explain how you would determine any properties included in your procedure, such as hardness and streak. [2]
   (b) State two physical properties that would be most useful in telling these two minerals apart. [2]

32. State one reason why color is frequently less valuable than cleavage in mineral identification. [1]

33. In one or more complete sentences, explain why minerals are considered a nonrenewable resource. [1]

34. Worldwide, there are copper reserves that could supply our current needs for an estimated 32 years.
   (a) State one way in which copper is important in our daily lives. [1]
   (b) State two ways in which the number of years the current reserves will last could be extended. [2]

EXTENDED CONSTRUCTED RESPONSE QUESTIONS

Base your answers to questions 35 through 38 on the tables below. Table 1 lists the projected lifetimes of selected mineral resources in the United States. Table 2 lists the percentages of recycling of selected metals in the United States.

TABLE 1 PROJECTED LIFETIMES OF U.S. MINERAL RESERVES

Material	Projected Lifetime (years)
Bauxite	5
Chromium	0
Cobalt	0
Copper	17.5
Iron ore	205
Lead	5.5
Manganese	0
Nickel	0.1
Tin	0.4
Zinc	11
Gold	72
Silver	*
Platinum group	*
Gypsum	27
Phosphate	27
Potash	13.4
Sulfer	10

*Accurate consumption data not available

TABLE 2 METAL RECYCLING IN THE UNITED STATES, 1986–1995 AS PERCENTAGES OF CONSUMPTION

Metal	1986	1989	1992	1995
Alumninum	15	22	30	25
Chromium	21	21	26	22
Cobalt	15	14	25	22
Copper	22	25	42	39
Lead	50	60	62	57
Manganese	0	0	0	0
Nickel	25	34	30	34
Zinc	7	10	39	25

35. In one or more complete sentences, state one reason why iron is not among the metals listed in Table 2. [1]

36. According to Tables 1 and 2, for which of these mineral resources is the United States most dependent on imports from other countries?
    (a) Identify the mineral resource. [1]
    (b) In one or more complete sentences, explain why the United States is so dependent on other countries for this resource. [1]

37. State two reasons why manganese is more expensive than zinc. [2]

38. Suppose chromium recycling is banned because it generates toxic pollutants. In one or more complete sentences, predict the effect such a ban would have on the price of chromium and support your prediction with evidence from Tables 1 and 2. [2]

CHAPTER 15
ROCKS

KEY IDEAS Rocks are the naturally formed, solid, coherent materials that comprise Earth. Most rocks are aggregates of minerals; that is, they are made of many grains stuck together. The grains may be mineral crystals or tiny bits of broken rock, or even solid parts of once-living organisms. The grains may be small or large, tightly or loosely bound together, and may all be made of one type of mineral or represent a variety of minerals.

Rocks are grouped into three "families" based upon how they were formed: igneous, metamorphic, and sedimentary. Igneous rocks form by the solidification of molten rock; metamorphic rocks form when heat, pressure, or chemical activity causes changes in existing rocks; and sedimentary rocks form by the chemical precipitation or the compaction and/or cementation of sediments. The composition and structure of a rock offer important clues as to its origin.

Of the three types of rock, no one type necessarily originates before another. Rock is constantly being created, destroyed and altered in internal and external Earth processes. The schematic model of the possible sequences of processes by which any one rock may have been transformed into another is known as the rock cycle.

KEY OBJECTIVES
Upon completion of this chapter, you will be able to:

- Explain that rocks are usually composed of one or more minerals.
- Describe how rocks are classified by their origins, mineral contents, and textures.
- Explain how the characteristics of rocks indicate the processes by which they formed and the environments in which these processes took place.
- Describe how the properties of rocks determine their uses and also influence land usage by humans.

WHAT IS A ROCK?

A *rock* is a naturally formed, nonliving, firm and coherent mass of solid Earth material. A pile of sand is not a rock because the grains are not bound together, so the sand is not firm and coherent. A tree is not a rock even though it is solid, because a tree is living. But coal, which has been compressed into a coherent mass of dead leaves, twigs, and other plant parts, *is* a rock. Most rocks, though, are aggregates of minerals.

There are three large "families" of rocks, each defined by the processes that form them. ***Igneous*** rocks (from the Latin *ignis*, meaning "fire") form by the cooling of molten rock. ***Sedimentary*** rocks form when sediments are compacted and/or cemented together, or by the precipitation of dissolved minerals. *Sediments* are small bits of matter deposited by water, ice, or wind. They can be fragments of rocks, shells, or the remains of plants or animals. ***Metamorphic*** rocks are formed when already existing rock is changed by great heat, great pressure, or chemical action. Let us now consider how each type of rock forms, and how the way in which it forms determines its characteristics.

TYPES OF ROCK

Igneous Rocks

Formation of Igneous Rocks

Igneous rocks form by the crystallization of a molten mixture of minerals and dissolved gases. This molten mixture is called *magma* while it is beneath Earth's surface and *lava* when it escapes onto the surface. Most magmas can be thought of as fiery soups of fast-moving ions, that is, electrically charged atoms or molecules. As the magma cools, these ions slow down, are attracted to one another, and settle into stable structures—molecules of mineral compounds joined in a structural array, or mineral crystals. See Figure 15.1.

The formation of mineral crystals is a process, not an instantaneous event. As magma or lava cools, the amount of time available for crystal structures to form determines how large they can become. Thus, the size of the mineral crystals in an igneous rock provides a clue to the speed at which the rock crystallized, and that in turn is a clue to its origin.

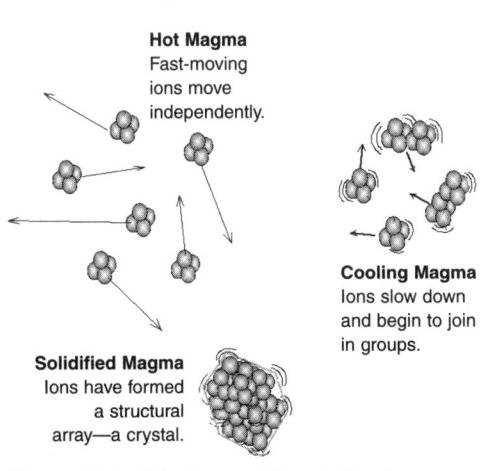

Figure 15.1 Cooling and Crystallization.

If cooling is very rapid, groups of atoms may solidify into a substance that lacks any distinctive structure—a glass. Such rapid cooling typically occurs at Earth's surface, where molten material comes into sudden contact with air or water at temperatures that are hundreds of degrees cooler than its temperature. Volcanic glasses, such as obsidian, can be observed forming when lava flows into the ocean and rapidly cools. From this type of observational data, we can infer the origin of glass found elsewhere, far from current volcanic activity or perhaps even buried beneath layers of other rock.

If cooling occurs more slowly, mineral crystals may begin to form, but may solidify while still microscopic in size. This, too, typically occurs at or near the surface of Earth. For example, the surface of lava, which may flow out of a fissure as a glowing liquid, can be observed to quickly cool and "skin over." This thin skin of rock shields the molten lava beneath it and allows the lava to cool more slowly. The lava inside may take days, weeks, or even months to solidify, thus allowing it to form crystals large enough to be seen with a microscope or magnifying glass.

Only very slow cooling allows the formation of crystals large enough to be visible to the unaided eye. Such slow cooling occurs underground, where the cooling magma is blanketed by the surrounding rock. Rock is a poor conductor of heat; and magma surrounded by rock, like hot coffee in a Thermos bottle, loses heat very slowly. Some underground pools of magma cool only a few degrees per century! Thousands of years may be required for the magma to totally crystallize. The result is large, visible crystals.

Characteristics of Igneous Rocks

Since most rocks are mixtures of minerals, rocks cannot be identified in the same way as minerals. A single rock may consist of minerals of several colors, hardnesses, and densities. For example, granite may contain white quartz with a hardness of 7, pink feldspar with a hardness of 6, and black mica with a hardness of 2–3. Thus, granite does not have a single characteristic color or hardness.

Two more general properties that are useful in identifying rocks are texture and mineral composition. *Texture* is the size, shape, and arrangement of the mineral crystals or grains in a rock. *Mineral composition* is simply the minerals that comprise the rock.

Igneous rocks are composed of randomly scattered, tightly interlocking mineral crystals. The main differences in the textures of igneous rocks involve the sizes and compositions of the mineral crystals. Most igneous rocks consist of a mixture of several of the following common minerals: quartz, feldspar, biotite, amphibole, pyroxene, and olivine.

Igneous rocks can be divided into two major types: intrusive and extrusive. Classification is based on whether the rocks formed from magma or lava. Magma and lava solidify into rock under very different conditions and thus produce rocks with different characteristics. Except for the volcanic glasses, all igneous rocks consist of tightly interlocking mineral crystals (see Figure 15.2). The texture of these crystals is a key to how a particular rock was formed.

Textures of Igneous Rocks

Intrusive Rocks

Magma, as previously defined, is molten rock inside Earth. Magma can move around by pushing its way into cracks or crevices in surrounding rock or by intruding into them. If the magma remains trapped underground and cools enough to solidify, an igneous rock will form. Igneous rock formed by the cooling of magma is called *intrusive*, or *plutonic*, rock.

Figure 15.2 Tightly Interlocking Crystals of an Igneous Rock.

As described earlier, magma generally cools slowly underground because it is blanketed by the surrounding rocks. Rocks are poor conductors of heat and do not allow the heat in the magma to escape very rapidly. When magma cools slowly, ions in the magma have time to align themselves in orderly structures called *crystals*. The slower the cooling, the more ions are able to align and the bigger the crystal. As a result, intrusive rocks generally have crystals large enough to be seen with the unaided eye. While you may not think a 1-millimeter crystal is large when you look at it, remember that millions upon millions of ions had to move into alignment with each other to reach that size. Rocks with large, visible crystals are said to have a coarse texture.

Granite, gabbro, and pegmatite are typical intrusive rocks.

Extrusive Rocks

When magma reaches the surface and pours out of Earth, it is called *lava*. Lava is usually pushed out, or extruded, from inside Earth in volcanoes or through large cracks in Earth's crust. Therefore, igneous rock formed from lava is called *extrusive*, or *volcanic*, rock.

Lava cools and solidifies into rock rapidly. When lava is extruded from a volcano or cracks in the crust, it is instantly exposed to an environment that may be 1,000°C colder than the lava's location just moments before. Since lava cools and solidifies rapidly, the ions in it have less time to align themselves into crystals, and the crystals that form are generally too small to be seen by the unaided eye. A magnifier or microscope may be needed to see the crystals in an extrusive rock. Rocks with crystals that can be seen only with magnification have a fine texture.

In some cases, magma cools so rapidly that no crystalline structures can form and the ions literally become frozen in place. When a rock of this type is examined with a microscope, no crystals can be seen. The physical characteristics of these unusual rocks are the same as those of glass, so they are called *volcanic glasses*. Their texture, in which no mineral crystals or grains are visible, is termed glassy. Obsidian is a common volcanic glass.

Rocks

Sometimes, as lava is being extruded, gases dissolved in the lava form tiny bubbles. A glassy, lava froth forms similar to the froth that overflows from a champagne bottle after the cork is popped. In thick, viscous lava, the rock solidifies around the bubbles before the gas can completely escape, and a type of rock called *pumice* is formed. Since most of the glassy bubbles remain sealed, pumice will often float on water. In thinner, runnier lavas, the gas tends to bubble away rather than to form pumice. However, rapidly cooling surfaces may become pockmarked with small openings, or ***vesicles***, made by the final escaping gas bubbles just as the lava solidified.

When lava is very forcibly extruded, as in an explosive volcanic eruption, a whole array of unusual rock materials form. These range from the thin, glassy strands ("Pele's hair") produced when very liquid lava is sprayed out, to large, streamlined globs ("volcanic bombs") that solidify while hurtling through the air. Such shapes are clues to the way in which the rock formed.

Basalt and rhyolite are typical extrusive rocks. The differences in the grain sizes of typical intrusive and extrusive rocks are shown in Figure 15.3.

|---1 mm---| |---1 mm---| |---1 mm---|
(a) Basalt: Fine texture   **(b) Diabase**   **(c) Gabbro:** Coarse texture

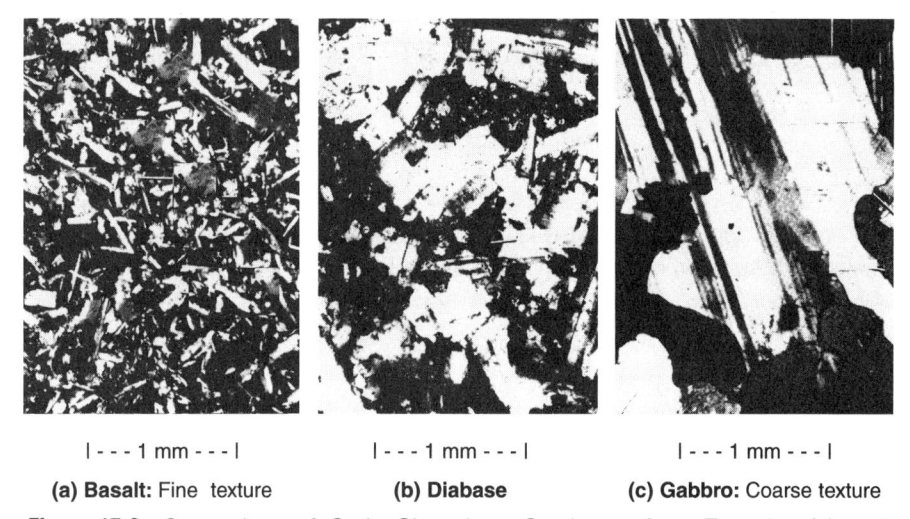

Figure 15.3 Comparison of Grain Sizes in a Continuum from Extrusive (a) and Intrusive (b and c) Rocks. Basalt (a) forms at the surface. Diabase (b) may form as an intrusion near the surface as in a dike or sill. Gabbro (c) cools very slowly deep underground. Source: *Physical Geology*, 2nd Ed., Richard Foster Flint and Brain J. Skinner, John Wiley, 1974, 1977.

Porphyrys

In most igneous rocks the crystal grains are all roughly the same size, all large or all small. One type, however, has an unusual texture. In porphyrys, coarse mineral grains are embedded in a mass of fine mineral grains (see Figure 15.4). The isolated, large crystals are called *phenocrysts*; the surrounding fine-grained material is the *groundmass*. Porphyrys are thought to form in two stages: (1) a magma begins to solidify slowly at depth, and (2) the magma rises rapidly to the surface, where it finishes solidifying. The crystals formed at depth are large because the rate of cooling is slow. Nearer the surface, cooling is more rapid and the crystals formed are tiny.

Mineral Composition of Igneous Rocks

The mineral composition of an igneous rock is the product of the magma from which the rock solidifies. All common igneous rocks are mixtures of two or more of the following minerals: quartz, potassium feldspar, plagioclase feldspar, biotite mica, amphibole, pyroxene, and olivine. All are silicates, but they vary in composition, color, density, and other properties.

Figure 15.4 Porphyritic Texture— Large Phenocrysts of Plagioclase Feldspar in a Basalt Matrix. Source: *Physical Geology*, 2nd Ed., Richard Foster Flint and Brian J. Skinner, John Wiley, 1974, 1977.

These silicate minerals are divided into two main groups: the *felsic* group, consisting of quartz and potassium and plagioclase feldspars, and the *mafic* group, consisting of silicates rich in magnesium and iron, such as olivine, pyroxene, amphibole, and biotite. The word *felsic* comes from *fel*dspar and *si*lica (the name of the compound SiO$_2$, of which quartz is composed). *Mafic* comes from *ma*gnesium and the symbol *Fe* for iron. Felsic lavas have a thick, pudding-like consistency. As a result, they clog up fissures, allowing pressure to build up and leading to explosive eruptions. Mafic lavas have a thinner, syrupy consistency. They tend to flow more freely and to erupt as a flowing river of lava or spurt out of fissures like a fountain. Also, felsic minerals are lighter in color and lower in density than mafic minerals. Felsic rocks are common in the continents, while mafic rocks are more frequently found in the ocean basins.

The Scheme for Igneous Rock Identification

There are more than 1,500 different igneous rocks, and the system for classifying them is very complex. Petrologists are trying to simplify the system. Figure 15.5 can be used to make a very basic classification of an igneous rock.

This scheme has two main parts: an upper half, in which some of the most common igneous rocks are arranged by characteristics, and a lower half, a diagram that shows the percent mineral compositions of these rocks.

In the upper half, rocks such as granite and rhyolite, which are composed of light-colored minerals, are on the left. Rocks such as basalt and gabbro, which are composed of dark-colored minerals, are on the right. Since dark-colored minerals tend to be denser than light-colored minerals, the dark-colored rocks on the right are denser than the light-colored rocks on the left. Glassy extrusive rocks (e.g., obsidian and pumice) are at the top. Fine-

Rocks

grained extrusive rocks (e.g., rhyolite and basalt) are in the middle. Coarse-grained, intrusive rocks (e.g., granite and gabbro) are on the bottom.

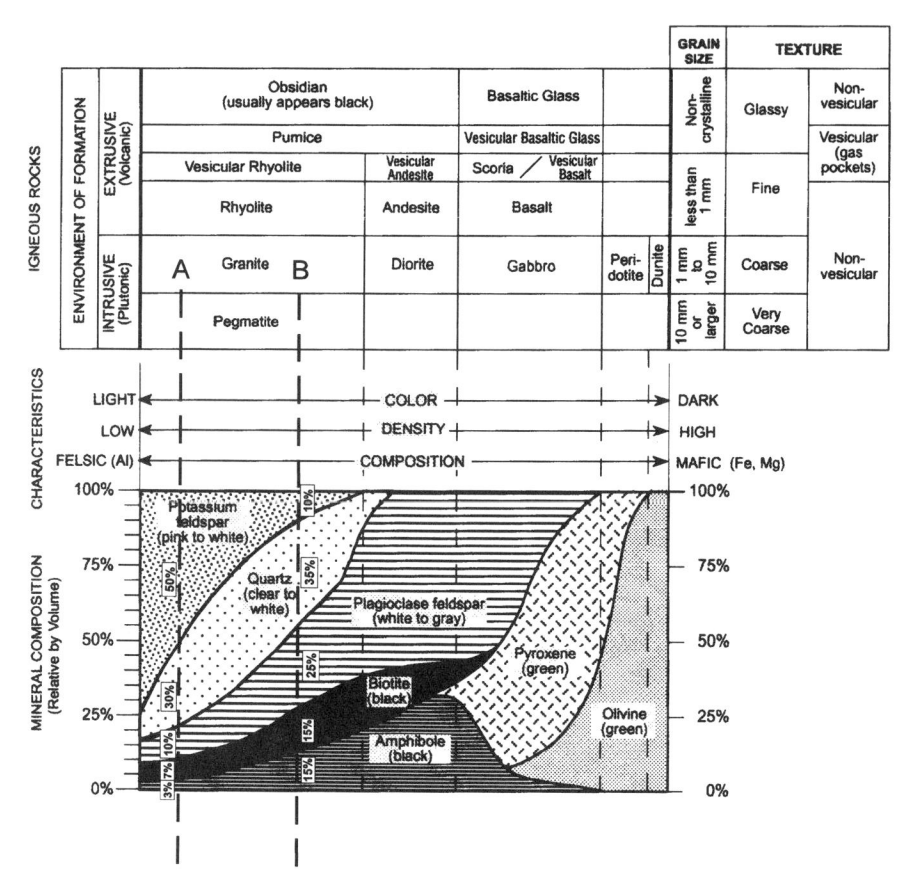

Figure 15.5 Scheme for Igneous Rock Identification. Source: The State Education Department, *Earth Science Reference Tables,* 2001 ed. (Albany, New York; The University of the State of New York).

In the lower half, the diagram shows the relative percents of the minerals of which the rocks are composed. As you can see, a particular type of rock can have a mineral composition that varies within a certain range. Look, for example, at the two different granites. Granite *A* consists of about 50% potassium feldspar, 30% quartz, 10% plagioclase feldspar, 7% biotite, and 3% amphibole. Granite *B* consists of only 10% potassium feldspar, 35% quartz, 25% plagioclase feldpsar, and 15% each of biotite and amphibole.

The following is a brief description of the rocks in Figure 15.5, starting at the top.

OBSIDIAN and BASALTIC GLASS form when lava cools so rapidly that crystals have no time to form before the lava solidifies. These rocks are very hard and brittle and break along shell-like depressions, forming sharp edges. Primitive peoples used them to make knives.

314

PUMICE hardens just as escaping gases whip the lava into a froth. The cooling rate is so fast that gases are trapped inside the frothy hardening rock. Thus, pumice is full of holes and looks somewhat like a rocky sponge. In fact, pumice is often used to scrub surfaces clean. Formerly, sailors used pumice stones to scrub the decks of wooden ships. Because of the holes, most pumice is so light it floats on water.

VESICULAR ROCKS form when gas escaping from lava bubbles away into the atmosphere rather than being trapped inside and forming pumice. Vesicles are small openings formed by escaping gas just as the lava flow hardens. In many flows the vesicles later become filled with calcite, quartz, or some other mineral deposited by heated groundwater. The type of lava determines which vesicular rock forms. For example, a rhyolite lava will form vesicular rhyolite. In scoria, the gases escape from the just-hardening lava as large bubbles, forming a texture full of cavities. Large bubbles are more likely to form in a thin, fluid basalt lava. Thus, scoria has a basaltic composition.

RHYOLITE contains the same minerals as granite, but it is fine grained and much rarer than granite. It is formed in lava flows and can be found in volcanic regions.

ANDESITE is named for the Andes Mountains, where it was first studied. Andesites are fine grained and usually gray or green in color. They have the same composition as diorites.

BASALT is the most abundant extrusive rock. Most lava flows form basalt, which is dark colored and fine grained. The ocean floor is mostly basalt. Sometimes basalt rock is crushed and used in road and railroad beds.

GRANITE consists mainly of quartz and feldspars. It is light colored and coarse grained and is thought to form by very slow cooling deep within Earth. Granite is the most abundant of all the igneous rocks. Most of the crust beneath the continents is made of granite. Granite is commonly used as a building stone and in making monuments.

DIORITE is coarse grained and contains about equal amounts of light- and dark-colored minerals, giving it a "salt and pepper" appearance.

GABBRO is a dark, coarse-grained rock. It is a common intrusive rock in New England and other mountain areas. It is sometimes used as a building stone.

PERIDOTITE is composed almost entirely of two minerals: olivine and pyroxene. It occurs widely, mostly in small plutonic bodies. It is a dark-colored rock with a high density—3.3 grams per cubic centimeter.

DUNITE is composed almost entirely of olivine. It is very rare and ranges in color from yellow to dark green, with a texture like hard brown sugar. It is named for the Dun Mountains of New Zealand, one of the few places it is found.

PEGMATITE has abnormally large crystals. It forms from a mixture of minerals that have relatively low melting temperatures, and are therefore the last to solidify from a cooling magma. It forms from a magma rich in silica and consists largely of quartz, potassium feldspar, and muscovite and biotite mica. Individual minerals grains several meters long have been found in some pegmatites, and it is mined for muscovite mica, which can sometimes be found in sheets up to a meter across.

Sedimentary Rocks

Sediments

All sedimentary rocks are composed of sediments. *Sediments* can be broadly defined as solid fragments of material that have been transported and then deposited by air, water, or ice. Most sediments are fragments of rock, but bits and pieces of plants and animals, or even molecules dissolved in water, can also be sediments.

When deposited, sediments typically form loose, unconsolidated layers on Earth's surface. These layers may then be transformed by physical and chemical changes into sedimentary rock.

Lithification

Lithification is the conversion of loose sediments into coherent, solid rock by processes such as cementation, compaction, dessication, and crystallization. This conversion may occur while the sediment is being deposited, soon after deposition, or long after deposition.

Cementation is the binding together of sediment particles by substances such as clay, carbonates, and hydrates of iron. These substances literally act as cement to hold particles together, in a solid mass, when they *crystallize* or otherwise accumulate between the loose sediment particles.

Compaction is the reduction in volume or thickness of a sediment layer due to the increasing weight of overlying sediments that are continually being deposited or to the pressure produced by movements of Earth's crust. Compaction usually results in a decrease in pore space within the sediments as they become more tightly packed. This, in turn, commonly leads to *desiccation*, the drying out or drying up of water, as it is forced out of the pore spaces during compaction. As the water dries up, minerals dissolved in the water crystallize and help to hold the rock together.

Figure 15.6 illustrates the compaction and cementation processes.

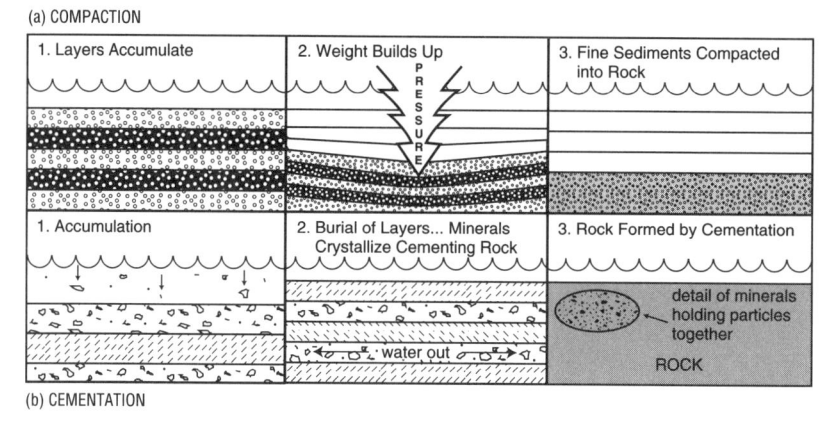

Figure 15.6 Compaction and Cementation.

Most sedimentary rocks form as a result of the compression and cementing of sediments. In most cases these sediments are rock fragments that range in size from tiny particles, such as clay and silt, to larger particles, such as sand and pebbles.

Some sedimentary rocks, however, result from the evaporation of seawater or from organic processes. Seawater has many substances dissolved in it; most are salts. When seawater evaporates, the salts are left behind. They form salt crystals, which settle out of the water in layers. Rock salt forms in this way as seawater evaporates.

Some organisms, such as corals, take substances dissolved in seawater into their bodies and form skeletons. When these organisms die, their skeletons settle to the ocean floor and can build up in layers. Biogenic limestones are formed in this way.

Plant materials that build up in layers can change drastically during compaction. They become desiccated, and the substances from which they were composed break down. Since carbon is the chief component of organic materials, it is often the end product of the lithification of plant parts. Coal typically forms in this way.

Characteristics of Sedimentary Rocks

Sedimentary rocks have three unique characteristics:

They *form at or near the surface of Earth at normal temperatures and pressures*. As a result, the environment in which sedimentary rocks form is quite different from the environments in which most other types of rock have their origins. Therefore, the minerals contained in sedimentary rocks are often quite different from those found in igneous or metamorphic rocks.

They *form layers*. Water is the chief transporter of sediments on Earth's surface. While being transported by water, most sediments become rounded. Sediments deposited in water settle into horizontal layers of rounded particles. This layering is preserved during lithification, so most sedimentary rocks usually contain rounded grains cemented in layers. These successive layers, known as strata, or beds, preserve evidence of past surface conditions. The individual layers can be easily distinguished because they usually differ in some way, such as color, thickness, size of particles, mineral composition, or degree of cementation.

The *surfaces of layers in sedimentary rock often contain distinctive features* identical to those found on modern sediments. Examples are ripple marks created by waves in shallow water, polygon-shaped cracks formed by fine sediment that was dried out by the Sun, footprints of walking animals, and the tracks and trails of crawling animals such as snails.

When sand and coarser sediments are moved along by air or water, they often form streamlined mounds such as dunes or ripples. Within these mounds, the layers of sediment are inclined in a down-wind or down-current direction. As successive layers of dunes or ripples pass over the surface, they leave behind layers of inclined sediments, called *cross-bedded sediments*, as shown in Figure 15.7.

Rocks

They *contain fossils*. Plants and animals often live in areas where sediments are being deposited. The remains of these plants and animals are often incorporated into layers of accumulating sediment and into the sedimentary rocks that form from them.

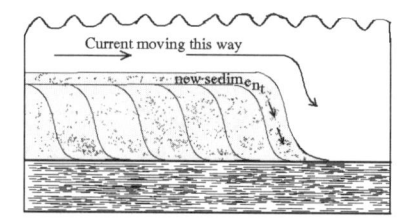

Figure 15.7 Diagram of Cross-Bedded Sediments.

Observations of current sediment layers and surfaces indicate that some characteristics enable us to infer which is the top, and which the bottom, of a series of sedimentary rock layers. This, in turn, allows us to infer the relative ages of the layers (bottom is oldest, and top is youngest). When a mixture of sediments settles in water, the larger particles reach the bottom first, and the smallest last. The result is graded bedding, as shown in Figure 15.8. When graded bedding is preserved in a sedimentary rock, we can infer that the original top was where the smaller particles are located. Similarly, cross-bedded sediments (Figure 15.8b) allow us to infer not only top and bottom, but also the direction in which the wind or water current was moving when the beds were deposited. If a rock layer contains ripple marks, mud cracks, footprints, or trails, we can infer the top of the layer from their position. Also, when certain shells settle in water (Figure 15.8c), they end up with one side facing upward most of the time, so their position in a sedimentary rock again allows us to infer top and bottom.

(a) Graded Bedding   (b) Cross Bedding   (c) Position of Fossils

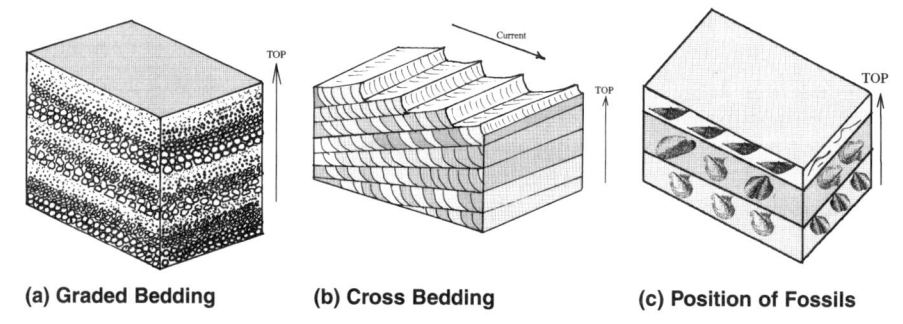

Figure 15.8 Diagram of Graded Bedding, Cross-bedding, and Fossil Position All Showing the Top of a Series of Layers. (a) Graded Bedding. Within each bed, particles are sorted and form layers that decrease in size from the bottom to the top of the bed. (b) Cross Bedding. Particles are well sorted and form inclined layers within each bed. The layers in the bed are steeply sloping at the top and gently sloping at the bottom. (c) Position of Fossils. The hard parts of many organisms that become fossilized have a preferred orientation after settling through water. For example, clam shells settle more often with the cupped side of the shell facing down.

Types of Sediment Particles

There are three major types of sediment particles: clastic, organic or biogenic, and chemical.

Clastic sediments are rock or mineral fragments formed by the breakdown of rock due to weathering. The vast majority of sediments are of this type. These particles are grouped and named by size according to the Wentworth scale, as shown in Table 15.1.

TABLE 15.1 THE WENTWORTH SCALE

Particle-Size Range (diameter, cm)	Particle Name	Sediment Composed of That Particle
Greater than 25.6	Boulder	Boulder gravel
6.4–25.6	Cobble	Cobble gravel
0.2–6.4	Pebble	Pebble gravel
0.006–0.2	Sand	Sand
0.0004–0.006	Silt	Silt
Less than 0.0004	Clay	Clay

Organic, or *biogenic*, *sediments* are particles produced by the life activities of plants or animals. The most abundant organic sediment consists of the shells of aquatic animals. These may range from the large calcium carbonate shells of clams or snails to the microscopic silica skeletons of amoebas and diatoms. There are even some algae, which absorb calcium carbonate from seawater and incorporate it into their skeletons. When these organisms die, their shells and skeletons settle to the bottom and accumulate in layers of mud.

In warm seas, coral reefs form when huge populations of little animals called corals secrete calcium carbonate. Wave action smashes the reef into pieces, along with the shells of other animals living on the reef, producing a coarse sand or granule-sized sediment.

Organic sediments also include the woody tissue of plants (tree trunks, branches, twigs, and leaves), as well as spores and pollen. *Peat* is an organic sediment formed from plant material that accumulates in swamps, where stagnant water conditions prevent the oxidation of the plant material. Over time, buried peat may be transformed into coal.

The fatty or waxy tissues of land plants and the soft tissues of aquatic plants and animals do not form particles. However, this soft organic matter is usually mixed in with sediment particles and is considered to be the raw material from which petroleum forms underground.

Fossils in sedimentary rocks provide evidence of the environment in which they formed. The fossil of an organism that is clearly a fish, together with fossilized clam shells, indicates an aquatic environment. Pollen from citrus or palm trees in a rock is evidence of a warm terrestrial environment.

Chemical sediments consist of particles (crystals) that have crystallized out of solutions at or near Earth's surface. All the water on Earth's surface has some material dissolved in it. Seawater, for example, contains over 80 elements and hundreds of different compounds. A number of conditions can cause some of the substances dissolved in water to crystallize.

Let's use a simple analogy. Think of a solution as a container in which a substance is being held. A glass of juice is a good example. The solvent is like the glass, and the solute is like the juice. If there is more juice than can fit into the glass, the juice will overflow and spill to the ground. Now think

Rocks

of seawater. Water is the solvent, or glass; salts are the solute, or juice. Overflow is like crystallization; spilling to the ground is like precipitation.

In our example, there are two things that can change—the size of the glass and the amount of juice. Having more solvent is like having a bigger glass. Similarly, if more water is added, seawater can hold more salt. On the other hand, removing water is like having a smaller glass. If water is removed, less salt can be held.

Evaporation removes water from seawater. When the water in a solution evaporates, the solution becomes more and more concentrated. When it reaches saturation, the substance dissolved in the solution begins to crystallize out. Sediments made of particles that crystallized as a result of evaporation are called *evaporites*. Halite crystals are a common chemical sediment that forms along the edges of bodies of water undergoing extensive evaporation.

If you have ever tried to dissolve sugar in iced tea, you know that less sugar dissolves in cold tea than hot tea. The temperature of a solution affects the amount of material that can be dissolved in it. As the temperature of a solution falls, its ability to hold dissolved materials decreases. Cooling a solution is like making the glass in the preceding example smaller without removing any of the juice. The result is overflow. For example, when the mineral-laden water of a hot spring cools, calcite or even opal may precipitate from the water.

Particles may also crystallize out of a solution if more solute is added than the solution can hold. If you add juice to an already full glass, the glass will overflow. For example, aquatic plants and animals can increase the amount of carbon dioxide (CO_2) dissolved in the water around them and thereby cause calcium carbonate to crystallize out and sink to the bottom, or precipitate. Also, if a chemical reaction in a solution produces a substance that doesn't dissolve in water, that substance will also precipitate.

The Scheme for Sedimentary Rock Identification

Sedimentary rocks are classified as *fragmental*, *organic*, or *chemical*, depending on the sediments from which they lithified. As with igneous rocks, texture and mineral composition are the main properties used to classify and name sedimentary rocks, as shown in Figure 15.9. In sedimentary rocks, the mineral composition of the sediment particles tells us where the particles came from, and the texture provides clues about the processes that deposited the sediments.

In Figure 15.9, sedimentary rocks are divided into two main groups. In the upper part of the chart are "Inorganic Land-Derived Sedimentary Rocks," that is, rocks formed from clastic sediments. In the lower part are "Chemically and/or Organically Formed Sedimentary Rocks," that is, rocks formed from chemical or organic sediments.

Clastic sedimentary rocks are classified on the basis of grain size. Nonclastic rocks are identified primarily by composition and texture.

Scheme for Sedimentary Rock Identification

INORGANIC LAND-DERIVED SEDIMENTARY ROCKS

TEXTURE	GRAIN SIZE	COMPOSITION	COMMENTS	ROCK NAME	MAP SYMBOL
Clastic (fragmental)	Pebbles, cobbles, and/or boulders embedded in sand, silt, and/or clay	Mostly quartz, feldspar, and clay minerals; may contain fragments of other rocks and minerals	Rounded fragments	Conglomerate	
			Angular fragments	Breccia	
	Sand (0.2 to 0.006 cm)		Fine to coarse	Sandstone	
	Silt (0.006 to 0.0004 cm)		Very fine grain	Siltstone	
	Clay (less than 0.0004 cm)		Compact; may split easily	Shale	

CHEMICALLY AND/OR ORGANICALLY FORMED SEDIMENTARY ROCKS

TEXTURE	GRAIN SIZE	COMPOSITION	COMMENTS	ROCK NAME	MAP SYMBOL
Crystalline	Varied	Halite	Crystals from chemical precipitates and evaporites	Rock Salt	
	Varied	Gypsum		Rock Gypsum	
	Varied	Dolomite		Dolostone	
Bioclastic	Microscopic to coarse	Calcite	Cemented shell fragments or precipitates of biologic origin	Limestone	
	Varied	Carbon	From plant remains	Coal	

Figure 15.9 Scheme for Sedimentary Rock Identification. Source: The State Education Department, *Earth Science Reference Tables*, 2001 ed. (Albany, New York; The University of the State of New York).

Metamorphic Rocks

Formation of Metamorphic Rocks

Metamorphic rocks get their name from the Greek words *meta*, meaning "change," and *morphs*, meaning "form." A metamorphic rock has literally been changed from its original form. A rock changes when it is exposed to an environment significantly different from the one in which it originally formed. In general, three factors cause metamorphic changes in rock: heat, pressure, and chemical activity.

Heat
Most minerals expand when heated, causing the atoms to move apart, and the bonds that hold them together to be stretched and weakened. If a mineral is heated enough, all of the bonds break, and the mineral melts. If a rock experiences such heating, a magma forms. In metamorphism, however, heating breaks *some* but not all of the bonds in the minerals of the rock. *Some* atoms break loose, migrate through the rock, and join with other atoms to form new minerals—metamorphic minerals—in the rock. The result is to change both the chemical composition and the structure of the rock, that is, to change the form of the rock. Such extreme heating of rock may be caused by deep burial (temperature increases with depth) or by contact with hot materials, such as magma.

Pressure

The effect of pressure on minerals is opposite to that of heat. Pressure forces the atoms in the mineral much closer together. The resulting stress on the bonds causes some of them to break. This, in turn, enables the atoms to rearrange into a more compact structure. The result is a significant change—a denser, harder rock.

Pressure exerted on rock may be due to deep burial or to movements in Earth's crust. Pressure due to deep burial is equal in all directions, but crustal movements may produce pressures that are greater in one direction than another, thereby deforming or flattening rocks. The effects on the particles in a rock are different in these two cases, as shown in Figure 15.10.

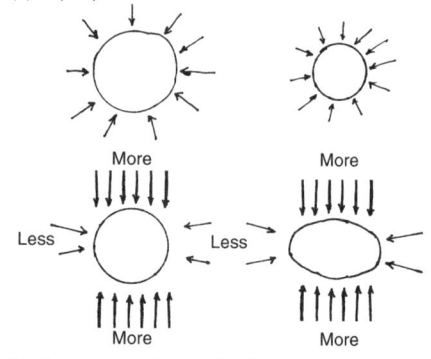

Figure 15.10 Equal Versus Directed Pressure on a Particle. Equal pressure (a) causes the particle to decrease in size, but the shape is unchanged. Unequal pressure (b) causes both a decrease in size and a change in shape.

Chemical Activity

A rock can also be changed when solutions rich in dissolved ions move through it. Such solutions are given off by cooling magmas; others come from metamorphism taking place deep underground. As these solutions move through pores in surrounding rock, they interact with minerals in the rock to form new minerals. Sometimes the water itself combines with minerals to form new ones. For example, olivine reacts with water to form talc or serpentine.

Metamorphism

The changes that occur in solid rock due to great temperature, great pressure, and chemical activity are known as *metamorphism*. Metamorphism occurs while rocks are still in the solid state. Changes produced by weathering, although they occur while rocks are solid, are not considered metamorphism. Neither are igneous processes such as the melting of rocks in high-temperature zones. True metamorphism occurs beneath Earth's surface at high temperatures and pressures. Under the conditions that produce metamorphism, ordinarily rigid, brittle rocks behave like soft modeling clay. The rocks can be bent and twisted without breaking. Rocks that originally had a clastic texture can recrystallize so that the metamorphic rock has a crystalline texture. Conversely, rocks that were originally crystalline can be crushed and broken into pieces and then relithified to form a clastic texture.

The preexisting rock from which a metamorphic rock forms is called the *parent rock*. By investigating the conditions that produce metamorphic changes, it is possible to infer the parent rock from the mineral compositions and structures of most metamorphic rocks.

Metamorphic rocks occur on a continuum from little alteration in the parent rock to major changes. Rarely are there sharp boundaries between parent rocks and metamorphic rocks; instead, gradations lead from one to the other. Limestone grades almost imperceptibly into marble. Shales grade into schists. Sandstones grade into quartzites. This characteristic has been helpful in determining the origins of certain metamorphic rocks. Metamorphic rocks are divided into groups based on the parent materials from which they were derived.

Types of Metamorphism

There are two major types of metamorphism: contact and regional. *Contact metamorphism* occurs when molten rock comes into contact with surrounding rocks. Heat given off by the cooling magma is conducted into the rock, where it causes changes in texture and mineral composition. Contact metamorphism is most intense at the contact between the magma and the surrounding rock, and decreases with distance from the magma until points are reached at which the effects of the heat are not felt. Generally, crystals are largest at the point of contact and become smaller with distance.

Regional metamorphism occurs over large areas and is generally associated with mountain building. Regional metamorphism takes its name from the large scale on which it occurs. For example, about 300 million years ago North America and Africa collided along a boundary between plates of Earth's crust. The result was the crushing of rocks from Alabama to New York and the formation of the Appalachian Mountains. During regional metamorphism, deeply buried sedimentary and igneous rocks are squeezed both by the weight of rock layers above and by movements of the crust, are intruded by igneous rocks, and are infused with chemically active fluids.

Characteristics of Metamorphic Rocks

When a rock is metamorphosed, nearly every characteristic can change, including texture and mineral composition. These changes can be so great that the metamorphic rock bears little, if any, resemblance to its original state. Therefore, to infer what a metamorphic rock once was, one needs to know what kinds of changes take place during metamorphism.

Changes in Texture

Crystalline Texture. During metamorphism the size, shape, and spacing of the grains (crystals or particles) in a rock are changed. Under both heat and pressure, the grains in a rock may partially melt and fuse together, forming larger grains (recrystallization); or, pressure may break brittle grains into smaller pieces, destroying the original texture of the rock. The heat and the pressure of metamorphism fuse any fragmental textured rock into a solid, crystalline mass. As a result, most metamorphic rocks have a crystalline texture.

Rocks

Increased Density. Pressure from deep burial closes pore spaces in sedimentary rocks and openings in igneous rocks. Pressure also compresses grains so that they are smaller and closer together. The rock becomes more compact. Its mass is forced into a smaller volume, and its density increases. Thus, the igneous or sedimentary rock is changed—it becomes metamorphic rock.

Foliation and Banding. A rock is said to be foliated when its crystals are arranged in layers or bands along which the rock breaks easily. These layers may be microscopic or thick enough to be easily seen. *Foliation* results when mineral grains recrystallize or are flattened under pressure. Even pressure compresses the grains so that they are smaller and closer together, but retain their original shape. Uneven pressure not only compresses grains but also changes their shape. Grains that may have been randomly oriented will form parallel layers (foliation) or become aligned when deformed under pressure.

Banding occurs when minerals of different densities recrystallize under pressure and separate into layers, like a mixture of oil and water. Since light-colored minerals tend to be less dense than dark ones, the layers that form are alternating light and dark.

Figure 15.11 shows the effects of pressure and recrystallization on rocks.

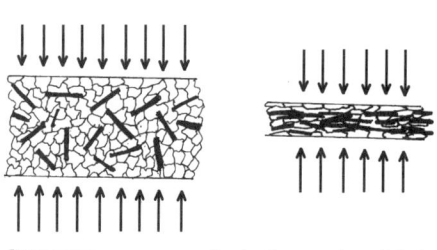

Compression causes randomly dispersed crystals to become oriented in a direction perpendicular to the pressure. The oriented crystals form thin sheets or layers called *foliation*.

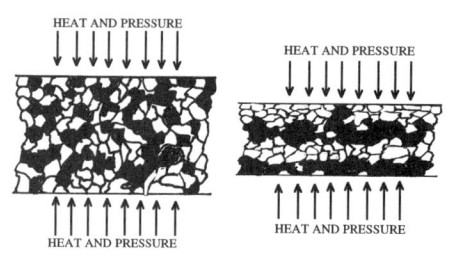

Migration of mineral crystals due to density differences produces *banding*—regions of light- and dark-colored minerals.

Figure 15.11 Random to Oriented Patterns Due to Pressure and Fusing of Grains, and Banding Caused by Density Separation.

Distortion of Layers. When sedimentary rocks are metamorphosed, the layers in the rock may be distorted. Softened by the heat and squeezed by the pressure, the once-flat layers become twisted and contorted. Such distortion is commonly seen in large-scale formations of metamorphic rock.

Changes in Mineral Composition

The changes described above can occur without altering the original mineral composition of the rock. When a rock is metamorphosed, however, the original minerals often disappear and new minerals form in their place. The underlying reason for this change is lack of stability. Minerals that are stable in one environment are not stable in another. Sugar is stable at room temperature; but when heated strongly, it becomes unstable and breaks down into carbon and water. Clay minerals, such as kaolinite, provide another good example. Kaolinite forms by the weathering of feldspar. It has an open, sheetlike molecular structure that is stable at the temperatures and pressures normally found at Earth's surface. Under the heat and pressure associated with deep burial, however, it cannot survive, and forms micas that have a more compact, sheetlike molecular structure.

Classification of Metamorphic Rocks

Metamorphic rocks are classified according to mineral composition and texture, including foliation and banding.

An interesting question associated with metamorphic rocks is, When has a rock changed enough to be called metamorphic? Rocks do not change instantaneously from one type to another; the changes occur along a continuum from little alteration to major changes. Generally, if a change can be recognized in a rock, the rock is considered metamorphic. Table 15.2 lists some common rocks and the rocks they can become when metamorphosed.

TABLE 15.2 DERIVATIONS OF COMMON METAMORPHIC ROCKS

Parent Rock	Metamorphic Rock
Shale	Compaction and reorientation → slate
	Slate → micas and chlorite form → phyllite
	Phyllite → grains recrystallize and get coarser → schist
	Schist → banding occurs → gneiss
Shale	Contact with magma → granofels
Sandstone	Quartzite
Limestone or dolostone	Marble
Basalt	Hydration reactions → greenschist
	Greenschist → dehydration → amphibolite
	Amphibolite → banding occurs → gneiss

The Scheme for Metamorphic Rock Identification

Metamorphic rocks are classified according to their textures (including foliation and banding) and compositions. In Figure 15.12 metamorphic rocks are divided into two main groups, foliated and nonfoliated, based on texture. Then, within each of these groups, the rocks are classified by grain size.

If you look at the column headed "Composition," you will see that the foliates share a common composition, while each of the nonfoliates has a distinctly different composition. The majority of foliated rocks are derived from clastic sedimentary rocks rich in clay minerals, such as shale and siltstone. The "Comments" column tell you what the parent rock was in each case.

Notice that slate, schist, and gneiss really form a continuum. Slate forms first as shale is slightly metamorphosed. The flakelike micas align because of compression and form surfaces along which the rock will split. With further metamorphosis, alignment of the mica becomes even more pronounced and phyllite is formed. As metamorphosis continues, chlorite changes to biotite, foliation becomes stronger, grains become coarser because of recrystallization, and a schist is formed. Still further metamorphism causes the crystals to grow even larger, alternating layers of light and dark minerals (bands) form as crystals migrate because of density differences, and a gneiss forms. The larger, interlocking crystals of a gneiss cause these rocks to break less regularly than schists, but they will often break completely across layers. In the highly metamorphic environment in which gneisses form, rock is plastic and will often be contorted into the weirdly twisted patterns seen in gneiss.

Texture		Grain Size	Composition	Type of Metamorphism	Comments	Rock Name	Map Symbol
FOLIATED	MINERAL ALIGNMENT	Fine	MICA	Regional	Low-grade metamorphism of shale	Slate	
FOLIATED	MINERAL ALIGNMENT	Fine to medium	MICA, QUARTZ, FELDSPAR, AMPHIBOLE, GARNET	(Heat and pressure increase with depth) ↓	Foliation surfaces shiny microscopic mica crystals	Phyllite	
FOLIATED	MINERAL ALIGNMENT	Fine to medium	MICA, QUARTZ, FELDSPAR, AMPHIBOLE, GARNET, PYROXENE	(Heat and pressure increase with depth) ↓	Platy mica crystals visible from metamorphism of clay or feldspars	Schist	
FOLIATED	BANDING	Medium to Coarse	MICA, QUARTZ, FELDSPAR, AMPHIBOLE, GARNET, PYROXENE	(Heat and pressure increase with depth) ↓	High-grade metamorphism; some mica changed to feldspar; segregated by mineral type into bands	Gneiss	
NONFOLIATED		Fine	Variable	Contact (Heat)	Various rocks changed by heat from nearby magma/lava	Hornfels	
NONFOLIATED		Fine to coarse	Quartz	Regional or Contact	Metamorphism of quartz sandstone	Quartzite	
NONFOLIATED		Fine to coarse	Calcite and/or dolomite	Regional or Contact	Metamorphism of limestone or dolomite	Marble	
NONFOLIATED		Coarse	Various minerals in particles and matrix	Regional or Contact	Pebbles may be distorted or stretched	Metaconglomerate	

Figure 15.12 Scheme for Metamorphic Rock Identification. Source: The State Education Department, *Earth Science Reference Tables*, 2001 ed. (Albany, New York; The University of the State of New York).

WHAT IS THE ROCK CYCLE?

Where do rocks come from? You might say igneous rock comes from magma, sedimentary rock comes from sediments, and metamorphic rock comes from other rocks that have changed. But consider that magma is melted rock, sediments are broken rock, and there had to be an already-existing rock to be metamorphosed. Basically, then, rocks come from other rocks.

Rocks are constantly changing from one type to another in a never-ending *rock cycle*, shown in Figure 15.13. The outside circle shows the different forms in which rock matter can exist: magma, igneous rock, sediments, and so on. Arrows leading from one form to another are labeled with the changes that are taking place. Arrows inside the circle show some alternative paths in the rock cycle. For example, an igneous rock does not always break down into sediments. If buried, it may be metamorphosed; therefore, there is an arrow inside the circle leading from igneous rock to metamorphic rock.

Let's follow a rock through one possible cycle, beginning with magma. A magma solidifies and forms an igneous rock. The igneous rock is uplifted and exposed at the surface, where weathering breaks it down into sediments. Rain washes the sediments into a stream. The stream carries them to the ocean, where they are deposited and buried. Compaction and cementation produce a sedimentary rock. The sedimentary rock is exposed to heat and pressure, perhaps as two of the plates in Earth's crust collide, and a metamorphic rock is formed. If the metamorphic rock is forced deep enough into the crust, it melts and forms magma.

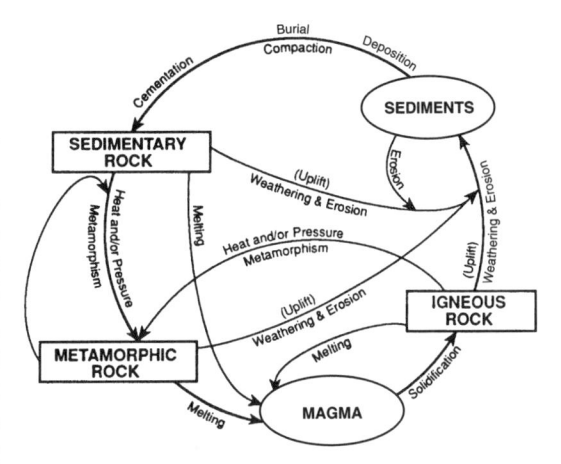

Figure 15.13 Rock Cycle in Earth's Crust. Source: The State Education Department, *Earth Science Reference Tables*, 2001 ed. (Albany, New York; The University of the State of New York).

Notice that we have come full circle back to where we started. This is an example of the rock cycle. It is, however, only one of many possible paths. Except for the stray meteorite that reaches the surface, Earth is essentially a closed system. Rock doesn't enter or leave, but is constantly recycling from one form to another.

The rock cycle diagram has no beginning and no end, but of course the rock cycle really did start somewhere. There is much evidence that Earth was originally totally molten. No solid rock existed; there was only magma. Therefore, it is thought that the original rocks formed as this magma cooled, and the rock cycle began with igneous rocks.

MULTIPLE-CHOICE QUESTIONS

In each case, write the number of the word or expression that best answers the question or completes the statement.

1. All rocks contain
   - (1) minerals
   - (2) sediments
   - (3) intergrown crystals
   - (4) fossils

Rocks

2. Although more than 2,000 minerals have been identified, 90 percent of Earth's lithosphere is composed of 12 minerals: feldspar, quartz, mica, calcite, amphiboles, kaolinite, augite, garnet, magnetite, olivine, pyrite, and talc.
   The best explanation for this fact is that most rocks
   (1) are monomineralic
   (2) are composed only of recrystallized minerals
   (3) have a number of minerals in common
   (4) have a 10% nonmineral composition

3. Rocks are classified on the basis of
   (1) the way in which they were formed
   (2) the shape of the sample
   (3) their age in millions of years
   (4) the mass of the sample

4. Which properties of a rock provide the most information about the environment in which the rock formed?
   (1) texture, composition, and structure
   (2) mass, composition, and color
   (3) density, volume, and mass
   (4) color, texture, and volume

5. Which would most likely occur during the formation of igneous rock?
   (1) compression and cementation of sediments
   (2) recrystallization of unmelted material
   (3) solidification of molten materials
   (4) evaporation and precipitation of sediments

6. According to the *Earth Science Reference Tables*, which property is most useful for identifying igneous rocks?
   (1) kind of cement   (3) number of minerals present
   (2) mineral composition   (4) types of fossils present

Base your answers to questions 7 and 8 on your knowledge of earth science and the diagrams below, which represent cross sections of four rock samples. Each cross section illustrates the sediments, minerals, or structural appearance of the rock sample.

(A)   (B)   (C)   (D)

7. Which rock sample is most likely monomineralic?
   (1) A
   (2) B
   (3) C
   (4) D

Note that question 8 has only three choices.
8. Which rock sample is most likely a nonsedimentary rock?
   (1) A
   (2) B
   (3) C

9. The diagrams below represent four rock samples. Which rock took the longest time to solidify from magma deep within Earth?

(1) Bands of alternating light and dark minerals	(2) Easily split layers of 0.0001-cm-diameter particles cemented together	(3) Glassy black rock that breaks with a shell-shape fracture	(4) Interlocking 0.5-cm-diameter crystals of various colors

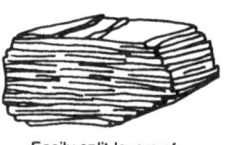

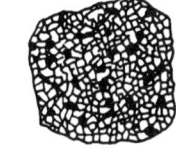

10. Extremely small crystal grains in an igneous rock are an indication that the crystals formed
    (1) under high pressure
    (2) over a short period of time
    (3) from an iron-rich magma
    (4) deep below the surface of Earth

11. Which igneous rock crystallized quickly near the surface of Earth, is light in color, and contains quartz and plagioclase feldspars?
    (1) granite
    (2) gabbro
    (3) basalt
    (4) rhyolite

Rocks

12. According to the *Earth Science Reference Tables*, which graph best represents the comparison of the average grain sizes in basalt, granite, and rhyolite?
Key to graph abbreviations:
B—Basalt G—Granite R—Rhyolite

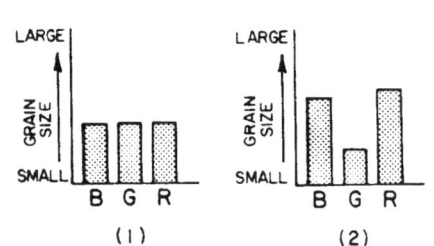

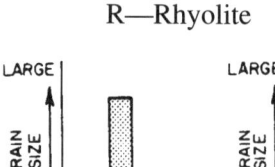

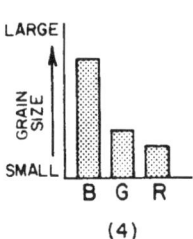

13. The diagram to the right represents a cross section of a coarse-grained nonsedimentary rock. According to the *Earth Science Reference Tables*, this rock is most likely to be
    (1) basalt
    (2) gabbro
    (3) rhyolite
    (4) granite

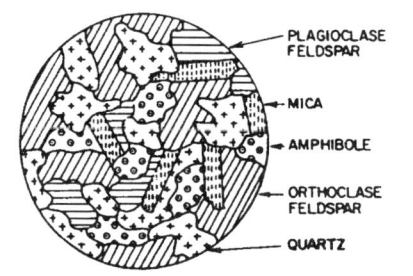

14. Sand collected at a beach contains a mixture of pyroxene, olivine, amphibole, and plagioclase feldspar. According to the *Earth Science Reference Tables*, the rock from which this mixture of sand came is best described as
    (1) dark-colored with a mafic composition
    (2) dark-colored with a felsic composition
    (3) light-colored with a mafic composition
    (4) light-colored with a felsic composition

15. According to the *Earth Science Reference Tables*, generally, as the percentage of felsic minerals in a rock increases, the rock's color will become
    (1) darker and its density will decrease
    (2) lighter and its density will increase
    (3) darker and its density will increase
    (4) lighter and its density will decrease

16. An igneous rock that crystallized deep below Earth's surface has the following approximate composition: 70 percent pyroxene, 15 percent plagioclase, and 15 percent olivine. According to the *Earth Science Reference Tables*, what is the name of this igneous rock?
    (1) granite   (3) gabbro
    (2) rhyolite   (4) basalt

330

17. A fine-grained rock has the following mineral composition: 30 percent plagioclase feldspar, 60 percent pyroxene, 5 percent olivine, and 5 percent amphibole. This rock is most likely
    (1) rhyolite
    (2) granite
    (3) gabbro
    (4) basalt

18. According to the *Earth Science Reference Tables*, rhyolite and granite are alike in that both are
    (1) fine-grained
    (2) dark-colored
    (3) mafic
    (4) felsic

19. According to the Scheme for Igneous Rock Identification in the *Earth Science Reference Tables*, which statement best describes the percentage of plagioclase feldspars in a sample of gabbro?
    (1) The percentage of plagioclase feldspars in gabbro can vary.
    (2) Gabbro always contains less plagioclase than pyroxene.
    (3) Plagioclase feldspars always make up 25% of a gabbro sample.
    (4) Gabbro contains no plagioclase feldspars.

20. Where are Earth's sedimentary rocks generally found?
    (1) in regions of recent volcanic activity
    (2) deep within Earth's crust
    (3) along the midocean ridges
    (4) as a thin layer covering much of the continents

21. Which process forms a sedimentary rock?
    (1) cooling of molten magma within Earth's crust
    (2) recrystallization of unmelted material within Earth's crust
    (3) cooling of a lava flow on Earth's surface
    (4) precipitation of minerals as seawater evaporates

22. Rock layers showing ripple marks, cross-bedding, and fossil shells were formed
    (1) from solidification of molten material
    (2) from deposits left by a continental ice sheet
    (3) by high temperature and pressure
    (4) by deposition of sediments in a shallow sea

23. Large rock-salt deposits in the Syracuse area indicate that the area once had
    (1) large forests
    (2) a range of volcanic mountains
    (3) many terrestrial animals
    (4) a warm, shallow sea

Rocks

24. According to the *Earth Science Reference Tables*, inorganic, land-derived sediments are classified on the basis of
    (1) composition
    (2) grain size
    (3) type of cement
    (4) rate of formation

25. Which sedimentary rock is most likely of bioclastic origin?
    (1) limestone
    (2) shale
    (3) halite
    (4) conglomerate

26. The diagrams below represent magnifications of rocks. Which is most likely a diagram of a nonsedimentary rock?

    KEY
    Quartz   Amphibole   Feldspar

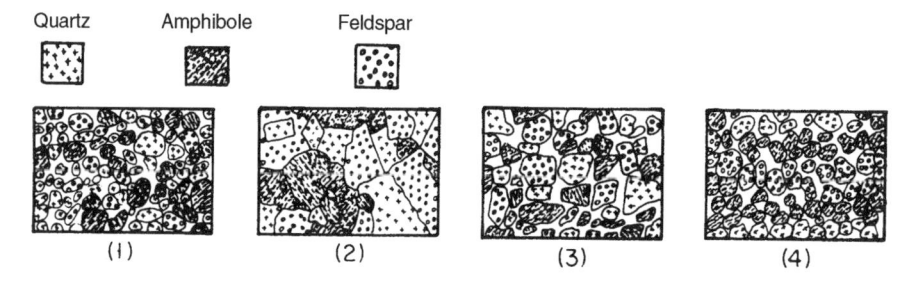

    (1)   (2)   (3)   (4)

27. According to the *Earth Science Reference Tables*, a sedimentary rock formed as the result of compaction and cementation of uniformly sized fine sediments with a diameter of 0.1 centimeter is called
    (1) conglomerate
    (2) shale
    (3) sandstone
    (4) slate

28. According to the *Earth Science Reference Tables*, which sedimentary rock could form as a result of evaporation?
    (1) conglomerate
    (2) shale
    (3) limestone
    (4) sandstone

29. According to the *Earth Science Reference Tables*, which characteristic determines whether a rock is classified as a shale, a siltstone, a sandstone, or a conglomerate?
    (1) the absolute age of the sediments within the rock
    (2) the mineral composition of the sediments within the rock
    (3) the particle size of the sediments within the rock
    (4) the density of the sediments within the rock

30. According to the *Earth Science Reference Tables*, which type of sedimentary rock contains the greatest range of particle sizes?
    (1) conglomerate
    (2) sandstone
    (3) shale
    (4) siltstone

31. According to the *Earth Science Reference Tables*, which sedimentary rock would be formed by the compaction and cementation of particles 1.5 centimeters in diameter?
    (1) shale
    (2) sandstone
    (3) conglomerate
    (4) siltstone

32. When dilute hydrochloric acid is placed on the sedimentary rock limestone and the nonsedimentary rock marble, a bubbling reaction occurs with both rocks. What does this indicate?
    (1) The minerals of these two rocks have similar chemical compositions.
    (2) The molecular structures of these two rocks have been changed by heat and pressure.
    (3) The physical properties of these two rocks are identical.
    (4) The two rocks originated at the same location.

33. Metamorphic rocks result from the
    (1) erosion of rocks
    (2) recrystallization of rocks
    (3) cooling and solidification of molten magma
    (4) compression and cementation of soil particles

34. Which characteristics indicate that a rock has undergone metamorphic change?
    (1) The rock shows signs of being heavily weathered and forms the floor of a large valley.
    (2) The rock is composed of intergrown mineral crystals and shows signs of deformed fossils and structure.
    (3) The rock becomes less porous when exposed at the surface and is finely layered.
    (4) The rock contains a mixture of different-sized rounded grains or both felsic and mafic silicate minerals.

35. Which characteristic of rocks tends to increase as the rocks are metamorphosed?
    (1) density
    (2) volume
    (3) permeability
    (4) number of fossils present

36. Which rocks would most likely be separated by a transition zone of altered rock (metamorphic rock)?
    (1) sandstone and limestone
    (2) granite and limestone
    (3) shale and sandstone
    (4) conglomerate and siltstone

Rocks

37. According to the *Earth Science Reference Tables*, which metamorphic rock has visible mica crystals and a foliated texture?
(1) pumice
(2) shale
(3) schist
(4) slate

38. Which statement is supported by information in the Rock Cycle in Earth's Crust diagram in the *Earth Science Reference Tables*?
(1) Metamorphic rock results directly from melting and crystallization.
(2) Sedimentary rock can be formed only from igneous rock.
(3) Igneous rock always results from melting and solidification.
(4) All sediments turn directly into sedimentary rock

39. According to the Rock Cycle in Earth's Crust diagram in the *Earth Science Reference Tables*, which type(s) of rock can be the source of deposited sediments?
(1) igneous and metamorphic rocks, only
(2) metamorphic and sedimentary rocks, only
(3) sedimentary rocks, only
(4) igneous, metamorphic, and sedimentary rocks

40. Which processes result in the formation of a nonsedimentary rock?
(1) compression and cementation
(2) solidification of molten material
(3) evaporation and precipitation
(4) biologic processes

41. According to the *Earth Science Reference Tables*, which sequence of events occurs in the formation of a sedimentary rock?

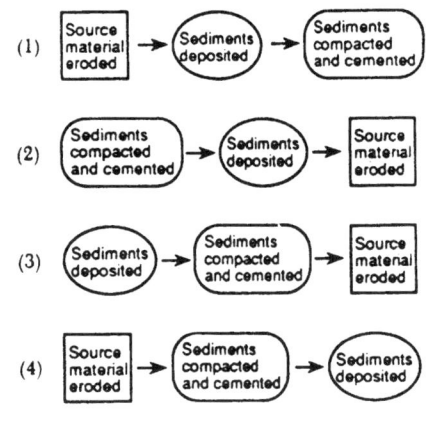

(1) Source material eroded → Sediments deposited → Sediments compacted and cemented

(2) Sediments compacted and cemented → Sediments deposited → Source material eroded

(3) Sediments deposited → Sediments compacted and cemented → Source material eroded

(4) Source material eroded → Sediments compacted and cemented → Sediments deposited

Base your answers to questions 42 through 46 on your knowledge of earth science, the *Earth Science Reference Tables*, and the diagrams of five rock samples (including metamorphic gneiss) shown below.

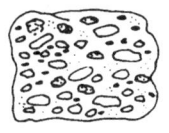

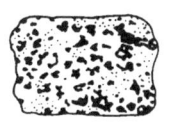

Basat   Gneiss   Conglomerate   Granite   Sandstone

42. Which rock is composed of sediments that have a range of sizes and that originate from different rock types?
    (1) basalt   (3) conglomerate
    (2) gneiss   (4) granite

43. Which rock shows banding that formed as a result of the recrystallization of unmelted material under high temperature and pressure?
    (1) gneiss   (3) granite
    (2) conglomerate   (4) sandstone

44. The granite most likely was formed by the processes of
    (1) erosion and deposition
    (2) compaction and cementation
    (3) heating and metamorphism
    (4) melting and solidification

45. The diagram to the right represents the percentage by volume of each mineral found in a sample of basalt. Which mineral is represented by the letter *X* in the diagram?
    (1) orthoclase feldspar
    (2) plagioclase feldspar
    (3) quartz
    (4) mica

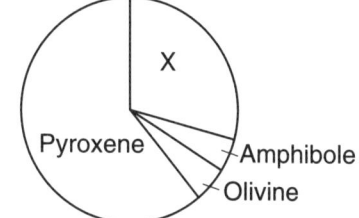

46. Which rock was formed by the compression and cementation of sediments with particle sizes ranging from 0.08 to 0.1 centimeter?
    (1) basalt   (3) granite
    (2) conglomerate   (4) sandstone

Rocks

Base your answers to questions 47 through 51 on your knowledge of earth science, the *Earth Science Reference Tables*, and the table below. The table provides data about the textures and mineral compositions of four different igneous rock samples having the same volume.

Rock	Texture	Orthoclase Feldspar	Quartz	Plagioclase Feldspar	Mica	Amphibole	Pyroxene
A	Coarse	62%	20%	7%	7%	4%	0%
B	Coarse	24%	40%	19%	10%	7%	0%
C	Fine	6%	16%	41%	14%	23%	0%
D	Fine	0%	0%	50%	0%	6%	44%

47. Which igneous rock sample contains the most quartz by volume?
(1) A
(2) B
(3) C
(4) D

48. According to the Rock Cycle in Earth's Crust diagram in the *Earth Science Reference Tables*, all four rock samples have undergone
(1) compaction and sedimentation
(2) volcanic eruption
(3) solidification from a molten state
(4) deposition and burial

49. Which two igneous rocks most likely formed closest to the surface of Earth?
(1) A and B
(2) C and D
(3) B and C
(4) A and D

50. Which rock sample is probably basalt?
(1) A
(2) B
(3) C
(4) D

51. Which rock sample has the greatest density and also contains the most magnesium?
(1) A
(2) B
(3) C
(4) D

Base your answers to questions 52 through 55 on the diagram below, which represents a scheme for classifying rocks. The letters *A, B, C* and *X, Y, Z* represent missing labels.

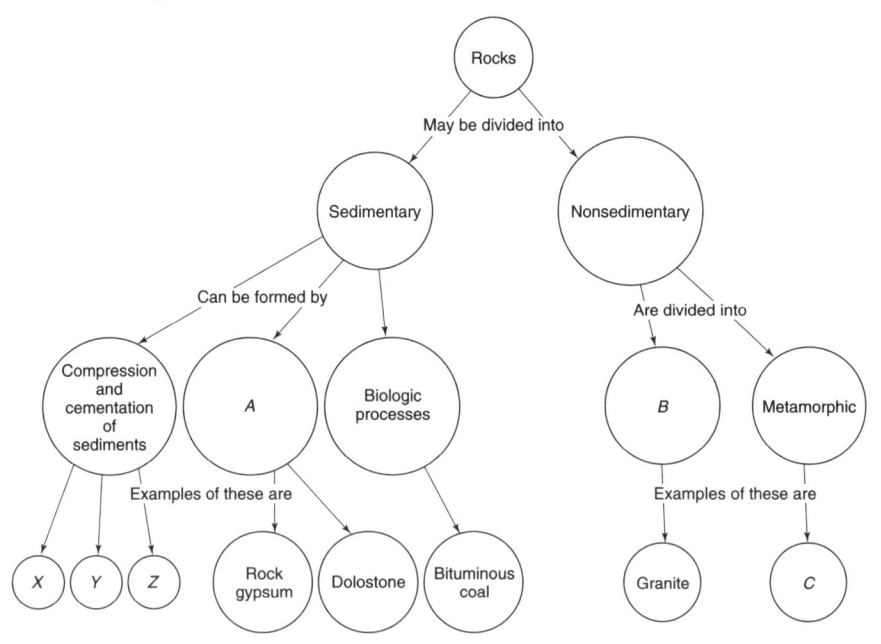

52. The classification of rocks into sedimentary or nonsedimentary groups is based primarily on the rocks'
(1) origin
(2) density
(3) color
(4) age

53. Which processes would form the type of rock that is represented by circle *B*?
(1) deposition and compaction
(2) weathering and erosion
(3) melting and solidification
(4) faulting and folding

54. If the rock in circle *C* formed from limestone, it would be called
(1) schist
(2) anthracite coal
(3) marble
(4) slate

55. Which rocks could be represented by circles *X, Y* and *Z*?
(1) shale, slate, and schist
(2) sandstone, shale, and siltstone
(3) anthracite coal, metaconglomerate, and rock salt
(4) breccia, gneiss, and rhyolite

Rocks

CONSTRUCTED RESPONSE QUESTIONS

The chart below shows the different rock families and their subdivisions. Circled letters *A*, *B*, and *C* indicate parts of the chart that have not been completed.

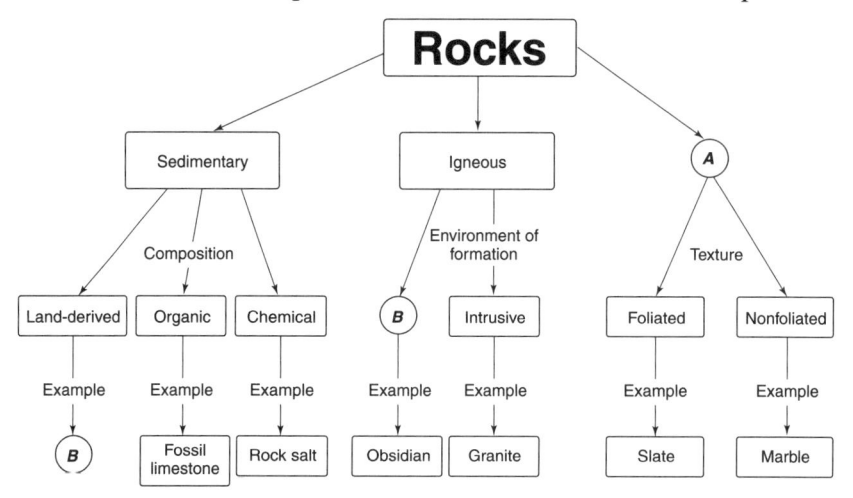

56. Complete the chart by writing the missing terms next to letters *A*, *B*, and *C* below.

A. _____[1]
B. _____[1]
C. _____[1]

Base your answers to questions 57 through 59 on the information, diagram, and data table below.

To sort a quartz sediment sample by particle size, a student shook the sample through a column containing screens *A* through *E*. The mesh of the screens (the open spaces between the wires) had different-sized openings, as represented by the diagram. The results are given below in the student's data table.

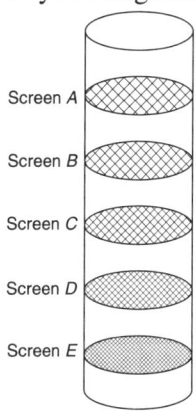

Student Data Table

Screen	Screen Mesh Ooening Size (cm)	Percentage of Particles Trapped by Screen (%)
A	0.1	0
B	0.05	30
C	0.025	45
D	0.0125	15
E	0.00625	10

(Screen mesh not drawn to scale)

338

57. Explain why screens *B* through *E* must be arranged in the order shown in the diagram to separate the sediment, as shown in the student data table. [1]

58. State two processes that must occur in nature to change a deposit of these sediments into clastic sedimentary rock. [1]

59. Which clastic sedimentary rock may be formed from particles of the same size as this quartz sediment sample? [1]

Base your answers to questions 60 through 62 on the Rock Cycle in Earth's Crust diagram in the *Earth Science Reference Tables*.

60. Construct two event sequences through which an igneous rock could be formed from a sedimentary rock. Represent each event sequence as shown in the sample below, which shows how an igneous rock could become a metamorphic rock.

SAMPLE: $\boxed{\text{Igneous rock}}$ → Heat and/or pressure → $\boxed{\text{Metamorphic rock}}$

61. In complete sentences, state all of the possible ways in which an igneous rock can undergo change according to the rock cycle diagram. [3]

62. The rock cycle diagram shows no arrow leading from "Magma" to "Metamorphic Rock."
  (a) In one or more complete sentences, explain why such a pathway is not possible. [1]
  (b) Construct an event sequence through which magma *could* theoretically be changed into metamorphic rock. [2]

Rocks

EXTENDED CONSTRUCTED RESPONSE QUESTIONS

Base your answers to question 63 on the cross section below, which shows the surface and subsurface rock near New York City.

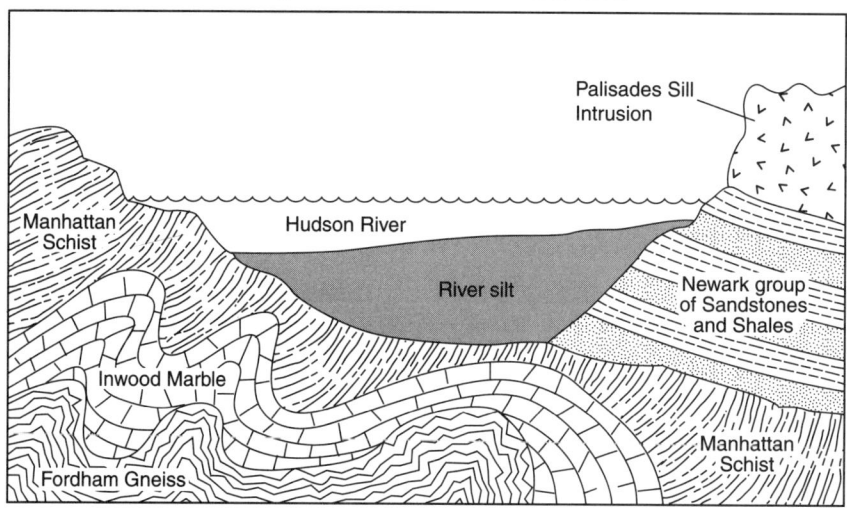

Geologic Section Across the Hudson River

(Not drawn to scale)

63. The Palisades Sill is an intrusive rock called diabase. A portion of the Palisades Sill intrusion contains large crystals and consists of 61 percent plagioclase feldspar and 33 percent pyroxene.
   (a) In one or more complete sentences, describe the conditions under which large crystals may form in an igneous intrusion. [1]
   (b) State another igneous rock that is similar in texture and composition to the diabase of the Palisades Sill. [1]
   (c) In one or more complete sentences, compare the origin of the rocks along the Hudson River's channel on the left side of this cross section with the origin of the rocks along the Hudson River's channel on the right side of this cross section. [1]
   (d) State the name of the rock formation that was originally limestone. [1]
   (e) Diabase is an intrusive igneous rock. Diabase forms underground from magma that cools over a long period of time. In one or more complete sentences, explain why the diabase of the Palisades Sill, which was not *formed* on the surface, is *found exposed* at the surface in this cross section. [1]

Unit Five: THE DYNAMIC EARTH

Chapter 16
EARTHQUAKES AND EARTH'S INTERIOR

KEY IDEAS With current technology, we cannot drill more than a few kilometers into the solid Earth. How, then, can we know anything of the internal structure of Earth? Earthquakes provide the key to inferring this internal structure. Earthquake waves literally travel around and through Earth, because rock is an excellent medium for conducting wave motion. Networks of seismometers, which detect earthquake waves, blanket Earth's surface. The records of earthquakes detected by seismometers can be analyzed to determine precisely when and where an earthquake occurred. By analyzing records of many earthquakes, as well as earthquake-wave behavior, scientists have been able to piece together a picture of Earth's internal structure. One of the most important results of such analysis has been the inference of a core that is chemically different from the rest of Earth and that is divided into a solid inner core and a liquid outer core.

KEY OBJECTIVES
Upon completion of this chapter, you will be able to:
- Describe the causes and effects of earthquakes.
- Explain how seismic records can be analyzed to locate the epicenter of an earthquake.
- Explain how earthquake waves can be used to create and refine a model of Earth's interior.
- Compare and contrast the composition, density, and phase of matter in the inner core, outer core, mantle, and crust.
- Identify regions of high earthquake activity, and relate them to the structure of Earth's crust.

EARTHQUAKES

An *earthquake* is a sudden trembling of the ground. Seismologists, scientists who study earthquakes, estimate that over one million quakes occur each year—about one every second! During most earthquakes, the shaking is so slight it is barely noticeable to the senses. During minor earthquakes, the ground feels a bit like a railway station when the train rumbles in. Major earthquakes cause violent shaking and lurching of the ground. The ground may split open, buildings may topple, pipes and power lines may be broken, and roadways may collapse. A destructive earthquake is a catastrophe, something to be planned for and guarded against.

Causes of Earthquakes

The major cause of earthquakes is *faulting*, the sudden movement of rock along planes of weakness, called faults, in Earth's crust. However, some earthquakes are caused by volcanic eruptions, and explosions set off by humans are responsible for a few. Nuclear testing is usually detected by the earthquakes it causes.

The *elastic rebound theory* explains the mechanism by which faulting causes earthquakes. At some places in Earth's crust, immense forces push or pull on the rock. Under this stress, rock can bend elastically, much as a wooden ruler can bend. However, if the stress increases, a point is reached where the rock can bend no farther without breaking; it is at its elastic limit. If stressed beyond this point, the rock snaps and the two broken edges whip back, or rebound. Great masses of rock suddenly scrape past each other, and the shock of this wrenching action jars the crust and sets an earthquake in motion. The point where the rock breaks is called the *focus* of an earthquake. The focus may be just beneath the surface or hundreds of kilometers down.

As the rock rebounds from this deformation, it pushes against surrounding rock, causing it to deform. The energy of this back and forth motion, or vibration, is transmitted as a series of deformations spreading out through the rock in all directions. The motion of a change in a medium is termed *wave motion*. The deformations of rock that spread out from the focus of an earthquake are therefore called *earthquake waves*, or *seismic waves*. Waves transport energy from one place to another through the motion of a change in the medium. Seismic waves transport energy from the focus, through the surrounding rock, to Earth's surface. The earthquake is first felt at the *epicenter*, a point on the surface directly above the focus. See Figure 16.1.

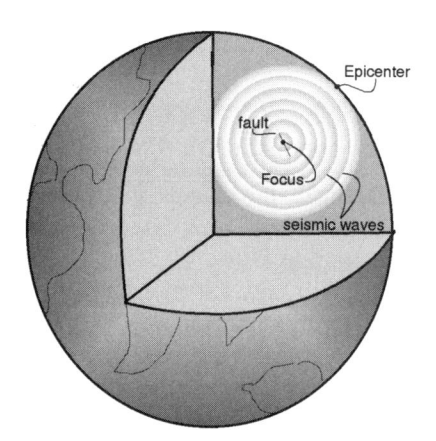

Figure 16.1 Fault, Focus, Epicenter, and Emanating Seismic Waves.

Earthquake Waves

To understand the behavior of earthquake waves, it is necessary to review a few basic concepts relating to wave behavior in general.

Wave Behavior

A wave in a rope is a simple analogy that will help you visualize what is going on when seismic waves travel through Earth. It will also help you to understand some of the inferences that can be drawn from the behavior of seismic waves.

If you give one end of a stretched rope a quick shake, a bump, or wave, travels down the rope at some speed. See Figure 16.2. If the rope is uniform and completely flexible, the wave keeps the same shape as it moves down the rope. This is what happens in an earthquake, when faulting gives the crust a quick shake that produces seismic waves.

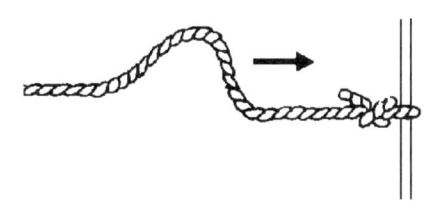

Figure 16.2 Wave Traveling Along a Rope.

As a wave travels down a rope, its forward part is moving upward and its rear part is moving downward. See Figure 16.3. The up and down motions of the string have kinetic energy. Work has to be done to produce the wave by pulling against the tension of the string, so the deformed string has potential energy. Thus, a wave contains both kinetic and potential energy as it travels along a rope. In like manner, seismic waves transfer energy as they travel through Earth. It is this energy that causes earthquake damage.

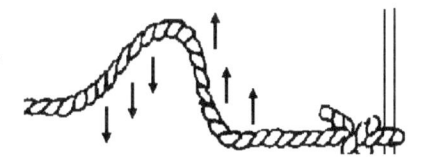

Figure 16.3 Movement of Rope as Wave Passes Through it.

The speed of the wave depends on the properties of the rope—how dense it is and how tightly it is stretched. Pulses move slowly down a loose, heavy rope because this rope has a lot of inertia and responds slowly to the forces acting on it. On the other hand, waves move rapidly down a taut, light rope because this rope has a greater tendency to straighten out. The same is true of seismic waves moving through Earth. The denser and more rigid the Earth material, the faster seismic waves travel through them.

When a wave reaches the end of

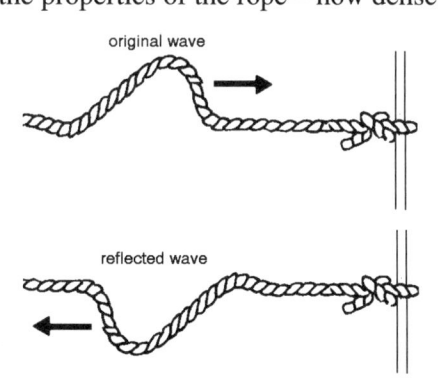

Figure 16.4 Wave Reflecting from a Fixed Support.

Unit Five **THE DYNAMIC EARTH**

Earthquakes and Earth's Interior

the rope, it may be reflected and travel back toward its source. Depending on how the end is held, the reflected wave may be upright, be inverted, or disappear completely. If the end is fixed to a support (see Figure 16.4), the wave exerts a force on the support, which then exerts an equal but opposite force on the rope. This opposite force causes the rope to be displaced in the opposite direction from the original wave. An inverted wave forms and travels back along the rope in the opposite direction from the original wave. If the end is looped around a support so that it is connected but still free to move, an upright wave is reflected. If the end is held in a state somewhere between totally fixed and totally free, the wave disappears completely.

Now, consider what happens if two different ropes, a heavy (more dense) rope and a thin (less dense) rope are spliced together. See Figure 16.5. One end is tied to a support, and the other end is given a shake so that a wave is produced. When the wave reaches the splice, the wave passes from the thin rope to the heavy rope and is said to be transmitted. However, the transmission is not total because a reflected wave also appears at the splice and travels back down the thin rope. This type of wave behavior is what we see when seismic waves cross a boundary between two distinctly different regions in Earth's interior. This wave behavior was an important clue to Earth's layered internal structure.

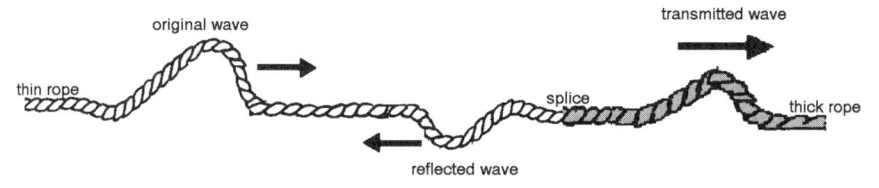

Figure 16.5 Transmitted and Reflected Waves at a Splice Between a Heavy and a Light Rope.

If one wave follows another in a rope, the result is a *periodic* wave. See Figure 16.6. In periodic waves a certain waveform, or shape, is repeated at regular intervals. Three quantities are often used to describe periodic waves: wave velocity, wavelength, and frequency. *Wave velocity* is the distance a wave moves each second; *wavelength* is the distance between adjacent crests or troughs; *frequency* is the number of waves that pass a given point in a second. Differences in wavelength, frequency, and the plane in which the waves oscillate are what distinguish different types of seismic waves.

If a periodic wave travels across a boundary between a more dense and a less dense medium at an angle, it is bent, or refracted. See Figure 16.7. This *refraction* occurs because, as the wave crosses the boundary, the front of the wave is moving at a different speed from the rear of the wave. The effect is somewhat like the result when the wheel at one end of an axle turns faster than the wheel at the other end—the path of the axle curves. Here, again, this is what we see when seismic waves cross a boundary in Earth's interior at an angle.

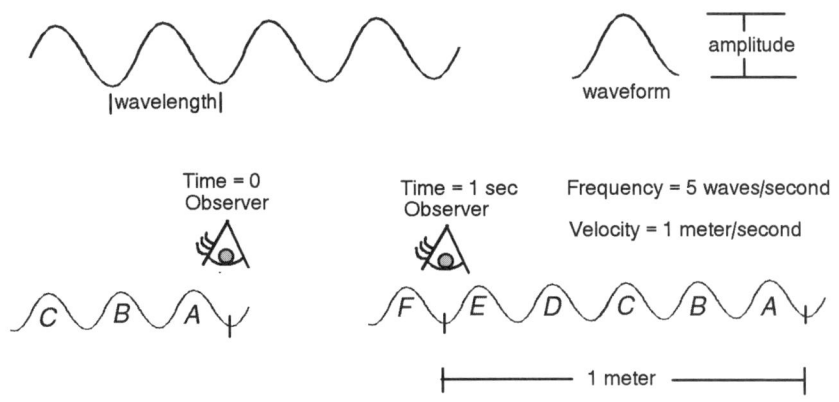

Figure 16.6 Periodic Wave, Showing Waveform, Wavelength, Wave Velocity, and Frequency.

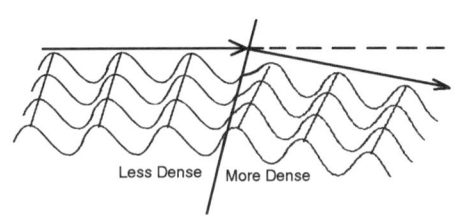

Figure 16.7 Periodic Wave Crossing a Boundary Between a More Dense and a Less Dense Medium and Being Refracted.

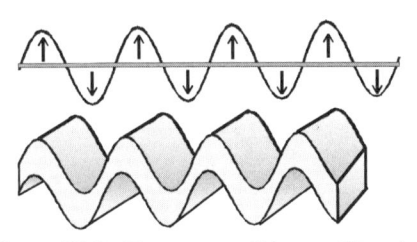

Figure 16.8 Transverse Waves, Showing Perpendicular Motion.

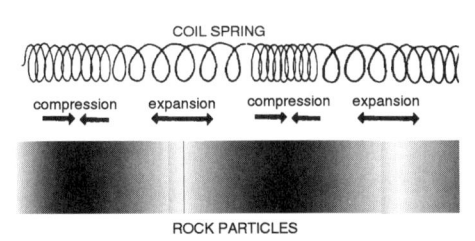

Figure 16.9 Longitudinal Waves in a Coil Spring and in Rock. As a longitudinal wave moves through a spring, its coils alternately compress and expand. In a rock, the individual particles are compressed and expanded.

Wave Types

In *transverse waves* the particles of the medium move back and forth in a direction perpendicular to the direction of wave motion. Waves in a stretched rope are transverse waves because any point on the rope moves side to side perpendicularly to the direction in which the wave is moving, as shown in Figure 16.8.

Longitudinal waves occur when the particles of the medium move back and forth in the same direction in which the waves travel. A good analogy is the movement of a compression down a coil spring when the end is pushed in and out. See Figure 16.9.

Seismic Waves

The complex motions that occur during faulting generate several types of periodic seismic waves. Some are transverse waves; others are longitudinal. Seismology took a giant step forward with the invention of the modern seismograph. A *seismograph* is a device that detects,

Earthquakes and Earth's Interior

measures, and records the motions of Earth that are associated with seismic waves.

A seismograph's operation is based upon the law of inertia: Objects that are at rest tend to remain at rest. A simple seismograph consists of a heavy, suspended weight that tends to remain at rest while Earth moves around it. A recording device (e.g., a pen) is attached to the weight. A recording medium, such as paper, is wound around a clockwork-driven drum mounted firmly in bedrock. When a series of seismic wave pulses causes the bedrock to move back and forth, the drum moves with it, but the heavy weight and its attached pen remain motionless for a long time. As the paper moves under the motionless pen, a line is drawn on the paper that records the back and forth motion of the bedrock, as shown in Figure 16.10a. The line recorded on paper by a seismograph is called a *seismogram* (see Figure 16.10b).

Modern seismographs use a laser beam reflected from a mirror on the weight instead of a pen. The laser beam exposes a track along photographic paper that is attached to the drum. When the bedrock vibrates, the track becomes a wavy line. In most earthquake recording stations, at least three seismographs are used, one to measure vibrations in each of three dimensions: x, y, and z (typically north-south, east-west, and up-down).

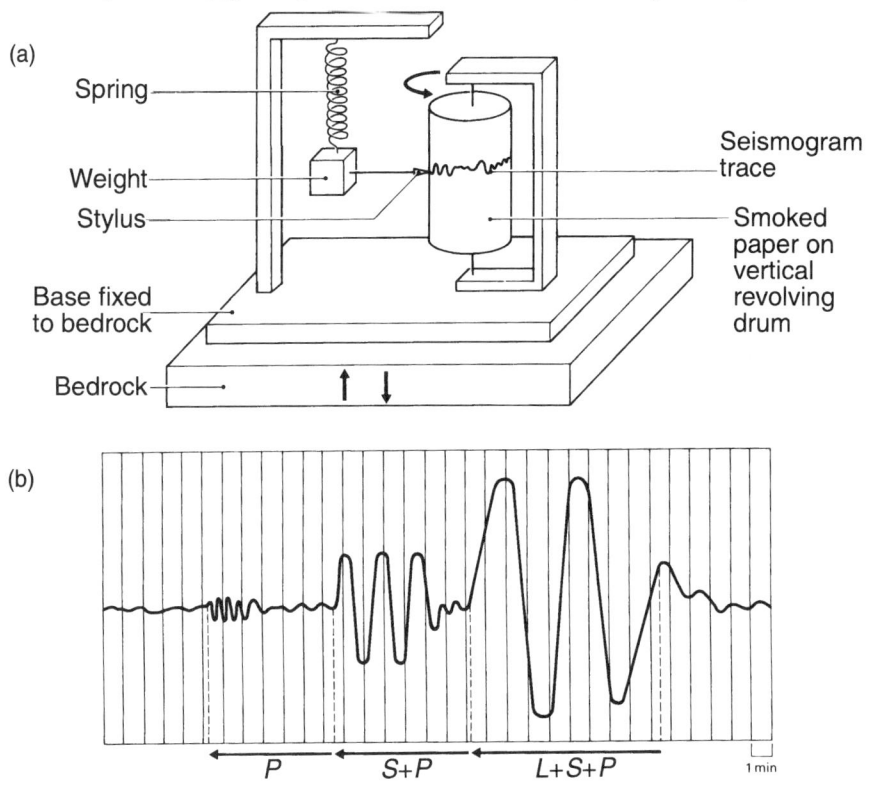

Figure 16.10 Simple Seismograph and Seismogram. P, S, and L represent P-, S-, and *Love* waves, respectively. Source: Earth Science on File. © Facts on File, Inc., 1988.

A careful study of seismograms reveals four different types of seismic waves. Two types, *P*-waves and *S*-waves, transport energy through the body of Earth and are therefore called *body waves*. Two other types, Love waves and Rayleigh waves, transport energy along the surface of Earth and are therefore called *surface waves*. Although surface waves play a major role in the damage caused by earthquakes, it is the body waves that reveal the structure of Earth's interior and make it possible to locate an earthquake's focus and epicenter. For this reason we will focus our attention on *P*-waves and *S*-waves.

P-waves

P-waves are longitudinal waves. They are alternating pulses of compression and expansion of rock in the same direction in which the wave travels. *P*-waves are the first (or **P**rimary) waves to reach a seismograph after an earthquake occurs because they travel fastest, at speeds of about 8 kilometers per second in the upper crust. *P*-waves can travel through solids, liquids, and gases because all three can be compressed and expanded. *P*-waves travel faster in these materials because these materials have stronger internal bonds; and, as occurs in a taut rope, the tendency to pull back into their original volume is greater. Since denser materials tend to be more rigid, *P*-wave speed increases in denser rock.

S-waves

S-waves are transverse waves. They twist rock back and forth, deforming its shape in a direction perpendicular to that of wave travel. *S*- (or **S**econdary) waves are the second type to arrive at a seismograph. *S*-waves arrive after *P*-waves because *S*-waves travel at a slower speed—about 4 kilometers per second in the upper crust. Like *P*-waves, *S*-waves travel faster in denser, more rigid rocks. *S*-waves can be transmitted only through solids. These waves cannot travel through liquids and gases because, when they are deformed, they do not return to their original shape.

Analyzing Seismograms

P-wave and *S*-wave recordings from a single event can be used to find the distance to the epicenter of an earthquake. With recordings from three or more seismic recording stations, the location of the epicenter can be determined. The basis for these determinations is the difference in the speeds of *P*-waves and *S*-waves.

When an earthquake occurs, all four types of seismic waves start moving outward from the focus at the same time. However, since they travel at different speeds, they do not all arrive at a seismograph at the same time. The *P*-waves, which travel the fastest, arrive first, followed by the *S*-waves some time later; the surface waves arrive last. The farther a seismograph is from the epicenter, the greater the difference between the arrival times of the *P*-waves and the *S*-waves.

For example, suppose *P*-waves travel 8 kilometers per second and *S*-

Earthquakes and Earth's Interior

waves travel 4 kilometers per second. The *P*-waves will arrive at a seismograph 80 kilometers away in 10 seconds. The *S*-waves will reach the same seismograph in 20 seconds. The arrival times at this first station will be 20 – 10 or 10 seconds *apart*. The same *P*-waves will reach a seismograph 200 kilometers away in 25 seconds, and the *S*-waves will arrive in 50 seconds. The arrival times at the farther station will be 50 – 25 or 25 seconds *apart*.

Determining Distance to the Epicenter of an Earthquake

For every distance that seismic waves travel, there is a corresponding difference in arrival times. This relationship between the difference in arrival time between *P*-waves and *S*-waves can be seen clearly in Figure 16.11. The travel time axis is marked off in minutes, and each minute is subdivided into three 20-second segments. The epicenter distance is marked off in thousands of kilometers ($10^3 = 1,000$) with each 1,000 kilometers subdivided into five 200-kilometer segments. The lines marked *S* and *P* show the times that *S*-waves and *P*-waves, respectively, take to travel a given distance. The vertical difference between the lines is the difference in arrival times at a given distance. It shows, for example, that *P*-waves travel 1,000 kilometers in 2 minutes while *S*-waves take about 4 minutes. The difference in arrival times is 2 minutes.

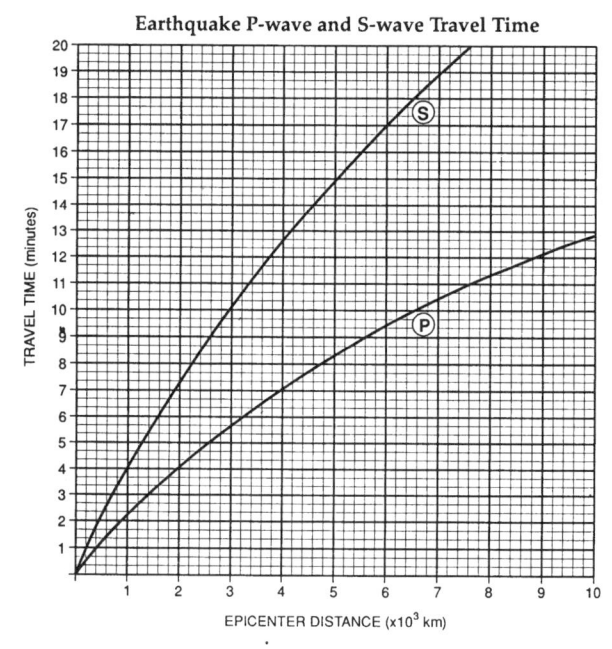

Figure 16.11 Earthquake *P*-wave and *S*-wave Travel Time. Source: The State Education Department, *Earth Science Reference Tables*, 2001 ed. (Albany, New York; The University of the State of New York).

Using Figure 16.11 and the relationship between difference in arrival times and distance, one can determine the distance between a seismograph and the epicenter of an earthquake as follows:

1. Determine the arrival times of the *P*-waves and the *S*-waves.
2. Calculate the *difference* in arrival times by subtracting the *P*-wave arrival time from the *S*-wave arrival time.
3. Using the travel time axis of the graph, mark off a distance along the edge of a piece of scrap paper equivalent to the *difference* in arrival times.
4. Keeping the edge of the paper vertical, find the point at which the marked-off difference in arrival times corresponds to the vertical distance between the *P*-wave line and the *S*-wave line.
5. Read the epicenter distance corresponding to that location on the graph.

Example:

The seismogram in Figure 16.12 is a record of an earthquake that was recorded by a seismograph. The seismogram does not tell how far away the earthquake occurred, but it does show when the *P*-waves and *S*-waves arrived. How far away from the seismograph was the epicenter of the earthquake?

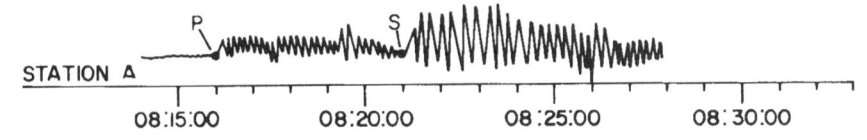

Figure 16.12 Seismogram.

To determine the epicenter, proceed as follows:

1. Note that the *P*-waves arrived at 08:16:00 and the *S*-waves arrived at 08:21:00.
2. Subtract: 08:21:00 − 08:16:00 = 00:05:00 or 5 min.
3. Using the travel time axis in Figure 16.11, mark off 5 min on the edge of a piece of paper.
4. Keeping the paper vertical, find the place where the marked-off portion of the paper just fits between the *S*-wave and *P*-wave lines.
5. Read the epicenter distance on the axis. It is 3,400 km.

The epicenter distance is 3,400 kilometers. These five steps are illustrated in Figure 16.13.

Determining the Location of the Epicenter

Knowing how far a seismograph station is from the epicenter does not pinpoint the location of the epicenter. In the example above, it could be 3,400 kilometers away in any direction. All of the points that are 3,400 kilometers from the seismograph in station *A* form a circle with a radius of 3,400 kilometers around that station. The epicenter is located somewhere on this circle. See Figure 16.14a.

Earthquakes and Earth's Interior

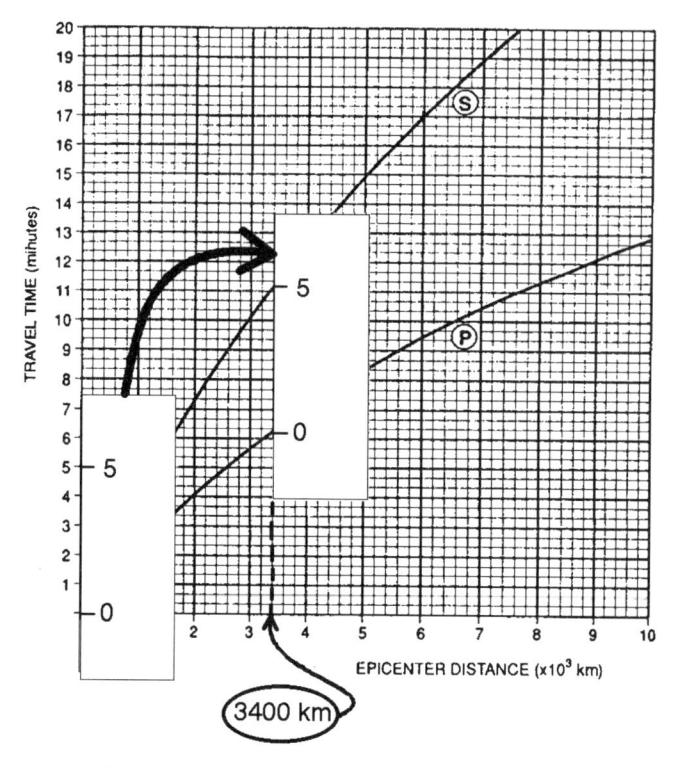

Figure 16.13 Five Steps in Determining Epicenter Distance.

However, if the same earthquake is 2,000 kilometers from a second station, station B, a circle can be drawn around that station too. The possible locations are now narrowed down to two points, as shown in Figure 16.14b. A third recording of the same earthquake, at station C, will eliminate one of these two points and pinpoint the location of the epicenter, as shown in Figure 16.14c.

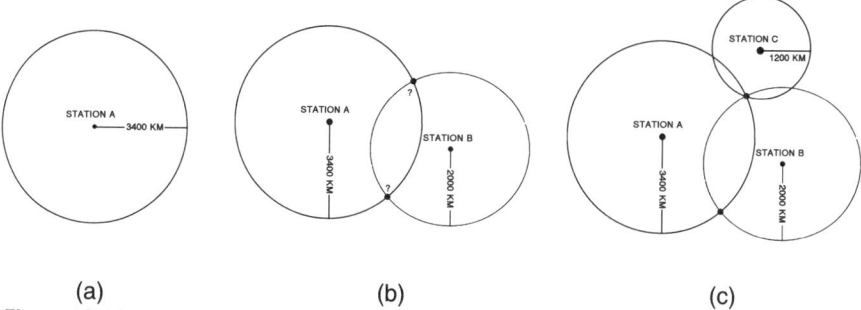

Figure 16.14 Information from Three Seismograph Stations Is Needed to Pinpoint the Location of the Earthquake Epicenter.

Example:

The seismograms shown in Figure 16.15 were recorded at three seismic recording stations for the same earthquake as in the preceding example. How can the epicenter of this earthquake be located?

The epicenter distance from station A is known to be 3,400 km. The difference in P-wave and S-wave arrival times at station B is 7 min, yielding an epicenter distance of 5,400 km. The difference in P-wave and S-wave arrival times at station C is 9 min, for an epicenter distance of 7,600 km.

Plotted on a map using the map scale, stations *A*, *B*, and *C* pinpoint the epicenter. See Figure 16.15.

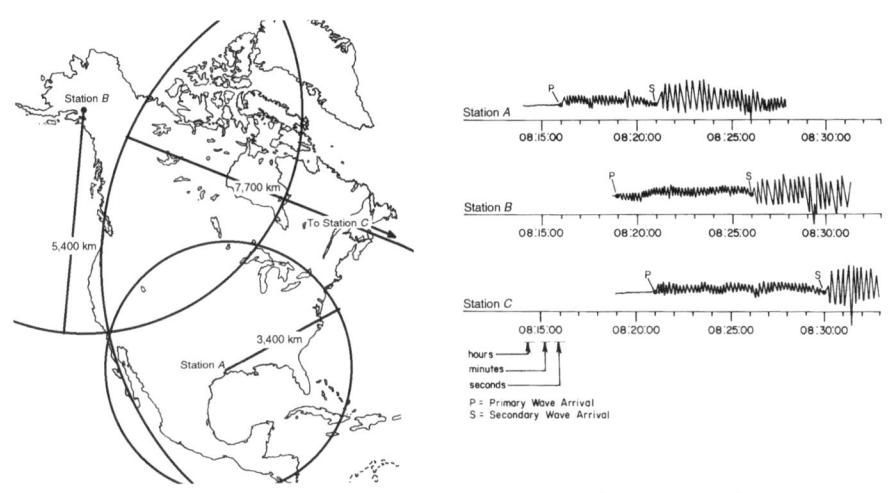

Figure 16.15 Locating the Epicenter. When the three seismograms above are analyzed, the distance from seismograph to the earthquake epicenter can be determined. Plotted on a map, only one place is the correct distance from all three stations.

Measuring Earthquakes

As seismic waves travel farther from the focus, they lose energy and their effects lessen. At a distance of 100 kilometers from the focus, the energy of an earthquake is only about 1/10,000 of what it was at 1 kilometer from the focus. The Richter and the Mercalli scales are used to measure the strengths of earthquakes.

The Richter Magnitude Scale (see Figure 16.16) is based upon the energy released by an earthquake, and energy is determined by direct measurements of the motions of the crust using seismic instruments. The greater the energy released by an earthquake, the larger the amplitude (height) of the earthquake waves. The Richter scale consists of numbers ranging from 0 to 8.6. The scale is logarithmic, meaning that the energy of a shock increases by powers of 10 in relation to the Richter magnitude numbers. In other words, each increase of 1 unit on the Richter scale corresponds to a tenfold increase in the amplitude (height) of the earthquake wave. Also, every tenfold increase in wave height corresponds to a hundredfold increase in energy! Thus, a magnitude-2 earthquake wave is 10 times higher than a magnitude-1 earthquake wave and releases 100 times as much energy.

The actual destructiveness of an earthquake, though, is influenced also by factors other than the amount of energy released. For example, destructiveness depends also upon nearness to the epicenter and the nature of the subsurface materials. Therefore, a magnitude-5 earthquake may cause vastly different amounts of damage in two different places. Since humans are concerned about damage to their land and property, a scale of earthquake damage intensity was developed. The modified Mercalli Intensity Scale (see Table 16.1) is based upon observations of earthquake damage and is concerned chiefly with the impact of an earthquake on structures made by humans and on human activities. Engineers and city planners often use this scale in reaching decisions about structural and land-use issues in earthquake-prone areas.

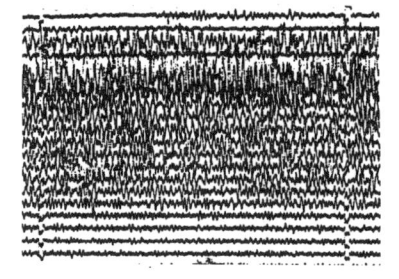

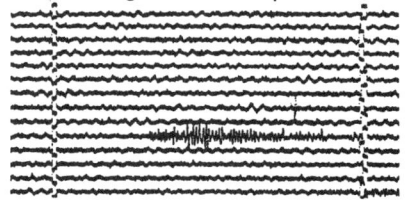

Figure 16.16 The Richter Magnitude Scale. This is based on the motion of bedrock as recorded by seismographs.

Effects of Earthquakes

The effects of earthquakes depend mainly upon the strength of the seismic waves and the nature of the material through which they pass. Earthquake effects include ground shaking and failure, surface faulting, changes in wells and geysers, and tsunamis.

Ground shaking affects solid rock the least, and loose, water-soaked ground the most. The same shock that causes solid bedrock to sway slightly may cause intense jolting in nearby wet clay or landfill. Tapping a bowl of Jello is a good analogy. The solid bowl vibrates but moves hardly at all, while the Jello jiggles and quivers noticeably. In ground failure, strong vibrations cause loose or water-soaked ground to break apart and settle, forming cracks and fissures. On hillsides these fissures can trigger landslides, slumping, and mudflows.

Surface faulting occurs when movement along a fault causes the surface to be lifted, lowered, or shifted sideways. The result is displacement of surface features and structures.

TABLE 16.1 MODIFIED MERCALLI INTENSITY SCALE

Intensity Value	Description of Effects
I	Not felt.
II	Felt by persons at rest; felt on upper floors because of sway.
III	Felt indoors. Hanging objects swing. Feels like a passing truck.
IV	Hanging objects swing. Feels like heavy truck passing, or a jolt is felt. Standing cars rock. Windows and dishes rattle. Glasses clink. In the upper range of IV, wooden frames and walls creak.
V	Felt outdoors and direction can be estimated. Sleepers are awakened. Liquids are disturbed, some spill. Small, unstable objects upset. Doors swing open and close. Shutters and pictures move.
VI	Felt by all. Persons walk unsteadily, and many are frightened and run indoors. Windows, dishes, and glassware are broken. Furniture is moved or overturned. Pictures fall from walls. Weak plaster cracks.Trees and bushes visibly shaken or heard to rustle.
VII	Difficult to stand. Noticed by drivers of cars. Hanging objects quiver. Furniture is broken. Weak chimneys break at roof line. Fall of plaster, loose bricks, masonry. Waves on ponds and water muddied. Small slides and cave-ins of sand and gravel banks.
VIII	Steering of moving cars is affected. Partial collapse of masonry structures. Chimneys and smokestacks twist and fall. Frame houses move on foundation if not bolted down. Branches broken from trees. Wet ground and steep slopes crack.
IX	General panic. Weak masonry destroyed, stronger masonry cracks, is seriously damaged. Frame structures not bolted shift off foundations. Frames cracked. Reservoirs seriously damaged. Underground pipes broken. Conspicuous cracks in ground.
X	Masonry and frame structures destroyed along with their foundations. Some bridges destroyed. Serious damage to reservoirs, dikes, dams, and embankments. Large landslides. Water thrown out of lakes, canals, rivers, etc. Rails bent slightly.
XI	Rails bent greatly. Underground pipelines completely destroyed and out of service.
XII	Damage nearly total. Large rock masses displaced. Objects thrown into the air. Lines of sight and level are distorted.

Modified from: H.O. Wood and F. Neumann, "Modified Mercalli Intensity Scale of 1931," *Bulletin of the Seismological Society of America*, Vol. 21, No. 4, pp. 277–288.

As seismic waves travel through water-saturated rock or soil, water is compressed and forced through pore spaces. Like a hand alternately squeezing and releasing a soaked sponge in a bucket, seismic waves force water in and out of rock and soil surrounding wells, causing their water levels to fluctuate. Seismic waves have a similar effect on rock fissures, causing geysers to behave erratically.

Tsunamis are immense sea waves caused by earthquakes beneath the ocean floor or by undersea landslides. Tsunamis may be only a few meters high, but they have very long wavelengths and travel much more rapidly than ordinary ocean waves. Tsunamis have been clocked moving faster than 500 kilometers per hour with wavelengths as great as 200 kilometers. When tsunamis reach shallow water, they are slowed by friction with the ocean bottom. In bays and narrow channels, however, their high speed and long wavelength may be funneled into huge breaking waves more than 20 meters high. The force exerted by such a huge, fast-moving mass of water can do extensive damage.

Earthquakes are geologic hazards to humans, but anticipating the hazards earthquakes pose and preparing to meet them can minimize loss of lives and property. Figure 16.17 shows areas of various degrees of seismic risk in the United States.

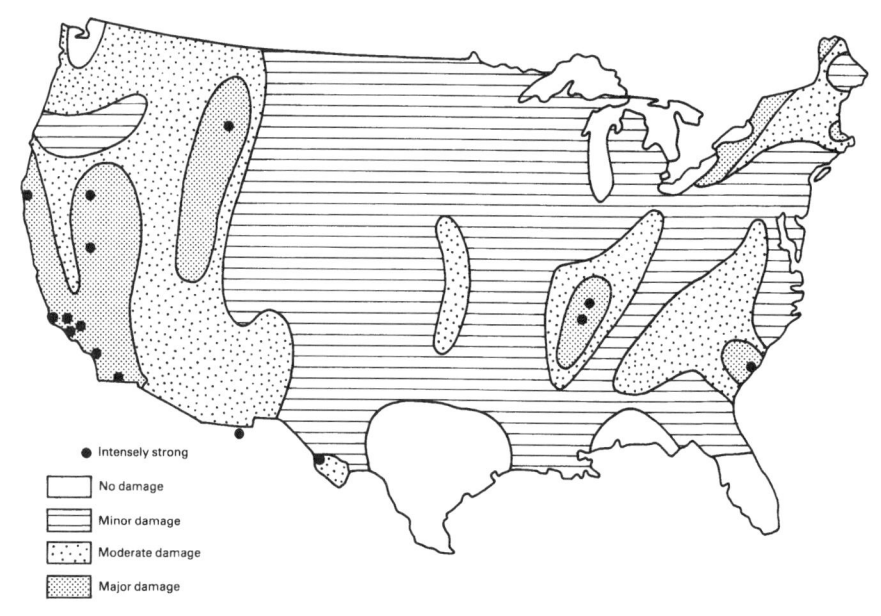

Figure 16.17 Seismic Risk Map of the United States. Source: Earth Science on FIle. © Facts on File, Inc., 1988.

Earthquake Zones

If the locations of earthquakes are plotted on a map of Earth, an interesting pattern emerges, as shown in Figure 16.18. Rather than being randomly spread over Earth, earthquakes occur in distinct, narrow zones. These zones include the mid-ocean ridges, the rim of the Pacific Ocean (called the *Ring of Fire* for its many active volcanoes), and the Mediteranean Belt. As you will discover in Chapter 18, the locations of earthquake epicenters reveal the locations of boundaries between the plates of Earth's crust.

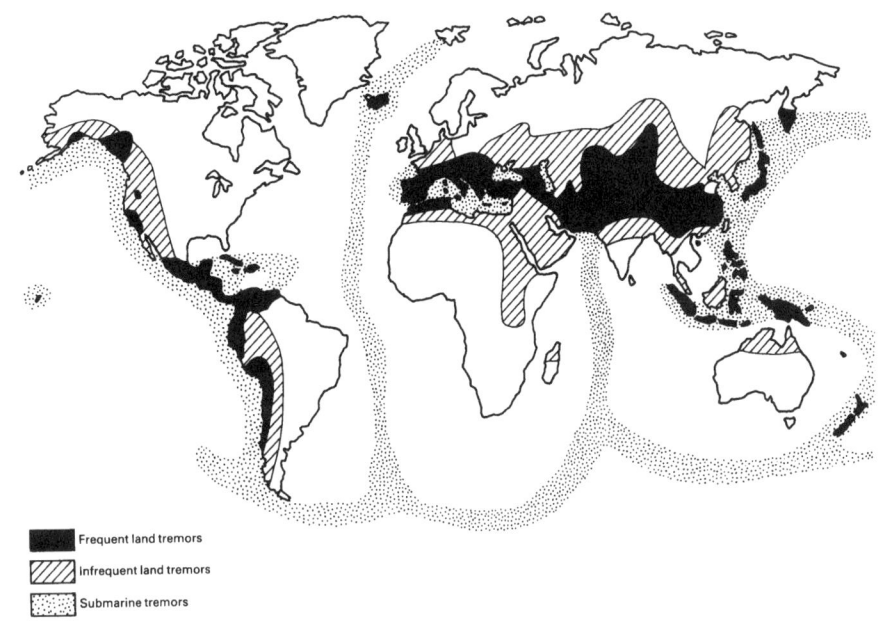

Figure 16.18 Worldwide Earthquake Distribution. Source: Earth Science on File. © Facts on File, Inc., 1988.

- ■ Frequent land tremors
- ▨ Infrequent land tremors
- ▨ Submarine tremors

THE INFERRED STRUCTURE OF EARTH'S INTERIOR

Whenever an earthquake occurs, its seismic waves are recorded by seismographs at hundreds of seismic recording stations throughout the world. Analysis of these seismograms reveals much about the structure of Earth's interior.

As was discussed earlier, seismic waves are reflected from the surfaces of materials and refracted as they move across boundaries between substances of different densities. If Earth had a homogeneous composition and density, seismic waves would move through it in simple, straight-line paths with reflections only when the surface was reached. They would also travel directly from the earthquake's focus to all points on Earth's surface. Furthermore, if the location of an earthquake's epicenter is known, the arrival times of P-waves and S-waves at all stations should be easy to predict.

This is not the case. P-waves are absent from seismographs at certain distances from the epicenter and are more highly concentrated elsewhere. Predictions of arrival times based upon straight-line travel differ markedly from actual measurements. Moreover, the paths of seismic waves curve. This gradual bending is due to the gradual increase in density with depth, caused by increasing pressure. There are many more reflected waves than would be expected from a homogeneous Earth. There are also places on Earth's

Earthquakes and Earth's Interior

surface where *P*-waves or *S*-waves, or both, do not arrive directly from the focus of the earthquake. These facts led seismologists to conclude that Earth's interior is not homogeneous.

The Crust

In 1910, Andrija Mohorovicic discovered that two sets of *P*-waves and *S*-waves arrived at locations close to an earthquake's epicenter, but only one set arrived farther away. Mohorovicic explained this phenomenon by hypothesizing a boundary between rocks of different densities that would cause the waves to refract. As shown in Figure 16.19, stations close to the epicenter of an earthquake can receive both direct and refracted seismic waves. Farther away, however, waves cannot reach the station without crossing the boundary, so only refracted waves arrive.

This boundary, called the *Moho* in honor of Mohorovicic, marks the boundary between the outermost layer of Earth, the crust, and the mantle beneath it. The density of the crust varies from about 2.7 grams per cubic centimeter near the surface to about 3.1 grams per cubic centimeter near the Moho. The thickness of the crust varies, but is usually greatest beneath mountains and least beneath oceans.

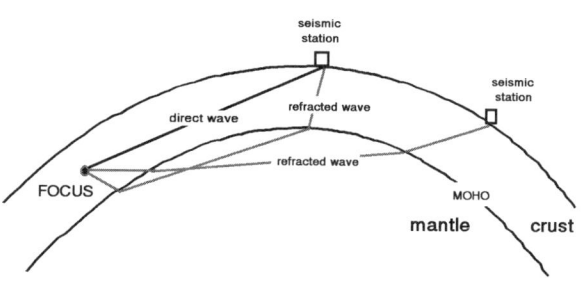

Figure 16.19 The Moho Boundary.

The Mantle and Core

Between the epicenter and an angle of 103° from the epicenter, *P*-waves and *S*-waves arrive directly from the focus, as would be expected. Beyond 103°, however, the *S*-waves *disappear*! This region is known as the *S*-wave shadow zone. Seismic stations in the *S*-wave shadow zone receive no *S*-waves from an earthquake.

Between 103° and 143° from the epicenter, the *P*-waves *also* disappear. But beyond 143°, the *P*-waves reappear. Therefore, the region between 103° and 143° from the epicenter is known as the *P*-wave shadow zone. Seismic stations located between 103° and 143° of an earthquake's epicenter fall within both the *S*-wave shadow zone and the *P*-wave shadow zone and record no seismic waves from the earthquake. Seismic stations beyond 143° in either direction from the epicenter record only *P*-waves. That this odd pattern of shadow zones can be explained by Earth's internal structure is shown in Figure 16.20.

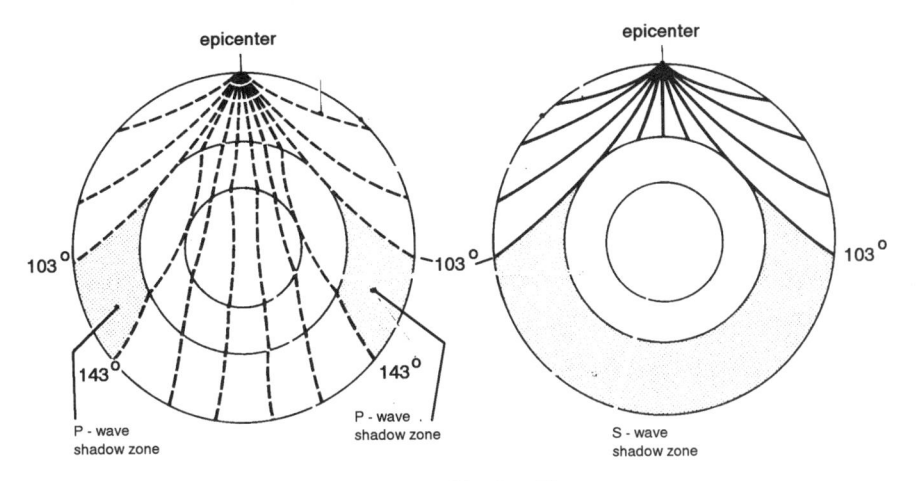

Figure 16.20 The *P*-Wave and *S*-Wave Shadow Zones.

A boundary at a depth of about 2,900 kilometers, between rocks of markedly different densities, produces a strong refraction of *P*-waves that cross it. Up to 103°, waves arrive directly from the epicenter. The strong refraction as soon as the boundary is encountered immediately angles the waves away to 143°, so no waves reach the region in between. Since *P*-waves continue to be received past 143°, they are able to pass through the material beneath the boundary. This boundary marks the bottom of the mantle; the layer beneath it is called the *core*.

The complete absence of *S*-waves beyond 103° indicates that they are unable to travel through the material beneath the mantle. Since transverse waves cannot travel through fluids, the disappearance of *S*-waves makes sense if the core is a fluid.

Careful analysis of refraction of *P*-waves that pass through the core and the presence of reflections from inside the core indicate that another boundary inside the core divides it into an inner core and an outer core. Only the outer core is thought to be a fluid.

Summary of Inferences about Earth's Interior

Figure 16.21 summarizes what has been inferred about Earth's interior based upon studies of seismic waves and other data. The diagram at the top shows a cross section of Earth with the interior layers labeled. Along the right edge, the density of each layer is shown. Notice that the mantle is divided into two regions, the asthenosphere, or plastic mantle, and the stiffer mantle. Notice, too, the sharp increase in density between the mantle and the core, which explains the strong refraction of *P*-waves. Directly beneath the cross section is a graph showing pressure in *millions of atmospheres* at different depths. The dotted lines indicate boundaries between crust, mantle, and outer and inner cores. On the bottom is a similar graph showing temperatures and

Earthquakes and Earth's Interior

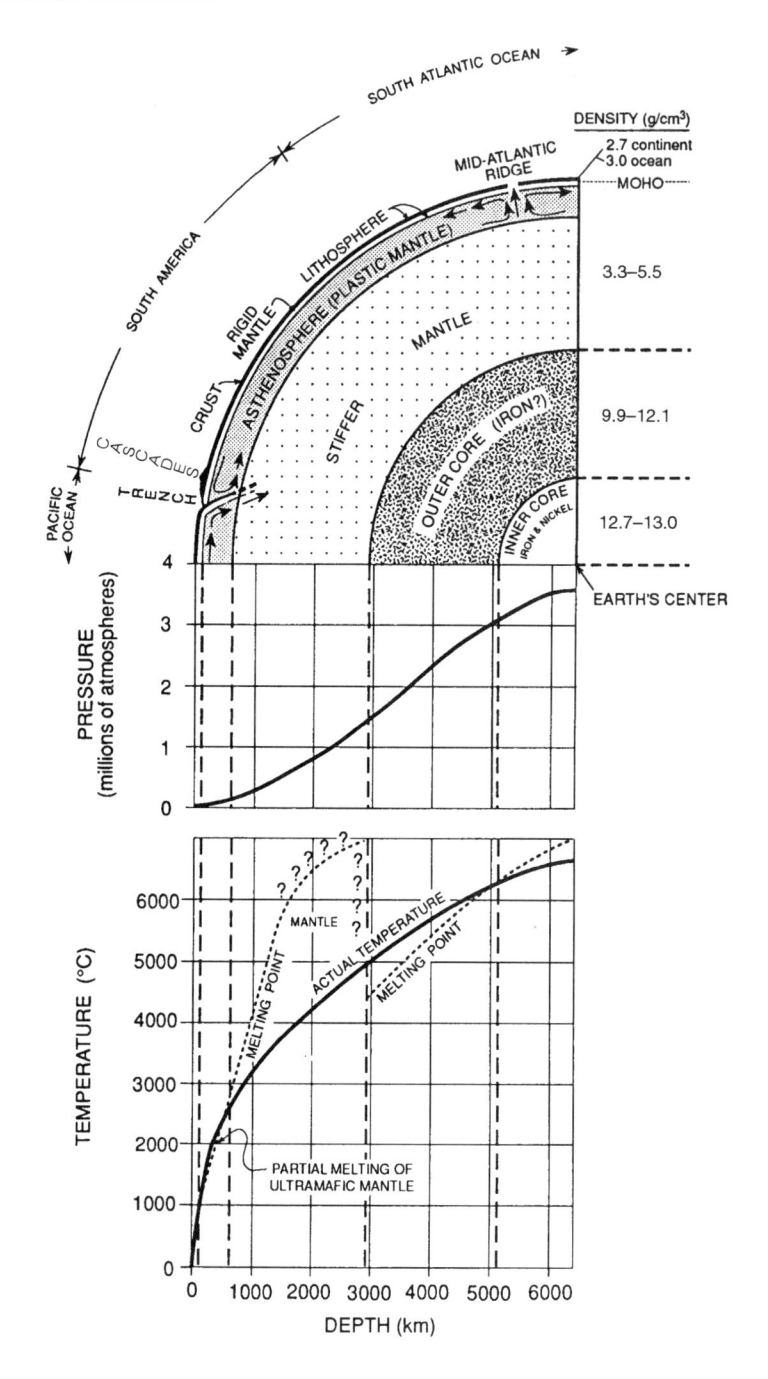

Figure 16.21 Inferred Properties of Earth's Interior. Source: The State Education Department, *Earth Science Reference Tables*, 2001 ed. (Albany, New York; The University of the State of New York).

melting points of rock at different depths. Notice that in the outer core the actual temperature is higher than the melting point of the rocks.

Meteorites are thought to be fragments of an Earth-like planet. Some are similar in composition to Earth rocks, but others are composed of iron-nickel alloy. An iron-nickel alloy would have the density and rigidity of the core as deduced from seismic wave studies. Therefore, Earth's core is believed to consist of an iron-nickel alloy. An iron-nickel core in which convection currents transfer heat outward (see Chapter 10) would also help to explain Earth's magnetic field.

MULTIPLE-CHOICE QUESTIONS

In each case, write the number of the word or expression that best answers the question or completes the statement.

1. The most frequent cause of major earthquakes is
    (1) faulting
    (2) folding
    (3) landslides
    (4) submarine currents

2. The immediate result of a sudden slippage of rocks within Earth's crust is
    (1) isostasy
    (2) erosion
    (3) an earthquake
    (4) the formation of convection currents

3. Where does most present-day faulting of rock occur?
    (1) in regions of glacial activity
    (2) in the interior areas of continents
    (3) at locations with many lakes
    (4) at interfaces between moving parts of Earth's crust

4. Where have earthquakes occurred most frequently during the last 100 years?
    (1) in the polar regions
    (2) in the interior of continental areas
    (3) along the Pacific Ocean coastlines
    (4) along the Atlantic Ocean coastlines

5. The lines on the map below represent faults and fractures in the bedrock of New York State.

Earthquakes and Earth's Interior

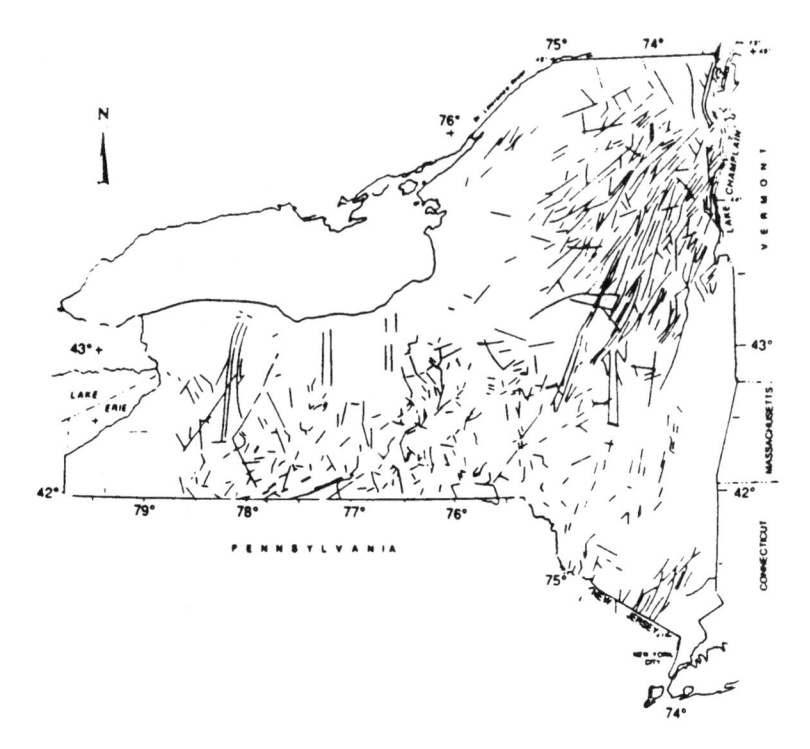

According to the *Earth Science Reference Tables*, at which location on this map are earthquakes most likely to occur?
(1) New York City
(2) Kingston
(3) Buffalo
(4) Mt. Marcy

6. A large belt of mountain ranges and volcanoes surrounds the Pacific Ocean. Which events are most closely associated with these mountains and volcanoes?
(1) hurricanes
(2) tornadoes
(3) sandstorms
(4) earthquakes

7. The diagram below represents a geologic cross section of a portion of Earth's crust. Which geologic event that is shown in the diagram suggests past crustal movement at that location?

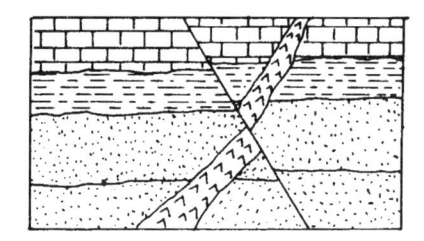

KEY:
Igneous Intrusion
Sedimentary Rocks

360

(1) deposition of sediments
(2) intrusion of molten material (magma)
(3) folding of rock layers
(4) faulting of rock layers

8. Which cross-sectional diagram below best represents a landscape region that resulted from faulting?

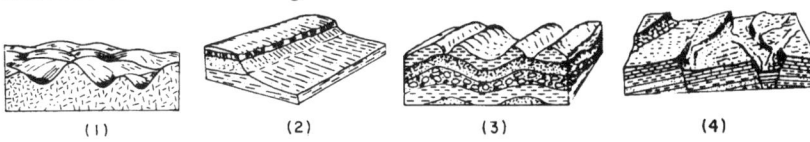

9. Which statement best describes the relationship between the travel rates and travel times of earthquake *P*-waves and *S*-waves from the focus of an earthquake to a seismograph station?
   (1) *P*-waves travel at a slower rate and take less time.
   (2) *P*-waves travel at a faster rate and take less time.
   (3) *S*-waves travel at a slower rate and take less time.
   (4) *S*-waves travel at a faster rate and take less time.

Note that question 10 has only three choices.

10. As the distance from the epicenter of an earthquake increases, the time between the arrival of the *P*-waves and *S*-waves at a seismograph station will
    (1) decrease
    (2) increase
    (3) remain the same

11. The distance between an earthquake epicenter and the location of a seismograph station can be calculated because
    (1) seismographs are sensitive to directions
    (2) earthquake waves decay at known rates
    (3) shear waves will not pass through liquids
    (4) shear waves and compression waves travel at different speeds

12. An earthquake occurred at 5:00:00 A.M. According to the *Earth Science Reference Tables*, at what time would the *P*-wave reach a seismograph station 3,000 kilometers from the epicenter?
    (1) 5:01:40 A.M.   (3) 5:05:40 A.M.
    (2) 5:04:30 A.M.   (4) 5:10:15 A.M.

Earthquakes and Earth's Interior

13. The difference in arrival times at a seismograph station for *P*- and *S*-waves from an earthquake is 5.0 minutes. According to the *Earth Science Reference Tables*, how far away is the epicenter of the earthquake?
    (1) 1.3×10^3 km
    (2) 2.6×10^3 km
    (3) 3.5×10^3 km
    (4) 8.1×10^3 km

14. A seismograph station records an earthquake with a difference of 9 minutes between the arrival times of initial *P*-waves and *S*-waves. According to the *Earth Science Reference Tables*, approximately how far from the epicenter is this station?
    (1) 2,500 km
    (2) 5,800 km
    (3) 7,700 km
    (4) 9,000 km

15. The epicenter of an earthquake is located near Massena, New York. According to the *Earth Science Reference Tables*, the greatest difference in arrival times of the *P*- and *S*-waves for this earthquake would be recorded in
    (1) Albany
    (2) Utica
    (3) Plattsburgh
    (4) Binghamton

16. Which graph best represents the relationship between differences in arrival times of *P*-waves and *S*-waves and distance from the earthquake epicenter?

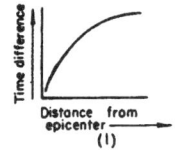

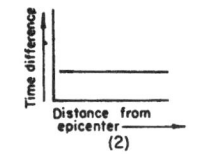

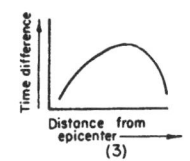

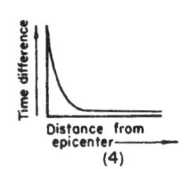

Base your answers to questions 17 and 18 on your knowledge of earth science, the *Earth Science Reference Tables*, and the chart below. This chart provides partial seismic data for an earthquake that was detected at two different seismographic stations. The earthquake occurred at 10 hr:00min:00 sec A.M. eastern standard time (e.s.t.).

	P-wave Arrival Time (e.s.t)	S-wave Arrival Time (e.s.t.)	Distance from Epicenter (km)
Triville	10 hr:07min:00 sec A.M.		
Endburg			9,000

17. At what time (e.s.t.) did the *S*-wave arrive at Triville?
    (1) 10 hr:01min:45 sec
    (2) 10 hr:04min:00 sec
    (3) 10 hr:07min:00 sec
    (4) 10 hr:12min:40 sec

18. At what time (e.s.t.) did the *P*-wave arrive at Endburg?
    (1) 10 hr:01min:45 sec
    (2) 10 hr:09min:00 sec
    (3) 10 hr:12min:10 sec
    (4) 10 hr:22min:15 sec

19. Four seismograph stations received data from the same earthquake. The table below shows the differences in travel times for the *P*- and *S*-waves recorded at each station.

Station	Travel-Time Difference
A	4 min 32 sec
B	3 min 52 sec
C	3 min 10 sec
D	4 min 17 sec

    Which station is the closest to the epicenter of the earthquake?
    (1) A   (3) C
    (2) B   (4) D

20. A seismograph station is 2,000 kilometers from an earthquake epicenter. According to the *Earth Science Reference Tables*, how long will an *S*-wave take to travel from the epicenter to the station?
    (1) 7 min 20 sec   (3) 3 min 20 sec
    (2) 5 min 10 sec   (4) 4 min 10 sec

21. Two different cities experience an earthquake with a magnitude of 5.5 on the Richter Magnitude Scale. However, on the Modified Mercalli Intensity Scale, the earthquake was rated V in one city and VII in the other city. The best explanation for this difference is that
    (1) one city is nearer the Equator and the other is near a pole
    (2) the earthquake occurred at 8 P.M. in one city and at 4 A.M. in the other
    (3) one city is built on bedrock while the other is built on loose sediments
    (4) one city is in a drier climate zone than the other

Earthquakes and Earth's Interior

Base your answers to questions 22 through 26 on your knowledge of earth science, the *Earth Science Reference Tables*, and the map below. The map shows the epicenters and intensities of recent earthquakes in New York State. The state has been subdivided into four regions (A,B,C,D). In the key, VIII represents the most intense earthquakes and IV represents the least intense.

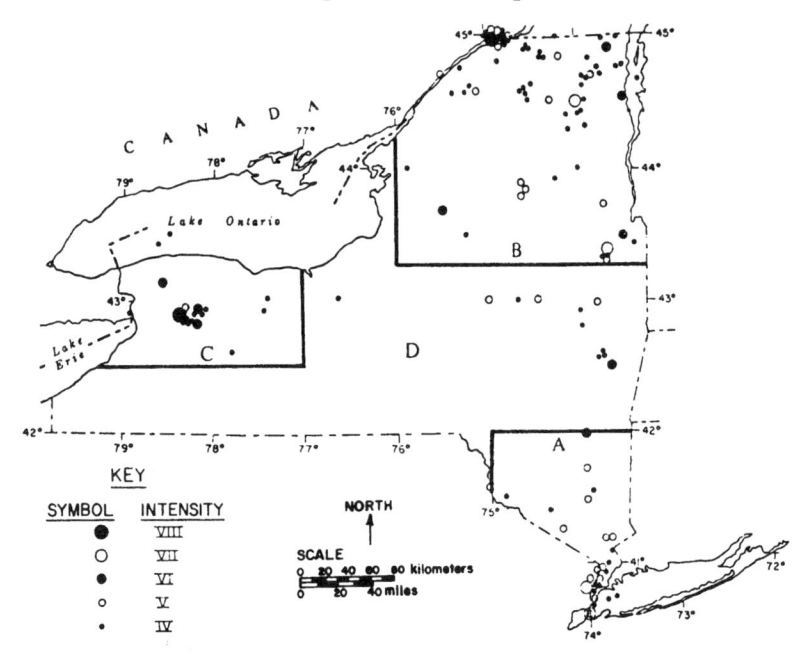

22. Which levels of earthquake intensity were experienced most frequently in New York State?
    (1) V and VII
    (2) VI and VII
    (3) IV and VIII
    (4) IV and V

23. Which is the best approximation of the length of the northern boundary line of zone *A*?
    (1) 80 km
    (2) 120 km
    (3) 100 km
    (4) 155 km

24. Which city is most likely to experience an earthquake at some future time?
    (1) Massena
    (2) Jamestown
    (3) Elmira
    (4) Utica

25. Which landscape region occupies most of earthquake region *B*?
    (1) Erie-Ontario Lowlands
    (2) Adirondack Highlands
    (3) Appalachian Uplands
    (4) New England Highlands

26. What type of rocks are found surrounding the epicenters of region *C*?
    (1) intensely metamorphosed
    (2) highly metamorphosed
    (3) igneous
    (4) sedimentary

Base your answers to questions 27 and 28 on the map below, which shows a portion of California along the San Andreas fault zone. The map gives the probability (percentage chance) that an earthquake strong enough to damage buildings and other structures will occur between the present time and the year 2024.

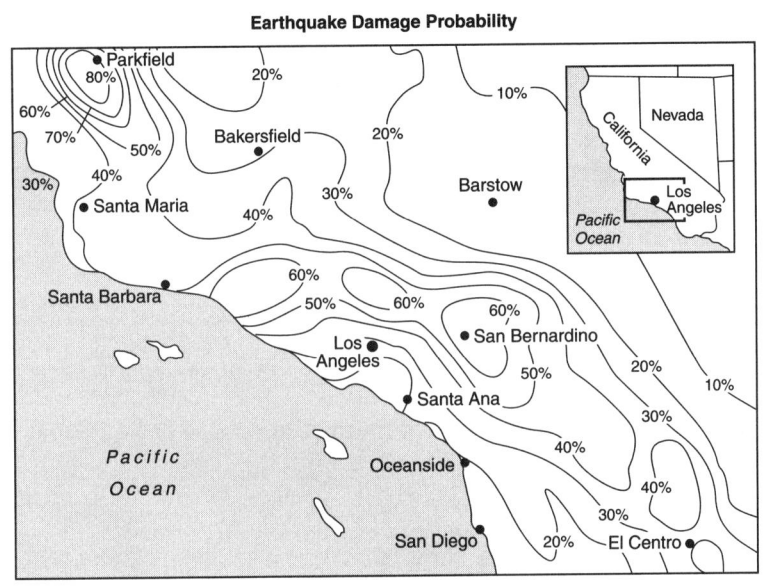

Earthquake Damage Probability

27. Which map represents the most likely location of the San Andreas fault line?

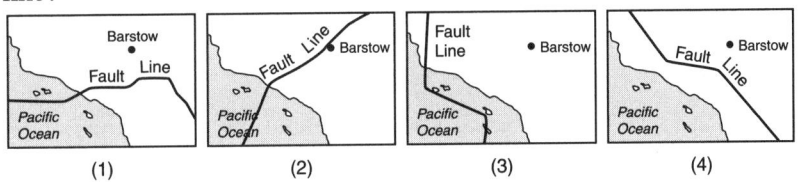

28. Which city has the greatest danger of damage from an earthquake?
    (1) Parkfield   (3) Santa Barbara
    (2) San Diego   (4) San Bernardino

Earthquakes and Earth's Interior

29. Useful information regarding the composition of the interior of Earth can be derived from earthquakes because earthquake waves
    (1) release materials from within Earth
    (2) travel through Earth at a constant velocity
    (3) travel at different rates through different materials
    (4) change the radioactive decay rates of rocks

30. According to the *Earth Science Reference Tables*, as the depth within Earth's interior increases, the
    (1) density, temperature, and pressure increase
    (2) density, temperature, and pressure decrease
    (3) density and temperature increase, but pressure decreases
    (4) density increases, but temperature and pressure decrease

31. The rock located between 2,900 kilometers and 5,200 kilometers below Earth's surface is inferred to be
    (1) an iron-rich solid   (3) a silicate-rich solid
    (2) an iron-rich liquid   (4) a silicate-rich liquid

32. Which has been studied to obtain information about Earth's inner core?
    (1) seismic waves   (3) index fossils
    (2) sedimentary rocks   (4) exposed bedrock

33. The theory that the outer core of Earth is composed of liquid material is best supported by
    (1) seismic studies that indicate that shear waves do not pass through the outer core
    (2) seismic studies that show that compressional waves can pass through the outer core
    (3) density studies that show that the outer core is slightly more dense than the inner core
    (4) gravity studies that indicate that gravitational strength is greatest within the core

34. Which is the most accurate statement about Earth's interior?
    (1) The temperature within the mantle decreases with depth.
    (2) The pressure within the mantle increases with depth.
    (3) The continental crust is thinnest under the mountain regions.
    (4) The two most common elements in Earth's crust are iron and nickel.

35. According to the *Earth Science Reference Tables*, which zone of Earth has the greatest density?
    (1) crust   (3) outer core
    (2) mantle   (4) inner core

Note that question 36 has only three choices.

36. Which of the following earthquake waves can travel through both solids and fluids?
(1) S-waves, only
(2) P-waves, only
(3) S-waves and P-waves

37. Earthquake S-waves do not travel through Earth's
(1) crust
(2) Moho
(3) mantle
(4) core

38. Which graph best represents the relationship between depth and density for Earth's interior? [Refer to the *Earth Science Reference Tables*.]

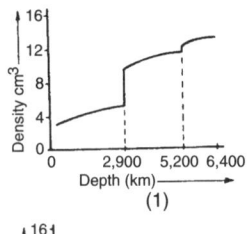

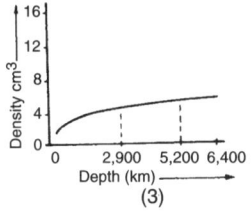

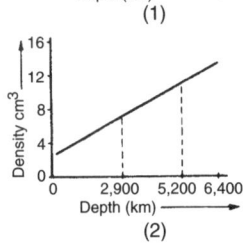

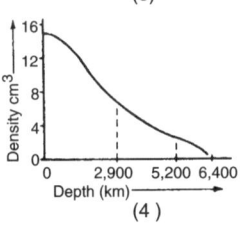

39. Theories about the composition of Earth's core are supported by meteorites that are composed primarily of
(1) oxygen and silicon
(2) aluminum and iron
(3) aluminum and oxygen
(4) iron and nickel

Earthquakes and Earth's Interior

Base your answers to questions 40–43 on your knowledge of earth science, the *Earth Science Reference Tables*, and the three seismograms shown below. The seismograms were recorded at earthquake recording stations A, B, and C. The letters P and S on each seismogram indicate the arrival times of the compressional (primary) and shear (secondary) seismic waves.

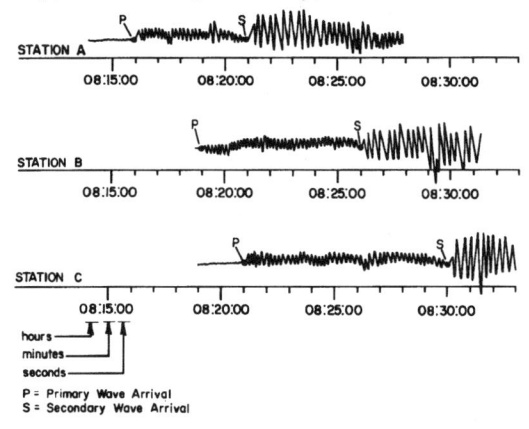

P = Primary Wave Arrival
S = Secondary Wave Arrival

40. At what time did the S-waves arrive at station A?
 (1) 08:16:00 (3) 08:27:00
 (2) 08:21:00 (4) 08:30:00

41. Approximately how far from station B is the earthquake epicenter located?
 (1) 3,500 km (3) 5,300 km
 (2) 4,300 km (4) 6,300 km

42. A fourth station recorded the same earthquake. The P-waves arrived, but the S-waves did not arrive. The best explanation for the absence of the S-wave is that it
 (1) was never transmitted by the earthquake
 (2) stopped when it reached the liquid part of Earth's interior
 (3) stopped when it reached a solid part of Earth's interior
 (4) traveled only on Earth's surface and did not penetrate Earth's interior

43. The radius of each circle in the diagrams below represents the distance from each seismograph recording station, A, B, C, to the epicenter. Which diagram correctly illustrates the positions of the three recording stations relative to the location of the earthquake epicenter?

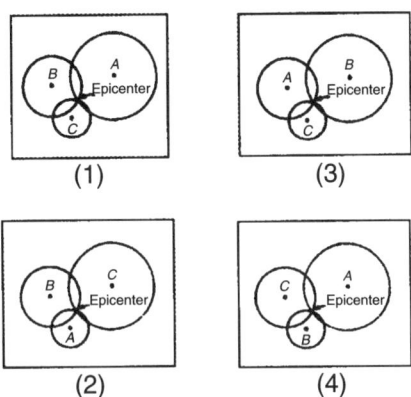

(1) (3)

(2) (4)

Base your answers to questions 44–48 on your knowledge of earth science, the *Earth Science Reference Tables*, and the diagram below. The diagram represents a cross section of Earth, showing the paths of earthquake waves from a single earthquake source. Seismograph stations are located at points A through F, and all are in the same time zone.

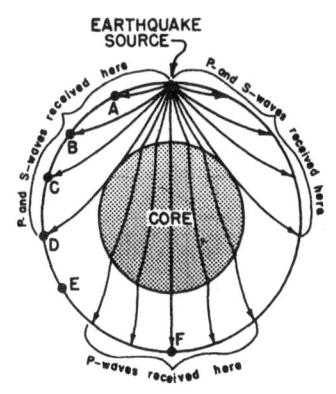

44. At which station is the difference in time between the arrivals of P- and S-waves the greatest?
    (1) A (3) C
    (2) B (4) D

45. Station E did not receive any P-waves or S-waves from this earthquake because P-waves and S-waves
    (1) cancel each other out
    (2) are bent, causing shadow zones
    (3) are changed to sound energy
    (4) are converted to heat energy

Earthquakes and Earth's Interior

46. What explanation do scientists give for the fact that station *F* did not receive *S*-waves?
    (1) Earth's inner core is so dense that *S*-waves cannot pass through.
    (2) Earth's outer core is liquid, which does not allow *S*-waves to pass.
    (3) *S*-waves do not have enough energy to pass completely through Earth.
    (4) *S*-waves are absorbed by Earth's crust.

47. Seismograph station *D* is 7,700 kilometers from the epicenter. If the *P*-waves arrived at this station at 2:15 P.M., at approximately what time did the earthquake occur?
    (1) 1:56 P.M.
    (2) 2:00 P.M.
    (3) 2:04 P.M.
    (4) 2:08 P.M.

48. Seismograph station *B* recorded the arrival of *P*-waves at 2:10 P.M. and the arrival of *S*-waves at 2:15 P.M. Approximately how far is station *B* from the earthquake epicenter?
    (1) 1,400 km
    (2) 2,400 km
    (3) 3,400 km
    (4) 4,400 km

Constructed Response Questions

49. Using one or more complete sentences, identify the two kinds of seismic body waves, and explain how they differ. [2]

50. Areas identified as high-risk on the seismic risk map shown in Figure 16.17 may not have had significant earthquake activity for a century or more. In one or more complete sentences, explain why they are mapped as high-risk regions. [1]

51. Using one or more complete sentences, state the difference between earthquake magnitude and earthquake intensity. [1]

52. The Richter Magnitude Scale is a logarithmic scale. In one or more complete sentences, explain the difference between a logarithmic scale and a linear scale. [1]

Base your answer to questions 53 through 55 on the two graphs below, which show the frequency of earthquakes in the Denver area and the disposal of liquid waste at the nearby Rocky Rountain Arsenal. In the mid-1960s the city of Denver began experiencing small earthquakes. A geologist hypothesized a connection between the earthquakes and the injection of liquid wastes into wells at the nearby Rocky Mountain Arsenal. His idea was that the liquid wastes were lubricating old faults near the arsenal, allowing them to slip and cause earthquakes.

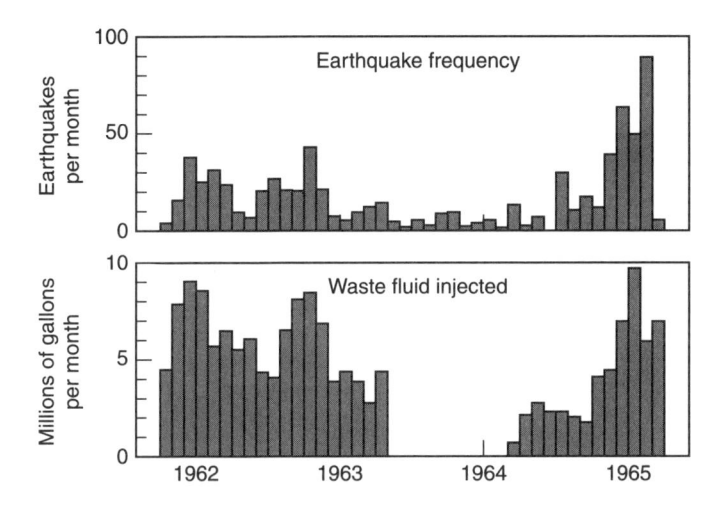

[Graphs showing Earthquake frequency (earthquakes per month, 0–100) and Waste fluid injected (millions of gallons per month, 0–10) from 1962 to 1965.]

53. In one or more complete sentences, state the relationship shown on the graphs, if any, between the frequency of earthquakes in the Denver area and the disposal of liquid waste at the Arsenal. [1]

54. Evaluate the data in the graphs, and in one or more complete sentences give your opinion as to whether or not the data support the geologist's hypothesis. Support your opinion with data from the graphs. [1]

55. Design an experiment that could be carried out to test the geologist's hypothesis. [2]

EXTENDED CONSTRUCTED RESPONSE QUESTIONS

Base your answers to questions 56 through 59 on the information and map below and your knowledge of earth science.

An earthquake occurred in the southwestern part of the United States. Mercalli-scale intensities were plotted for selected locations on the map, as shown below. (As the numerical values of Mercalli ratings increase, the damaging effects of the earthquake waves also increase.)

Earthquakes and Earth's Interior

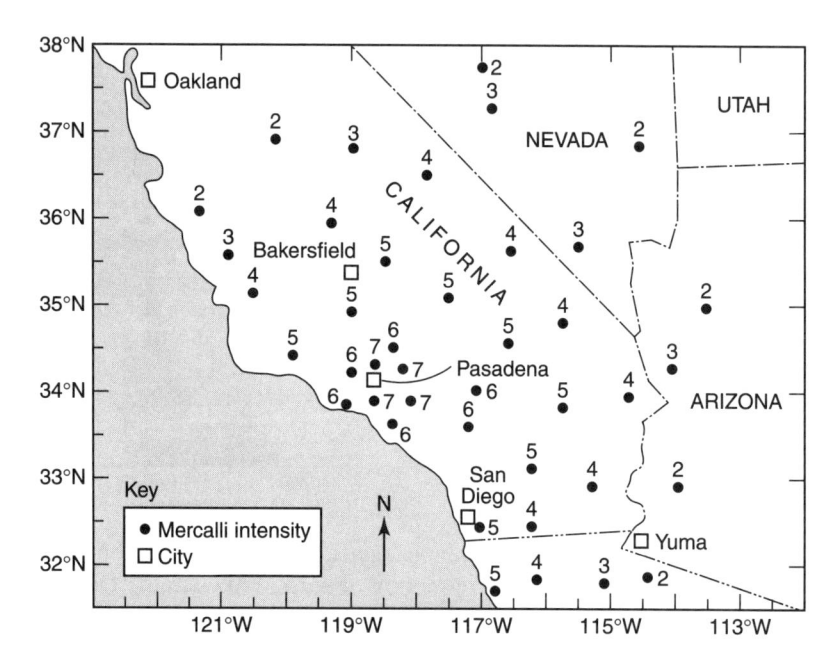

56. Using an interval of 2 Mercalli units and starting with an isoline representing 2 Mercalli units, draw an accurate isoline map of earthquake intensity. [4]

57. State the name of the city that is closest to the earthquake epicenter. [1]

58. State the latitude and longitude of Bakersfield. [2]

59. Using one or more complete sentences, identify the most likely cause of earthquakes that occur in the area shown on the map. [2]

CHAPTER 17
VOLCANOES AND EARTH'S INTERNAL HEAT

> **KEY IDEAS** Earth's internal heat engine is powered by heat from the decay of radioactive materials and residual heat from Earth's formation. Measurements of heat flow at and near Earth's surface indicate a steady increase in temperature with depth, reaching temperatures of 6,000°C or more at the core. Heat flowing outward from this fiery interior is largely responsible for a variety of phenomena ranging from Earth's magnetic field to crustal motion and volcanic activity.

KEY OBJECTIVES
Upon completion of this chapter, you will be able to:

- Explain that Earth has internal sources of energy that create heat.
- Explain how the transfer of heat energy within Earth's interior results in the formation of regions of different densities, and how these density differences lead to motion.
- Describe the origin of magma, and identify regions of frequent volcanic activity.
- Compare and contrast intrusive and extrusive volcanic activity.

EARTH'S INTERNAL HEAT ENGINE

Beneath its cool outer crust, Earth is glowing hot. Evidence for a hot interior comes from exploration of the crust, both above and below the surface. At the surface, we find numerous examples of heat escaping from Earth's interior: molten rock pours from volcanoes, and boiling hot water and steam rise from hot springs and jet out of geysers.

Further evidence comes from mines and deep wells. As we drill deeper into the crust, the temperature of the rock gets higher and higher. The deepest oil wells go down about 8 kilometers, and the temperature of the oil from these wells is 150°C. Measurements of rock temperature in mines and boreholes show an average temperature rise of 1°C per 30 meters; this increase in temperature with depth is known as the *geothermal gradient*. The rate of temperature increase, however, falls off rapidly with depth. See Figure 17.1. Although the rate of temperature increase in the mantle is lower, considering

Volcanoes and Earth's Internal Heat

that Earth's center is more than 6,000 *kilometers* down, it is not unlikely that temperatures there exceed 6,000°C.

Heat flows continuously from Earth's fiery depths toward its cool surface. The flow of heat from Earth's interior is roughly 50 calories (209 joules) per square centimeter per year. This is extremely small compared to the amount of heat that Earth's surface receives from the Sun each year—an average 179,000 calories (749,000 joules) per square centimeter. Therefore, Earth's heat flow from its interior contributes little to its heat balance or to the powering of convection in the atmosphere or oceans. However, this heat does power convective motions within Earth's mantle and core.

Sources of Internal Heat

Original Heat

Some of the heat inside Earth was trapped when it was formed. As discussed in Chapter 10, the heat produced by accretion caused Earth to reach internal temperatures high enough to melt rock. Over time, the outer surface cooled, forming a solid crust. Solid rock is a poor conductor of heat; therefore, once the crust formed, the heat beneath

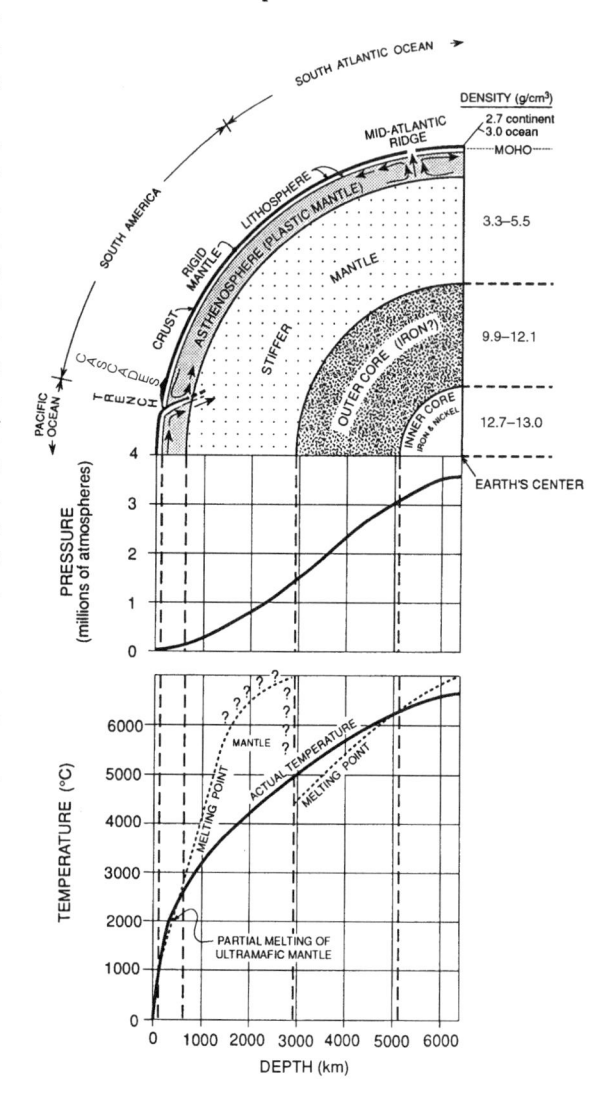

Figure 17.1 The Geothermal Gradient. The solid black line on the graph in the lower part of the diagram is the geothermal gradient, the actual change in temperature with increasing depth. The dotted line indicates the changing melting temperature of rock due to changes in composition and pressure. Source: The State Education Department, *Earth Science Reference Tables*, 2001 ed. (Albany, New York; The University of the State of New York).

374

it was trapped. Measurements of the rate at which Earth is losing heat indicate, however, that if this internal store were the only source of heat, Earth would already have totally solidified.

Radioactive Decay

The bodies that accreted to form Earth probably had a composition similar to that of stony meteorites known as chondrites. Among the elements found in chondrites are uranium, thorium, and potassium, each of which has radioactive isotopes. Since these elements have big ions, they don't fit into crystalline structures easily. When Earth's outer layers melted after accretion, these elements tended to stay in the molten magma and floated to the surface. As Earth cooled, they became concentrated in the rocks of the crust, particularly in granite.

The radioactive decay of these elements produces heat energy. It is estimated that the total mass of granite in the crust emits about 20 *trillion* watts of power per year. Volcanic rock derived from the mantle indicates that the concentration of radioactive elements in the mantle is lower. Nevertheless, the mantle is so much larger and more massive than the crust that it supplies about 10 trillion watts of power per year from radioactive decay alone. Thirty trillion watts per year is about 33 calories (140 joules), or more than half of the total heat flow from Earth's interior of 50 calories (209 joules) per square centimeter per year. Thus, more than half of the heat flowing out of Earth is due to radioactive decay within the crust and mantle.

Tidal Friction

Another possible source of ongoing heat production is tidal friction. Gravitational attraction between Earth and the Moon and between Earth and the Sun exert unequal forces on Earth that tend to deform it. The ever-changing configuration of the Earth-Moon-Sun system produces a rhythmic pattern of deformations—the tides. Not only Earth's oceans are affected; the atmosphere and solid earth show detectable tides as well. Measurements made in the 1970s showed that at middle latitudes a surface point on the crust is 30 centimeters farther from the center of Earth at high "earth tide" than at low "earth tide." In other words, tidal forces cause solid rock to bend upward, forming a tidal bulge. The tidal bending and unbending of rock layers in the crust releases heat, just as bending a wire coat hanger back and forth causes it to become hot to the touch.

Internal Heat Transfer

Heat tends to move from areas of high concentration to areas of low concentration. Simply stated, heat moves from materials at a higher temperature to materials at a lower temperature. Heat is transferred in three ways: conduction, radiation, and convection. See Figure 17.2.

Volcanoes and Earth's Internal Heat

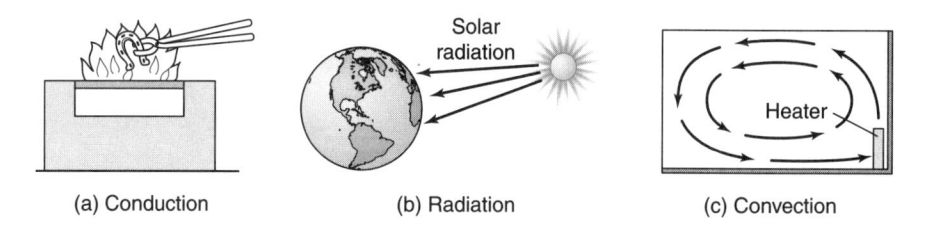

Figure 17.2 Mechanisms of Heat Transfer. Source: *Physics*, 2nd Ed., Arthur Beiser, Benjamin Cummings, 1978.

(a) Conduction (b) Radiation (c) Convection

Conduction (Figure 17.2a) is the transfer of heat energy from one molecule to another through collision. Conduction involves actual contact between two materials. Molecules with a lot of heat energy are fast moving. When they collide with cooler, slower moving molecules, an energy transfer occurs. The faster moving molecule slows down, and the slower moving molecule speeds up. Think of a bowling ball hitting the pins at the end of the alley. The ball slows down when it hits the pins, but the pins speed up and go flying. Rock is a poor conductor of heat because its molecules are bound in crystalline structures and cannot move about freely. As a result, heat moves very slowly through rock by conduction.

Radiation (Figure 17.2b) is the transfer of energy by electromagnetic waves. As atoms and molecules vibrate, they send out waves of energy that travel through space. Within rock, however, surrounding atoms and molecules block and trap that energy. Thus, loss of Earth's internal heat by radiation occurs almost entirely at the surface.

Convection (Figure 17.2c) is the actual motion of a volume of hot fluid from one place to another. Heated fluids are less dense than cooler fluids. Thus, when a portion of a fluid is heated, it becomes less dense than the surrounding cooler fluid and rises upward. As the heated fluid rises, it carries upward the heat energy within it. The upward motion of the hot fluid displaces cold fluid in its path, thereby setting up convection currents.

Seismic evidence shows that the outer core of Earth is a fluid. Therefore, convection currents in the outer core can carry heat outward. Enough heating occurrs in Earth's mantle to soften the rock and allow convection to remove the heat there as well. Thus, convection is occurring in both the core and the mantle. Plumes of hot material driven by heat from the core interrupt simple convection currents in the mantle. As a result, the nature of convection in the mantle is complicated. As these complex currents of hot material rise upward against the crust, they spread out and move sideways. As you will learn in Chapter 18, it is inferred that the force of these currents dragging along the base of the crust drives the motion of crustal plates.

VOLCANOES

Origin of Magma

Molten rock inside Earth is called *magma*. Melting occurs only in places where heat is concentrated or pressure is reduced. Measurements made on lavas show that some magmas are fluid at about 1,000°C. As shown in Figure 17.1, temperatures high enough to melt rock and form magma probably occur in the upper mantle at depths of 70–200 kilometers. This is the root zone of volcanoes.

Why is Earth's mantle not entirely molten? The answer is that pressure influences melting temperature. Although temperatures within the mantle are high enough to melt rock, the great pressures there hold crystals together and thus prevent the rock from turning into a liquid. For example, the mineral albite melts at 1,104°C at Earth's surface; but at a depth of 100 kilometers, where the pressure is 35,000 times greater, it melts at 1,440°C. Thus, the depth at which a rock melts depends on its composition and on the temperature and pressure at that depth.

The composition of continental crust and of oceanic crust is not the same. Continental crust is granitic, consisting of minerals with rather low melting points. Granitic magma can begin to form at depths of 30–60 kilometers. Ocean crust is basaltic, consisting of minerals with higher melting temperatures. Thus, basaltic magmas begin to form between depths of 100 and 350 kilometers. In either type of rock, some minerals melt at lower temperatures than others, so the magma that forms is a slush of hot, viscous liquid surrounding hot but still solid crystals. Geologists call this magma a "zone of partial melt." Until the liquid portion exceeds half of the melt, this magma behaves more like a solid than a liquid.

Once a body of magma has formed, it will start to rise because it is less dense than the surrounding rock. The rising magma works its way upward through small cracks and fissures. As magma rises into fissures, it widens them and can cause new ones to develop, resulting in earthquakes. Rising also decreases the pressure on the magma, allowing more minerals to remain molten at lower temperatures. Near Earth's surface, magmas can melt surrounding rock, forming underground pools of magma known as *magma chambers*. From these magma chambers, the molten rock may reach the surface through cracks and fissures to erupt, or may cool and solidify underground.

Basaltic magma is more fluid than granitic magma. Therefore, it rises more quickly; and, because of dropping pressure, its melting temperature falls faster than the magma is cooling. Thus, a lot of basaltic magma makes it all the way to Earth's surface as a fluid and erupts as lava. Granitic magma, on the other hand, is much more viscous, so its rate of ascent is slowed. It often cools faster than dropping pressure lowers its melting temperature. Therefore, most granitic magmas solidify underground rather than reaching the surface and forming lavas.

The Formation of Volcanoes

A *volcano* is both the opening in the crust (the vent) through which magma erupts and the mountain built by the erupted material. See Figure 17.3. Volcanoes form where cracks in Earth's crust lead to a magma chamber. Since liquid magma is less dense than the surrounding solid rocks and is under pressure, it rises toward the surface through these cracks. As magma rises, dissolved gases in it expand and are released, giving an upward boost to the magma. The closer to the surface, the less confining pressure there is to overcome and the faster the magma and gases move. Depending upon its viscosity, magma may pour out quietly onto the surface as a flood of molten rock, or spurt out explosively, showering the surrounding terrain with solid rock and globs of molten rock. Once magma emerges from the surface, it is called *lava*.

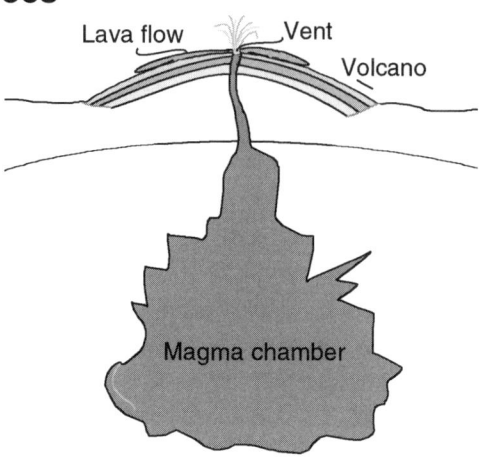

Figure 17.3 A Cross Section of a Volcano. Magma rises from chambers in the upper mantle until it exits the crust through an opening called the *vent*. Thereafter, it is called *lava*, and accumulates in a mound called a *volcano*.

Volcanic Structures

Erupted materials spread out in all directions around the vent, or central opening, of a volcano. With each new eruption, more material piles up around the vent and a cone-shaped mound forms. The shape of the cone depends upon the viscosity of the magma.

Fluid lava tends to flow quietly through the vent, spreading out in wide, thin sheets. The result is a flat, wide cone called a *shield volcano* (for its resemblance to a shield when viewed from overhead).

Thick, pasty lava containing a lot of dissolved gas does not flow easily through the vent, which tends to become clogged or completely blocked. As a result, gases cannot escape and pressure builds up until a violent explosion takes place. During an explosive eruption, rock and soil surrounding the vent are thrown upward, along with chunks and droplets of lava blown out of the vent. The lava cools as it hurtles through the air and is generally solid by the time it reaches the ground. Collectively, all of the fragments of solidified lava ejected during a volcanic eruption are called *tephra*. This "rain" of solid particles produces a steep, narrow cone called a *cinder cone volcano* (for the particles' resemblance to cinders).

Alternating explosive and quiet eruptions give rise to large, symmetrical *composite cones* made of alternating layers of solidified lava and volcanic rock particles such as ash and cinders.

In some cases, magma emerges through long, open cracks called *fissures*. Lava pouring out of fissures spreads out in wide, thin sheets. Instead of forming a cone as it would around a central vent, it forms a sheet, called a *lava plateau*, that blankets the surrounding land.

The four volcanic structures discussed above are illustrates in Figure 17.4.

(a) Shield Volcano

(b) Cinder Cone Volcano

(c) Composite Cone

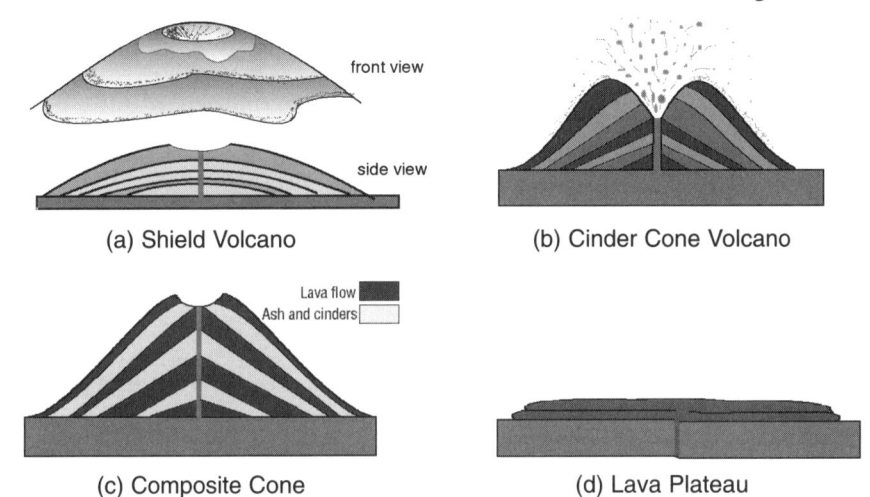

(d) Lava Plateau

Figure 17.4 Typical Volcanic Structures. (a) Broad, gently sloping shield volcanoes form from successive lava flows. (b) Steep-sided cinder cones form from the buildup of ash and cinders. (c) Composite cones of intermediate slope form when periods of lava flows are interspersed with eruptions of ash and cinder. (d) Very fluid lava that erupts along a fissure spreads out to form a lava plateau.

Intrusive Activities and Structures

Some magma moves around within Earth but never reaches the surface. Underground flows of magma, or *intrusions*, move into and through the countless underground cracks in the rocks of Earth's crust. There, intrusions may solidify into rock structures called *plutons*. See Figure 17.5. Plutons are named for their shapes and include dikes, sills, laccoliths, stocks, and batholiths. *Dikes* are flat, slablike structures that cut across layers of rock like a wall. *Sills* have a similar shape, but lie parallel to layers of rock (as a win-

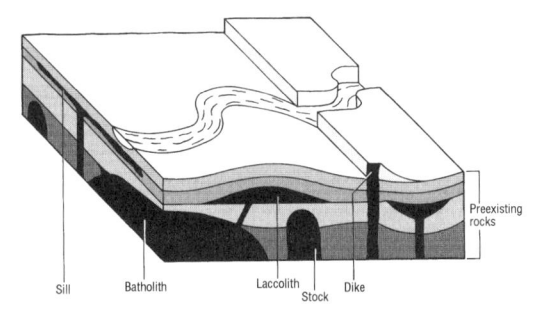

Figure 17.5 Some Types of Plutons.

dow sill is parallel to the ground). *Laccoliths* are large structures with flat bottoms and arched tops. The word *laccolith* means "lake-rock," and these structures resemble inverted lakes. Laccoliths form when magma is forced between rock layers faster than it can spread out, causing it to push the overlying layers of rock upward to form a dome mountain. Large, irregularly shaped plutons are called *stocks* if they cover less than 75 square kilometers, and *batholiths* if they cover a larger area.

Zones of Volcanic Activity

If the locations of active volcanoes are plotted on a map of Earth, the interesting pattern shown in Figure 17.6 emerges.

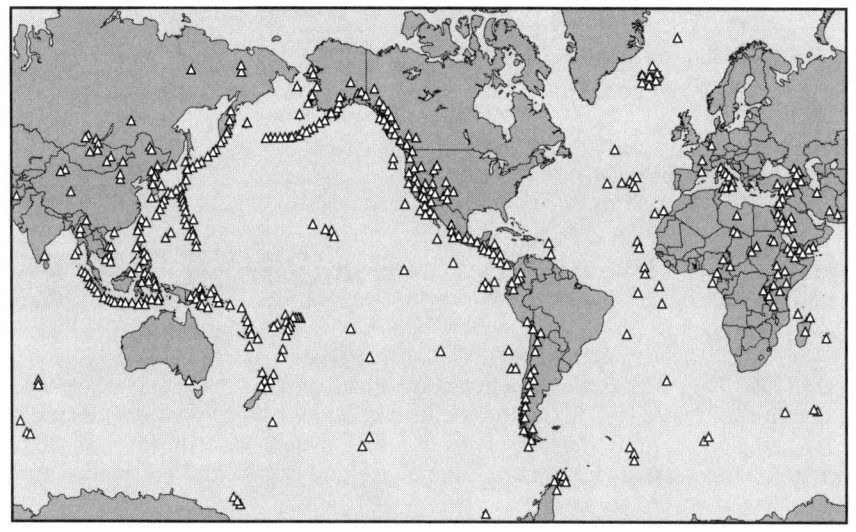

Figure 17.6 Earth's Active Volcanoes. Source: Adapted from image at http://nmnhwww.si.edu/gvp/volcano/index.htm, website of the Global Volcanism Program, National Museum of Natural History, E-121, Smithsonian Institution, Washington, D.C.

Active volcanoes are not distributed evenly on Earth. Rather, they occur in narrow, elongated zones in some areas and are completely absent from others. For example, active volcanoes surround the Pacific Ocean Basin, a zone known as the Ring of Fire, but none of the edges of continents bordering the Atlantic Ocean are volcanic. A narrow zone of volcanoes can also be found near the centers of most oceans, producing the vast chain of undersea mountains known as the mid-ocean ridge. A third zone runs along the northern edge of the Mediterranean Sea; known as the Mediterranean Belt, it includes historic volcanoes such as Mt. Etna and Mt. Vesuvius in Italy.

You may have noticed by now a remarkable similarity between the pattern of volcanic activity shown here, and the pattern of earthquake activity described in Chapter 16. As you will discover in Chapter 18, earthquake and volcano zones define the edges of the huge, platelike slabs of rock that make up Earth's crust.

MULTIPLE-CHOICE QUESTIONS

In each case, write the number of the word or expression that best answers the question or completes the statement.

1. The source of energy for the high temperatures found deep within Earth is
   (1) tidal friction
   (2) incoming solar radiation
   (3) decay of radioactive materials
   (4) meteorite bombardment of the Earth

2. According to the *Earth Science Reference Tables*, as depth within Earth's interior increases,
   (1) density, temperature, and pressure increase
   (2) density, temperature, and pressure decrease
   (3) density and temperature increase, but pressure decreases
   (4) density increases, but temperature and pressure decrease

3. Which observation provides the strongest evidence for the inference that convection cells exist within Earth's mantle?
   (1) Sea level has varied in the past.
   (2) Marine fossils are found at elevations high above sea level.
   (3) Displaced rock strata are usually accompanied by earthquakes and volcanoes.
   (4) Heat-flow readings vary at different locations in Earth's crust.

4. The accompanying diagram shows a cross-sectional view of Earth's interior. The motion represented by the arrows indicates that Earth's mantle
   (1) has the properties of a fluid
   (2) is composed of solid metamorphic rocks
   (3) is not affected by the heat from Earth's core
   (4) is denser than the core

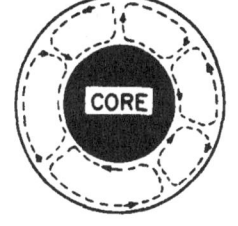

5. Which set of Earth processes are thought to be most closely related to each other because they normally occur in the same zones?
   (1) mountain building, earthquakes, and volcanic activity
   (2) mountain building, shallow-water fossil formation, and rock weathering
   (3) volcanic activity, rock weathering, and deposition of sediments
   (4) earthquakes, shallow-water fossil formation, and shifting magnetic poles

Volcanoes and Earth's Internal Heat

6. The primary cause of convection currents in Earth's mantle is believed to be the
   (1) differences in densities of Earth materials
   (2) subsidence of the crust
   (3) occurrence of earthquakes
   (4) rotation of Earth

7. An intrusion of igneous rock that cuts across the rock layers is called a
   (1) dike   (3) laccolith
   (2) fault   (4) sill

8. The greatest evidence of recent volcanic activity in the continental United States can be found in the vicinity of the
   (1) Pacific Northwest   (3) east coast
   (2) Great Lakes   (4) Adirondack Mountains

9. Successive horizontal lava flows produce regions of high relief called
   (1) mountains   (3) volcanoes
   (2) plains   (4) plateaus

10. Explosive volcanoes are usually characterized by
    (1) steeply sloped lava cones
    (2) steeply sloped cinder cones
    (3) gently sloped lava cones
    (4) gently sloped cinder cones

11. Which statement concerning a sill is true?
    (1) It may be vertical.
    (2) It must be vertical.
    (3) It may cut across sedimentary strata.
    (4) It must be horizontal.

12. Which statement applies equally to batholiths, laccoliths, dikes, and sills?
    (1) They create folded mountains.
    (2) They are composed of intrusive rocks.
    (3) They are flat sheets of metamorphic material.
    (4) They are composed largely of obsidian.

13. A study of rocks in an extinct volcano indicates a composition largely of cinders and other igneous fragments. This extinct volcano was probably formed as a result of
    (1) lava flows
    (2) igneous intrusions
    (3) explosive eruptions
    (4) magma solidifications

14. Which rock is most likely to be found as part of a dike?
    (1) granite
    (2) obsidian
    (3) pumice
    (4) scoria

15. Which diagram best represents the transfer of heat by convection in a liquid?

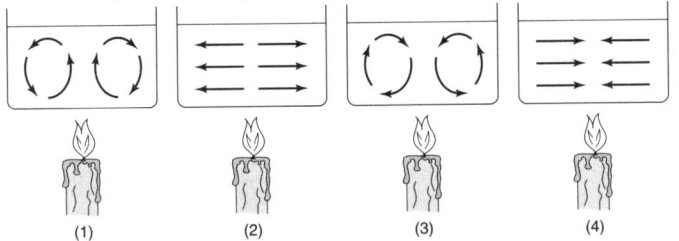

Base your answer to question 16 on your knowledge of earth science and on the diagram below, which represents a cross-sectional view of a volcanic island that has not erupted during recorded history.

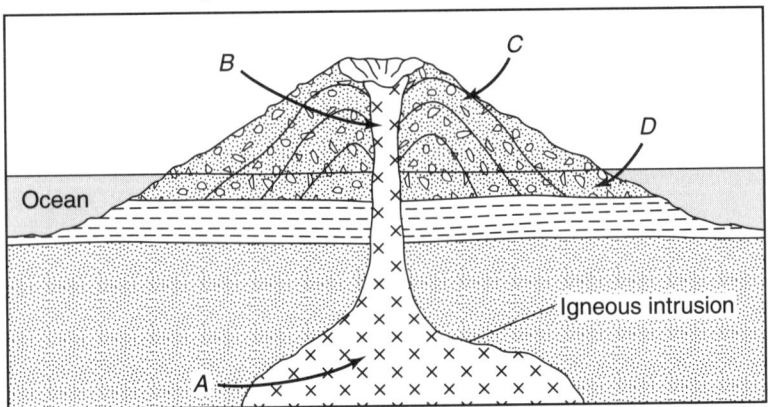

16. Where would rocks containing the largest mineral crystals most likely be found?
    (1) deep in the interior (A)
    (2) near the surface of the cooled lava (B)
    (3) in the cinders forming the cone (C)
    (4) in the solidified lava flow in the ocean (D)

17. In which method is energy transferred by density differences?
    (1) absorption
    (2) conduction
    (3) convection
    (4) radiation

18. Thick, pasty lava with a high gas content tends to erupt
    (1) quietly
    (2) explosively
    (3) from fissures in a broad sheet
    (4) from undersea vents

Volcanoes and Earth's Internal Heat

19. An igneous intrusion loses heat to its surroundings primarily by
    (1) conduction
    (2) convection
    (3) radiation
    (4) absorption

20. A flow of magma cools and hardens into rock before it reaches the surface. Which of the following structures most likely formed from the magma?
    (1) a cinder cone
    (2) a basalt plateau
    (3) a geyser
    (4) a dike

21. Magma chambers tend to form in areas beneath Earth's surface where
    (1) heat builds up and pressure is reduced
    (2) pressure builds up and heat is reduced
    (3) pores in Earth are filled with groundwater
    (4) there are extensive deposits of coal, oil, or natural gas

22. Rock ejected from a volcano during an explosive eruption will most likely consist of
    (1) rounded rock particles with frosted surfaces
    (2) rock particles that are mostly monomineralic
    (3) a wide range of rock-particle shapes and sizes
    (4) large, individually formed mineral crystals

23. A lava flow that has cooled and solidified into rock is observed to have a surface pockmarked by numerous spherical cavities. These cavities are most likely the result of
    (1) rainwater dissolving out water-soluble mineral crystals
    (2) the corrosive effects of sustained periods of acid rain
    (3) rounded pebbles pried out of the surface by frost action
    (4) gas bubbling out of the lava just as it hardened

24. As depth beneath the surface increases, the temperature at which a particular mineral will melt
    (1) increases
    (2) decreases
    (3) remains the same
    (4) changes from Fahrenheit to Celsius degrees

25. According to the *Earth Science Reference Tables*, the melting point of materials in the outer core is
    (1) lower than their actual temperature
    (2) higher than their actual temperature
    (3) lower than that of materials in the crust
    (4) higher than that of materials in the inner core

Base your answers to questions 26 through 29 on your knowledge of earth science, the *Earth Science Reference Tables*, and the map and information below.

One eruption of Mt. St. Helens in Washington State resulted in the movement of volcanic ash across the northwestern portion of the United States. The map shows the movement of the ash along paths at three different altitudes above sea level: path *A* at 1.5 kilometers, path *B* at 3.0 kilometers, and path *C* at 5.5 kilometers. Above each lines across each path, the time interval between the eruption and the position of the leading edge of the ash is indicated. Points *X* and *Y* are places on Earth's surface.

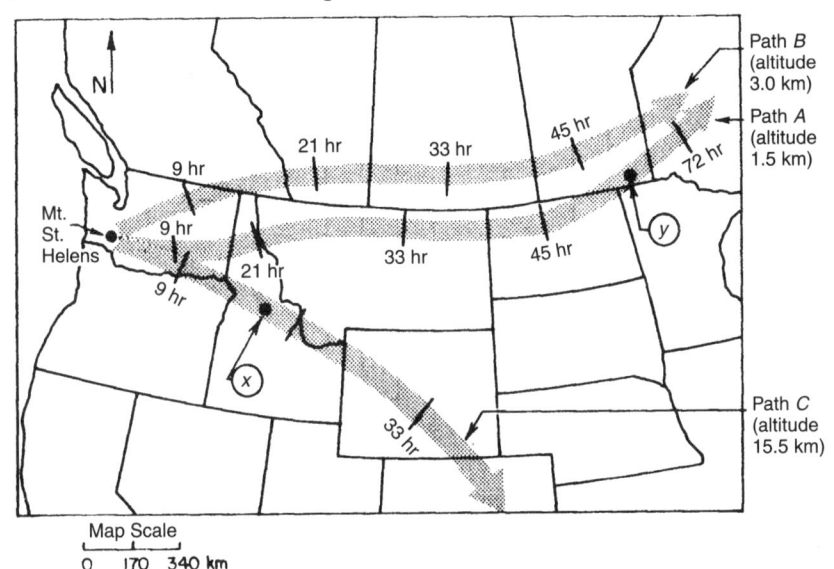

26. The data for paths *A*, *B*, and *C* were collected in which region of the atmosphere?
    (1) troposphere   (3) stratosphere
    (2) mesosphere   (4) thermosphere

Note that question 27 has only three choices.
27. How would the amount of insolation reaching point *X* 24 hours after the eruption compare with the amount of insolation this point usually received?
    (1) The amount of insolation would be less.
    (2) The amount of insolation would be greater.
    (3) The amount of insolation would be the same.

Volcanoes and Earth's Internal Heat

28. Approximately how many hours did it take for the ash front of path A to arrive at point Y?
    (1) 39   (3) 45
    (2) 59   (4) 72

29. Why did the general direction of the ash front along path C not follow the general direction of the ash fronts along path A and path B?
    (1) The eruption threw material toward the southeast.
    (2) Northwest winds existed at the 5.5-km level.
    (3) Mountain ranges southeast of the volcano are 1.5 km high.
    (4) Only the ash along path C was affected by Earth's rotation.

CONSTRUCTED RESPONSE QUESTIONS

30. State three types of evidence that support the statement "There is a tremendous amount of heat within Earth." [3]

31. In one or more complete sentences, explain the difference between magma and lava. [1]

32. The table below lists the melting points and boiling points of several Earth materials at 1 atmosphere of pressure (the average air pressure at sea level).

Substance	Melting Point	Boiling Point
Diamond	3,550°C	4,826°C
Iron	1,535°C	2,750°C
Quartz	1,610°C	2,230°C

(a) State the depth at which the temperature of Earth's interior is inferred to be at the melting point listed in the table for each of the three substances. [3]
(b) State the depth at which the temperature of Earth's interior is inferred to be at the boiling point listed in the table for each of the three substances. [3]
(c) In one or more complete sentences, explain why quartz does not melt or boil in the stiffer mantle even though temperatures there are high enough. [1]

33. Identify three types of volcanic structures. In one or more complete sentences, explain how each structure is formed. [3]

34. In one or more complete sentences, explain why the rock that forms from igneous intrusions typically consists of large mineral crystals. [1]

35. State two differences between a quiet eruption and an explosive eruption. [2]

EXTENDED CONSTRUCTED RESPONSE QUESTIONS

Base your answer to questions 36 through 40 on the diagram below, which represents a cross section of a portion of Earth's crust. Letters *A* through *M* identify rock layers, structures and boundaries within the cross section.

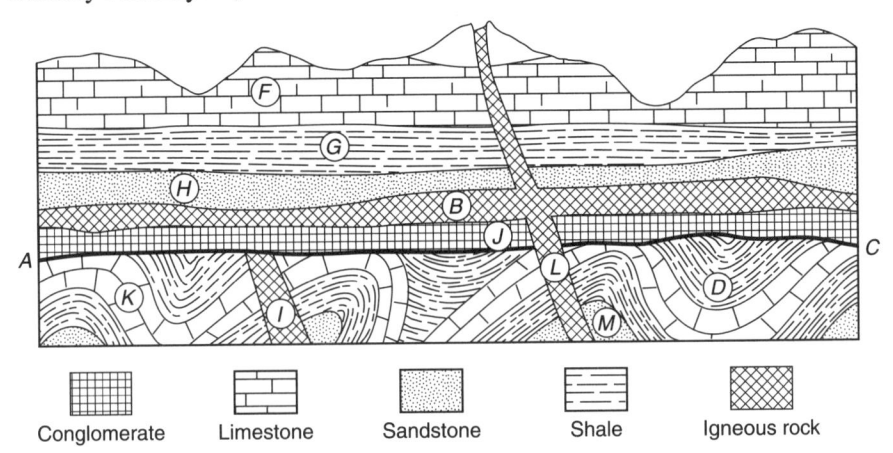

36. State the rock layer that best represents a dike. [1]

37. State the rock layer that best represents a sill. [1]

38. Place a circle around the rock structure that best represents a volcano. [1]

39. In one or more complete sentences, explain how the igneous intrusion labeled *B* most likely affected the rock adjacent to it in rock layer *H*. [1]

40. State two likely sources of the heat originally contained in the igneous rock that formed the intrusions in the cross section. [2]

CHAPTER 18
PLATE TECTONICS

KEY IDEAS There is much evidence that Earth is a dynamic geologic system whose crust is in nearly constant motion. The theory of plate tectonics is an excellent example of how our ideas about Earth change as new discoveries are made. Building upon earlier theories of continental drift, isostasy, and ocean floor spreading, plate tectonics was able to explain observations that could not be accounted for by these earlier theories.

According to the plate tectonics theory, Earth's crust, or lithosphere, consists of separate plates that rest on the more fluid asthenosphere and move slowly in relationship to one another, creating convergent, divergent, and transform plate boundaries. The outward transfer of Earth's internal heat drives convective circulation in the mantle that moves the lithospheric plates comprising Earth's surface. Plate boundaries are the sites of most earthquakes, volcanoes, and young mountain ranges.

Many of Earth's surface features, such as mid-ocean ridges/rifts, trenches/subduction zones/island arcs, mountain ranges (folded, faulted, and volcanic), hot spots, and the magnetic and age patterns in surface bedrock, are a consequence of forces associated with plate motion and interaction.

KEY OBJECTIVES
Upon completion of this chapter, you will be able to:

- Describe evidence of crustal movements.
- Explain how the theories of continental drift, isostasy, and ocean floor spreading were combined in a single unifying theory—the theory of plate tectonics.
- Explain how differences in density resulting from heat flow within Earth's interior cause movement of the lithospheric plates.
- Relate specific forms of crustal activity, such as earthquakes, volcanoes, and the deformation and metamorphism of rocks during the formation of mountains, to the various types of plate boundaries.
- Describe how many of the processes of the rock cycle are consequences of plate dynamics, including the production of magma, regional meta-

morphism, and the creation of major depositional basins through downwarping of Earth's crust.
- Explain how plate motions have resulted in global changes in geography, climate, and the patterns of organic evolution.

THE DYNAMIC CRUST

Earth's crust, or *lithosphere*, is almost constantly in motion. This motion may be abrupt, as in an earthquake, or so gradual that it is imperceptible to the senses. Movements of the lithosphere exert forces on rock, causing them to be deformed. Evidence that the lithosphere is moving ranges from direct observation of motion to inferences based upon displacement of structures and deformation of rock.

Evidence of Crustal Movements

Earthquakes are unmistakable evidence of crustal movement. During an earthquake, movement of the crust occurs along faults. Every one of the nearly 1 million earthquakes that occur annually involves movement of Earth's crust.

Volcanic eruptions also involve movements of the crust. As magma moves beneath and then out of a volcano, the crust around the volcano rises and sinks measurably, often causing the brittle crust to crack and form fissures. These breaks are direct evidence of movement.

Displaced structures are also evidence of crustal movement. When the crust moves, structures built on its surface move along with it.

Bench marks are permanent metal plaques set in the ground giving the exact locations and precisely determined elevations of points. See Figure 18.1a. Bench marks are used as references in geologic surveys and tidal observations. Measurements of bench-mark elevations for more than 100 years reveal that in large areas of the United States the ground is slowly moving upward or downward. See Figure 18.1b. Crustal motion can also be detected and measured using satellite laser technology.

Tilted or folded rock layers are also evidence of crustal movement. Sedimentary strata, lava flows, and tephra are originally deposited in horizontal layers. Therefore, where such materials are observed to be steeply tilted, or bent and folded, it can be inferred that they have been moved from their original positions.

Sedimentary rock layers at high elevations are further evidence of crustal movement. Most layers of sediments are deposited at or below sea level, and may be buried deeper before changing into sedimentary rock. In many places, however, layers of sedimentary rock are found high atop mountains or plateaus. Some are located as much as 2 kilometers above current sea level, far higher than any known body of water, past or present. From this it can be inferred that these rock layers have been uplifted.

Plate Tectonics

Figure 18.1 (a) A United States Geological Survey Bench Mark. A bench mark is a small plaque, usually mounted in rock or on a concrete base, marked with the precise location and elevation of a point. Source: United States Geological Survey. (b) Changes in Bench Mark Elevations over a 100-Year Period. Source: *Physical Geology*, 2nd Ed., Richard Foster Flint and Brian J. Skinner, John Wiley, 1974, 1977.

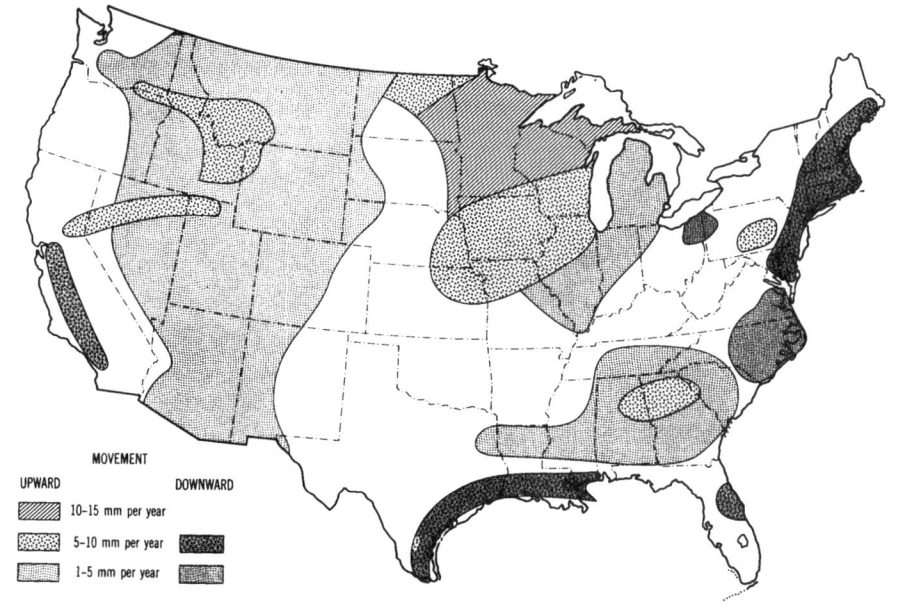

Fossils provide striking evidence of crustal movement. Fossils of marine life, such as corals and clams, found at mountaintops in the Himalayas are evidence of uplift. Fossils of terrestrial or shallow-water organisms beneath the deep ocean bottom indicate sinking, subsidence of the crust.

Thick layers of shallow-water sediments, as much as 15 kilometers thick in some places, are interesting evidence of crustal movement. The average depth of the ocean is 3.8 kilometers with several isolated points having depths of about 10 kilometers. How can 15 kilometers of sediment accumulate in a body of shallow water without filling it completely? One probable inference is that the crust beneath the sediment is sinking while the sediment is accumulating, so sediment layers build up while the water depth remains fairly constant.

Causes and Effects of Crustal Movements

Causes

Unbalanced forces acting on Earth's crust cause crustal movement. Some of the forces acting on the crust are gravity (including gravitational attraction by the Sun and Moon) and forces produced by Earth's rotation, by the expansion and contraction of rock material due to heating and cooling, and by density currents in Earth's mantle.

Stress, Tension, Compression, and Shear

The many forces acting on rock are called *stress*. Stress can act upon rock in several different ways. A rock is under *uniform* stress when the stress in all directions is equal. *Tension* stresses act in opposite directions, pulling rock apart or stretching it. *Compression* stresses act toward each other, pushing or squeezing rock together. *Shear* stresses may act toward or away from each other, but they do so along different lines of action, causing rock to twist or tear.

Effects

When stress acts upon a rock, the rock *strains*, or changes in size, shape, or both. Uniform stress causes rock to change in size, but not in shape. Tension, compression, and shear stresses cause a change in shape and may also cause a change in size. See Figure 18.2.

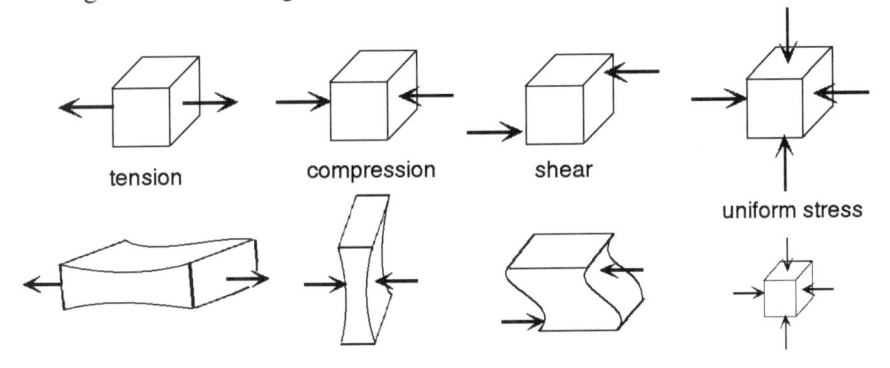

Figure 18.2 Changes in Rock Shape and/or Size over Time Due to Tension, Compression, Shear, and Uniform stress.

Deformation and Fracture

When rock is stressed, it goes through a series of changes. The first step is elastic deformation, in which the rock strains, but the change is not permanent. If the stress is removed, the rock will return to its original shape and size. The next step is ductile deformation, which begins when stress reaches a point called the *elastic limit*. At the elastic limit, the stress exceeds the strength of the rock's internal bonding and permanent changes occur. If

Plate Tectonics

stress is then removed, the deformation is permanent and the rock no longer returns to its original shape or size. Ductile behavior is similar to what happens when you squeeze or stretch modeling clay without breaking it. The final step is fracture, in which the stress actually causes the rock to break, or fracture. In general, high temperatures and pressures favor ductile behavior and make fracture less likely to occur. See Figure 18.3.

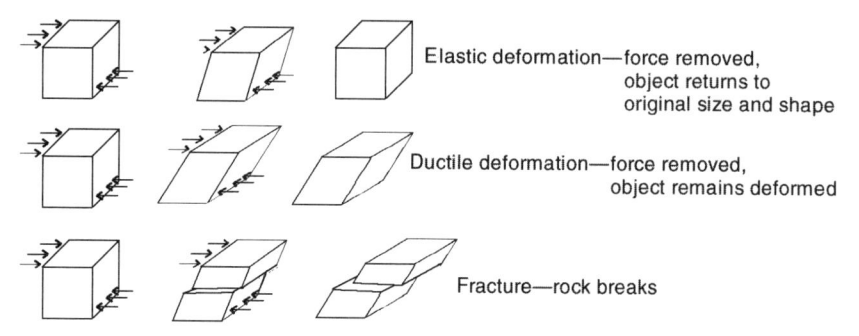

Figure 18.3 Elastic Deformation, Ductile Deformation, and Fracture.

Folds

Ductile deformation of layered rock forms bends or warps called *folds*. Folding is due to compression stresses. An upward-arched fold is an *anticline*; a downward, valleylike fold is a *syncline*. The sides of a fold are called its *limbs*. Folds range from simple *monoclines*, in which only one limb is bent, to *overturned* folds, in which both limbs are tilted in the same direction, to *recumbent* folds, which are bent back on themselves almost horizontally. See Figure 18.4.

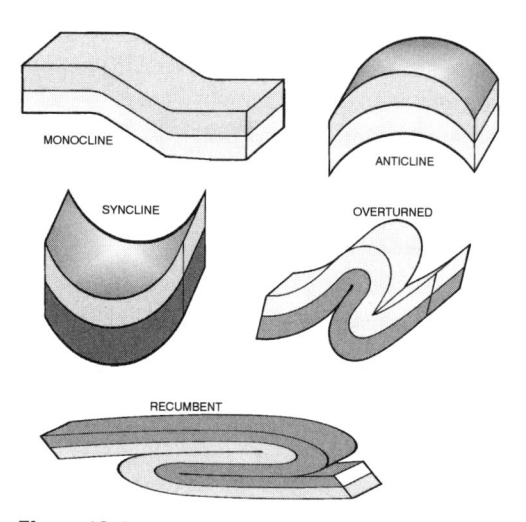

Figure 18.4 Various Types of Folds.

Joints and Faults

Joints and faults are fractures that form when rock is stressed until it breaks. *Joints* are fractures along which there has not been any movement of the rock. *Faults* are fractures along which the rock has moved and one side of the fracture is displaced relative to the other side. Faulting is always associated with earthquakes. A fault is classified by the angle of the fracture and the direction of relative movement along it. Five types of faults are illustrated in Figure 18.5.

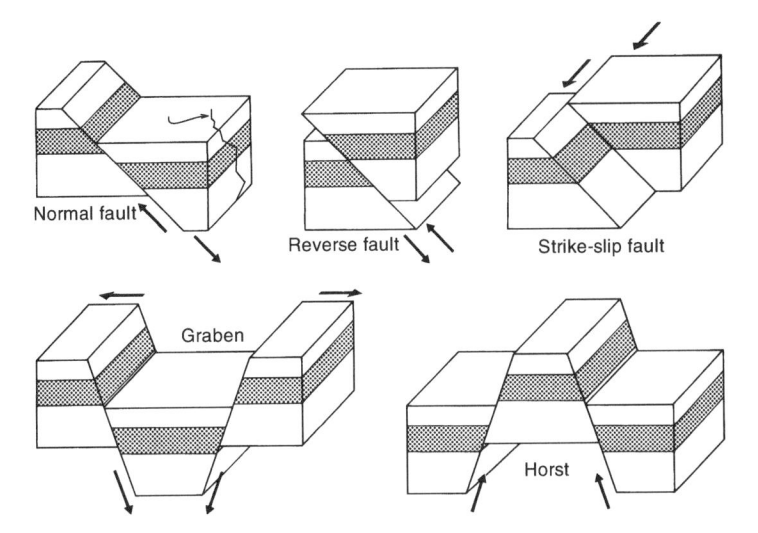

Figure 18.5 Various Types of Faults.

PLATE TECTONICS

The development of the plate tectonics theory is an example of how a theory gains acceptance based upon the evidence that supports it and its ability to explain puzzling observations. Plate tectonics makes sense of such a large range of phenomena that it has become a unifying principle in geology. Two earlier ideas—continental drift and seafloor spreading—were based upon evidence that was later incorporated into a single unifying theory—plate tectonics.

The Theory of Continental Drift: Supporting Evidence

Matching Shorelines The shorelines on both sides of the Atlantic match strikingly, especially the shapes of the Atlantic coasts of Africa and South America. The likeness is even clearer if the edges of the continental shelves are used rather than present shorelines. This match gave rise to the idea that the continents were once joined together but then drifted apart. At first the idea was rejected because it seemed ridiculous to think that anything as large as a continent could move around. In 1912, however, Alfred Wegener reintroduced the idea of *continental drift*—the concept that continents drift slowly across Earth's surface, sometimes colliding and sometimes breaking into pieces. Wegener and his colleagues presented striking evidence that the world's landmasses had once been joined together, but over time had broken apart into the continents of today and slowly drifted to their present positions. The original landmass was dubbed Pangaea, meaning "all land." See Figure 18.6.

Plate Tectonics

Continental and Ocean Crust Statistical studies of elevations and depths indicate that there are two distinct levels for the world's surface—the continents and the ocean floors. These levels alternate and exist side by side with almost no transition between the two, suggesting a series of side-by-side blocks. Some blocks are continents; some are ocean floors.

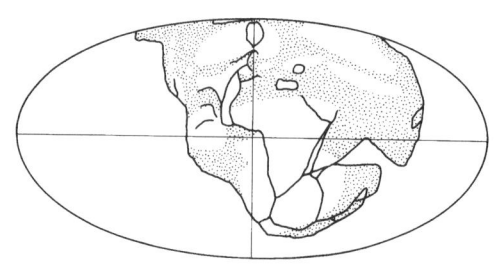

Figure 18.6 Pangaea as Proposed by Wegener. Source: Earth Science on File. © Facts on File, Inc., 1988.

The compositions and thicknesses of the ocean floors and the continents differ. The ocean floors are thin and made of dense basalt, while the continents are thick and consist of less dense granite. See Figure 18.7.

Wegener used a still earlier idea—isostasy—to explain the two levels of the crust. Isostasy stated that the crust was floating on hot, fluid rock in the mantle and offered two possible explanations for continents being higher than ocean floors: continents are higher because less dense materials float higher than more dense materials in the same fluid (Pratt's hypothesis), and continents are higher because thick objects float higher than thin objects in the same fluid (Airy's hypothesis). See Figure 18.8. The existence of blocks would explain how continents could "drift"—the blocks simply moved sideways.

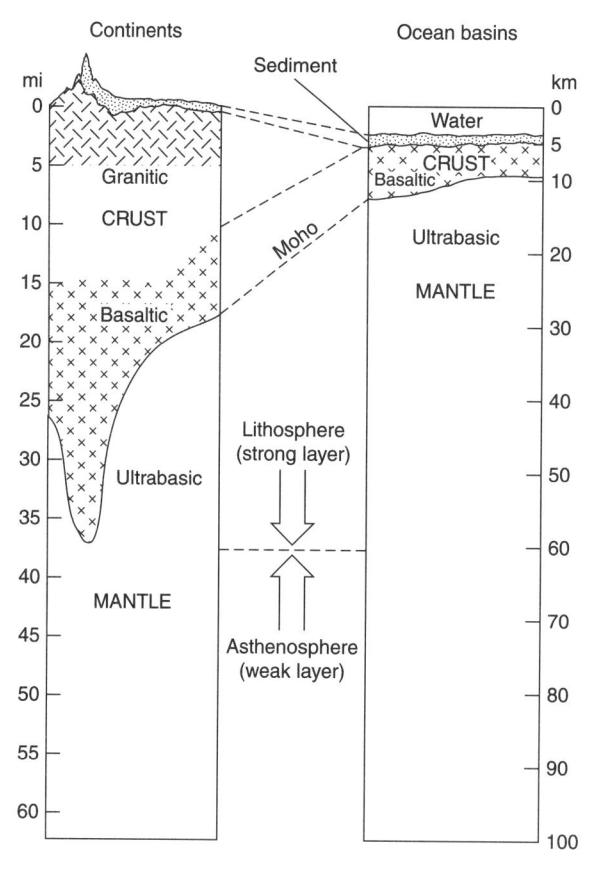

Figure 18-7 Compositions and Thicknesses of Continents and Ocean Floors. Source: *The Earth Sciences*, Arthur N. Strahler, Harper & Row, 1971.

394

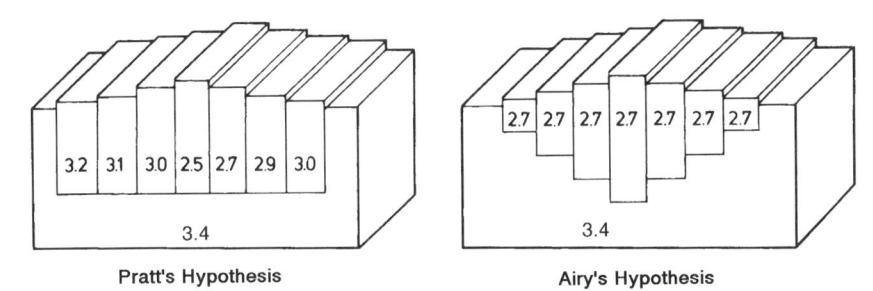

Figure 18.8 Isostasy. In both hypotheses, the rocks of the crust float on a denser (3.4 g/cm^3) mantle. In Pratt's hypothesis, mountains are the result of less dense rock (2.5 g/cm^3) floating higher than dense rock (3.2g/cm^3). In Airy's hypothesis, all the rocks of the crust have the same density (2.7 g/cm^3). Mountains occur where the crust is thicker, and ocean basins occur where it is thinner. Source: Earth Science on File. © Facts on File, Inc., 1988.

Geological Structures Comparison of geological structures on the two sides of the Atlantic Ocean shows a remarkable correspondence. For example, the Sierras near Buenos Aires contain a succession of beds very like those of the Cape Mountains in South Africa. The large gneiss plateau of Africa is strikingly similar to that of Brazil, and matching pockets of igneous rocks and sedimentary rocks can be found in both. The sequence of rock types—older granite, younger granite, alkaline rocks, Jurassic age volcanic rocks, and kimberlite—also matches. The diamond fields of both South Africa and Brazil occur in the kimberlite beds. The Falkland Islands contain rock layers almost indistinguishable from those of the African Cape yet markedly different from layers in nearby Patagonia. Figure 18.9 shows some of the matches of rock types in South America and Africa. Similar matches of North America with Europe, Greenland with North America and Norway, and Madagascar with Africa, as well as between other locations, can also be found.

Fossils and Contemporary Organisms Fossils of identical land animals (Mesosauroidae) are found nowhere else but in southern Africa and South America. Fossils of identical trees (*Glossopteris*) are found in Australia, India, and South America. Sixty-four percent of all fossil reptiles from the Carbonaceous Period are identical in Europe and North America.

Contemporary organisms also show a corresponding pattern of distribution. For example, *Lumbricus* earthworms are found from Japan to Spain in Eurasia, but only in the eastern United States. Freshwater perch are found in Europe and Asia, but only in the eastern United States. Similar patterns exist for pearl mussels, mud minnows, and garden snails. Common heather is found only in Europe and Newfoundland. European and eastern North American eels both spawn in the Sargasso Sea off the North American coast. Again, similar patterns can be found between other now-separated continents.

Plate Tectonics

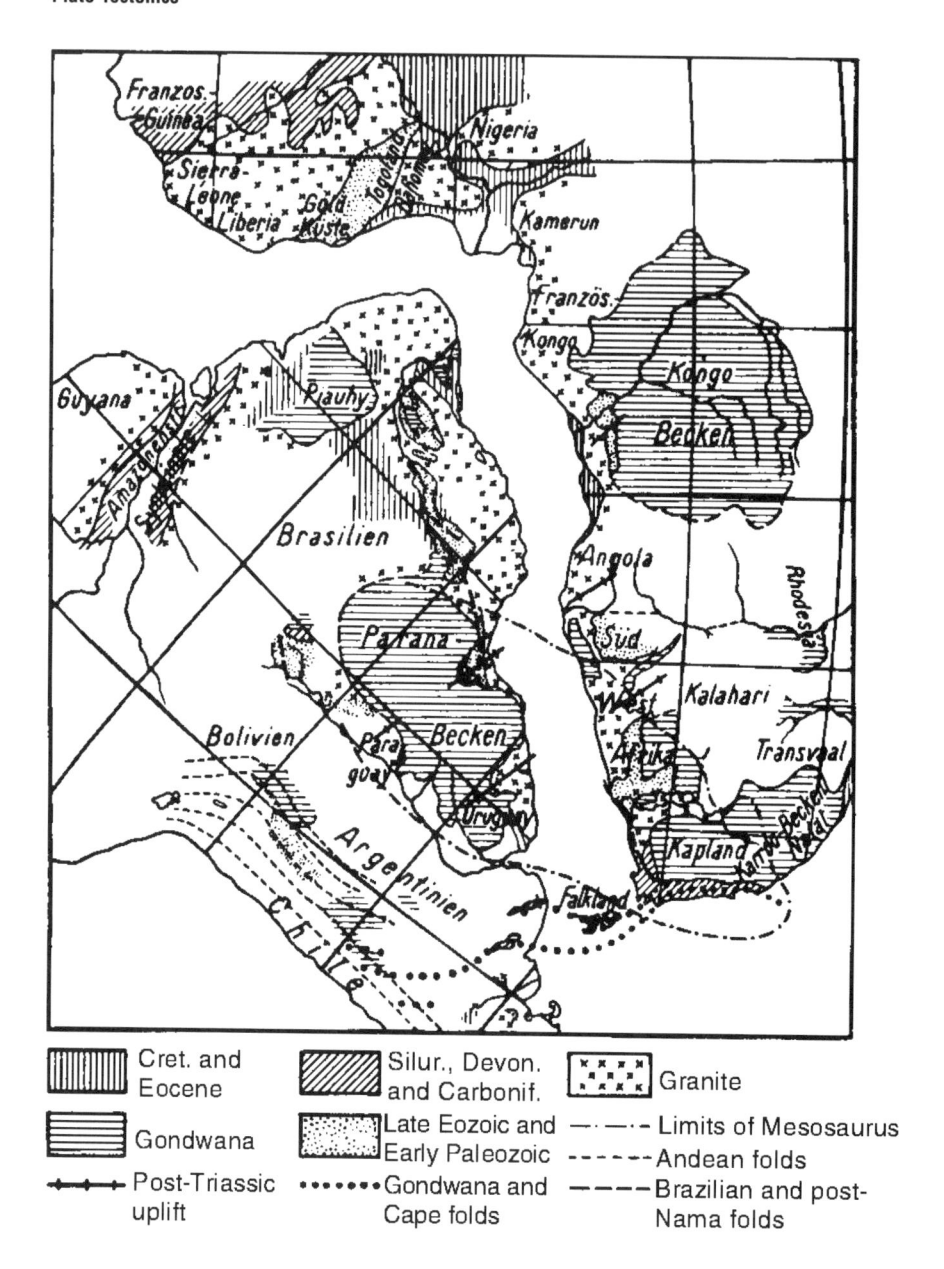

Figure 18.9 Evidence of Continental Drift. Alfred Wegener's correlation of rock type and surface features along the margins of Africa and South America. Source: *The Origins of Continents and Oceans*, Alfred Wegener, by permission of Dover Publications, Inc., 1966.

Past Climates There is also unmistakable evidence that 300 million years ago an ice sheet covered South America, southern Africa, India, and southern Australia. This ice sheet was huge, similar to the one covering Antarctica today. Glacial striations even indicate the direction of ice flow on all four continents. This is difficult to explain without continental drift. If the continents were in their present positions, the ice sheet would have covered all the southern oceans and in places crossed the Equator. In order for this to happen, Earth would have had to be very cold, yet there is no evidence of glaciation at the same time in the Northern Hemisphere. If the continents drifted, though, then 300 million years ago they all could have been joined and some could have been located adjacent to the South Pole, as shown in Figure 18.6. Similarly, places that are today located near the poles contain deposits of coal, which could have formed only in equatorial rain forests. This, too, can be explained by drifting continents.

All of the evidence presented above supports the idea that continents have moved over time. However, it does not suggest a mechanism that would move the continents.

The Theory of Ocean Floor Spreading: Supporting Evidence

In the 1950s the ocean floors were explored extensively by oceanographic research vessels. Much information was gathered and led Professor Harry Hess of Princeton University to propose the idea that the ocean floor is spreading sideways away from the mid-ocean ridges. The evidence supporting this idea suggests a mechanism that could have moved the continents apart.

Mid-Ocean Ridges New instruments enabled scientists to survey the ocean floor with unprecedented accuracy. The ocean basins, once thought to be flat plains, were shown to contain a chain of undersea mountains running down the center of almost every ocean in the world. The mountain chain was split by a deep rift, and the surface around it was riddled with faults. This rugged terrain suggested that movement is occurring on the ocean floors.

Young Rock Surveys of the ocean floors showed that the rocks beneath the oceans are generally younger than the rocks of the continents. Most ocean rocks are only a fraction of the age of some continental rocks. In addition, the age of the rocks on the ocean bottoms increases with distance away from the mid-ocean ridges, suggesting that rocks form at the ridges and then move sideways away from them.

Paleomagnetism Basalt, the major rock of the ocean floors, is rich in iron compounds. Magnetite, pyrrhotite, and hematite are strongly affected by magnetism. When lava is extruded in the mid-ocean rifts, crystals of minerals affected by magnetism align with Earth's magnetic field while the lava cools. When solidified, the igneous rock contains a record of Earth's magnetic field locked in its crystals. Studies of ancient rocks show that Earth's magnetic field has reversed many times in the past. The reversals seem to take place at irregular intervals ranging from 20,000 to several million years. Research ships carrying sensitive instruments that can measure small changes in magnetism have discovered that magnetism in rocks on the ocean floors reveals a striking pattern of strips of normal and reversed fields. See Figure 18.10. The fact that the pattern is identical on either side of the mid-ocean ridges also suggests that rocks form at the ridges and then move sideways away from them.

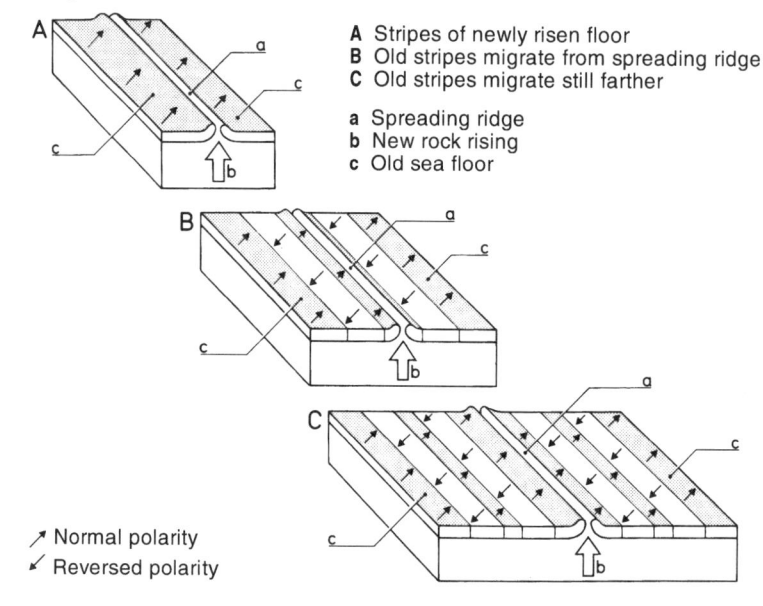

A Stripes of newly risen floor
B Old stripes migrate from spreading ridge
C Old stripes migrate still farther

a Spreading ridge
b New rock rising
c Old sea floor

↗ Normal polarity
↙ Reversed polarity

Figure 18.10 Paleomagnetism Patterns on the Ocean Floor. Source: *Earth Science on File.* Copyright © 1988 by Diagram Visual Information Ltd. Reprinted by permission of Facts on File, Inc.

The proposed mechanism for ocean floor spreading is as follows. Lava is extruded in the rift. The lava hardens to form new ocean floor with Earth's magnetic field "frozen" in its crystals. New lava erupts, splitting the just-formed rock and pushing it aside. The new lava hardens. As this process is repeated, new ocean floor is constantly being formed, and the floor on either side of the mid-ocean ridge is pushed sideways.

Trenches If new ocean floor is constantly being created along the ridges, Earth should be getting larger, but it is not. Therefore, ocean floor is being destroyed at the same rate at which it forms. But *where* is it being destroyed? The answer lies in the trenches. ***Trenches*** are deep crevices in the ocean floor where it bends downward sharply. Beneath the trenches, earthquakes occur frequently and the positions of their foci show a pattern of increasing depth with distance from one side of the trench. The plunging ocean floor and deepening pattern of earthquake foci beneath the trenches are consistent with a situation where sinking convection currents in the mantle are pulling the ocean floor downward into the mantle, where it melts and is destroyed in a process called *subduction*. See Figure 18.11. With this final piece of evidence the picture was complete, and scientists had not only an explanation for the features and formation of the ocean floor, but also a mechanism for continental drift. Shortly thereafter, continental drift and ocean floor spreading were unified into the theory of plate tectonics.

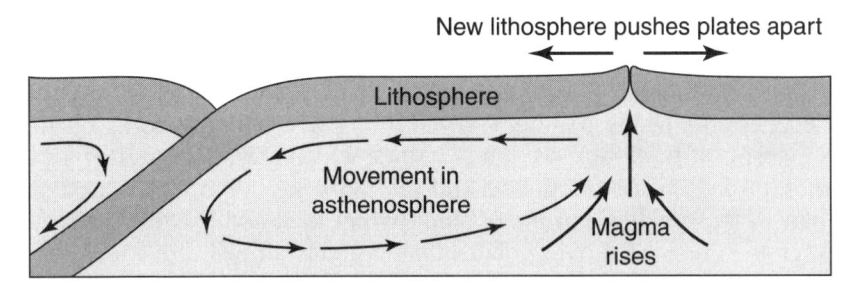

Figure 18.11 Ocean-floor is created along the ridges, but is destroyed as it plunges into the mantle (forming trenches) and melts. Source: *The Dynamic Earth*, Brian J. Skinner and Stephen C. Porter, John Wiley, 1989.

The Plate Tectonics Theory

By the 1960s, scientists had enough evidence to believe that continents and seafloors were indeed moving. The evidence showed not only motion, but also motion in many directions. This motion did not agree with the existing model of Earth, which considered the continents and seafloors part of a single, unbroken shell of rocky crust. Clearly, a solid shell cannot move in many different directions at once!

The Lithosphere and the Asthenosphere

As stated earlier, analysis of earthquake waves led to the discovery that Earth's interior consists of layers that have different properties. Based upon density differences, Earth can be divided into the inner and outer core, the mantle and the crust. If, instead of density, the *rigidity* of rock is considered, another pattern emerges. The crust and upper mantle to a depth of 100 kilometers is strong and rigid, forming a layer called the *lithosphere*. The region of the upper mantle

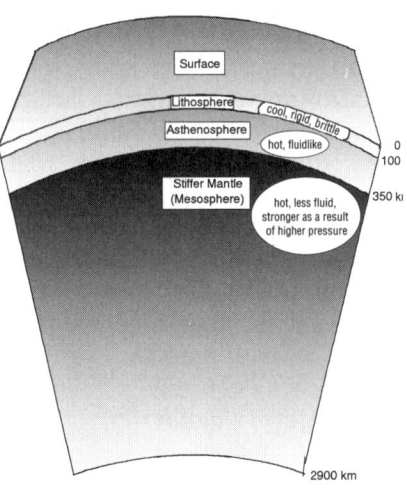

Figure 18.12 Lithosphere and Asthenosphere.

between 100 and 350 kilometers in depth is weak and behaves like a viscous fluid; this layer is called the *asthenosphere*. See Figure 18.12.

Scientists used the lithosphere/asthenosphere pattern to devise a new model of Earth's structure that fit the evidence. They combined aspects of continental drift, isostasy, and ocean floor spreading into a single unifying theory—***plate tectonics***. (*Tectonics* is the branch of geology dealing with the forces affecting the structure of Earth's crust.) In this new model, the lithosphere is seen, not as an unbroken solid, but fragmented and consisting of several huge pieces or *plates*—hence the name *plate* tectonics. The plates are envisioned as "floating" on a layer of fluid rock set in motion by huge convection currents in the mantle. Carried along like bobbing corks on these fiery currents of fluid rock, the plates collide with one another, jostle past one another, or drift apart. Continents and seafloors are in motion because they are part of these plates and move along with them, and it is the interaction between the edges of moving plates that explains all of Earth's features, as well as earthquakes, volcanic activity, and crustal movements.

In simple terms, then, plate tectonics embodies two key ideas: the rigid lithosphere consists of great slabs called *plates*; and these plates move sideways around Earth, sliding on the fluidlike asthenosphere. As the plates move laterally around Earth, their edges interact in one of three ways: they spread apart, they collide, or they slide past each other. Let us now consider these interactions and see how they account for a wide range of observations.

Interactions at Plate Boundaries

There are three types of plate boundaries. See Figure 18.13.

Divergent boundaries are places where adjacent plates are moving apart. Divergent boundaries create tension stresses, causing rock to fracture. Fractures result in earthquakes and open rifts through which magma can rise to the surface and solidify to form new lithosphere. The mid-ocean rifts are divergent boundaries, and the mid-ocean ridges are volcanic mountains formed by magma emerging from the rifts.

Convergent boundaries are places where adjacent plates are moving toward each other and colliding. When plate edges collide, compression and shear stresses cause rocks to fold, fracture, and move along faults. Plates consisting of ocean crust and plates of continental crust interact differently along divergent boundaries. Since ocean crust is denser than continental crust, ocean crust tends to plunge under continental crust when plates collide, forming a *subduction zone*. In subduction zones, the plunging plate is forced into the mantle and melts. If two plates carrying continental crust collide, they both crumple and fold. The Andes and Himalayan mountains formed along convergent boundaries.

Transform boundaries are places where plates move past each other in strike-slip motions. Shear stresses cause rock to fracture, forming numerous faults. The plates on either side of transform fault margins smash and rub against each other, as two ships in a near-collision grind their sides together.

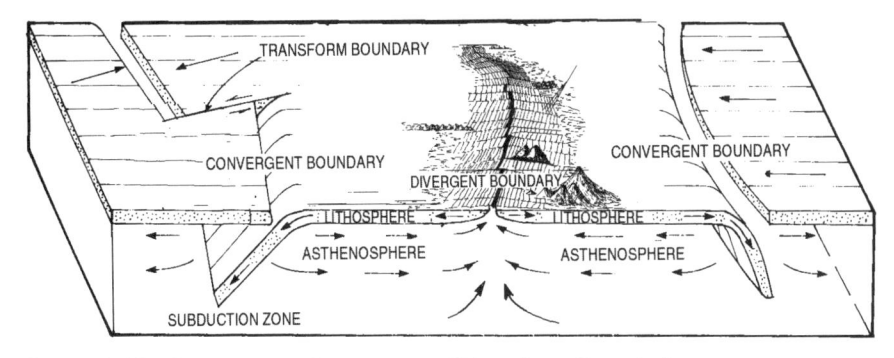

Figure 18.13 Convergent, Divergent, and Transform Boundaries.

The rock along these faults is intensely shattered, and the sliding motion of the plates causes many earthquakes. If the plates separate a little, some magma may "leak" through the boundary, causing small-scale volcanism. The San Andreas fault is part of a huge transform boundary along which the Pacific plate is moving past the North American plate.

Hot Spots Some volcanoes occur far from plate edges. For example, the

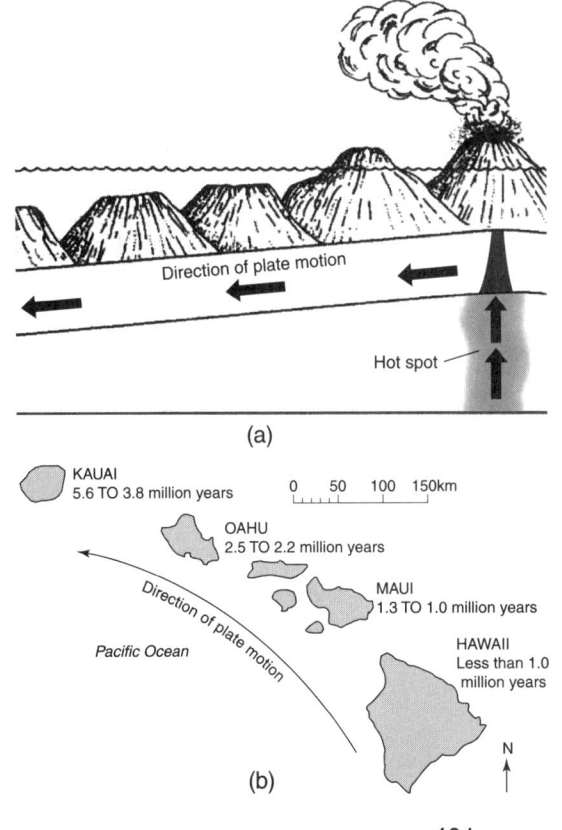

Figure 18.14 (a) Cross-section Showing a Series of Volcanoes Formed by a Plate Moving over a Hot Spot. The extinct volcanoes are eroded to nearly flat surfaces at sea level and are then submerged, forming flat-topped seamounts called *guyots*. Source: *Mountains of Fire*, Robert W. Decker and Barbara B. Decker, Cambridge University Press, 1991. (b) The Progressive Increase in Age of Hawaiian Islands with Distance Away from Currently Active Hawaiian Volcanoes. This supports the hot-spot hypothesis and indicates plate movement toward the northwest.

Plate Tectonics

Hawaiian volcanoes are located 4,000 kilometers from the nearest plate edge. How does plate tectonics account for volcanic activity far from plate boundaries? To explain the unusual position of these volcanoes and relate them to plate tectonics, Canadian geophysicist J. Tuzo Wilson proposed the concept of hot spots.

According to this concept, there are long-lasting zones of rising hot magma, or *hot spots*, deep within Earth at several places beneath moving plates. Large batches of magma rise from these hot spots and work their way upward through the moving plate. The magma rises because it is more buoyant than the surrounding rock, wedges apart cracks in the plate, and erupts, forming a volcano. As the plate moves along, older volcanoes are carried away in the direction of plate motion and new ones form over the hot spot. See Figure 18.14a.

Wilson's idea explained something that had puzzled scientists for years—why the Hawaiian Islands have active volcanoes only at the southeastern edge of the chain. Furthermore, knowing the rate of plate motion, one should be able to predict the ages of the older volcanic islands. Radioactive dating of the rocks found on various Hawaiian Islands confirmed this hypothesis. See Figure 18.14b.

The geothermal activity—hot springs, geysers, and volcanic rocks—at Yellowstone National Park is the result of another hot spot. The Galapagos, the Azores, and the Society Islands are also examples of volcanic islands formed by hot spots. Thus, while plate tectonics explains the location of belts of volcanic activity, hot spot volcanoes reveal the direction and rate of plate motion.

Key Ideas of the Plate Tectonics Theory

The plate tectonics theory, then, consists of the following ideas.

1. Earth's crust consists of a series of rigid slabs of rock called *lithospheric plates*. Each plate is about 100 kilometers thick. Earth's surface consists of about six major plates and several smaller pieces. See Figure 18.15.
2. The plates rest on a relatively fluid layer of the upper mantle and lower crust called the *asthenosphere*. This fluid layer lets the plates move independently of the deeper rocks in the mantle. Boundaries between plates are regions of volcanic and earthquake activity because they are where the plates collide and scrape against each other. The interiors of plates are relatively quiet.
3. The type of crust atop each plate determines whether the plate carries an ocean floor, a continent, or both. The continents and ocean floors are merely passengers on the moving plates. When the plates change position, the continents and ocean floors move along with them. Past positions of the plates can be inferred from evidence such as fossils, paleomagnetism, and past climates.

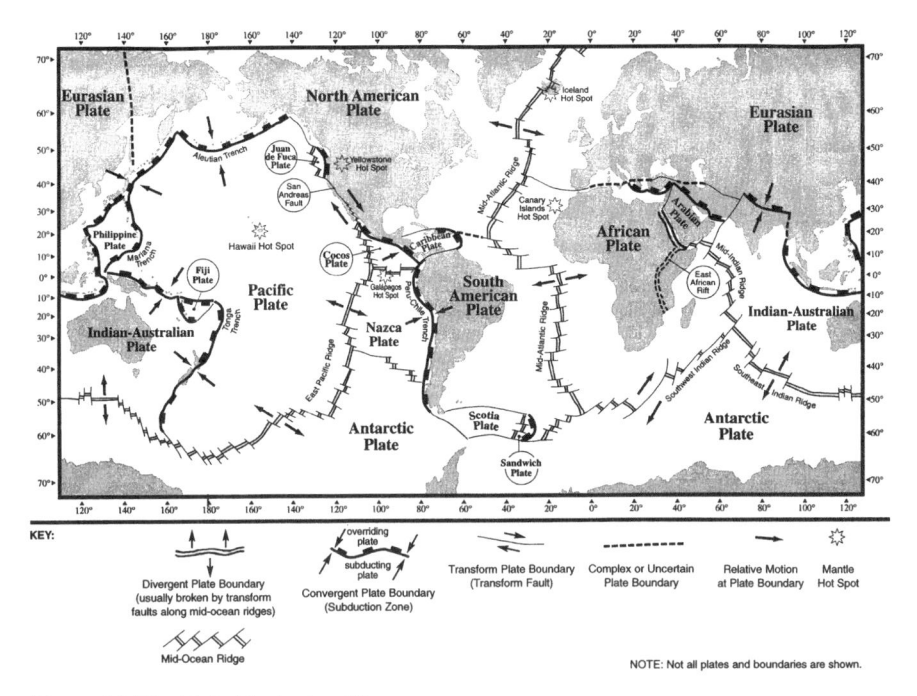

Figure 18.15 Plate Tectonics Diagram.

4. There are three types of plate boundaries: convergent, divergent, and transform. Plates collide at convergent boundaries, separate at divergent boundaries, and slide past each other at transform boundaries. When an ocean plate collides with a continental plate, the denser ocean plate is pushed under the continental plate, or subducted. Most mountains form at divergent and convergent boundaries.
5. In subduction zones, ocean floor plates are consumed as they are pushed into the hot mantle and melt. As the plate plunges into the mantle, it remains rigid and produces deep-focus earthquakes. Friction between the subducted plate and the adjacent plate melts rock along the interface and produces volcanic activity directly above and parallel to the trench. For this reason most trenches are bordered by volcanic island arcs or volcanic mountain chains.
6. The creation of ocean floor in the mid-ocean rifts and its destruction in the trenches are in equilibrium. Therefore, Earth remains the same size.
7. Forces exist within Earth that are powerful enough to move the lithospheric plates. Two hypotheses have been proposed to explain plate motion: the plates are pushed by convection currents in the mantle, and they are pulled downward by gravity in subduction zones. Observations of heat flow support the concept of convection in the mantle. However, the scientific community is not yet unified in identifying the specific forces that propel the plates.

Plate Tectonics

8. Convection in the mantle may produce "hot spots" over plumes of rising heat. As plates move over a hot spot, the magma works its way up through the plate and a volcano forms. Volcanic chains in the center of a plate, such as the Hawaiian Islands, are thought to have formed in this way.

Figure 18.16 is a simplified cross section of Earth showing the structures described above.

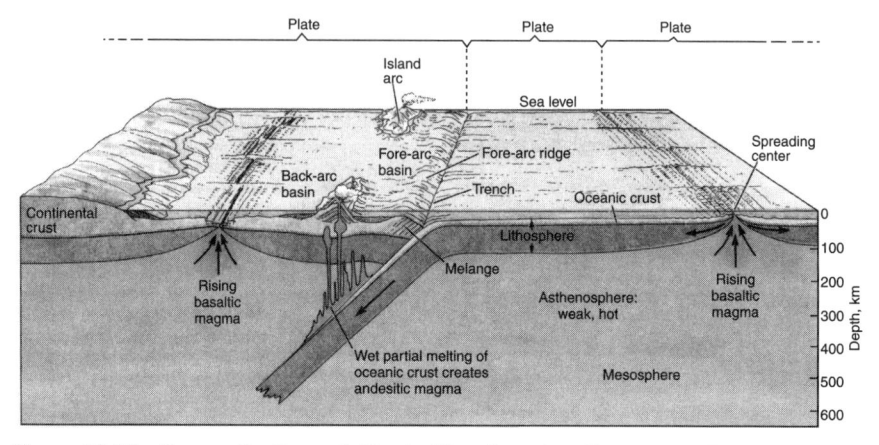

Figure 18.16 Cross Section of Earth Showing the Structures Found in Plate Tectonics. Source: *The Dynamic Earth*, 2nd Ed., Brian J. Skinner and Stephen C. Porter, John Wiley, 1992.

MULTIPLE-CHOICE QUESTIONS

In each case write the number of the word or expression that best answers the question or completes the statement.

1. The diagrams below show cross sections of exposed bedrock. Which cross section shows the *least* evidence of crustal movement?

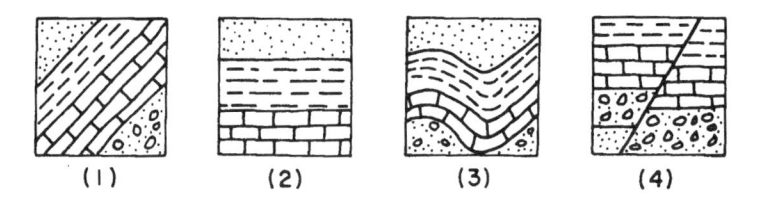

2. A line of former beaches along a coast, all 50 meters above sea level, is evidence of
   (1) present erosion
   (2) the present melting of polar icecaps
   (3) land uplift
   (4) a decrease in the deposition of marine fossils

3. Fossils of organisms that lived in shallow water can be found in horizontal sedimentary rock layers at great ocean depths. This fact is generally interpreted by most earth scientists as evidence that
   (1) the cold water deep in the ocean kills shallow-water organisms
   (2) sunlight once penetrated to the deepest parts of the ocean
   (3) organisms that live in deep water evolved from species that once lived in shallow water
   (4) sections of Earth's crust have changed their elevations relative to sea level

4. The best evidence of crustal movement is provided by
   (1) dinosaur tracks found in surface bedrock
   (2) marine fossils found on a mountaintop
   (3) weathered bedrock found at the bottom of a cliff
   (4) ripple marks found in sandy sediment

5. Two geologic surveys of the same area, made 50 years apart, showed that the area had been uplifted 5 centimeters during the interval. If the rate of uplift remains constant, how many years will be required for this area to be uplifted a total of 70 centimeters?
   (1) 250   (3) 350
   (2) 500   (4) 700

6. Which observation provides the strongest evidence for the inference that convection cells exist within Earth's mantle?
   (1) Sea level has varied in the past.
   (2) Marine fossils are found at elevations high above sea level.
   (3) Displaced rock strata are usually accompanied by earthquakes and volcanoes.
   (4) Heat-flow readings vary at different locations in Earth's crust.

7. How was the valley shown in the diagram below most likely formed?

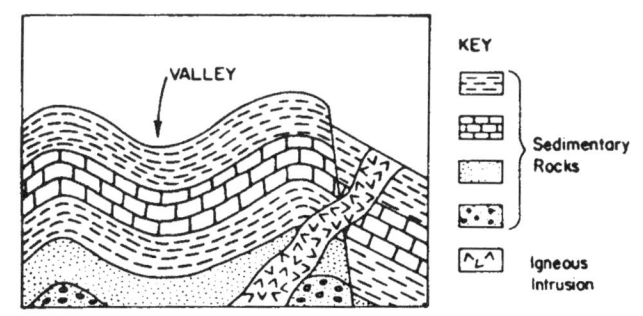

   (1) by the deposition of sediments
   (2) by the extrusion of igneous material
   (3) by the faulting of rock layers
   (4) by the folding of rock layers

Plate Tectonics

8. A sandstone layer is found tilted at an angle of 75° from the horizontal. What probably caused this 75° tilt?
   (1) The sediments that formed the sandstone layer were originally deposited at a 75° tilt.
   (2) The sandstone layer has changed position because of crustal movement.
   (3) The sandstone layer has recrystallized because of contact metamorphism.
   (4) Nearly all sandstone layers are formed from wind-deposited sands.

9. Which Earth process most likely formed the depression now occupied by the lake shown in the diagram below?

   (1) glaciation   (3) climate changes
   (2) erosion   (4) faulting

10. The diagram below represents a geologic cross section of a portion of Earth's crust. Which geologic event shown in the diagram would suggest past crustal movement at that location?

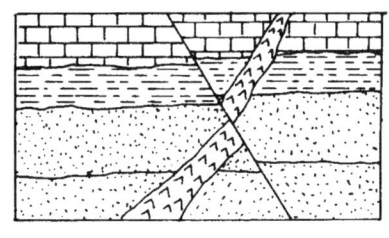

    KEY:
    [▲▲▲] Igneous Intrusion
    [▓▓▓] Sedimentary Rocks

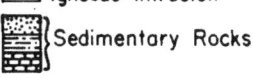

    (1) deposition of sediments
    (2) intrusion of molten material (magma)
    (3) folding of rock layers
    (4) faulting of rock layers

11. The diagram below represents a cross section of a portion of Earth's crust.

    Which past activity in this region is suggested by the shape of these sedimentary rock layers?
    (1) widespread volcanic activity
    (2) horizontal sorting
    (3) glacial deposition
    (4) crustal movements

12. Folded sedimentary rock layers are usually caused by
    (1) deposition of sediments in folded layers
    (2) differences in sediment density during deposition
    (3) a rise in sea level after deposition
    (4) crustal movement occurring after deposition

13. Which statement best supports the theory that all the continents were once a single landmass?
    (1) Rocks of the ocean ridges are older than those of the adjacent seafloor.
    (2) Rock and fossil correlation can be made where the continents appear to fit together.
    (3) Marine fossils can be found at high elevations above sea level on all continents.
    (4) Great thicknesses of shallow-water sediments are found at interior locations on some continents.

14. The theory of continental drift does *not* explain the
    (1) matching of rock features on continents thousands of kilometers apart
    (2) melting of glacial ice at the close of the Pleistocene Epoch
    (3) apparent fitting together of many continental boundaries
    (4) fossils of tropical plants in Antarctica

15. According to the *Earth Science Reference Tables*, during which geologic period were the continents all part of one landmass, with North America and South America joined to Africa?
    (1) Tertiary   (3) Triassic
    (2) Cretaceous   (4) Ordovician

Plate Tectonics

16. Which statement best supports the theory of continental drift?
    (1) Basaltic rock is found to be progressively younger at increasing distances from a mid-ocean ridge.
    (2) Marine fossils are often found in deep-well drill cores.
    (3) The present continents appear to fit together as pieces of a larger landmass.
    (4) Areas of shallow-water seas tend to accumulate sediment that gradually sinks.

17. According to the Inferred Positions of Earth Landmasses diagram in the *Earth Science Reference Tables*, on what other landmass would you most likely find fossil remains of the late Paleozoic reptile called Mesosaurus, shown to the right?
    (1) North America
    (2) Africa
    (3) Antarctica
    (4) Eurasia

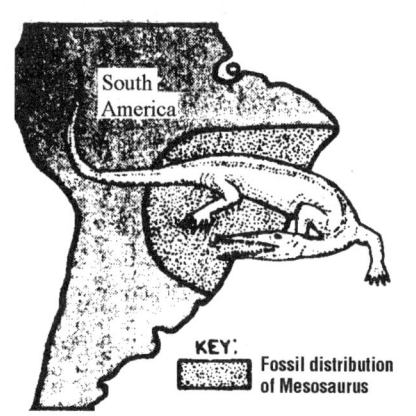

18. Which is the best evidence supporting the concept of ocean floor spreading?
    (1) Earthquakes occur at greater depths beneath continents than beneath oceans.
    (2) Sandstones and limestones can be found both in North America and in Europe.
    (3) Volcanoes appear at random within the oceanic crust.
    (4) Igneous rocks along the mid-ocean ridges are younger than those farther from the ridges.

19. Igneous materials found along oceanic ridges contain magnetic iron particles that show reversal of magnetic orientation. This is evidence that
    (1) volcanic activity has occurred constantly throughout history
    (2) Earth's magnetic poles have exchanged positions
    (3) igneous materials are always formed beneath oceans
    (4) Earth's crust does not move

Note that question 20 has only three choices.
20. As the distance from the mid-ocean ridges increases, the age of the ocean basin rock will
    (1) decrease
    (2) increase
    (3) remain the same

21. Which statement best describes the continental and oceanic crusts?
    (1) The continental crust is thicker and less dense than the oceanic crust.
    (2) The continental crust is thicker and more dense than the oceanic crust.
    (3) The continental crust is thinner and less dense than the oceanic crust.
    (4) The continental crust is thinner and more dense than the oceanic crust.

22. Which evidence does *not* support the theory that Africa and South America were once part of the same large continent?
    (1) correlation of rocks on opposite sides of the Atlantic Ocean
    (2) correlation of fossils on opposite sides of the Atlantic Ocean
    (3) correlation of coastlines on opposite sides of the Atlantic Ocean
    (4) correlation of living animals on opposite sides of the Atlantic Ocean

23. Igneous rock along oceanic ridges is younger than igneous rock farther from the ridges. This evidence supports the theory that
    (1) the ocean floor is stable
    (2) the ocean floor is spreading
    (3) volcanoes once existed on both sides of oceanic ridges
    (4) oceanic ridges are areas of subsidence

24. The drawing below represents the ocean floor between North America and Africa.

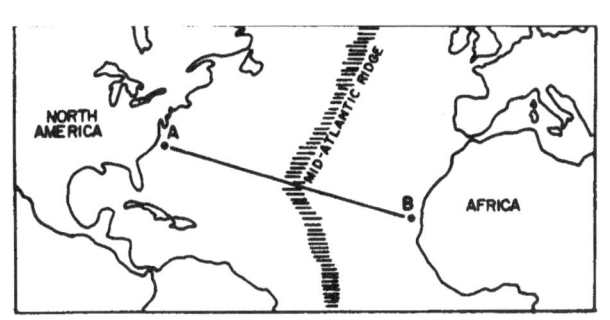

Plate Tectonics

Which graph best represents the age of the bedrock in the ocean floor along line *AB*?

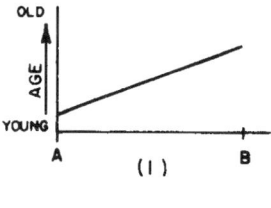

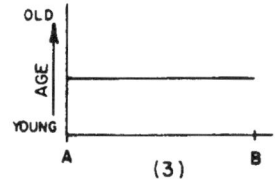

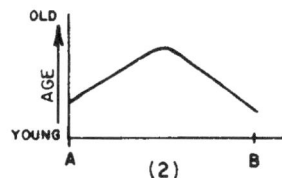

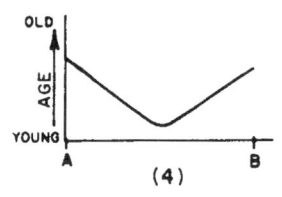

(1) Age increases from A to B (linear increase)
(2) Age peaks in the middle between A and B
(3) Age remains constant from A to B
(4) Age is lowest in the middle between A and B (V-shape)

25. According to the *Earth Science Reference Tables*, during which geologic time period were the continents of North America, South America, and Africa closest together?
(1) Tertiary
(2) Cretaceous
(3) Triassic
(4) Ordovician

26. What is the primary force in the theory of ocean floor spreading?
(1) Density differences in the mantle cause convection cells.
(2) Planetary winds in the atmosphere blow against landmasses.
(3) Earth rotates on its axis.
(4) Earth revolves around the Sun.

27. Which is suggested by the occurrence of higher than average temperature below the surface of Earth in the area of the mid-Atlantic Ridge?
(1) the existence of convection cells in the mantle
(2) the presence of heat due to orographic effect
(3) a high concentration of magnetism in the mantle
(4) the existence of a thinner crust under mountains

28. In which set are the Earth processes thought to be most closely related to each other because they normally occur in the same zones?
(1) mountain building, earthquakes, and volcanic activity
(2) mountain building, shallow-water fossil formation, and rock weathering
(3) volcanic activity, rock weathering and deposition of sediments
(4) earthquakes, shallow-water fossil formation, and shifting magnetic poles

410

29. The primary cause of convection currents in Earth's mantle is believed to be the
    (1) differences in densities of Earth materials
    (2) subsidence of the crust
    (3) occurrence of earthquakes
    (4) rotation of Earth

Base your answer to the following question on your knowledge of earth science and the diagrams below.

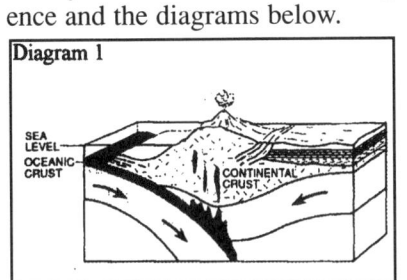

Diagram 1

Diagram 2

Arrows show the direction of plate movement.

30. At which location in Diagram 2 would the plate boundary shown in Diagram 1 most likely be found?
    (1) A
    (2) B
    (3) C
    (4) D

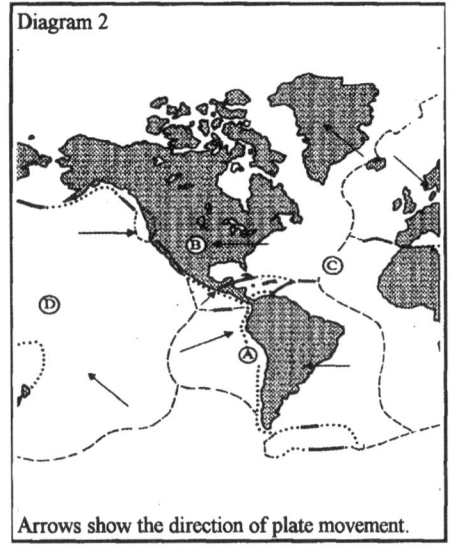

31. The diagram below represents a partial cross section of a model of Earth. The arrows show inferred motions within Earth.

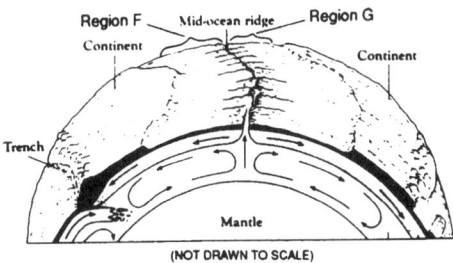

(NOT DRAWN TO SCALE)

Which property of the oceanic crust in regions F and G is a result of these inferred motions?
(1) The crystal size of the rock decreases constantly as distance from the mid-ocean ridge increases.
(2) The temperature of the basaltic rock increases as distance from the mid-ocean ridge increases.
(3) Heat-flow measurements steadily increase as distance from the mid-ocean ridge increases.
(4) The age of the igneous rock increases as distance from the mid-ocean ridge increases.

Plate Tectonics

32. According to the *Earth Science Reference Tables*, which of the following locations is the site of a convergent plate boundary?
    (1) the mid-Atlantic Ridge
    (2) the Aleutian trench
    (3) the Southeast-Indian Ridge
    (4) the Pacific/North American plate boundary

Base your answers to questions 33 through 37 on your knowledge of earth science, the *Earth Science Reference Tables*, and the map information below. The map shows the present-day relative positions of South America and Africa and the age of the rocks composing the two continents. Letters *A–H* indicate specific rock units. The apparent close correlation between rocks on the two continents provided early evidence for the theory of continental drift.

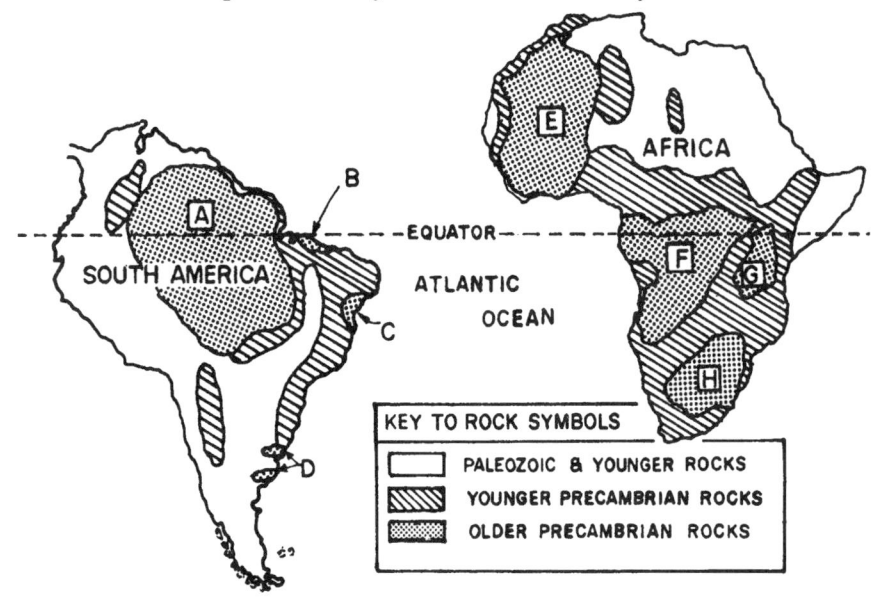

33. According to the *Earth Science Reference Tables*, when did Africa and South America completely separate and move away from each other into two distinct landmasses as shown above?
    (1) before the Cambrian Period
    (2) during the Carboniferous Period
    (3) during the Triassic Period
    (4) after the Cretaceous Period

34. Which Precambrian rock unit in South America is most probably a former section of rock unit *F* in present-day Africa?
    (1) *A*   (3) *C*
    (2) *B*   (4) *D*

35. What present-day Atlantic Ocean feature is part of the evidence that these two continents are still drifting apart?
    (1) a magnetic pole
    (2) land sediments
    (3) a deep ocean trench
    (4) a mid-ocean ridge

36. Which model best indicates the overall direction of movement of the African landmass relative to the position of South America during this time of continental drift?

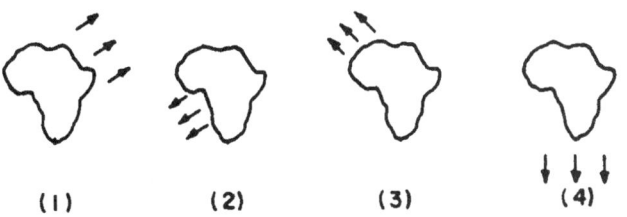

Note that question 37 has only three choices.
37. Compared to the age of the rocks composing these continental landmasses, the age of the oceanic crust between them is mostly
    (1) younger
    (2) older
    (3) approximately the same

Base your answers to questions 38 through 42 on your knowledge of earth science, the *Earth Science Reference Tables*, and the diagram below. The diagram represents the age of the basaltic ocean crust in the Atlantic Ocean between the United States and Africa. Line *AB* is drawn for reference purposes only.

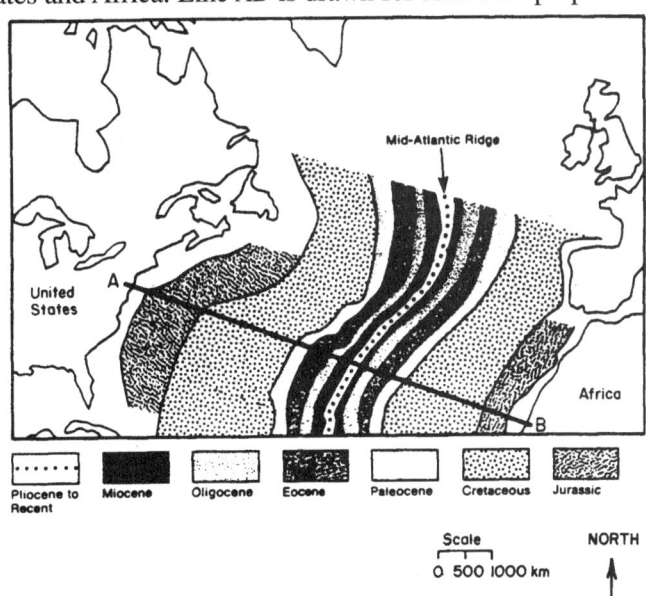

413

Plate Tectonics

38. Which statement is best supported by the diagram?
  (1) The ocean crust is the same age along line *AB*.
  (2) The oldest ocean crust is located near the continents.
  (3) The age of the ocean crust increases from point *A* to point *B*.
  (4) Most of the ocean crust along line *AB* formed in the Paleozoic Era.

39. The age of formation of the ocean crust along line *AB* suggests that the United States and Africa are moving
  (1) eastward   (3) closer together
  (2) westward   (4) farther apart

40. Which diagram most closely represents the cross section of the ocean floor along line *AB*?

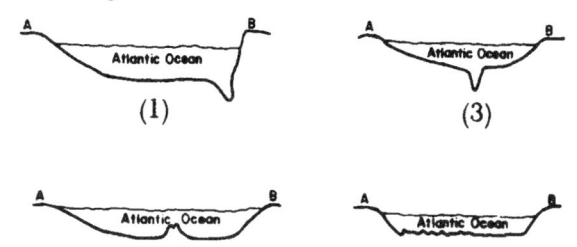

41. According to the diagram, the width of the Cretaceous rock east of the mid-Atlantic Ridge along line *AB* is approximately
  (1) 1,000 km   (3) 1,600 km
  (2) 1,200 km   (4) 4,000 km

42. On which map do the marks best represent the distribution of the frequency of earthquakes for this area?

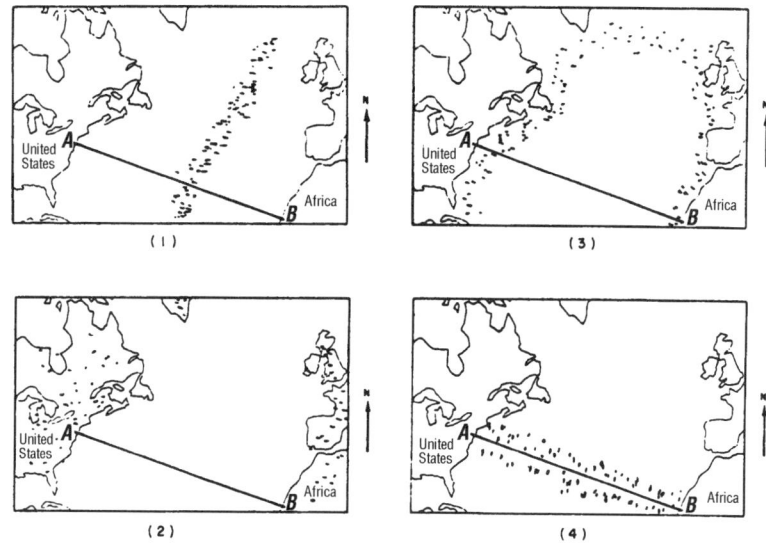

414

Base your answers to questions 43 through 46 on your knowledge of earth science, the *Earth Science Reference Tables*, and the diagrams below. The diagrams represent geologic cross sections of the upper mantle and crust at four different Earth locations. In each diagram, the movements of the crustal sections (plates) are shown by arrows and the locations of frequent earthquakes are indicated by symbols as shown in the key. Diagrams are not drawn to scale.

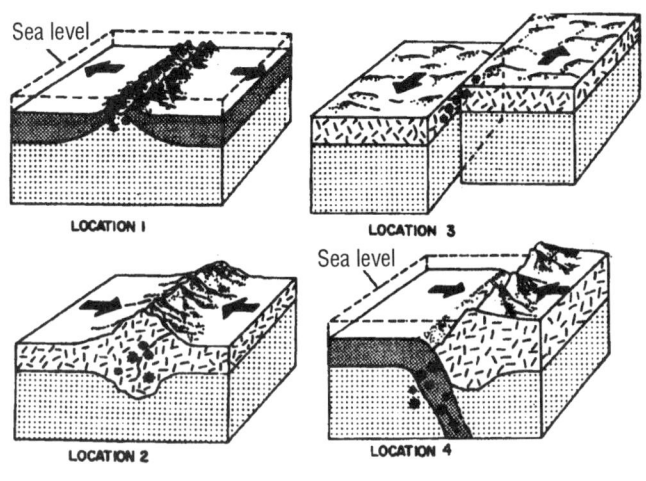

KEY
- CONTINENTAL CRUST (GRANITE)
- OCEANIC CRUST (BASALT)
- MANTLE
- ✱ EARTHQUAKE FOCUS
- ➡ DIRECTION OF PLATE MOVEMENT

43. According to the diagrams, the most probable cause of the shallow earthquakes at location 3 is
    (1) the collision of two oceanic plates
    (2) the spreading of two oceanic plates
    (3) the sinking of an oceanic plate under a continental plate
    (4) the horizontal shifting of two continental plates

44. Compared to the oceanic crust, the continental crust at location 4 is
    (1) thicker and less dense
    (2) thicker and more dense
    (3) thinner and less dense
    (4) thinner and more dense

45. Which location could be found along the west coast of South America?
    (1) 1   (3) 3
    (2) 2   (4) 4

Plate Tectonics

46. At a depth of 2,500 kilometers below location 1, what is the actual temperature of the rock ?
    (1) 1,000° C
    (2) 6,800° C
    (3) 4,600° C
    (4) 1,000,000°C

CONSTRUCTED RESPONSE QUESTIONS

Base your answers to questions 47 through 49 on the diagram below. The diagram represents the supercontinent Pangaea, which began to break up approximately 220 million years ago.

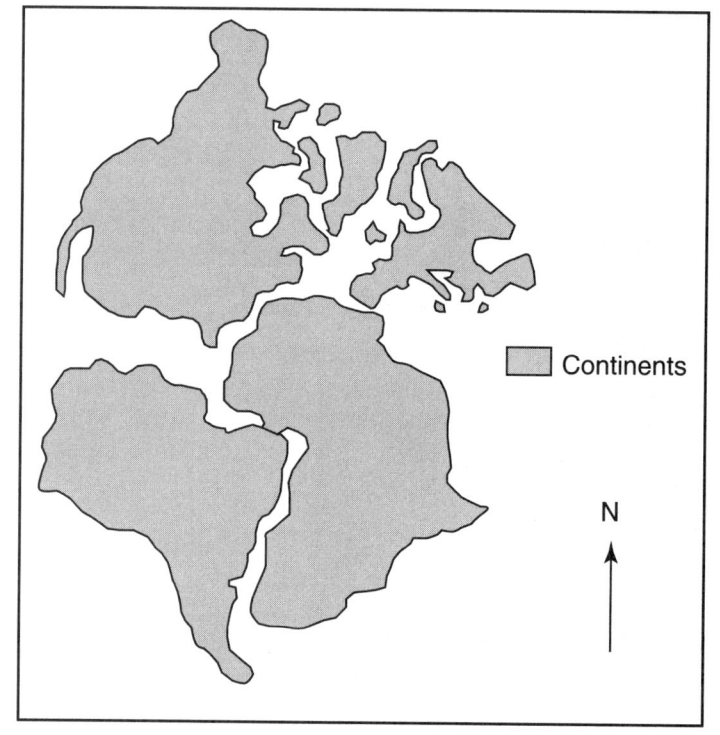

47. During which geologic period within the Mesozoic Era did the supercontinent Pangaea begin to break apart? [1]

48. State one form of evidence that supports the inference that Pangaea existed. [1]

49. State the compass direction toward which North America has moved since Pangaea began to break apart. [1]

Base your answers to questions 50 through 52 on the cross section of a portion of Earth's interior below. The cross section shows the focal depths of some earthquakes that occurred west of the Tonga Trench. Data were collected along the 22°S parallel of latitude.

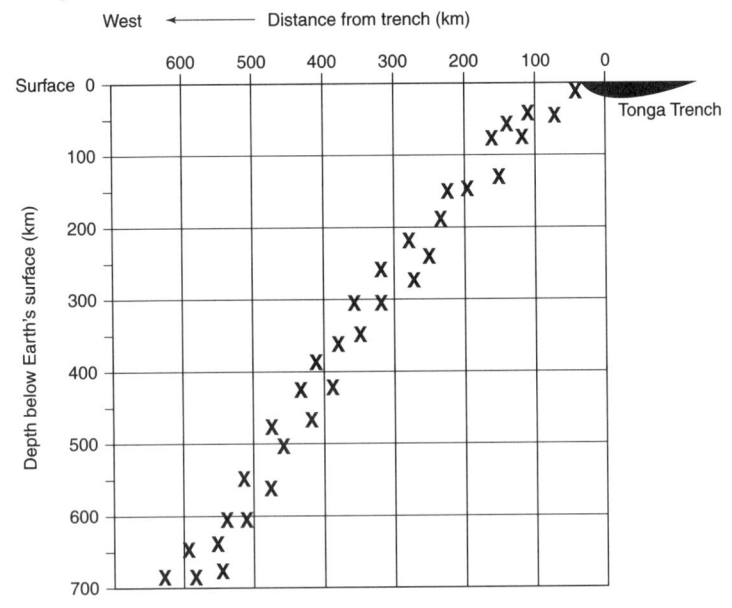

50. State the relationship between the depth of an earthquake's focus and the earthquake's distance from the Tonga Trench. [1]

51. The Tonga Trench is the crustal surface boundary between two tectonic plates. State the names of the two plates. [1]

52. The focal depth pattern shown on the cross section represents the location of the subsurface boundary between two tectonic plates. Describe the relative motions of the plates along this boundary. [1]

Plate Tectonics

EXTENDED CONSTRUCTED RESPONSE QUESTIONS

Base your answers to questions 53 through 57 on your knowledge of earth science, the *Earth Science Reference Tables*, and the diagrams below. Diagram I is a map showing the locations and bedrock ages of some of the Hawaiian Islands. Diagram II is a cross section of an area of Earth illustrating a stationary magma source and the process that could have formed the islands.

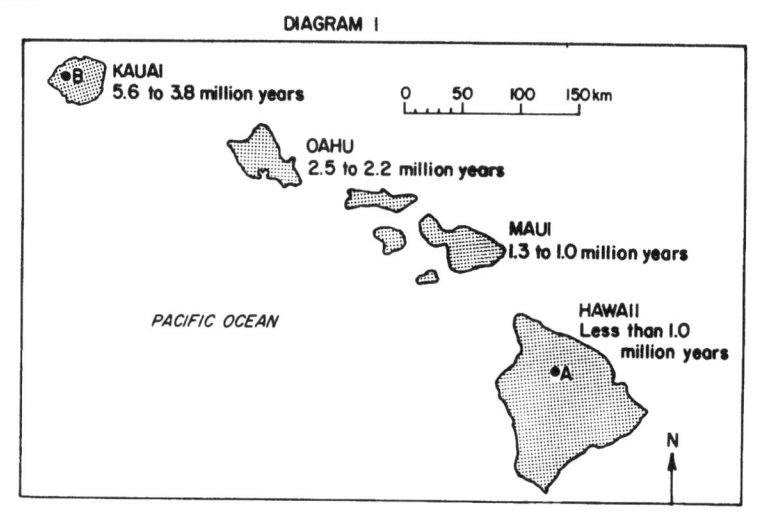

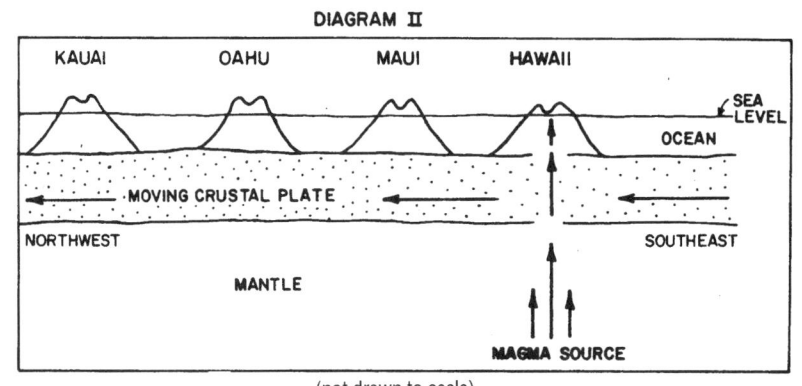

DIAGRAM I

KAUAI
5.6 to 3.8 million years

0 50 100 150 km

OAHU
2.5 to 2.2 million years

MAUI
1.3 to 1.0 million years

PACIFIC OCEAN

HAWAII
Less than 1.0 million years

N

DIAGRAM II

KAUAI   OAHU   MAUI   HAWAII

SEA LEVEL
OCEAN

MOVING CRUSTAL PLATE

NORTHWEST   SOUTHEAST

MANTLE

MAGMA SOURCE

(not drawn to scale)

53. On Diagram I, draw an arrow indicating the direction of motion of the Pacific Plate. [1]

54. (a) Using the ages and map scale given in the diagram, calculate the rate, in kilometers per million years, at which the plate is moving over the stationary magma source between points A and B. [Hint: Use the formula for rate of change given in the *Earth Science Reference Tables*.]

418

(b) What would this rate of plate motion be in millimeters per year? [Hint: 1 kilometer = 1 million millimeters] [2]

55. If each island formed as the crustal plate moved over the magma source in the mantle as shown in Diagram II, describe the most likely location in which the next volcanic island would form. [1]

56. Which of the Hawaiian Islands has the greatest probability of experiencing a volcanic eruption? Support your answer with evidence from the diagrams. [2]

57. Most Hawaiian volcanoes erupt quietly, spewing forth rivers of molten lava that run down their slopes toward the ocean. On the basis of this information, choose three types of igneous rock listed in the *Earth Science Reference Tables* that you would expect to find on the Hawaiian Islands. Justify each of your three choices. [3]

Unit Six: WEATHERING, EROSION, AND DEPOSITION

Chapter 19
WEATHERING AND SOIL FORMATION

> **KEY IDEAS** In preceding units, we explored the mechanisms by which the materials that comprise Earth's crust were formed. In this unit we will discuss the processes that shape the material exposed at the surface of Earth's crust; a dynamic, changing system of minerals, rocks, weather, water, ice, wind, and living organisms.

KEY OBJECTIVES

Upon completion of this chapter, you will be able to:

- Describe the weathering processes that result in the physical and chemical breakdown of crustal material and describe the products of weathering.
- Explain how weathering and biological activity over long periods of time create soils.

WEATHERING

The surface of Earth's crust, or lithosphere, is constantly exposed to the changing weather of the atmosphere. This environment is very different from the one in which most rocks and minerals were formed. As these materials adjust to their new environment, they change. They break down into smaller pieces, and chemical reactions change them into new substances. Since the processes that produce these changes result from exposure to the weather, they are called *weathering*. **Weathering** *is the breakdown of rocks into smaller particles by natural processes*. Whenever rocks are exposed to the air, water, and living things at or near Earth's surface, weathering occurs. Weathering processes are divided into two general types: physical and chemical.

Physical Weathering

Some of the changes caused by weathering involve form only. Weathering may break a large, solid mass of rock into loose fragments varying in size and shape but identical in composition to the original rock. Processes that break down rocks without changing their chemical compositions are called *physical weathering*.

Frost Action

Frost action is the result of an unusual property of water. Most materials expand when heated and contract when cooled. This is true of water except that, when water is cooled from 4°C to 0°C, it *expands*. It expands most when, at 0°C, it solidifies into ice; then its volume increases by 9%. The expansion of water as it cools and solidifies can exert huge forces on anything confining it—forces measuring tens of thousands of pounds per square inch!

Water, in the form of rain, melting snow, or condensation, seeps into any cracks or pores in rock. When the temperature drops below the freezing point of the water in these cracks and pores, the water changes into ice. The expanding ice exerts tremendous pressure against the confining rock. Acting like a wedge, it widens and extends the opening. Then, when the ice thaws, the water seeps deeper into the opening. When the water refreezes, the process is repeated. In this way, the alternate freezing and thawing of water, or *frost action*, breaks rocks apart (see Figure 19.1).

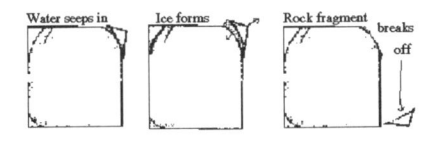

Figure 19.1 Frost Action

Frost action is particularly effective where bedrock is directly exposed to the atmosphere, moisture is present, and the temperature fluctuates frequently above and below the freezing point of water. These conditions often exist during the winter in temperate climates such as New York State's, and can occur also on mountain tops and at high elevations in spring or fall. Daytime temperatures rise above freezing, causing snow and ice to melt, only to drop below freezing again at night, producing frost action.

Frost action on cliffs of bare rock breaks loose fragments that fall to the base of the cliff. When this takes place rapidly, a pile of fragments called a *talus slope* accumulates at the base of the cliff (see Figure 19.2).

The potholes that threaten motorists in many of our northern states are caused by frost action on exposed road surfaces. Farmers in New England have to clear their fields every spring of rocks and boulders pushed up out of the ground by frost action in soil. As you might expect, frost action is almost nonexistent in states such as Florida and Hawaii.

Figure 19.2 A Talus Slope Forms at the Base of a Cliff. The angle of repose is the steepest slope at which the fragments remain stable.

421

However, through much of the northern United States, it is probably the primary weathering process, and, throughout the world, probably the most significant physical weathering process.

Abrasion

Rocks can also be broken down by *abrasion*, that is, by rubbing against each other. Rock abrasion occurs mainly when fragments are being carried along by agents of erosion. A typical example occurs in streams. As the fragments are carried along by the water, they bounce off and rub against each other. This abrasion breaks smaller pieces off the surface, and the fragments become still smaller and also more rounded (see Figure 19.3). The fragments abrade the bedrock beneath the stream as well. Wind also weathers rock by abrasion. Anyone who has sat on a sandy beach on a windy day can attest to the abrasive power of wind-driven sand. Abrasion is a major physical weathering process.

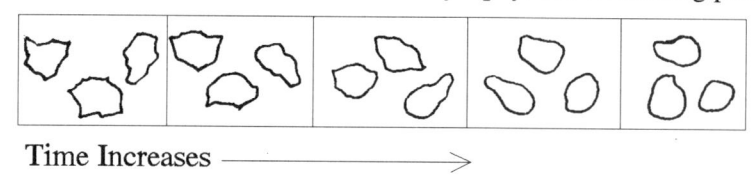

Time Increases ————→

Figure 19.3 Rounding of Particles Due to Abrasion

Exfoliation

Exfoliation is the scaling off, or peeling, of successive shells from the surface of rocks. Exfoliation generally occurs in coarse-grained rocks that contain the mineral feldspar. Whenever the surface of such rocks becomes wet, moisture penetrates pores and crevices between the mineral grains and reacts with the feldspar. A chemical change occurs, producing a new substance: kaolin. This clay has a greater volume than the feldspar it replaces. The expansion pries loose the surrounding mineral grains, and a thin shell of surface material flakes away. (Note that this is a physical process caused by a chemical change.)

With successive wettings of the rock surface, the process is repeated. Since more surface is exposed at the corners of cracks or joints, these split off at the fastest rate. The result is a rounding in the shape of the rock as seen in Figure 19.4. Feldspar is a common rock-forming mineral; therefore, exfoliation is a significant weathering process.

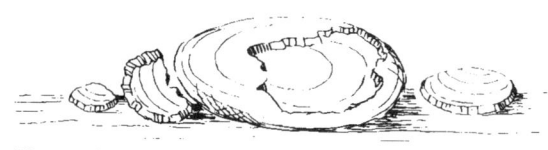

Figure 19.4 Exfoliation

Plant and Animal Action

Plants and animals interact with rock in a number of ways that cause the rock to break down into smaller pieces. When a rock develops cracks, small particles of rock and soil are washed into the cracks by rain or blown in by wind. If a seed finds its way into a crack, it can germinate and begin to grow. As the plant grows, it sends tiny rootlets deeper into the crack in search of water. The growing rootlets thicken and press against the sides of the crack. Like the expansion of ice in frost action, the growing roots widen and extend the crack (see Figure 19.5a). Eventually the rock is broken apart. The roots of

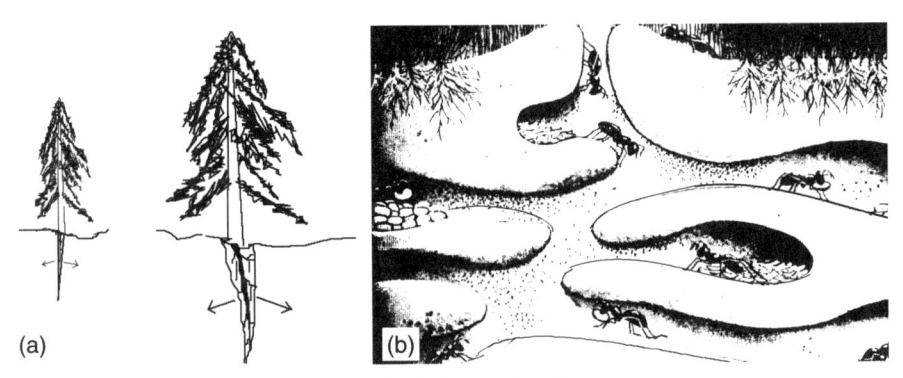

Figure 19.5 Plant (a) and Animal (b) Action on Rocks.

tiny plants like mosses and lichens produce a rock-dissolving acid as they grow and decay, thereby further accelerating the breakdown of the rocks.

Although animals (with the exception of humans) do not directly attack rock, they contribute indirectly to its weathering in many ways. More than 100 years ago, Charles Darwin calculated that on 1 acre of land earthworms bring as much as 10 metric tons of particles to the surface each year. Exposed to the atmosphere again, these particles are subjected to further breakdown by weathering processes.

Ants (see Figure 19.5b), termites, woodchucks, moles and other burrowing animals also play a role in weathering. Furthermore, the burrows they create allow air and water to penetrate deeper beneath the surface and to weather the underlying bedrock.

Humans, however, have probably contributed more to the physical weathering of rock than any other single species. Roadcut building, rock quarrying, and strip mining are just a few examples of the human activities that break up rock. In addition, these activities expose vast quantities of fresh rock to other weathering processes.

The contribution of each individual organism to the weathering of Earth's crust may seem small. However, consider the number of living things on Earth and the length of time they have been at work. Taken collectively, the total amount of rock they have affected is tremendous.

Changes in Temperature

Rocks are often exposed to large temperature changes. As rocks heat up during the day, they expand; as they cool off at night, they contract. You might expect this constant change to cause rocks to crack and break up, but experiments have indicated that this is not generally the case. Only extreme temperature changes, such as those resulting from forest and brush fires, cause rocks to crack or flake off at the surface.

Pressure Unloading

Some rocks form deep beneath Earth's surface. There they are under great pressure—millions of pounds per square inch. As the rocks form, stresses build up inside them that cannot be released because of the pressure. When forces within Earth bring these rocks to the surface, the pressure is reduced. The resultant expansion and release of stress cause the rocks to develop large cracks, or joints, at weak points in their structure—a phenomenon known as *pressure unloading*. Pressure unloading can also occur when glaciers melt away and the pressure exerted by the weight of the ice on underlying rock is released.

Chemical Weathering

Chemical weathering breaks down rocks by changing their chemical compositions. Most rocks form in an environment that is very different from that at the surface of Earth. Many of the substances in the atmosphere, for example, were not present in the environment where the rocks were formed. When the minerals in a rock are exposed to these substances, they may react with them to form new compounds with properties different from those of the original minerals. Such changes almost always weaken the structure of the rock, so that it either falls apart or is more easily broken down by physical weathering. Oxygen, water, and carbon dioxide are chiefly responsible for the chemical weathering of rocks.

Oxidation

The atmosphere is about 21 percent oxygen. Oxygen combines with many substances in a reaction called *oxidation*, which is an important chemical weathering process. Oxygen reacts most readily with minerals containing iron, such as magnetite, pyrite, amphibole, and biotite. Oxidation of iron produces compounds of iron and oxygen, iron oxides, of which hematite (Fe_2O_3) and magnetite (Fe_3O_4) are common examples. Since the iron in hematite and magnetite can be extracted economically, these compounds are also considered iron ores.

If water is present during oxidation, another reaction can occur; a compound of iron, oxygen, and water called *goethite* can form. Goethite has a yellowish brown color. If goethite is dehydrated, hematite forms. Many reddish or yellow-brown soils and weathered rocks get their color from the hematite or goethite they contain.

Oxidation of iron oxide in the presence of water to form goethite:
4FeO + 2H$_2$O + O$_2$ → 4(FeO-OH)
iron oxide   water   oxygen   goethite

Dehydration of goethite to form hematite:
2(FeO-OH) → Fe$_2$O$_3$ + H$_2$O
goethite   hematite   water

How does oxidation break down or weather rocks? When oxygen combines with iron, the chemical bonds between the iron and other substances in the rock are broken, thereby weakening its structure. Compare the strength of a rusty nail with that of a new one. The effect on rock is similar. In addition to iron, other elements in rock, such as aluminum and silicon, can combine with oxygen. The effects are much the same: an oxide is formed, the structure is weakened, and the rock disintegrates.

Hydration, Hydrolysis, and Solution

Most of Earth's surface is covered with water. Trillions of gallons fall on Earth as rain every day. With such abundance it is not surprising that water is an important agent of chemical weathering.

When water combines with another substance, the reaction is called *hydration*. For example, the hydration of anhydrite forms gypsum.

Hydration of anhydrite to form gypsum:
CaSO$_4$ + 2H$_2$O → CaSO$_4$ • 2H$_2$O
anhydrite   water   gypsum

Water can also form hydrogen ions (H+) and hydroxide ions (OH–). When these ions replace the ions of a mineral, the reaction is called *hydrolysis*. Feldspar, amphibole, and biotite are common minerals that can undergo hydrolysis, which causes them to swell and crumble. The product is a powder of clay minerals that are insoluble in water. For example, the hydrolysis of feldspar forms kaolinite, a clay mineral.

Hydrolysis of feldspar to form kaolinite:
4 KAlSi$_3$O$_8$ + 4 H$^+$ + 2H$_2$O → Al$_4$Si$_4$O$_{10}$(OH)$_8$ + 8SiO$_2$
K-feldspar   hydrogen ions   water   kaolinite   silica

(Note that the symbol K does not appear on the right. The potassium is released as a positive ion into the soil water.)

Water also weathers rock simply by dissolving it. This process is called *solution*. So many materials dissolve, at least to some extent, in water that water is often called the *universal solvent*. Halite (rock salt) and gypsum are good examples of water-soluble minerals. Water will slowly dissolve these minerals out of a rock, thereby exposing the surrounding minerals to further weathering. In some cases the rock's structure may be so weakened by the empty spaces created that the rock just crumbles. In solution, the dissolved minerals may react with each other to form new substances. If these substances are insoluble in water, they precipitate out.

Carbonation

Carbon dioxide (CO_2) is a colorless, odorless gas that comprises 0.03–0.04 percent of Earth's atmosphere. The chemical combining of carbon dioxide with another substance is called *carbonation*. As a gas, carbon dioxide has almost no effect on rocks. When carbon dioxide comes in contact with water, though, the following carbonation reaction takes place:

Production of carbonic acid by solution of carbon dioxide:
$$CO_2 + H_2O \rightarrow H_2CO_3$$
carbon dioxide water carbonic acid

Although carbonic acid is a fairly weak acid, it attacks common rock minerals. Carbonic acid reacts readily with minerals containing the elements sodium, potassium, magnesium, and calcium. The compounds formed when these elements react with carbon dioxide are called *carbonates*.

Carbonic acid is most destructive to the mineral calcite, which is completely dissolved by carbonic acid. Rocks, such as limestones, that are almost entirely composed of calcite are totally dissolved away by carbonic acid in rainwater and groundwater. Spectacular caverns can form as groundwater containing carbonic acid seeps through bedrock composed of calcite and eats huge holes in it (see Figure 19.6).

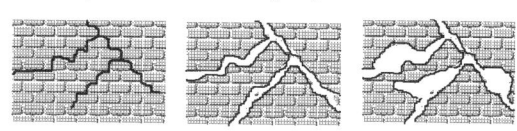

Figure 19.6 Carbonation Caverns Form as Rainwater Containing Carbonic Acid Seeps into Cracks in Limestone.

Other Chemical Factors

In addition to carbonic acid, other naturally occurring acids attack rocks and minerals. Some of these acids are produced by lightning or the decay of organic material; others, as waste products of certain plants and animals. These acids dissolve in rainwater as it falls through the atmosphere and seeps through the soil. Upon reaching bedrock, they dissolve some rocks and cause others to crumble.

Primitive plants such as lichens can grow on bare rock surfaces. They have limited nutrient needs and can obtain these directly from the rock. Lichens grow when the rock surface is wet and lie dormant when it is dry. Secretions from the lichens eat away at the rock's surface, dissolving out mineral nutrients and loosening mineral particles. These mineral particles accumulate in rock crevices, along with dust from the atmosphere and debris from dead lichens. Seeds and spores may then take hold in these tiny pockets of soil and grow, further increasing the amount of chemical and physical weathering the rock undergoes.

An increasingly important source of acids that weather rock is human activity. Factories, homes, and automobiles release vast quantities of waste gases and other pollutants into the atmosphere. Many, such as the oxides of nitrogen and sulfur, react with water to form very strong, reactive acids.

In some areas surrounding large cities and industrial complexes, the concentration of these acids has reached alarming levels. Rainfall in these areas contains so much acid that it is called *acid rain*. Acid rain accelerates the weathering of rock and the breakdown of structures built by humans, and it damages plant and animal life.

Factors Affecting Weathering

How long does it take for a rock to be broken down by weathering processes? The answer is complex since many factors influence the rate at which a rock will weather. Among the more important factors affecting weathering are climate, particle size, exposure, mineral composition, and time.

Climate

Climate, which is the single most important factor affecting weathering, is the average condition of the atmosphere in a region over a long period of time. It is typically expressed in terms of two factors: temperature and precipitation (see Figure 19.7). Both factors influence the type and the rate of weathering. Warm climates favor chemical weathering; cold climates, physical weathering, principally frost action. In both cases, the more moisture present, the more pronounced the weathering.

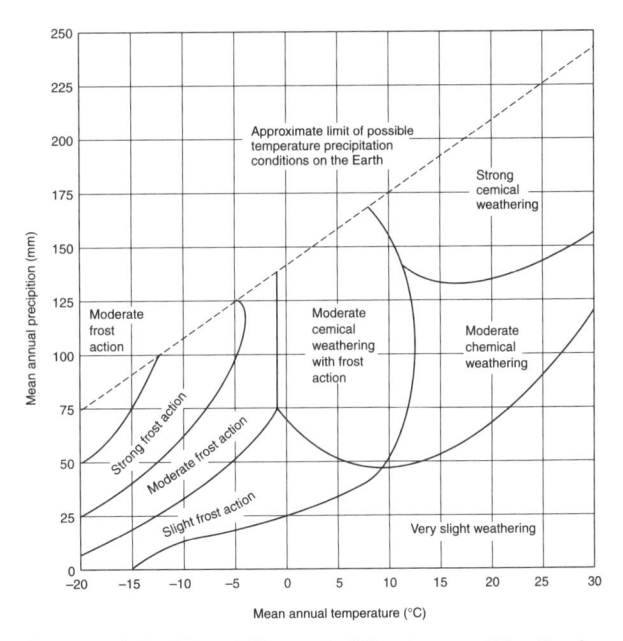

Figure 19.7 The Effect of Climate on Weathering Processes. Physical weathering is dominant in areas where rainfall and temperature are both low. High temperature and precipitation favor chemical weathering.

Weathering and Soil Formation

Chemical reactions tend to occur at a faster rate as temperature increases. Many of these reactions require water, which is a reactant in hydration and carbonation and provides a medium in which acid reactions can occur. Hot, moist climates also support increased biological activity, ranging from burrowing to the production of humic acid as plant matter decomposes. Thus, a hot, moist climate is the ideal environment for rapid chemical weathering.

In cold climates, physical weathering by frost action predominates. It is most effective in cold climates that alternately freeze and thaw. Here again, precipitation is a key factor. Without water, ice cannot form and frost action cannot occur. A cold, moist climate will therefore experience strong frost action.

Temperate climates, such as New York's, combine aspects of both extremes. Hot, moist summers followed by moist winters with many freeze-thaw cycles result in an environment in which rocks weather rapidly. A classic example of the destructiveness of this climate is its effect on Cleopatra's Needle, a granite obelisk with hieroglyphics cut into its surface that was moved from Egypt to Central Park in 1880. In Egypt's hot, dry climate the obelisk stood almost unchanged for 3,000 years. In the moist, changing climate of New York City, however, it began to rapidly deteriorate. Today, its surface cracked, worn, and discolored, the hieroglyphics almost unreadable, it stands as mute testimony to the effect of climate on weathering.

Particle Size

The size of rock particles greatly affects the rate at which chemical weathering occurs. Under the same conditions, the smaller the pieces of a particular rock, the faster they will weather. The reason is that a given volume of small particles has more surface area than the same volume of large particles (see Figure 19.8).

A reaction can take place only when the rock comes into contact with the chemical weathering agent. The more rock surface exposed, the more rock is able to react, and the faster the rate of the reaction.

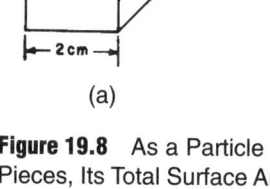

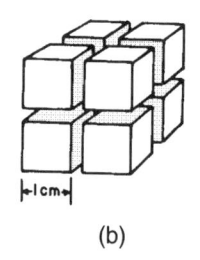

Figure 19.8 As a Particle Is Broken into Smaller Pieces, Its Total Surface Area Increases. (a) This cube has a surface area of 24 square centimeters. (b) Each of these eight small cubes has a surface area of 6 square centimeters, for a total of 48 square centimeters.

Size is also a factor in what happens to rock particles after they are produced by weathering. Small particles may be transported to a new location that has a different climate, or to a body of water that contains a variety of reagents.

Exposure

Exposure determines the degree to which a rock comes into contact with weathering agents. Soil, ice, and vegetation can cover a rock and thereby decrease its contact with weathering agents. Rocks thus protected tend to weather more slowly than those completely exposed at the surface. Also, the weathered surface of a rock can itself shield fresh material underneath, thereby slowing the rate of weathering. Another factor that affects exposure is the slope of the land. On steep slopes, loose materials move downhill because of gravity or are carried downhill by erosion, thus continually exposing fresh rock.

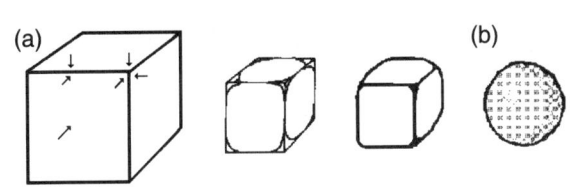

Figure 19.9 Results of Exposure. (a) Corners and edges weather fastest, resulting in a rounded particle (b).

Exposure also has its effects on each individual particle. As you can see in Figure 19.9a, the corners of a cube are exposed to weathering on three sides, the edges on two sides, and the faces on one side. The net effect is that the corners of a particle weather away most quickly and the faces most slowly. Over time the result is a spherical particle, as shown in Figure 19.9b.

Mineral Composition

The mineral composition of a rock determines its physical and chemical properties and thus its susceptibility to weathering. Mineral composition is of greatest importance in chemical weathering. Rocks composed of minerals that react readily with acids, water, or oxygen will weather more rapidly than those composed of less reactive minerals. For example, limestone, which is mostly calcite, is dissolved by even mildly acidic rainwater, while granite, which is mostly silicates, is almost unaffected.

Mineral composition also affects physical weathering (see Table 19.1). Softer rocks will abrade more readily than harder rocks. Solid, crystalline rocks have fewer openings into which water can penetrate than rocks composed of cemented-together particles.

TABLE 19.1 RELATIVE RESISTANCES OF THE MAJOR ROCK-FORMING MINERALS TO WEATHERING

Mineral	Resistance to Weathering
Olivine	Least resistance
Pyroxene	
Amphibole	
Biotite	↓
Plagioclase feldspars	
Muscovite	
Orthoclase	
Quartz	Most resistant

Time

Weathering is a slow process. The 3.5-billion-year-old gneisses in Greenland are scarcely weathered. Even easily weathered rocks may take hundreds of years to be completely broken down. The longer a rock is exposed to weathering processes, however, the more it deteriorates and disintegrates. Eventually, all rocks exposed at Earth's surface are completely broken down.

THE PRODUCTS OF WEATHERING

Since Earth formed, the rocks exposed at its surface have been attacked by forces that disintegrate them. Whether these forces are physical or chemical, the result is the same—solid rock is broken into fragments. These fragments range from the tiniest particle dissolved in water to the largest boulders. The fragments produced form sediments and soils.

Sediments

The fragments or particles of rock produced by weathering are called *sediments*. Sediments are named according to their sizes (Table 19.2).

TABLE 19.2 SEDIMENT SIZES AND NAMES

Particle Diameter (cm)	Name
<0.0004	Clay
0.0004–0.006	Silt
0.006–0.2	Sand
0.2–6.4	Pebbles
6.4–25.6	Cobbles
>25.6	Boulders

Soils

Soil is the accumulation of loose, weathered material that covers much of the land surface of Earth. Soil varies in depth, composition, age, color, and texture. Although its chief component is weathered rock, a true soil also contains water, air, bacteria, and decayed plant and animal material (humus).

The rock from which a soil forms is called the *parent material*. Soil that forms directly from the bedrock beneath it is *residual soil* (see Figure 19.10). If the soil forms from material that was transported to the location by erosion, it is *transported soil*. Transported soil may have a different mineral composition from the underlying bedrock.

As a soil forms, the processes of weathering and plant growth develop recognizable layers, or horizons, in the soil. These horizons differ in structure, composition, color, and texture. A soil that has been forming long enough to have developed distinct horizons is called a *mature soil*. In *immature soils*, horizons are indistinct or altogether lacking.

A *soil profile* is a cross section of a soil from surface to bedrock. Figure 19.11 shows the profile of a typical mature soil with a description of each horizon.

Parent material greatly influences the type of soil that forms, especially when the soil is just starting to form or is immature. Over a long period of time, however, climate has a greater influence on the type of soil that forms. Whatever the parent material, the profiles of mature soils that formed in similar climates are very much alike.

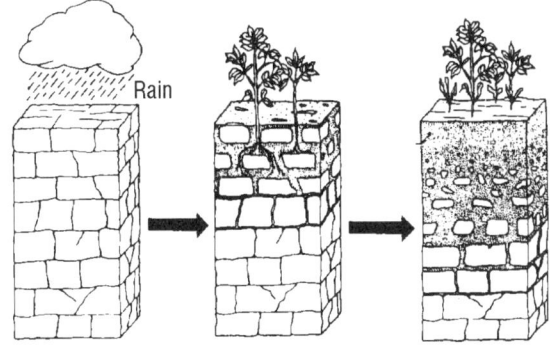

Figure 19.10 The Development of a Residual Soil. Weathering opens bedrock to air, water, and dust. Plant and animal life accelerate the breakdown of rock and allow air and water to penetrate ever more deeply. Eventually a deep layer of soil develops atop the bedrock.

In the United States, two general types of soils have developed (see Figure 19.12). In the western half of the country, where rainfall is less than 63 centimeters yearly, the soils that formed are mainly a type called *pedocals*. Pedocals are rich in calcium compounds and tend to be slightly alkaline, a

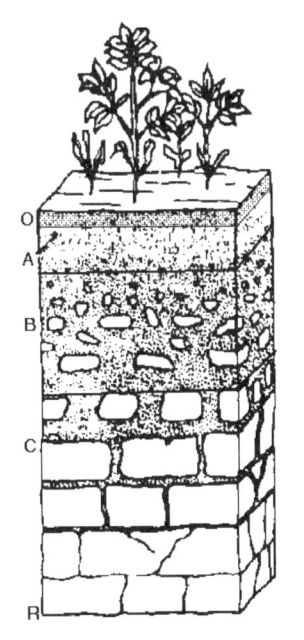

Horizon O (organic horizon)
The material in this horizon is commonly called topsoil and supports plant life. This horizon contains fresh to partly decomposed organic matter. Its color varies from dark brown to black.

Horizon A (leached horizon)
The top of this horizon is highly decomposed organic matter mixed with minerals. Its particles, exposed since the soil began to form, are sand sized or smaller. In a process called *leaching*, soluble minerals and tiny clay particles are carried down to lower layers as water seeps downward through this layer. Leaching is one of the processes that cause horizons to form. This horizon ranges from brown to gray in color.

Horizon B (accumulation horizon)
Materials leached out of horizon A are deposited here. This horizon is richer in clay, has less organic material, and has more and larger particles of bedrock than horizon A. As a result it is less fertile. The clay in this horizon gives it a clumpy or blocky consistency compared to loose, sandy horizon A. The leached materials deposited here (clays and iron oxides) give this horizon a reddish brown or tan color.

Horizon C (partly weathered horizon)
This horizon consists of partly weathered bedrock. Sometimes referred to as subsoil, it is the cracked, broken surface of the bedrock. It marks the final transition from topsoil to unaltered bedrock or parent material.

Parent material (R)
Unaltered bedrock

Figure 19.11 Profile of a Mature Soil.

Weathering and Soil Formation

perfect growing environment for grasses. In the eastern United States, where rainfall exceeds 63 centimeters yearly, the soils are mainly *pedalfers*. Pedalfers are rich in aluminum and iron compounds produced when water and oxygen react with common rock-forming minerals, and soluble calcium compounds are washed away by successive rainfalls.

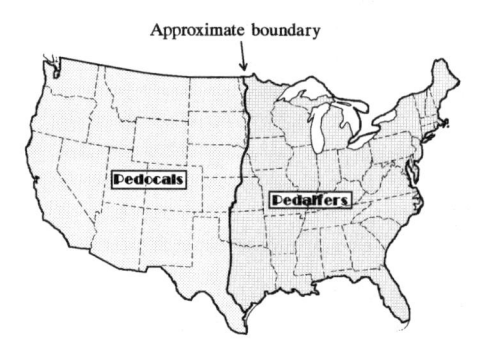

Figure 19.12 Two General Types of Soil in the United States.

MULTIPLE-CHOICE QUESTIONS

In each case, write the number of the word or expression that best answers the question or completes the statement.

1. At high elevations in New York State, which is the most common form of physical weathering?
   (1) abrasion of rocks by the wind
   (2) alternate freezing and melting of water
   (3) dissolving of minerals into solution
   (4) oxidation by oxygen in the atmosphere

2. Which property of water makes frost action a common and effective form of weathering?
   (1) Water dissolves many Earth materials.
   (2) Water expands when it freezes.
   (3) Water cools the surroundings when it evaporates.
   (4) Water loses 80 calories of heat per gram when it freezes.

3. Which is the best example of physical weathering?
   (1) the cracking of rock caused by the freezing and thawing of water
   (2) the transportation of sediment in a stream
   (3) the reaction of limestone with acid rainwater
   (4) the formation of a sandbar along the side of a stream

4. Which weathering process tends to form spherical boulders?
   (1) carbonation   (3) frost action
   (2) exfoliation   (4) oxidation

5. Which substance has the greatest effect on the rate of weathering of rock?
   (1) nitrogen   (3) water
   (2) hydrogen   (4) argon

6. Which type of climate causes the fastest chemical weathering?
   (1) cold and dry
   (2) cold and humid
   (3) hot and dry
   (4) hot and humid

7. The weathering of Earth materials is most affected by
   (1) topography
   (2) longitude
   (3) altitude
   (4) climate

8. Which factor has the *least* effect on the weathering of a rock?
   (1) climatic conditions
   (2) composition of the rock
   (3) exposure of the rock to the atmosphere
   (4) the number of fossils found in the rock

9. Chemical weathering will occur most rapidly when rocks are exposed to the
   (1) hydrosphere and lithosphere
   (2) mesosphere and thermosphere
   (3) lithosphere and atmosphere
   (4) hydrosphere and atmosphere

10. The four limestone samples illustrated below have the same composition, mass, and volume. Under the same climatic conditions, which sample will weather fastest?

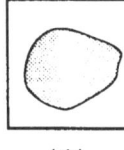

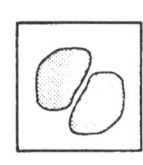

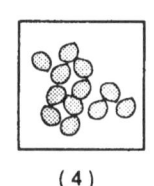

11. Two different kinds of minerals, *A* and *B*, were placed in the same container and shaken for 15 minutes. The accompanying diagrams represent the sizes and shapes of the various pieces of mineral before and after shaking. What caused the resulting differences in shapes and sizes of the minerals?
    (1) Mineral *B* was shaken harder. 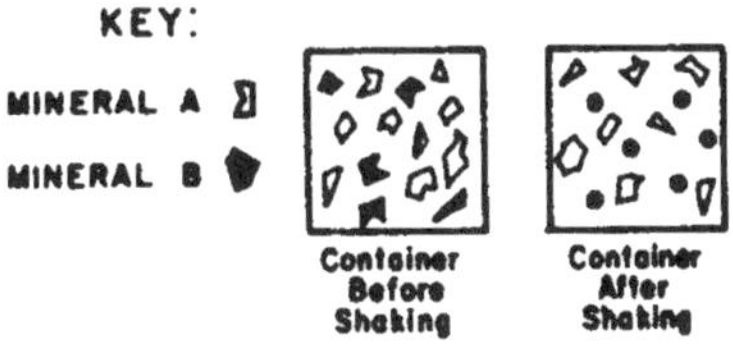
    (2) Mineral *B* had a glossy luster.
    (3) Mineral *A* was more resistant to abrasion.
    (4) Mineral *A* consisted of smaller pieces before shaking began.

Weathering and Soil Formation

12. For a given mass of rock particles, which graph best represents the relationship between the size of rock particles exposed to weathering and the rate at which weathering occurs?

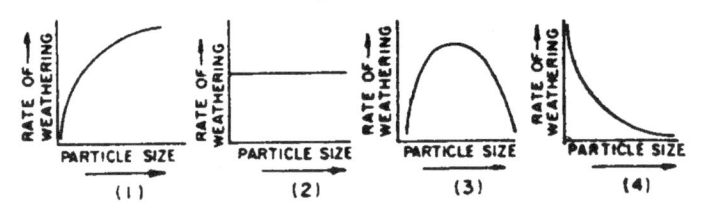

13. The diagram below represents a sedimentary rock outcrop.

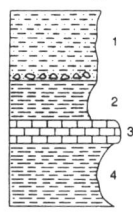

    Which rock layer is the most resistant to weathering?
    (1) 1 (3) 3
    (2) 2 (4) 4

14. Two tombstones, *A* and *B*, which have the same size and style of lettering, have been standing in a cemetery for 100 years. The lettering is clear on *A* but not on *B*. Which is the most probable reason for the difference?
    (1) *B* was more protected from the atmosphere.
    (2) The minerals in *A* are more resistant to weathering than those in *B*.
    (3) *A* is more porous than *B*.
    (4) *B* is smaller than *A*.

15. Why will a rock weather more rapidly if it is broken into smaller particles?
    (1) The mineral structure of the rock has been changed.
    (2) The smaller particles are less dense.
    (3) The total mass of the rock and the particles is reduced.
    (4) More surface area is exposed.

16. Which factors most directly control the development of soils?
    (1) soil particle sizes and method of deposition
    (2) bedrock composition and climate characteristics
    (3) direction of prevailing winds and storm tracks
    (4) earthquake intensity and volcanic activity

17. A variety of soil types are found in New York State primarily because areas of the state differ in their
    (1) amounts of insolation
    (2) distances from the ocean
    (3) underlying bedrock and sediments
    (4) amounts and types of human activities

18. Which change would cause the topsoil in New York State to increase in thickness?
    (1) an increase in slope
    (2) an increase in biologic activity
    (3) a decrease in rainfall
    (4) a decrease in air temperature

19. The chemical composition of a residual soil in a certain area is determined by the
    (1) method by which the soil was transported to the area
    (2) slope of the land and the particle size of the soil
    (3) length of time since the last crustal movement in the area occurred
    (4) minerals in the bedrock beneath the soil and the climate of the area

20. Particles of soil often differ greatly from the underlying bedrock in color, mineral composition, and organic content. Which conclusion about these soil particles is best drawn from these differences?
    (1) The particles are residual sediments.
    (2) The particles are transported sediments.
    (3) The particles are uniformly large-grained.
    (4) The particles are soluble in water.

21. Which is indicated by a deep residual soil?
    (1) resistant bedrock
    (2) a large amount of glaciation
    (3) a long period of weathering
    (4) a youthful stage of erosion

Base your answers to questions 22 through 26 on your knowledge of earth science. the *Earth Science Reference Tables*, and the graph below, which was prepared from the results of a study of four different types of cemetery stones. The graph shows, for each stone, the relationship between the age of the stone and the percentage that had weathered away.

Weathering and Soil Formation

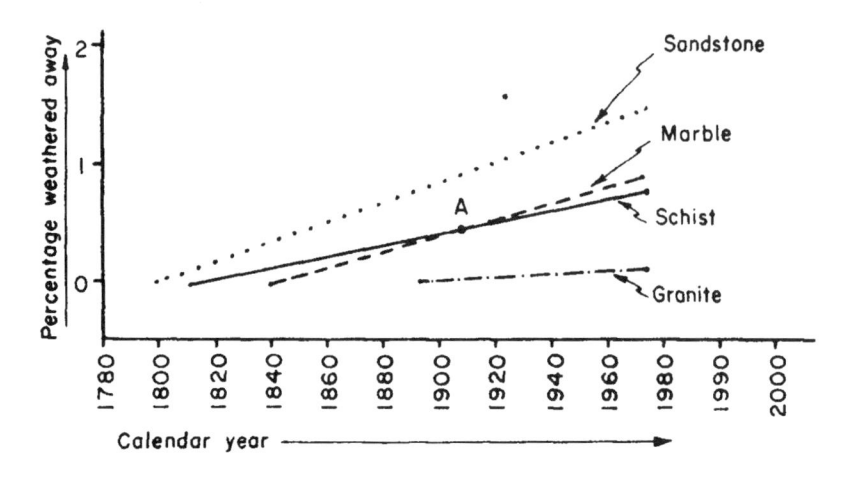

22. Which stone was found to have been exposed to weathering for the greatest number of years?
    (1) granite
    (2) schist
    (3) marble
    (4) sandstone

23. Which stone was the most resistant to weathering?
    (1) marble
    (2) schist
    (3) granite
    (4) sandstone

24. What total percentage of the schist should have weathered away by the year 2000?
    (1) 1.0%
    (2) 2.0%
    (3) 0.5%
    (4) 1.5%

25. Point A on the diagram represents the time at which
    (1) equal percentages of the marble and the schist had weathered away
    (2) the marble and schist weathered away at the same rate
    (3) climatic conditions changed the weathering rates
    (4) industrial pollutants changed the weathering rates

Note that question 26 has only three choices.
26. Studies have shown that pollutants added to the atmosphere in recent years are accumulating to cause an increase in the rate of weathering of marble. This factor should cause the line in the graph for marble in the future to
    (1) decrease in slope (curve downward)
    (2) increase in slope (curve upward)
    (3) remain at the same slope

Base your answers to questions 27 through 31 on your knowledge of earth science and the diagram below, which represents the dominant types of weathering for various climatic conditions.

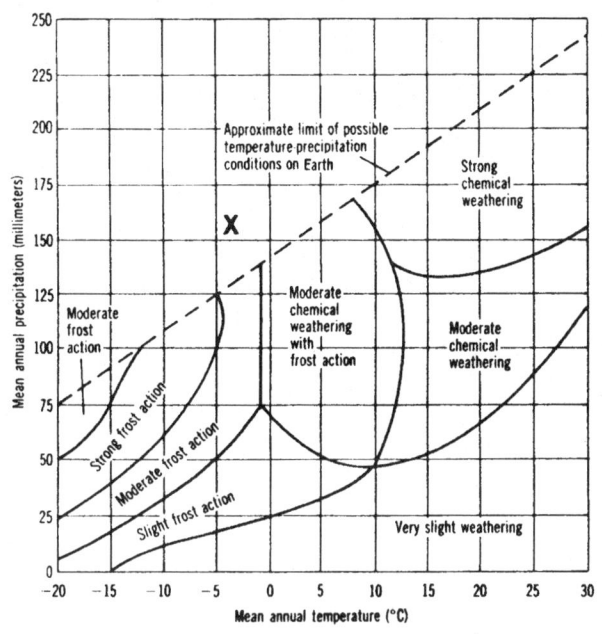

27. Which climatic conditions produce very slight weathering?
    (1) a mean annual temperature of 25°C and a mean annual precipitation of 100 mm
    (2) a mean annual temperature of 15°C and a mean annual precipitation of 25 mm
    (3) a mean annual temperature of 5°C and a mean annual precipitation of 50 mm
    (4) a mean annual temperature of –5°C and a mean annual precipitation of 50 mm

28. Why is no frost action shown for locations with a mean annual temperature greater than 13°C?
    (1) Very little freezing takes place at these locations.
    (2) Large amounts of evaporation take place at these locations.
    (3) Very little precipitation falls at these locations.
    (4) Large amounts of precipitation fall at these locations.

Weathering and Soil Formation

29. No particular type of weathering or frost action is given for the temperature and precipitation values at the location represented by the letter *X*. Why is this the case?
    (1) Only chemical weathering would occur under these conditions.
    (2) Only frost action would occur under these conditions.
    (3) These conditions create both strong frost action and strong chemical weathering.
    (4) These conditions probably do not occur on Earth.

30. Four samples of the same material with identical composition and mass were cut as shown in the diagrams below. When the samples are subjected to the same chemical weathering, which sample will weather at the fastest rate?

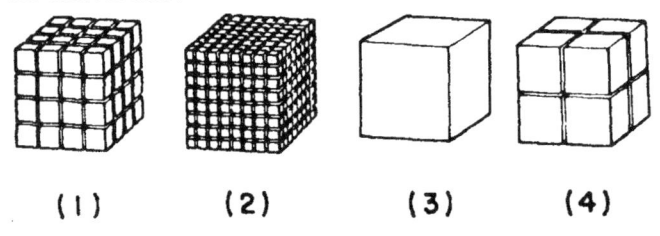

    (1)   (2)   (3)   (4)

31. Assume that the rate of precipitation throughout the year is constant. Which graph would most probably represent the chemical weathering of most New York State bedrock?

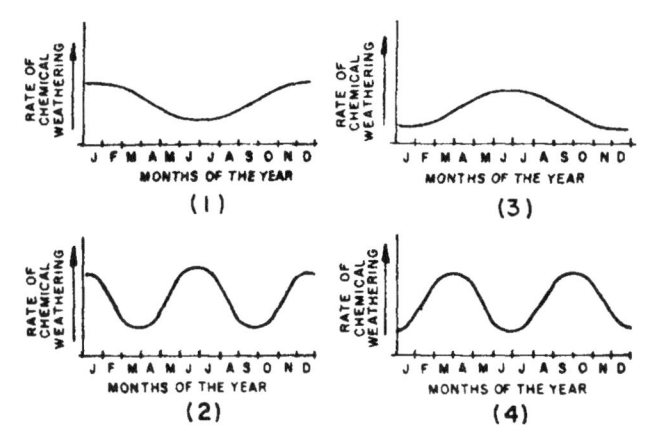

CONSTRUCTED RESPONSE QUESTIONS

Base your answers to questions 32 through 34 on the graph below, which gives the composite results of investigations by four different classes of the weathering of rock by abrasion. Each class used a different rock (A, B, C, or D) as shown. The period of shaking was the same for each class, and each class started with the same mass of rock.

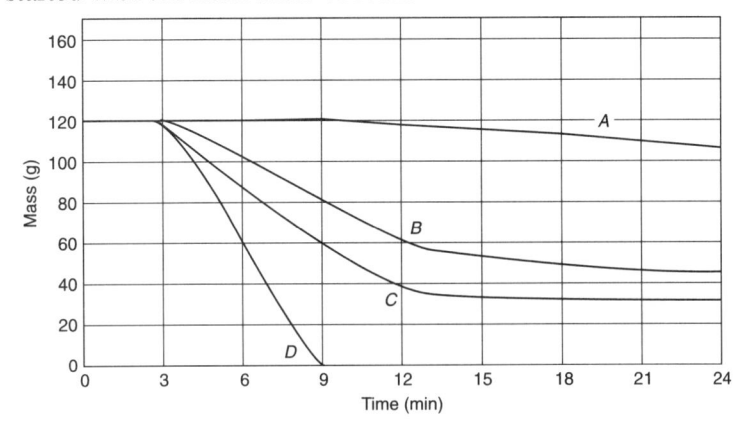

32. Which rock is most likely the hardest? In one or more complete sentences, justify your choice with evidence from the graph. [2]

33. State a likely reason for the flattened portion of the curves for rocks B and C between the 12-minute and 24-minute intervals. [1]

34. A town is building a canal to divert water from a stream through a new park. The town intends to line the canal with rock to keep the stream from eroding the valuable parkland. The town purchasing agent has chosen rock D because it is slightly less expensive than rock A. In one or more complete sentences, evaluate his choice of building material in view of the data in the graph. [2]

Base your answers to questions 35 through 37 on the diagram below, which shows the soil profile formed in an area of granite bedrock. Four different soil horizons, A, B, C, and D, are shown.

Weathering and Soil Formation

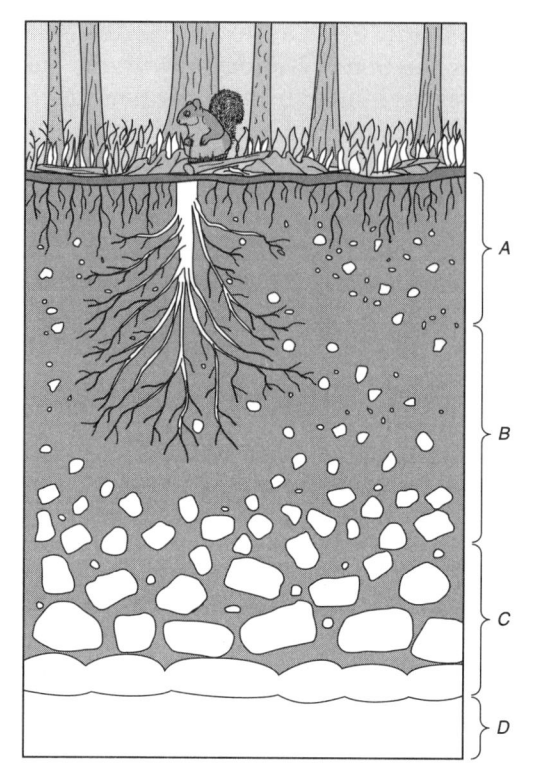

35. Identify the soil horizon containing the highest percentage of organic material. [1]

36. Soil horizon *B* is rich in clay and soluble minerals.
    (a) State the process by which this soil horizon has become enriched in these materials. [1]
    (b) In one or more complete sentences, compare the fertility of soil in this layer with that of horizon *A*. [1]
37. State one factor that affects the type of soil that forms in a region. [1]

EXTENDED CONSTRUCTED RESPONSE QUESTIONS

Base your answers to questions 38 through 41 on the given information and the data table below. Samples of three different rock materials, *A*, *B*, and *C*, were placed in three containers of water and shaken vigorously for 20 minutes. At 5-minute intervals, the contents of each container were strained through a sieve. The masses of the materials remaining in the sieve were measured and recorded as shown in the data table below.

MASSES OF MATERIAL REMAINING IN SIEVE

Shaking Time (min)	Rock Material A (gr)	Rock Material B (gr)	Rock Material C (gr)
0	25.0	25.0	25.0
5	24.5	20.0	17.5
10	24.0	18.5	12.5
15	23.5	17.0	7.5
20	23.5	12.5	5.0

38. Using the information in the data table, construct a line graph on the grid provided below, following the directions in parts (a) through (c) below.

    (a) Plot the data for rock sample *A* for the 20 minutes of the investigation. Surround each point with a small circle, and connect the points. [1]

    Example:

    (b) Plot the data for rock sample *B* for the 20 minutes of the investigation. Surround each point with a small triangle, and connect the points. [1]

    Example:

    (c) Plot the data for rock sample *C* for the 20 minutes of the investigation. Surround each point with a small square, and connect the points. [1]

    Example:

Weathering and Soil Formation

Mass of Rock Versus Shaking Time

○ Rock sample A
△ Rock sample B
▪ Rock sample C

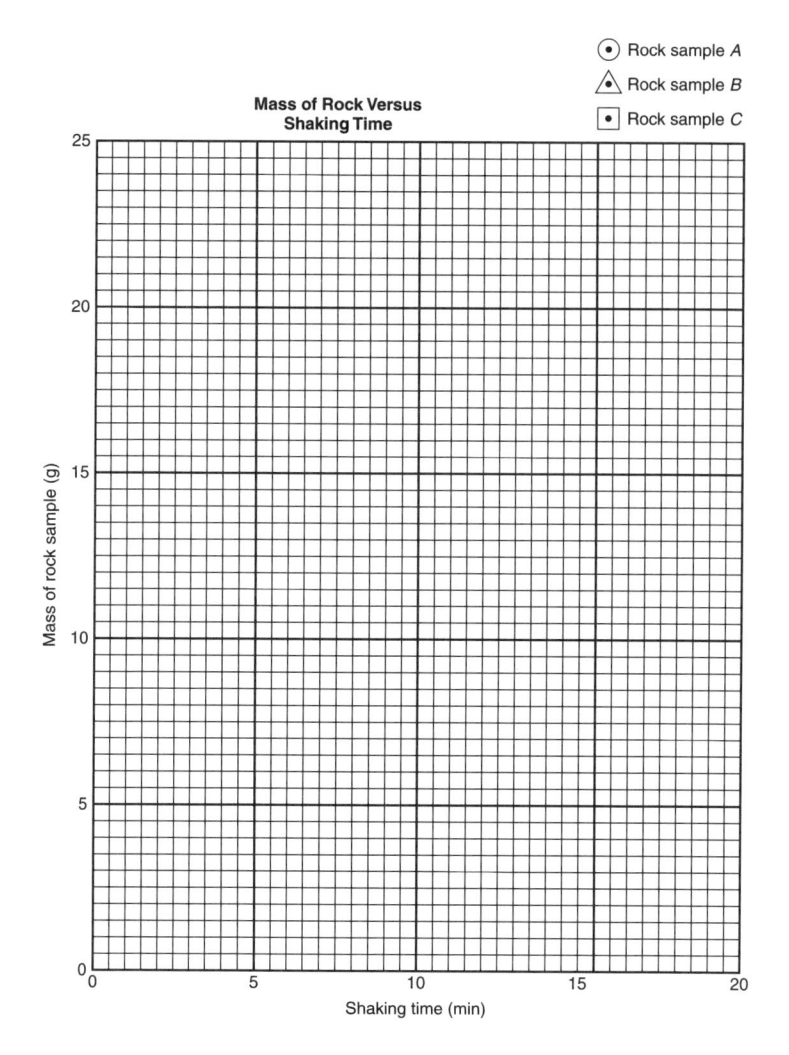

(Graph: Mass of rock sample (g) vs. Shaking time (min), x-axis 0–20 min, y-axis 0–25 g)

39. Using one or more complete sentences, state the most likely reason for differences in the weathering rates of the three rock materials. [2]

40. Using the directions in parts (a) through (c) below, calculate the average rate of change in the mass of rock material *C* for the 20 minutes of shaking.
    (a) Write the equation for the rate of change in mass. [1]
    (b) Substitute date into the equation. [1]
    (c) Calculate the rate of change in mass, and label your answer with proper units. [1]

41. Using one or more complete sentences, describe the most likely appearance of the corners and edges of rock material *C* at the end of the 20 minutes. [2]

CHAPTER 20
EROSION

> **KEY IDEAS** Once weathering has broken rock down into smaller particles, the natural agents of erosion—generally driven by gravity—remove, transport, and deposit them. The natural agents of erosion include streams, ocean currents, and wave action, (moving water), glaciers (moving ice), wind (moving air) and mass movements. Each of these agents of erosion produces distinctive changes in the material that it transports and creates characteristic surface features and landscapes. In certain erosional situations, destruction of property, personal injury, and loss of life can be reduced by effective emergency preparedness.

KEY OBJECTIVES
Upon completion of this chapter, you will be able to:

- Explain how natural agents of erosion, generally driven by gravity, remove, transport, and deposit weathered rock particles.
- Compare and contrast the distinctive changes each agent of erosion makes in the material it transports, as well as the characteristic surface features and landscapes each produces.
- Recognize the factors that affect erosion and relate them to landforms produced by erosion.
- Cite examples of the impacts of various types of erosion on human activities.

EROSION

What happens to the sediments produced by weathering? In most cases they are moved, often great distances from where they originated. Any process that moves sediments from one place to another on Earth's surface is called *erosion*. As you read these words, erosion is changing Earth in countless ways. Waves crashing against shores are scouring away sand, reshaping the coastline. In deserts, hot winds are moving towering dunes of sand grain by grain. On cold, barren mountaintops fragments of rock broken loose by weathering are tumbling to the ground. Glaciers creeping downhill are tearing huge boulders from the ground and carrying them along. Streams, from tiny trickles to raging torrents, are carving their way into the crust.

Erosion

Together, weathering and erosion wear away Earth's crust. Weathering breaks down solid rock; erosion carries away the pieces. In this way fresh rock is exposed to weathering, and the cycle repeats itself. Sometimes erosion is rapid and unmistakeable, as in a landslide. Usually, though, the process is gradual, and a long time passes before it is evident that erosion is taking place.

Evidence of Erosion

Any sediment moved from its source is evidence of erosion. At times, this evidence is striking. The owner of a house returning after a mudslide knows that the house was not built with one end hanging over the edge of a cliff. The lack of soil under the house is evidence of erosion. In much the same way, valleys and canyons are evidence of erosion. The Grand Canyon is one of the most spectacular examples of erosion in the United States. Imagine how much sediment had to be carried away to form it!

Quite often, though, the evidence of erosion is subtle. A boulder sitting on bedrock may not immediately seem to indicate erosion. If the boulder is granite and the bedrock is limestone, however, the boulder certainly wasn't derived from the bedrock; it had to be carried there from some other source. Similarly, layers of sediment can often be found overlying bedrock that has an entirely different composition. This, too, is evidence that erosion has occurred. Since the sediment did not come from the bedrock, it must have been carried there from some other source.

AGENTS OF EROSION

Erosion moves sediments. To move anything, a force is needed. The primary driving force of erosion is **gravity**. Gravity can move sediments by acting on them directly. On cliffs and steep slopes, sediments broken loose by weathering move downhill under the direct influence of gravity. Gravity can also move sediments by acting on them indirectly, through *agents of erosion*. For example, water runs downhill under the direct influence of gravity. The running water, in turn, can exert a force on sediments in its path, causing them to move. Thus the running water is an agent of erosion. Some other agents of erosion are waves, currents, winds, and glaciers.

An agent of erosion, together with the driving force (usually gravity) that sets in motion the agent that picks up and transports sediment, comprises an *erosional system*. Erosional systems are like natural conveyor belts, moving sediments from higher to lower elevations.

Mass Wasting

Mass wasting is the downhill movement of sediments under the direct influence of gravity. On most slopes, some kind of downhill movement of sediments is proceeding all the time because, on a slope, gravity acts as if it has two parts, or components. One part, which we will call the normal force (F_N), pulls downward perpendicular to the surface. The other part, which we will call the downhill force (F_D), pulls downhill parallel to the surface. The downhill force gives rise to a frictional force (F_F) that opposes the motion of the object. The frictional force depends on two factors: the normal force holding the two surfaces together and the nature of the two surfaces. The more tightly the two surfaces are pressed together by the normal force, the greater the frictional force between the two. (This explains why an empty box is easier to push across a floor than a similar box filled with something heavy.) As long as the frictional force is greater than the downhill force, the object will remain stationary.

As the slope gets steeper, more of the gravity acts in a downhill direction and less in a direction perpendicular to the surface (i.e., the downhill force increases and the normal force decreases). When the downhill force is greater than the frictional force, the object moves downhill. See Figure 20.1.

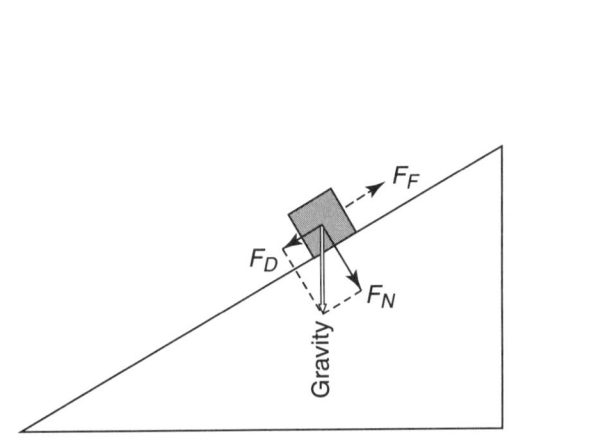

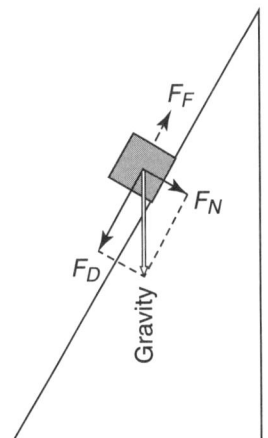

Figure 20.1 Forces Acting on an Object Resting on a Slope. In the diagram, F_N = normal force, F_D = downhill force, and F_F = frictional force. As a slope steepens, the downhill force increases while the normal force and frictional force decrease. When the downhill force exceeds the frictional force, the object moves downhill. Source: *Environmental Geology*, Carla W. Montgomery, McGraw-Hill, 1997.

The steepest slope angle at which a particular sediment remains stable is called its *angle of repose* (see Figure 20.2). The size, shape, and density of a rock particle affect its angle of repose. Thus, sand, gravel, and clay have different angles of repose. When their angles of repose are exceeded, sediments move downhill and mass wasting occurs. Any perceptible downslope move-

Erosion

ment of rock, soil, or a mixture of the two is commonly referred to as a *landslide*. This is a very general term, however, and may involve many different types of movement of many different types of material. Depending on the slope and the angles of repose of the sediments, mass-wasting processes may be rapid or slow.

Figure 20.2 Angles of Repose in a Pile of Dry Sand.

Rapid Mass Wasting

Rockfalls (see Figure 20.3a) occur when rock fragments broken loose by weathering fall from cliffs or bounce by leaps down steep slopes. Rockfalls are the most rapid of all mass-wasting processes. "Fallen Rock Zone" is a common sign near roadcuts on many highways or in mountainous areas where rockfalls occur frequently. In populated areas or on heavily traveled roads, rockfalls can be very dangerous and have resulted in loss of life. Some localities have spent much money cutting back the rocky slopes of roadcuts or covering them with steel cable nets to prevent rockfalls.

Rockslides (see Figure 20.3b) occur on less steep slopes when rock masses

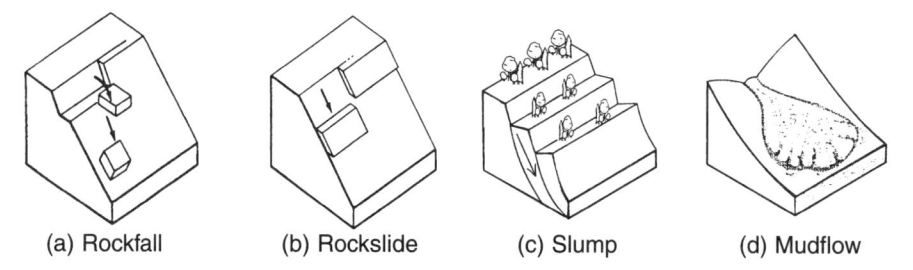

(a) Rockfall   (b) Rockslide   (c) Slump   (d) Mudflow

Figure 20.3 Four Rapid Mass-Wasting Processes.

or debris slide downhill. Most rockslides are triggered by heavy rains or earthquakes. Water acts as a lubricant between particles. The shock of an earthquake knocks particles apart, decreasing the friction between them and the underlying surface. The downhill component of gravity is then greater than the friction holding the particles in place, and the particles move downhill. Rockslides can move enormous amounts of material, often millions of cubic meters. In 1959, twenty-seven people were killed when an earthquake caused an entire mountainside to slide into the Madison River gorge in Montana.

Slump (see Figure 20.3c) is a mass-wasting process in which a huge mass of bedrock or soil slides downward from a cliff in one piece. The mass, or slump block, slides along a curved plane of weakness as shown in the diagram. The slump block rotates and comes to rest with its upper surface tilting toward the cliff. Slump is common where ocean waves or streams undercut cliffs.

A *mudflow* (see Figure 20.3d) is the rapid, downhill flow of a fluid mixture of rock, soil, and water. Mudflows generally occur after heavy rains saturate soil that has no protective covering of vegetation, and usually occur in semiarid regions or on slopes denuded by construction or lumbering. The rain mixes with the soil to form a thin mud, which flows downhill, thickening as more soil and debris are picked up and increasing in speed. Since the mudflow is much heavier and thicker than water, the impact when it hits something in its path is devastating. Entire houses, cars, and trees have been swept away by mudflows. A mudflow comes to rest when it reaches the bottom of the slope or when it has picked up enough material so that it thickens to a point where it can no longer move.

Slow Mass Wasting

In humid regions, slopes are usually covered with vegetation. The vegetation protects the surface from the impact of raindrops, and its root system holds the soil together, inhibiting downhill movement. However, mass wasting occurs even on slopes covered with vegetation, albeit slowly.

On vegetated slopes, rainwater entering the soil loosens the particles and makes them slippery, decreasing their angles of repose. In this situation, an earthflow may occur. In an *earthflow* (see Figure 20.4), a shallow layer of soil and vegetation, saturated with water, slowly slides downhill. An earthflow may take several hours to ooze its way down a slope. Earthflows are a common cause of road and rail blockages. While they are usually not life threatening since they move at a snail's pace, they can cause considerable property damage.

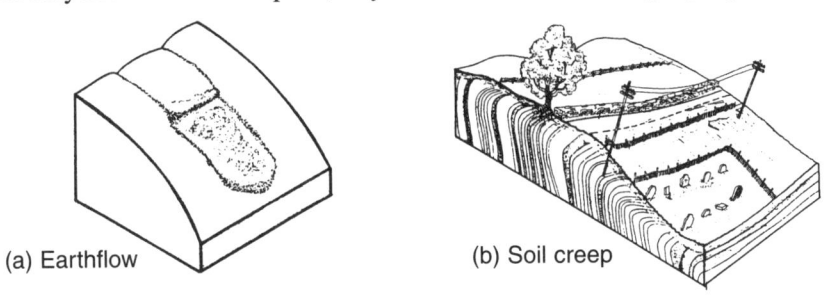

(a) Earthflow   (b) Soil creep

Figure 20.4 Two Slow Mass-Wasting Processes.

The slowest of all mass-wasting processes is soil creep. *Soil creep* (see Figure 20.4b) is the invisibly slow, downhill movement of soil, carrying vegetation with it. The tilting of old poles, fenceposts, or tombstones and bulging or broken retaining walls are all evidence of creep.

Any disturbance of the soil on a slope causes creep. Freezing and thawing, wetting and drying, and trampling or burrowing by animals are just a

Erosion

few of the things that can disturb the soil on a slope. When disturbed, the soil particles shift and settle. As they do so, gravity moves them downhill. Creep is common in regions where the soil alternately freezes and thaws frequently.

Large objects on or in a creeping slope are carried along by the moving soil. Boulders that have crept down a hillside may accumulate at the base of the slope, forming a boulder field. Trees growing on creeping slopes are often misshapen. The tree grows straight up, but the soil that its roots are in slowly creeps downhill.

Mass wasting is usually only the first step in the eroding of Earth. It delivers sediments to the base of slopes. There, agents of erosion, such as streams, can pick up the sediments and carry them farther.

Water in Motion: Runoff and Streams

Raindrops and Runoff

Raindrops may fall for thousands of feet before they hit the ground, and the force of falling raindrops can move sediments. When raindrops strike the ground, their impact causes a geyserlike splashing of loose soil. In a process called *splash erosion*, each raindrop impact lifts some soil and drops it in a new position. On one field observed during a rainstorm, splash erosion moved 91 metric tons of soil per acre.

Runoff is precipitation that does not evaporate or sink into the ground. Under the influence of gravity, runoff flows downhill. On even the gentlest slopes, runoff may flow downhill in a thin sheet called *overland flow*. Overland flow exerts a dragging force on the ground. Depending on its speed, it can exert a force great enough to move sediments ranging from fine clay to coarse sand or gravel. Erosion by overland flow is called *sheet erosion*.

Vegetation greatly decreases both splash and sheet erosion. The leaves and stems of plants absorb the impact of falling raindrops, and the network of plant roots anchors the soil in place. Plant stems also block and slow the downhill movement of water, thereby decreasing the force it exerts on sediments in its path. In arid regions or on slopes denuded of vegetation, however, splash and sheet erosion can remove vast quantities of loose soil.

Splash erosion and sheet erosion are serious problems on farms. They carry away the rich topsoil essential for healthy plant growth. Contour plowing, terracing, strip cropping, and crop rotation help lessen soil erosion.

On steep, unprotected slopes, erosion by runoff can become intense. Numerous tiny grooves, or rills, form as the water runs downhill, carrying away sediment. If erosion continues, the rills may widen and deepen into a gully. Gullies act as funnels for runoff. They concentrate the force of the flowing water, thus increasing the rate of erosion.

Streams

A *stream* is any body of running water that moves under gravity, in a relatively narrow but clearly defined channel, to progressively lower levels. A stream may be small enough to step across, or so wide that the other side is barely visible. Streams are the most important agent of erosion because they affect more of Earth's surface than any other agent. Most of the sediment carried downhill by runoff ends up in streams.

Streams form as runoff from several slopes that drain into low-lying areas between the slopes. These natural depressions fill with water, forming a puddle, a pond, or even a lake. As runoff continues, they may eventually overflow. Then the overflow continues flowing downhill, along natural passageways or depressions in the surface, to lower and lower levels. As the water flows along these passageways, it cuts into the surface, forming a clearly defined path, or *channel*, along which it flows. Once established, the same channel provides a pathway for all later runoff.

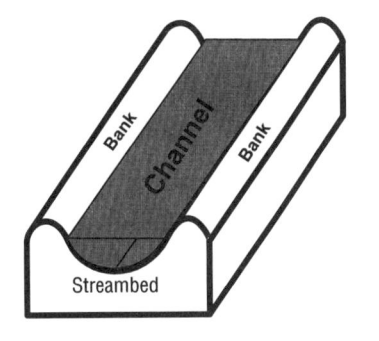

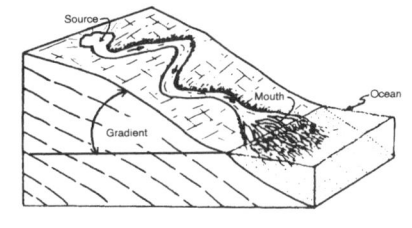

Figure 20.5 A Stream and Its Parts. A stream begins at its **source**, flows along a path called a **channel**, and ends at its **mouth**. The channel consists of sides, called **banks**, and a bottom, called the **streambed**, whose slope is the stream's **gradient**.

Figure 20.5 shows a typical stream and its parts.

Stream Transport

Water flowing downhill can be very powerful. Have you ever seen whitewater rafters shooting a rapids? The water flowing downhill exerts quite a force! Imagine trying to paddle upstream through the rapids. It would be almost impossible because the force of the flowing water is so great. Now imagine a grain of sand or a pebble in the path of such a stream. Any loose sediments are likely to be carried away.

Material carried by a stream is called its *load*. A stream's load consists mainly of sediments, which can vary in amount and type, depending on the speed and volume of the stream. The steeper the slope, or gradient, of a stream's channel, the faster the water flows. Water can carry more and larger sediments as its speed increases. At any speed, the more water there is in a stream, the more load it can carry.

Streams transport different sediments in different ways. See Figure 20.6. Large sediments, such as pebbles, cobbles, and boulders, are heavy. Even the fastest flowing streams may not be able to pick them up. But sediments can be moved without picking them up! Do you have to pick up a car that has run

Erosion

out of gas in order to move it? Of course not; you can push or drag the car so that it rolls along the road. In much the same way, large particles can be carried downhill by streams. The force of the flowing water in a stream can push or drag large particles downhill by rolling or sliding in a motion called *traction*. If the sediment particles move in a series of bounces, hops, or leaps along the streambed, the motion is called *saltation*.

Smaller particles, such as silt, sand, and clay, are lighter in weight and therefore require less force to move. Although they are heavier than water, the flowing water can pick up these small sediments; and, whenever they begin to sink, the force of the turbulent water pushes them back up again—they are carried downhill *suspended* in the water.

Some sediments dissolve in water. Water that has combined with carbon dioxide or other gases in the atmosphere to become acidic (see Chemical Weathering, page 426) is especially effective at dissolving sediments. Sediments that are dissolved are carried downhill *in solution*.

Stream Erosion

Streams not only transport sediments but also change the sediments they carry, wear away the surfaces over which they flow, and create new sediments. Pure water flowing over rock wears it away very little, but water carrying sediment acts like a cutting tool. When the sediments being carried hit rock in the stream channel, chips of rock break off. Rolling and sliding pebbles, cobbles, and boulders collide and knock chips off each other. They crush and grind up smaller particles between them. The process of crushing, grinding, and wearing away rock by the impact of sediments is called *abrasion*. Abrasion wears away a stream's channel and causes the sediments carried by the stream to become rounded.

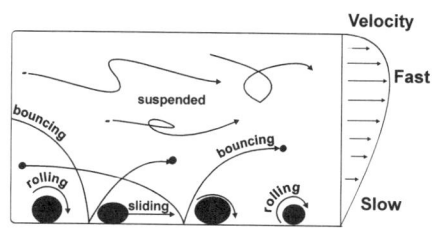

Figure 20.6 Methods of Particle Transport. Larger sediments are transported in streams by rolling, sliding, and bouncing along the streambed. Smaller sediments are carried in suspension or may even be dissolved in the water of the stream.

By abrasion, streams can cut through bedrock over which they flow. The potholes shown in Figure 20.7 were carved out of bedrock by sediments in a swiftly flowing stream. The waterfall formed where a stream flowed over hard bedrock onto soft bedrock. The soft bedrock wore away faster, lowering the streambed, and a waterfall was formed. Prolonged erosion can carve deep canyons and valleys into Earth's surface. The Grand Canyon is an amazing example of what sediment-laden water can do.

A Stream's Life Cycle

As a stream cuts into the surface and carries away sediment, its channel becomes wider, deeper, and longer. Over time the shape and behavior of the stream change as the surface over which it flows is changed.

Youth. Newly formed streams generally flow down steep slopes. The water flows quickly, often forming rapids. See Figure 20.8a. Over time, the stream wears a long, deep groove, or valley, in the surface. Stream-cut valleys typically have steep sides that form a distinctive V-shape. A young stream valley is shown in Figure 20.8b.

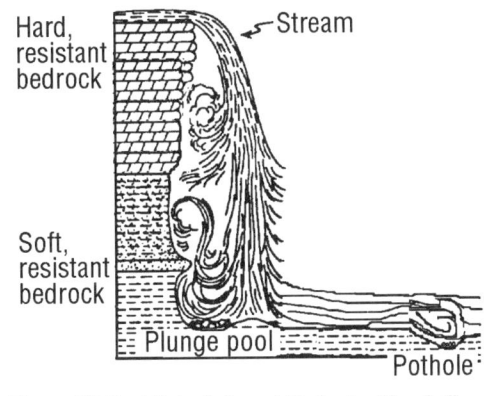

Figure 20.7 Waterfall and Pothole. The falling water creates turbulence that swirls sediments against the streambed, forming potholes.

Maturity. As a stream erodes the slope over which it flows, the slope becomes less steep. As a result, the water flows more slowly and sediment accumulates at the bottom of the valley, making it flat. As mass

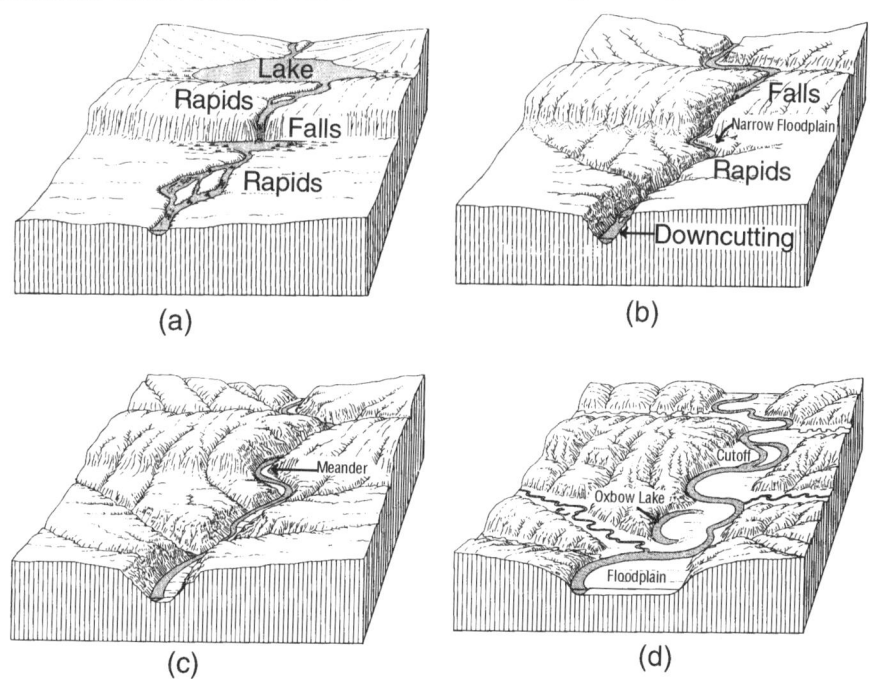

Figure 20.8 The Three Stages in the Life Cycle of a Stream. (a) The relief of the land is steep, and the stream has lakes, rapids, and falls. (b) Continued downcutting of the streambed eliminates falls, and cutting back of the valley walls allows a narrow floodplain to form. (c) Further downcutting and erosion of the valley walls enlarges the floodplain and allows the stream to meander widely between the valley walls. (d) Shifting meanders result in cutoffs and oxbow lakes. Source: *The Earth Sciences*, Arthur N. Strahler, Harper & Row, 1971.

Erosion

wasting and runoff wear away the sides of the valley, the V-shape becomes wider and less steep. The slower moving water flows around obstacles instead of tumbling over them, and the stream's path forms curving loops called *meanders*. When runoff increases suddenly, as after a torrential downpour or sudden spring thaw, the stream overflows. Spreading out over the valley floor, the stream slows down and deposits sediments to form a broad, flat *floodplain* adjacent to the stream. Figure 20.8c shows a typical stream at the mature stage.

Old Age. Eventually, the slopes around a stream are worn away almost completely. A flat lowland of gently rolling hills, called a *peneplain*, is formed. Since the slope of the land is very gentle, the water in the stream flows slowly. Its meanders become highly curved and winding and at times may actually loop over one another. During floods, water may gush over the land between meanders, forming a *cutoff*. If the new channel cuts deep enough, part of the meander may become isolated, forming an *oxbow lake*. Figure 20.8d shows a typical stream at the old stage.

Whatever the stage of development, the water in most streams eventually reaches the oceans.

Water in Motion: the Oceans

The ocean never rests. Anyone who has swum in, sailed on, or just sat on the beach beside the ocean knows that it is constantly moving. Waves, currents, and tides all move ocean waters; and, once in motion, ocean water can transport sediments and erode the land.

Ocean Currents

Ocean currents are mainly the result of two interactions with the atmosphere: heating of the ocean and wind. The ocean is able to store more heat than the atmosphere or the land because of water's high specific heat. Differences in the intensity of solar radiation received at the Equator and near the poles results in an uneven distribution of heat in the oceans. The oceans contain more heat near the Equator than near the poles.

Convection Currents
The heat in the oceans near the Equator is transferred toward the poles by convection, or the movement of seawater due to density differences. Dense, cold water formed near the poles sinks and slowly flows toward the Equator. This cold water is oxygen-rich; without it, bottom waters would quickly become oxygen depleted by the oxidation of organic matter that sinks through it. As the cold water moves toward the Equator, it displaces warmer water upward, causing *upwelling*. Upwelling brings nutrients and oxygen to the surface, thereby supporting a rich growth of plants and animals.

Wind-Driven Currents

Uneven heating of Earth's surface and the air above it creates similar circulations in the atmosphere. As warm air at the Equator rises, air from adjacent areas gets sucked in to replace it, creating wind. Winds do not move straight toward the Equator, but curve because of the Coriolis effect. Thus, the winds near the Equator, called the *trade winds*, approach the Equator at about a 45° angle. Over the ocean, with little to block their path, the trade winds are among the steadiest winds on Earth. At middle latitudes, the more variable *prevailing westerlies* move in the opposite direction. At high latitudes are the most variable winds, the *polar easterlies*. See Figure 20-9.

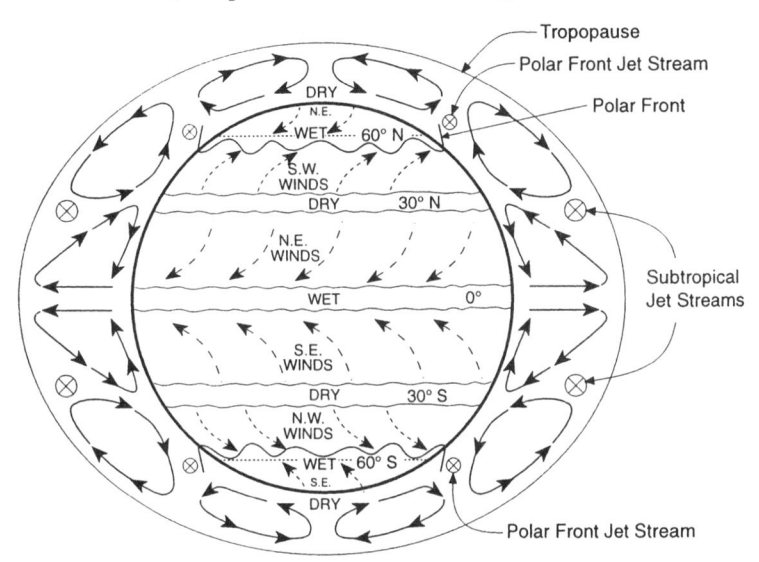

Figure 20.9 Planetary Wind and Moisture Belts in the Troposphere. The drawing shows the locations of the belts near the time of an equinox. The locations shift somewhat with the changing latitude of the Sun's vertical ray. In the Northern Hemisphere, the belts shift northward in summer and southward in winter. Source: The State Education Department, *Earth Science Reference Tables*, 2001 ed. (Albany, New York; The University of the State of New York).

Winds exert a force on the ocean surface and produce wind-driven currents. All of the major surface ocean currents are driven by the wind and follow roughly the pattern of surface winds. When pushed by wind, however, the surface water does not move in the same direction as the wind but, because of the Coriolis effect, curves off at about a 45° angle. As a result, the surface currents driven by the trade winds do not approach the Equator at the same 45° angle as the winds that are pushing them, but are themselves bent 45° from the wind direction. Thus, throughout Earth, the trade winds form westward-moving currents, known as *equatorial currents*, that are parallel to the Equator.

In the Atlantic and Pacific oceans, these westward-moving equatorial currents are obstructed by landmasses and are deflected north and south. These deflected currents, which run north and south along the western boundary of

Erosion

the oceans, are among the largest and strongest ocean currents. The Gulf Stream, a western boundary current in the Atlantic Ocean, runs north along the eastern coast of the United States and transports more than 100 times the output of all the rivers in the world.

When the western boundary currents reach the midlatitudes, the prevailing winds, which blow from the west, push them back east across the ocean. The result is a large, circular pattern of motion called a *gyre*. The North and South Atlantic and Pacific, as well as the Indian Ocean, all have large-scale gyres. In addition, the northern polar regions of the Atlantic and Pacific have smaller gyres. See Figure 20.10.

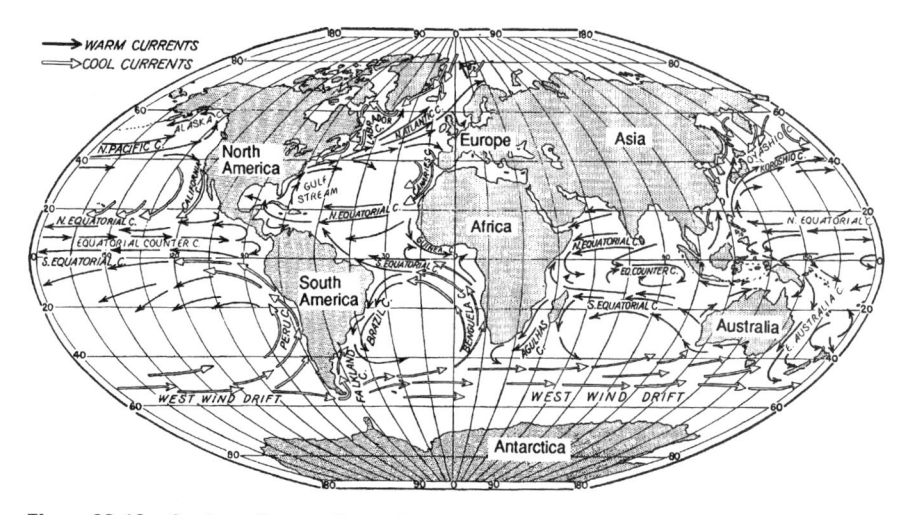

Figure 20.10 Surface Ocean Currents. Source: The State Education Department, *Earth Science Reference Tables*, 2001 ed. (Albany, New York; The University of the State of New York).

One unfortunate effect of the large-scale gyres is that they carry pollutants from the continental coastlines far out into the oceans. For example, many of the largest cities in the United States are located along the east coast. The sewage, garbage, and industrial waste from these cities have long been dumped into the ocean. The Gulf Stream then carries these pollutants far out into the mid-Atlantic. These practices may have had little impact when populations were small, but today they represent serious threats to the health of ocean life and the future well-being of our oceans.

Waves

Wind-Generated Waves

Wind not only drives surface currents but also creates waves. Wind forms waves on the surface of the ocean by transferring energy from the moving air to the water. The transfer of energy is not completely understood, but appears to involve pressure differences caused by the deflection of wind over the wave (much as air flowing over a wing provides lift). It also seems to be related to a resonating movement of the surface caused by wind pushing against the surface in turbulent conditions (like a hand pushing down repeatedly on the surface of the water). The result in either case is a transverse wave that moves through the surface water of the ocean.

The highest part of a wave is called the *crest;* the lowest part, the *trough*. The size of an ocean wave is usually expressed as the **wave height**—the vertical distance between the crest and the trough. Wave crests and troughs can be closer together or farther apart. The distance from crest to crest, or from trough to trough is called the *wavelength*. See Figure 20.11.

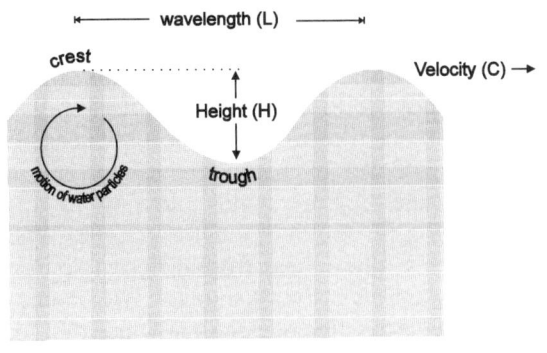

Figure 20.11 A Typical Transverse Wave. Source: *An Introduction to Oceanography*, David A. Ross, Prentice-Hall, 1970.

When a wave moves through the ocean, the water is carried up and forward by the crest and then down and back by the trough. As a result, the water doesn't really get anywhere as a wave passes through it, but instead moves in a circular pattern. Though the crest carries water up and forward, the trough carries it down and back, so the water ends up right where it started.

Near the surface, the movement of water particles is almost perfectly circular, but the diameter of the circular motion decreases quickly with depth. This decline is due to a loss of energy during transfer to deeper water. At a depth of one-fourth the wavelength, the circle is only one-fifth of its diameter near the surface. At greater depths the movement is little more than a back and forth rocking motion. Below 100 meters, little if any motion is felt from most wind-generated waves.

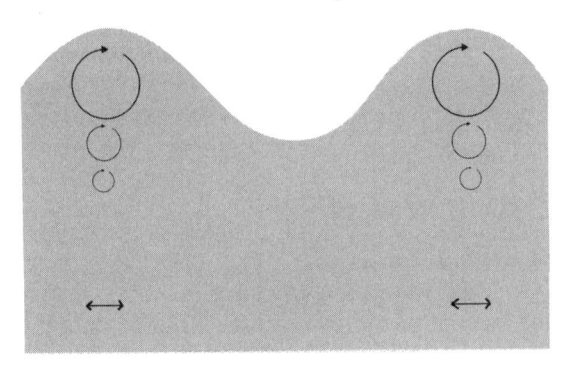

Figure 20.12 Movement of a Wave Through Seawater.

The height and the period of wind-generated waves are affected by three factors: the speed of the wind, the length of time it blows, and the *fetch*, or distance of water over which it blows. The higher the wind speed and the longer the wind blows, the higher the wave height. Fetch is important in determining wavelength. The longer the fetch, the longer the wavelength.

Wind-generated waves can be divided into three categories: seas, swells, and surf.

Erosion

Seas are waves directly affected by the wind. While the wind is blowing, it pushes the wave crests up into sharp peaks and stretches out the troughs. The surface of the ocean is often a confused jumble of seas of varying heights moving in different directions, rather than a set of regular waves moving in one direction. This occurs because the ocean's surface at any one point is affected by a mixture of waves coming in from many different places, generated by winds blowing in different directions, at different speeds, and for different lengths of time. When the ocean is choppy, we are seeing a mixture of seas from different sources.

Swells are waves with smoothly rounded crests and troughs. Swells form as seas leave the area of turbulent winds that formed them because waves with a long wavelength travel faster than those with a short wavelength (velocity = wavelength/period). Thus, over time, the long waves outdistance the short ones and the waves sort out into regions of waves of similar wavelengths. The waves in these uniform patterns are called *swells*. As waves travel farther from their source, their height decreases but their length remains constant. Waves showing such patterns can travel across entire oceans.

Surf occurs near shore when a wave breaks. The difference between surf, or breaking waves, and seas or swells is that the water in breaking waves is not moving in a circular pattern but is actually moving toward the shore. As a result, a lot of energy is directed at the beach.

Tsunamis

Tsunamis are waves caused by undersea movements of the crust due to earthquakes, slumping, or volcanic eruptions. The sudden up-and-down movement of the bottom creates a wave that has a long wavelength and travels at high speed. In deep water, tsunamis may have wavelengths up to 700 kilometers and travel at speeds over 350 kilometers per hour, yet may be only several centimeters high. When tsunamis reach a coastline, though, they form huge waves that break against the coast and cause great destruction.

Coastal Processes

Breaking Waves

Breaking waves, or surf, forms as follows. When waves enter shallow water, the bottom of the wave comes into contact with the ground at a depth equal to one-half the wavelength. This contact slows the bottom of the wave and forces the wave upward, while the top continues at its original speed. When the top of the wave outpaces the rest of the wave, it is no longer supported and falls over. See Figure 20.13.

Wave Refraction

Waves that enter shallow water at an angle to the beach are refracted; their direction of travel is changed. This refraction occurs because one part of the wave reaches shallow water and slows down while the rest of the wave is still in deep water and moving faster. Like a rolling log whose one end hits a tree, causing the whole log to swing around, the faster moving end of the wave swings around when the end in shallow water slows down.

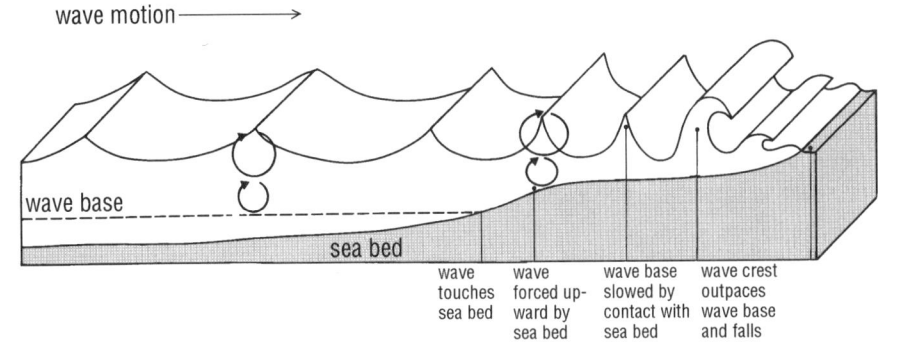

Figure 20.13 Mechanics of a Breaking Wave. Source: *Earth Science on File.* © Facts on File, Inc., 1988.

On irregularly shaped coastlines, waves converge on headlands and diverge at depressions in the shoreline, as shown in Figure 20.14. The results are erosion of the headland by the force of the waves directed against it and the deposition of sediments in the calmer zone of divergence in inlets. Over time, the irregularly shaped shoreline is straightened out.

Longshore Currents

When waves break, the water is carried into the surf zone and travels toward the beach. The water that washes up against the beach runs back along the incline under the surf zone. When waves approach the beach at an angle, the water moves upward at an angle and falls straight back. The result is a movement of water parallel to the beach called a *longshore current*. See Figure 20.15.

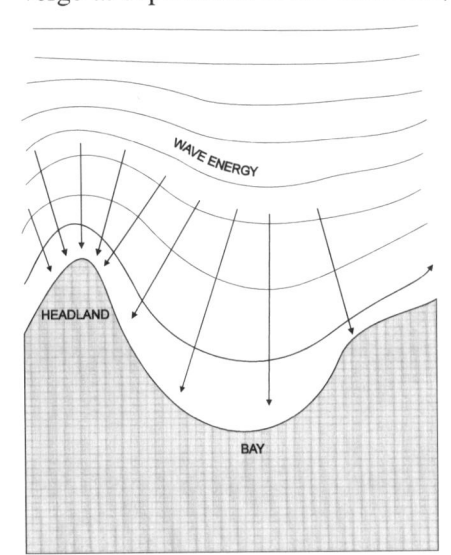

Figure 20.14 Wave Refraction at Headlands and Depressions.

The longshore current and the up-and-back motion of water washing along the beach move sand along the shore. The net result is to transport sediment along the beach and erode it inland. In populated areas, people try to impede this erosion by building structures in the surf zone to block the longshore current and trap beach sediments from being washed away. See Figure 20.16. These attempts are only moderately successful in the long run. Also, the structures often cause increased erosion down-current as the longshore current carries away sediment but brings none in from up-current.

Erosion

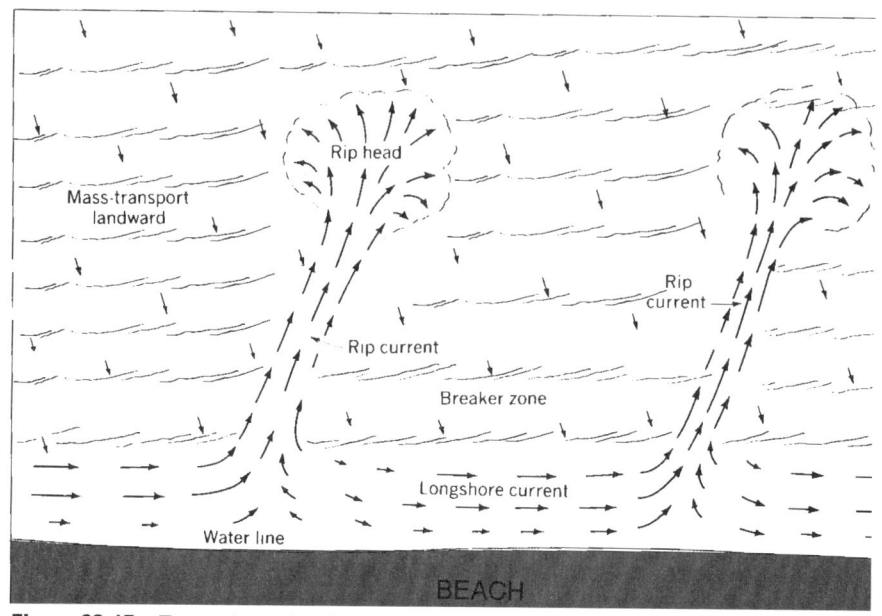

Figure 20.15 Formation of a Longshore Current. Source: *The Earth Sciences*, Arthur N. Strahler, Harper & Row, 1971.

Shoreline Erosion

Ocean water is constantly in motion. Waves, currents, and tides are examples of the movement of ocean water. Moving ocean water can transport sediments and erode the land.

Waves

Waves can be very powerful. A single wave can send tons of water crashing against a shore. The force of waves can break up rocks on the shore. Loose fragments are stirred up and carried along in the turbulence of breaking waves.

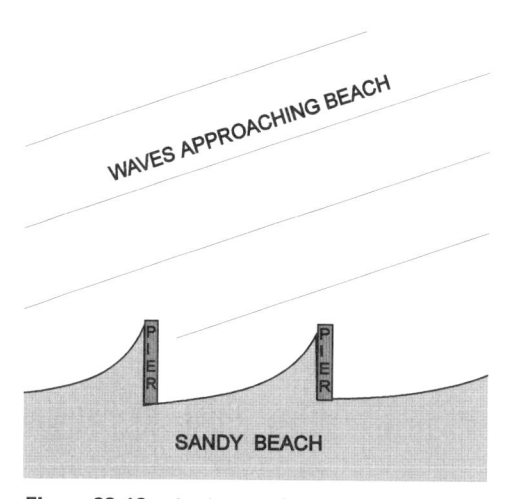

Figure 20.16 Jetties or Groins Built To Trap Sediment.

Wave erosion produces a number of shoreline features, some of which are shown in Figure 20.17. When waves erode steep, rocky shores, they remove materials from the base of the slopes. The result is a *wave-cut notch* that undercuts the slope. Overhanging rocks and soil break away, leaving a steep-sided *sea cliff*. When wave erosion penetrates deeply into a cliff, a large

458

hole or *sea cave* may form. As weaker rock is worn away from the cliff, pillars of resistant rock, called *stacks*, may be left behind. As the erosion continues, fragments broken from the cliff are ground up by abrasion in the turbulent waves. This buildup of material forms a flat platform, or *sea terrace*, at the base of the cliff.

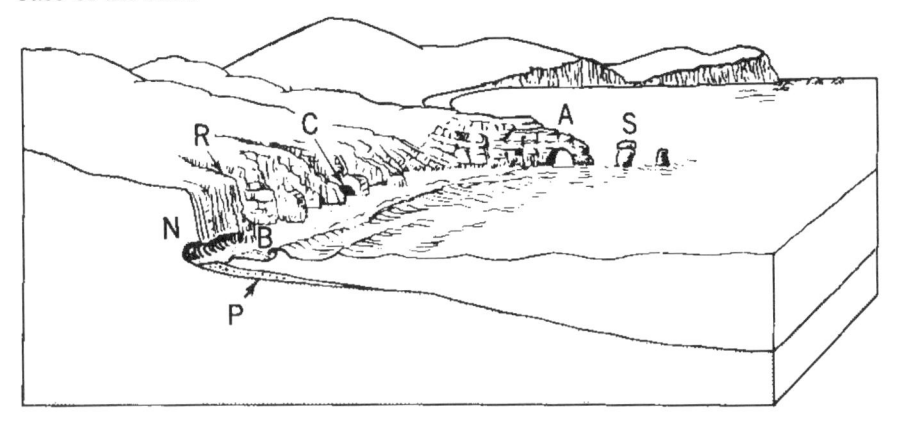

Figure 20.17 Shoreline Features Formed by Wave Erosion of a Rocky Coast. *P*—platform of abraded sediments, *N*—wave-cut notch at the base of a cliff, *R*—crevice eroded back into the rock, *B*—beach of sediments eroded out of the cliff, *C*—sea cave, *A*—arch, *S*—sea stack. Source: *The Earth Sciences*, Arthur N. Strahler, Harper & Row, 1971.

Currents

Currents are movements of water. On gently sloping, sandy shores waves mainly shift sediment around. When waves strike perpendicularly to the shore, sediments are moved in and out with the waves. Waves that strike the shore at angles other than 90°, however, are reflected at an angle and interfere with each other, forming a longshore current parallel to the shore. See Figure 20.15. Longshore currents can move sediments along just as streams do. Along some coasts, tides flowing into and out of narrow openings form swift currents. These, too, can move large amounts of sediments.

Wind Erosion

Wind that blows fast enough can pick up and carry loose particles of rock. Wind is also an agent of erosion, though not an important one in most regions. Where rainfall is abundant, plants growing on the surface protect underlying sediments from the full force of the wind. Where the ground is moist, sediment particles tend to stick together and are more difficult for the wind to dislodge and move. (Think of the difference between blowing on a handful of dry, powdery clay and blowing on a handful of wet, sticky clay.) In arid regions, however, there is little plant cover, and the soil, loose sediments, and weathered bedrock are dry and exposed. Under such conditions, wind becomes an important agent of erosion.

Deflation and Abrasion

Wind erodes the land in two ways: deflation and abrasion. In *deflation*, the wind picks up and carries away loose particles much as a stream carries sediments. In *abrasion*, exposed rock is worn away by wind-driven particles. The rock particles worn away by abrasion are then carried away by the wind.

Wind is like a stream of air, and wind carries sediments in much the same way that a stream of water does. However, winds are usually able to carry only small particles, such as sand, silt, clay, and dust, because the winds' medium, air, is not very dense. It exerts less force on a particle than does a denser medium, such as water, moving at the same speed. Imagine the difference between being hit by a balloon filled with air and a balloon filled with water!

Wind carries very small particles, such as clay and dust, in suspension. The finest particles may be carried many meters up into the atmosphere. When volcanoes erupt, dust and ash are often spewed high into the atmosphere, where winds carry them for hundreds or even thousands of miles before they finally settle to the ground. Some particles of dust and ash may reach the jet streams of the stratosphere and remain suspended for years after they are picked up by these fast-moving winds.

Wind carries larger particles closer to the ground. They slide, roll, and bounce, skipping along the surface. As they do so, they collide with exposed bedrock or other particles. In the process of abrasion, these collisions scratch the surface of the exposed rock and cause tiny chips to break off. Like a sandblaster, the wind-borne particles cut and polish the rock surfaces they impact. Over time, rock exposed to wind-driven particles is worn away. The surfaces of both the exposed rock and the particles take on a frosted appearance because of the many tiny scratches inflicted on them.

The amount and the type of erosion caused by wind depend on three factors: the speed of the wind, the size of the particles it carries, and the length of time it blows. The faster the wind blows, the larger the particles it can transport. The larger the particles carried by the wind, the greater the abrasion of exposed surfaces. The longer the wind blows, the more deflation and abrasion that can occur.

Features Produced by Wind Erosion

As deflation removes layer after layer of loose surface materials, a shallow depression, known as a *deflation hollow*, is formed. The semiarid Great Plains of the midwestern United States are dotted with tens of thousands of deflation hollows. In wet years they are covered by grass; and when rainfall is unusually high, they may fill with runoff, forming shallow lakes. In dry years, however, the grass dies off and the wind continues to deflate the bare soil.

When deflation lowers the land surface to the level of the water table, erosion is stopped. The exposed groundwater holds the soil particles together and enables plants to grow. The plants' roots help hold the loose soil in place, and their leaves and stems block the wind, ending any further erosion. A patch of vegetation surrounded by arid wasteland is called an *oasis*.

Deflation usually removes only fine particles. If the soil is composed of a mixture of particle sizes, the larger particles, which are too heavy for the wind to move, are left behind. These larger particles shield the finer particles beneath them from deflation. As exposed fine particles are removed, coarse particles form an increasing percentage of the particles left at the surface. Eventually, a continuous layer of pebbles, gravel, and other coarse particles, called a *desert pavement*, may form, as shown in Figure 20.18.

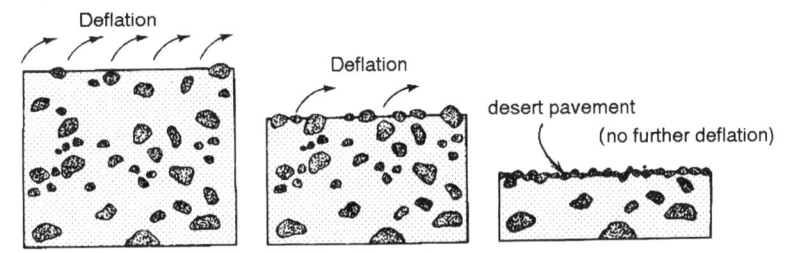

Figure 20.18 Formation of a Desert Pavement. As deflation removes finer sediments, larger sediments that are left behind eventually form a continuous layer called a desert pavement, which blocks further deflation. Source: *Physical Geology*, 2nd Ed., Richard Foster Flint and Brian J. Skinner, John Wiley, 1977.

Abrasion by wind-blown particles produces spectacular features in arid regions. Caves, arches, and bridges may be cut into exposed bedrock. Unusually shaped rock formations may result when rocks of different hardnesses are exposed to abrasion simultaneously.

Loose rock particles, too heavy to be transported by the wind, can also be abraded by wind-blown particles. Over time, they develop flat surfaces on the side facing the prevailing winds (see Figure 20.19). These unusual particles are called *ventifacts*, from the Latin word *ventus*, meaning "wind."

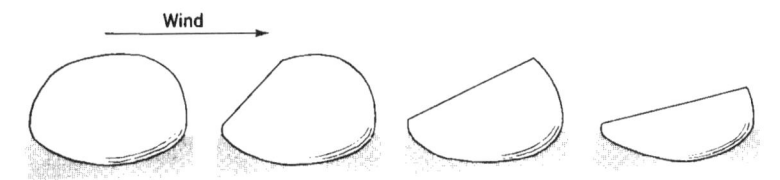

Figure 20.19 Ventifact Formation. Abrasion of the exposed surface of a rock fragment by wind-born particles forms a flat-faced ventifact. Source: *Physical Geology*, 2nd Ed., by Richard Foster Flint and Brian J. Skinner, John Wiley, 1977.

Glaciers

A *glacier* is a large mass of ice, formed on land by the compaction and recrystallization of snow, that is moving downhill or outward under the force of gravity. Modern glaciers are found only at high elevations and in polar regions where snowfall exceeds melting over an extended period of time.

Erosion

Formation of Glaciers

For a glacier to form, the average amount of incoming snowfall must be greater than the average amount lost by melting or evaporation year after year.

When incoming snowfall exceeds snow loss, the net amount of snow remaining on the ground increases each year. Freshly fallen snow is a fluffy mass of delicate ice crystals that consists mostly of pockets of trapped air and therefore has a low density. The situation quickly changes, however, as snow piles up on the ground and sits there for weeks or months. Because of evaporation, the ice crystals become smaller and rounder and form a denser, *granular snow*. As more and more snow builds up, alternate melting and refreezing, together with the weight of overlying layers, compact the granular snow into an even denser mass called *firn*. Firn is usually a year or more old and has little or no pore spaces. Eventually, the ice crystals in firn grow together into a solid mass of ice (see Figure 20.20). When snow is converted to ice in this way, some of the dust and gases of the atmosphere are trapped in the ice. Scientists can sample the trapped air and compare it with the current atmosphere to determine what, if any, changes have occurred.

When ice has accumulated to a thickness such that its own weight deforms the ice mass and it moves slowly under pressure, a glacier has formed. Usually this stage is reached when the ice is 100 meters or more thick.

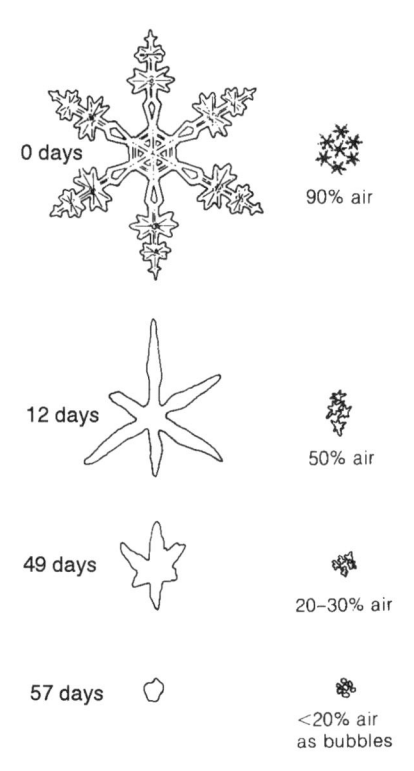

0 days — 90% air

12 days — 50% air

49 days — 20–30% air

57 days — <20% air as bubbles

Figure 20.20 Transformation of Snow Crystals into Glacial Ice. Because of greater surface area, a snowflake's intricate edges melt and evaporate faster than its center. Over time, the snowflake becomes a rounded, compact pellet of ice.
Sources: *Earth*, 4th Ed., Frank Press and Raymond Seiver, W.H. Freeman, 1986.

Movement of Ice

Under pressure, ice behaves like a very thick fluid. The ice *flows* downhill over Earth's surface, like thick syrup poured on a tilted plate. If there is just a little syrup, it spreads out into a thin layer and moves very slowly. The rate of flow of syrup can be increased in two ways: by adding more syrup, so that the layer gets thicker and flows more rapidly, and by increasing the tilt of the plate. In much the same way, glaciers flow faster where the glacial ice is thickest and where the slope is steepest.

Variations in the rate of flow of ice cause a glacier to thicken and also to thin out. Where the slope is steep, the ice thins out because it is flowing downslope faster than it is being replenished. The rapid downhill flow exerts a tension on the ice that causes cracks called *crevasses* to form. On extreme slopes the ice may break apart into blocks that cascade downslope in a rock fall. On gentler slopes the ice slows down. The ice from upstream pushes against the slower moving ice, and the compression causes the glacier to bulge and thicken.

These variations in the rate of flow of ice are illustrated in Figure 20.21.

Because the ice in a glacier is flowing, the glacier moves in a unique way. Glaciers do *not* move over Earth's surface like a brick sliding down a board. A brick slides down an incline along its base as a single unit; nothing moves around *inside* the brick. In ice, however, both sliding, along the base of the glacier, and flowing, throughout the body of the ice, occur. See Figure 20.22 for a diagrammatic comparison of the two types of motion.

The amount of a glacier's movement that is due to the base sliding along the ground, as compared to the amount caused by internal flow, depends on the temperature of the glacier. Temperature affects the behavior of ice in two key ways:

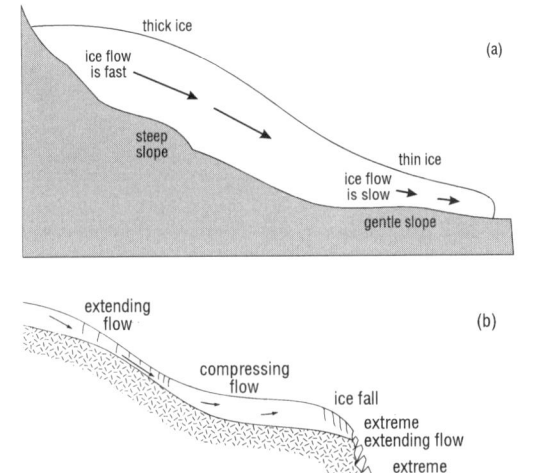

Figure 20.21 Glacial Flow Rates. (a) Ice flows fastest on steep slopes and in places where the ice is thickest. (b) On steep slopes, the ice thins out and may even split, forming crevasses, because it is flowing downhill faster than it is being replenished from farther upstream. On gentle slopes, the ice thickens as ice from upstream piles up behind slower moving ice.
Source: *Living Ice: Understanding Glaciers and Glaciation*, Robert P. Sharp, Cambridge University Press, 1988.

- Ice that is near its melting point is less rigid and is therefore more easily deformed than ice that is well below its melting point. For example, ice at 0°C deforms 10 times faster than ice at −22°C.
- Increasing the pressure on matter causes an increase in temperature. Pressure exerted on ice that is near its melting point may produce enough warming to result in melting.

Erosion

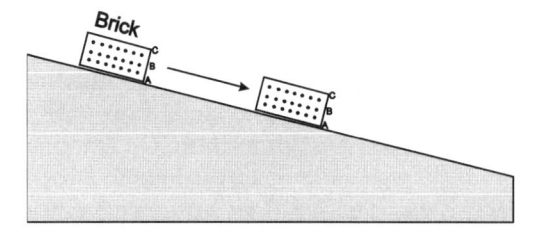

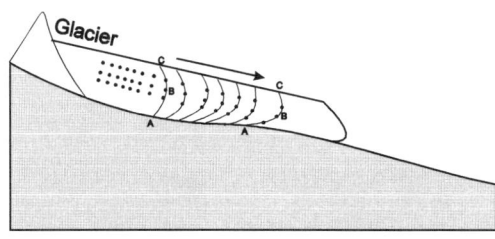

Figure 20.22 A Brick and a Glacier in Motion. A glacier both slides along its base and flows like a fluid. The relative positions of particles inside the body of the ice change as the glacier flows downhill.

Warm glaciers, whose ice is near its melting point, slip along their bases more readily than cold glaciers. When warm glaciers encounter an obstacle, the pressure of the ice against the obstacle causes the ice to deform and flow around the obstacle. The glacier may also bypass the obstacle by melting on one side and refreezing on the other. The pressure on the upstream side of the obstacle may cause the ice to melt and then refreeze on the downstream side, where pressure is lower. Meltwater produced by pressure can also fill in cavities and act as a lubricant, allowing the ice to move over the ground more easily. See Figure 20.23.

In cold glaciers moving over bedrock, almost none of the movement is due to slipping of the base along the ground. The bond between the ice and the rock to which it is frozen is greater than the internal strength of the ice. Therefore, under pressure, the ice will deform before it breaks loose from the bedrock.

Types of Glaciers

There are two main types of glaciers: valley glaciers and the much larger continental glaciers. *Valley glaciers* are constrained by topography, and their form and flow are strongly influenced by the shape of the land. *Continental glaciers* submerge the land to the extent that the size and shape of the glacier, rather than the shape of the land, control glacial form and flow.

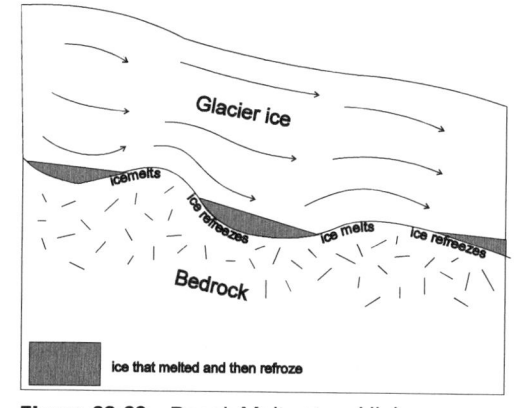

Figure 20.23 Basal Meltwater. High pressure on the upstream side of an obstacle causes ice to melt. The meltwater refreezes on the downstream side of the obstacle, where pressures are lower. Source: Adapted from *Glaciers and Landscape: A Geomorphological Approach*, David E. Sugden and Brian S. John, John Wiley, 1976.

Valley Glaciers

Valley glaciers (see Figure 20.24) form between mountain slopes at high elevations. They start as snowfields that accumulate in bowl-shaped hollows called *cirques*. As the snowfield deepens and ice forms from firn, the ice flows out of the cirque and down the valley, confined by the surrounding slopes.

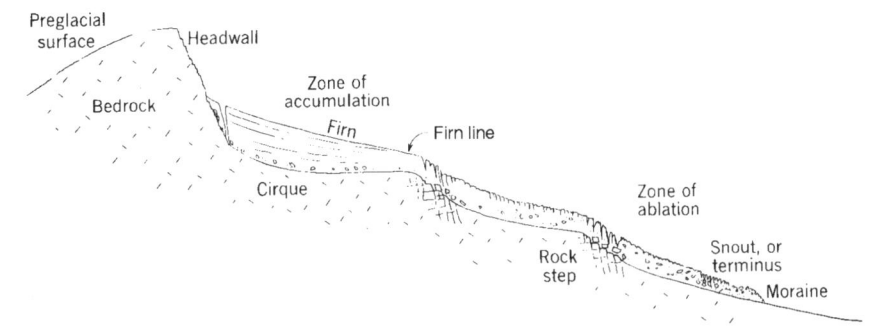

Figure 20.24 Side View of a Typical Valley Glacier. Snow accumulates in a bowl-shaped cirque and is compacted into firn. The firn recrystallizes into ice and moves downhill; the firn line marks the beginning of the glacial ice. When the glacier reaches warmer temperatures, ablation consumes the ice. The glacier ends in a thin snout edged by a pile of debris called a *moraine*. Source: *The Earth Sciences*, 2nd Ed., Arthur N. Strahler, Harper & Row, 1971.

As the glacier moves down the valley, it may meet other glaciers from adjacent valleys and the glaciers may merge, forming a larger glacier. Unlike streams, which mix when they merge, glaciers do not mix but rather "weld" together and flow downslope side by side. A large valley glacier may have been fed by dozens of smaller tributary glaciers flowing from high mountain valleys.

Valley glaciers end at a point where there is a balance between ablation and accumulation. This point can be a sensitive indicator of changes in climate. A general warming trend will cause the end of the glacier to melt back up the valley, or retreat. A cooling trend will cause the end of the glacier to move down the valley, or advance.

When a glacier ends on land, streams of meltwater flowing from the glacier often form a river. Large blocks of ice that fall off the face of the glacier may be buried by sediment. A glacier that ends in a body of water forms steep cliffs of ice. Chunks of ice that break off and fall into the water will float since ice is less dense than liquid water. These floating masses of ice, or icebergs, then drift with the currents and gradually melt.

Continental Glaciers

Continental glaciers form at high latitudes where temperatures are always cold. These glaciers are immense masses of ice that spread over the entire land surface, rather than being confined to valleys. Two continental glaciers exist today: one covers Greenland, and the other Antarctica. The Greenland glacier (see Figure 20.25) covers more than 1.7 million square kilometers and is 3.2 kilometers thick at its center. The Antarctic glacier covers more than 12

Erosion

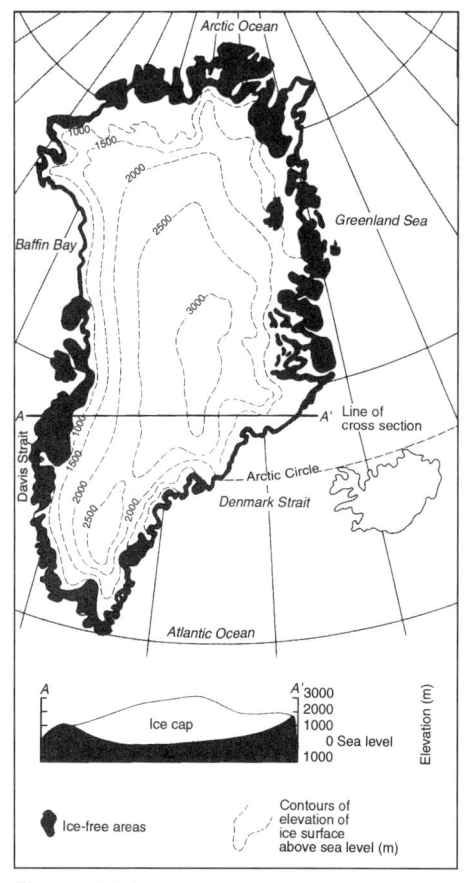

Figure 20.25 The Greenland Glacier. Contour lines show the elevations of the ice surface in meters, and a cross-sectional view reveals the lenslike shape of the glacier. Source: *Earth,* 4th Ed., Frank Press and Raymond Seiver, W.H. Freeman, 1986.

million square kilometers (about 1.5 times the size of the 48 mainland U.S. states) and is 4 kilometers thick in places. It even extends over the ocean along the margins of the continent of Antarctica, forming several ice shelves.

A continental glacier has a lenslike shape; it is thickest at the center and thinner at the edges. The thick central area is where snow accumulates in a huge firn field and slowly turns to ice. The ice then flows outward in all directions under the pressure of its own weight. The huge weight of a continental glacier actually depresses the surface of the land, and the topography of the land is extensively changed by the movement of the glacier over its surface.

Processes of Glacial Erosion

Glaciers erode the surfaces over which they move by abrasion and plucking. *Abrasion* is the wearing, grinding, scraping, or rubbing away of rock surfaces by friction. *Plucking* is the process by which rock fragments are loosened, picked up, and carried away by glaciers.

Abrasion

Abrasion is accomplished mainly by fragments of rock frozen in the ice at the bottom of a glacier. These rub and scrape against the underlying bedrock as the base of the glacier slips downslope. Sharp, angular fragments of hard materials such as quartz are very effective abraders. Clay is a good polishing material but doesn't abrade the bedrock very much.

Plucking

Plucking occurs in several ways. The water that forms when pressure on the ice causes melting can seep into cracks and pores in the bedrock. When this water refreezes, it shatters the bedrock and the broken pieces are picked up and carried away by the glacier. Plucking that occurs in this way is probably the most important because it goes on continuously over much of the glacier's bed as it moves along.

Plucking can also occur when glaciers move over bedrock that is broken into large pieces by jointing. Joints are fractures in rock and usually occur in three planes, forming large, rectangular blocks. These blocks freeze to the base of the glacier and are pushed out of the ground and carried along as the ice moves downslope.

A third type of plucking may occur where the glacier is thick and the bedrock is brittle. In such places, the pressure exerted by a glacier may be great enough to crush, shear off, or shatter brittle bedrock, forming fragments that are carried away by the glacier.

Features Resulting from Glacial Erosion

The processes of abrasion and plucking leave behind unmistakable evidence of a glacier's passage over bedrock. Numerous features, ranging from scratched and polished rock surfaces to basins excavated by plucking to streamlined hills of solid rock, are formed by glacial erosion.

Glacial Polish

Fine materials such as clays and silt that are frozen into the ice at the base of a glacier rub against bedrock as the glacier moves. These particles smooth the rock surface as sandpaper smooths wood. The smooth rock surfaces produced by this type of abrasion are said to be *glacially polished*.

Glacial Striations

Glacial striations (see Figure 20.26) are small scratches in the surface of the bedrock. They are created when fragments frozen into the bottom of a glacier scrape against the bedrock as the glacier slides over the bedrock. On smooth surfaces the striations are usually straight and parallel, but on uneven surfaces with protrusions the striations may curve because the ice deformed as it flowed around these obstacles. A striation may also thicken or thin out, depending on what happened to the abrading particle as it scraped against the bedrock. The long axis of a glacial striation indicates the direction of glacial movement.

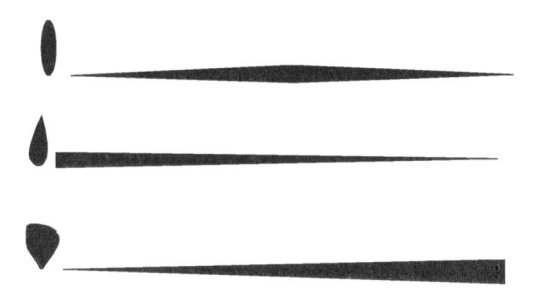

Figure 20.26 Glacial Striations and the Particles That Made Them. Striations vary in size and shape, depending on the nature of the rock particles that formed them and the way that the particles were dragged over the bedrock.

Glacial Grooves

Glacial grooves are long, deep, U-shaped depressions in bedrock with

Erosion

smooth bottoms and sides and rounded edges. Most grooves are 10–20 centimeters deep, twice as wide, and tens of meters long. Really large grooves, however, can be more than 1 meter deep and hundreds of meters long. Grooves are shaped by abrasion and often form along fractures, soft layers, or other weak zones in the bedrock. Once started, a groove acts as a channel for further ice flow and concentrates the abrasive force of the glacier in this channel.

Rock Drumlins
Drumlins are hills of loose rock and soil molded into a streamlined shape by glaciers moving over them. Rock drumlins are large outcrops of rock that have been abraded into a streamlined shape by glaciers. Rock drumlins are large, often measuring 100 meters in length.

U-Shaped Valleys
Glaciers move from high elevations to low elevations along the easiest paths. Those paths are usually stream valleys (see Figure 20.27a) that were eroded before the glacier formed. Stream valleys in mountainous regions tend to be steep-sided and V-shaped and to follow a narrow, twisting path between adjacent slopes (see Figure 20.28a). When a glacier moves down a stream valley, it alters the valley's shape. The ice does not meander back and forth as readily as water, so it often breaks through obstacles to form a straighter path. The ice also cuts back the walls of the valley, making them steeper. The result is a wider, straighter, steeper-sided, *U-shaped valley* (Figure 20.27b).

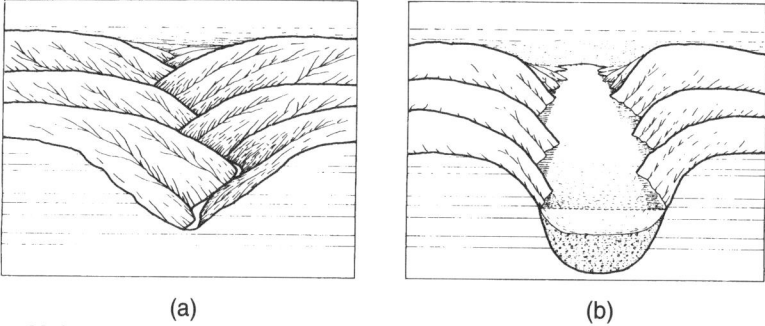

(a)   (b)

Figure 20.27 (a) A V-shaped Stream Valley and (b) a U-shaped Glacial Valley.
Source: *Geology of New York: A Simplified Account*, Y. Isachsen et al., eds., New York State Museum Geological Survey, The State Education Department, The University of the State of New York, 1991.

Hanging Valleys
A *hanging valley* (Figure 20.28c) is a valley of a tributary glacier whose floor is above, or hanging over, the floor of the main glacier into which it feeds. The difference in the levels of the valley floors can be hundreds of meters; and if a stream flows down the tributary valley, it may form a spectacular waterfall as it plunges to the floor of the main valley. Hanging valleys may be formed if the main glacier cuts back the valley of a tributary, or if it

simply erodes faster and deeper than the tributary.

Cirques, Arêtes, and Horns

A *cirque* is the open, half-bowl-shaped hollow high on a mountainside at the head of a valley glacier. Cirques form when snowfields on mountain slopes melt and refreeze during spring and fall. The repeated freezing and thawing breaks apart the rocks. The rock fragments creep downslope or, when broken into small enough pieces, are carried away by meltwater. In this way, the snowfield creates a hollow in the land on which it rests. As the hollow grows, so does the depth of the snowfield, eventually becoming large and deep enough to form a glacier. Once formed, the glacier plucks rock from the base of the cirque, further deepening it.

Where cirques from adjacent slopes meet, a narrow, knife-edged ridge called an *arête* is formed between them. Where three or four cirques on the flanks of a mountain peak meet, the central peak takes on a pyramidal shape called a *horn*.

These three features—cirque, arête, and horn—are shown in Figure 20.28b.

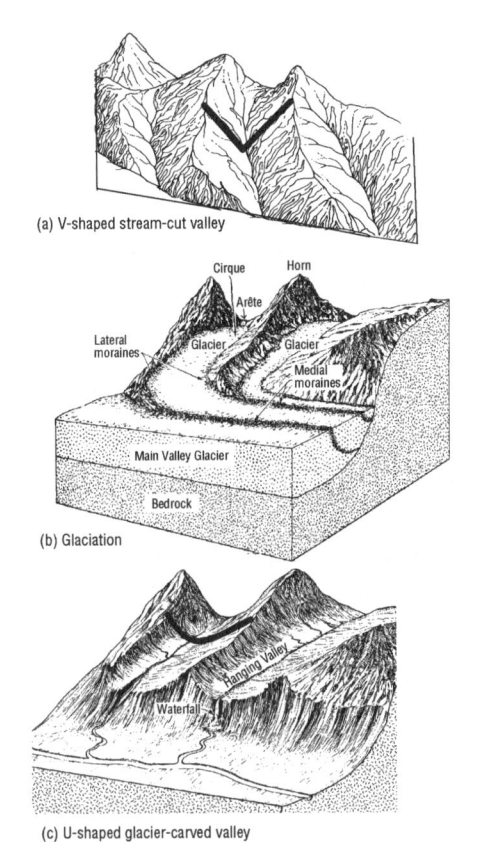

Figure 20.28 Landscape Before, During, and After Glaciation. (a) Valley before glaciation. (b) Glacial erosion forms cirques in hollows between slopes. Erosion in adjacent cirques forms sharp ridges called arêtes and converts rounded peaks into sharp horns. Eroded material carried along the ice margins forms moraines. (c) Hanging valleys form where tributary glaciers flowed into the main glacier. Source: *Earth*, 4th Ed., Frank Press and Raymond Seiver, W.H. Freeman, 1986.

Some Environmental and Human Impacts of Erosion

Mass Wasting

The sudden failure of slopes usually involves some triggering event. It can be the vibrations of an earthquake, the addition of water by intense or prolonged rainfall, an increase in the steepness of the slope (due to natural processes or human activities) or the loss of stabilizing vegetation.

The movement of rock material down a slope by mass-wasting processes can have a striking impact on the environment and on human activities. For example, a stream that is cutting a valley can create unstable slopes. The resulting landslides into the valley can dam up the stream, creating a natural reservoir that floods surrounding land. In 1983, a landslide in Utah blocked Spanish Fork Canyon, creating a large lake that cut off a transcontinental railway line and a major highway. Debris flows can bury or sweep away large areas of vegetation, leaving massive scars on the landscape.

Unstable slopes are a hazard to human structures built on or below them. Every year, rockfalls along highway cuts cause motorist fatalities. Housing developments built atop slopes are frequently undermined and destroyed by mass wasting. However, with careful planning, building regulations, and zoning laws, and the use of stabilization techniques, the impact of mass wasting on human activities can be reduced.

Hazardous slopes can be stabilized by changing the steepness of the slope, reducing the weight acting on the slope, removing water from the rocks and soil, and planting vegetation. Some commonly used stabilization techniques include (1) the use of retaining devices (steel nets, concrete walls, rock bolts), (2) the removal of water by installing drainage systems or by covering the slope with an impermeable material, (3) grading the slope to reduce its steepness, and (4) building walls to divert flows. All of these preventive measures, however, are costly. The best strategy is to avoid building on unstable slopes in the first place.

Streams

Erosion by moving water affects human lives in many ways. Overland flow may wash away the topsoil your family just bought and spread on the lawn. Splash and sheet erosion of topsoil on farmland decreases fertility and lowers crop yields, resulting in higher prices at the supermarket. Erosion and flooding by streams leave thousands homeless annually. Nevertheless, there are benefits to living near running water, such as fertile soils, access to water, food from marine life, energy, waste absorption and dispersal, and possibly a transportation route.

Flooding is *the* major stream hazard, causing loss of life and structural damage. Flooding is a natural response to an unusually high input of water over a short time, but human activities can influence flooding in many ways. Paving surfaces or constructing buildings changes runoff patterns and infiltration rates. Changes to river channels and dam construction alter a stream's flood patterns.

The primary impacts of flooding are due to actual contact with the flowing water. They include death by drowning, damage to structures, crop loss, and erosion of flooded areas. Floods also impact humans and the environment indirectly by interrupting services, changing river channels, and destroying wildlife habitat.

Regions at risk of flooding and the degree of risk each faces can be determined if accurate maps and records of past floods have been kept. These allow predictions to be made based upon the frequency and extent of past flooding. However, records may not go back long enough to permit precise predictions about rare, severe floods.

The recurrent hazard of flooding has prompted efforts to reduce risk in flood-prone areas by limiting the human population there. Some communities now have floodplain zoning, which prohibits new construction. The Federal Emergency Management Agency (FEMA) requires that, if the owner of a flood-damaged property gets more than 50 percent of its value in compensation, the structure must be demolished and the site rezoned to prohibit future construction. Many lenders place restrictions on mortgages for properties in floodplains, allowing only appropriate activities such as recreation or certain types of farming to be carried out there. In some places there are calls for the buyout of all properties in flood-prone areas and the return of the land to its original wetland status. For urban areas that already exist in floodplains, however, flood hazard will continue to be an issue.

Waves and Currents

Erosion is one of the most important problems in coastal zones, mainly because 85 percent of shorelines is privately owned and rising sea levels are eroding all shorelines. Erosion is most pronounced on beaches and coastal cliffs, where smaller sediments are easily moved by waves and currents. It tends to be less of a problem on rocky coastlines or beaches made of pebbles, which are harder to move.

There are really only three ways to fight erosion of shoreline areas: dredge sediments to fill in where erosion has occurred, dam rivers or build coastal structures such as jetties and breakwaters to decrease the force of eroding waves and currents, and develop coastal dune areas as barriers to inland erosion.

Unfortunately, some of these strategies have proved ineffective. For example, *beach replenishment* involves replacing eroded sand with sand brought in from someplace else. This is an expensive undertaking considering the huge amount of sand needed. Also, most beach replenishment doesn't last very long—40 percent of all replenished beaches are gone in less than 2 years, less than half last for 2–5 years, and only 12 percent last more than 5 years. Beach stabilization is expensive and, as you can see, usually ineffective in the long term. Furthermore, it often changes the nature of the shoreline, and this can have negative effects on the organisms that inhabit it.

Of course, the best solution would be to recognize that coastal erosion is a natural process and allow it to happen without interference. Perhaps the coastal zone is not an environment that is favorable for private ownership.

Wind

Wind erosion presents humans with many problems. In arid regions, poles carrying electric power and telephone lines have been worn through by abrasion, causing them to fall and interrupt service. In times of drought, the precious topsoil of farmland stripped of vegetation must be protected from deflation. Vehicles used in arid regions require special features to keep abrasive particles out of their moving parts and to prevent particles suspended in the air from clogging fuel, air, and cooling lines.

Awareness of wind erosion and its causes enables humans to guard against it. Farmers plant windbreaks of drought-resistant trees. In many arid regions, off-road vehicles, such as dirt bikes and dune buggies, are banned because they destroy the fragile vegetation that protects the land from wind erosion.

Glaciers

Glaciers are responsible for much of North America's topography. Thus, many familiar landforms are the products of glacial erosion. Glacial erosion sculpted the rugged peaks of the Adirondacks and created the beautiful fjord known as the Hudson River Valley. The Great Lakes occupy basins carved out and deepened by glaciers and were once filled with glacial meltwater. Erosion by meltwater from the last glacial period formed most of the Mississippi River's drainage basin.

The modern-day burning of fossil fuels has increased the amount of carbon dioxide in the air. The result has been greenhouse-effect heating that has begun to melting Earth's remaining glaciers and ice sheets; this in turn has led to a rise in sea level and to coastal flooding. The global melting of ice may have other consequences, such as changes in global weather patterns that could have serious effects on agriculture.

MULTIPLE-CHOICE QUESTIONS

In each case, write the number of the word or expression that best answers the question or completes the statement.

1. On Earth, which agent of erosion is responsible for moving the largest amount of material?
   (1) groundwater
   (2) glaciers
   (3) running water
   (4) wind

2. Which is the best evidence that erosion has occurred?
   (1) a soil rich in lime on top of limestone bedrock
   (2) a layer of basalt on the floor of the ocean
   (3) sediments found in a sandbar of a river
   (4) a large number of fossils embedded in limestone

3. For which movement of Earth materials is gravity *not* the main force?
   (1) sediments flowing in a river
   (2) boulders carried by a glacier
   (3) snow tumbling in an avalanche
   (4) moisture evaporating from an ocean

4. Which is the best explanation for the shape of the cliff in the diagram?

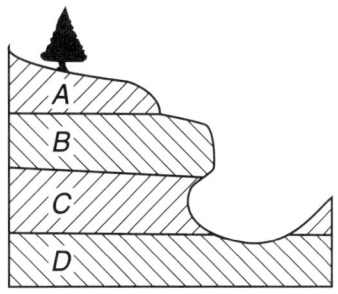

   (1) Rocks A and C are made of larger particles than rock B.
   (2) The particles in rocks A and C are more firmly cemented than those in rock B.
   (3) The minerals in rocks A and C erode faster than those in rock B.
   (4) Rocks A and C have not been exposed to weathering as long as rock B.

Note that question 5 has only three choices.
5. If the rate of erosion in a particular landscape on Earth's surface increases and the uplifting forces remain constant, the elevation of that landscape will
   (1) decrease
   (2) increase
   (3) remain the same

6. Which erosional force acts alone to produce avalanches and landslides?
   (1) gravity   (3) running water
   (2) winds   (4) sea waves

7. Abrasive action is *least* effective in the process of erosion caused by
   (1) glaciers   (3) rivers
   (2) groundwater   (4) waves

8. Which characteristic of a transported rock is most helpful in determining its agent of erosion?
   (1) age   (3) composition
   (2) density   (4) physical appearance

Note that question 9 has only three choices.
9. As the amount of plant life in a region is decreased by the activities of humans, the amount of erosion will probably
   (1) decrease
   (2) increase
   (3) remain the same

Erosion

10. In which area will surface runoff most likely be greatest during a heavy rainfall?
    (1) a sandy desert
    (2) a wooded forest
    (3) a level grassy field
    (4) a paved city street

11. Which occurs as a stream is gradually uplifted?
    (1) Its ability to erode will probably increase.
    (2) Its potential energy will probably decrease.
    (3) Its amount of streambed deposits will probably increase.
    (4) Its stream discharge will probably decrease.

12. Which graph best shows the relationship between the slope of a streambed and the velocity of stream flow?

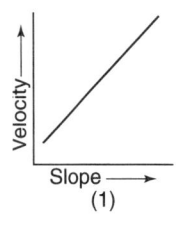

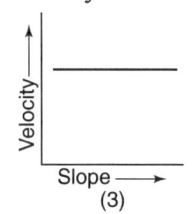

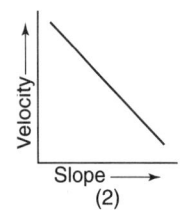

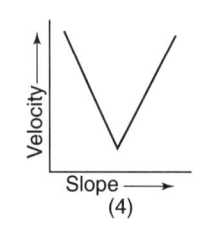

13. Which agent of erosion forms U-shaped valleys?
    (1) groundwater
    (2) winds
    (3) rivers
    (4) glaciers

14. Which agent of erosion forms water gaps?
    (1) glaciers
    (2) ocean waves
    (3) rivers
    (4) wind

15. The diagram to the right represents a winding stream. At which location is stream erosion most likely to be greater than stream deposition?
    (1) A
    (2) B
    (3) C
    (4) D

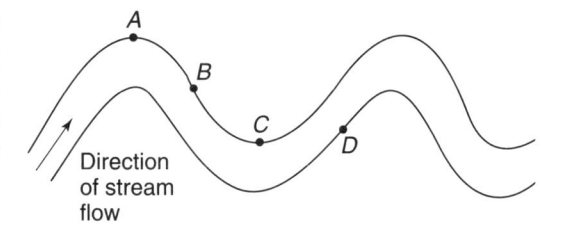

16. In the two diagrams below, the lengths of the arrows represent the relative velocities of stream flow at various places in a stream. Diagram I shows the different water velocities across the surface, Diagram II shows the different water velocities (V) at various depths.

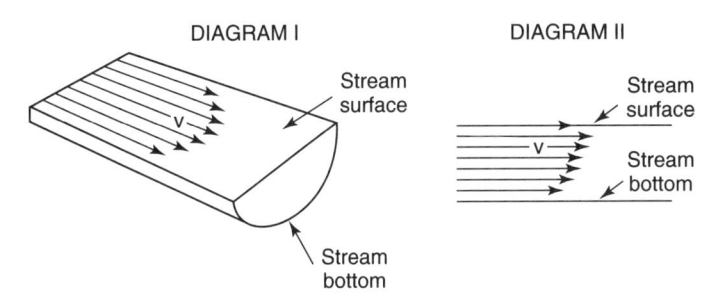

DIAGRAM I

Stream surface

Stream bottom

DIAGRAM II

Stream surface

Stream bottom

At which location in the stream is the water velocity greater?
(1) at the center along the bottom
(2) at the center near the surface
(3) at the sides along the bottom
(4) at the sides near the surface

Note that questions 17 and 18 have only three choices.

17. As the volume of water in a section of a stream channel increases, the average velocity of the water
(1) decreases
(2) increases
(3) remains the same

18. As the velocity of a stream decreases, the rate of downcutting by the stream will probably
(1) decrease
(2) increase
(3) remain the same

19. According to the *Earth Science Reference Tables*, what is the approximate minimum stream velocity needed to keep a 6.4-centimeter particle moving?
(1) 100 cm/s   (3) 225 cm/s
(2) 175 cm/s   (4) 315 cm/s

20. According to the *Earth Science Reference Tables*, which stream velocity transports cobbles, but not boulders?
(1) 50 cm/s   (3) 200 cm/s
(2) 100 cm/s   (4) 400 cm/s

Erosion

Note that question 21 has only three choices.

21. As the volume of a stream increases, the amount of material that can be carried by the stream generally
    (1) decreases
    (2) increases
    (3) remains the same

22. Which of the rock fragments shown in the diagrams below shows the *least* evidence of erosion?

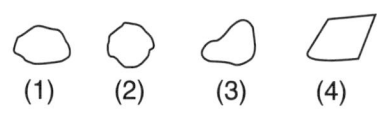

    (1)   (2)   (3)   (4)

23. Which material could best be carried in solution by a stream?
    (1) granite   (3) gabbro
    (2) quartz   (4) salt

24. A river transports material by suspension, rolling, and
    (1) solution   (3) evaporation
    (2) sublimation   (4) transpiration

25. Which graph best shows the relationship between the size of a rock particle transported by a stream and the stream velocity?

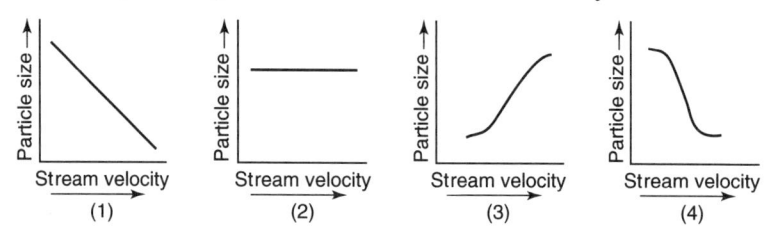

26. Which rock material is most likely to be transported by wind?
    (1) large boulders with sets of parallel scratches
    (2) jagged cobbles consisting of intergrown crystals
    (3) irregularly shaped pebbles that contain fossils
    (4) rounded sand grains that have a frosted appearance

27. Which landscape features are primarily the result of wind erosion and deposition?
    (1) U-shaped valleys containing unsorted layers of sediment
    (2) V-shaped valleys containing well-sorted layers of sediment
    (3) terraces of gravel containing unsorted layers of sediment
    (4) cross-bedded sand deposits containing finely sorted layers of sediment

28. What is the major cause of convection currents in the ocean ?
    (1) rotation of Earth
    (2) dilution of surface water by rainfall
    (3) friction between water and air
    (4) density differences due to changing temperature and salinity

29. What is the chief cause of the surface currents of the ocean ?
    (1) tides
    (2) the Moon
    (3) prevailing winds
    (4) rivers

Base your answers to questions 30 through 34 on the diagram below. The diagram represents a shoreline in New York State along which several general features have been labeled. Letter *B* identifies a location on the shoreline.

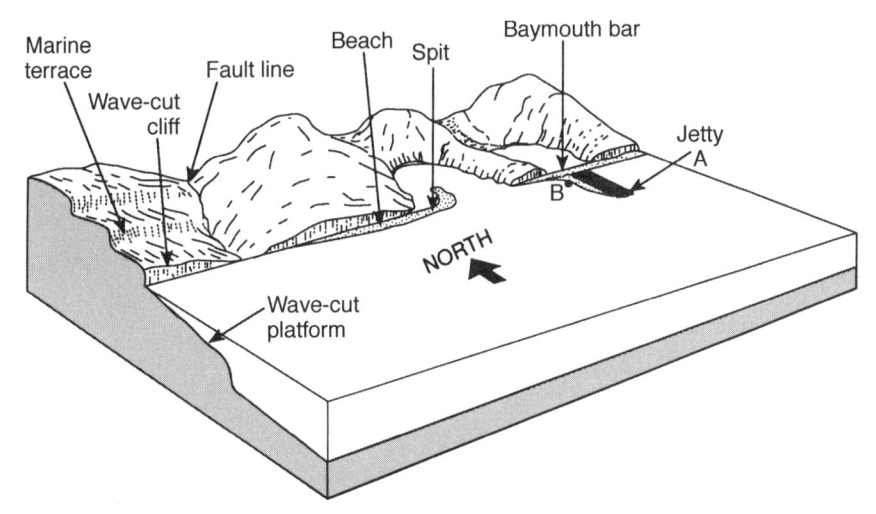

30. After the formation of the baymouth bar, the jetty (a structure made of rocks that extends into the water), labeled *A*, was constructed perpendicularly to the shoreline. Which statement best describes the result of the construction of this jetty?
    (1) Water-current velocity at location *B* decreased.
    (2) Water-current velocity at location *B* increased.
    (3) Sand deposition at location *B* decreased.
    (4) Sand deposition was not affected.

Erosion

31. What is the most likely source of the waves approaching this coastline?
    (1) variations in water temperature
    (2) density differences within the water
    (3) the rotation of Earth
    (4) surface winds

32. Past movements along the fault line most likely caused the formation of
    (1) the baymouth bar   (3) longshore currents
    (2) tsunamis   (4) tidal currents

33. Which statement best describes the longshore current that is modifying this coastline?
    (1) The current is flowing northward at a right angle to the shoreline.
    (2) The current is flowing southward at a right angle away from the shoreline.
    (3) The current is flowing eastward parallel to the shoreline.
    (4) The current is flowing westward parallel to the shoreline.

34. The marine terraces represent former positions of the wave-cut platform. These terraces indicate that
    (1) crustal uplift has occurred
    (2) crustal sinking has occurred
    (3) the coastline was subjected to many tidal waves
    (4) the coastline was affected by strong winter storms

Base your answers to questions 35 and 36 on the diagram below, which shows the movement of water particles in ocean waves. The particles are represented by black dots. Letters *A*, *B*, *C* and *D* are points of reference.

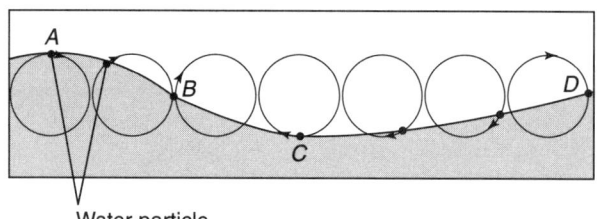

35. The passage of a wave will cause a particle of water at the surface to
    (1) move horizontally, only   (3) move in a circular pattern
    (2) move vertically, only   (4) remain stationary

36. The wave pattern shown in the diagram would occur most often in
    (1) a longshore current parallel to a coastline
    (2) shallow water in the breaker zone of a beach
    (3) a turbidity current
    (4) deep water

Base your answers to questions 37 through 40 on the profile below. Four zones within the profile are labeled. The profile shows ocean waves approaching a beach.

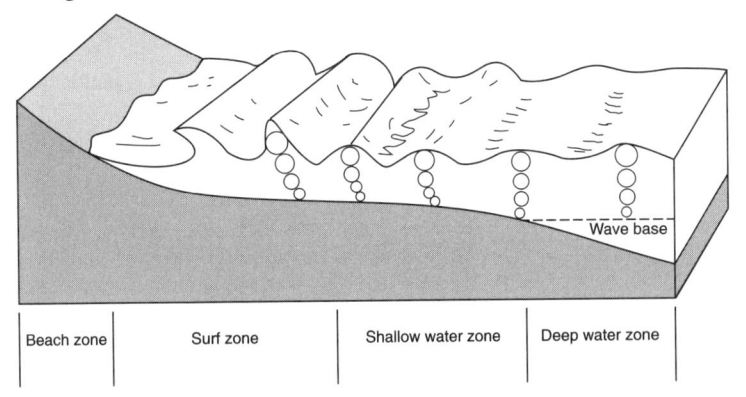

37. In which zone is sand most likely to be moving along the shore?
    (1) deep-water zone
    (2) shallow-water zone
    (3) surf zone
    (4) beach zone

38. Which statement best describes the waves in the surf zone?
    (1) The waves are unaffected by the ocean bottom.
    (2) The wave height is decreasing.
    (3) The speed of the wave bottom is decreasing.
    (4) The waves collapse as their heights increase.

39. The sand on this beach has its origin in the weathering of granite bedrock. According to the *Earth Science Reference Tables*, a mineral that is likely to be found in this sand is
    (1) quartz
    (2) calcite
    (3) halite
    (4) olivine

Note that question 40 has only three choices.
40. If a strong coastal storm with high wind speeds moves into the region, the wave heights are likely to
    (1) increase
    (2) decrease
    (3) remain the same

Erosion

41. Which geologic evidence best supports the inference that a continental ice sheet once covered most of New York State?
    (1) polished and smooth pebbles; meandering rivers; V-shaped valleys
    (2) scratched and polished bedrock; unsorted gravel deposits; transported boulders
    (3) sand and silt beaches; giant swamps; marine fossils found on mountaintops
    (4) basaltic bedrock; folded, faulted, and tilted rock structures; lava flows

42. Which is the best evidence that more than one glacial advance occurred in a region?
    (1) ancient forests covered by glacial deposits
    (2) river valleys buried deeply in glacial deposits
    (3) scratches in bedrock that is buried by glacial deposits
    (4) glacial deposits that overlay soils formed from glacial deposits

43. Which landscape characteristic best indicates the action of glaciers?
    (1) few lakes
    (2) deposits of well-sorted sediments
    (3) residual soft soil covering large areas
    (4) polished and scratched surface bedrock

44. The formation of the Finger Lakes of central New York State and the formation of Long Island are both examples of
    (1) climatic changes resulting in a modification of the landscape
    (2) uplifting and leveling forces in dynamic equilibrium
    (3) soil associations differing in composition depending upon the bedrock composition
    (4) human activities altering the landscape

The diagram below represents a landscape area.

45. Which process is primarily responsible for the shape of the surface shown in the diagram?
    (1) crustal subsidence
    (2) wave action
    (3) glacial action
    (4) stream erosion

46. Which diagram best represents a cross section of valley that was glaciated and then eroded by stream?

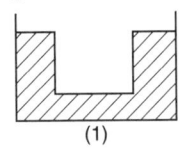

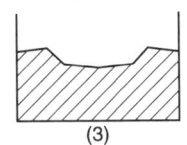

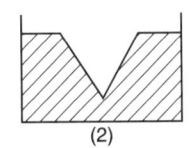

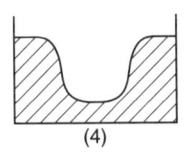

47. Many elongated hills, each having a long axis with a mostly north-south direction, are found scattered across New York State. These hills contain unsorted soils, pebbles, and boulders. Which process most likely formed these hills?
    (1) stream deposition
    (2) wind deposition
    (3) wave deposition
    (4) glacial deposition

48. Which rock material was most likely transported to its present location by a glacier?
    (1) rounded sand grains in a river delta
    (2) rounded grains in a sand dune
    (3) residual soil on a flat plain
    (4) unsorted loose gravel in hills

Base your answers to questions 49 through 51 on your knowledge of earth science, the *Earth Science Reference Tables*, and the diagram below. The diagram represents two branches of a valley glacier. Points *A*, *B*, *G*, and *H* are located on the surface of the glacier. Point *X* is located at the interface between the ice and the bedrock. The arrows indicate the general direction of ice movement.

Erosion

49. Which force is primarily responsible for the movement of the glacier?
 (1) groundwater
 (2) running water
 (3) gravity
 (4) wind

50. Metal stakes were placed on the surface of the glacier in a straight line from position *A* to position *B*. Which diagram best shows the positions of the metal stakes several years later?

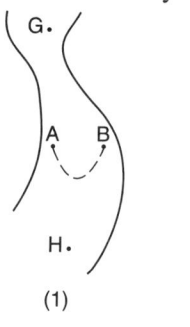

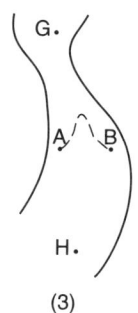

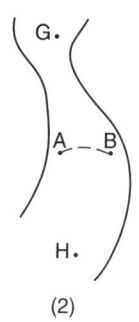

51. Which cross section best represents the valley shapes of this landscape area after the glacier melts?

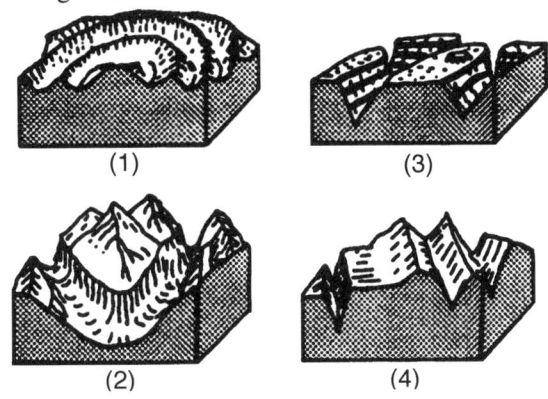

    (1)   (3)
    (2)   (4)

52. The direction of movement of a glacier is best indicated by the
    (1) elevation of erratics
    (2) alignment of grooves in bedrock
    (3) size of hanging valleys
    (4) amount of deposited sediments

53. Which force is primarily responsible for the movement of a glacier?
    (1) groundwater
    (2) running water
    (3) gravity
    (4) wind

54. If the front of an active glacier is observed to be stationary, it is correct to infer that the ice in the glacier is
    (1) not advancing
    (2) melting as fast as it advances
    (3) advancing faster than it melts
    (4) melting faster than it advances

CONSTRUCTED RESPONSE QUESTIONS

Base your answers to questions 55 and 56 on the map and cross sections below. The map shows measured changes in the position of Niagara Falls since 1678. The cross sections show two parts of Niagara Falls: Horseshoe Falls and American Falls. Letters *A* through *D* represent the rock layers at both locations.

Erosion

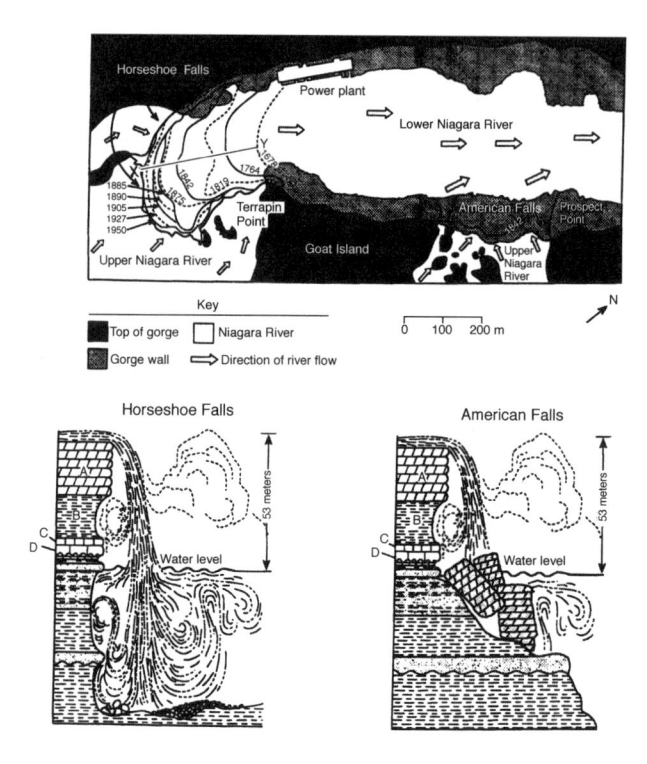

55. Calculate the rate of erosion for Horseshoe Falls along line *X-Y* between 1678 and 1950 using the following directions:
    (a) State the equation for rate of change. [1]
    (b) Substitute data into the equation. [1]
    (c) Calculate the rate of erosion, and record your answer using the correct units. [1]

56. Two students have proposed the following hypotheses to explain why Horseshoe Falls has eroded more than American Falls since 1842.
    Hypothesis 1: American Falls erodes more slowly than Horseshoe Falls because the rock layers at Horseshoe Falls are made of a different material that is much less resistant to erosion.
    Hypothesis 2: American Falls erodes more slowly than Horseshoe Falls because less water flows over American Falls.
    (a) Evaluate these hypotheses in view of the information in the map and cross sections, and state which hypothesis is best supported by the evidence. [1]
    (b) State one form of evidence in the map and cross sections that supports the hypothesis you identified in (a). [1]

57. In one or more complete sentences, explain what is meant by an "agent" of erosion. [1]

58. In one or more complete sentences, explain why water is often responsible for triggering mass-wasting movements. [1]

59. Complete the table below by placing a check (√) in the appropriate column.

Mass-Wasting Process	Slow	Rapid
Creep		
Earthflow		
Rockslide		
Mudflow		
Slump		
Rockfall		

60. State three ways in which streams transport sediments. [1]

61. Draw a correctly labeled diagram of an ocean wave that includes the following features: wavelength, wave height, crest, and trough. [2]

62. In one or more complete sentences, explain how wind erosion forms a desert pavement. [2]

63. State two differences between a valley glacier and a continental glacier. [2]

64. In one or more complete sentences, explain how it is possible for a glacier to be constantly moving although its terminus (end) is not observed to change position. [1]

EXTENDED CONSTRUCTED RESPONSE QUESTIONS

Base your answers to questions 65 and 66 on the paragraph and drawings below. The drawings show artists' conceptions of two planned structures.

Before structures such as roads, buildings and bridges are built; an earth scientist is often consulted. Earth scientists understand the natural processes of erosion and their potential hazards. Therefore, earth scientists can help engineers plan and build structures in such a way that they will not be endangered by these processes.

Erosion

Structure 1   Structure 2

Source: *Macmillan Earth Science: Teacher's Resource Book*, Eric Danielson and Edward J. Denecke, Jr., Macmillan, 1986

65. Identify the erosional process that would endanger Structure 1, and in one or more complete sentences, explain how that structure would be endangered. [1]—each process [1]—how it endangers each structure.

66. Identify the erosional process that would endanger Structure 2, and in one or more complete sentences, explain how the structure would be endangered. [1]—process [1]—how it endangers the structure.

Base your answers to questions 67 through 70 on the following experiment devised by a student.

The student wants to determine whether particle shape affects the angle of repose of Earth materials and proposes the following experimental procedure:

1. Obtain about 500 cubic centimeters of five types of Earth materials.
2. Use a hand lens to observe and record the shape of the particles in each Earth material.
3. Pour a sample of material 1 into a pan, forming a small pile.
4. Place a ruler along the edge of the pile as shown in the diagram below.

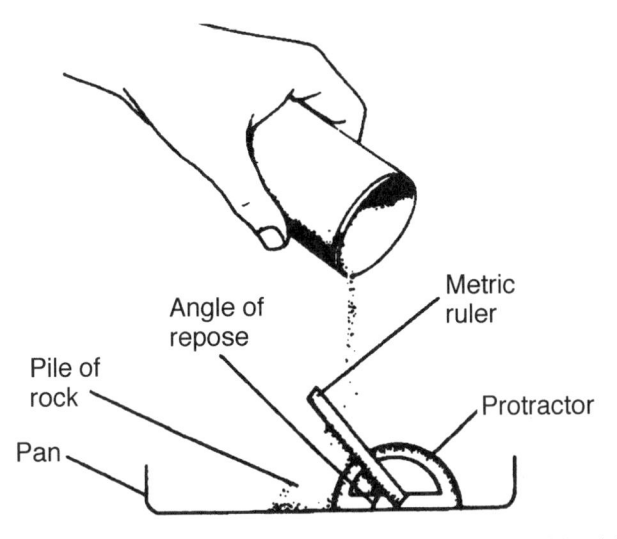

Source: *Macmillan Earth Science: Teacher's Resource Book*, Eric Danielson and Edward J. Denecke, Jr., Macmillan, 1986

5. Using a protractor, measure and record the angle of repose for the material.
6. Repeat steps 3 through 5 for each Earth material.

67. State a hypothesis about the relationship between particle shape and angle of repose. [1]

68. In one or more complete sentences, explain why this experimental procedure is flawed. [2]

69. Identify three variables in this experiment, and state how each could be controlled. [1] for each variable [1] for method of controlling the variable

70. Write an experimental procedure to test the following hypothesis: If the particle size of an Earth material decreases, the angle of repose of the material will also decrease. [4]

CHAPTER 21
DEPOSITION

KEY IDEAS Patterns of deposition result from a loss of energy within the transporting system and are influenced by the size, shape, and density of the transported particles. Sediment deposits may be sorted or unsorted.

Sediments of inorganic and organic origin often accumulate in depositional environments. Sedimentary rocks form when sediments are compacted and/or cemented after burial or as the result of chemical precipitation from seawater.

Landscapes are regions in which the physical features of Earth's surface are related by their structure, the processes that formed them, and the way in which they developed. They can be classified by characteristics such as relief, stream patterns, and soil associations. Landscapes with similar characteristics can be grouped into distinctive landscape regions.

KEY OBJECTIVES
Upon completion of this chapter, you will be able to:
- Explain how the characteristics of sediments and the media carrying them affect their deposition.
- Describe the processes by which the various agents of erosion deposit the sediments they are transporting.
- Describe some of the common landforms resulting from erosion and deposition by mass movements, moving water, glaciers, and wind.
- Identify and describe landscape regions.

DEPOSITION

At some point, all agents of erosion begin to deposit sediment. *Deposition* is the process by which transported sediment is dropped in a new place. The same agents that erode sediment also deposit it. Several factors influence when deposition will begin.

Speed of Medium

The *medium* is the substance that carries sediment. In streams the medium is water, in winds it is air, and in glaciers it is ice. Generally, the slower a medium is moving, the less its carrying power. If the carrying power drops below the level needed to transport a particle, that particle will drop to the ground. Less carrying power results in deposition.

You can see this for yourself. Imagine you are flying a kite and the wind slows down. The slower the wind blows, the less carrying power it has. The kite will begin to drop. If the wind stops blowing, the kite will settle to the ground. Deposition occurs in much the same way. As the medium slows, sediment begins to settle out.

Figure 21.1 shows the sediment that water carries at different speeds. Notice that water carries more sediment sizes when it is moving fast. As water slows down, some sediment is deposited. Large, dense particles will settle first, and light, small particles last. The result is a separation of different particles into layers.

The separation of particles during deposition is called *sorting*. Figure 21.2 shows the sorting that occurs when a stream flows into a larger body of water, such as a lake or an ocean. As the stream flows into the larger body, it slows down. Coarse gravel and pebbles settle first. Farther from the shore, the movement of the water is slower and sand settles out. Still farther from shore, in very slow moving water, the finest particles of silt and clay settle. Sorted layers of particles result.

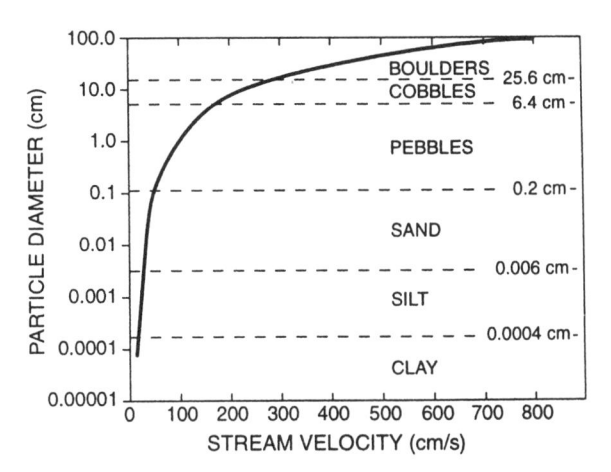

Figure 21.1 Relationship of Transported Particle Size to Water Velocity. This generalized graph shows the water velocity needed to maintain, but not start, movement. Variations occur because of differences in particle density and shape. Source: The State Education Department, *Earth Science Reference Tables*, 2001 ed. (Albany, New York; The University of the State of New York).

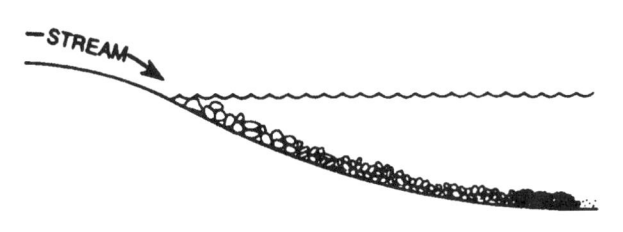

Figure 21.2 Horizontal Sorting of Particles Carried by a Stream. As a stream decreases in velocity, the heavier particles settle first; the lighter particles are carried farther downstream before settling.

Deposition

Characteristics of Particles

Particles may vary in size, shape, and density. Each of these characteristics influences the rate at which settling occurs. To learn how particle size influences settling rate, look at Figure 21.3a. Notice that, when all the rock particles are deposited, sorting has occurred. How can that be? The answer is that some particles must settle before others. The particles on the bottom settled fastest, and those on top slowest. Since the larger particles are on the bottom, they settled fastest. The smaller ones, on top, settled slowest.

In general, the larger the particle, the faster it will settle. Very small particles, such as fine clay, take a long time to settle. The muddy appearance of many rivers and lakes is due, in part, to particles that have not yet settled. To see how particle shape affects settling rate, look at Figure 21.3b. Notice that, when all the particles are deposited in water, the bottom layer consists of round particles. These particles must have settled first. Also notice in the figure a top layer of flat particles. These particles must have settled last.

In general, round particles settle faster than flat ones. To understand how particle shape affects settling rate, imagine dropping a crumbled paper and a flat paper to the ground at the same time. The crumbled paper, like a rounded particle, lands faster. Friction between the large surface area of the flat paper and the air slows that paper's fall. A flat shape in rock particles has the same effect in water. Friction between the flat particle and the water will decrease the rate at which the particle settles.

Finally, to see how the density of a particle affects settling, look at Figure 21.3c, Notice, once again, that sorting has taken place. The densest particles form the bottom layer; the least dense particles, the top layer, Thus it can be concluded that the densest particles settled fastest.

To summarize, several factors affect the deposition of sediment. The speed of the medium determines when particles will be deposited. The size, shape, and density of a particle determine the rate at which it will settle. Differences in the settling rates of sediments result in sorting during deposition.

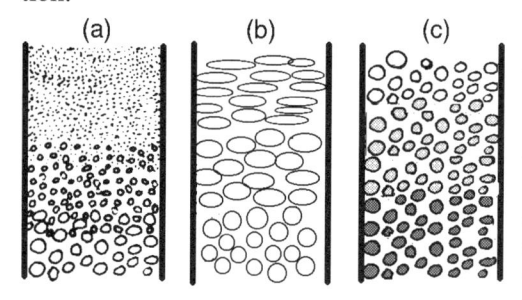

(a) All other factors being the same, large particles settle faster than smaller ones.
(b) All other factors being the same, particles with rounded shapes settle faster than flat particles.
(c) All other factors being the same, particles composed of denser materials settle faster than those composed of less dense materials.

Figure 21.3 Settling of Different Particles in a Column. When sediments settle in a quiet medium, they form a series of layers. The fastest settling particles reach the bottom first and form the lowest layer. They are followed by slower settling particles, which form the next layer; the slowest settling particles form the top layer.

Bedding of Sediment

As a result of sorting during deposition, sediment forms layers consisting of different types of particles. Sediment deposited in a quiet body of water will usually form layers similar to those shown in Figure 21.4. Each layer of sediment is called a *bed*, and each bed usually represents a period of deposition. Beds of sediment are commonly found in sedimentary rock.

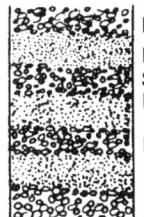

Hurricane, September
No rainfall, August
Summer storm, July
Periodic gentle rain, May-June
Heavy rains, May
No rainfall, April
Heavy rainfall and spring thaw, March

Figure 21.4 Bedding of Sediment. Rapid deposition results in *graded bedding*, a vertical sorting. Each sequence represents a depositional event.

DEPOSITION BY MASS MOVEMENTS

Deposition by Gravity

During mass movements, sediment is pulled downhill by gravity and is deposited, when it stops moving, at the base of a slope. Sediment deposited by gravity does not settle through a medium, as does sediment deposited by wind or water. Thus, little, if any, sorting takes place. Most deposits resulting from mass movements are unsorted and do not show any distinct layering. In unsorted deposits, all types of particles are arranged without any order, as shown in Figure 21.5.

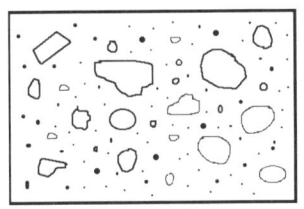

Figure 21.5 Unsorted Deposit. Unsorted sediment contains rock fragments of all sizes randomly mixed together.

Deposits by Rapid Mass Movements

Rapid mass movements deposit sediment. As deposition takes place, landforms develop. These landforms are evidence that deposition has occurred.

Rockfalls
During rockfalls and rockslides, fragments slide downhill until they reach the base of a slope, forming a talus. A *talus* is a pile of broken rock that builds up at the base of a cliff or steep slope. Boulders may roll down a talus and onto the flat ground at the base of the slope. As the talus weathers, soil may form on the surface. Then, if the talus is not being constantly covered by new rock fragments, plants may take root and grow. See Figure 21.6a.

Deposition

Mudflow
During a mudflow, mud moves downhill until it reaches the base of a slope. The mud then spreads out in a thin, wide sheet. See Figure 21.6b. Huge boulders carried by the mudflow may be left to stand isolated on the gently rolling land.

Deposits by Slow Mass Movements
Slow mass movements also deposit sediment. Earthflows, slump, and soil creep all produce sediment deposition from which landforms may develop.

Earthflow
An earthflow deposits a thick pile of sediment. See Figure 21.6c. Earthflow deposits are generally thicker than deposits produced by mudflows. The difference between the two can be demonstrated by spilling a jar of thick honey and a jar of thin tomato sauce next to each other. The thick honey, like an earthflow, produces a small, thick deposit. The thinner sauce, like a mudflow, produces a thin spread. Compare Figures 21.6b and 21.6c.

Slump
Slump forms a deposit that looks like an apron spread on the ground. Notice in Figure 21.6d the steplike appearance of the deposit. The steps are the tops of blocks of material that broke loose from the cliff. Very often, earthflows begin as slump.

Soil Creep
As stated earlier, soil creep is a very slow process. Material deposited by soil creep is often hard to recognize because it is usually covered with plants, and looks like part of the hill on which it formed. Only a person trained to recognize deposition will perceive the series of ripples at the base of the gentle slope as evidence of deposition by soil creep. See Figure 21.6e.

Effects of Mass Movements
Deposition by mass movement along roadsides can cause serious problems. Poorly planned roads

(a) Talus

(b) Mudflow

(c) Earthflow

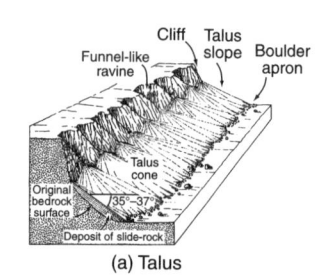

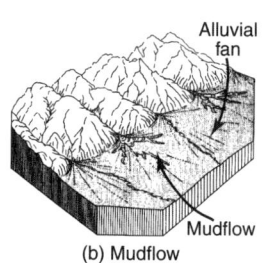

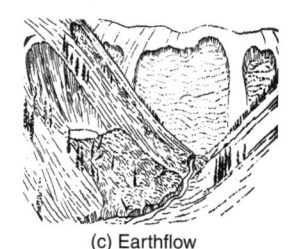

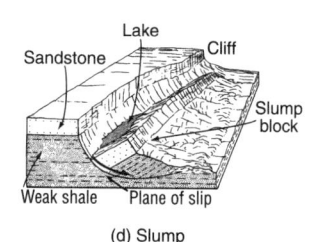

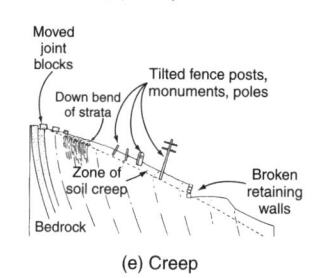

(d) Slump

(e) Creep

Figure 21.6 Deposits Formed by Mass Movements. Source: *The Earth Sciences*, Arthur N. Strahler, Harper & Row, 1971.

may be blocked by deposits. In fact, material deposited by slides, flows, and slump may make a road impassable. Interstate 40, near Rockwood, Tennessee, is one such road; one stretch has been blocked more than 20 times by major slides. Clearing such roads costs taxpayers millions of dollars each year. In addition, these poorly planned roads pose dangers to motorists.

DEPOSITION BY MOVING WATER

More than 70 percent of Earth's surface is covered by water. Therefore, it is not surprising that most deposition occurs in water. In this section, you will learn how sediment carried by moving water is deposited and how deposition changes the shape of the land.

Deposition by Streams

Rivers and other streams carry large amounts of sediment. Water from the Colorado River is used to irrigate crops; canals carry the water to the fields. In just one of these canals, up to 5,000 metric tons of sediment has to be removed from the water every day.

Streams deposit sediment when they slow down or decrease in volume. Such slowdown or decrease in volume occurs when the slope decreases, runoff and seepage from groundwater decrease, or the stream enters a standing body of water, such as a lake or an ocean. Deposition by streams results in layers that are sorted according to particle size, shape, and density. In addition, these deposited sediments tend to be smoothed and rounded as a result of abrasion while being carried in the stream.

Many landforms develop as a result of stream deposition. These landforms are found in almost all parts of the United States.

Oxbow Lakes

In a preceding chapter you learned how meanders are formed. As meanders become more curved, their ends move closer together. Eventually, the stream erodes its way through the land, separating the meanders (see Figure 21.7). Most of the water then flows through this steeper, straighter shortcut. The slower moving water in the old meander deposits its load. When deposition blocks off the entrance to the old meander and it becomes separated from the stream, an oxbow lake is formed.

Floodplains and Levees

Because of its increased speed and volume, a stream in flood carries much more than the normal amount of sediment. Also, when a stream is in flood, it overflows its banks. Outside the channel, the water is slower moving and shallower. As soon as the flood water leaves the channel, it begins to deposit its sediment load over the surrounding land to form a *floodplain*. The larger

Deposition

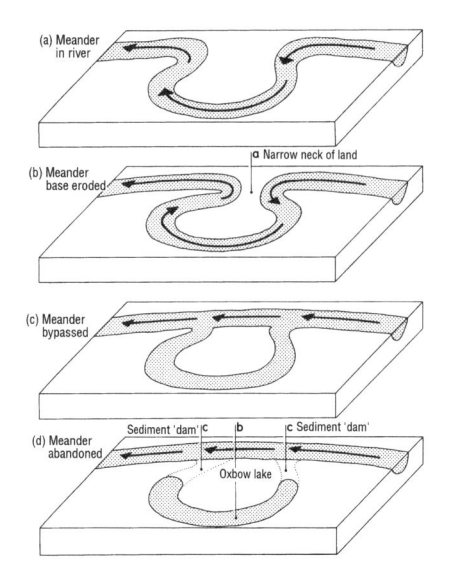

Figure 21.7 Formation of an Oxbow Lake. Water flows faster along the outside curve of a meander than it flows along the inside curve because it has farther to travel in order to make the turn. The faster moving water erodes the bank on the outside curve until just a narrow neck of land (a) separates the loops of the meander. When the meanders erode through the narrow neck, most of the stream flow bypasses the meander. Because less water flows through the meander, water velocity decreases and deposition builds sediment "dams" (c) that separate the meander from the stream, forming an oxbow lake (b). Source: *Earth Science on File.* © Facts on File, Inc., 1988.

particles settle out first, building a *levee*, that is, a low, thick, ridgelike deposit along the banks of both sides of the stream (see Figure 21.8).

Floodplains make good farmland; sediment deposited after each flood renews the soil. For this reason, farmers have been able to raise crops on the floodplains of the Nile River for over 4,000 years!

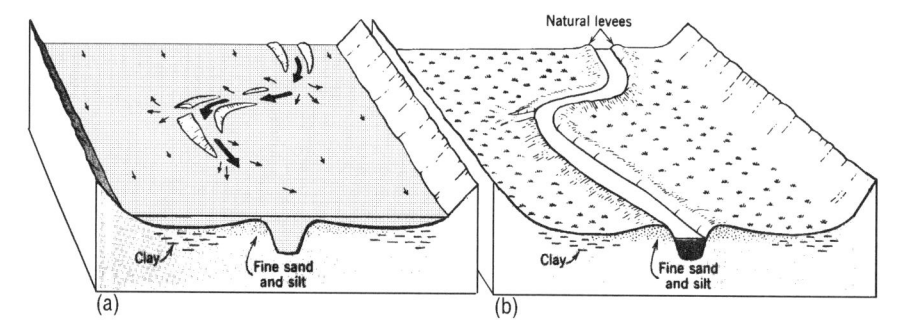

Figure 21.8 Formation of a Floodplain (a) and Levee (b). During a major flood, the stream's floodplain becomes a lake. Water in the main channel flows rapidly. However, water escaping the channel immediately slows down and deposits fine sand and silt, forming a natural levee. As the water continues to spread out and slow down, clay is deposited on the surrounding lower land. When the flood ends, the levees remain as low ridges along the sides of the channel and the floodplain is covered with wet clay and swampy areas.
Source: *Physical Geology*, 2nd Ed., Richard Foster Flint and Brian J. Skinner, John Wiley, 1977.

Deltas and Alluvial Fans

When a stream flows into a quiet body of water such as an inland sea, ocean, or lake, it stops moving. The point at which this occurs is called the *mouth* of the stream. Here most of the stream's sediment is deposited. The resulting

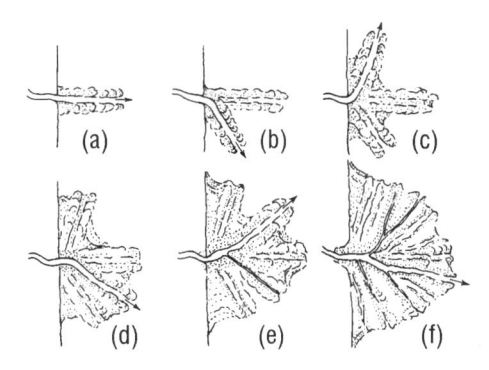

Figure 21.9 Stages (a)–(f) in the Formation of a Delta. A delta is formed as a stream flows into a standing body of water such as a lake or an ocean. Source: *The Earth Sciences*, by Arthur N. Strahler, Harper & Row, 1971.

landform is a *delta*, as seen in Figure 21.9. A delta is a large, flat, fan-shaped pile of sediment at the mouth of a stream. It is composed of sand, silt, and clay and is shaped like a triangle. It gets its name from the Greek letter delta, which is written as Δ.

When streams flowing down steep mountain valleys reach flat, open land, they slow down and deposit much of their coarser load. The resulting landform is an *alluvial fan*, that is, a large mound of coarse sediment deposited by a stream onto open, flat land (see Figure 21.10).

Although an alluvial fan is similar in appearance to a delta, there are some differences. An alluvial fan has steeper sides and is made of coarser sediment than a delta.

Deposition by Oceans

Oceans also deposit sediment. Whenever waves and currents slow down, they deposit their loads. Deposition by waves and currents builds new landforms that change the shape of shorelines.

Beaches

A *beach* is a deposit of sediment along the shoreline of a body of water. Beaches are formed as waves wash up against the land and slow down. Beaches constitute an almost continuous fringe around the continents and most islands. Although most people think of a beach as a strip of fine, white sand along the seashore, beaches may consist of any earth material from fine clay to boulders.

Figure 21.10 Formation of an Alluvial Fan. An alluvial fan forms where a stream flows from a steep gradient onto a gentle gradient. Source: *The Earth Sciences*, Arthur N. Strahler, Harper & Row, 1971.

The nature of beach sediments depends on the source material and the weathering processes at work in the area. Beaches are shaped by waves and are changed daily and monthly by tides. Figure 21.11 shows the general features of a beach. *Berms* are flat areas formed by wave action. The breaking wave loosens sediments upslope and drags them back toward the water as the wave recedes.

Deposition

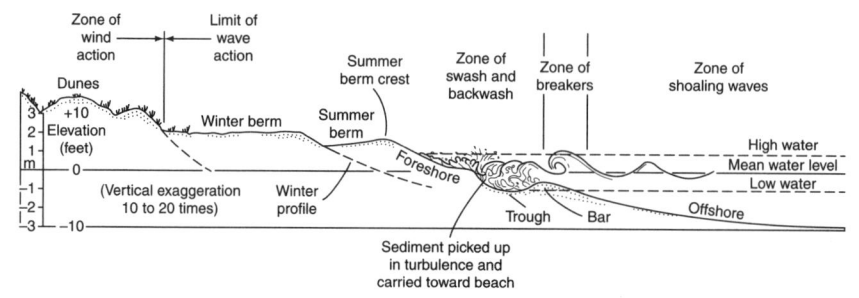

Figure 21.11 Beach Features. The usual direction in which breaking waves carry sand is toward the beach. This sand, together with sediments eroded from coastal cliffs, builds a flat area, or a berm. Source: *The Earth Sciences*, Arthur N. Strahler, Harper & Row, 1971.

Figure 21.12 shows the changing shoreline conditions in winter and summer. During stormy winter months, waves tend to be high and to have short periods. The sand picked up by one wave does not have a chance to settle to the bottom before the next wave sets the sand in motion again. As a result, the sand stays in suspension and is carried away from the beach by the backrush of water until it reaches deeper, calmer water and can settle, forming an offshore sand bar. See Figure 21.12a.

In summer months, the weather is calmer and the waves that reach the shore tend to be swells of long periods. The breaking waves pick up sand in shallow water and carry it up onto the beach. Most of the sand is then carried back toward the sea by the backrush of water along the beach. There is time during the backrush, however, for some particles to settle before the next wave again carries sand toward the beach. The overall result is a shifting of sand toward the beach. See Figure 21.12b.

Sandbars and Spits

A *sandbar* is a long, narrow pile of sand deposited in open water. Sandbars may form where receding waves wash beach sediments into deeper, quieter waters and also where the shoreline curves away from a longshore current. As the current curves away from the shoreline, it flows into deeper, quieter water and deposits its sediments. Most bars formed in this way are attached at one end to the mainland and are called *spits*. *Tombolos* are sandbars that connect an island to the mainland. See Figure 21.13.

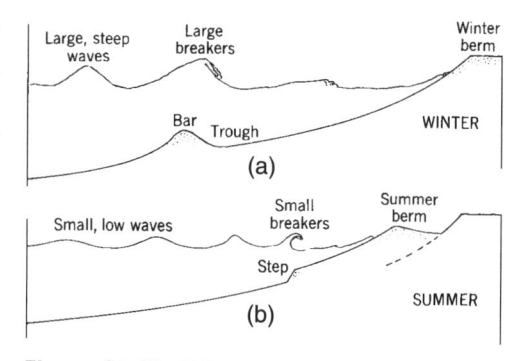

Figure 21.12 Winter (a) and Summer (b) Beach Profiles. High waves of short period generally occur in winter months and drag sediments offshore, forming a sandbar. In summer, waves generally have long periods and move sediment onto the beach. Source: *The Earth Sciences*, Arthur N. Strahler, Harper & Row, 1971.

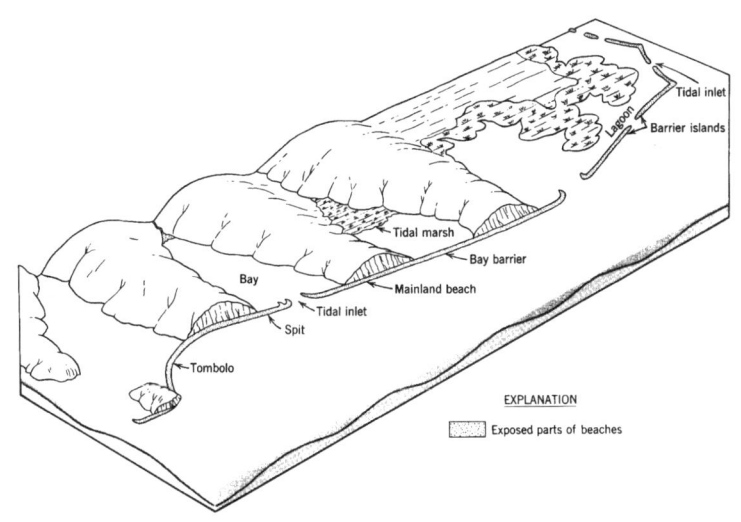

Figure 21.13 Some Coastal Landforms Created by Deposition of Sediments by Ocean Waves and Currents. Source: *Physical Geology*, 2nd Ed., Richard Foster Flint and Brian J. Skinner, John Wiley, 1977.

Barrier Bars and Lagoons

Eventually, waves deposit enough sand on a bar so that it is above water. Large waves continue to add material, and in time the bar is built up well above sea level. Such large bars are known as *barrier bars*. Barrier bars may completely block off a bay, forming a *tidal marsh*, or may parallel the coast, broken here and there by tidal inlets into elongated *barrier islands*. The calm, protected bodies of water between barrier bars and the mainland are called *lagoons*. See Figure 21.13.

Problems Associated with Deposition by Moving Water

Deposition by moving water can create problems in harbors, which must be deep and have clear openings to the sea to enable large ships to pass through. Occasionally, sandbars and spits block a harbor's entrance. In addition, sediment from streams and rivers deposited in harbors can create areas of shallow water. Ships are then in danger of running aground. To prevent this from happening, sediment must be removed from harbors by dredging.

DEPOSITION BY WIND

Wind is moving air. When a wind slows or stops moving, the particles it is carrying settle to the ground and are deposited. Wind deposits generally contain fine sediments and result in the formation of certain surface features.

Deposition

Loess

Loess is a thick, unlayered deposit of very fine, buff-colored sediment deposited by wind. Loess deposits may vary in thickness from a few centimeters to several hundred meters. Loess is found throughout the world. There are large deposits in the central United States, Europe, and parts of China. Loess is a source of excellent soil because of its ability to hold large amounts of water.

Dunes

Dunes are mounds of sand deposited by wind. They are often found in barren, desert regions and can also occur on large, sandy beaches. A dune often forms when windblown sand meets an obstacle such as a rock or a bush.

Figure 21.14 shows how a dune develops. On the windward side of the dune, that is, the side facing the wind, sand that strikes an obstacle falls to the ground. In time, the sand forms a mound that slopes gently up toward the tip of the obstacle. Wind then pushes sand grains up this slope to the crest, or top, of the dune.

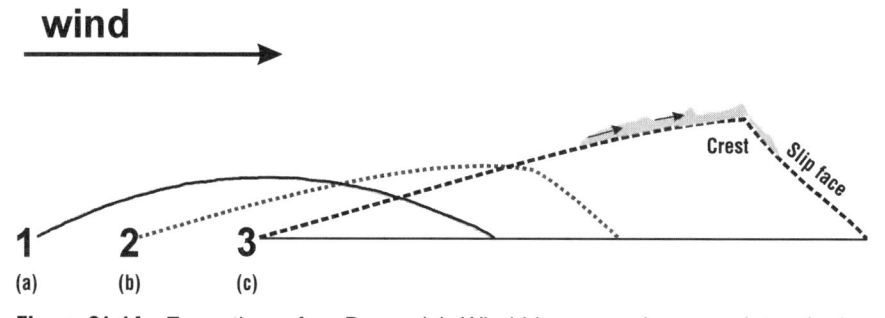

Figure 21.14 Formation of a Dune. (a) Wind-blown sand accumulates in low mounds. (b) Steady winds carry sand up the side of the dune, and it rolls over the crest onto the sheltered side. (c) As sand is picked up on the gently sloping side facing the wind and deposited on the steep slip face, the entire dune moves.

On the leeward side of the dune, that is, the side sheltered from the wind, winds are very slow. As a result, the sand grains rolling over the crest fall to the ground because the winds can no longer support them. A deposit of sand forms, creating a steep-sided mound. As sand continues to be deposited, this mound may become unstable. Indeed, sand may slip or slide downhill. For this reason, the steep, leeward side of a dune is called the *slip face*.

As winds blow sand from the windward side of a dune and deposit it on the leeward side, the entire dune moves in the direction in which the wind is blowing. In this way, dunes travel along a desert. Some desert dunes may move from 10 to 25 meters in a year. Moving dunes have been known to engulf forests, farmlands, and buildings.

Sand dunes have a variety of heights, shapes, and patterns. Dunes may reach heights of 30–100 meters, depending on wind speed and the amount of sand in the area. Three common shapes and patterns of sand dunes are shown in Figure 21.15. The type of dune that forms in an area depends on the prevailing wind patterns and the amount of sand.

In the United States, the dunes of our coastal areas are an important line of defense for inland properties from storm waters. From 1936 to 1940, the National Park Service built nearly 1,000 kilometers of fencing along the coast of North Carolina. The fencing formed an obstacle to the windblown sand, trapping it and thereby forming dunes. These dunes were then planted with grass, shrubs, and over two million trees! Today, the resulting barrier of dunes protects inland regions from serious damage by hurricanes.

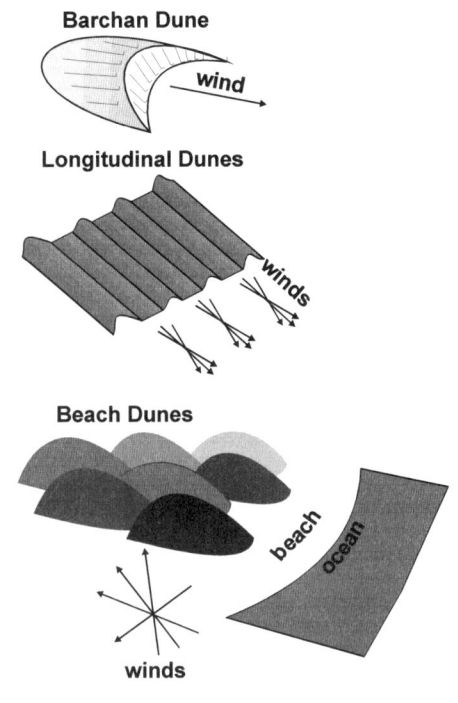

Figure 21.15 Common Shapes and Patterns of Sand Dunes.

DEPOSITION BY GLACIERS

The way in which glaciers transport and deposit sediments is very different from that of any other agent of erosion. To fully understand glacial deposition, let us first consider how glaciers transport their sediment load.

Glacial Transport

Glaciers transport materials suspended in the ice, along the sides and bottom of the ice, on the surface of the ice, and in the area directly alongside the ice. Although it is tempting to picture a glacier as a huge block of ice pushing a pile of debris ahead of it, most glaciers transport very little material by "bulldozing" ahead of the ice. Most of the material carried by a glacier is suspended in the ice.

One of the characteristics that affect the ability of any medium to carry materials in suspension is the viscosity, or resistance to flow, of the medium. The higher the viscosity, the greater is the medium's ability to hold materials in suspension. Ice is millions of times more viscous than liquid water and can carry just about anything in suspension. A boulder that couldn't be

Deposition

budged by the fastest moving stream of water, no less remain suspended in the water, can easily be carried along suspended in the ice of a glacier. See Figure 21.16.

Debris can become suspended in glacial ice in several ways. Many glaciers originate in snowfields high in mountainous regions. Landslides and rockfalls from the steep slopes that surround the glacier deposit debris on top of the snow. When more snow accumulates, the debris is buried and eventually encased in the ice that forms. Also, rock protrusions around which a glacier flows shed fragments that become incorporated into the ice. Where two glaciers merge, the debris along the edges of the glaciers becomes incorporated into the body of the larger glacier that forms. Material also becomes incorporated into the bottom and sides of a glacier as the ice melts and refreezes around loose rock fragments. The 10- to 100-centimeter layer along the base of a glacier is filled with suspended debris.

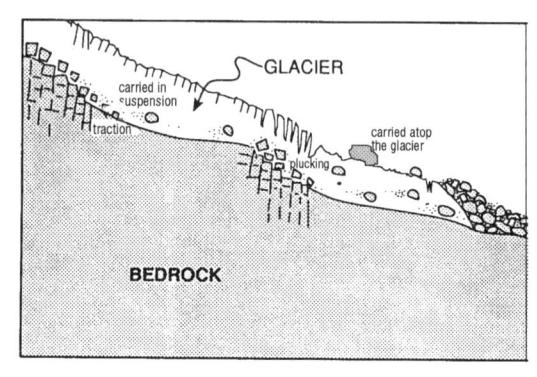

Figure 21.16 Transport of Sediment by a Glacier. A glacier carries sediments on its surface, suspended in the ice, and dragged along its sides and bottom.

Material along the bottom of the glacier is carried along by traction. In *traction*, the particles are not encased in the ice but are dragged along, slipping, sliding, and rolling beneath the moving ice. Material along the sides of the glacier is also moved by traction, and rock fragments may slide or roll off the ice onto ground next to the glacier. Meltwater-soaked debris with the consistency of wet concrete also flows off the sides of the glacier.

Most material carried on the top of a glacier was originally suspended in the ice but has been exposed as melting occurred. Thick debris, however, shields the ice beneath it from melting and results in high spots on the ice surface. Blocks of rock that fall onto a glacier from surrounding slopes can be carried for great distances, conveyor-belt style, atop the glacier.

Types of Glacial Deposits

Part of a glacier's load is carried frozen within the ice or scattered on its surface. The rest is pushed up in piles around the edges of the glacier and carried along as the glacier moves over Earth's surface. When the glacier reaches a warm region, or the climate changes and becomes warm enough, the ice begins to melt. As the glacier melts, the sediments it was carrying are deposited. These deposits may be sorted or unsorted, and each type is formed differently.

Unsorted Deposits

At the melting edge of the glacier, sediments within the ice, as well as those carried on top of it, are released. Particles of all shapes and sizes fall to the ground in a confused jumble, forming an unsorted deposit. In addition, as the ice melts back, piles of sediment that were pushed up around the edges of the glacier are left behind. These sediments, too, are unsorted. They were tumbled about and mixed as the ice scraped them from the ground and pushed them along. See Figure 21.17a.

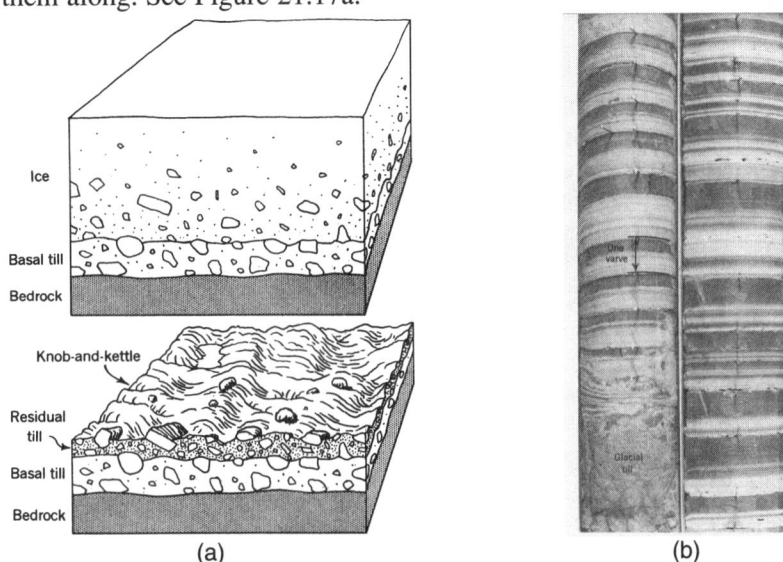

Figure 21.17 Glacial Deposits. (a) Glacial till. As glacial ice melts, the sediments are released and drop to the ground in a confused jumble, forming an unsorted deposit. (b) Sorted varved clays from a glacial meltwater lake deposit. Meltwater streams may sort glacial sediments; or, as in this photograph, the sediments may settle out in sorted layers at the bottom of lakes fed by meltwater. Source: *The Earth Sciences*, Arthur N. Strahler, Harper & Row, 1971.

Sorted Deposits

Some glacial sediment is deposited into streams that originated from a melting glacier. This material is sorted by the running water and then deposited in layers. Fine sediments carried into lakes by meltwater lakes may settle out in sorted layers on the lake bed. See Figure 21.17b.

How Glaciers Deposit Sediments

Some deposition takes place beneath the ice. The base of a warm glacier is continuously melting because of heat radiating from the ground and heat created by friction with the ground. As particles in the base of the glacier are freed by melting, some are rolled or dragged along until they become lodged in the ground. Refreezing of large particles, which jut out more than finer ones, also leaves behind deposits enriched in fine particles.

Materials on top of a glacier are deposited in one of two ways: either meltwater and debris slides carry them off the glacier, or they are set down on the ground as the ice beneath them melts. When meltwater deposits these materials, they are well sorted and somewhat rounded. Materials that slide or fall off the glacier, however, are jumbled masses of large and small angular fragments. By the time a glacier reaches its end, it has already undergone much melting. The remaining material on its surface is dropped off the end of the glacier as the ice beneath it melts, forming a pile called an *end moraine*.

Meltwater carries debris from the glacier beyond the end of the glacier and deposits it over the land. In summer, glacial meltwater streams are in flood and carry huge amounts of sediment, including abundant fine particles. These fine sediments are formed by the crushing pressure and abrasion between particles as they are carried along inside the glacier. If numerous streams spread out over the land and deposit glacial material in a wide sheet, the result is an *outwash plain*. If the meltwater streams are confined in a valley, they may deposit their load on the floor of the valley.

Meltwater often fills closed depressions, forming glacial lakes. If the lake forms along the edge of the glacier, debris can fall directly into the lake. If the water undercuts the ice face, icebergs fall into the lake and drift across its surface. As an iceberg melts, it rains *ice-rafted* debris onto the lake bed. During the winter, the lake freezes over and deposition ceases except for the settling out of the very finest particles of clay. In the spring when the ice thaws, meltwater carries in coarser material, which is deposited atop the clay. The result is a pair of layers known as a *varve*. Each varve represents one seasonal cycle—a year. See Figure 12.17b.

Landforms Resulting from Glacial Deposition

Glacial drift is any material deposited as a result of glacial activity, including material deposited by meltwater and debris slides. *Glacial till* is material deposited directly from ice. The main difference between the two is sorting; drift may be sorted because of the action of meltwater, whereas till is unsorted.

When glacial drift and till are deposited, they form a number of distinctive landscape features, as shown in Figure 21.18.

Moraines

Moraines are deposits of till. A moraine that forms a thin, widespread layer of till is called a *ground moraine*. Ground moraines form gently rolling hills and valleys. Piles of till deposited around the side of a glacier are called *lateral moraines*. Deposits of till at the front of the glacier are *terminal moraines*. Terminal and lateral moraines form long, parallel ridges. The ridges mark the boundaries where a glacier once existed. See Figure 21.19.

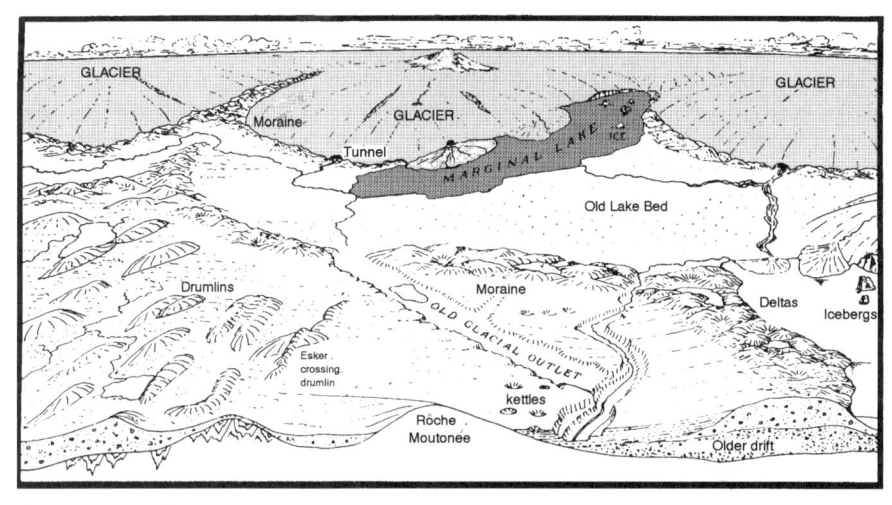

Figure 21.18 Landforms Resulting from Glacial Deposits. Source: Educational Leaflet #28, The New York State Museum, Albany, New York.

Drumlins

Drumlins are long, low mounds of till that have a rounded, teardrop shape, as seen in Figure 21.18. By looking at the figure, you can see the direction in which the glacier moved. The rounded part of the drumlin points in the direction from which the glacier advanced. Drumlins are thought to be molded by the ice of a glacier as it slides over previously deposited piles of sediment.

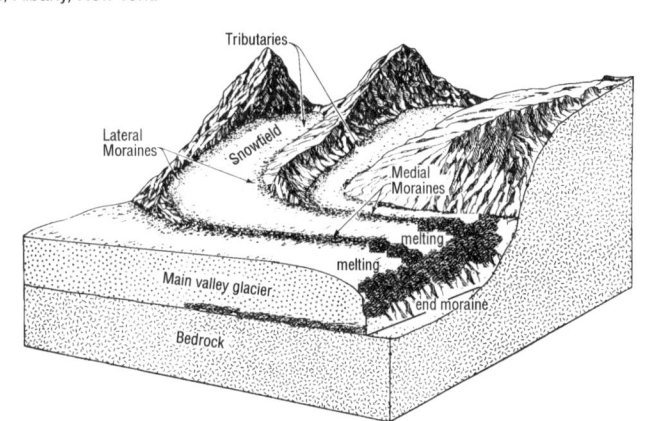

Figure 21.19 How Moraines Form. Lateral moraines consist of materials carried along the sides of a glacier. When two glaciers merge, their lateral moraines are welded together between the glaciers, forming a medial moraine. End moraines form where debris melts out of the ice at the end of the glacier. Source: *Earth*, 4th Ed., Frank Press and Raymond Seiver, W.H. Freeman, 1986.

Erratics

Erratics are large, isolated boulders deposited by a glacier. Many erratics are more than 3 meters in diameter and weigh thousands of metric tons. Such boulders are much too large to be carried by wind or water, but glaciers often transport them far from their sources. When the ice melts, the large boulder comes to rest on a surface having a composition different from its own.

Deposition

Outwash Plains

At the leading edge of a glacier, some melting is almost always taking place. Streams of glacial meltwater carry sediments out beyond the glacier and deposit them in sorted layers. Over time, a broad plain is built up in front of the glacier by deposition of sediments from glacial meltwater. Such a landform is called an *outwash plain* because the sediments it is made of were "washed out" beyond the glacier by meltwater streams.

Kettles and Kettle Lakes

A *kettle* is a pit found in a glacial deposit. There are several steps in the formation of a kettle. First, a chunk of glacial ice breaks loose. Next, sediment from the glacier covers the chunk of ice. After a while, the ice melts, the overlying sediment sinks, and the kettle is formed. See Figure 21.20. A kettle may later fill with glacial meltwater, rainwater, or groundwater, forming a *kettle lake*.

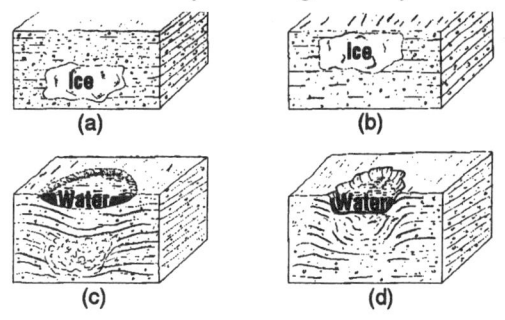

Figure 21.20 Formation of a Kettle. Kettles form when buried blocks of ice melt and the overlying sediments collapse, forming a hole. The melting of deeply buried blocks of ice forms rounder and shallower kettles. In (a) and (a'), the depression formed by melting a deeply buried ice block is seen, whereas (b) and (b') shows the depression formed by melting a shallowly buried ice block. Source: *Living Ice: Understanding Glaciers and Glaciation*, Robert P. Sharp, Cambridge University Press, 1988.

Glacial Lakes

In addition to kettle lakes, glaciers may form lakes in two other ways. A glacier may gouge out a large depression in Earth's surface that may fill with water. Alternatively, a *glacial lake* may form when ice or moraines dam the course of a stream; the flow of water backs up behind the moraine, creating a lake. Long, narrow glacial lakes, such as the ones found in central New York, are known as *finger lakes*.

Ice-dammed lakes can cause disastrous floods when the ice dam melts and releases the lake water. There is much evidence that catastrophic floods caused by ice damming occurred repeatedly during the last ice age, 10,000–20,000 years ago.

Glacial lakes impact society in many ways. They form some of this country's most spectacular scenery. In addition to being beautiful, glacial lakes have both recreational and commercial uses. New York's Finger Lakes are used extensively for fishing, boating, and swimming. The Great Lakes, a system of five huge glacial lakes, are the largest repositories of fresh water on Earth. Native Americans and early explorers used them as key transportation routes. Later, they served as major shipping routes for getting the bountiful harvests of our country's heartland to major population centers back east. Large cities, such as Chicago and Detroit, grew from port cities along this route. The building of the Erie Canal created the link needed for a water route from the Great Lakes to New York City, and catapulted New York City from a small coastal port to the major international center of trade it is today.

LANDSCAPE DEVELOPMENT

Throughout this unit, you have learned about geologic processes that change the surface. Crustal movements cause the surface to be uplifted in some places and to subside in others. Volcanic activity injects new rock beneath the surface in some places while bringing new rock to the surface in others. Weathering and erosion reshape the land by wearing away rock exposed at the surface and transporting sediments downslope. When these sediments are deposited, characteristic features are formed. Together, all of these geologic processes create the *topography* of the land, that is, the landforms that shape the surface of Earth. ***Landforms*** are physical features of Earth's surface that have characteristic shapes and are produced by natural processes. Landforms include both major forms, such as mountains, plateaus, and plains, and minor forms, such as hills, valleys, drumlins, dunes, and many others.

Characteristics of Landscapes

A ***landscape*** is region in which the landforms are related by their structures, the processes that formed them, and the ways in which they developed. Landscapes are the product of the interaction of geological processes, climate, and human activities on the land over a long period of time. Landscapes can be classified by characteristics that can be readily observed and measured, such as relief, stream patterns, and soil associations.

Relief

Relief refers to the physical shape or general unevenness of a part of Earth's surface, such as variations in slope or elevation. It is expressed in terms of the vertical difference in elevation between the highest and lowest points in a given region. A region showing great variations in elevation has "high relief," and a region showing little variation in elevation has "low relief." Regions with high relief typically have steep slopes; those with low relief, gentle slopes. Most landscapes can be classified as mountains, plateaus, or plains based upon their reliefs.

   Mountains are parts of Earth's crust that project at least 300 meters above the surrounding land. A mountain generally has steep sides, a restricted summit area, and considerable bare-rock surface. Most mountains are formed by crustal movements. Two common types are fold mountains and fault-block

Deposition

mountains. Fold mountains form when layers of rock are folded by compression, wrinkling Earth's crust. Fault-block mountains form when tension forces cause blocks of crust to move along faults. On one side of the fault, rocks rise; on the other side, rocks subside. See Figure 21.21.

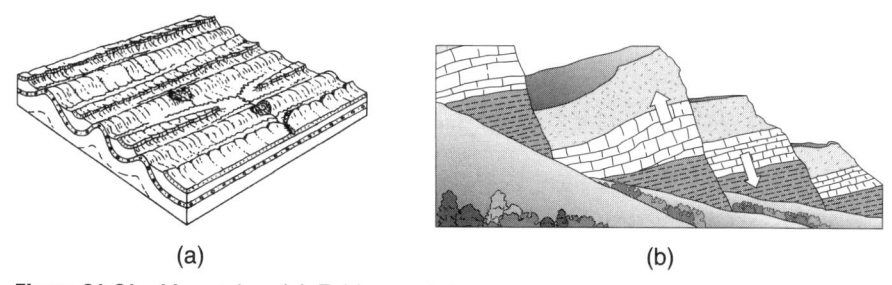

(a)   (b)

Figure 21.21 Mountains. (a) Fold mountains. Source: *The Earth Sciences*, Arthur N. Strahler, Harper & Row, 1971. (b) Fault-block mountains. Source: *Macmillan Earth Science*, Eric Danielson and Edward J. Denecke, Jr., Macmillan, 1989.

Plateaus are large areas of flat land at high elevations. A plateau generally has an underlying structure made of horizontal layers of rock that were gently uplifted. Plateaus are often found next to mountain ranges and were probably raised by the same forces that formed the mountains but were not faulted or folded as greatly as the rocks of the mountains. Although made of horizontal rock layers, the surface of a plateau is often not level. Streams and other agents of erosion can cut deep valleys and canyons into the surface of the plateau.

Plains are large areas of flat land at low elevations. Plains are usually formed by the deposition of sediments in horizontal layers at or below sea level. Thus, the underlying structure of a plain consists of horizontal layers of rock. Many plains have undergone uplifting, which has raised their surfaces above sea level, but have not been uplifted as much as plateaus.

Stream Patterns

Stream patterns refer to the patterns formed by the system of streams in an area as they flow down slopes and join with other streams. A stream that flows into and joins another stream is called a *tributary*. The area from which precipitation drains into a stream or system of streams is called a ***drainage basin***. See Figure 21.22. Higher areas, called *divides*, separate drainage basins. Streams from neighboring drainage basins may join into larger streams, which, in turn, may join with others to form even larger streams. Together, all of the streams in an area and their tributaries form a ***stream drainage system***. Drainage systems often show distinct patterns that reflect the geology of the region. Figure 21.23 shows some of the more common stream drainage patterns that develop on different geological structures.

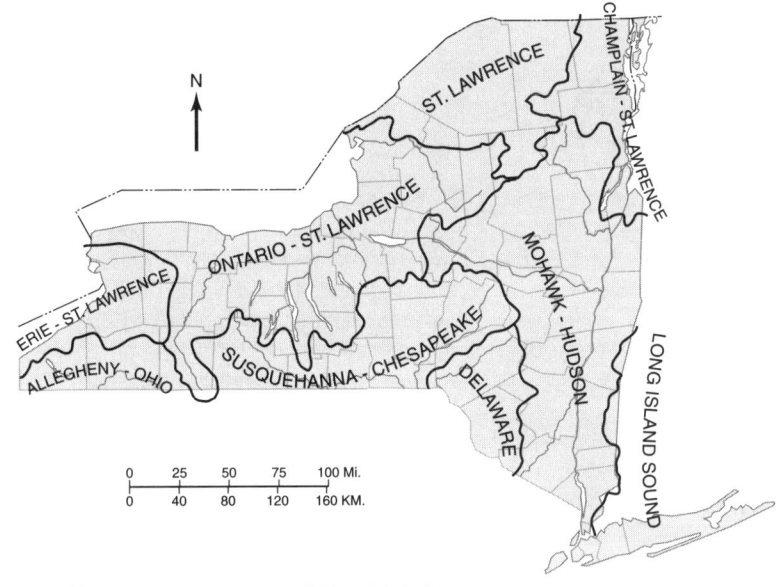

Figure 21.22 Drainage Basins of New York State.

Stream Pattern	Geologic Structures
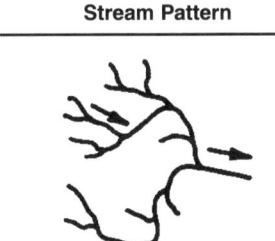 (a) Dendritic	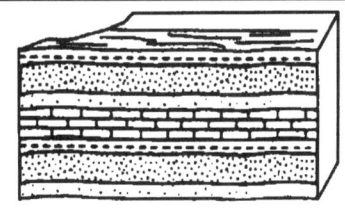 Form where underlying rocks have no pronounced differences in erosion rate, so stream flow is equal in all directions.
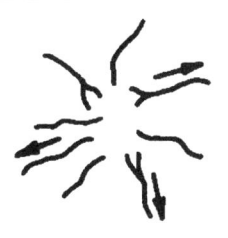 (b) Radial	 Form on rounded surfaces, such as conical volcanoes or dome mountains, where stream flow away from a central high point.

Deposition

Stream Pattern	Geologic Structures
 (c) Annular	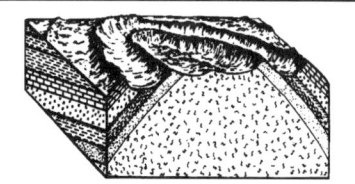 Form around dome mountains with tilted rock layers on their flanks. Differential erosion causes concentric ridges to form. Streams flow in low areas between these ridges.
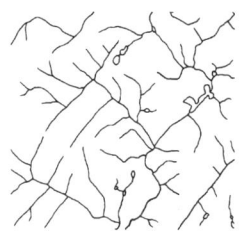 (d) Rectangular	 Form is folded, faulted, or tilted strata with long strips of rock of unequal resistance to erosion. Major streams flow along belts of weak rock, making many right-angle turns.
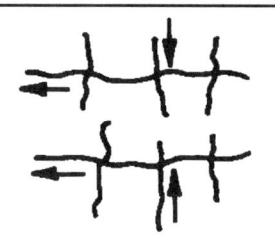 (e) Trellis	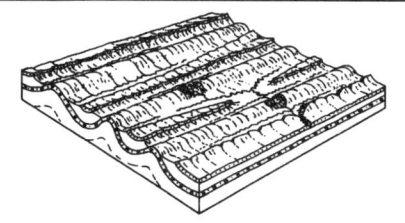 Form in tilted, folded, or faulted strata with long strips of rock of unequal resistance to erosion. Major streams run through long valleys, following belts of weak rock between parallel ridges of stronger rock. Tributary streams enter major streams at right-angles.

Figure 21.23 Stream Drainage Patterns. (a) Dendritic (b) Radial (c) Annular (d) Rectangular (e) Trellis.

Soil Associations

Soil associations are groups of two or more soils, occurring together in a characteristic pattern in a given geographical area. Soils are grouped according to characteristics such as composition, organic content, particle size and shape, porosity and permeability, and maturity (i.e., the degree to which the soil has developed horizons). Soil associations form patterns that reflect the soil parent material, underlying bedrock, slope, vegetative cover, and plant litter in a region.

Climate and Landscapes

Climate refers to the characteristic weather of a region over an extended period of time, particularly the region's precipitation and temperature patterns. Temperature and precipitation affect the development of a landscape by influencing its relief, stream patterns, and soil associations.

Climate influences relief by controlling such factors as the type and rate of weathering, the amount and duration of the runoff that removes and transports weathered materials, and the amount, type, and distribution of plant cover.

The development of a landscape's relief depends on the balance between weathering and the removal of weathered materials by erosion. Temperature and precipitation strongly influence the intensity of weathering. Hot, arid regions and cold, arid regions undergo minimal weathering. Chemical weathering dominates in hot, humid regions, while physical weathering (mainly frost action during colder months) dominates in cold, humid regions. Thus, bedrock tends to be broken down at a faster rate in humid climates than in dry regions. Since running water is the predominant agent of erosion on Earth, precipitation is essential to the removal and transport of weathered materials. Thus, precipitation influences the rate at which the relief of the landscape changes. Over the long term, the more abundant the precipitation, the more rapidly the landscape's relief changes.

Climate also influences relief by controlling the amount, type, and distribution of plant cover. In humid climates, abundant rainfall promotes a protective cover of vegetation that protects the soil from rapid erosion by runoff and produces gently rounded slopes. In arid climates, with little protective vegetation, erosion by rapid runoff leads to steep slopes and exposed bedrock.

You may be surprised to learn that, while arid regions may erode more slowly than humid regions in the long term, the lack of plant cover in arid regions allows the exposed soil to be very rapidly eroded during the few periods of heavy rainfall that do occur. Thus, some of the most rapid erosion rates occur in deserts after a storm! However, over the long term, the brief periods of intense erosion in arid regions are outweighed by the prolonged periods of more moderate erosion in humid regions.

Climate influences stream patterns by controlling such factors as the amount and duration of precipitation, differences between the windward and leeward sides of mountains, and the relationship between precipitation and evaporation. For example, in arid climates, smaller streams tend to remain dry for most of the year, while humid climates tend to have more permanent streams. Thus, humid climates often have a denser network of tributary streams than arid climates. The leeward sides of mountains tend to have less annual precipitation than the windward sides. Thus, the network of streams may vary from one side of the mountains to another.

Soil associations are strongly influenced by climate. Arid climates tend to have soils that are thinner and contain less organic matter than those in humid climates. Since physical weathering is dominant, soil particles in arid climates tend to be larger and more angular than those in humid climates The more abundant rainfall in humid climates results in greater infiltration, which, in turn,

Deposition

speeds up the rate at which soil horizons form. Thus, humid climates develop mature soils much more rapidly than arid climates. Moisture and higher temperatures also favor chemical weathering processes, which also speed the breakdown of weathered materials and the development of soil horizons.

Landscape Regions

Landscapes with similar relief, stream patterns, and soil associations can be grouped into distinctive *landscape regions*. Landscape regions typically have distinctive physical features that set them apart from one another. On a large scale, such as the continental United States, regions with similar landscape regions are grouped into *physiographic provinces* that are classified mainly as mountains, plateaus, and plains. See Figure 21.24.

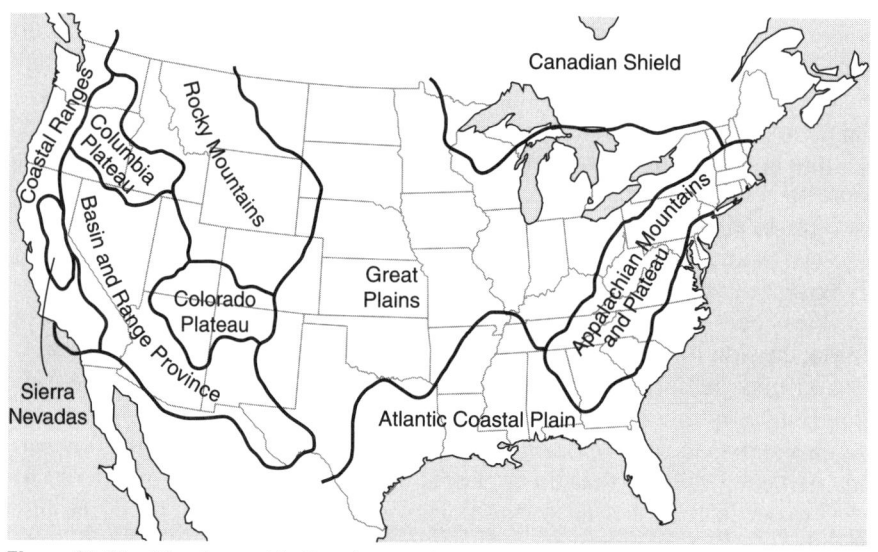

Figure 21.24 Physiographic Provinces of the United States. Source: *UPCO's Review of Earth Science*, Robert B. Sigda, United Publishing Co., 1995.

New York State is divided into a variety of smaller landscape regions based on relief, stream patterns, and the type and structure of the underlying bedrock. See Figure 21.25. The accompanying table gives a brief description of each of the landscape regions in New York State.

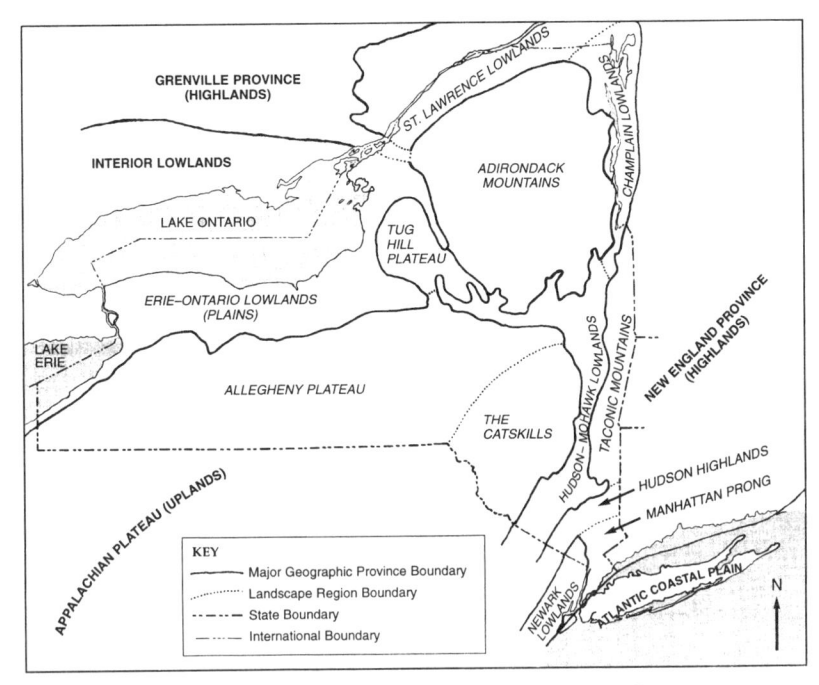

Figure 21.25 Generalized Landscape Regions of New York State. Source: The State Education Department, *Earth Science Reference Tables*, 2001 ed. (Albany, New York; The University of the State of New York).

LANDSCAPE REGIONS OF NEW YORK STATE

Adirondack Mountains. A circular region that is part of the Grenville Province, a large belt of deeply eroded metamorphic bedrock of Proterozoic age. Erosion has stripped away most of the overlying rock, and it is criss-crossed with faults. The resistant rocks of the Adirondacks eroded more slowly than the rocks of the St. Lawrence and Champlain lowlands around them. The Adirondacks have long, straight valleys that formed along faults, gently curved ridges of resistant rock, and a radial drainage pattern.

St. Lawrence and Champlain lowlands. The St. Lawrence and Champlain lowlands are part of the Interior Lowlands that extend west through the Great Plains. They consist of layers of sedimentary rock that once covered the Adirondacks but were later eroded away. They now form the low-lying plains occupied by the St. Lawrence River and Lake Champlain.

Taconic Mountains. The Taconic Mountains are a region of intensely folded and faulted metamorphic rocks that were thrust into that area from the east by the collision between a volcanic island arc and the North American continent about 450 million years ago. Ridges and valleys generally run north-south because softer rocks have worn away to produce valleys, while more resistant rocks form the ridges.

Alleghany Plateau. The Alleghany Plateau forms the northern end of the Appalachian Plateau to the southwest. It consists of flat-lying layers of sedimentary rock that were deposited in a warm, shallow sea that covered much of New York State during the Late Silurian and Devonian. Later the entire area was uplifted to form the plateau. The plateau surface rises to the east until it becomes the Catskills. Streams and their tributaries have cut the surface of the plateau into hilly uplands.

Deposition

The Catskills. The Catskills are the deeply eroded remains of a huge, apronlike delta formed during the Devonian from sediments that eroded from the ancient Acadian Mountains along the shore of the warm, shallow sea that covered much of New York State. Since then, streams have cut deeply into the sedimentary layers, forming steep-sided valleys that separate rounded "mountains."

Erie-Ontario Lowlands. The Erie-Ontario Lowlands are low, flat areas to the north and west of the Alleghany Plateau, and separated from it by escarpments, or cliffs. Though they are plains, the Erie-Ontario Lowlands are covered with hills of unsorted glacial till and sorted meltwater deposits.

Tug Hill Plateau. East of Lake Ontario, the land rises steadily to form the Tug Hill Plateau, a relatively flat, rocky area separated from the Adirondack Mountains by the Black River Valley. The plateau stands in the path of winter storms that pick up moisture from Lake Ontario. Annual snowfall averages 20 feet, so the plateau is one of the snowiest regions east of the Rocky Mountains.

Hudson Highlands. The Hudson Highlands consist of metamorphic rocks that were originally deposited as sedimentary and volcanic rocks 1.3 billion years ago. They were then metamorphosed into gneiss and marble during a collision of continents 1.1 billion years ago that thrust up an ancient mountain chain—the Grenville Mountains. The Grenville Mountains have since been eroded flat, exposing their roots—the rocks of the Hudson Highlands.

Hudson-Mohawk Lowlands. This region covers most of the Hudson River Valley and the Mohawk River Valley. The bedrock consists of relatively soft sedimentary rocks that are easily eroded. The surrounding highlands are made of more resistant rocks. These lowlands provide the nation's only natural, navigable waterway through the Appalachian Mountains and serve as an important transportation route between the Atlantic and the Great Lakes.

Newark Lowlands. The Newark Lowlands form a gently rolling surface broken by ridges, between the Hudson Highlands and the Manhattan Prong. Formed on layers of igneous and sedimentary rock dating from the Age of Dinosaurs (Triassic-Jurassic), the surface of these lowlands is broken by ridges of igneous rock that is more resistant to erosion. The Palisades Sill intruded into the rocks of the Newark Lowlands 195 million years ago.

Manhattan Prong. The Manhattan Prong consists of metamorphic rock that was folded and faulted by the same collision that formed the Taconic Mountains and was later eroded by glacial ice into a landscape of rolling hills and valleys. Resistant gneiss, schist, and quartzite form the hills, and less resistant marble underlies the valleys.

Atlantic Coastal Plain. The Atlantic Coastal Plain is a flat, low-lying area that slopes toward the Atlantic Ocean. In New York State, it consists of Staten Island and Long Island, both of which are important residential areas. Long Island is composed mainly of glacial sediments. Two terminal moraines form its hilly northern edge while a broad, flat glacial outwash plain stretches from the terminal moraine to the Atlantic Ocean. Long offshore barrier islands with broad beaches make Long Island a popular recreational area.

MULTIPLE-CHOICE QUESTIONS

In each case, write the number of the word or expression that best answers the question or completes the statement.

1. Why do the particles carried by a river settle to the bottom as the river enters the ocean?
   (1) The density of the ocean water is greater than the density of the river water.
   (2) The kinetic energy of the particles increases as the particles enter the ocean.
   (3) The velocity of the river water decreases as it enters the ocean.
   (4) The large particles have a greater surface area than the small particles.

2. A mixture of sand, pebbles, clay, and silt, of uniform shape and density, is dropped from a boat into a calm lake Which material most likely will reach the bottom of the lake first?
   (1) sand   (3) clay
   (2) pebbles   (4) silt

3. Which rock particles will remain suspended in water for the longest time?
   (1) pebbles   (3) silt
   (2) sand   (4) clay

Base your answers to questions 4 and 5 on your knowledge of earth science and on the diagrams below, which represent four columns of sedimentary deposits found at different locations.

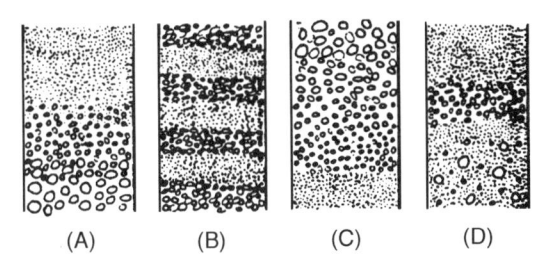

4. Which column best represents deposits from a stream that had a continuously decreasing water velocity?
   (1) A   (3) C
   (2) B   (4) D

5. Which column most clearly indicates that there were cyclic changes in the environment?
   (1) A   (3) C
   (2) B   (4) D

Deposition

Note that question 6 has only three choices.
6. As sediment deposition increases, the pressure on lower layers will
   (1) decrease
   (2) increase
   (3) remain the same

7. The rate at which particles are deposited by a stream is *least* affected by the
   (1) size and shape of the particles
   (2) velocity of the stream
   (3) stream's elevation above sea level
   (4) density of the particles

8. The diagram below represents a top view of a river emptying into an ocean bay. A-B is a reference line along the bottom of the bay. Which characteristics would most likely decrease along the line from A to B?

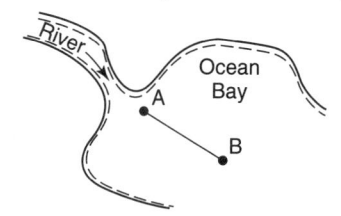

   (1) the amount of salt in solution
   (2) the size of the sediments
   (3) the density of the water
   (4) the depth of the water

9. A river carrying pebbles, sand, silt, and clay flows into the ocean. The sediments are sorted by size as they are deposited at different distances from shore. Which sedimentary rock will most likely form from the sediment deposited farthest from shore? [Refer to the *Earth Science Reference Tables*.]
   (1) conglomerate
   (2) sandstone
   (3) siltstone
   (4) shale

10. Which deposit is formed along rivers?
    (1) coastal plain
    (2) floodplain
    (3) lake plain
    (4) till plain

11. How will the formation of a cutoff change a meandering stream?
    (1) It will reduce the stream's velocity.
    (2) It will straighten the stream's course.
    (3) It will widen the stream's course.
    (4) It will increase the stream's rate of deposition.

12. The particles in a sand dune deposit are small and very well sorted and have surface pits that give them a frosted appearance. This deposit most likely was transported by
    (1) ocean currents
    (2) glacial ice
    (3) gravity
    (4) wind

13. The map below represents a winding stream. At which location is stream deposition most likely to be greater than stream erosion?

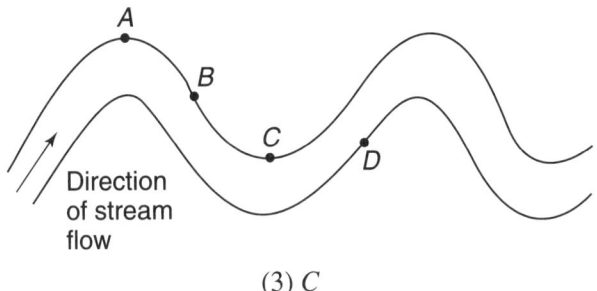

    (1) A
    (2) B
    (3) C
    (4) D

14. Granite pebbles are found on the surface in a certain area where only sandstone bedrock is exposed. Which is the most likely explanation for the presence of these pebbles?
    (1) The granite pebbles were transported to the area from a different region.
    (2) Some of the sandstone has been changed into granite.
    (3) The granite pebbles were formed by weathering of exposed sandstone bedrock
    (4) Groundwater tends to form granite pebbles within layers of sandstone.

15. The diagram below shows a cross section of soil from New York State containing pebbles, sand, and clay.

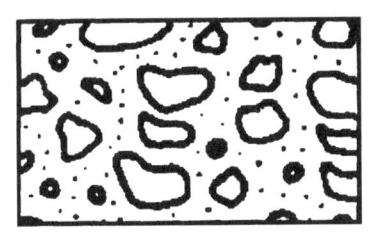

    The soil was most likely deposited by
    (1) an ocean current
    (2) the wind
    (3) a river
    (4) a glacier

Deposition

16. Which agent was responsible for the deposition of fine-grained, angular particles found in a sandstone ?
    (1) wave action
    (2) ground water
    (3) running water
    (4) wind

17. Which landscape features are primarily the result of wind erosion and deposition?
    (1) U-shaped valleys containing unsorted layers of sediment
    (2) V-shaped valleys containing well-sorted layers of sediment
    (3) terraces of gravel containing unsorted layers of sediment
    (4) cross-bedded sand deposits containing finely sorted layers of sediment

Base your answers to questions 18 through 22 on the *Earth Science Reference Tables*, the information and diagrams below, and your knowledge of earth science.

A mixture of colloids, clay, silt, sand, pebbles, and cobbles is put into stream I at point A. The water velocity at point A is 400 centimeters per second. A similar mixture of particles is put into stream II at point A. The water velocity in stream II at point A is 80 centimeters per second.

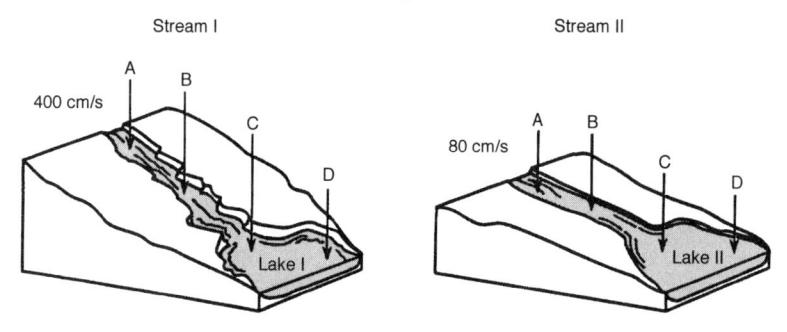

18. Which statement best describes what happens when the particles are placed in the streams?
    (1) Stream I will move all particles that are added at point A.
    (2) Stream II will move all particles that are added at point A.
    (3) Stream I cannot move sand.
    (4) Stream II cannot move sand.

19. Which statement is the most accurate description of conditions in both streams?
    (1) The greatest deposition occurs at point B.
    (2) Particles are carried in suspension and by bouncing along the bottom.
    (3) The particles will have a greater velocity than the water in the stream.
    (4) The velocity of the stream is the same at point B as at point C.

20. If a sudden rainstorm occurs at both streams above point A, the erosion rate will
    (1) increase for stream I, but not for stream II
    (2) increase for stream II, but not for stream I
    (3) increase for both streams
    (4) not change for either stream

21. What will most likely occur when the transported sediment reaches lake II?
    (1) Clay particles will settle first.
    (2) The largest particles will be carried farthest into the lake.
    (3) The sediment will become more angular because of abrasion.
    (4) The particles will be deposited in sorted layers.

Note that question 22 has only three choices.
22. In lake I, as the stream water moves from point C to point D, its velocity
    (1) decreases
    (2) increases
    (3) remains the same

Base your answers to questions 23 through 27 on the information in the table below, which shows the distribution of particles in the load of a certain stream before the load reaches the stream's mouth.

Particles	Miles from Stream's Source		
	100	300	500
Gravel	25%	5%	0%
Coarse sand	30%	18%	1%
Medium sand	30%	50%	5%
Fine sand	14%	20%	25%
Silt	0.8%	5%	50%
Clay	0.2%	2%	19%

23. The decrease in the sizes of the particles carried by the stream as it flows away from its source is most probably caused by the
    (1) decreasing velocity of the stream
    (2) decreasing amounts of material carried in solution
    (3) increasing volume of the stream
    (4) increasing gradient of the stream

24. How is most of the load probably carried at a point 500 miles from the stream's source?
    (1) in suspension
    (2) in solution
    (3) by action of wind and convection currents
    (4) by rolling action of the particles along the bottom of the streambed

Deposition

25. If the sediments were produced from granite, which mineral would most likely form the coarsest material in the load at 500 miles from the source?
   (1) biotite
   (2) amphibole
   (3) feldspar
   (4) quartz

26. What kind of rock will be formed from the sediment deposited in the area between 300 and 500 miles from the source?
   (1) conglomerate
   (2) limestone
   (3) sandstone
   (4) shale

27. The graph below represents the distribution of which type of particle?

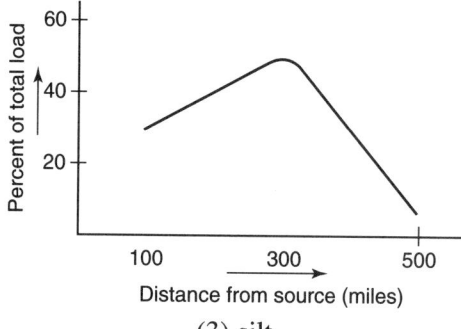

   (1) gravel
   (2) medium sand
   (3) silt
   (4) clay

28. Many elongated hills, each having a long axis with a mostly north-south direction, are found scattered across New York State. These hills contain unsorted soils, pebbles, and boulders. Which process most likely formed these hills?
   (1) stream deposition
   (2) wind deposition
   (3) wave deposition
   (4) glacial deposition

29. Which rock material was most likely transported to its present location by a glacier?
   (1) rounded sand grains found in a river delta
   (2) rounded grains found in a sand dune
   (3) residual soil found on a flat plain
   (4) unsorted loose gravel found in hills

30. Often, sediments that are carried down submarine canyons by turbidity currents settle out over the ocean floor. Which cross section represents the pattern of deposition of these sediments?

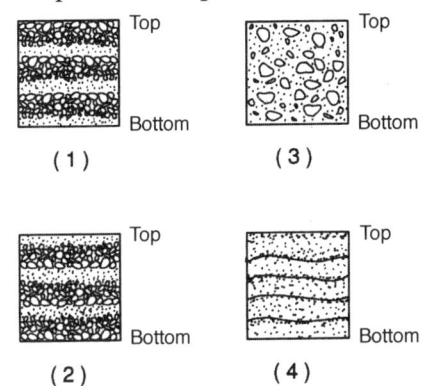

(1) (2) (3) (4)

Base your answers to questions 31 through 34 on the *Earth Science Reference Tables*, the diagram below, and your knowledge of earth science. The diagram represents a glacier moving out of a mountain valley. The water from the melting glacier is flowing into a lake. Letters *A* through *F* identify points within the erosional/depositional system.

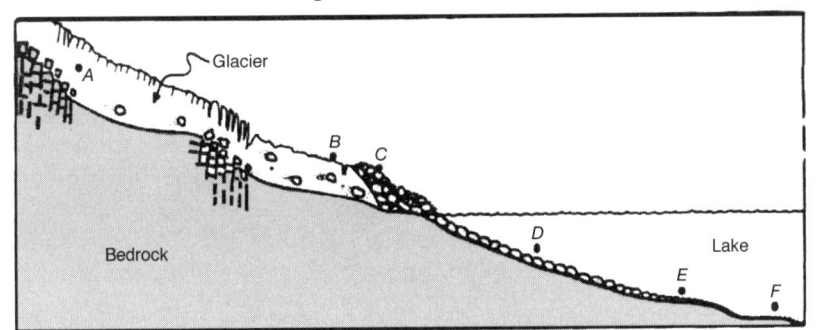

31. Deposits of unsorted sediments would probably be found at location
 (1) *E* (3) *C*
 (2) *F* (4) *D*

32. An interface between erosion and deposition by the glacier is most likely located between points
 (1) *A* and *B* (3) *C* and *D*
 (2) *B* and *C* (4) *D* and *E*

33. Colloidal-sized sediment particles carried by water are most probably being deposited at point
 (1) *F* (3) *C*
 (2) *B* (4) *D*

Deposition

34. Which graph best represents the speed of a sediment particle as it moves from point D to point F?

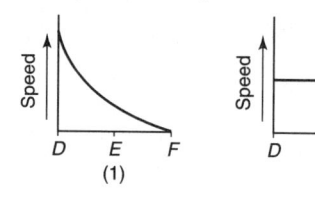

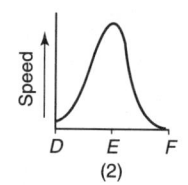

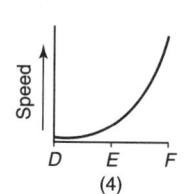

(1) [graph: speed decreasing from D to F]
(2) [graph: speed peaks at E]
(3) [graph: speed constant from D to F]
(4) [graph: speed increasing from D to F]

35. A glacial deposit would most likely consist of
    (1) particles in a wide range of sizes
    (2) particles the size of pebbles and larger
    (3) sediments in flat, horizontal layers
    (4) sediments found only in the bottoms of stream valleys

36. Which statement presents the best evidence that a boulder-sized rock is an erratic?
    (1) The boulder has a rounded shape.
    (2) The boulder is larger than surrounding rocks.
    (3) The boulder differs in composition from the underlying bedrock.
    (4) The boulder is located near potholes.

37. Lateral moraines of a valley (alpine) glacier are composed principally of rock debris
    (1) deposited by glacial meltwater
    (2) freed from the enclosing ice by frost heaving
    (3) derived from the weathering of the valley walls
    (4) deposited by streams flowing along the ice margins

Base your answers to questions 38 through 42 on the diagrams below. Diagram I represents a section of the northeastern United States and Canada. Five different source regions, A through E, are shown along with the positions of glacial deposits containing boulders that originated from each location. Diagram II represents the appearance of the surface of a typical boulder from any of the deposit locations.

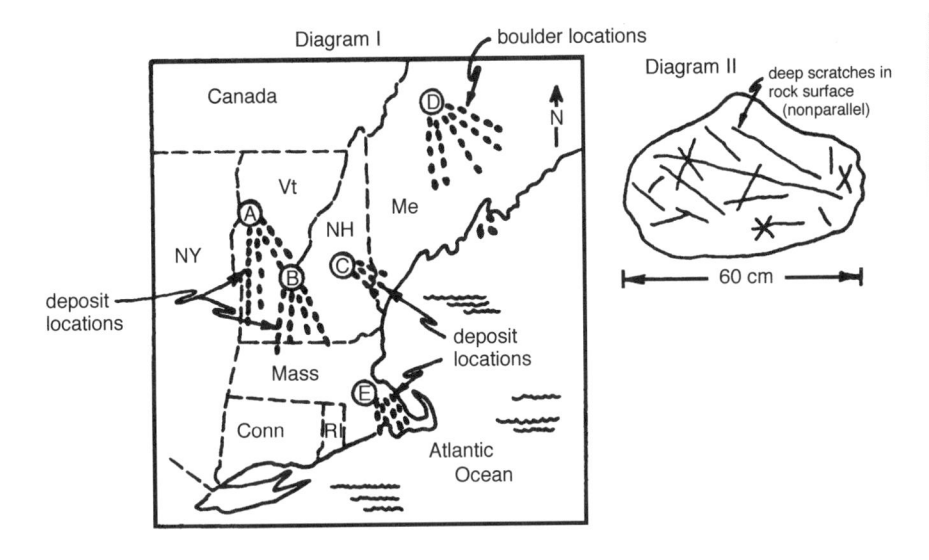

38. The force that caused the deposits to be distributed in the pattern shown in Diagram I most likely came from which general direction?
(1) northwest
(2) northeast
(3) southwest
(4) southeast

39. Which characteristic do all of the deposits most likely have in common?
(1) They have the same chemical composition.
(2) They were eroded from source region *A*.
(3) They are composed of unsorted sediments.
(4) They are found at the ends of large rivers.

40. According to the *Earth Science Reference Tables*, during which geological epoch were the deposits most likely transported to the locations shown in Diagram I?
(1) Mesozoic
(2) Pleistocene
(3) Jurassic
(4) Paleocene

41. The scratches in the boulder shown in Diagram II were most likely caused by the
(1) internal arrangement of the minerals in the boulder
(2) splitting of a large boulder into two smaller boulders
(3) erosion of the boulder by running water
(4) movement of the boulder over bedrock

42. About how many times larger is the actual boulder than the model shown in Diagram II?
(1) 5
(2) 14
(3) 30
(4) 60

Deposition

Base your answers to questions 43 through 46 on the *Earth Science Reference Tables*, the maps and cross section below, and your knowledge of earth science. The map shows the stages of growth of a stream delta. Point *X* represents a location in the stream channel. The side view of the stream shows rock particles transported in the stream at a point close to its source.

Maps: Stages in Growth of a Stream Delta

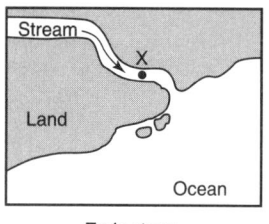

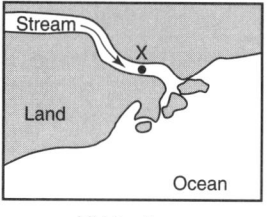

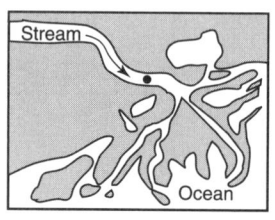

Early stage   Middle stage   Late stage

Side View of a Stream

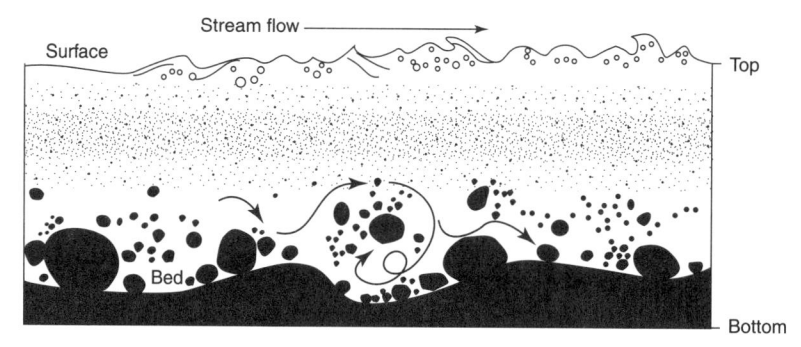

43. The rock materials transported in the stream are most likely transported
    (1) in solution, only
    (2) in suspension, only
    (3) in solution and suspension, only
    (4) in solution, in suspension, and by rolling

44. The velocity of the stream at location *X* is controlled primarily by the
    (1) amount of sediment carried at location *X*
    (2) distance from location *X* to the stream source
    (3) slope of the stream at location *X*
    (4) temperature of the stream at location *X*

45. A decrease in the velocity of the stream at location *X* will usually cause an increase in
    (1) downcutting by the stream
    (2) deposition within the stream channel
    (3) the size of the particles carried by the stream
    (4) the amount of material carried by the stream

46. Which most likely describes the sediments in the delta?
   (1) jagged fragments deposited in elongated hills
   (2) unsorted, mixed sizes deposited in scattered piles
   (3) large cobbles deposited in parallel lines
   (4) round grains deposited in layers

47. The diagrams below represent two different landscape regions, *A* and *B*.

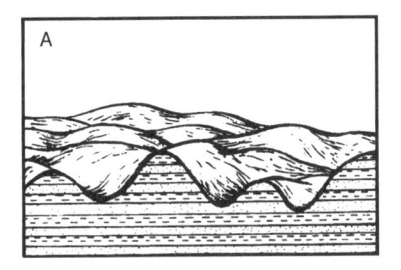

   Which factor was probably most important in causing the two landscapes to develop differently?
   (1) type of bedrock
   (2) amount of folding
   (3) time
   (4) climate

48. Which diagram shows the stream drainage pattern that usually develops on the surface of horizontal rock?

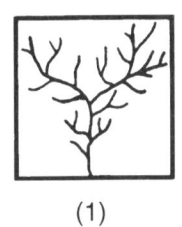

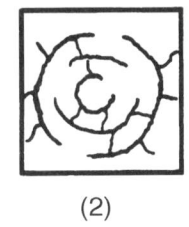

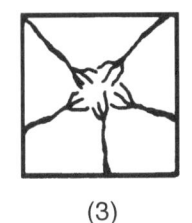

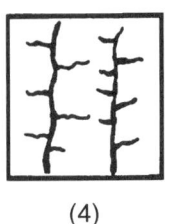

   (1)   (2)   (3)   (4)

49. Which landscape probably resulted from the erosion of folded rock layers?

Deposition

50. Which cross section best represents the general bedrock structure of New York's Alleghany Plateau?

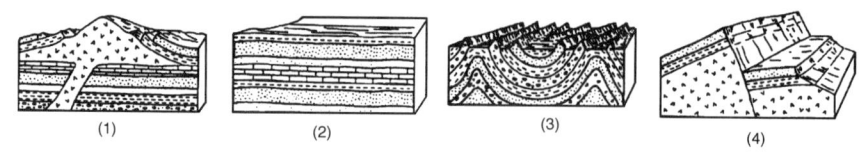

　　(1)　　　　(2)　　　　(3)　　　　(4)

51. According to the *Earth Science Reference Tables*, much of the surface bedrock of the Adirondack Mountains consists of
    (1) slate and dolostone
    (2) gneiss and quartzite
    (3) limestone and sandstone
    (4) conglomerate and red shale

Base your answers to questions 52 through 56 on the *Earth Science Reference Tables*, the map below, and your knowledge of earth science. The map shows the major landscape regions of New York State.

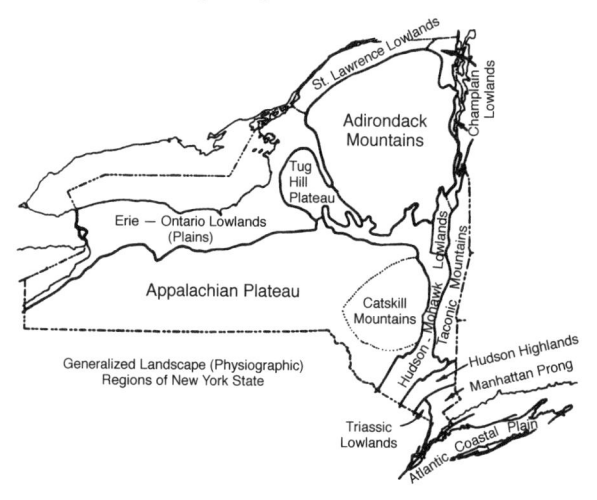

Generalized Landscape (Physiographic) Regions of New York State

52. Which major landscape region covers the greatest surface area in New York State?
    (1) Hudson Highlands
    (2) Appalachian Plateau
    (3) St. Lawrence Lowlands
    (4) Atlantic Coastal Plain

53. Recent measurements of elevation in New York State seem to indicate that the land is very slowly rising in the Adirondack Mountain region. Which statement best explains this change?
    (1) The Adirondack Mountains are in a zone of crustal uplift.
    (2) The Adirondack Mountains are in a zone of very few earthquakes.
    (3) The rocks in the Adirondack Mountains are younger than those in other regions of New York.
    (4) The gravitational attraction of the Moon is greater in the higher Adirondack region.

54. During which period was most of the surface bedrock that separates the Adirondacks from the Catskills formed?
    (1) Precambrian   (3) Jurassic
    (2) Ordovician   (4) Triassic

55. Salt deposits are often found in bedrock of Silurian age. Which landscape region in New York State most likely contains salt deposits?
    (1) Adirondack Mountains
    (2) Taconic Mountains
    (3) Atlantic Coastal Plain
    (4) Erie-Ontario Lowlands

56. Which landscape region contains the location 44°00N and 74°30W?
    (1) Adirondack Mountains
    (2) Hudson Highlands
    (3) Erie-Ontario Lowlands
    (4) Tug Hill Plateau

Base your answers to questions 57 through 59 on the *Earth Science Reference Tables* and the map below which shows present drainage basins in New York State.

Deposition

57. In the Susquehanna-Chesapeake drainage basin, rivers flow over bedrock composed primarily of
   (1) gneisses, quartzites, and marbles
   (2) shales, slates, and diabases
   (3) conglomerates, granites, and rhyolites
   (4) limestones, shales, and sandstones

58. In which drainage basin are the Finger lakes located?
   (1) Mohawk-Hudson
   (2) Susquehanna-Chesapeake
   (3) Ontario-St. Lawrence
   (4) Champlain-St. Lawrence

59. In which drainage system are the Newark Lowlands located?
   (1) Mohawk-Hudson
   (2) Susquehanna-Chesapeake
   (3) Ontario-St. Lawrence
   (4) Champlain-St. Lawrence

CONSTRUCTED RESPONSE QUESTIONS

Base your answers to questions 60 through 62 on the information below.

A mountain is a landform with steeply sloping sides whose peak is usually thousands of feet higher than its base. Mountains often contain a great deal of nonsedimentary rock and have distorted rock structures caused by faulting and folding of the crust.

A plateau is a broad, level area at a high elevation. It usually has an undistorted, horizontal rock structure. A plateau may have steep slopes as a result of erosion.

60. State why marine fossils are not usually found in the bedrock of the Adirondack Mountains. [1]

61. State the agent of erosion that is most likely responsible for shaping the Catskill Plateau so that it physically resembles a mountainous region. [1]

62. State the approximate age of the surface bedrock of the Catskills. [1]

Base your answers to questions 63 and 64 on the *Earth Science Reference Tables*, your knowledge of earth science, and the diagrams below, which represent stream drainage patterns.

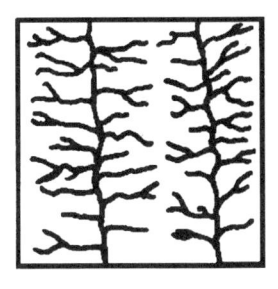

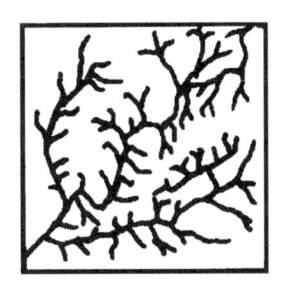

63. Label each diagram as "dendritic," "trellis," or "radial." [3]

64. State the major landscape region of New York State that displays a radial drainage pattern. [1]

65. In one or more complete sentences, explain the difference between horizontal and vertical sorting. [2]

66. In one or more complete sentences, explain why glacial deposits are often composed of unsorted sediments, while stream deposits are composed of sorted sediments. [2]

67. State three characteristics of a particle that affects its settling rate. [3]

68. A student has collected samples of two different types of gravel. Devise an experimental procedure the student could follow to determine which gravel would abrade faster in a stream. [4]

69. State two types of observational evidence that could be used to determine the direction from which a glacier advanced. [2]

EXTENDED CONSTRUCTED RESPONSE QUESTIONS

Base your answers to questions 70 through 74 on your knowledge of Earth science, the *Earth Science Reference Tables*, and the newspaper article below.

Deposition

Clash over Hudson Cleanup*

Scientists and professionals involved in environmental cleanups said yesterday that it is technically possible—though extremely difficult—to safely and completely dredge 2.65 million cubic yards of contaminated sediment from a 35-mile stretch of the upper Hudson River.

Mike O'Toole, director of environmental remediation for the NYS Department of Environmental Conservation said, "PCBs (*poly*-*c*hlorinated *b*iphenyls) are primarily attached to cohesive sediments like clay, and can be removed safely by using silt screens and dams." Environmental dredging, he said, is different from dredging for navigational purposes, in which open buckets are used to remove large amounts of sediment and debris from the bottom of a waterway. In an environmental dredging operation, he said, the site is physically secured from the rest of the waterway.

If there is little or no current, O'Toole said, dredgers employ a silt screen that catches any particles stirred up by the dredging operation and prevents them from floating downstream. If there is a current, "you use sheets of steel locked together and pounded into the sediment around the work site. This physically separates the site from the rest of the water, and any deposits which are stirred up fall back into the site, rather than migrating downstream."

O'Toole acknowledged that environmental dredging operations are expensive, "but are well worth the cost in terms of protecting the future health of the river."

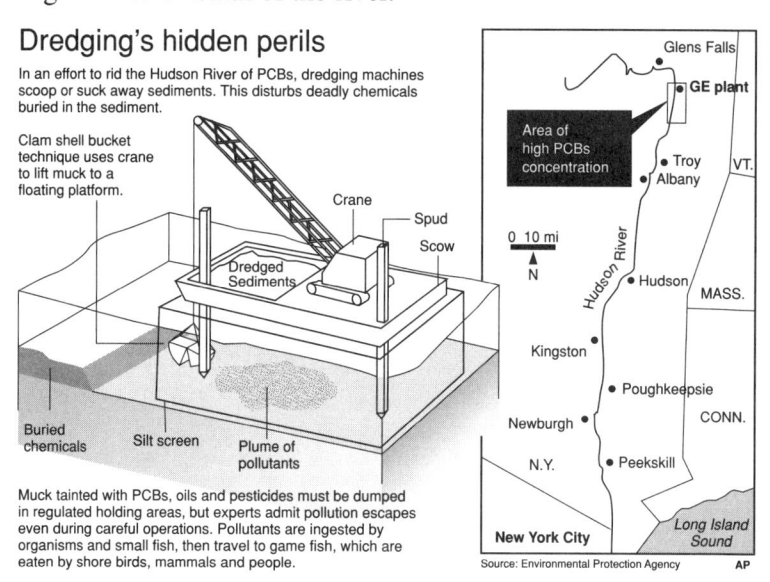

Dredging's hidden perils

In an effort to rid the Hudson River of PCBs, dredging machines scoop or suck away sediments. This disturbs deadly chemicals buried in the sediment.

Clam shell bucket technique uses crane to lift muck to a floating platform.

Crane — Spud — Scow — Dredged Sediments

Buried chemicals — Silt screen — Plume of pollutants

Muck tainted with PCBs, oils and pesticides must be dumped in regulated holding areas, but experts admit pollution escapes even during careful operations. Pollutants are ingested by organisms and small fish, then travel to game fish, which are eaten by shore birds, mammals and people.

Glens Falls • GE plant
Area of high PCBs concentration
• Troy • Albany VT.
0 10 mi N
Hudson River
• Hudson MASS.
• Kingston
• Poughkeepsie CONN.
Newburgh •
N.Y. • Peekskill
New York City
Long Island Sound

Source: Environmental Protection Agency AP

*Source: Roger Witherspoon, "Clash over Hudson Cleanup," in *The Journal News*, Gannett Satellite Information Network, Inc., Thursday, December 7, 2000, page 3A.

70. A dredging company proposes using a silt screen with openings 0.0006 centimeters in diameter when dredging the Hudson as part of the cleanup. In one or more complete sentences, explain
   (a) the purpose of a silt screen in environmental dredging, [1]
   (b) why the proposed silt screen would or would not be effective. [1]

71. According to the *Earth Science Reference Tables*, in which landscape region of New York State is the area of high PCB concentration, shown in the map in the article, located? [1]

72. The diagram in the article shows a plume of pollutants extending downstream from the silt screen. State one way in which the pollutant plume would be affected by an increase in the speed of the river current. [1]

73. According to the article, under what conditions would sheets of steel locked together be used to isolate the worksite instead of a silt screen? [1]

74. State one way in which pollutants that escape from the worksite could endanger humans. [1]

Unit Seven: THE ATMOSPHERE, WEATHER, AND CLIMATE

Chapter 22

THE ATMOSPHERE

> **KEY IDEAS** The atmosphere can be thought of as a huge heat engine that converts heat energy into mechanical energy. Earth's atmospheric heat engine is powered primarily by solar energy and is influenced by gravity.
>
> When the various electromagnetic waves that emanate from the Sun strike the atmosphere, only a small percentage are directly absorbed, especially by gases such as ozone, carbon dioxide, and water vapor. Clouds and Earth's surface reflect some energy back into space, and Earth's surface absorbs some. The energy absorbed is then transferred between Earth's surface and the atmosphere by conduction, convection, radiation, and evaporation.
>
> The transfer of heat energy from solar radiation, and of water vapor, into and out of the atmosphere causes regions of different densities to form. The rise or fall of regions of different densities due to the action of gravitational force produces atmospheric circulation, which is affected by Earth's rotation. Atmospheric circulation distributes solar energy over the whole Earth. The interaction of these processes results in the complex atmospheric occurrence known as weather.

KEY OBJECTIVES

Upon completion of this unit, you will be able to:

- Explain why the Sun emits different types of electromagnetic radiation.
- Compare and contrast the different forms of electromagnetic radiation shown in the *Earth Science Reference Tables*.
- Describe the mechanisms by which solar energy is absorbed by the atmosphere and the factors that determine the amount of solar energy an area receives.
- Explain how density differences in the atmosphere cause atmospheric circulation.
- Explain how Earth's rotation influences atmospheric circulation (a phenomenon known as the Coriolis effect), and interpret a diagram of the planetary wind and pressure belts.
- Explain how energy can be stored or released during phase changes.

SOLAR RADIATION: THE ATMOSPHERE'S PRIMARY ENERGY SOURCE

The Sun is the atmosphere's major source of energy. Energy from the Sun reaches Earth in the form of electromagnetic waves.

Energy from Electromagnetic Waves

A magnet can make a compass needle move from a distance because the magnet is surrounded by an invisible magnetic field. If you move the magnet, the magnetic field will move and the moving magnetic field, in turn, will cause the compass needle to move. In much the same way, a statically charged balloon held near your head will attract your hair from a distance because the balloon is surrounded by an invisible electric field. If you move the balloon, the invisible electric field moves with it and you can feel the effect of the moving field on your hair. Such fields extend outward infinitely in all directions from their sources, but they weaken with distance from these sources (see Figure 22.1).

All matter is composed of atoms. Every atom consists of electrically charged particles such as protons and electrons, which are surrounded by an electric field. Whenever a charged particle moves, a magnetic force is produced and the particle is then also surrounded by a magnetic field. These two fields—an electric field and a magnetic field—exist simultaneously around all electrically charged particles that are moving. Together, they are called an *electromagnetic field*, and they extend outward infinitely in all directions around the particles.

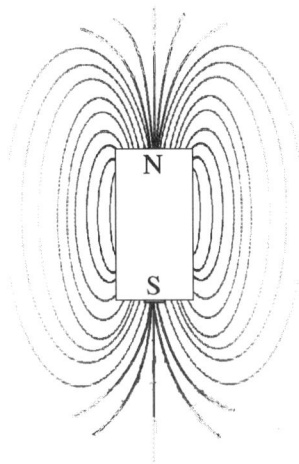

Figure 22.1 An Invisible Magnetic Field Surrounding a Magnet. The lines represent magnetic force; closely spaced, the lines mean stronger magnetic force. Note that the strength of the magnetic field decreases with distance.

When a particle moves back and forth, its electromagnetic field moves with it (see Figure 22.2a). Like a ripple in a pond, this movement spreads out through the field in the form of a transform wave. The way that the wave moves through the electromagnetic field depends on the way in which the particle moves. The faster the particle oscillates, the shorter the wavelength, or distance between successive "ripples" in the electromagnetic field. Every frequency of oscillation produces a wave of a different wavelength (see Figure 22.2b).

A moving wave contains energy. It can exert forces on matter with which it interacts. Since an electromagnetic field does not require a medium to exist, it can extend through space. Disturbances in electromagnetic fields

The Atmosphere

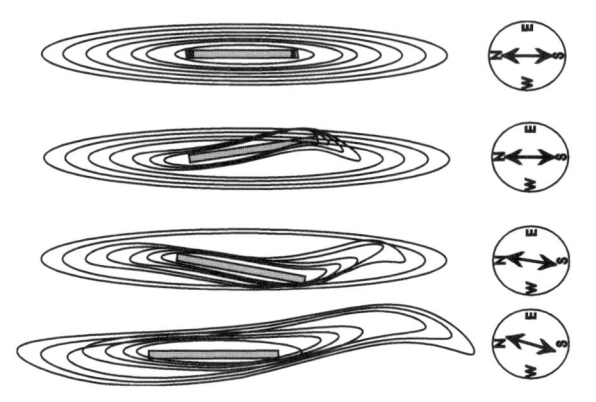

(a) As a magnet moves, its magnetic field moves too. Like a ripple in a pond, the movement of the field spreads outward in all directions at a rate of 3×10^8 meters per second—the speed of light. When the moving field reaches *another* field (such as the one around a compass needle), it exerts a force of attraction or repulsion that can cause motion of the field and even of the object that produced the field. A similar disturbance would travel through the electric field around an electrically charged particle in motion.

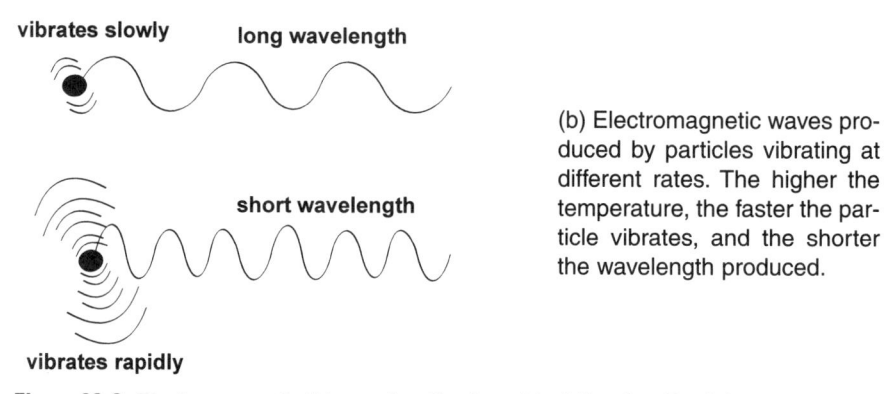

(b) Electromagnetic waves produced by particles vibrating at different rates. The higher the temperature, the faster the particle vibrates, and the shorter the wavelength produced.

Figure 22.2 Electromagnetic Waves Are Produced by Vibrating Particles.

around particles can travel through space as they move outward, or radiate, through the field. In this way, energy is transferred from the Sun to Earth without the existence of a physical medium between the two.

The Electromagnetic Spectrum

Electromagnetic waves are classified by their lengths and range from short waves, such as X rays, to long waves, such as radio waves. The ***electromagnetic spectrum*** (see Figure 22.3) is a continuum in which electromagnetic waves are arranged in order, from longest to shortest wavelength. Infrared (heat) waves and visible-light waves fall roughly in the middle range of wavelengths. The wavelength, or distance between "ripples" of visible light waves, falls between 10^{-6} and 10^{-7} meter.

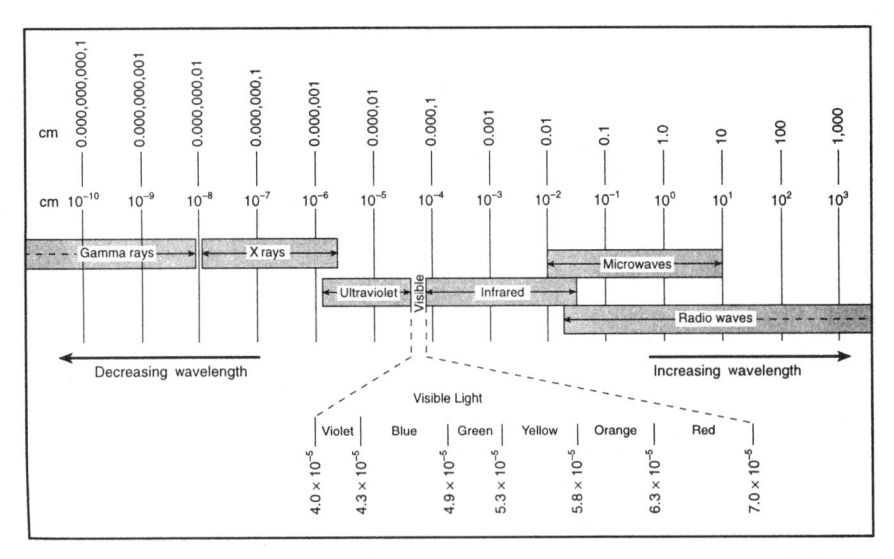

Figure 22.3 Electromagnetic Spectrum. This shows the categories into which electromagnetic waves are classified according to their lengths. Source: *Earth Science Reference Tables*—2001 Edition.

Many different waves emanate from the Sun (see Figure 22.4) because it contains particles moving at many different speeds. However, most waves coming from the Sun are in the visible-light and infrared ranges. To move at the speed needed to emit waves of this length, particles at the Sun's surface must have temperatures between 5,000K and 7,000K. Earth's magnetic field and the Van Allen belts of charged particles in the upper atmosphere deflect many of the waves with short lengths. This circumstance is fortunate, since most short-wavelength radiation is harmful to living things.

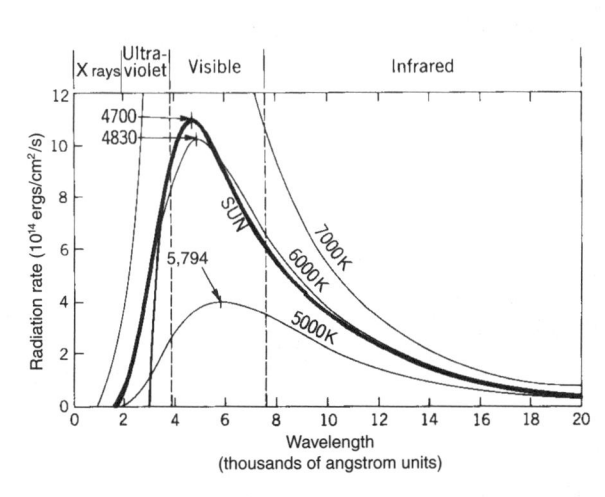

Figure 22.4 Makeup of Solar Radiation. The Sun emits electromagnetic radiation ranging in wavelength from infrared to X rays. The majority of the waves are in the visible-light range with a peak at 4,700 angstrom units (1 angstrom unit = 1×10^{-9} meter—one billionth of a meter). This is very close to the theoretical output of a body at a temperature at 6,000K, which is why the Sun's surface temperature is thought to be between 5,000 and 7,000K.

The Global Radiation Budget

When Earth intercepts electromagnetic waves from the Sun, the moving waves exert a force on the fields surrounding particles of matter in the atmosphere, hydrosphere, and lithosphere, causing the particles to move. This increased motion shows up as an increase in the level of Earth's heat energy. At the same time, Earth's matter radiates energy in the form of electromagnetic waves back out into space, causing a decrease in the level of Earth's heat energy. Thus, both incoming solar radiation and Earth's outgoing, or terrestrial, radiation pass through the atmosphere. Table 22.1 shows the global radiation budget.

TABLE 22.1 GLOBAL RADIATION BUDGET*

Incoming Solar Radiation	Gain (%)	Loss (%)
Reflection from clouds to space		21
Diffuse reflection (scattering) to space		5
Direct reflection from Earth's surface		6
Net energy loss back to space		32
Absorbed by clouds	3	
Absorbed by molecules, dust, water vapor, and CO_2	15	
Absorbed by Earth's surface	50	
Net incoming energy gained by Earth-atmosphere system	**68%**	

Outgoing Terrestrial Radiation	Gain (%)	Loss (%)
Total infrared radiation emitted by Earth's surface	98	
Absorbed by atmosphere	90	
Lost to space		8
Total infrared radiation emitted by the atmosphere	137	
Absorbed by Earth's surface	77	
Lost to space		60
Net outgoing energy lost from entire Earth-atmosphere system		**68%**
Net energy leaving *Earth's surface* (98% emitted − 77% reabsorbed)		21
Net energy leaving the *atmosphere* (137% emitted − 90% reabsorbed)		47

*Source: *The Earth Sciences*, Arthur N. Strahler, Harper & Row, 1971.

Radiative Balance

Earth's average global temperature depends upon the balance between the energy gained by absorbing sunlight and the energy lost by radiating heat. As you can see in Table 22.1, the Earth-atmosphere system's net energy gain from incoming solar radiation is the same as its net energy loss by terrestrial radiation—68 percent. Over long periods of time, then, the average level of Earth's heat energy remains fairly constant, and the average temperature of Earth's surface hardly changes. Therefore, we say that Earth is in *radiative balance*; it is reradiating as much energy as it absorbs.

This fact, however, *does not mean that all points on Earth's surface are in radiative balance at all times*. Most places on Earth go through temperature changes on both a daily and a yearly basis. These temperature changes indicate that there are times when a place is gaining more energy than it is losing and times when it is losing more energy than it is gaining.

There is also much evidence that over long periods of time Earth experiences warming and cooling trends. These may be triggered by changes in Earth's tilt due to precession as it spins on its axis and to increases or decreases in certain gases in the atmosphere.

Let us now consider some of the factors that influence the way in which Earth absorbs incoming solar radiation.

FACTORS AFFECTING INSOLATION

The Sun provides nearly all of the energy received at Earth's surface. Solar radiation reaches the upper atmosphere at a fairly constant rate of about 200 kilocalories per minute per square meter. About one-third of this radiation is reflected back into space, mostly by clouds. As the remaining radiation passes through the atmosphere, some is absorbed by gas molecules. Some is refracted as it crosses boundaries between layers of differing densities. Some is scattered by particles of dust or aerosols in the air. The radiation that reaches Earth's surface, called *insolation*, short for *in*coming *sol*ar radi*ation*, is either reflected or absorbed.

A number of factors control the amount of solar energy that an area absorbs or reflects, including the angle at which insolation strikes the surface, the length of time each day that insolation is received (the duration of insolation), and the nature of the surface.

Angle of Insolation

Latitude

Since Earth is a sphere, insolation does not strike all points on Earth's surface at the same angle. Near the Equator insolation strikes the surface almost vertically, but near the poles it strikes at a more glancing angle. Therefore, the insolation reaching the tropics is more concentrated than that reaching polar regions (see Figure 22.5).

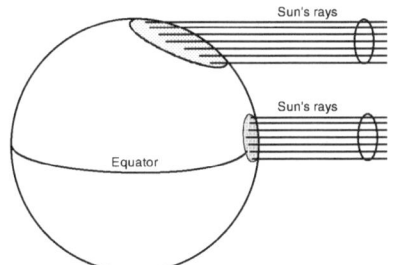

Figure 22.5 Insolation near the Equator and the North Pole. Note that the two beams of sunlight approaching Earth are identical, but the angle at which they strike Earth's surface causes insolation received near the Equator to be more concentrated than that received near the poles. The result is greater heating at the Equator.

Daily and Annual Cycles

The angle at which insolation strikes Earth's surface at any location also varies in both daily and annual cycles. The daily cycle starts at dawn with insolation striking the surface at a very low angle; it then becomes increasingly direct throughout the morning, reaches its greatest directness at noon,

535

The Atmosphere

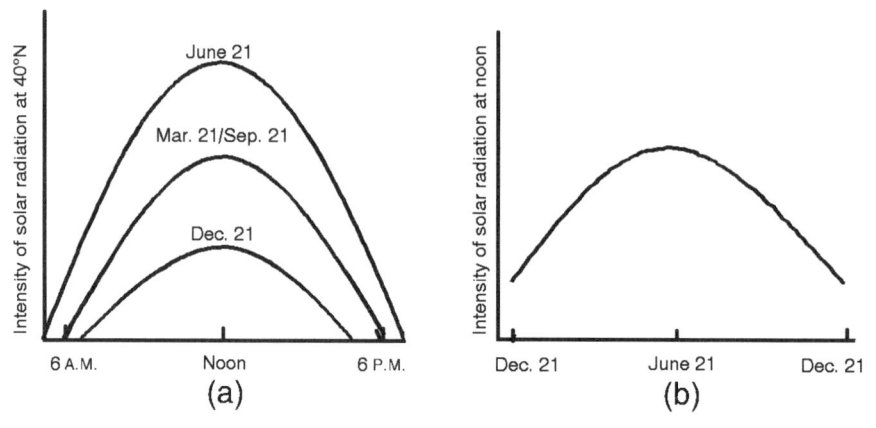

Figure 22.6 Daily and Yearly Cycles of Solar Radiation Intensity. (a) Daily cycle of intensity of solar radiation from 6 A.M. to 6 P.M. at 40°N on the equinox and solstice days. Note that for each day there is a cyclic change from low intensity in early morning to higher intensity at noon and then back to low intensity in late afternoon. (b) Yearly cycle of intensity of solar radiation at noon for an entire year. Note the cyclic change from low intensity at noon on the winter solstice to high intensity at noon on the summer solstice and back to low intensity at the next winter solstice.

and becomes increasingly indirect again throughout the afternoon until sunset.

The annual cycles vary with latitude and Earth's position in its orbit around the Sun. In the United States, insolation is most direct on June 21, the summer solstice, and least direct on December 21, the winter solstice.

Figure 22.6a shows the daily cycle of intensity of solar radiation; Figure 22.6b, the yearly cycle of intensity of solar radiation at noon.

Duration of Insolation

The length of time that the surface receives insolation each day, or the *duration of insolation*, depends on the season and the latitude of the location.

Season

The length of daylight varies in an annual cycle with the season (see Figure 22.7). In the United States, the greatest number of hours of insolation is received on June 21, the summer solstice. Each day thereafter, the number of daylight hours decreases, reaching a minimum on December 21, the winter solstice. Then the number of daylight hours increases daily until it peaks again the following June 21. At the fall and spring equinoxes, the number of daylight hours equals the number of hours of darkness. New York State varies from roughly 15 hours of daylight at the summer solstice to only 9 hours at the winter solstice.

Latitude

The length of daylight on any given day also varies with location north or south of the Equator (see Figure 22.8). Since Earth's axis of rotation is tilted, the circles that form the parallels of latitude are also tilted. At different latitudes, different fractions of the parallels are in daylight and darkness. Thus, observers at different latitudes spend different fractions of each 24-hour rotation in daylight and in darkness. When the Northern Hemisphere is tilted toward the Sun, points between the North Pole and the Arctic Circle rotate through an entire 24-hour day without ever entering into darkness. With movement southward, the number of daylight hours decreases.

Although the maximum duration of insolation occurs on June 21 in the Northern Hemisphere, maximum temperatures are reached sometime after this date. Temperatures continue to increase after June 21 because the number of daylight hours still exceeds the number of nighttime hours for many weeks after this date. As long as more energy is received each day than is lost overnight, the temperature will continue to increase.

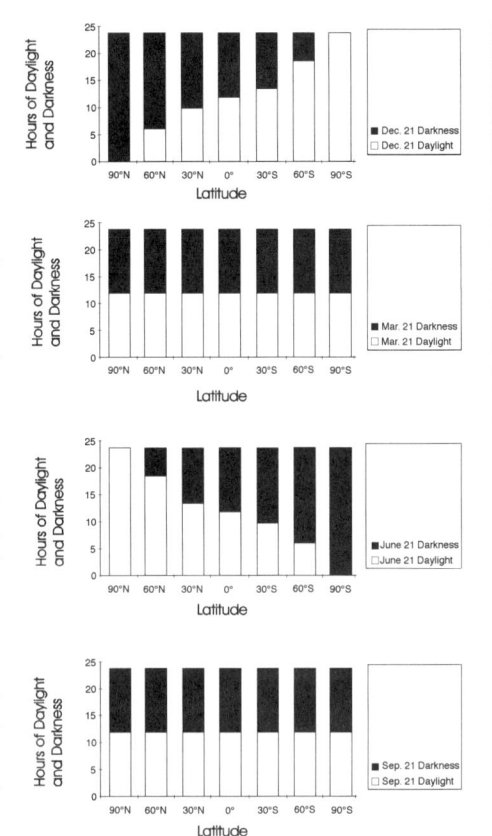

Figure 22.7 Duration of Day and Night at Different Locations on Key Dates Throughout the Year.

Nature of Earth's Surface

Earth's surface consists of a wide variety of substances, including ice, water, soil of many types, and vegetation. All interact differently with the insolation that strikes them.

Color and Texture

Just as a ball is more likely to bounce off a concrete wall than off a pillow, insolation striking different substances will have a greater or lesser tendency to be absorbed or reflected. Which process will predominate depends on the molecular structure of the substance. The color of a substance is a fairly good indicator

The Atmosphere

of whether the substance absorbs more insolation than it reflects or vice-versa.

The lighter the color of a substance, the more light is being reflected by it; the darker the color, the more light is being absorbed. When insolation strikes substances such as sand and snow, their light color indicates that much of the insolation is being reflected. On the other hand, when insolation strikes rich black soil or deep green leaves, their dark color indicates that much of the insolation is being absorbed. In general, dark-colored substances absorb more insolation than light-colored ones, and the more insolation a substance absorbs the more it is heated.

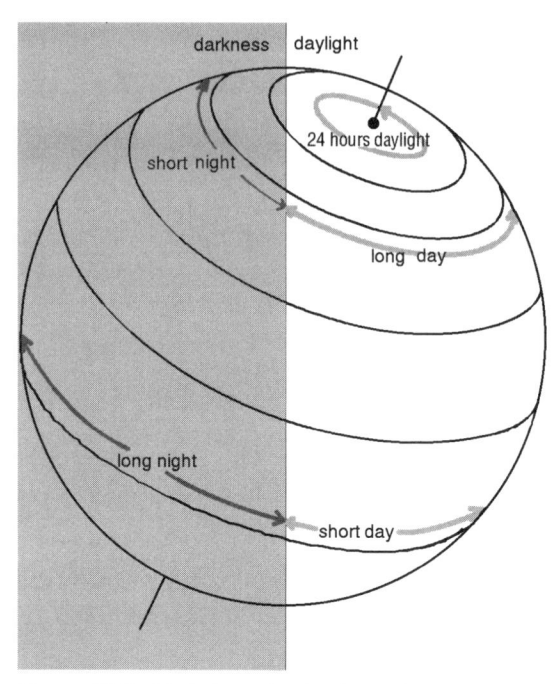

Figure 22.8 Variations in the Lengths of Day and Night Because of Earth's Tilted Axis of Rotation.

The word *texture* refers to the smoothness or roughness of a surface. On a smooth surface, insolation is more likely to be reflected away from the surface than on a rough surface. The irregularities and indentations on a rough surface cause some of the insloation to be reflected in such a direction that it hits the surface a second, or even a third, time before leaving (see Figure 22.9). Each time the insolation strikes the surface, a little more of its energy is absorbed. Therefore, if all other characteristics of two surfaces are the same, a rough surface will absorb more insolation than a smooth one.

Specific Heat Capacity

When different substances absorb equal amounts of heat energy, they do not all change temperature by the same number of degrees. The difference can be thought of as the result of a kind of molecular inertia. Heavy molecules and tightly held molecules require more energy to get them moving than light and loosely held molecules. The temperature of a substance is strictly a result of how fast the molecules are moving. If you put the same amount of energy into two substances, one with tightly held molecules and one with loosely held molecules, the loosely held molecules would move faster and that substance would register a higher temperature.

Table 22.2 shows the specific heat capacities of different Earth materials

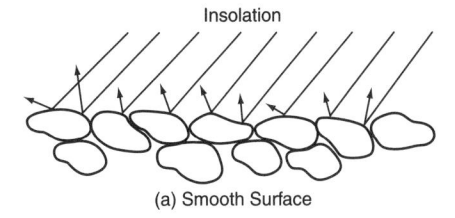

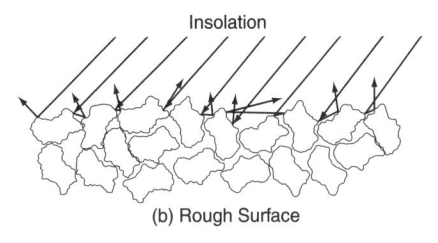

Figure 22.9 Insolation Striking (a) a Smooth Surface and (b) a Rough Surface.

and also the number of degrees each substance will increase in temperature if 1 calorie of heat energy is added to it. Notice that water increases in temperature much less than rock material such as granite or basalt with the addition of the same amount of heat. Therefore, if the same amount of insolation is absorbed by an ocean and by the sand-sized particles of broken rock that make up the adjoining beach, the sand will become much hotter than the water. (This is the reason why we seek out bodies of water to swim in when temperatures on land are high.) In general, land increases in temperature more than water when exposed to the same insolation.

TABLE 22.2 SPECIFIC HEAT CAPACITIES OF COMMON EARTH MATERIALS

Material	Specific Heat Capacity (cal/g • °C)	Temperature Increase if 1 Calorie of Heat Is Added to 1 Cubic Centimeter of Material (°C)
Water		
Solid	0.5	2
Liquid	1.0	1
Vapor	0.5	2
Dry air	0.24	4.2
Granite	0.19	5.3
Basalt	0.20	5
Marble	0.21	4.8
Iron	0.11	9.1
Copper	0.09	11.1
Lead	0.03	33.3

Latent Heat of Water

When water is melting or evaporating, it absorbs energy with no resulting change in temperature. The energy is used to break internal bonds rather than to increase the speed at which water molecules are moving. Since this heat causes no change in temperature, it is called *latent heat* (*latent* means "hidden"). As a result, ice that is melting or water that is evaporating from ocean surfaces absorbs insolation without increasing in temperature. Therefore, ice-covered land and oceans remain cooler than adjacent land areas.

Latent heat is not an issue on land because the substances of which land is composed do not melt or evaporate in the normal range of temperatures at Earth's surface. However, since water is so abundant and exists in all three phases on Earth, latent heat has an important effect on the temperature changes that occur when insolation strikes water.

HOW ENERGY ENTERS THE ATMOSPHERE

As you have learned, the atmosphere does not absorb much solar energy directly. How, then, is the atmosphere heated? Most insolation passes right through the atmosphere to Earth's surface, where it is absorbed and changed into forms of energy that the atmosphere can absorb. Earth's surface, then, heats the atmosphere. Let us now consider how energy moves from Earth's surface into the atmosphere and, once there, how it moves within the atmosphere.

Conduction

When Earth's surface absorbs solar energy, the absorbed energy causes the molecules at the surface to move more rapidly. We perceive this absorbed energy as heat and measure it in terms of the temperature of the substance as shown by a thermometer. Since Earth's surface is in continuous contact with the base of the atmosphere, its surface molecules strike molecules of gases in air. During these collisions, some heat energy from Earth's surface molecules is transferred to the gas molecules of the atmosphere. This kind of energy transfer, in which heat energy moves from molecule to molecule by collisions, is known as *conduction*. Conduction is important because it is the chief method by which energy moves from Earth's surface to the atmosphere.

Convection

Within the atmosphere, heat spreads very slowly by conduction. Therefore, the air touching the ground becomes warmer and warmer. As this air warms, it expands and becomes less dense than the surrounding air. Finally, a bubble of warm air rises, and cooler air settles to the ground in its place, a type of movement known as *convection*. Since the rising bubble of warm air carries the heat it contains upward with it, convection transfers heat energy *within* the atmosphere. Convection is important because it is the chief method by which energy conducted into the atmosphere from Earth's surface is then carried throughout the atmosphere.

Radiation

Energy from the Sun reaches Earth by *radiation*, the release and transfer of energy in the form of electromagnetic waves. As described earlier, all moving atoms and molecules emit electromagnetic radiation, and the wavelength emitted depends upon how fast these particles are moving. However, since both Earth and its atmosphere are much cooler than the Sun, the energy they radiate has much longer wavelengths. Most of the energy radiated by Earth's surface is infrared radiation.

This fact is important because, although the atmosphere is transparent to most of the Sun's short-wavelength radiation, it is not as transparent to infrared radiation. Ozone, carbon dioxide, and water vapor in the atmosphere absorb or reflect most of Earth's infrared radiation; the rest goes through the atmosphere and out into space. Thus, short wavelengths can readily enter the atmosphere, but long wavelengths cannot readily escape, a phenomenon known as the *greenhouse effect* (see Figure 22-10).

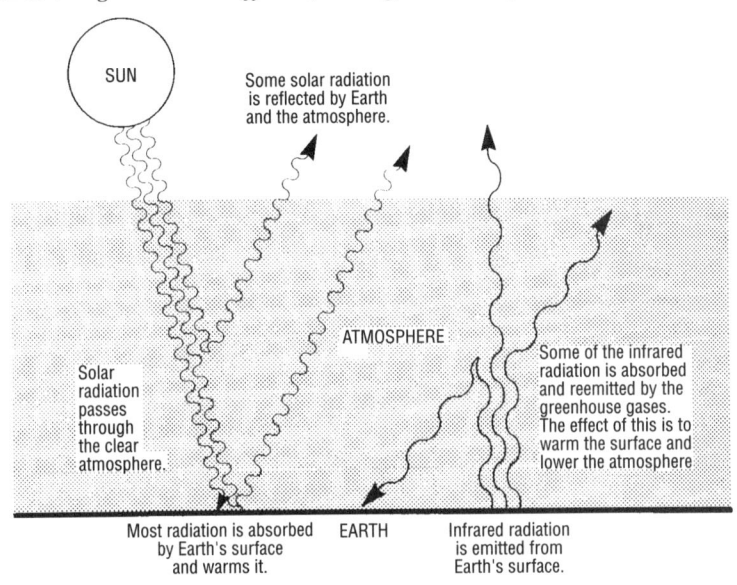

Figure 22.10 The Greenhouse Effect. This sequence of events is called the *greenhouse effect* because the glass in a greenhouse allows light to enter but blocks heat from escaping. Source: Climate Change, The Intergovernmental Panel on Climate Change, World Meteorological Organization/UN Environmental Programme, Cambridge University Press.

This term is appropriate because, just as glass surrounds a greenhouse, the atmosphere surrounds Earth. Glass, like the atmosphere, allows visible light rays to pass through it but blocks infrared rays. Similarly, a greenhouse lets sunlight in but does not allow heat to escape. The greenhouse effect keeps the greenhouse—and Earth—warmer than they would otherwise be.

Latent Heat

Heat also enters the atmosphere as energy stored in molecules of water that have evaporated. Try wetting your finger and waving it in the air. Notice that it feels cooler? The reason is that water is evaporating from it. Each molecule of liquid water has to absorb a certain amount of heat from your finger before it can evaporate. Then, when it evaporates and becomes water vapor, each water vapor molecule carries with it into the air the heat energy it absorbed from your finger. The energy stored in water vapor molecules in the air moves along with the air as it circulates in the atmosphere.

Usually, when matter gains heat energy, its temperature increases; and when matter loses heat energy, its temperature decreases. However, when matter changes phase, heat is absorbed or given off in the process of forming or breaking bonds between molecules, rather than in causing molecules to move faster or slower. Thus, as mentioned earlier, the heat energy gained or lost during a phase change does not cause a change in temperature and is therefore latent heat.

The energy *gained* when ice melts into liquid water amounts to about 80 calories per gram of water. The same amount of energy, 80 calories per gram of water, is *released* when liquid water freezes into ice. The energy *gained* when liquid water evaporates into water vapor amounts to about 540 calories per gram of water. Here, too, the same amount of energy, 540 calories per gram of water, is *released* when the water vapor condenses back into liquid water. The changes of phase that water undergoes and the heat given off or absorbed in each process are summarized in Figure 22.11a.

Water covers more than 70 percent of Earth's surface, so most of the lower atmosphere is in contact with water. Each day solar energy absorbed by this water causes evaporation, sending vast amounts of water vapor into the atmosphere. Every gram of water vapor in the atmosphere contains the 540 calories of latent heat energy that was gained during the change of water from a liquid to a gas.

Remember, though, that latent heat stored during evaporation does not show up as a temperature change. The water vapor starts out no warmer than the water from which it formed. How, then, is the atmosphere warmed by latent heat? The heat stored in water molecules when they evaporated is released when the water changes back to water or ice. At the moment the water vapor condenses, it releases its latent heat into the air. As a result, condensation is a warming process that causes the temperature of air to rise as droplets of water in clouds grow in size. This increase in temperature causes the air to expand, become less dense, and rise. Thus, latent heat is an important energy source for violent storms. The release of latent heat during large-scale condensation produces localized heating, which causes updrafts. At the same time, precipitation associated with the heavy condensation causes downdrafts. The result is the violent circulations of air that occur in storms such as hurricanes and tornadoes.

Circulation of the Atmosphere

The transfer of heat energy from solar radiation, and water vapor, into and out of the atmosphere causes regions of different densities to form. The equatorial regions of the Earth are predominantly water and receive more solar energy per year than the polar regions. Air over the Equator becomes warm and moist through contact with the Earth's surface. Air that is warm and moist is less dense than cooler, drier air. Thus, the air at the Equator is surrounded by air of higher density that pushes inward toward the Equator and displaces the warm, moist equatorial air upward. As this air is forced upward it expands outward and cools, resulting in condensation and precipitation. The end result is cooler, dryer air that is denser than the air beneath it and

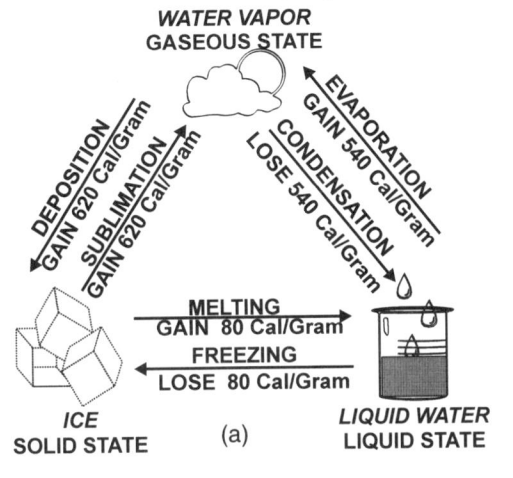

(a) Water gains or loses energy when it undergoes a change in state of matter. Since this gain or loss of heat energy does not result in a change in temperature, it is hidden or *latent heat*.

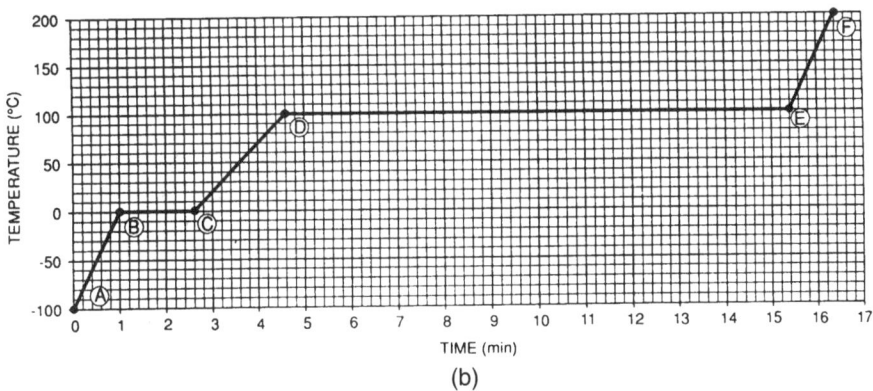

(b) The heating curve for a sample of water heated from a starting temperature of −100°C to a final temperature of +200°C. The same amount of heat is added to the sample every minute. From A to B the ice increases in temperature until it reaches its melting point. From B to C there is a change in state from solid ice to liquid water with no increase in temperature (the curve is flat, like a plateau). From C to D liquid water increases in temperature until it reaches its boiling point. From D to E there is a change in state from liquid water to water vapor, again with no increase in temperature. From E to F water vapor is increasing in temperature. Note that the flat area from B to C is shorter than the flat area from D to E since less energy is required to change water from a solid to a liquid than to change water from a liquid to a gas.

Figure 22.11 Changes in State of Water (a) and Heating Curve for Water (b).

543

The Atmosphere

begins to sink back toward the surface. Together, the rising warm, moist air and sinking cool, dry air form a circular pattern of motion called a *convection cell*. See Figure 22.12. A similar convection cell forms over the poles, although it begins with sinking air, which spreads southward when it reaches the Earth's surface, warms, and rises as shown in Figure 22.13. In between the polar and equatorial convection cells lies a third cell, which is set in motion by westerly winds in the middle latitudes. Thus, three convection cells girdle both Earth's Northern and Southern Hemispheres. This *three-cell theory* of global atmospheric circulation explains how the atmosphere distributes solar energy over the whole Earth. See Figure 22.14. The interaction of these processes results in the complex atmospheric occurrence known as weather.

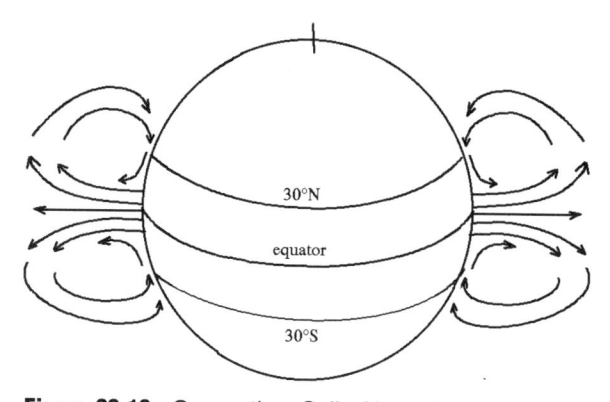

Figure 22.12 Convection Cells Near the Equator. Air near the Equator is heated by more intense insolation, becomes less dense, and floats upward in the surrounding denser air. As the air rises, it expands and cools, causing it to sink back toward the surface. The result is a circular movement of air, or convection cell.

The Coriolis Effect and Global Atmospheric Circulation

The rise or fall of regions of different densities as a result of the action of gravitational force produces atmospheric circulation, which is affected by Earth's rotation. One of the consequences of Earth's rotation is a tendency of all matter that is in motion on Earth's surface to be

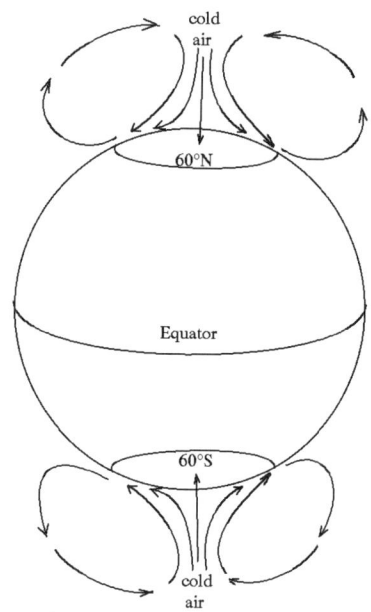

Figure 22.13 Convection Cells Near the Poles. Air near the poles cools because of less intense insolation, becomes more dense, and sinks downward. The sinking air spreads outward at the surface, is warmed as it moves away from the poles, and rises. The result is a circular movement of air, or convection cell.

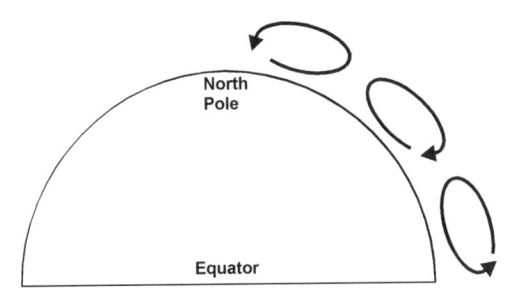

Figure 22.14 The Three-Cell Theory of Convection in the Atmosphere.

deflected to the right from its point of origin in the Northern Hemisphere and to the left in the Southern Hemisphere. This is called the Coriolis effect, after the French mathematician Gustave Gaspard Coriolis, who first analyzed it in the nineteenth century. The Coriolis effect is not an actual force but rather is the *apparent* effect of a number of different forces acting on any particles in motion.

The main elements of the Coriolis effect are the curvature of Earth's surface, the rotation of Earth, and the tendency of objects in motion to remain in motion in a straight line (Newton's first law of motion). The east-west path between any two points on Earth's surface is not a straight line, but a curve—a segment of a parallel (line of latitude). As a result, any object following a straight path will appear, to an observer on Earth, to curve from its path. We say "appear" because actually it is the observer who is traveling a curved path while the object travels in a straight line. In Figure 22.15 the solid line represents the actual path of an observer on the surface of a spherical Earth as it rotates. The dotted line represents a straight path for a moving object. The Coriolis effect is most pronounced where there is the greatest difference between the curvature of a parallel and a straight-line path—in other words, near the poles.

A similar deflection in the north-south direction is due to the difference in the speeds at which points on Earth's surface are moving because of rotation. As Earth rotates, every object on its surface rotates with it. In one 24-hour rotation, objects near the poles travel less distance than objects near the Equator and thus are moving slower (see Figure 22.16).

Now, let's suppose a rocket at the Arctic Circle (60°N) is aimed due south

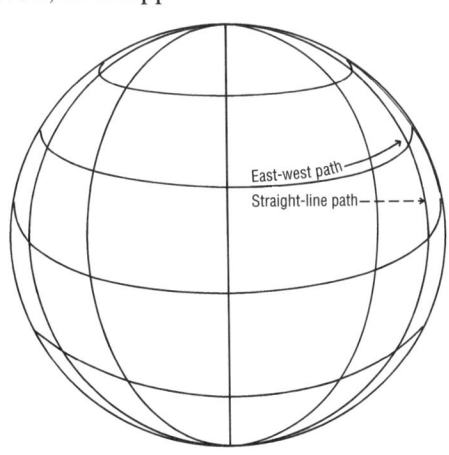

Figure 22.15 Deflection Due to Curvature of Earth's Surface. An object moving in a straight line travels along the dashed line. An observer on Earth's surface moves due east-west as Earth rotates, following a path that curves. To an observer on Earth's surface, the straight-line path seems to veer off, or be deflected, to the right in the Northern Hemisphere. In the Southern Hemisphere, the deflection is to the left.

The Atmosphere

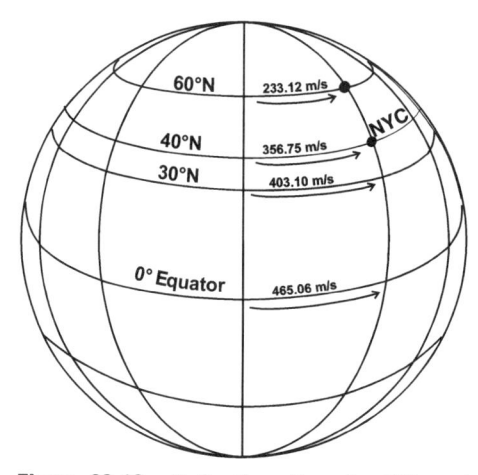

Figure 22.16 Deflection Due to Different Velocities at Different Latitudes. As Earth rotates, objects on its surface move at different speeds, depending on their latitudes. Objects at the Equator move faster because they cover more distance in one 24-hour rotation than objects near the poles.

at New York City (40°N). Before the rocket is even launched, it is moving west to east at the speed of Earth's surface at the Arctic Circle—233 meters per second. The target, New York City, is also moving west to east at the speed of Earth's surface at New York City—356 meters per second. Note that the target is moving west to east faster than the rocket! When the rocket is fired, it moves due south and west to east; but since the target is moving west to east faster than the rocket, the rocket lags behind the target and actually hits the ground behind the target. To an observer, the rocket appears to follow a path that curves to the *right* of the target. Similarly, a rocket fired from New York City toward the Arctic Circle would land ahead of the target because New York City is moving west to east faster than the target. Nevertheless, the deflection would still be to the *right*. As seen in Figure 22.17, because of the Coriolis effect no matter what direction the motion, deflection is always to the right in the Northern Hemisphere.

If we repeat this experimentation in the Southern Hemisphere, we find that the direction of deflection is reversed—the rocket appears to curve to the *left* of the target. Thus, deflection is to the right of the wind's point of origin in the Northern Hemisphere and to the left of its point of origin in the Southern Hemisphere, as shown in Figure 22.17. (A good way to recall the direction of deflection is to

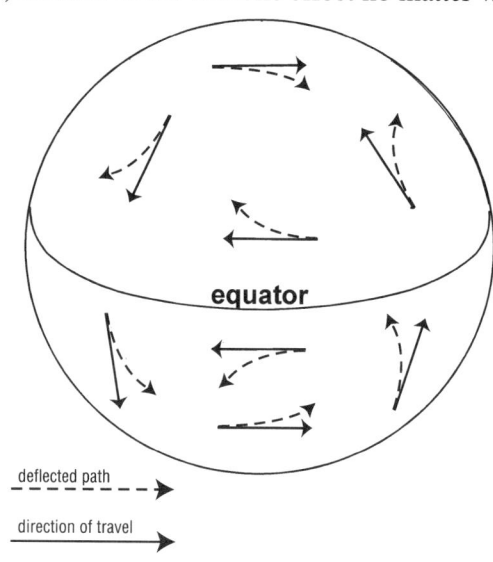

Figure 22.17 The Coriolis Effect. Objects are deflected to the right from their points of origin in the Northern Hemisphere and to the left from their points of origin in the Southern Hemisphere.

remember that someone who is *left*-handed is called a *south*paw.)

For objects following paths that fall somewhere between due north-south and due east-west, the total deflection is a combination of deflection due to curvature and deflection due to differences in rotational speed. Of course, frictional forces within the atmosphere and with Earth's surface modify the Coriolis effect, but it still persists. Over short distances, however, the Coriolis effect is imperceptible, so don't worry about veering off to the right when you walk home from school or drive to the mall in the family car. The Coriolis effect becomes a factor only over great distances, such as those covered by planetary winds.

MULTIPLE-CHOICE QUESTIONS

In each case, write the number of the word or expression that best answers the question or completes the statement.

1. At which temperature would an object radiate the *least* amount of electromagnetic energy?
   (1) the boiling point of water (100°C)
   (2) the temperature at the stratopause (0°C)
   (3) the temperature of the North Pole on December 21 (-60°F)
   (4) room temperature (293K)

2. The graph below represents the relationship between the intensity and the wavelength of the Sun's electromagnetic radiation. Which statement is best supported by the graph?

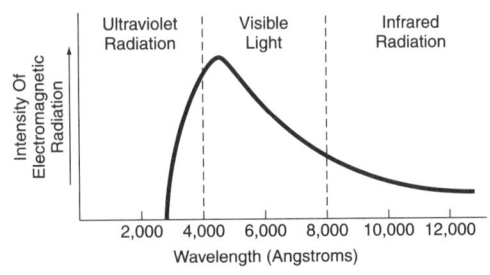

   (1) The infrared radiation given off by the Sun has a wavelength of 2,000 angstroms.
   (2) The maximum intensity of radiation given off by the Sun occurs in the visible region.
   (3) The infrared radiation given off by the Sun has a shorter wavelength than ultraviolet radiation.
   (4) The electromagnetic energy given off by the Sun consists of a single wavelength.

The Atmosphere

3. An object that is a good radiator of electromagnetic waves is also a good
   (1) insulator from heat
   (2) reflector of heat
   (3) absorber of electromagnetic energy
   (4) refractor of electromagnetic energy

4. Most of the energy in Earth's atmosphere comes from
   (1) the rotation and revolution of Earth
   (2) the rotation of Earth and wind from Earth
   (3) radioactive decay of elements and radiation from Earth
   (4) radiation from Earth and insolation from the Sun

5. Earth's surface temperatures are due chiefly to energy received from
   (1) insolation from the Sun
   (2) radioactivity within Earth's crust
   (3) fusion in Earth's core
   (4) friction between the crust and the mantle of Earth

6. By which process does most of the Sun's energy travel through space?
   (1) absorption   (3) convection
   (2) conduction   (4) radiation

7. Which color is the best radiator of electromagnetic energy?
   (1) red   (3) black
   (2) white   (4) yellow

The diagram below shows part of the electromagnetic spectrum.

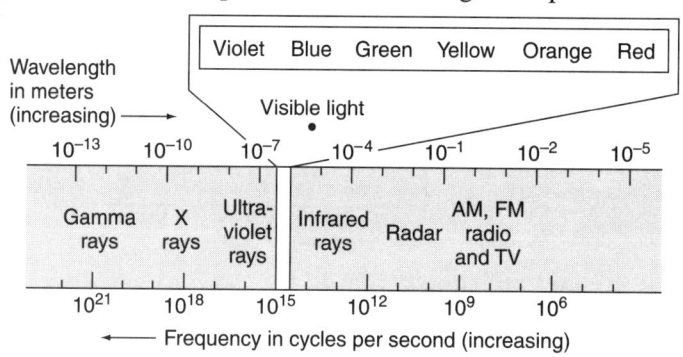

8. Which form of electromagnetic energy shown on the diagram has the lowest frequency and longest wavelength?
   (1) AM radio   (3) red light
   (2) infrared rays   (4) gamma rays

9. In which region of the electromagnetic spectrum is most of the outgoing radiation from Earth?
   (1) infrared
   (2) visible
   (3) ultraviolet
   (4) X ray

10. When is an object in radiative balance?
    (1) when the radiation emitted by the object is equal to that absorbed by the object
    (2) when the radiation emitted by the surroundings is equal to that absorbed by the object
    (3) when the radiation emitted by the object is equal to that absorbed by the surroundings
    (4) when the wavelength of the radiation emitted by the object is equal to that absorbed by the object

Base your answers to questions 11 through 14 on your knowledge of earth science, the *Earth Science Reference Tables*, and the diagram and graph below. In the diagram, equal masses of water and soil are located at identical distances from the lamp. Both were heated for 10 minutes, and then the lamp was removed. The water and soil were then allowed to cool for 10 minutes. The graph shows the temperature data obtained during the investigation.

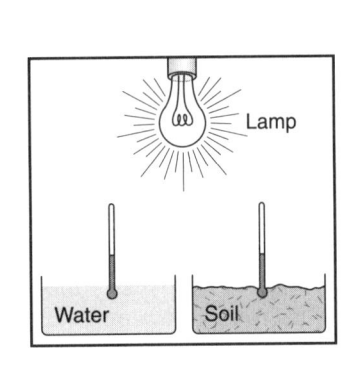

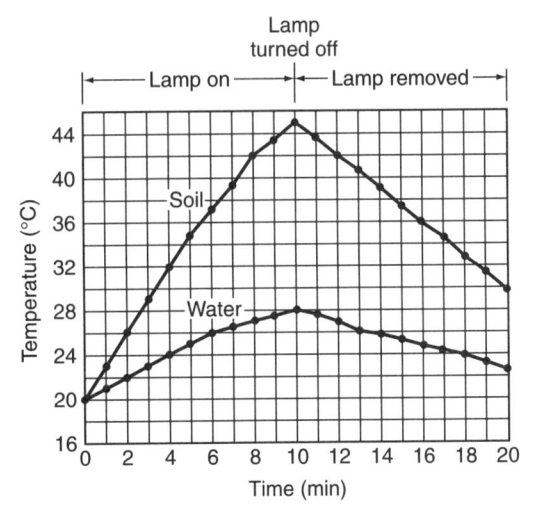

11. What were the temperature readings of the water and soil at the time the lamp was turned off?
    (1) water: 20°C, soil: 20°C
    (2) water: 23°C, soil: 30°C
    (3) water: 28°C, soil: 45°C
    (4) water: 45°C, soil: 28°C

The Atmosphere

12. By which process was most of the energy transferred between the lamp and the water during the first 10 minutes of the investigation?
    (1) conduction
    (2) convection
    (3) reflection
    (4) radiation

13. What was the rate at which the soil temperature changed during the first 10 minutes of the investigation?
    (1) 0.8 C°/min
    (2) 2.5 C°/min
    (3) 8 C°/min
    (4) 25 C°/min

14. Compared to the water, the soil became warmer during the heating period because the soil
    (1) has a lower specific heat
    (2) was closer to the lamp
    (3) reradiated less heat
    (4) has a lower density

15. Which model best represents how a greenhouse remains warm as a result of insolation from the Sun?

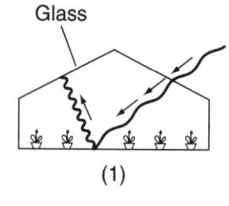

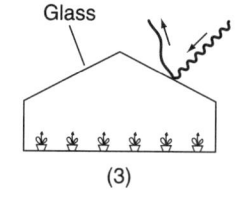

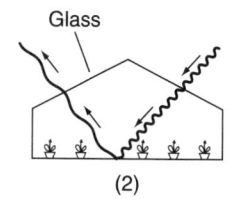

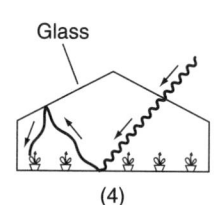

    Key
    ～～～ Short waves
    —— Long waves

16. Which latitude on Earth would receive the highest average yearly insolation per square meter of surface if the atmosphere were completely transparent at all locations?
    (1) 90°N
    (2) 23½°N
    (3) 0°
    (4) 23½°S

17. Which graph best illustrates the relationship between the angle of insolation and the time of day at a location in New York State?

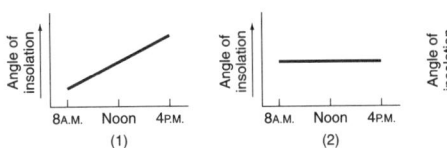

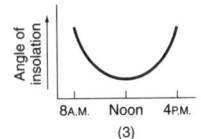

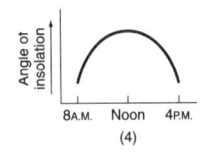

18. Which graph best represents the relationship between latitude north of the Equator and the length of the daylight period on March 21?

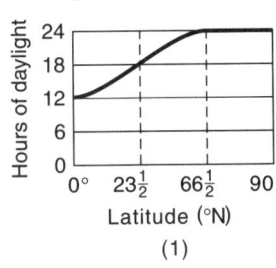

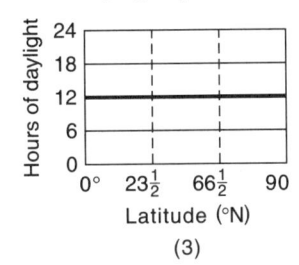

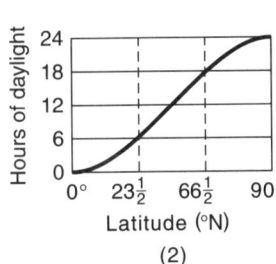

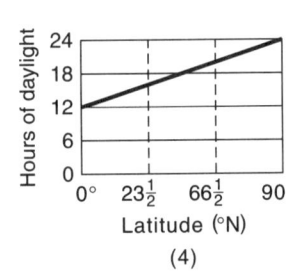

19. What is the usual cause of the drop in temperature that occurs between sunset and sunrise at most New York State locations?
    (1) strong winds
    (2) ground radiation
    (3) cloud formation
    (4) heavy precipitation

20. The diagram to the right represents a portion of Earth's surface that is receiving insolation. Positions A, B, C, and D are located on the surface. 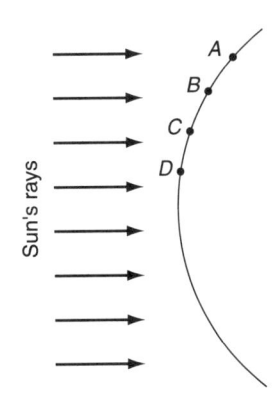
    At which position would the intensity of insolation be greatest?
    (1) A
    (2) B
    (3) C
    (4) D

551

The Atmosphere

Base your answers to questions 21 through 27 on the *Earth Science Reference Tables*, the diagrams and graph below, and your knowledge of earth science. The diagrams show the general effect of Earth's atmosphere on insolation from the Sun at middle latitudes during both clear-sky and cloudy-sky conditions. The graph shows the percentage of insolation reflected by Earth's surface at different latitudes in the Northern Hemisphere in winter.

Insolation in the Atmosphere

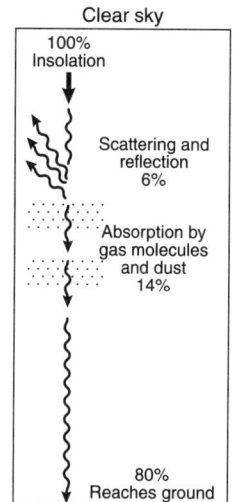

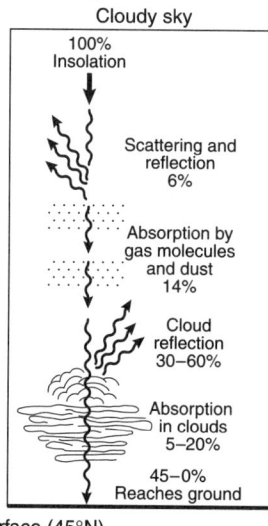

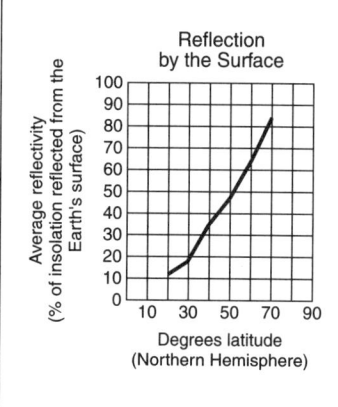

Clear sky:
- 100% Insolation
- Scattering and reflection 6%
- Absorption by gas molecules and dust 14%
- 80% Reaches ground

Cloudy sky:
- 100% Insolation
- Scattering and reflection 6%
- Absorption by gas molecules and dust 14%
- Cloud reflection 30–60%
- Absorption in clouds 5–20%
- 45–0% Reaches ground

Earth's surface (45°N)

Reflection by the Surface (graph: Average reflectivity (% of insolation reflected from the Earth's surface) vs. Degrees latitude (Northern Hemisphere))

21. Approximately what percentage of the insolation actually reaches the ground at 45°N on a clear day?
    (1) 100%   (3) 60%
    (2) 80%   (4) 45%

22. Which factor keeps the greatest percentage of insolation from reaching Earth's surface on cloudy days?
    (1) absorption by cloud droplets
    (2) reflection by cloud droplets
    (3) absorption by clear-air gas molecules
    (4) reflection by clear-air gas molecules

23. According to the graph, on a winter day at 70°N, what approximate percentage of the insolation is reflected by Earth's surface?
    (1) 50%   (3) 85%
    (2) 65%   (4) 100%

24. Which statement best explains why, at high latitudes, reflectivity of insolation is greater in winter than in summer?
    (1) The North Pole is tilted toward the Sun in winter.
    (2) Snow and ice reflect almost all insolation.
    (3) The colder air holds much more moisture.
    (4) Dust settles quickly in cold air.

25. The radiation that passes through the atmosphere and reaches Earth's surface has the greatest intensity in the form of
    (1) visible-light radiation
    (2) infrared radiation
    (3) ultraviolet radiation
    (4) radio-wave radiation

26. Electromagnetic energy that reaches Earth from the Sun is called
    (1) insolation   (3) specific heat
    (2) conduction   (4) terrestrial radiation

27. Which two factors determine the number of hours of daylight at a particular location?
    (1) longitude and season
    (2) longitude and Earth's average diameter
    (3) latitude and season
    (4) latitude and Earth's average diameter

The Atmosphere

Base your answers to questions 28 through 32 on your knowledge of earth science and the graph below, which shows measurements of the insolation above Earth's atmosphere and at Earth's surface on a clear day. [Note that the graph does not show the entire solar spectrum at the longer wavelengths.]

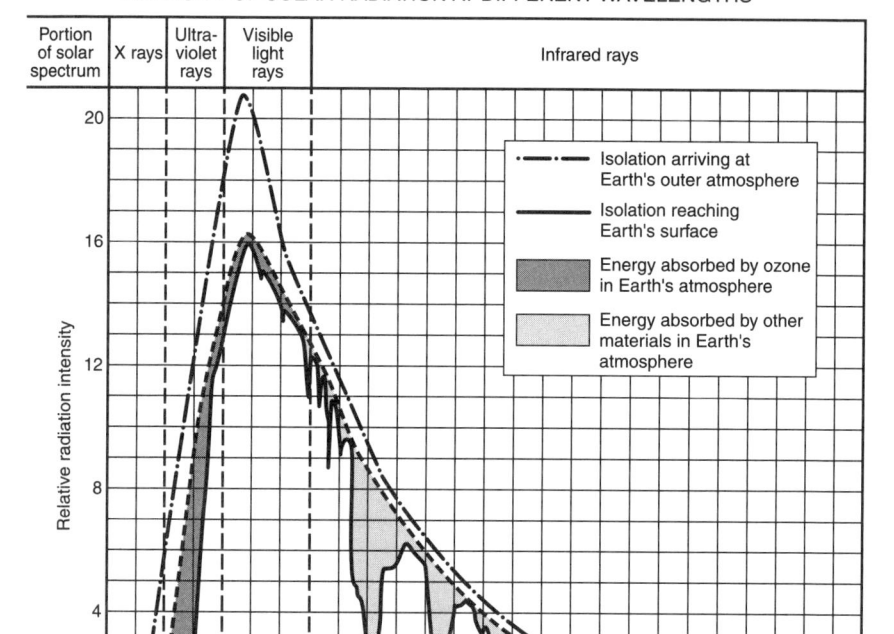

INTENSITY OF SOLAR RADIATION AT DIFFERENT WAVELENGTHS

28. Which of the following types of solar radiation has the longest wavelength?
    (1) X rays
    (2) ultraviolet rays
    (3) visible light rays
    (4) infrared rays

29. The greatest intensity of energy reaching the outer atmosphere of Earth from the Sun has a wavelength of approximately
    (1) 4.5×10^0 angstroms
    (2) 4.5×10^3 angstroms
    (3) 3.0×10^3 angstroms
    (4) 9.1×10^2 angstroms

30. In which portion of the solar spectrum does ozone absorb the greatest amount of energy?
    (1) X rays
    (2) ultraviolet rays
    (3) visible light rays
    (4) infrared rays

31. What quantity is most likely represented by the area between the "dot-dash" curve (- ·· -) and the "dash" curve (- - -)?
    (1) the amount of radiation given off by the Sun
    (2) the amount of radiation absorbed in outer space
    (3) the amount of insolation reflected by the atmosphere back into space
    (4) the amount of insolation absorbed by Earth's surface

32. According to the graph, in which portion of the solar spectrum is the greatest total amount of energy absorbed by ozone and other materials in Earth's atmosphere?
    (1) ultraviolet rays and infrared
    (2) X rays and visible light
    (3) visible light and infrared
    (4) X rays and infrared

33. Which factor most affects the total amount of solar energy received by a location on Earth's surface during a period of 1 year, assuming no cloud cover?
    (1) climate
    (2) longitude
    (3) latitude
    (4) distance from a body of water

34. Earth's axis of rotation is tilted 23½° from a line perpendicular to the plane of its orbit. What would be the result if the tilt was only 13½°?
    (1) shorter days and longer nights at the Equator
    (2) colder winters and warmer summers in New York State
    (3) less difference between winter and summer temperatures in New York State
    (4) an increase in the amount of solar radiation received by Earth

35. For which location and date will the duration of insolation be shortest?
    (1) 60°N on June 21
    (2) 23½°N on June 21
    (3) 60°N on December 21
    (4) 23½°N on December 21

36. A surface will most effectively reflect insolation if it is
    (1) rough and light-colored
    (2) smooth and light-colored
    (3) rough and dark-colored
    (4) smooth and dark-colored

The Atmosphere

37. When visible light strikes a snow-covered, flat field at a low angle, most of the energy will be
    (1) absorbed by the snow
    (2) refracted by the snow
    (3) reflected by the snow
    (4) radiated by the snow

38. Bodies of water cool slower than land areas because
    (1) some insolation is converted into potential energy as water evaporates
    (2) bodies of water have a lower density than land areas
    (3) water has a higher specific heat than land
    (4) water is a better reflector of sunlight than land

Base your answers to questions 39 through 44 on your knowledge of earth science and the graph below. This graph shows the varying amounts of insolation received at Brockport, New York, on three different dates under clear or partly cloudy skies.

Insolation Received at Brockport, N.Y. (43°13'N, 77°56'W)

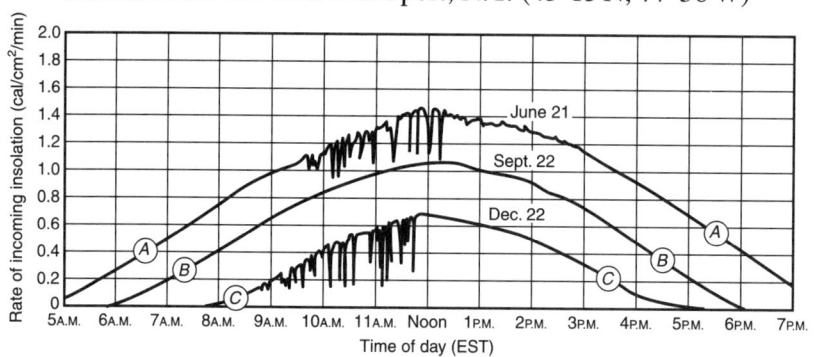

39. The duration of insolation at Brockport on September 22 was approximately
    (1) 6 hr
    (2) 9 hr
    (3) 12 hr
    (4) 15 hr

40. Which graph most nearly represents the maximum rate of insolation on the three dates plotted above?

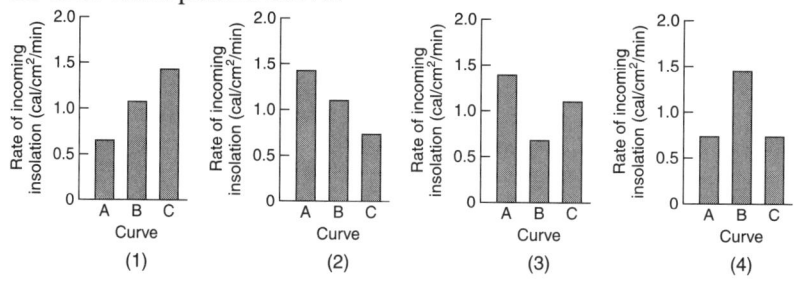

556

41. Why is there a difference in the times of sunrise and sunset for each of the three curves?
    (1) The duration of insolation varied for each of the three days.
    (2) The amount of cloud cover varied for each of the three days.
    (3) The daily temperature varied for each of the three days.
    (4) The total energy output of the Sun varied for each of the three days.

42. What most likely caused the irregular changes in insolation shown on curves *A* and *C* between 9 A.M. and noon?
    (1) temperature changes within the atmosphere
    (2) variations in the Sun's energy output
    (3) instrument error
    (4) changes in the amount of cloud cover

43. How would insolation curves for the same three dates in Rome, New York (43°13' N, 75°27' W) compare to the curves for Brockport, New York. (43°13' N, 77°56' W), assuming the same type of sky and cloud conditions?
    (1) The Rome curves would appear essentially the same.
    (2) The Rome curves would be higher and longer.
    (3) The Rome curves would be lower and shorter.
    (4) Rome curves *A* and *B* would be higher and longer, but curve *C* would be lower and shorter.

44. The change from the vapor phase to the liquid phase is called
    (1) evaporation   (3) precipitation
    (2) condensation   (4) transpiration

45. A sample of water undergoes the phase changes from ice to vapor and back to ice as shown in the model below. During which phase change does the sample gain the greatest amount of energy?

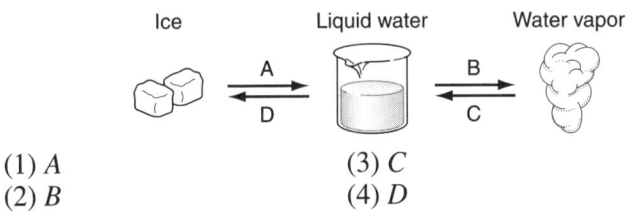

    (1) *A*   (3) *C*
    (2) *B*   (4) *D*

Note that question 46 has only three choices.

46. If equal masses of water in various phases (states) are compared, which phase will contain the greatest amount of stored energy (latent heat)?
    (1) solid ice
    (2) liquid water
    (3) water vapor

The Atmosphere

47. Why is the condensation of water vapor considered to be a process that heats the air?
(1) Liquid water has a lower specific heat than water vapor.
(2) Energy is released by water vapor as it condenses.
(3) Water vapor must absorb energy in order to condense.
(4) Air can hold more water in the liquid phase than in the vapor phase.

48. An ice cube is placed in a glass of water at room temperature. Which heat exchange occurs between the ice and the water within the first minute?
(1) The ice cube gains heat, and the water loses heat.
(2) The ice cube loses heat, and the water gains heat.
(3) Both the ice cube and the water gain heat.
(4) Both the ice cube and the water lose heat.

49. Which process results in a release of latent heat energy?
(1) melting of ice
(2) heating of liquid water
(3) condensation of water vapor
(4) evaporation of water

50. The diagram below shows a container of water that is being heated.

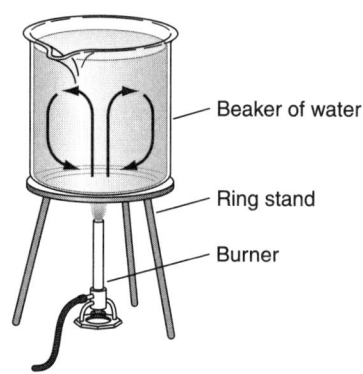

The movement of water shown by the arrows is most likely caused by
(1) density differences
(2) insolation
(3) the Coriolis effect
(4) Earth's rotation

Note that question 51 has only three choices.
51. As a sample of very moist air rises from sea level to a higher altitude, the probability of condensation occurring in that air sample will
    (1) decrease
    (2) increase
    (3) remain the same

52. Between the years 1850 and 1900, records indicate that Earth's mean surface temperature showed little variation. This would support the inference that
    (1) Earth was in radiative balance
    (2) another ice age was approaching
    (3) more energy was entering than was leaving Earth
    (4) the Sun was emitting more energy

53. An observer noted that, even though the ice on a pond was melting, the temperature of the liquid water in the pond remained constant until all the ice melted. Which statement best explains this observation?
    (1) The angle of insolation was not great enough.
    (2) The daylight hours are longer than the nighttime hours.
    (3) The nighttime hours are longer than the daylight hours.
    (4) The heat energy was being used to change the solid ice to liquid water.

Base your answers to questions 54 through 57 on your knowledge of earth science and on the diagram below. The diagram shows the apparatus used as a model to study large-scale motions within the Earth's atmosphere. The water was heated for several minutes.

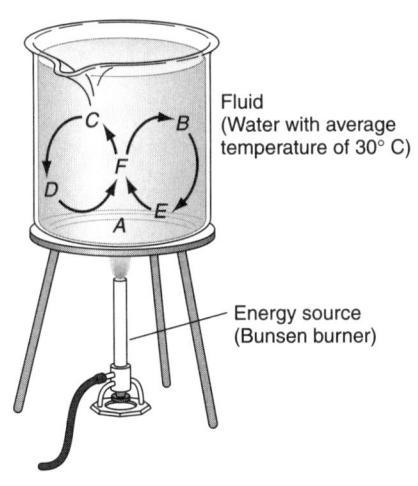

The Atmosphere

54. What type of energy transfer is indicated by the arrows in the diagram?
    (1) conduction   (3) radiation
    (2) convection   (4) insolation

55. The density of the water in the beaker is least at point
    (1) A   (3) C
    (2) B   (4) D

56. What primary source of energy causes Earth's atmosphere to move in a manner similar to the motion of the water in the beaker?
    (1) ocean currents   (3) the Sun
    (2) the Moon   (4) tides

Note that question 57 has only three choices.
57. As water moves from A to F to B, the water's kinetic energy will generally
    (1) decrease
    (2) increase
    (3) remain the same

58. The Coriolis effect would be influenced most by a change in Earth's
    (1) rate of rotation
    (2) period of revolution
    (3) angle of tilt
    (4) average surface temperature

59. The diagram below represents an activity in which an eye dropper was used to place a drop of water on a spinning globe. Instead of flowing due south toward the target point, the drop appeared to follow a curved path and missed the target.

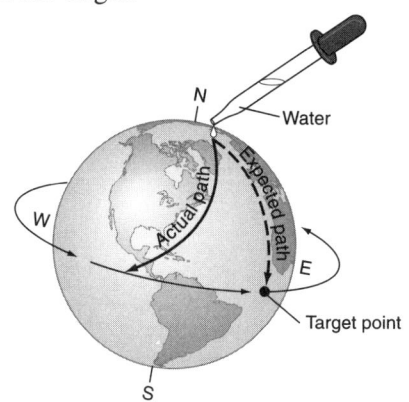

   The curved-path phenomenon most directly affects Earth's
   (1) tilt   (3) atmospheric circulation
   (2) Moon phases   (4) tectonic plates

560

CONSTRUCTED RESPONSE QUESTIONS

60. Base your answers to (a) through (c) on the graph below, which shows the average daily temperatures and the duration of insolation for a location in the midlatitudes of the Northern Hemisphere during a year.

Surface Temperature (—) versus Duration of Insolation (- - -)

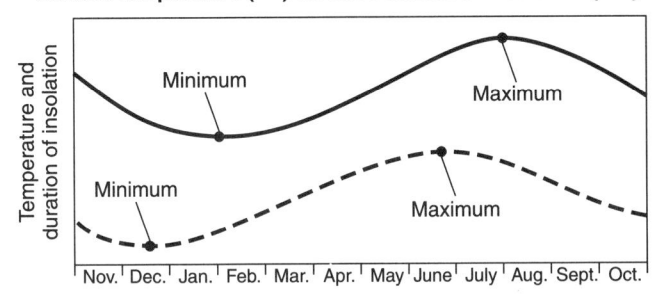

(a) State the month of the year during which insolation reaches its minimum at this location. [1]
(b) State how the relationship between surface temperature and duration of insolation during the months of January and July differs from their relationship during most other months of the year. [2]
(c) In one or more complete sentences, explain why surface temperatures continue to increase between late June and early August even though the duration of insolation is decreasing during the same time period. [2]

61. In one or more complete sentences, explain why the formation of ice crystals in the atmosphere (snow) has a warming effect on the atmosphere. [2]

62. State the relationship between the angle of insolation and the intensity of the radiation striking the ground. [2]

63. In one or more complete sentences, explain why Earth is not constantly increasing in temperature even though the planet is constantly receiving energy from the Sun. [2]

64. In one or more complete sentences, explain why, even though the North Pole receives 24 hours of insolation for part of the year, it is not the warmest place on Earth during that time. [2]

65. In one or more complete sentences, explain which of the following would be more effective in cooling a drink: adding 10 grams of water at 0°C or adding 10 grams of ice at 0°C. [2]

The Atmosphere

66. Base your answers to parts (a) through (d) on the graph below, which shows the temperatures recorded when a sample of water was heated from −100°C to +200°C. The water received the same amount of heat every minute.

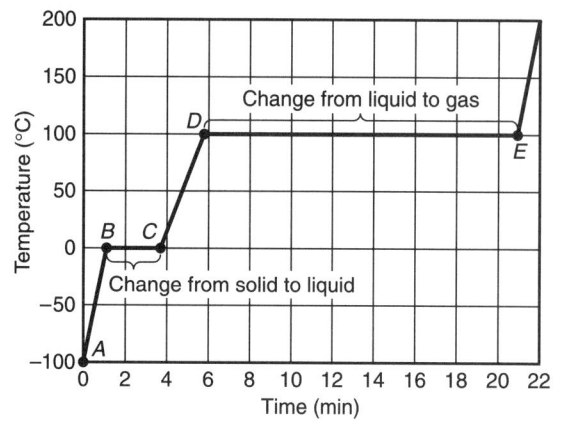

(a) In one or more complete sentences, explain why the water did not change temperature between points *B* and *C* or between points *D* and *E*, even though the water received the same amount of heat energy each minute. [1]
(b) State the lettered points between which the water gained the greatest amount of heat energy. [1]
(c) State the lettered points between which the water was changing from a solid to a liquid. [1]
(d) The water received the same amount of heat every minute. In one or more complete sentences, explain why the water temperature remained unchanged between points *B* and *C* for less than 2 minutes, but remained unchanged between points *D* and *E* for more than 10 minutes. [1]

EXTENDED CONSTRUCTED RESPONSE QUESTIONS

67. Base your answers to parts (a) and (b) on the table below, which provides information about several different gases and their relationships to the greenhouse effect.

Name of Gas	Symbol	Concentration in Troposphere (parts per billion)	Relative Greenhouse Effect per Kilogram of Gas (in equal concentrations)	Decay Time (years)
Carbon dioxide	CO_2	353,000	1	120
Methane	CH_4	1,700	70	10
Ozone	O_3	10–50	1,800	0.1
Chloro-fluorocarbon	CFC-11	0.28	4,000	6.5
	CFC-12	0.48	6,000	120

(a) In one or more complete sentences, state one reason why a scientist would recommend restricting the emission of chlorofluorocarbon CFC-12 into the atmosphere, based on the information in the table. [2]

(b) Which of the following would contribute more to the greenhouse effect? State a reason for your answer. [2]

Releasing 1 kilogram of methane directly into the atmosphere

or

Burning 1 kilogram of methane, resulting in the release of about 3 kilograms of carbon dioxide into the atmosphere

68. Base your answers to parts (a) through (c) on the diagram below, which shows identical solar electric cells placed on the roofs of two adjacent homes, *A* and *B*, in New York City.

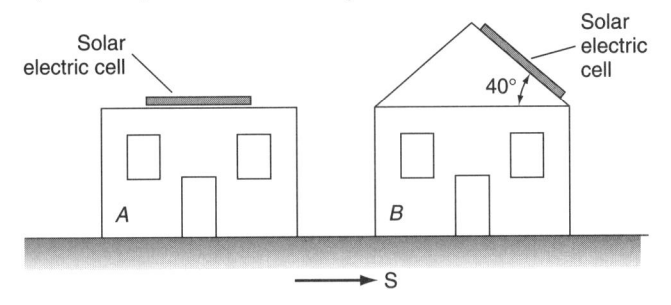

(a) State on which home the roof-mounted solar cell will produce more electricity on March 21 at solar noon. [1]
(b) In one or more complete sentences, explain why the home you chose will produce more electricity on March 21 at solar noon. [1]
(c) Explain how the output of home *B*'s solar cell would change if it was relocated to the north side of the roof. [1]

69. Base your answers to parts (a) through (c) on the chart below, which shows the amount of solar energy reflected back into space by various surfaces.

Type of Surface	Solar Energy Reflected Back Into Space (%)
Thick clouds	75-90
Thin clouds	30-50
Water	10
Sand	15-45
Grassy field	10-30
Fresh snow	75-95
Forest	3-10

(a) In one or more complete sentences, explain why water along a shore heats up less than the adjacent sand on a summer day even though sand reflects more solar energy back into space. [1]

(b) State one reason why a grassy field reflects less solar energy than a snow-covered field. [1]
(c) Scientists have determined that Earth is currently experiencing a period of global warming that is believed to be the result of an increased greenhouse effect. Higher temperatures favor increased evaporation and cloud formation. According to the chart, how would an increase in cloud cover affect the rate of global warming? [1]

CHAPTER 23
WEATHER

KEY IDEAS Weather, the present condition of the atmosphere at any given location, is described by the physical characteristic of the atmosphere. Weather patterns become evident when atmospheric variables are observed, measured, and recorded. Atmospheric variables include air temperature, air pressure, moisture (relative humidity and dew point), precipitation (such as rain, snow, hail, and sleet), wind speed and direction, and cloud cover. Atmospheric variables are measured using instruments such as thermometers, barometers, psychrometers, precipitation gauges, anemometers, and wind vanes.

Atmospheric variables can be represented in a variety of formats including radar and satellite images, synoptic weather maps, atmospheric cross-sections, and computer models. Analysis of atmospheric variables reveals that they are interrelated. Atmospheric moisture, temperature and pressure distributions, winds and jet streams, air masses, frontal boundaries, and the movement of cyclonic systems and associated tornadoes, thunderstorms, and hurricanes occur in observable patterns. Loss of property, personal injury, and loss of life can be reduced by effective emergency preparedness.

KEY OBJECTIVES
Upon completion of this unit, you will be able to:
- Define weather, and identify the atmospheric characteristics used to describe it.
- Describe the relationships between important atmospheric characteristics.
- Explain how clouds and precipitation form.
- Construct weather maps, given weather information at numerous locations, and identify the positions of air masses and fronts.
- Predict weather and the movement of weather systems based on current weather maps.
- Describe the weather conditions associated with the various types of frontal boundaries.

- Explain how weather information can be used to make forecasts.
- Identify the conditions that typically lead to hazardous weather such as hurricanes and tornadoes.

HOW CAN WEATHER BE DESCRIBED?

Weather is the present condition of the atmosphere at any location. Weather changes are mainly the result of unequal heating, by solar radiation, of Earth's landmasses, oceans, and atmosphere. As heat energy is transferred at the boundaries between the atmosphere, oceans, and landmasses, layers of different temperatures and densities form in the atmosphere. When gravity acts on these layers of different densities, they rise and fall, producing circulation of air in the atmosphere. This circulation causes the air at any one location to constantly be moved away and replaced by air from a different location—with a resulting change in weather.

The constantly changing weather is described in terms of the characteristics of the atmosphere that change, or atmospheric variables. The *atmospheric variables* used to describe weather are measured with weather instruments and include the temperature, pressure, humidity, movements, and transparency of the air. Atmospheric variables are interrelated, so that a change in one is likely to cause a change in another. This interrelatedness makes weather prediction a complex task. However, measurement of atmospheric variables and the creation of field maps based on these data reveal large-scale patterns that can be used to make predictions.

Air Temperature

The air temperature at any given location is related to the heat energy present in the atmosphere at that location. Solar radiation is the main source of heat energy in the atmosphere; therefore anything that influences solar radiation will ultimately influence air temperature. Although the radiation emitted by the Sun is fairly constant, the amount that reaches Earth's surface and is converted to heat energy in the atmosphere is influenced by many factors. These include the angle at which solar radiation strikes Earth's surface; the number of hours of solar radiation per day; the reflection, refraction, or absorption of solar radiation by the atmosphere (e.g., cloud cover); and the nature of the surface materials absorbing solar radiation. In general, the more direct the solar radiation and the longer it occurs, the higher the temperature.

The angle at which solar radiation strikes Earth's surface varies during the course of a day. Because of Earth's rotation, radiation is received during the day but not at night. Therefore, air temperature often varies in a daily cycle. Similar changes occur as Earth moves through its orbit around the Sun, resulting in seasonal temperature variations.

Temperature is measured with thermometers. A ***thermometer*** works on a simple principle—matter expands when heated and contracts when cooled, and the amount of expansion or contraction depends on the amount of heat gained or lost. A typical thermometer measures the height of a column of alcohol or mercury in a glass tube. Some modern devices, however, use the change in electrical conductivity of certain metals when heated or cooled to measure temperature. Several scales, named for their creators, have been developed over the years to measure temperature: Fahrenheit, Celsius, and Kelvin. Meteorologists in the United States still use the Fahrenheit scale but are changing over to the Celsius.

Meteorologists use several types of specialized thermometers when measuring air temperature. Simple thermometers are mounted in shaded, vented boxes so that they will measure the temperature of the air and not be heated additionally by absorption of sunlight. Maximum-minimum thermometers have pinched tubes and markers that are pushed or dragged back as the fluid in them rises and falls. This feature enables them to record the highest and lowest temperatures during a time period by the positions of the markers. Continuous temperature readings are made with a thermograph, a thermometer in which a coil of heat-sensitive metal moves a pen against a rotating drum.

Air Pressure

Air is a mixture of gases. A gas consists of many tiny individual molecules that are far apart and moving rapidly. As they whiz to and fro, they are kept from escaping into space by the walls of a container or, in the case of the atmosphere, by Earth's gravity.

Gases exert pressure on surfaces with which they come into contact. Pressure is a force that acts on a surface area. The pressure that a gas exerts on a surface is the result of gas particles colliding with the surface. Since gas particles move randomly in all directions, they exert pressure equally in all directions. Figure 23.1 shows gas molecules speeding around in a container and striking surfaces held at different angles.

Air pressure is not an atmospheric variable that you can sense, in the way that you feel temperature or see cloud cover. By itself, knowledge of the air pressure in one location is of little use in weather forecasting. However, *field maps of air pressure in many locations* reveal distinct regions within the atmosphere, and *changes* in air pressure signal move-

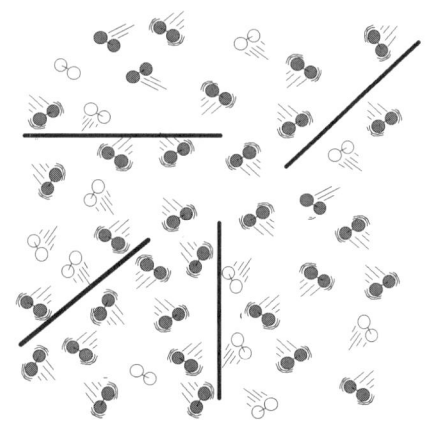

Figure 23.1 Gas Molecules in a Container. Regardless of the angle at which a surface is held, air molecules collide with it, exerting the same pressure in all directions.

Weather

ments of these large masses of air and their associated weather conditions. Rising pressure usually signals the approach or continuation of fair weather; falling pressure, the approach of a storm.

Air pressure is measured with a barometer. Two commonly used barometers are the mercury barometer and the aneroid barometer.

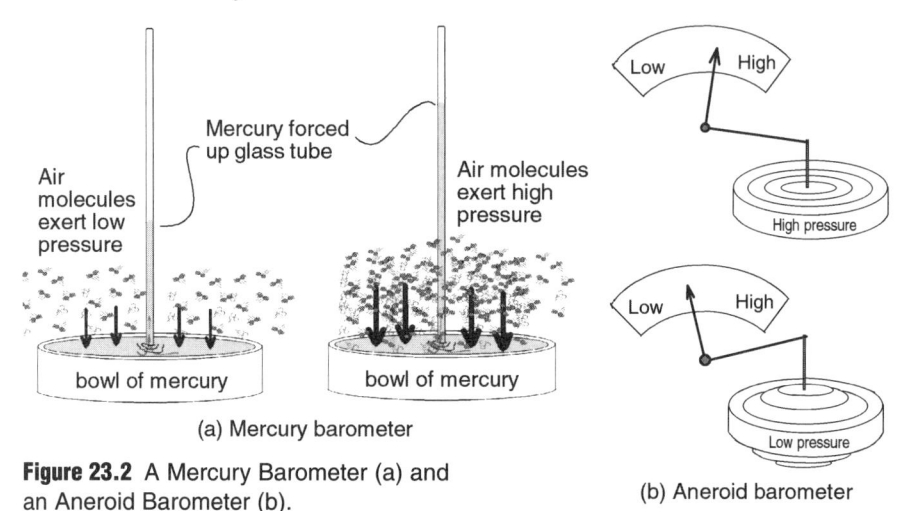

Figure 23.2 A Mercury Barometer (a) and an Aneroid Barometer (b).

(a) Mercury barometer

(b) Aneroid barometer

In a mercury barometer (see Figure 23.2a), air pressure forces liquid mercury up a tube. The mercury rises in the tube because the tube contains a vacuum and no pressure is exerted on the mercury inside the tube. The higher the air pressure outside the tube, the higher the mercury is pushed up in the tube. Since pressure is measured by the height to which the mercury is pushed, pressure is often described as "inches or millimeters of mercury." The word *aneroid* means no liquid, and an aneroid barometer (see Figure 23.2b) is a can with no liquid inside it. The can contains a spring scale. As air pressure on the sides of the can increases or decreases, the spring compresses and expands and the scale records the magnitude of the force.

Pressure is measured in units of force per unit area. A variety of units are used to describe atmospheric pressure. The ***atmosphere*** (atm) is the average pressure exerted by Earth's atmosphere at sea level and is equal to 1,013.2 millibars or 14.7 pounds per square inch. The bar is equal to 1×10^5 newtons per square meter, and the *millibar* is 1/1,000 of a bar or 100 newtons per square meter. (1 newton = 0.225 lb.) In meteorology, atmospheric pressure is measured in millibars. Normal atmospheric pressure is 1,013.2 millibars.

Air pressure is often misunderstood because much of our personal experience with pressure involves solids pressing against us, or our bodies pressing against something because of gravity acting on us. Since the molecules of solids are bonded together in a single unit, they act as a single unit. Under the influence of gravity the entire object is pulled downward and exerts a downward pressure. In gases, on the other hand, each tiny molecule acts independently. Gravity pulls the molecule down; but since it is hurtling through space, the effect is like that of gravity acting on a baseball speeding toward home. The ball sinks in a downward trajectory but still exerts a sizable horizontal force, as any professional catcher will attest. Think of gases as millions of tiny baseballs flying in all directions at the same time. Anything that affects the strength and the number of collisions that gas molecules will make with a surface also affects the pressure they exert.

Air Humidity

Humidity is the amount of moisture in the air. Humidity is not liquid water that is suspended in the air. Most of the moisture in the air is in the form of water vapor, a colorless, odorless gas that enters the atmosphere when liquid water evaporates or ice sublimes (changes directly from a solid to a vapor). Humidity is an important weather factor because the water vapor in the atmosphere is the source of the water that condenses to form clouds, fog, and precipitation.

Water molecules are always changing back and forth between phases (the three states of matter: vapor, liquid, and solid). Most water molecules enter the atmosphere from liquid water by evaporation and from the surface of ice and snow by sublimation (molecules leave a solid and enter directly into the vapor phase). Water leaves the atmosphere by condensation or deposition (molecules of water vapor form solid crystals without first becoming a liquid). This constant flow of water molecules between phases results in constantly changing levels of humidity in the atmosphere. By far, the greatest amount of activity occurs at the boundary between the atmosphere and the hydrosphere. Molecules of liquid water in the hydrosphere that gain energy from sunlight and their surroundings may become energetic enough to escape the liquid phase and enter the atmosphere as water vapor. When water vapor molecules in the atmosphere come into contact with a liquid water surface, they may be trapped there, so there is a constant two-way traffic of molecules to and from the liquid. The flow of molecules leaving the surface of the liquid and entering the atmosphere as water vapor (*evaporation*) is strongly influenced by temperature. At higher temperatures, molecules have more energy, so a greater number are likely to escape and enter the vapor phase. The flow of molecules arriving at the surface to exit the atmosphere as a liquid (*condensation*) depends on how densely crowded with molecules the water vapor above the liquid becomes. The more densely crowded the molecules, the more likely they are to come into contact with the surface and be trapped.

If more water molecules are leaving the surface than are arriving, there is net evaporation. If more water molecules are arriving at the surface than leaving, there is net condensation. If there is net evaporation, the amount of water vapor in the atmosphere increases. However, as you just learned, if the amount of water vapor in the atmosphere increases, molecules are more likely to come into contact with the surface and be trapped, so condensation also increases. Eventually, a point is reached where condensation and evaporation are in equilibrium, and the amount of water vapor above the surface remains unchanged. See Figure 23.3. The higher the temperature, the greater the amount of water vapor present in the atmosphere before equilibrium is reached. At 0°C, equilibrium is reached when air contains 5 grams of water vapor per cubic meter; at 20°C the amount is 17 grams per cubic meter, and at 100°C it is 598 grams per cubic meter! If for some reason (such as a sudden drop in temperature) the amount of water vapor exceeds the equilibrium amount, condensation will occur at a faster rate than evaporation until equilibrium is reestablished.

Net evaporation   Equilibrium   Net condensation

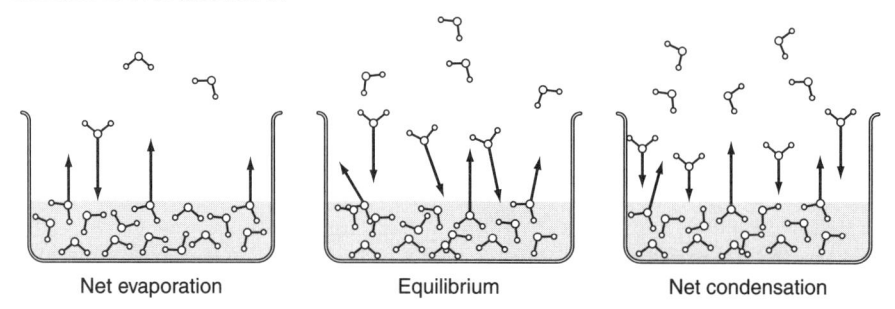

Figure 23.3 Net Evaporation and Net Condensation in Equilibrium. Equilibrium occurs when there are equal flows of molecules leaving and arriving at the surface of the liquid. Net evaporation occurs when the water-vapor content of the atmosphere is less than the equilibrium amount. Net condensation occurs when the water-vapor content of the atmosphere is greater than the equilibrium amount.

The amount of water vapor in the atmosphere is often expressed as absolute humidity or relative humidity. *Absolute humidity* is the number of grams of the water vapor molecules in 1 cubic meter of air. Since it is difficult to isolate the water vapor in air and to measure its mass, absolute humidity is seldom directly measured.

Relative humidity is the ratio of the water vapor now present in the atmosphere to the water vapor present when evaporation and condensation are in equilibrium at that temperature, times 100 percent.

$$\text{Relative humidity} = \frac{\text{Water vapor in atmosphere}}{\text{Water vapor in atmosphere at equilibrium}} \times 100\%$$

Relative humidity is a way of comparing the flows of water molecules leaving and arriving at a surface, and is useful in predicting how the air will feel to a person and whether condensation or evaporation is more likely to occur in the atmosphere.

A relative humidity of 20 percent means that the atmosphere currently contains only 20 percent of the water vapor it contains at equilibrium. As you have learned, at a temperature of 0°C, condensation and evaporation reach equilibrium when the amount of water vapor in the air reaches 5 grams per cubic meter. A relative humidity of 20 percent means that the air currently contains only 1 gram per cubic meter (20% of 5 g/m^3 = 1 g/m^3.) Therefore, there will be net evaporation until the air gains an additional 4 grams per cubic meter and reaches equilibrium (at which point there is no net evaporation).

A relative humidity of 100 percent means that the atmosphere contains 100 percent of the water vapor it contains at equilibrium. Therefore, there will be no net evaporation *or* net condensation. The higher the relative humidity, the more uncomfortable people feel because less moisture evaporates from their bodies before equilibrium is reached. Since less water evaporates from the skin, a wet and "sticky" feeling results.

It is important to remember, however, that a relative humidity of 100 percent does not mean that the air is 100 percent water vapor. The air still mostly consists of nitrogen and oxygen. The weight of the water vapor represents only a small fraction of the total weight of all the gases in the air. For example, the weight of water vapor in very warm, moist air may represent only 1/25 of the total weight of all the gases in the air. Nevertheless, if that is all of the water vapor the air contains at equilibrium, the relative humidity will be 100 percent. Similarly, the weight of water vapor in frigid arctic air may be as little as 1/10,000 of the total weight of the gases in the air; yet, if that is all of the moisture the air contains at equilibrium, the relative humidity will also be 100 percent.

High relative humidity also means that the water vapor in the atmosphere is close to its equilibrium amount. A decrease in temperature, which, in turn, decreases the equilibrium amount, could trigger condensation, which forms clouds, fog, or precipitation. A commonly used value in this connection is the ***dew point***, that is, the temperature at which the water vapor in the air will begin to condense into liquid water. This is the temperature at which the amount of water vapor currently in the air equals the equilibrium amount of water vapor.

Humidity is measured with a hygrometer or a psychrometer. A hygrometer consists of strands of human hair attached to a pointer. Human hair lengthens slightly as humidity increases. As the hair lengthens and shortens because of changing humidity, the pointer changes position.

A psychrometer consists of two thermometers, one whose bulb is kept dry and another whose bulb is kept wet by covering it with a cloth wick soaked with water. Evaporation from the wet wick causes cooling because the fastest molecules in the liquid water are the ones with enough energy to escape. The slower molecules left behind have a lower average energy; hence the liquid's temperature decreases as measured on the wet-bulb thermometer. See Figure 23.4. This cooling effect is the key to measuring relative humidity. The lower the relative humidity, the more water can evaporate from the wet-bulb thermometer and the more the thermometer is cooled. The dry-bulb thermometer, however, remains unchanged and registers the temperature of the air. Therefore, the difference in temperature between the wet-bulb thermometer and the dry-bulb thermometer is directly related to the relative humidity of the air.

Weather

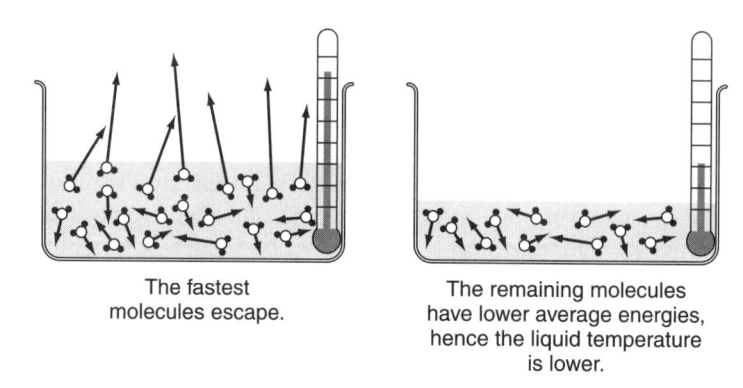

Figure 23.4 After evaporation, the remaining liquid is cooler than before.

The relationship among dry-bulb temperature, wet-bulb temperature, and relative humidity is shown in Table 23.1. To find the relative humidity, first calculate the difference between the wet-bulb and dry-bulb temperatures. Then find on the table the point where the dry-bulb temperature and the difference between the two temperatures intersect. The number at the intersection is the percent relative humidity.

Example:

Find the relative humidity if the dry-bulb temperature is 10°C and the wet bulb temperature is 5°C.

To find the relative humidity you need the *dry-bulb temperature* and the *difference between the dry-bulb and wet bulb temperatures.*

The dry-bulb temperature is given as 10°C.

The difference between the dry-bulb and wet-bulb temperatures must be calculated. It is 10°C − 5°C = 5°C.

The number on the table where these two temperatures intersect is 43. The relative humidity is 43%.

TABLE 23.1 RELATIVE HUMIDITY (%)

Dry-Bulb Temperature (°C)	\multicolumn{16}{c}{Difference Between Wet-Bulb and Dry-Bulb Temperatures (C°)}
	0	1	2	3	4	5	6	7	8	9	10	11	12	13	14	15
−20	100	28														
−18	100	40														
−16	100	48	0													
−14	100	55	11													
−12	100	61	23													
−10	100	66	33	0												
−8	100	71	41	13												
−6	100	73	48	20	0											
−4	100	77	54	32	11											
−2	100	79	58	37	20	1										
0	100	81	63	45	28	11										
2	100	83	67	51	36	20	6									
4	100	85	70	56	42	27	14									
6	100	86	72	59	46	35	22	10	0							
8	100	87	74	62	51	39	28	17	6							
10	100	88	76	65	54	(43)	33	24	13	4						
12	100	88	78	67	57	48	38	28	19	10	2					
14	100	89	79	69	60	50	41	33	25	16	8	1				
16	100	90	80	71	62	54	45	37	29	21	14	7	1			
18	100	91	81	72	64	56	48	40	33	26	19	12	6	0		
20	100	91	82	74	66	58	51	44	36	30	23	17	11	5	0	
22	100	92	83	75	68	60	53	46	40	33	27	21	15	10	4	0
24	100	92	84	76	69	62	55	49	42	36	30	25	20	14	9	4
26	100	92	85	77	70	64	57	51	45	39	34	28	23	18	13	9
28	100	93	86	78	71	65	59	53	47	42	36	31	26	21	17	12
30	100	93	86	79	72	66	61	55	49	44	39	34	29	25	20	16

Source: The State Education Department, *Earth Science Reference Tables*. 2001 ed. (Albany, New York; The University of the State of New York).

The difference between the wet-bulb temperature and the dry-bulb temperature is also directly related to the dew point of the air. The higher the relative humidity, the closer the air is to being filled to its capacity. Cooling air decreases its capacity to hold moisture. The closer the air is to being filled to capacity, the less it has to be cooled to reach the point of maximum moisture. Table 23.2, which shows the relationship among wet-bulb temperature, dry-bulb temperature, and dew point, can be used to find the dew point. The number on the table at the intersection of the dry-bulb temperature and the difference between the wet-bulb and dry-bulb temperatures is the dew point in degrees Celsius.

Example:

Find the dew point if the dry-bulb temperature is 10°C and the wet-bulb temperature is 5°C.

The dry-bulb temperature is given as 10°C.

The difference between the dry-bulb and wet-bulb temperatures must be calculated. It is 10°C − 5°C = 5°C.

The number on the table where these two temperatures intersect is −2. The

Weather

dew point is –2°C.

TABLE 23.2 DEW POINT TEMPERATURES (°C)

Dry-Bulb Tempera-ture (°C)	\multicolumn{16}{c}{Difference Between Wet-Bulb and Dry-Bulb Temperatures (C°)}

Dry-Bulb Temp (°C)	0	1	2	3	4	5	6	7	8	9	10	11	12	13	14	15
–20	–20	–33														
–18	–18	–28														
–16	–16	–24														
–14	–14	–21	–36													
–12	–12	–18	–28													
–10	–10	–14	–22													
–8	–8	–12	–18	–29												
–6	–6	–10	–14	–22												
–4	–4	–7	–12	–17	–29											
–2	–2	–5	–8	–13	–20											
0	0	–3	–6	–9	–15	–24										
2	2	–1	–3	–6	–11	–17										
4	4	1	–1	–4	–7	–11	–19									
6	6	4	1	–1	–4	–7	–13	–21								
8	8	6	3	1	–2	–5	–9	–14								
10	10	8	6	4	1	–2	–5	–9	–14	–28						
12	12	10	8	6	4	1	–2	–5	–9	–16						
14	14	12	11	9	6	4	1	–2	–5	–10	–17					
16	16	14	13	11	9	7	4	1	–1	–6	–10	–17				
18	18	16	15	13	11	9	7	4	2	–2	–5	–10	–19			
20	20	19	17	15	14	12	10	7	4	2	–2	–5	–10	–19		
22	22	21	19	17	16	14	12	10	8	5	3	–1	–5	–10	–19	
24	24	23	21	20	18	16	14	12	10	8	6	2	–1	–5	–10	–18
26	26	25	23	22	20	18	17	15	13	11	9	6	3	0	–4	–9
28	28	27	25	24	22	21	19	17	16	14	11	9	7	4	1	–3
30	30	29	27	26	24	23	21	19	18	16	14	12	10	8	5	1

Source: The State Education Department, *Earth Science Reference Tables*. 2001 ed. (Albany, New York; The University of the State of New York).

Air Movements

In the atmosphere, air circulates because of density differences. The vertical movements of rising and sinking air are called *air currents*. When sinking air reaches Earth's surface, it spreads out horizontally. When rising air expands, it also spreads out horizontally. Horizontal movements of air parallel to Earth's surface are called *winds*. Wind is described by both its speed and its direction.

Wind speed is measured with an anemometer. An anemometer consists of three or four cups mounted on a vertical axis and driven by the wind, causing the axis to spin. The speed of rotation of the axis varies with wind speed. Wind speed is measured in knots, a nautical measure of speed; 1 knot = 1.85 kilometers per hour, or 1.15 miles per hour.

Wind direction is determined with a wind vane. A wind vane is a pointer mounted on an axis that is attached to a compass rose. The tail of the pointer is larger in surface area than the tip. Thus, the wind exerts more pressure on the tail, causing it to swing around so that the tip points in the direction from which the wind is blowing. A wind is named according to the direction from which it blows. Just as a person who comes from the South is called a Southerner, a wind that blows from the south is called a south wind.

Atmospheric Transparency

All of the gases of which air is composed are transparent. However, there are many substances that become suspended in the atmosphere and decrease its transparency. These substances include dust, volcanic ash, smoke, salt particles, aerosols (droplets of liquids), ice particles, and water droplets. By far, water droplets have the greatest effect on atmospheric transparency. Clouds and fog consist of tiny, but visible, droplets of water. Atmospheric transparency is expressed in three ways: visibility, cloud ceiling, and cloud cover.

Visibility is the horizontal distance through which the eye can distinguish objects. It is usually expressed in miles.

Cloud ceiling is the base height of cloud layers. It is measured with a ceilometer, which uses pulses of light and a photoelectric telescope.

Cloud cover is the fraction of the sky that is obscured by clouds.

WHAT ARE THE RELATIONSHIPS AMONG THE SEVERAL ATMOSPHERIC VARIABLES?

Atmospheric variables are interrelated, so that a change in one is likely to cause a change in another. Since the atmosphere is gaseous, this interrelatedness is best understood in light of the kinetic theory of gases. According to the kinetic theory, gases consist of many tiny, individual molecules that do not interact with each other except when they collide. The molecules are far apart and are in constant random motion, and the temperature of the gas is proportional to the speed at which they are moving.

Air Pressure and Air Temperature

As explained earlier, air pressure is the result of the forces exerted by gas molecules colliding with a surface or each other. Air temperature depends on the speed of the gas molecules. Air pressure and temperature are related because both involve the motion of molecules. The higher the temperature, the faster the gas molecules are moving. The faster the gas molecules are moving, the more force they exert in a collision. Thus, you would expect that an increase in temperature would result in an increase in pressure, but in the atmosphere this is not the case. The reason is that the atmosphere is not contained by rigid walls but is an open system free to expand.

When the temperature of the atmosphere increases, its molecules move

Weather

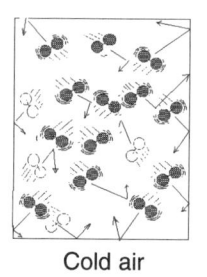

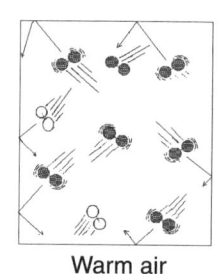

Figure 23.5 Relationship Between Air Temperature and Air Pressure. Even though the molecules are moving slower and collide with less force, the cold air on the left exerts more pressure because its closely spaced molecules collide much more frequently.

Cold air   Warm air

faster and collide more vigorously. As a result, the molecules bounce farther apart and the atmosphere expands. Even though the molecules are moving faster, and each individual impact exerts greater force, the total number of collisions decreases because the molecules are spread farther apart. The decrease in collisions far exceeds any increase in force due to faster moving molecules, so the net effect is a *decrease* in air pressure when air temperature increases. See Figure 23.5.

| Air temperature ↑, Air pressure ↓ | Air temperature ↓, Air pressure ↑ |

Now let us consider what happens to temperature if pressure is changed. See Figure 23.6. The main cause of pressure changes in the atmosphere is the rising or sinking of air in convection currents. When a parcel of air rises, the surrounding air exerts less confining pressure on the parcel and it expands. Conversely, when a parcel of air sinks, the surrounding air exerts more pressure on the parcel and it is compressed.

Any gas that is allowed to expand because the surrounding pressure is reduced will become cooler, and any gas compressed into a smaller volume will become warmer. Such temperature changes, which occur without heat being added from outside sources or heat being lost to the surroundings, are known as ***adiabatic*** changes.

In adiabatic temperature changes, the change in the temperature of the air reflects a change in the degree of crowding of its molecules. If a gas is allowed to expand into a larger volume, its molecules spread farther apart, and collide with one another less frequently. If a thermometer is placed in such air, fewer molecules will col-

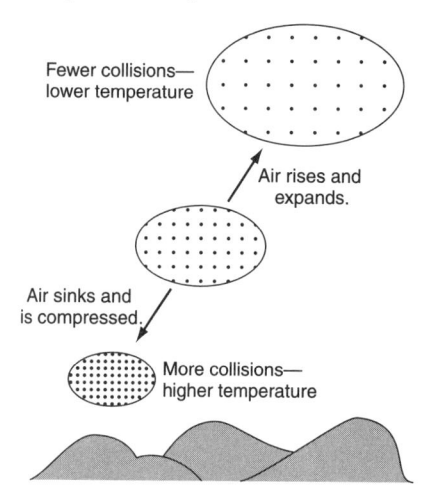

Figure 23.6 Relationship Between Air Pressure and Air Temperature. As air rises, pressure decreases, air molecules spread out and collide less frequently, and temperature decreases. As air sinks, pressure increases, air molecules crowd closer together and collide more frequently, and temperature increases.

lide with it. Fewer collisions means less energy transferred to the thermometer, causing less expansion of the fluid in the thermometer, and the thermometer registers a lower temperature. Thus, a decrease in pressure causes a decrease in temperature.

Compression forces the air molecules to crowd more closely together and causes more frequent collisions. If a thermometer is placed in the compressed air, more molecules will collide with it, and it will register a higher temperature. Thus an increase in pressure causes an increase in temperature.

| Air pressure ↑, Air temperature ↑ | Air pressure ↓, Air temperature ↓ |

Air Pressure and Humidity

The greater the mass of the molecules, the more force they exert during a collision. The mass of a molecule in air depends on the element(s) of which it is composed. Nitrogen and oxygen molecules have more mass than water molecules (molecular mass of N_2 = 28, of O_2 = 32, of H_2O = 18). As water vapor builds up in the atmosphere, lighter water molecules displace heavier nitrogen or oxygen molecules. Since the lighter water molecules exert less force during a collision than the heavier nitrogen and oxygen molecules, air pressure decreases as humidity increases. (Don't be confused by thinking of moist air as containing *liquid* water, which is denser than any gas in air; moist air contains water *vapor*, which is less dense than most gases in air.) Figure 23.7 shows the displacement of heavier gas molecules by lighter water vapor molecules as humidity increases.

| Humidity ↑, Air pressure ↓ | Humidity ↓, Air pressure ↑ |

Total Weight of All Molecules = 352
Dry Air

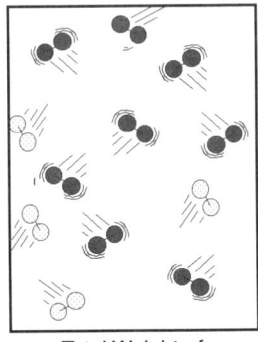

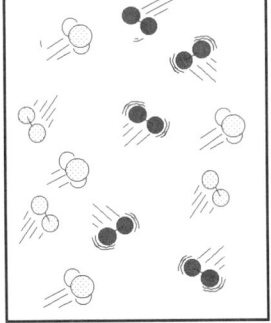

Total Weight of All Molecules = 308
Moist Air

Figure 23.7 Relationship Between Molecular Mass and Density. A volume of dry air has more mass than an equal volume of moist air because water molecules have less mass than the nitrogen and oxygen molecules they displace. Therefore, moist air is less dense than dry air.

Oxygen M.W. = 32
Nitrogen M.W. = 28
Water M.W. = 18

Weather

Air Temperature and Humidity

The higher the temperature, the greater the amount of water vapor present in the atmosphere before equilibruim is reached. Thus, warm air contains more water vapor at equilibrium than cool air. Why is this so? Imagine a sealed jar of dry air half filled with water and kept at a certain temperature. All along the water surface, water molecules evaporate and enter the air as water vapor. At the same time, some of the molecules of water vapor in the air lose energy, drop back to the water surface, and return to the liquid. At first, more water molecules leave by evaporation than return to the liquid, and the amount of water vapor in the jar increases. However, over time, the number of water vapor molecules returning to the liquid balances the number of water molecules leaving the liquid. Then, the amount of water vapor in the jar no longer increases. At any given temperature, a fixed amount of water vapor will be present in the air when equilibrium is reached. See Figure 23.8a.

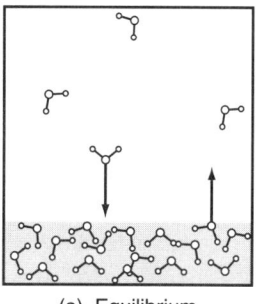

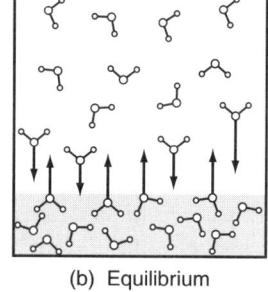

(a) Equilibrium at low temperature
(b) Equilibrium at high temperature

Figure 23.8 At lower temperatures (a) less water vapor is present in the atmosphere when equilibrium is reached than is present at higher temperatures (b).

Now let's consider what happens if we heat the sealed jar. In order for molecules of water in the liquid to break free and become water vapor, they require energy. At a higher temperature, more of the molecules in the liquid water will have enough energy to break free and become a gas. As the water vapor becomes more crowded with molecules, it becomes more likely that some will be trapped at the liquid surface and condense. Thus, the amount of water vapor in the warmed jar increases until a new balance between water molecules leaving and water molecules returning to the liquid is reached. The amount of water vapor in the jar at the higher temperature is greater than it was at the lower temperature. The same volume of space contains more water vapor at equilibrium! See Figure 23.8b. Roughly speaking, for every 10°C increase in temperature, the amount of water vapor in a given volume of air at equilibrium is doubled.

Since the amount of water vapor in the air at equilibrium changes with temperature, the relative humidity is also influenced by temperature. If the temperature increases, causing the equilibrium amount to increase, but the moisture in the air remains the same, the relative humidity will decrease. Conversely, if the temperature decreases, causing the equilibrium amount to decrease, but the moisture in the air remains the same, the relative humidity will increase.

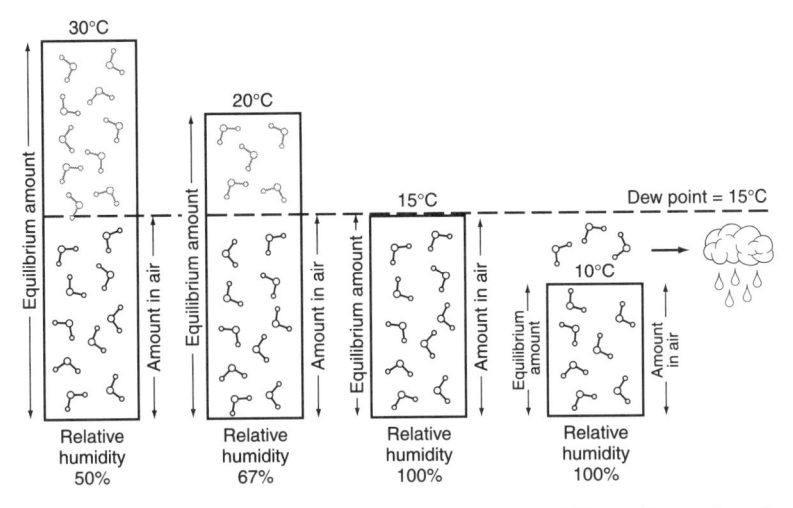

Figure 23.9 Cooling Decreases the Equilibrium Amount of Water Vapor, Leading to Condensation. As air is cooled, the amount of water vapor the air contains approaches the equilibrium amount. At the dew point, evaporation and condensation are in equilibrium. If the air is cooled below the dew point, net condensation results in formation of clouds and precipitation.

As air temperature decreases, it eventually reaches the **dew point**, the temperature at which the moisture the air contains equals the equilibrium amount. At that point, the relative humidity is 100 percent. If the temperature decreases further, moisture in the air will exceed the equilibrium amount, and condensation will occur at a faster rate than evaporation until equilibrium is reestablished. The condensed moisture forms droplets of water or directly forms ice crystals, which, in turn, may form clouds or precipitation. Therefore, as air temperature approaches the dew point, precipitation becomes more likely. See Figure 23.9.

Air Pressure and Winds

Winds blow from regions of high air pressure to regions of low air pressure. This is easy to understand if you think of two people pushing against each other. The person pushing harder will advance against the person pushing with less force. In much the same way, air exerting high pressure will advance against air exerting low pressure. The net movement of the air is from high pressure toward low pressure. See Figure 23.10.

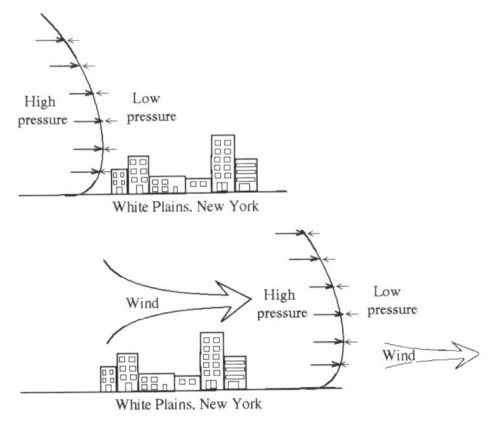

Figure 23.10 Air Pressure and Winds. Winds Blow from High to Low. High-pressure air exerts more force than low-pressure air and pushes the low-pressure air ahead of it, resulting in a horizontal movement of air—a wind.

Weather

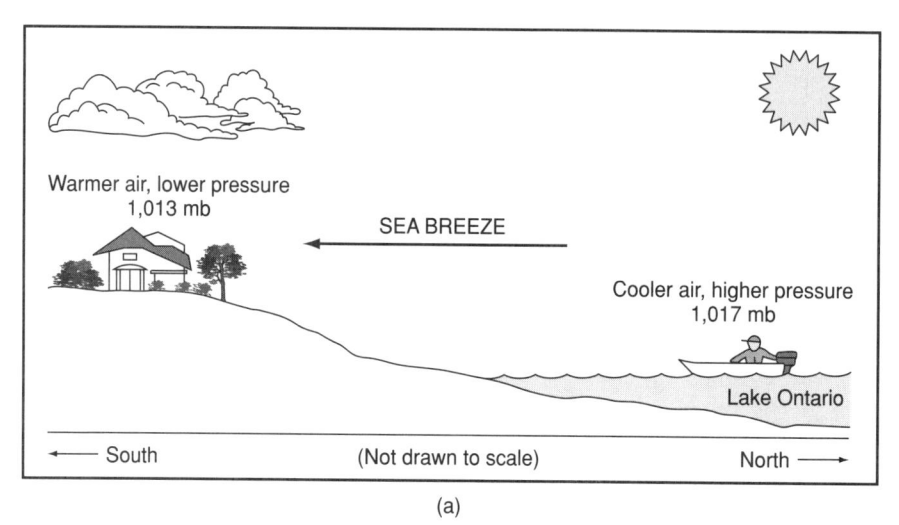

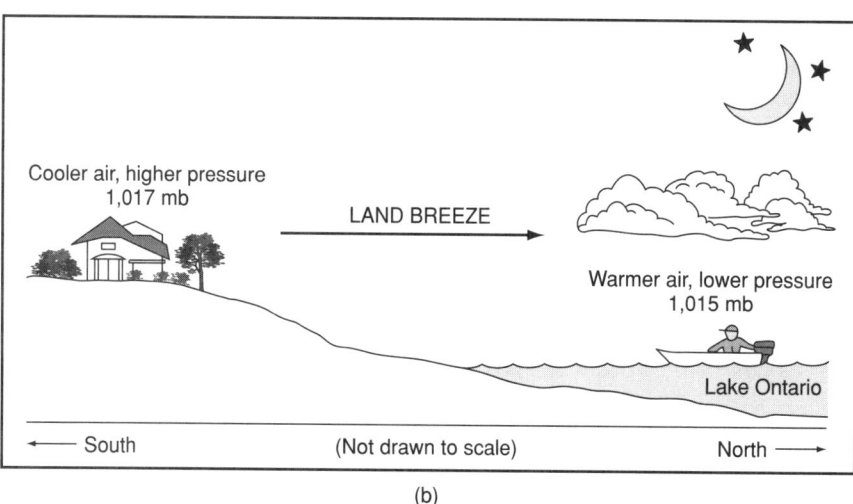

Figure 23.11 Land and Sea Breezes. (a) During the day, land surfaces heat up faster than water. The higher temperatures over the land result in lower air pressures, while the lower temperatures over the water result in higher air pressures. A wind develops that blows from the high-pressure area over the water toward the low-pressure area over the land. (b) At night, land surfaces cool off faster than water and the situation reverses, causing a wind to blow from the land toward the water.

Land and Sea Breezes

Land and sea breezes illustrate this air movement from high to low pressure regions very clearly. Differences in air pressure between adjacent regions occur wherever land and water meet. During the day, when exposed to sunlight, land surfaces increase in temperature more than water surfaces because water has a higher specific heat than soil. As a result, air over land surfaces will have a higher daytime temperature than adjacent air over water surfaces. Since warm air exerts less pressure than cool air, the pressure over the land will be lower and the pressure over the water will be higher. The result is a movement of air from the water toward the land. Since a wind is named for the direction or place from which it is blowing, this wind is called a *sea breeze*. See Figure 23.11a.

At night, though, land cools off faster than water. The air over the land becomes cooler than the air over the water. Thus, the air pressure is greater over the land than over the water, and air moves from the land toward the water. This wind is called a *land breeze* because it comes from the land. See Figure 23.11b.

Global Wind Belts

Differences in pressure between adjacent regions of air also occur on a global scale. In Chapter 22, you learned that unequal heating of the atmosphere at different latitudes causes three large atmospheric convection cells to form in the Northern and Southern Hemispheres. Rising air in these convection cells produces low-pressure belts that circle Earth at the Equator (0°), the Arctic Circle (60°N), and the Antarctic Circle (60°S). Sinking air produces high-pressure belts at the Poles (90°N, 90°S) and at the Tropics of Cancer (30°N) and Capricorn (30°S). Thus, a series of alternating high- and low-pressure belts circle Earth, giving rise to a series of global wind belts as air moves from regions of high pressure toward regions of low pressure. See Figure 23.12. The global winds that blow from the high pressure belts at 30° North and South latitude towards the low-pressure belt at the Equator are known as the *trade winds*. The Coriolis effect causes these large-scale winds to curve to the right in the Northern Hemisphere (so they blow from the northeast) and to the left in the Southern Hemisphere (so they blow from the southeast).

Global winds also blow from the high-pressure belts at 30° North and South latitudes toward the low-pressure belts at 60° North and South latitudes and are known as the *prevailing winds*. The

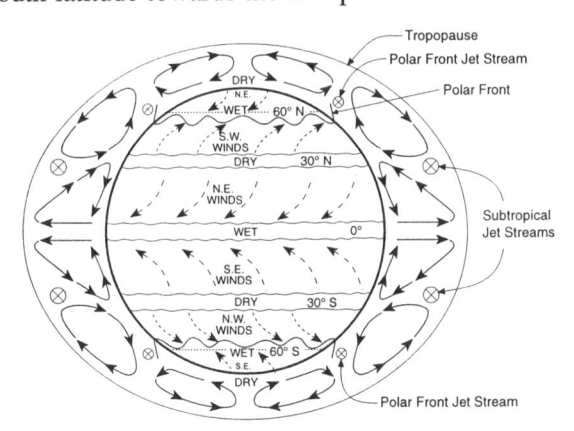

Figure 23.12 Planetary Wind and Moisture Belts in the Troposphere. The drawing shows the locations of the belts near the time of an equinox. The locations shift somewhat with the changing latitude of the Sun's vertical ray. In the Northern Hemisphere, the belts shift northward in summer and southward in winter. Source: The State Education Department, Earth Science Reference Tables, 2001 ed. (Albany, New York; The University of the State of New York).

Weather

Coriolis effect causes these large-scale winds to blow from the southwest in the Northern Hemisphere and from the southeast in the Southern Hemisphere.

The global winds that blow form the high-pressure regions over the poles toward the low pressure regions at 60°N and 60°S are known as the *polar easterlies*. The Coriolis effect causes these winds to blow from the northeast in the Northern Hemisphere and from the southeast in the Southern Hemisphere.

HOW DO CLOUDS AND PRECIPITATION FORM?

Clouds and precipitation form when air is cooled below its dew point and water vapor condenses into tiny water droplets or ice crystals. The conditions under which this cooling takes place determine what type of cloud or precipitation will form.

Condensation is the change of phase from gas to liquid. To condense into a liquid, a gas must have a surface to condense on. In the atmosphere, this surface is provided by particles suspended in the air, such as salt or ice crystals, dust, or smoke, called *condensation nuclei*. Water vapor in the air condenses when the air is cooled to the dew point and there are condensation nuclei on which condensation can form. *Sublimation* is the change of phase from a gas directly to a solid. Water vapor may sublime to form ice crystals if moisture is released from the air when its temperature is below the freezing point of water.

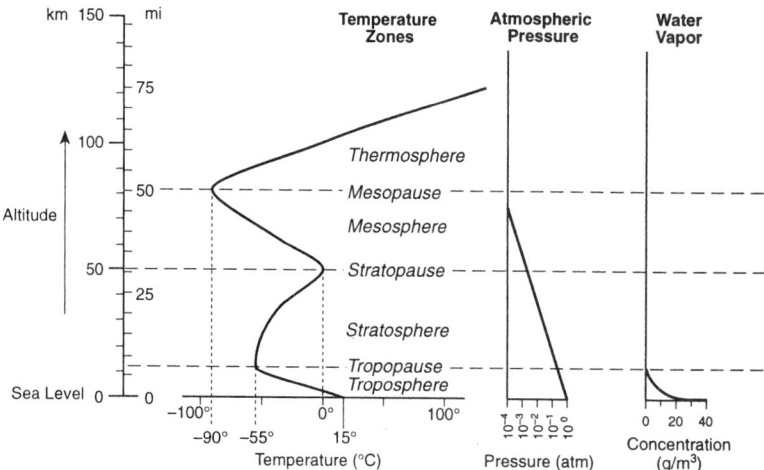

Figure 23.13 Physical Properties of the Atmosphere. Source: The State Education Department, *Earth Science Reference Tables*, 2001 ed. (Albany, New York; The University of the State of New York).

As you can see from Figure 23.13, temperature drops steadily with altitude in the troposphere. In the upper portion of the troposphere, temperatures below –40°C are the norm year-round. For this reason, even in the summer, cirrus clouds are composed of tiny ice crystals, and thunderstorms whose cloud tops extend into the upper troposphere may produce hail. Clouds are important components of the atmosphere for three reasons. They are the producers of precipitation, such as rain, snow, sleet, and hail. They are also major reflectors of solar radiation, and they serve as key indicators of overall weather conditions.

As shown in Figure 23.14, clouds are categorized by the altitudes at which they exist and their degrees of vertical development.

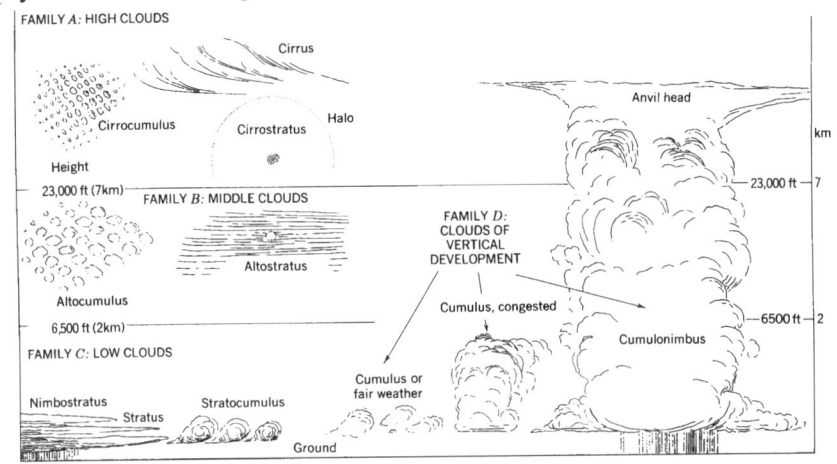

Figure 23.14 Cloud-Form Classification. Clouds are classified into families according to altitude and degree of vertical development. Source: Arthur N. Strahler, *The Earth Sciences*, Harper & Row, 1971.

Air that is warmed by contact with the Sun-heated surface of Earth expands, becomes less dense, and floats upward in the denser surrounding air. As the air rises, it continues to expand, causing a decrease in pressure and temperature. Warm air will also expand and cool if it is forced upward as wind pushes it over a natural obstacle, such as a mountain, or by cooler, denser air pushing its way underneath it. These three conditions that cause air to rise are diagrammed in Figure 23.15.

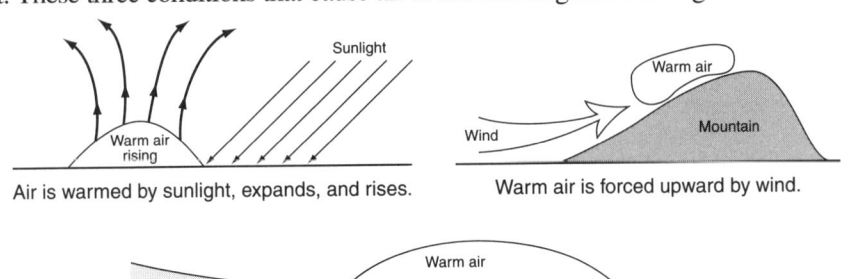

Figure 23.15 Three Conditions That Cause Air To Rise. Air rises when it is heated and becomes less dense, when it is forced upward by winds over rising land surfaces, and when it is forced upward by colder, denser air pushing under it.

Weather

The cooling of air as it rises results in condensation when the air temperature drops to the dew point and condensation nuclei are present. This condensation generates a *cloud* made of a great many tiny but visible water droplets or ice crystals. The *cloud base* indicates the elevation at which the rising air reached its dew point. Since this depends on the amount of moisture in the air and the starting temperature, clouds may form at many different elevations.

Fog is a cloud whose base is at ground level. It forms when moist air at ground level is cooled below its dew point. When moist air comes into contact with a surface such as grass, or with a car or other object that has cooled below the dew point because heat has radiated out of it overnight, droplets of water called *dew* condense on the surface. If the temperature is below freezing, tiny ice crystals called *frost* will form instead of dew.

The water droplets or ice crystals in clouds are very tiny (1/50 millimeter) and often remain suspended in the air. Thus, all clouds do not form precipitation. Precipitation occurs only when cloud droplets or ice crystals join together and become heavy enough to fall. Table 23.3 summarizes the different types of precipitation and their origins.

TABLE 23.3 TYPES OF PRECIPITATION AND ORIGINS

Name	Description	Origin
Rain	Droplets of water up to 4 mm in diameter	Cloud droplets coalesce when they collide.
Drizzle	Very fine droplets falling slowly and close together	Cloud droplets coalesce when they collide.
Sleet	Clear pellets of ice	Raindrops freeze as they fall through layers of air at below-freezing temperatures.
Glaze	Rain that forms a layer of ice on surfaces it touches	Supercooled raindrops freeze as soon as they come in contact with below-freezing surfaces.
Snow	Hexagonal crystals of ice, or needlelike crystals at very low temperatures	Water vapor sublimes, forming ice crystals on condensation nuclei at temperatures below freezing.
Hail	Balls of ice ranging in size from small pellets to as large as a softball, with an internal structure of concentric layers of ice and snow	Again and again hailstones are hurled up by updrafts in thunderstorms and then fall through layers of air that alternate above and below freezing. Each cycle adds a layer to the hailstone. The more violent the updrafts, the larger and heavier the hailstone can become before falling.

As precipitation falls through the atmosphere, condensation nuclei and other suspended material that adhere to it are brought down to Earth's surface. In this way dust and many pollutants are removed from the atmosphere by precipitation. However, some pollutants react with the water in precipitation to form harmful solutions such as acid rain, which has had many negative effects on the environment.

WHAT INFORMATION IS SHOWN BY WEATHER MAPS?

Synoptic Weather Maps

One of the most widely used methods of forecasting weather is known as *synoptic forecasting*. Synoptic forecasting is based on examination of a synopsis, or summary of the total weather picture at a particular time.

A synoptic weather map is made by measuring atmospheric variables at thousands of weather stations around the world four times a day. These data are then used to create field maps, which reveal large-scale weather patterns. By looking at a sequence of synoptic weather maps, meteorologists can perceive the developments and movements of weather systems and make predictions.

Synoptic weather maps use a symbol called a *station model* to show a summary of the weather conditions at a particular weather station. Figure 23.16 shows a typical station model and explains what each of its elements represents.

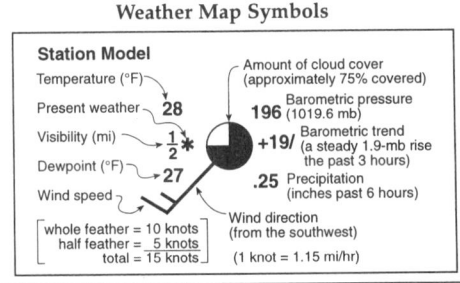

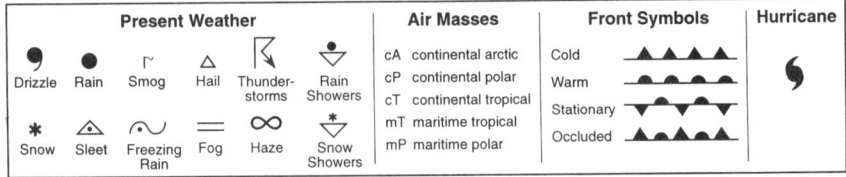

Figure 23.16 A Station Model. The symbol uses shorthand notation to summarize atmospheric variable for a particular location on a weather map. Source: The State Education Department, *Earth Science Reference Tables*, 2001 ed. (Albany, New York; The University of the State of New York).

Field Maps

Station models plotted on a map summarize a wide range of weather data and can be used to create many different field maps. A *field* is defined as a region of space that has a measurable quantity at every point. Field maps can be used to represent any quantity that varies in a region of space. One way to represent field quantities on a two-dimensional field map is to use *isolines*, which connect points of equal field value. For example, a temperature field

Weather

map contains lines connecting points of equal temperature, or *isotherms*. A pressure field map contains lines connecting points of equal pressure, or *isobars*. Field maps clearly show the weather patterns in the atmosphere.

Figure 23.17 shows a synoptic weather map, followed by the same map with isotherms and isobars drawn in. (Figures 23.17b and 23.17c).

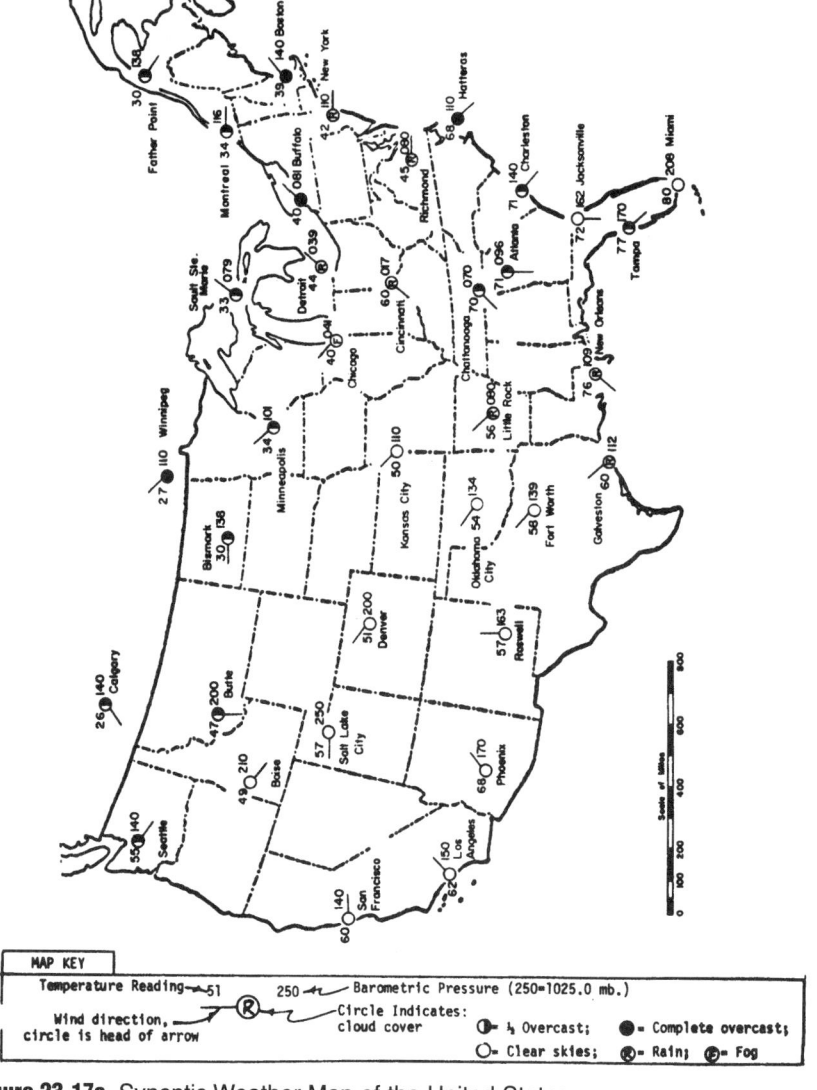

Figure 23.17a Synoptic Weather Map of the United States.

As you can see in Figure 23.17a, there are distinct regions in the atmosphere that have similar conditions. For example, a region of high pressure, cool temperature, and clear skies is centered near Salt Lake City, Utah. You can also see a region of low pressure, slightly warmer temperature, and rain centered near Cincinnati, Ohio.

586

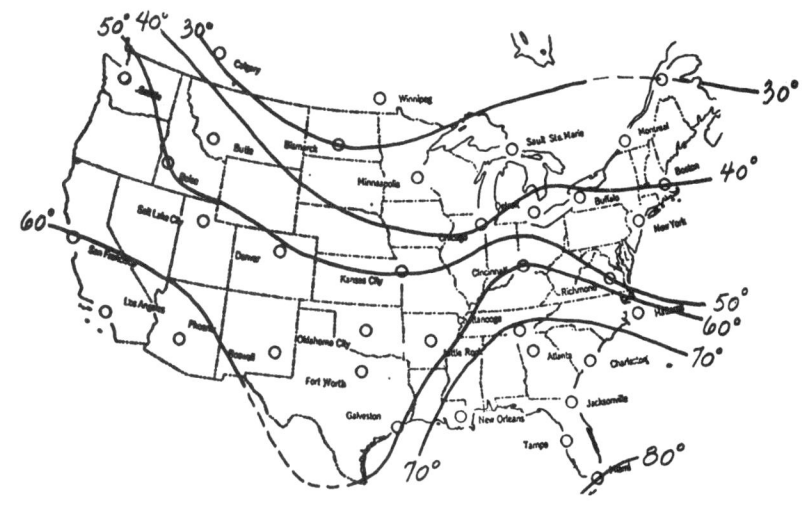

Figure 23.17b Isotherms.

Within a field, a field value changes as you move from place to place. The rate at which the field value changes is called its *gradient*. Rapid changes in field values, or steep gradients, show up as closely spaced isolines on a field map. In Figure 23.17c, you can see that the isobars are widely spaced within each of the two air masses, but are closely spaced between them. This means that there is little change in pressure within each of the regions, but there is a boundary between them where the pressure changes rapidly. You can see the same pattern repeated for temperature: there are regions of the atmosphere with relatively uniform characteristics, which have boundaries separating them from adjacent regions.

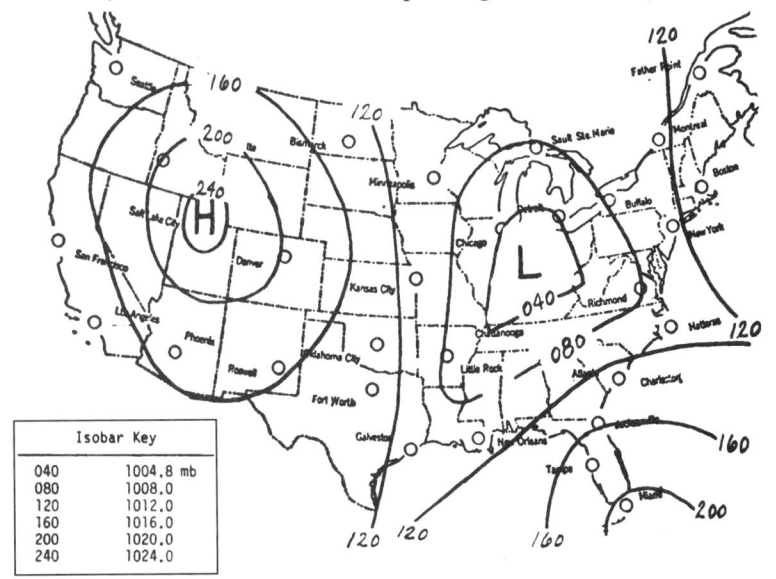

Isobar Key	
040	1004.8 mb
080	1008.0
120	1012.0
160	1016.0
200	1020.0
240	1024.0

Figure 23.17c Isobars.

Weather

HOW CAN WEATHER INFORMATION BE USED TO MAKE FORECASTS?

Most weather forecasts are based on the movements of large regions of air with fairly uniform characteristics, called *air masses*. By examining a series of synoptic weather maps, meteorologists can track the current movements of air masses and predict future movements.

Air masses

As stated above, a region of the atmosphere in which the characteristics of the air at any given level are fairly uniform is called an *air mass*. An air mass is identified by its average air pressure, air temperature, moisture content, and winds. The boundaries between air masses, where rapid changes occur, are called *frontal boundaries*, or *fronts*.

The characteristics of an air mass are the result of the geographical region over which it formed, or its *source region*. Air resting on, or moving very slowly over, a region tends to take on the characteristics of that region. For example, air that sits over the Gulf of Mexico in the summer is resting on very warm water. The air is warmed by contact with the water, and evaporation causes much humidity to enter the air. As a result the air becomes warm and moist, just like the Gulf of Mexico. In general, air masses that form near the poles are cold, and those that form near the Equator are warm. Air masses that form over water are moist; those that form over land, dry. The longer the air remains stationary over a region, the larger it becomes and the closer it matches the characteristics of the region. On weather maps, air masses are named for their source regions and are designated by a letter or letters as shown in Table 23.4.

TABLE 23.4 AIR-MASS NAMES AND SOURCES

	Arctic A	Polar P	Tropical T
	Formed over extremely cold, ice-covered regions	Formed over regions at high latitudes where temperatures are relatively low	Formed over regions at low latitudes where temperatures are relatively high
maritime m Formed over water, moist		mP—cold, moist Formed over North Atlantic, North Pacific	mT—warm, moist Formed over Gulf of Mexico, middle Atlantic, Caribbean, Pacific south of California
continental c Formed over land, dry	cA—dry, frigid Formed north of Canada	cP—cold, dry Formed over northern and central Canada	cT—warm, dry Formed over south-western United States in summer

Air masses are moved by planetary winds and jet streams (study Figure 23.12 carefully). In general, arctic and polar air masses are moved south and west over the United States by the northeast wind belt near the North Pole, and tropical air masses are moved to the north and east by the southwest wind belt between 30° and 60°N. As the air masses are carried over Earth's surface, they slowly change and take on the characteristics of the regions over which they are moving. For example, a cP air mass moving south out of Canada will gradually become warmer as it moves farther and farther south.

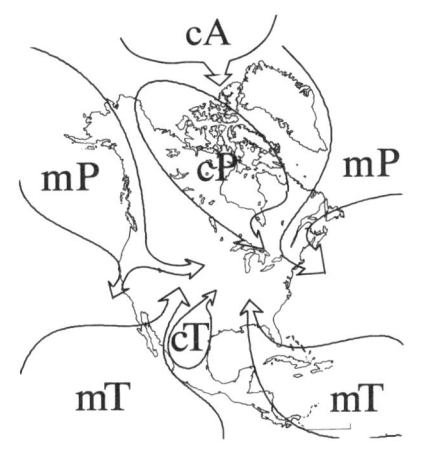

Figure 23.18 Air Masses That Influence Weather in the United States and Their Source Regions.

Air masses that affect U.S. weather and the regions in which they form are shown in Figure 23.18.

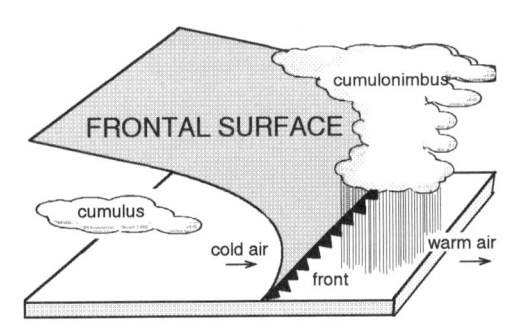

Figure 23.19 Perspective View of a Frontal Surface and a Front Line on the Ground. A *frontal surface* forms along the boundary between two different air masses. The line along which the boundary touches the surface is the *front*.

Weather Fronts

At any given point in time, several air masses will be moving across the United States. They generally move from west to east, driven by the prevailing southwesterlies. When different air masses meet, very little mixing of the air takes place and a sharp transition zone forms between them. This zone is termed a *weather front* because it marks the leading edge, or front, of an air mass that is pushing against another air mass. The whole surface along which the air masses meet is called the *frontal surface*, and the line on the ground marking the transition from one air mass to the other is the *front*. See Figure 23.19. Fronts are areas of rapid changes in weather conditions and are often sites of unsettled and rainy weather. Different types of frontal surfaces, such as cold, warm, stationary, and occluded fronts, will form, depending on which type of air mass is advancing against another.

Cold Fronts

A cold front forms along the leading edge of a cold air mass that is advancing

Weather

against a warmer air mass. The cold air mass is denser than the warmer air ahead of it, so it pushes against and under it like a wedge. This *forces* the warmer air upward rapidly and results in turbulence, rapid condensation forming heavy, vertically developed clouds, and heavy precipitation or thunderstorms. Cold fronts are often preceded by a zone of thunderstorms in the summer and of snow flurries in the winter. After the front passes, the temperature drops sharply and the pressure rises rapidly. Figure 23.20 shows a typical cold front.

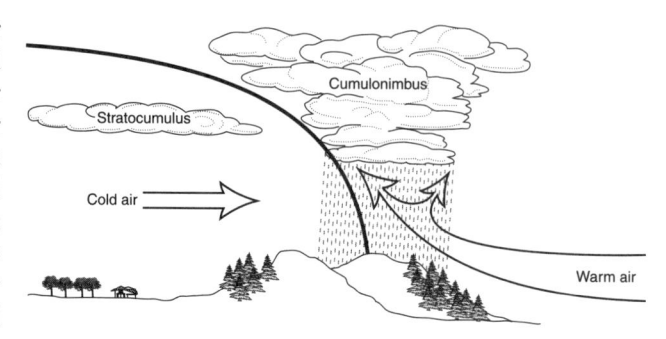

Figure 23.20 A Typical Cold Front.

Warm Fronts

A warm front forms where a warm air mass overrides the trailing edge of a cold air mass ahead of it. The less dense warm air mass rides up and over the denser cold air. As the warm air mass rises above the denser cold air mass, it expands and cools, causing condensation to occur over the wide, gently sloping boundary. The results are thickening, lowering clouds and widespread precipitation. Figure 23.21 shows a typical warm front.

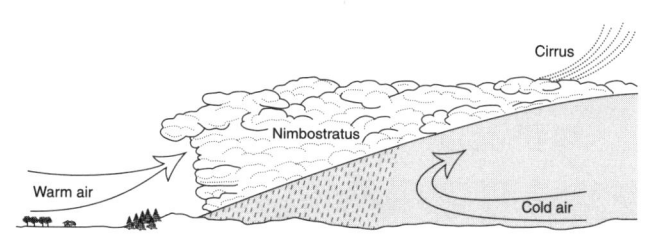

Figure 23.21 A Typical Warm Front.

Stationary Fronts

A stationary front forms along the boundary between a warm air mass and a cold air mass when neither moves appreciably in any direction. A stationary front slowly takes on the shape and characteristics of a warm front as the denser cold air slowly slides beneath the less dense, warmer air. Although stationary fronts develop the same widespread rain and cloudiness as do warm fronts, the rain and clouds may persist for a longer period, perhaps many days, until another air mass comes along with enough impetus to get the stalled air masses moving.

Occluded Fronts

Cold air is denser and exerts more pressure than warm air. Therefore, cold air masses and the cold fronts associated with them tend to move faster than warm air masses and warm fronts. If a slow-moving warm front is followed by a fast-moving cold front, the cold front will sometimes overtake the warm front. Then the cold air lifts the warm air entirely aloft, forming an occluded front. The lifting of the warm air mass causes large-scale condensation and precipitation, resulting in widespread rain and thunderstorms. Then, depending on whether the cold air mass coming from behind is colder or warmer than the cold air mass now ahead, a cold-front occlusion or a warm-front occlusion may form, as shown in Figure 23.22.

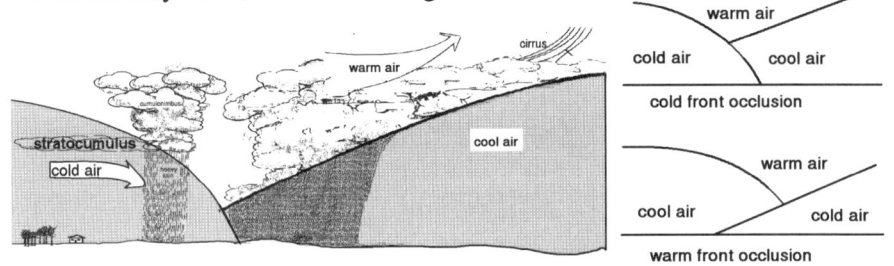

Figure 23.22 An Occluded Front. Two types of occlusions can form, depending on whether the advancing cold air mass is warmer or cooler than the cold air mass ahead of the warm air mass.

Cyclones, Anticyclones, and Wave Cyclones (Frontal Cyclones)

Regions of low pressure surrounded by higher pressure form winds that blow toward the center of the warm air mass but, because of the Coriolis effect, are deflected to the right. The result is a *cyclone*, a counterclockwise flow of air, that moves in a curved path toward the center of the low pressure.

Similarly, regions of high pressure surrounded by lower pressure form winds that blow outward from the center. The winds are deflected by the Coriolis effect into a flow of air, an *anticyclone*, that is clockwise and outward from the center.

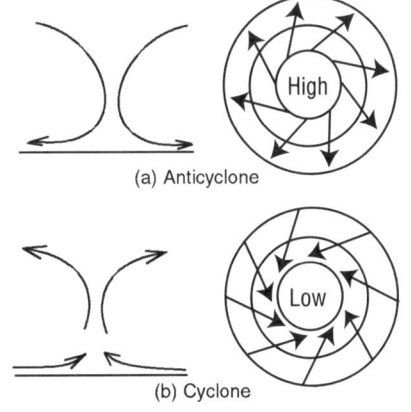

Figure 23.23 An Anticyclone (a) and a Cyclone (b).

A cyclone and an anticyclone are diagrammed in Figure 23.23.

Most of the weather systems that move across the United States are *wave cyclones*, or *frontal cyclones*. Wave cyclones are warm and cold fronts that move in a counterclockwise direction around a low-pressure center. They

Weather

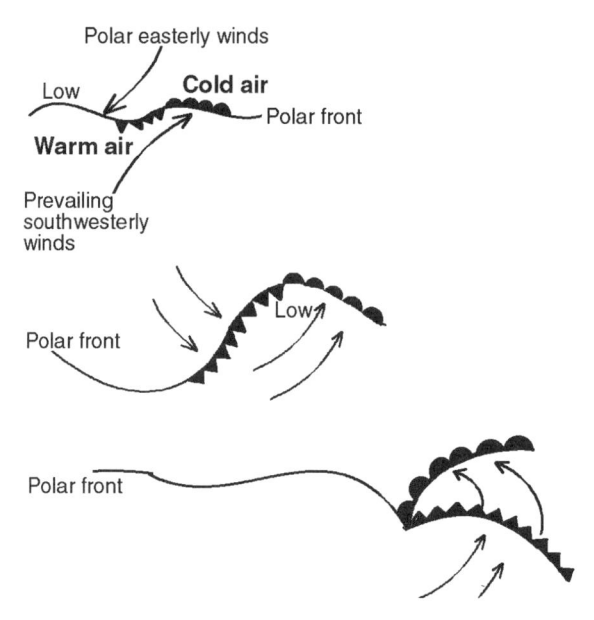

Figure 23.24 Formation of a Wave Cyclone.

typically form in the westerly wind belts along the boundary between polar and tropical air, called the *polar front* (see Figure 23.12). In the United States, this boundary falls just about along our northern border with Canada.

A wave cyclone forms when southwesterly winds push warm, low-pressure air against cold air along the polar front (see Figure 23.24). The result is to lower the pressure along the polar front, and the boundary bends to form a wave of cold air advancing from west to east against the warm air—a cold front. The cold air pushes the warm air ahead of it into the cold air on the other side of it, forming a warm front. The end result is a wavelike structure consisting of a cold front overtaking a warm front. The counterclockwise flow in the low-pressure center causes the cold front to swing around the equatorial side of the low-pressure center as the wave moves from west to east. The wave cyclone may even break loose from the polar front and be carried, swirling along, by the jet stream. In the United States, wave cyclones generally move from west to northeast and get carried out over the north Atlantic. The stormy weather associated with a wave cyclone may affect millions of square miles and last for several days.

HOW CAN WE FORECAST THE WEATHER?

Until relatively recently, weather forecasts were based on local observations made directly by human senses. Through first-hand experience and passed-down weather lore, people learned to recognize certain patterns of weather changes associated with cloud types or shifts in wind direction. It was not until the invention of the barometer and the thermometer in the 1600s, however, that accurate measurements of pressure and temperature changes could be made. Even then, forecasting was not practicable until communication among many points by telegraph made possible the rapid collection and sharing of weather data. The first government weather forecasts were issued in the United States in 1870.

Synoptic Weather Forecasting

The basis of most weather forecasting is the systematic collection of weather data from a network of weather-reporting sites that blanket the globe. At the simplest level, these data are summarized on a synoptic weather map using shorthand symbols called *station models*. At their most sophisticated, computer programs compile, store, analyze, and plot weather data, plot isolines for various weather factors, and can display weather data in a variety of graphic formats.

Synoptic weather maps are made four times a day—at midnight, noon, 6 A.M., and 6 P.M. Greenwich Mean Time. By looking at a sequence of these maps, a meteorologist can detect the development and movement of weather systems. The movements of these systems are projected into the future, and predictions of weather for various locations are made. Computer-drawn maps show wind, temperature, air pressure, and humidity patterns at many different atmospheric levels. Statistical analysis is then used to map projections of weather factors.

Synoptic forecasting is fairly accurate for large scale, short-term forecasts. However, there are always locally unique conditions that modify the general behavior of weather systems. For example, cities tend to absorb and reradiate more heat energy than rural areas, creating urban "heat islands." Local meteorologists must take these specific conditions into account when making forecasts, for they have a significant effect on such factors as the exact path a storm will take locally, the times when rain will begin and end, and the extremes that high and low temperatures will reach. For this reason, most local meteorologists have a network of weather hobbyists who feed them weather data in addition to those collected by official Weather Service stations.

Statistical Weather Forecasting

Statistical weather forecasting is based on the past behavior of the atmosphere. Mathematical equations are used to predict the probability that the atmosphere will behave in a certain way. Basically the system works something like this: Suppose you have math first period on Mondays, Tuesdays, and Wednesdays, and science first period on Thursdays. A person observing you would soon see a pattern in which three days of first-period math are followed by a first period of science. However, there would be exceptions. There might be a holiday on a Monday, so the sequence would be science after *two* maths. Or Thursday might be a holiday, so the sequence would be six maths before the next science. However, such deviations in the schedule would probably be the exception rather than the rule. Let's assume that in 10 weeks there was one deviation. Statistical forecasting would then say that for the next week there was a 9 in 10, or 90 percent, chance that three maths would be followed by a science.

Now let's apply this reasoning to weather. Suppose we observed that, nine out of the past ten times the wind shifted to blow from the northeast, we had rain within 6 hours. Then, if the wind shifted to the northeast, statistical forecasting would say that there was a 90 percent chance of rain in the next 6 hours.

The process of statistical forecasting is threefold: maintaining detailed historical records, identifying all previous occurrences of a pattern, and then calculating the probabilities. However, there is no way to anticipate unusual behavior, nor is past behavior a guarantee of future behavior.

Numerical Weather Forecasting

Numerical forecasting is based on fluid dynamics. Theoretically, if the initial states of the atmosphere, land surfaces, and oceans are known, we should be able to predict the behavior of the atmosphere based on the laws of physics governing heat and moisture transfer and fluid behavior. However, such complete information is not currently available.

The sheer quantity of data, coupled with the number and complexity of the required calculations, made numerical forecasting impracticable until computers capable of high-speed calculations were developed in the 1940s. Now, computers solve six equations simultaneously—one for each of the three dimensions of motion, and one each for the conservation of moisture, heat, and mass. These equations are solved for thousands of points on a map grid, and changes at each point are projected. These changes are then fed back into the loop for another round of calculations, producing a projection further into the future. As new actual data are entered into the equations throughout the day, the revised calculations serve to correct the predictions. Calculations for several levels in the atmosphere, projecting changes over the next 2 days, are made every 12 hours. A 5-day forecast is calculated once a day.

Long-Range Weather Forecasting

No matter what method is used, however, forecasting reliability decreases as the time frame increases. Short-term forecasts (1–3 days) of weather are usually far more accurate than long-term forecasts (a week or more). The decrease in forecast reliability over time is due to a number of factors: the wide spacing of initial data points, the unreliability in measurements taken over many areas, and a lack of understanding of the behavior of the atmosphere. A small error in an initial measurement can cause errors in computer-calculated forecasts.

For example, let's make a simple mathematical model that assumes that temperature increases or decreases in direct proportion to the ratio of day to night:

$$\text{Temperature today} \times \frac{\text{Length of night}}{\text{Length of day}} = \text{Temperature tomorrow}$$

Let's compare the temperatures predicted by this model, using a correct temperature reading and one in which an error was made:

Starting Temperature	Prediction When Daylight = 13 hours, Night = 11 hours	Prediction When Daylight = 14 hours, Night = 10 hours
21°C (correct)	21 × 13/11 = 24.8°C	24.8 × 14/10 = 34.7°C
22°C (incorrect)	22 × 13/11 = 26°C	26 × 14/10 = 36.4°C

As you can see, an initial error of 1°C has almost *doubled* as it was compounded in the calculations. Errors like these multiply as forecasts are extended farther into the future, and at some point the forecast becomes useless.

Chaos theory shows that small, unpredictable changes can result in large-scale changes in a system over time. For example, a forest fire set by lightning may cause a localized increase in temperature that may set in motion large-scale changes in air-flow patterns. For this reason, chaos theory seems to indicate that long-term forecasts may never be reliable.

WHERE AND HOW ARE THE MOST HAZARDOUS WEATHER SITUATIONS LIKELY TO OCCUR?

Thunderstorms

One of the most common weather hazards is a thunderstorm (see Figure 23.25). A thunderstorm begins as a convection cell in the atmosphere. Warm air rises, expands, cools, and then sinks back to Earth's surface. If the convection in the cell is strong, because of intense heating, the air may rise very rapidly, reaching heights of many kilometers in just a few minutes. A cloud forms in the updraft of the convection cell and grows as more and more warm, moist air is carried upward. The strong updrafts in the rising air support water droplets and ice crystals in the cloud so that they grow in size. When the updrafts can no longer support the moisture, it falls as rain or even hail. The falling rain sets up downdrafts that cause internal friction with updrafts, and the internal friction builds up static electric charges that may discharge as lightning. Thunderstorms are likely to occur wherever and whenever there is strong heating of Earth's surface.

Weather

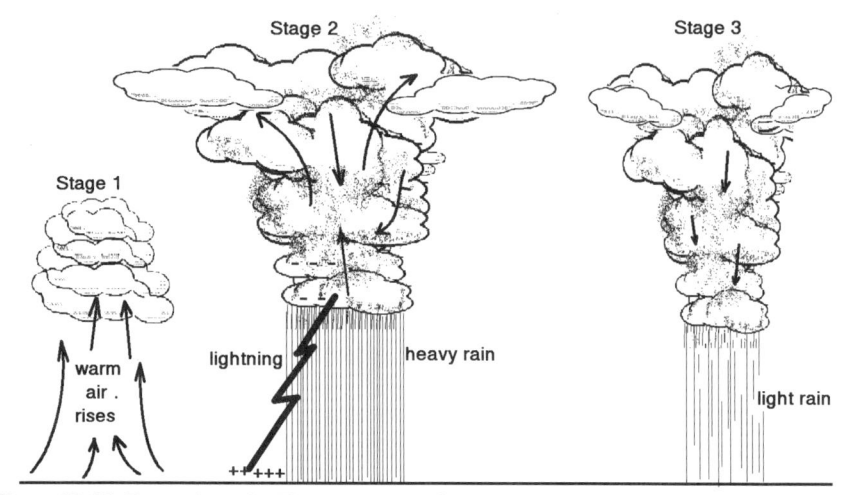

Figure 23.25 Formation of a Thunderstorm. Stage 1: Intense heating causes warm air to rise rapidly, forming tall, vertically developed clouds. Stage 2: Strong updrafts keep precipitation aloft until it is heavy. Falling rain creates downdrafts, and internal friction with updrafts causes buildup of static charge, leading to lightning and thunder. Stage 3: Downdrafts cause air to cool and updrafts subside, so rain becomes lighter.

Tornadoes

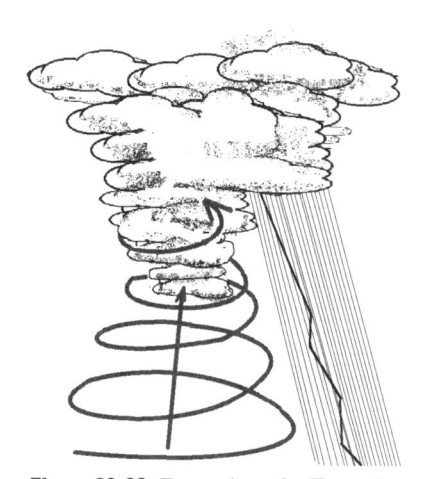

Figure 23.26 Formation of a Tornado.

Late in the day, when Earth's surface is warmest and convection is strongest, a tornado may form (see Figure 23.26). Tornadoes are small (most are less than 100 meters in diameter), brief (most last only a few minutes) disturbances that usually develop over land from intense thunderstorms. When heating is very intense, warm air rises in strong convection currents. The upward movement of the air causes a sharp decrease in pressure. Air rushes into the low-pressure region from the sides and, deflected by the Coriolis effect, is given a spin. As air moves toward the center of the updraft, it reaches high speeds because of the big difference in pressure. The rapidly moving air decreases the pressure even more, further feeding the updraft. The whole process spirals upward in intensity, and a funnel forms that eventually touches the ground. Wind speeds near the center of a tornado may reach 500 kilometers per hour or more.

Hurricanes

When air over warm oceans is heated by solar radiation, the warming of the air causes a decrease in pressure and the evaporation of water from the ocean adds humidity to the air, decreasing pressure even further. The result is numerous convection cells and thunderstorms. Widespread thunderstorm activity can merge into a large updraft that is part of a huge convection cell. The rising air further decreases the pressure, as does latent heat released when the moisture in the air condenses. As a result a region of very low pressure is created. Winds begin to blow toward the low-pressure center as air moves in from surrounding areas of higher pressure. The Coriolis effect deflects the winds, and a cyclone forms. Heat, released when fresh moisture brought in by winds condenses, feeds the convection cell with new energy, and it grows larger and stronger. When the winds in the cyclone exceed 119 kilometers per hour, we have a hurricane (see Figure 23.27). Fully formed hurricanes are huge cyclones, often exceeding 500 kilometers in diameter.

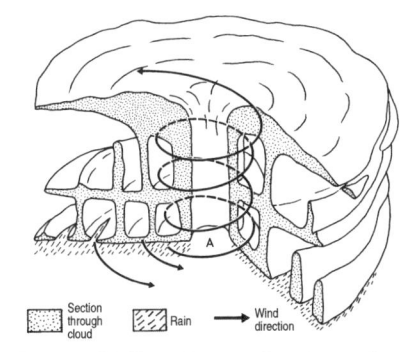

Figure 23.27 Cross Section of a Hurricane. Source: Earth Science on File. © Facts on File, Inc., 1988.

Hurricanes tend to form over oceans near the Equator during summer months, when the ocean surface is the warmest. Once the hurricane has formed, heat and moisture from the ocean provide the energy that maintains it. Anything that cuts off the supply of heat or moisture will diminish the strength of the hurricane, so hurricanes weaken as they move over land or cool water. Therefore a hurricane is usually most destructive when it first moves over land. However, even after moving over land, hurricanes can last for many days and cause much damage because of high winds and flooding.

Planetary winds tend to move hurricanes to the west until they reach 30° N; then the prevailing southwesterlies move them from southwest to northeast (see Figure 23.28). However, air masses that a hurricane encounters may deflect its path.

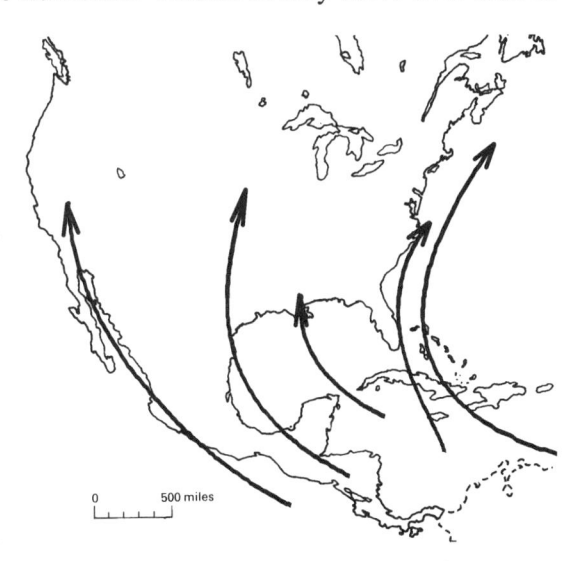

Figure 23.28 Typical Tracks of Tropical Cyclones—Hurricanes—near the United States.

Emergency Preparedness for Hazardous Weather

Hazardous weather, such as tornadoes, severe thunderstorms, and hurricanes, claims many lives and causes much property damage each year in the United States. Hazardous weather happens; there's not much humans can do about that. Measures can be taken, however, to reduce the risk of damage and loss of life. Reducing the risk is called *mitigation*. Some forms of mitigation are expensive and complicated, such as moving to a safer location or making structural changes to a house. Some measures are easier, such as developing an emergency plan, having disaster supplies on hand, protecting windows with shutters, and securing loose outdoor items that could cause damage when hurled by strong winds.

Emergency preparedness is an important responsibility of local government. Communities can help reduce risks by taking such actions as developing disaster plans, establishing evacuation routes and public shelters, enacting zoning regulations that limit development of high-risk areas, and enforcing building codes that require structures to withstand hazardous weather forces.

Watches and Warnings

Advance planning and quick response are the keys to surviving hazardous weather. Toward this end, the National Weather Service issues two levels of alerts about life-threatening weather such as tornadoes, severe thunderstorms, and hurricanes: watches and warnings. *Watch*es are issued for tornadoes and severe thunderstorms when weather conditions are such that these hazards are likely to develop in your area. Hurricane watches are issued when there is a threat of hurricane conditions within 24–36 hours. *Warnings* are issued when a tornado or severe thunderstorm has been sighted or indicated on radar in your area, or when hurricane conditions are expected in 24 hours or less.

When a severe thunderstorm or tornado is coming, there is very little time to make life-or-death decisions, so the National Weather Service issues a warning as soon as they are spotted. On the other hand, hurricanes are easier to track because they are so large and last so long. A 1-mile-wide tornado is huge and may last for minutes or hours; a 100-mile-wide hurricane is small, and hurricanes usually last for a week or more. Although hurricane warnings are easier to issue, the decisions involved are more complicated. An average hurricane warning along several hundred miles of United States coastline costs tens of millions of dollars spent for boarding up homes, closing down schools, businesses, and manufacturing plants, and evacuating people. Then, just a small deviation in a hurricane's path may mean the difference between a disaster and a very rainy day for a city under a hurricane warning. Forecasting is improving, however, and in the next few years meteorologists hope to make 30-hour forecasts as accurate as current 24-hour forecasts.

What to Do During Watches and Warnings

If a watch is issued for your area, you should stay alert to the weather by listening to the radio or television and be prepared to take shelter. This is the time to check emergency supplies and to remind family members of disaster plans and ways to respond during an emergency. Check for hazards outdoors and secure your home.

If a warning is issued, the danger is immediate and very serious, and everyone should go to a safe place. Once there, listen to a battery-operated radio or television for further instructions and for an official "all clear" before resuming normal activities. The Federal Emergency Management Agency has developed severe thunderstorm, tornado, and hurricane fact sheets that describe specific steps to be taken when watches and warnings are issued for these weather hazards. These fact sheets are available online at *www.fema.gov*.

MULTIPLE-CHOICE QUESTIONS

In each case write the number of the word or expression that best answers the question or completes the statement.

1. The diagram to the right represents a map of western and central New York State on a day in August. The location of an 18°C isotherm is shown. Why is the 18°C isotherm line farther north over the land than it is over Lake Ontario?
   (1) There is a high plateau at the eastern end of Lake Ontario.
   (2) The air is warmer over the land than over Lake Ontario.
   (3) Isotherms are the same shape as latitude lines.
   (4) The prevailing winds alter the path of the isotherms.

Note that question 2 has only three choices.

2. In New York State, when do the highest air temperatures for the year usually occur?
   (1) a few weeks before maximum insolation is received
   (2) a few weeks after maximum insolation is received
   (3) at the time that maximum insolation is received

Weather

3. The table below shows the noontime data for air pressure and air temperature at a location over a period of 1 week.

WEATHER DATA RECORDED AT NOON

Date	Nov. 9	Nov. 10	Nov. 11	Nov. 12	Nov. 13	Nov. 14	Nov. 15
Air temperature (°C)	1	6	0	−2	−4	5	10
Air pressure (mb)	1,024	998	1,015	1,021	1,030	1,013	?

Based on the data provided, which air pressure would most likely occur at noon on November 15?
(1) 987 mb
(2) 1,015 mb
(3) 1,017 mb
(4) 1,022 mb

Note that question 4 has only three choices.

4. Two weather stations are located near each other. The air pressure at each station is changing so that the difference between the pressures is increasing. The wind speed between these two locations will probably
(1) decrease
(2) increase
(3) remain the same

Base your answers to questions 5 through 7 on your knowledge of earth science and on the diagram below, which shows air temperatures and dew-point temperatures recorded during a 24-hour period in New York State.

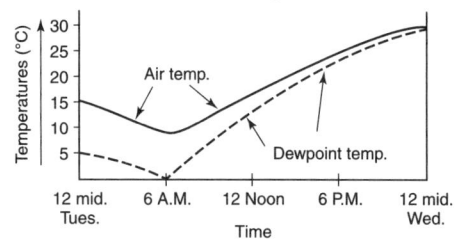

5. The probable changes in pressure for this period are best represented by which graph?

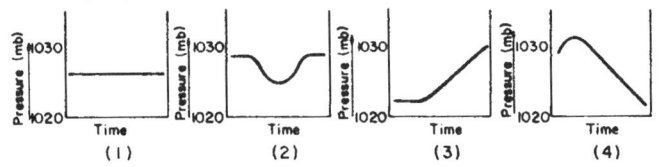

6. Which is the best explanation for the continual increase in temperature after 6:00 P.M. in the evening?
   (1) The amount of insolation increased at that time.
   (2) A warm air mass moved into the region.
   (3) The amount of cloud cover decreased, causing more radiative cooling of Earth.
   (4) The burning of fossil fuels to heat homes warmed the night air.

7. According to the graph, the probability of precipitation was greatest at
   (1) 6 A.M. Wednesday   (3) 12 noon Wednesday
   (2) 6 P.M. Wednesday   (4) 12 P.M. Wednesday

8. What would an air mass that forms over the land in central Canada most likely be labeled? [Refer to the *Earth Science Reference Tables*.]
   (1) cP   (3) mT
   (2) cT   (4) mP

9. The diagram below represents a cross-sectional view of air masses associated with a low-pressure system. The cold frontal interface is moving faster than the warm frontal interface. What usually happens to the warm air that is between the two frontal surfaces?

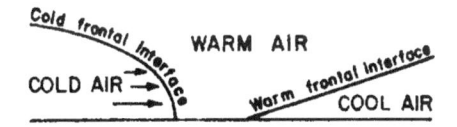

   (1) The warm air is forced over both frontal interfaces.
   (2) The warm air is forced under both frontal interfaces.
   (3) The warm air is forced over the cold frontal interface but under the warm frontal interface.
   (4) The warm air is forced under the cold frontal interface but over the warm frontal interface.

10. According to the *Earth Science Reference Tables*, which graph best represents the relationship between the altitude and the amount of water vapor in the atmosphere?

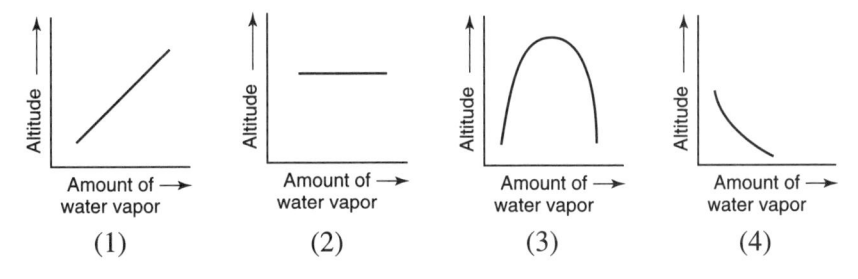

Weather

11. The characteristics of an air mass depend mainly upon the
    (1) rotation of Earth
    (2) cloud cover within the air mass
    (3) wind velocity within the air mass
    (4) surface over which the air mass was formed

12. How does air circulate within a cyclone (low-pressure area) in the Northern Hemisphere?
    (1) counterclockwise and toward the center of the cyclone
    (2) counterclockwise and away from the center of the cyclone
    (3) clockwise and toward the center of the cyclone
    (4) clockwise and away from the center of the cyclone

13. A high-pressure center is generally characterized by
    (1) cool, wet weather   (3) warm, wet weather
    (2) cool, dry weather   (4) warm, dry weather

14. An air mass from the Gulf of Mexico. moving north into New York State, has a high relative humidity. What other characteristics will it probably have?
    (1) warm temperatures and low pressure
    (2) cool temperatures and low pressure
    (3) warm temperatures and high pressure
    (4) cool temperatures and high pressure

15. Why do clouds usually form at the leading edge of a cold air mass?
    (1) Cold air contains more water vapor than warm air does.
    (2) Cold air contains more dust particles than warm air does.
    (3) Cold air flows over warm air, causing the warm air to descend and cool.
    (4) Cold air flows under warm air, causing the warm air to rise and cool.

16. Which diagram best represents the air circulation as seen from above in a high-pressure center (anticyclone) in the Northern Hemisphere?

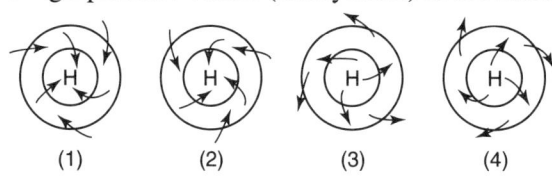

17. Precipitation often occurs along a frontal surface because the air along a frontal surface
    (1) has a high density
    (2) contains condensation nuclei
    (3) is rising
    (4) is low in humidity

18. According to the *Earth Science Reference Tables*, what is the air temperature shown on the station model below?

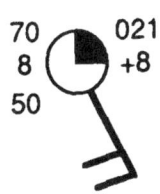

(1) 8°F
(2) 21°F
(3) 50°F
(4) 70°F

19. Which weather station model indicates the greatest probability of precipitation?

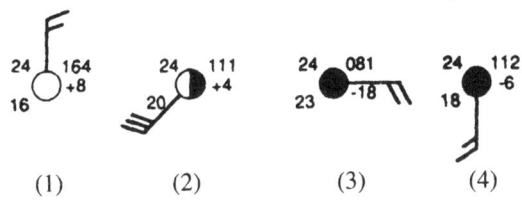

(1) (2) (3) (4)

Base your answers to questions 20 and 21 on the weather map shown below.

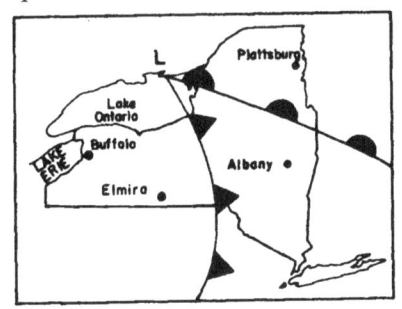

20. Which city is located in the warmest air mass?
(1) Buffalo
(2) Elmira
(3) Albany
(4) Plattsburg

21. Which city has most recently experienced a change in wind direction, brief heavy precipitation, and a rapid drop in air temperature?
(1) Buffalo
(2) Plattsburg
(3) Albany
(4) Elmira

Weather

Base your answers to questions 22 and 23 on the *Earth Science Reference Tables* and the diagram below of a weather station model,

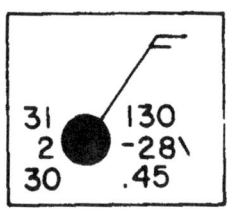

```
31      130
 2  ●  -28\
30      .45
```
(with E flag at top right)

22. The weather forecast for the next 6 hours at this station most likely would be
    (1) overcast, hot, unlimited visibility
    (2) overcast, hot, poor visibility
    (3) overcast, cold, probable snow
    (4) sunny, cold, probable rain

23. The barometric pressure at this station is
    (1) 1,013.0 mb   (3) 130.0 mb
    (2) 913.0 mb   (4) 10.28 mb

24. Which process most directly results in cloud formation?
    (1) condensation   (3) precipitation
    (2) transpiration   (4) radiation

25. Why is it possible for no rain to be falling from a cloud?
    (1) The water droplets are too small to fall.
    (2) The cloud is water vapor.
    (3) The dew point has not yet been reached in the cloud.
    (4) There are no condensation nuclei in the cloud.

26. Which atmospheric variable usually increases before precipitation occurs?
    (1) insolation   (3) visibility
    (2) air pressure   (4) relative humidity

27. Which is a form of precipitation?
    (1) frost   (3) dew
    (2) snow   (4) fog

28. Wind moves from regions of
    (1) high temperature toward regions of low temperature
    (2) high pressure toward regions of low pressure
    (3) high precipitation toward regions of low precipitation
    (4) high humidity toward regions of low humidity

29. A strong surface wind is blowing from city *A* toward city *B*. City *A* has a barometric pressure of 1,020 millibars. The barometric pressure of city *B* could be
    (1) 996 mb
    (2) 1,020 mb
    (3) 1,025 mb
    (4) 1,038 mb

30. Present-day weather predictions are based primarily upon
    (1) land and sea breezes
    (2) cloud heights
    (3) ocean currents
    (4) air-mass movements

31. What form of precipitation results from rain that falls through a freezing air mass ?
    (1) fog
    (2) snow
    (3) sleet
    (4) frost

32. Which natural process best removes pollutants from the atmosphere?
    (1) evaporation
    (2) precipitation
    (3) transpiration
    (4) convection

33. People sometimes release substances into the atmosphere to increase the probability of rain by
    (1) raising the air temperature within the clouds
    (2) providing condensation nuclei
    (3) lowering the relative humidity within the clouds
    (4) increasing the energy absorbed during condensation and sublimation

34. Which graph best represents the relationship between water-droplet size and the chance of precipitation?

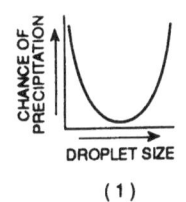

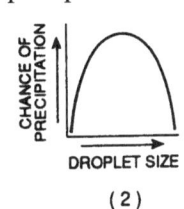

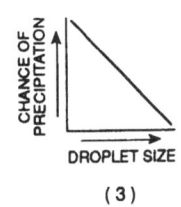

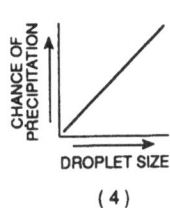

35. Condensation of water vapor in the atmosphere is most likely to occur when a condensation surface is available and
    (1) a strong wind is blowing
    (2) the temperature of the air is below 0°C
    (3) the air is saturated with water vapor
    (4) the air pressure is rising

Weather

36. At which location will a low-pressure storm center most likely form?
(1) along a frontal surface between different air masses
(2) near the middle of a cold air mass
(3) on the leeward side of mountains
(4) over a very dry, large, flat land area

37. During winter, New York City frequently receives rain when locations just north and west of the city receive snow. Which statement best explains this difference?
(1) The snow in the clouds has been depleted by the time the storm reaches New York City.
(2) The ocean modifies New York City's temperatures.
(3) New York City usually receives its weather from the south.
(4) New York City has a higher elevation.

Base your answers to questions 38 through 42 on your knowledge of earth science, *Earth Science Reference Tables*, and the diagrams below. Diagram I shows weather systems over a part of North America that includes five weather stations: A, B, C, D, and E. Solid lines represent isobars; dashed lines, X and Y, represent the horizontal locations of frontal surfaces. Diagram II shows the same weather system in cross-sectional view through stations A, B, C, D, and E. The relative temperatures of the different air masses are shown along with their directions of movement. The dashed lines represent the vertical locations of frontal surfaces.

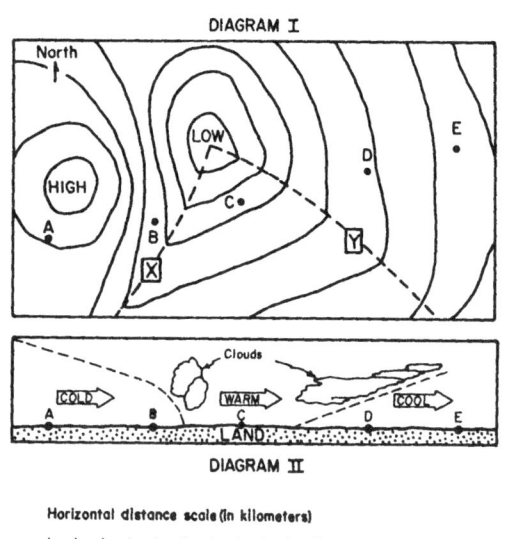

38. According to the *Earth Science Reference Tables*, which symbol should be used on a weather map to represent frontal surface (from) *X*?

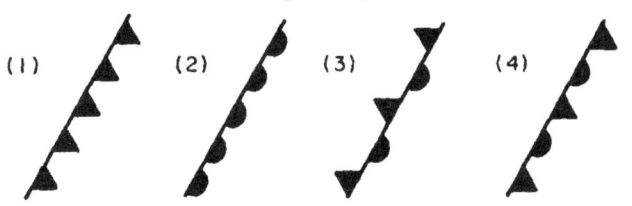

    (1)   (2)   (3)   (4)

39. Which location is probably experiencing the warmest air temperature?
    (1) *A*   (3) *C*
    (2) *B*   (4) *D*

40. Wind velocity is probably greatest at which location?
    (1) *A*   (3) *C*
    (2) *B*   (4) *D*

41. Which diagram best represents the direction of the winds that would normally be associated with these weather systems?

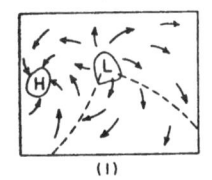

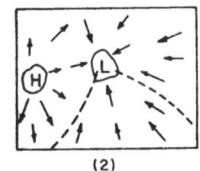

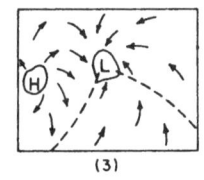

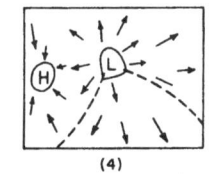

42. Which shaded area best represents the portion of Earth's surface that is most probably experiencing precipitation?

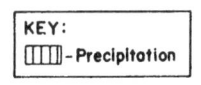

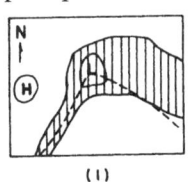

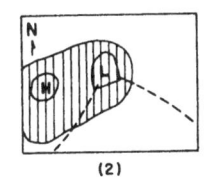

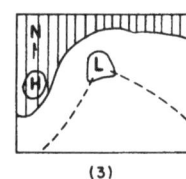

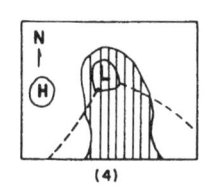

    KEY: ▦ - Precipitation

Weather

Base your answers to questions 43 through 47 on your knowledge of earth science, the *Earth Science Reference Tables*, and the diagram below. The diagram shows a common weather condition approaching a section of New York State.

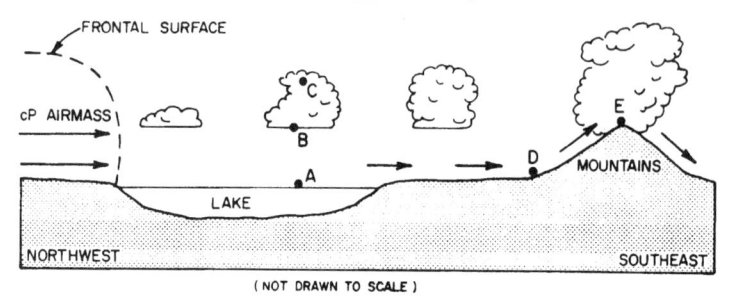

43. Which geographic area is the source of the cP air mass shown in the diagram?
(1) Caribbean Sea
(2) central Canada
(3) North Atlantic Ocean
(4) southwestern United States

44. Which surface weather map below best represents the frontal system shown in the diagram?

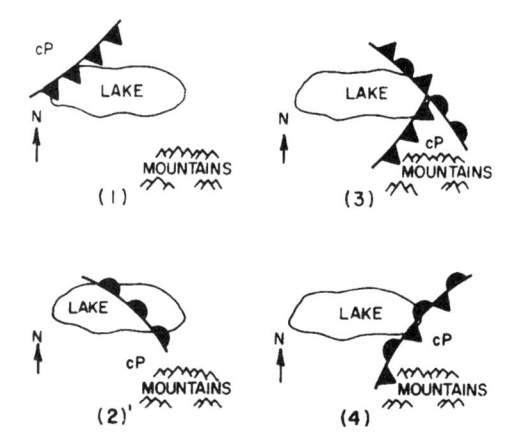

45. As the air moves from point *D* to point *E*, it will be
(1) warmed by compression
(2) warmed by expansion
(3) cooled by compression
(4) cooled by expansion

46. In October, the surface water temperature at this location is higher than the surface land temperature because the water
(1) is heated by the air above it
(2) is located farther south than the land
(3) has a higher specific heat than the land
(4) gains energy during a phase change

47. A strong wind blowing from the northwest toward the southeast would be caused primarily by differences in
(1) elevation
(2) cloud cover
(3) air pressure
(4) dew-point temperature

Base your answers to questions 48 through 52 on your knowledge of earth science and the diagram below. The diagram represents the general circulation of Earth's atmosphere and Earth's planetary wind and pressure belts. Points *A* through *F* represent locations on Earth's surface.

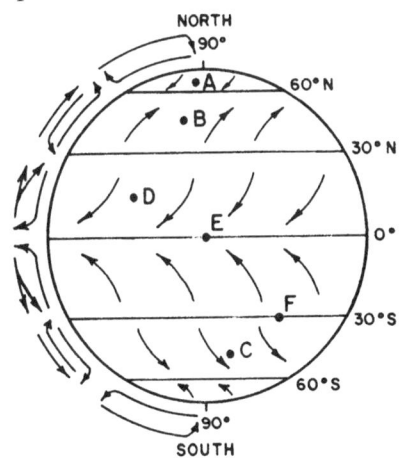

48. The curving paths of the surface winds shown in the diagram are caused by Earth's
(1) gravitational field
(2) magnetic field
(3) rotation
(4) revolution

49. Which location might be in New York State?
(1) *A*
(2) *B*
(3) *C*
(4) *D*

50. Which location is experiencing a southwest planetary wind?
(1) *A*
(2) *B*
(3) *C*
(4) *F*

Weather

51. Which location is near the center of a low-pressure belt where daily rains are common?
(1) E
(2) B
(3) F
(4) D

52. The arrows in the diagram represent energy transfer by which process?
(1) conduction
(2) radiation
(3) convection
(4) absorption

53. According to the *Earth Science Reference Tables*, an air temperature of 15°C is equal to
(1) −10°F
(2) 41°F
(3) 59°F
(4) 78°F

54. According to the *Earth Science Reference Tables*, an atmospheric pressure of 1,019 millibars is equal to
(1) 31.05 in. of mercury
(2) 30.15 in. of mercury
(3) 30.09 in. of mercury
(4) 30.00 in. of mercury

55. Which graph represents the greatest rate of temperature change?

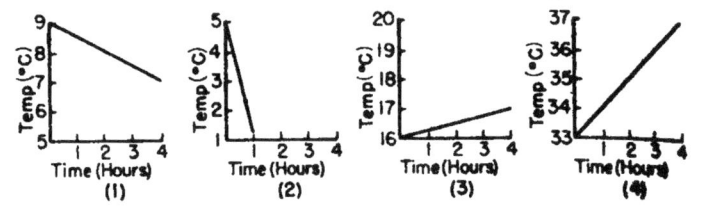

Base your answers to questions 56 and 57 on your knowledge of earth science, the *Earth Science Reference Tables*, and the diagram below, which shows the readings on a psychrometer (wet- and dry-bulb thermometers) after it has adjusted to atmospheric temperatures and humidity.

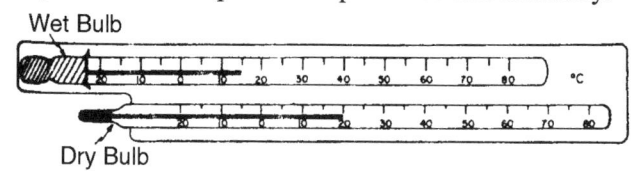

56. What is the approximate dew-point temperature?
(1) 5°C
(2) 12°C
(3) 15°C
(4) 20°C

57. Which statement best explains the difference in the readings on the two thermometers?
 (1) Evaporation removes heat energy from the wet bulb.
 (2) Evaporation absorbs heat energy from the surrounding air.
 (3) Condensation removes heat energy from the wet bulb.
 (4) Condensation absorbs heat energy from the surrounding air.

58. A container of water is placed in an open area outdoors so that the evaporation rate may be observed. The water will probably evaporate fastest when the air is
 (1) cool and humid
 (2) cool and dry
 (3) warm and humid
 (4) warm and dry

59. If the air temperature is 20°C, which dew-point temperature indicates the highest water-vapor content?
 (1) 18°C
 (2) 15°C
 (3) 10°C
 (4) 0°C

60. According to the *Earth Science Reference Tables*, what is the approximate dew-point temperature if the dry-bulb temperature is 24°C and the wet bulb temperature is 19°C?
 (1) 50°C
 (2) 16°C
 (3) 18°C
 (4) 22°C

61. There will most likely be an increase in the rate at which water will evaporate from a pond if there is a decrease in the
 (1) wind velocity
 (2) temperature of the air
 (3) moisture content of the air
 (4) altitude of the Sun

62. Rapidly falling barometric pressure readings usually indicate
 (1) clearing conditions
 (2) approaching storm conditions
 (3) decreasing humidity
 (4) decreasing temperatures

63. Air pressure is usually highest when the air is
 (1) cool and dry
 (2) cool and moist
 (3) warm and dry
 (4) warm and moist

64. Clouds usually form when moist air rises because
 (1) the air pressure increases
 (2) the dew point decreases
 (3) the air is cooled to its dew point
 (4) additional water vapor is added to the air

Weather

65 Which graph best represents the effect that heating has on air density in the troposphere?

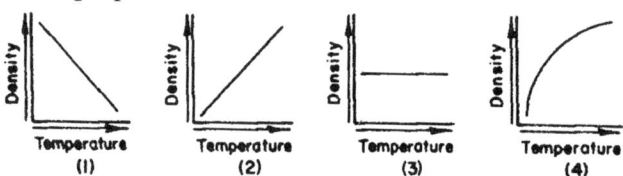

(1) Density decreasing with Temperature (downward line)
(2) Density increasing with Temperature (upward line)
(3) Density constant with Temperature (horizontal line)
(4) Density increasing and leveling off with Temperature (curve)

Note that questions 66 and 67 have only three choices.

66. Moisture is evaporating from a lake into stationary air at a constant temperature. As more moisture is added to this air, the rate at which water will evaporate will probably
  (1) decrease
  (2) increase
  (3) remain the same

67. As a stationary air mass is heated, its density will generally
  (1) decrease
  (2) increase
  (3) remain the same

68. Which statement best explains why the air pressure is usually greater over the ocean than over the land during daytime hours in the summer?
  (1) Air temperature is lower over the ocean.
  (2) Air is more humid over the ocean.
  (3) Prevailing winds are usually blowing from the land.
  (4) Water absorbs heat from the land.

CONSTRUCTED RESPONSE QUESTIONS

69. Base your answers to parts a through d on the diagram below, which represents a vertical cross section of a frontal system moving from west to east in an area of New York State. Temperatures for six weather stations, A through G, are shown on the diagram.

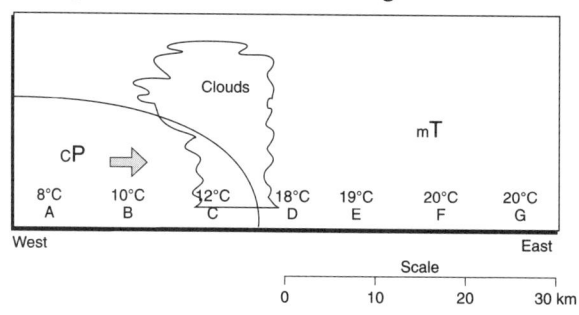

Diagram shows: cP air mass on west side with Clouds, mT air mass on east side.
Stations: A 8°C, B 10°C, C 12°C, D 18°C, E 19°C, F 20°C, G 20°C
Scale: 0, 10, 20, 30 km
West ← → East

612

(a) How will the cloud cover change at station *E* as the front approaches *and* after the front passes? [2]
(b) What barometric pressure change will occur as the front passes station *E*? [1]
(c) What change in temperature will occur after the front passes station *E*? [1]
(d) Describe the motion of the air masses that causes clouds to form along the front. [1]

70. Base your answers to parts (a) and (b) on the data table below, which shows the air temperatures and dew points over a 24-hour period for a particular location in New York State.

Time of Day	Air Temperature (°C)	Dew Point (°C)
12:00 midnight	19	12
2:00 A.M.	17	11
4:00 A.M.	14	10
6:00 A.M.	13	12
8:00 A.M.	15	11
10:00 A.M.	17	10
12:00 noon	18	9
2:00 P.M.	21	7
4:00 P.M.	23	6
6:00 P.M.	21	8
8:00 P.M.	19	10
10:00 P.M.	18	12
12:00 midnight	17	13

(a) Use the data to construct a graph on the grid, following the directions below. [5]
   1. Mark an appropriate scale on the axis labeled "Temperature."
   2. Plot a line graph for air temperature and label the line "Air temperature."
   3. Plot a line for the dew point and label the line "Dew point."
(b) Based on your graph, state the hour of the day when the relative humidity was lowest. [1]

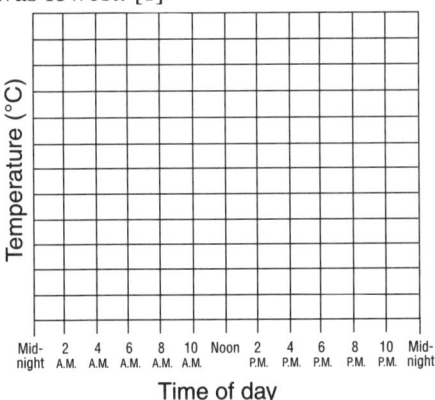

Weather

71. Base your answers to parts (a) through (c) on the weather maps below, which show the positions of a tropical storm at 10 A.M. on July 2 and on July 3.

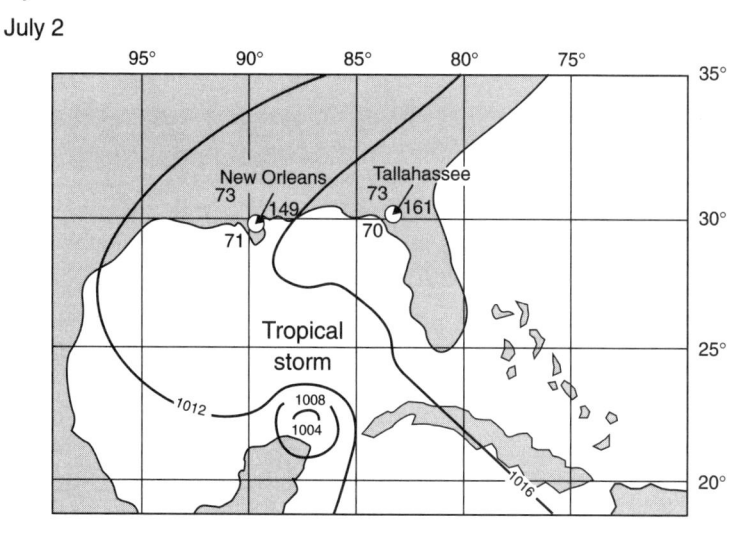

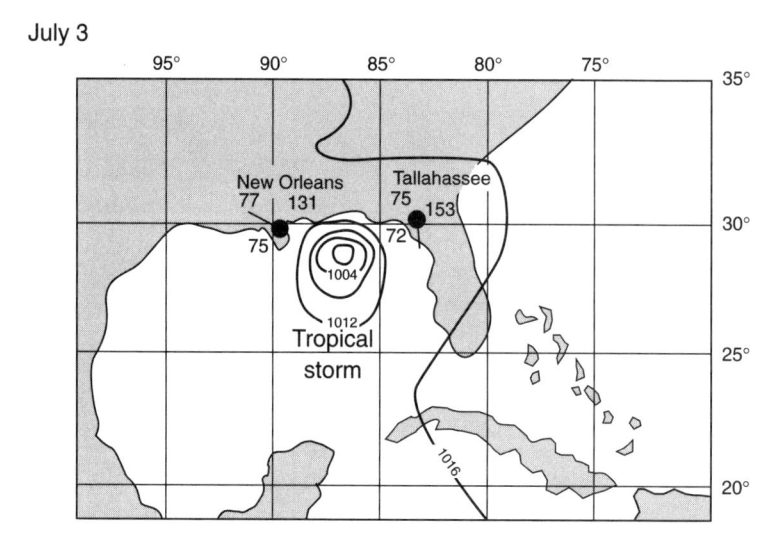

July 2

July 3

(a) State the dew-point temperature in Tallahassee on July 2. [1]
(b) Wind speed has been omitted from the station models. In one or more complete sentences, state how an increase in the storm's wind speed from July 2 to July 3 can be inferred from the maps. [2]
(c) The storm formed over warm tropical water. State what will most likely happen to the wind speed when the storm moves over land. [1]

72. Base your answers to parts (a) and (b) on the meteorological conditions shown in the table and the partial station model below, as reported by the weather bureau in the city of Oswego, New York.

Air temperature: 65°F
Wind direction: from the southeast
Wind speed: 20 knots
Barometeric pressure: 1017.5 mb
Dewpoint: 53°F

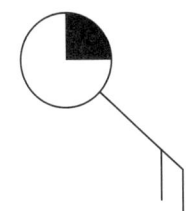

(a) Using the meteorological conditions given, complete the station model by recording the air temperature, dew point, and barometric pressure in the proper format. [2]
(b) State the sky conditions or amount of cloud cover over Oswego as shown by the station model. [1]

EXTENDED CONSTRUCTED RESPONSE QUESTIONS

73. Base your answers to parts (a) through (c) on the weather map below, which shows temperature readings at weather stations in the continental United States.

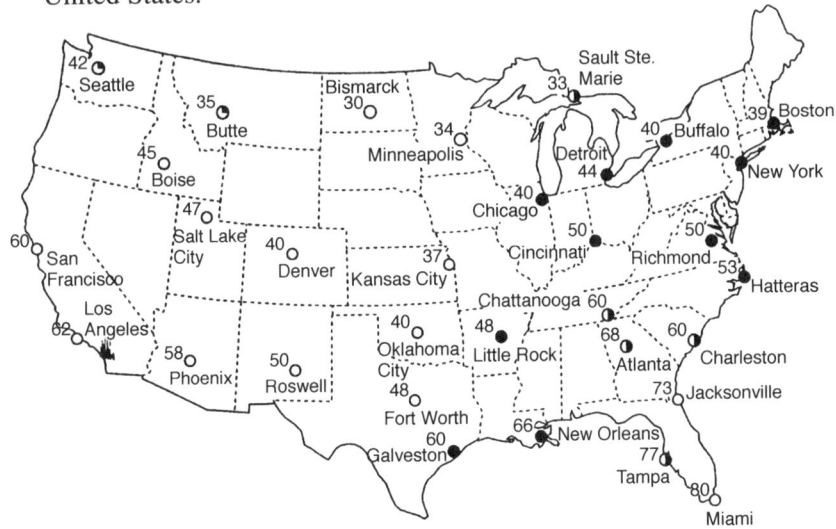

(a) On the weather map, draw three isotherms: the 40°F isotherm, the 50°F isotherm, and the 60°F isotherm. [2]

Weather

(b) In Richmond, Virginia, the wind direction is from the east at a speed of 20 knots. On the station model below, draw the correct symbols for wind direction and wind speed. [2]

50 ●

Richmond

(c) In addition to temperature, one other weather variable for each weather station is shown on the map. State the other weather variable. [1]

74. Base your answers to parts (a) through (e) on the information and data table below.

In August 1992, Hurricane Andrew, the most costly natural disaster in U.S. history, hit southern Florida. The data table below shows the location and classification of Hurricane Andrew on 7 days in August 1992.

DATA TABLE

Day	Latitude	Longitude	Storm Classification
August 18	13°N	46°W	Tropical storm
August 20	19°N	59°W	Tropical storm
August 22	25°N	66°W	Hurricane
August 24	25°N	78°W	Hurricane
August 26	28°N	90°W	Hurricane
August 27	32°N	91°W	Tropical storm
August 28	34°N	86°W	Tropical storm

(a) On the hurricane tracking map below, plot the locations of Hurricane Andrew given in the data table, following these directions:
 1. Use an **X** to mark each location on the grid. [1]
 2. Label each **X** with the appropriate date. The data for August 18 has been plotted on the map as an example. [1]
 3. Connect the **X**'s with a line to show the hurricane's path. [1]

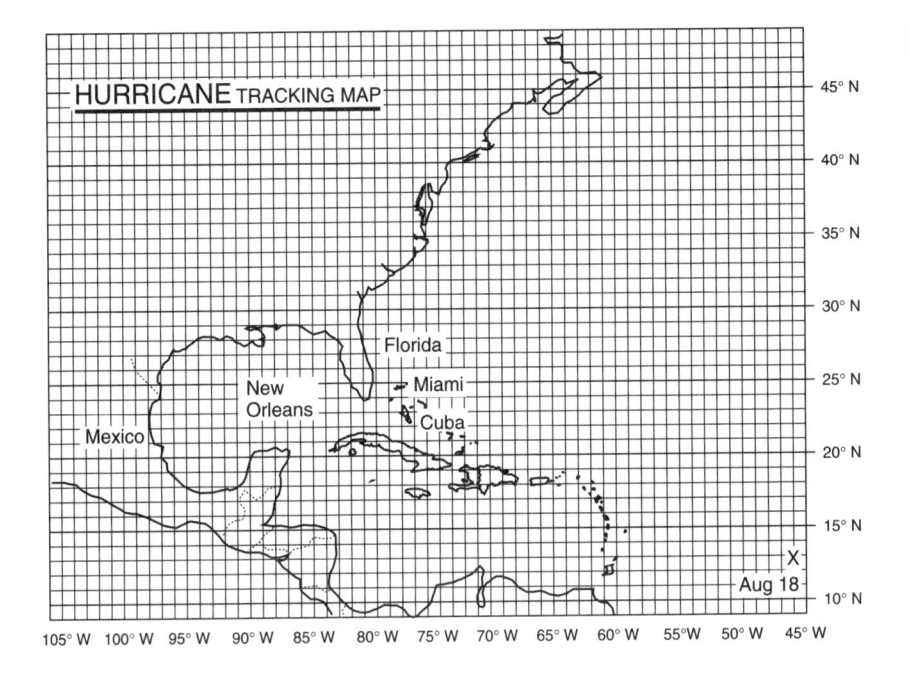

(b) As Hurricane Andrew approached Miami, Florida, cloudiness and precipitation increased dramatically. State how the air pressure at Miami was changing at this time. [1]

(c) By August 27, Hurricane Andrew was downgraded from a hurricane to a tropical storm because its wind speed decreased. State *one* reason why Hurricane Andrew's wind speed decreased at this time. [1]

(d) Describe one threat to human life and property that could have been caused by the arrival of Hurricane Andrew in southern Florida. [1]

(e) State three specific actions a person in southern Florida could take to increase safety or reduce injury or damage from a hurricane. [3]

CHAPTER 24
CLIMATE

> **KEY IDEAS** *Climate* is the term used to describe the general weather conditions in a given area over a long period of time. Global climate is determined by the interaction of solar energy with Earth's surface and atmosphere. Dynamic processes such as cloud formation and Earth's rotation, as well as the positions of mountain ranges and oceans, influence this energy transfer. Climate is important in determining the habitability of a location and the most reasonable land uses within the region. One of the most important effects of climate is the availability of water at the surface and within the ground.

KEY OBJECTIVES
Upon completion of this unit, you will be able to:

- Differentiate between weather and climate.
- Identify the various factors that control climate.
- Describe the role that the cycling of water and energy in and out of the atmosphere plays in determining climatic patterns.
- Describe several patterns of global climatic change.

WHAT IS CLIMATE?

Definition of Climate

Climate is the characteristic weather of a region, particularly temperature and precipitation, averaged over an extended period of time. We recognize that weather phenomena, when viewed against the long-term, stable backdrop of global climate, are temporary, local changes.

The climate of a region cannot be determined by looking at the weather over a single year. In 1988, the Midwest suffered a nearly rainless summer. Fields of grain in the nation's heartland turned brown and shriveled, water levels dropped in the Mississippi River, and wildfires blazed through millions of acres of forested land. Yet, in the summer of 1993, many of these same places were drenched by torrential rains, farmers watched floodwaters wipe out their fields, and the Mississippi River flooded communities all along its course. Many people wondered, are we seeing Earth's global climate change? One or two extreme summers can't answer that question. Our picture of climate develops slowly as we watch scores of seasons pass. Some winters are warmer than others, some summers dryer, some falls colder. We can gain a sense of the shifting patterns of climate only by comparing measurements taken over many years and even decades.

At any given point in time, a wide variety of climates can exist on Earth simultaneously. Earth's climates range from steamy rain forests to icy glaciers, hot deserts, and cool forests of conifers. However, as shown by the summers of 1988 and 1993 in the Midwest described above, any given region may also be subject to wide seasonal changes in temperature and rainfall. One of the challenges in studying climate is deciding just how to measure it.

The Main Elements of Climate: Temperature and Precipitation

Climate is the result of the interplay of a number of factors. One of the most important elements in determining the climate of a region is energy. As you know, Earth's main source of energy is sunlight, which warms the land, which, in turns, heats the atmosphere. Globally, the amount of heating that occurs depends on the amount of sunlight reaching Earth's surface. A key way of tracking energy flow in a region is by monitoring temperatures. Therefore, scientists who study climate keep track of air temperatures over land and sea, air temperatures at various altitudes, and ocean water temperatures around the globe.

Another major climate element is water. Water is important to all living things, and controls to a large extent the types of plant and animal life that can inhabit a region. It also plays an important role in weathering and is the main agent of erosion and deposition on Earth. Thus, the distribution of atmospheric moisture and the amount of precipitation are important factors in characterizing climates. Some other measurements used to monitor climates and climate changes are the amount of snow and ice cover on land, the extent of sea ice, and the concentrations of various gases in the atmosphere.

CLIMATE FACTORS

You have learned that the main elements of a region's climate are its temperature and precipitation patterns. However, various factors influence the cycling of water and energy in Earth's atmosphere and thereby produce different climates. Temperature and precipitation patterns are controlled by such factors as latitude, the distribution of land and water, high- and low-pressure belts and prevailing winds, monsoons, ocean currents, vegetative cover, elevation, mountain ranges, clouds, and cyclonic storm activity.

Climate

Latitude

Latitude is the primary factor of temperature control in climate because it determines both the angle and duration of insolation. Much more insolation is received in low latitudes near the Equator than in high latitudes near the poles.

Only the areas that lie between 23½° N (the Tropic of Cancer) and 23½° S (the Tropic of Capricorn) ever receive the vertical rays of the Sun. The average angle of insolation decreases toward the poles, causing the insolation to be spread over a larger area and therefore to be less intense. As the angle of insolation decreases, the solar energy passes through more of the atmosphere, so more is reflected or absorbed and insolation is further decreased.

While the average length of daylight, or duration of insolation, is 12 hours everywhere on Earth at the time of the equinoxes, at other times it varies. It is always 12 hours at the Equator; but as one moves north or south of the Equator, the duration of insolation varies in a cyclic manner. It increases toward the pole tilted toward the Sun, ranging from 12 hours at the Equator to 24 hours at the pole. It decreases toward the pole tilted away from the Sun, ranging from 12 hours at the Equator to 0 hours at the pole. The result is that the higher the latitude, the greater the cyclic change in duration of insolation over the course of a year. The poles range from 24 hours of insolation in the summer to no insolation in the winter. At the latitudes of New York State, insolation ranges from about 15 hours in the summer to about 9 hours in the winter.

Differences in the angle and duration of insolation with latitude result in three main temperature zones: the always-hot *torrid zone* near the Equator, the always-cold *frigid zone* near the poles, and the seasonally changing, intermediate *temperate zone* in between.

Distribution of Land and Water

The irregular distribution of land and water surfaces on Earth is another major factor controlling climate. As discussed earlier, land surfaces increase in temperature more than water surfaces when insolation strikes them, and they also cool off more rapidly than water surfaces. As a result, air temperatures are warmer in the summer and cooler in the winter over landmasses than they are over oceans at the same latitude. Large bodies of water tend to moderate the temperatures of nearby landmasses by warming them in winter and cooling them in summer. For this reason cities in the interior United States have a greater annual temperature range than coastal cities (see Figure 24.1).

	J	F	M	A	M	J	J	A	S	O	N	D
San Francisco, CA	49	51	53	56	58	61	63	63	64	61	55	50
St. Louis, MO	32	35	43	55	64	74	78	77	70	58	44	35

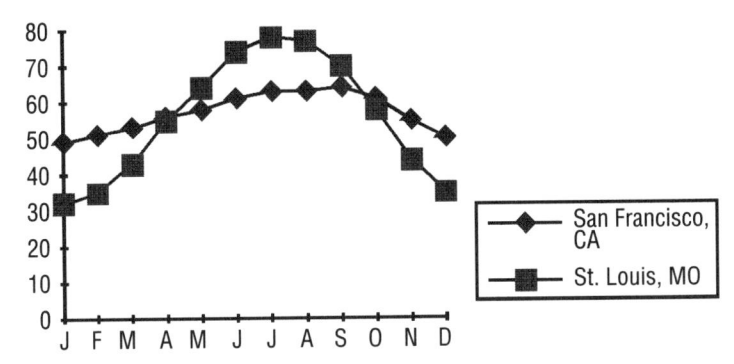

Figure 24.1 Average Monthly Temperatures at St. Louis, Missouri, an Inland City, and San Francisco, California, a Coastal City, Both at Approximately the Same Latitude.

Large areas permanently covered by ice also affect climate. The light color, high specific heat capacity, and latent heat of fusion of ice combine to keep it from melting completely in certain areas. The air over ice surfaces is chilled by contact with the ice, and this cold air, in turn, chills the land and water surrounding these surfaces.

Pressure and Wind Belts

The huge convection cells that form in the atmosphere because of unequal heating give rise to global pressure and wind belts. A belt of low pressure caused by rising warm air lies over the Equator. Low pressure favors evaporation, so the air there tends to be both warm and moist. The rising air spreads outward from the Equator, cools, loses moisture because of condensation, and then sinks back to Earth. This descending air creates a belt of high pressure on either side of the equatorial low between 25° and 30° north or south latitude. As the air sinks toward the ground, it is compressed and warms slightly. This warm, dry air is able to absorb moisture, and land surfaces in these regions tend to be dry. Most of Earth's deserts are located in these subtropical, high-pressure belts.

The adjacent high- and low-pressure belts give rise to winds that carry warm, moist air outward from the Equator. These are known as the tropical easterlies or *trade winds*.

The poles are regions of

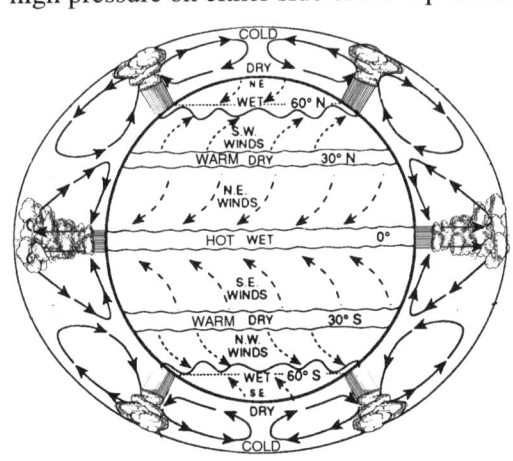

Figure 24.2 The Climatic Effects of Convection Cells in the Atmosphere.

high pressure because of their frigid, dry air. As the frigid air moves outward from the poles, it warms up and begins to rise, creating the subpolar low-pressure belts at 60° north and south latitude. The 30° high-pressure belts and the 60° low-pressure belts give rise to the moist but cooler prevailing westerlies.

In general, the high-pressure belts create regions of dry air, and the low-pressure belts create regions of moist air. The planetary wind belts carry moist or dry air over landmasses, greatly influencing their climates. These winds also transfer heat energy from the Equator toward the poles, modifying temperature patterns.

Figure 24.2 shows the planetary wind and pressure belts and their climatic effects.

Monsoons

Some landmasses are so large that they create their own seasonal convection cells. A *monsoon* is a seasonally reversing system of surface winds caused by temperature differences between land and ocean. A monsoon is a large-scale version of the type of circulation seen in land and sea breezes. The largest monsoon circulation occurs in southern Asia.

In land and sea breezes, we see *daily* changes in wind direction along a coastline. During the day, land heats up more quickly than water. Air above the land warms up and rises, forming a convection cell that draws cooler air from over the sea inland—a sea breeze. During the night, the land cools off more quickly than the water. The air above the warmer water rises, forming a convection cell that draws cooler air from over the land out to sea—a land breeze. (Refer to Figure 23.11: Land and Sea Breezes.)

In a monsoon, we see *seasonal* changes in wind direction between a large landmass and surrounding oceans. During the summer, the large Asiatic landmass warms up and a low-pressure center develops in the southern part of the continent. This low-pressure center results in winds that blow inland from the surrounding oceans. These moist summer monsoon winds bring clouds and heavy rains. During the winter, intense cooling produces a mass of very cold, very dry air over Siberia. The result is a high-pressure center in the northern part of the landmass that causes the winds to reverse direction and blow from the continental interior outward to the coast. Although these winter monsoon winds may warm up as they descend toward the coast, they pick up little moisture over the land and they bring clear, dry weather. See Figure 24.3.

A monsoon wind system also affects northern Australia. Since the seasons in Australia are opposite to those in Asia, however, winds blow in from the sea in January and outward from the continental interior in July. There is also some indication of a seasonal monsoon in the central part of the United States. During the summer, moist air generally flows from the Gulf of Mexico northward into the central plains. In the winter, cool, dry winds from Canada reverse this flow most of the time.

(a) Summer Monsoon Winds

(b) Winter Monsoon Winds

Figure 24.3 A Monsoon. A monsoon acts like a large-scale sea breeze in the summer (a) and a land breeze in the winter (b). Source: *The Earth Sciences*, Arthur N. Strahler, Harper & Row, 1971.

Climate

Ocean Currents

Ocean currents control climates by transferring heat from the Equator toward the poles, thereby cooling the equatorial regions, warming the polar regions, and influencing the temperature of adjacent landmasses these currents pass. The major surface ocean currents (see Figure 24.4) follow the prevailing winds that blow out of the subtropical high-pressure belts, giving rise to a clockwise flow in the Northern Hemisphere and a counterclockwise flow in the Southern Hemisphere.

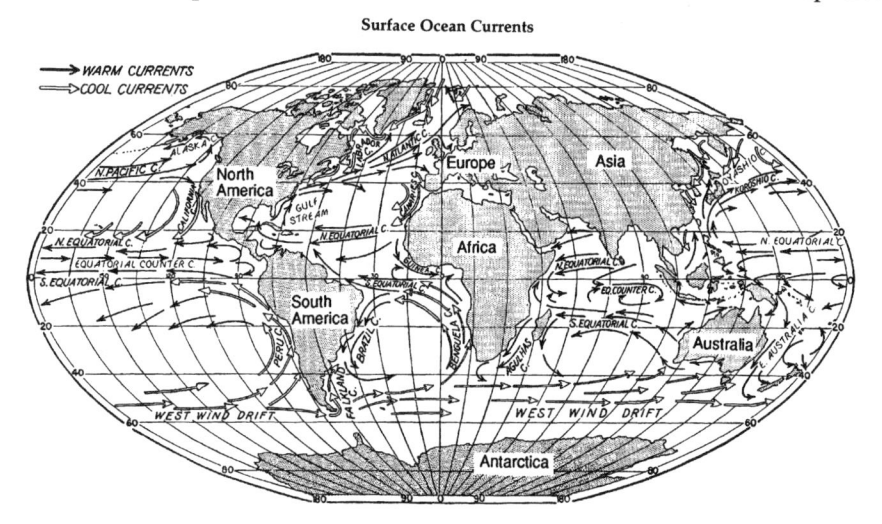

Figure 24.4 The Major Surface Ocean Currents of the World. Source: The State Education Department, *Earth Science Reference Tables*, 2001 ed. (Albany, New York: The University of the State of New York).

Wherever the ocean currents flow toward the poles, they carry warm water from the Equator. Wherever they flow toward the Equator, they carry cold water from the polar regions. Ocean currents such as the Gulf Stream and the North Atlantic Current carry warm water northward and eastward. The warm air over these ocean currents moderates the climate of the eastern coast of the United States, the Azores, and western Europe. The cold air over the icy waters of the Peru Current modifies the climate all along the western coast of South America, causing it to be much cooler and dryer than places with the same latitudes on the eastern coast of South America, where the Brazil Current and South Equatorial Current bring warm, moist air with their waters.

Vegetative Cover

Climate influences a region's natural vegetation; conversely, natural vegetation influences climate by affecting rainfall and temperature. Vegetation affects the water cycle by influencing the processes of evapotranspiration and surface runoff, which, in turn, influence rainfall. *Evapotranspiration* is a process that combines evaporation and transpiration, the two mechanisms by which liquid water is returned to the atmosphere as water vapor. ***Transpiration*** is the process by which plants release water vapor into the atmosphere through their leaves. It has been estimated that approximately half the rainfall in the Amazon Basin is derived from local evapotranspiration. Without the vegetation, the region would have a much drier climate.

Vegetation also influences temperature. Vegetation reflects less insolation back into space than bare surfaces, which tend to warm the climate. This effect is small, however, compared to the cooling that results when vegetation absorbs atmospheric carbon dioxide and thereby decreases the greenhouse effect. Thus, the net effect of deforesting land is to warm the climate.

Elevation

There is a gradual decrease in average temperature with elevation. This is due to the decrease in air pressure with elevation, which causes air to expand and cool. The fact that temperature decreases approximately 1°C for every 100-meter rise in elevation explains why high mountains may have tropical vegetation at their bases but permanent ice and snow at their peaks.

Mountain Ranges

Relief, or the differences in the heights of landforms in an area, is another factor that controls climate. For example, mountain ranges serve as barriers to outbreaks of cold air. In this way, the Alps protect the Mediterranean coast and the Himalayas protect India's lowlands.

In a process called the *orographic effect*, the windward side of a mountain facing winds carrying moisture-laden air usually receives much more precipitation than the leeward side. Mountain ranges force the air that is blown over them by winds to rise, and cool by expansion. This decreases the air's capacity to hold moisture and causes condensation, which forms clouds and precipitation. By the time the air reaches the top of the mountain, it has lost much of its moisture. When the air descends on the leeward side of the mountain, it is warmed by compression, its capacity to hold moisture is decreased, and precipitation is less likely. This orographic effect explains why cities along the Oregon coast west of the Cascades (e.g., Tillamook) have much precipitation while cities in the state's interior (e.g., Bend) have a dry climate (see Figure 24.5).

Clouds and Cyclonic Storm Tracks

Clouds control climate by reducing the amount of insolation gained or lost by a region. During daylight, clouds block sunlight and keep the temperature from rising as high as it would if the skies were clear. At night, the water droplets and water vapor in clouds absorb long-wavelength heat being radiated into space and thus keep the region from getting as cold as it would without them.

Climate

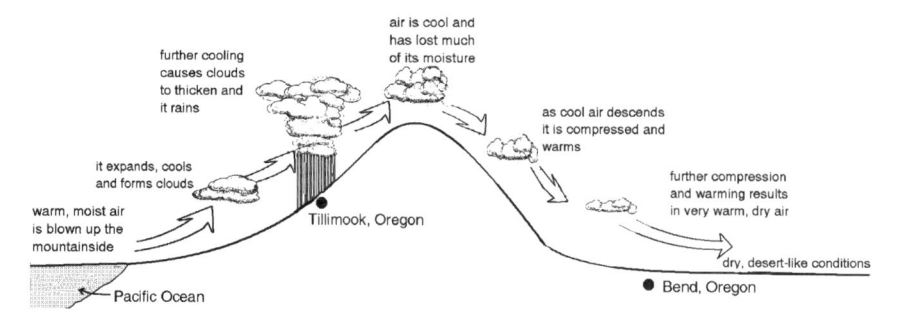

Figure 24.5 The Orographic Effect. On the windward side of the mountains, moist air is forced to rise and then cools by expansion, resulting in clouds and precipitation. On the leeward side, the cool air, which has lost much of its moisture, descends and warms by compression, resulting in warm, dry air that creates desertlike conditions.

The warm and cold fronts of wave cyclones that form along the polar front produce the changing weather of middle latitudes. These fronts produce precipitation and carry it over a wide area, causing changes in temperature as both warm and cold fronts pass through.

TYPES OF CLIMATES: THREE PATTERNS OF CLASSIFICATION

Climates are classified by moisture, temperature, and vegetation patterns. There are several systems in use today. The accompanying table summarizes terms often used to describe different patterns of moisture, temperature, and vegetation.

TABLE 24.2 DESCRIPTIVE TERMS FOR CLIMATIC PATTERNS

Moisture	Temperature	Vegetation
Arid	Polar	Desert
Semiarid	Subpolar	Grassland, steppe, taiga
Subhumid	Subtropical	Deciduous forest
		Coniferous forest
Humid	Tropical	Rain forest

Figure 24.6 is a map of the major climates of the world.

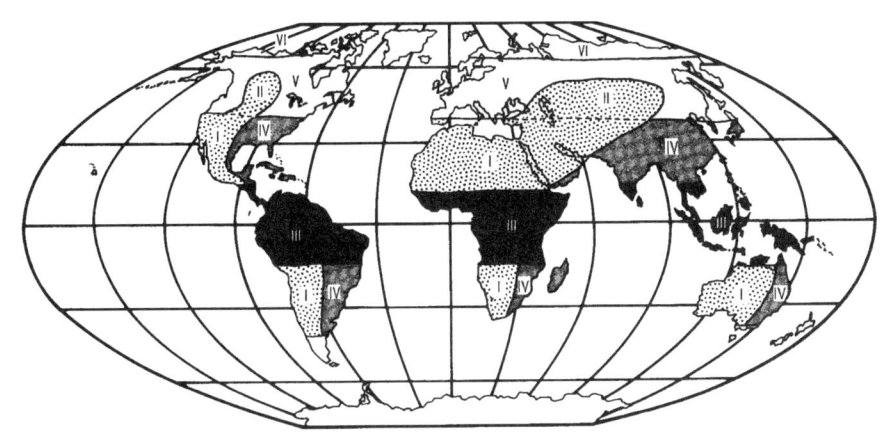

Figure 24.6 Major Climates of the World. I. Dry—hot summers, warm or cool winters. II. Dry—hot summers, cold winters. III. Humid—hot all year. IV. Humid—hot summers, cool winters. V. Humid—warm summers, cold winters. VI. Humid—cold summers, cold winters. Source: Part I: *Teacher's Guide—Investigating the Earth*, Earth Science Curriculum Project, Sponsored by the American Geological Institute and the National Science Foundation, Copyright © 1967, American Geological Institute.

EFFECTS OF CLIMATE

Climate has its greatest effect on vegetation. Individual plant species can generally survive only in a certain range of sunlight, temperature, precipitation, humidity, soil type, and wind. There is such a strong relationship between climate types and vegetation that climate regions are often named for their dominant vegetations.

Climate also affects the type of soil that will develop in an area. Warm, wet climates break down rock into soil faster than cool, dry climates. Soils in wet climates tend to be less fertile, though, because heavy precipitation and infiltration dissolve out nutrients needed by plants. Climate also affects soil by determining the types of plants that will become part of the soil when they eventually decay.

Another effect of climate is the shape into which landforms are eroded. In humid climates, where abundant rainfall makes running water the primary agent of erosion, landforms tend to have rounded contours. In arid climates, where wind is dominant and water is likely to erode Earth's surface in violent bursts, landforms are likely to be more jagged and angular. See Figure 24.7.

Climate

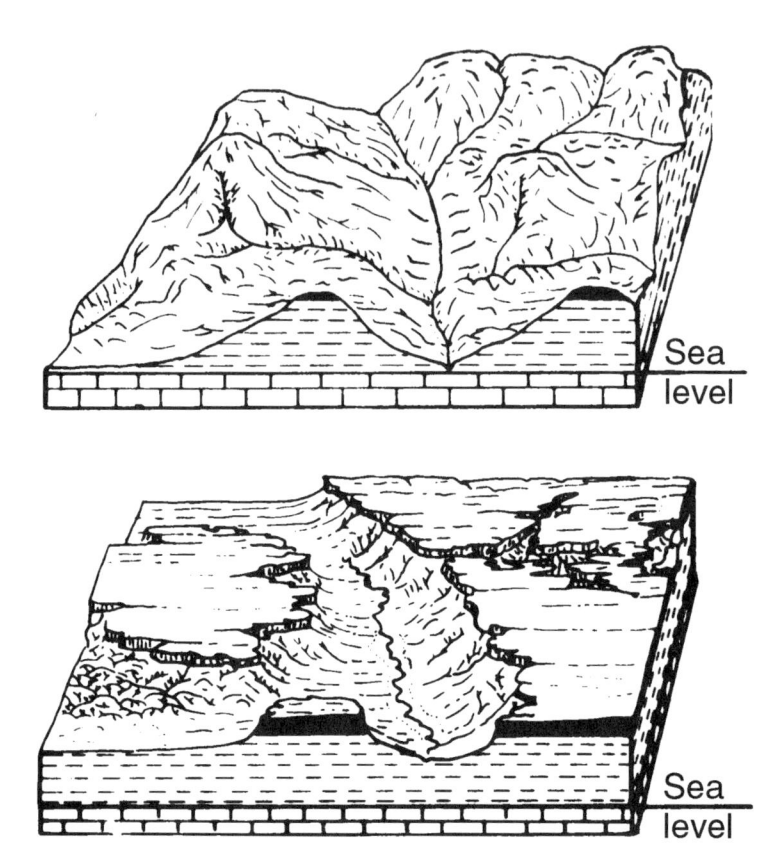

Figure 24.7 The Effects of Climate on Landforms. (a) In humid climates, abundant rainfall promotes a vegetative cover that protects the soil from rapid erosion by runoff and produces gently rounded slopes. (b) In arid climates, with little protective vegetation, erosion by rapid runoff and wind leads to steep slopes and exposed bedrock.

GLOBAL CLIMATIC TRENDS

Earth is an active planet. Plate motions move landmasses to different latitudes and change topography. Global temperatures change because of a variety of factors, including changes in the output of energy by the Sun, variation in the tilt of Earth's axis of rotation, and the blocking of incoming solar radiation by some pressure in the atmosphere. As a result, climatic zones shift over time. Some of the climatic trends noted in recent years include desertification, global warming, and the cyclic climate changes associated with El Niño.

Desertification

Desertification is the rapid development of deserts caused by the impact of human activities. Desertification does not occur because of forces originating within the desert. Rather, this patchy changing of dry, habitable land to unhabitable desert is the result of unwise land use by humans, accelerated by natural factors such as drought. Vegetation in dry lands is naturally scarce. Nevertheless, it is a valuable resource that protects the soil from erosion and provides food for people and for livestock. It may also be used for shelter and fuel.

When the sparse natural vegetation of dry land is severely disturbed by overgrazing, clearing for crops, or drought, the land deteriorates. Erosion removes fertile topsoil, leading to a drop in soil fertility. With fewer plant roots to break it up, the soil crusts over in the baking Sun and becomes less permeable to water. When precipitation does fall, less infiltrates and more runs off, leading to increased erosion and a decrease in soil moisture and groundwater. The result is a permanent conversion of mariginal dry lands to desert. Natural cycles of drought do play a role in desertification; but without human influence, the arid land is less severely damaged and its natural systems generally recover when the drought ends. Some projections suggest that within the next 20 years human activities will have caused one-third of the world's once-arable land to become useless for growing food crops.

Global Warming

Modern burning of fossil fuels has been increasing the amount of carbon dioxide in the air. In the past century, carbon dioxide levels in the atmosphere have increased by an estimated 25 percent, and these levels continue to climb. It is feared that, if the heat trapped by carbon dioxide increases with its concentration, rising levels of this gas will cause increased greenhouse-effect heating of the atmosphere. Global temperatures have shown an upward trend since 1850; it is difficult to tell, however, whether this increased warming is due to human impact or is part of the natural fluctuations in global temperatures.

One of the best known changes in global temperatures in modern times was the "little Ice Age," which lasted from 1450 to 1850. Paintings from this time show alpine glaciers in the Alps advancing dramatically. The high mountains of Ethiopia were covered in snow. There were times when a person could walk across New York Harbor from Manhattan to Staten Island. Then, about 1850, steady warming began.

Only in the mid-1990s did climatologists believe that they had enough evidence to state that greenhouse-effect heating has occurred and that the warming trend has moved beyond the probable range of normal changes in global temperatures. Global warming has probably accelerated desertification. It also has the potential to severely affect agriculture worldwide, increase sea level as glacial ice caps melt, and change global weather patterns. Recent projections suggest that soil-moisture levels in U.S. farmlands could drop by 40 percent if carbon dioxide levels double—an alarming prospect for farmers in areas that get barely enought rainfall now.

The El Niño-Southern Oscillation

The water in the ocean can be thought of as a series of thin layers, almost like pages in a book. When a steady wind blows over the surface of the ocean, the wind pushes the top layer, which begins to move. Due to the Coriolis effect, the top layer of water moves 45° to the right of the wind direction in the Northern Hemisphere (or to the left in the Southern Hemisphere). The top layer then pushes on the layer below it; and again the Coriolis effect comes into play, causing the second layer to move slightly, and a bit slower, to the right. As each layer pushes the one below it the direction of motion of the underlying layer continues to be successively further to the right. With increasing depth, each layer also moves more slowly until, at a depth of no more than a few hundred meters, the wind's effects are not felt at all. Though each layer moves in a slightly different direction, the net effect is that the upper part of the ocean is transported 90° to the wind direction. See Figure 24.8. When global winds blow steadily parallel to a coastline, this process carries warm surface waters offshore. As this warmer water is carried off, cold, nutrient-rich water rises to the surface to replace it—a process called upwelling. See Figure 24.9.

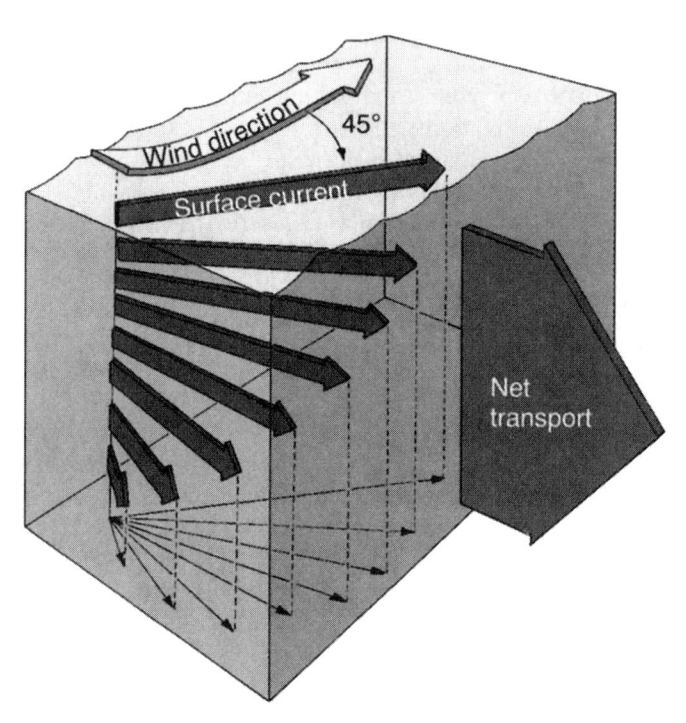

Figure 24.8 The Effects of Winds and Surface Currents. The net effect of winds blowing over surface waters is a current flowing at 90° to the wind direction.

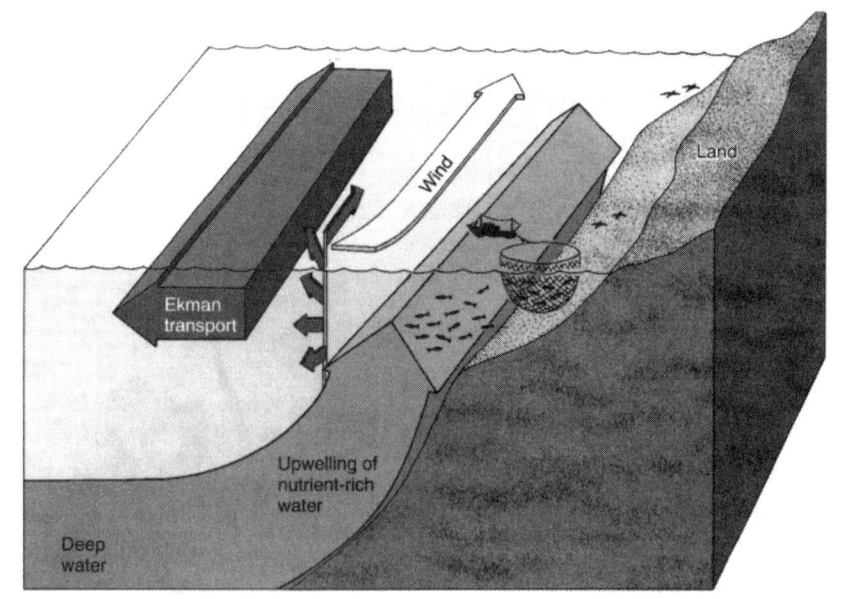

Figure 24.9 When prevailing winds blow parallel to a coast, warm surface water is carried offshore. Deep water that is rich in nutrient rises to the surface. This is called coastal upwelling. Source: *Marine Biology*, 3rd Ed., Peter Castro and Michael Huber, McGraw-Hill, 2000.

Along the eastern sides of many ocean basins, the global trade winds blow parallel to the coast and warm surface waters are carried offshore, producing strong coastal upwelling. As a result of the nutrient-rich water brought to the surface by this upwelling, these coastal areas rank among the ocean's richest fishing grounds.

The term *El Niño* originally referred to the changes in surface currents along the coasts of Peru and Chile. Trade winds blowing north along these coasts produce upwelling, making these waters rich fisheries. Every year, usually in December, the trade winds subside, upwelling slacks off, and the water becomes warmer. Local fisherman have known for centuries about this change in currents, which marks the end of the peak fishing season. Since the change comes around Christmas, the call it El Niño—"The Child." Every few years, however, the change is greater than usual. The surface water gets much warmer, the upwelling of nutrients ceases, and fish that depend on these food sources disappear.

The warming of the surface waters during El Niño causes a large region of low pressure to form over the southeastern Pacific Ocean. This, in turn, causes changes in the jet stream winds, which profoundly affect global weather patterns.

People in India depend on the summer monsoon to water their crops. When the monsoon fails, famine is widespread. In the early 1900s, analysis of weather records in an attempt to predict when India's summer monsoon would fail led to an important discovery: the air pressure over the Indian Ocean fluctuates. When it is extremely low, the normal monsoon flow car-

Climate

rying moist air from the ocean over the land is disrupted. Also, when the air pressure is unusually high over the Pacific Ocean, it tends to be unusually low over the Indian Ocean, and vice-versa. Over time, the air pressure seesaws back and forth over these two regions, a phenomenon called the **Southern Oscillation**. The changes in air pressure bring dramatic changes in wind and rainfall, and failure of the summer monsoon in India. Exactly what triggers the process is not yet clear; but when the air pressure starts to seesaw, failure of the monsoon can be expected.

By the 1950s it was clear that El Niño and the Southern Oscillation were not isolated events, but part of a complex interaction between ocean and atmosphere that spans half the globe. See Figure 24.10. The El Niño-Southern Oscillation, or ENSO, is not really an abnormal event, but simply one extreme of a regular climate cycle. Nevertheless, the weakening or reversal of wind and ocean current patterns drastically changes normal regional weather systems.

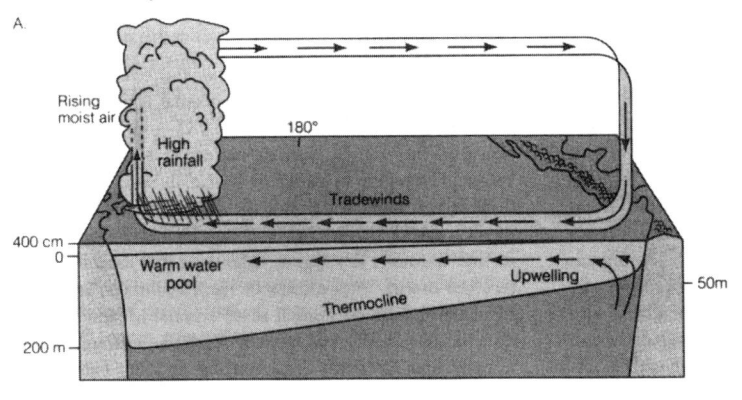

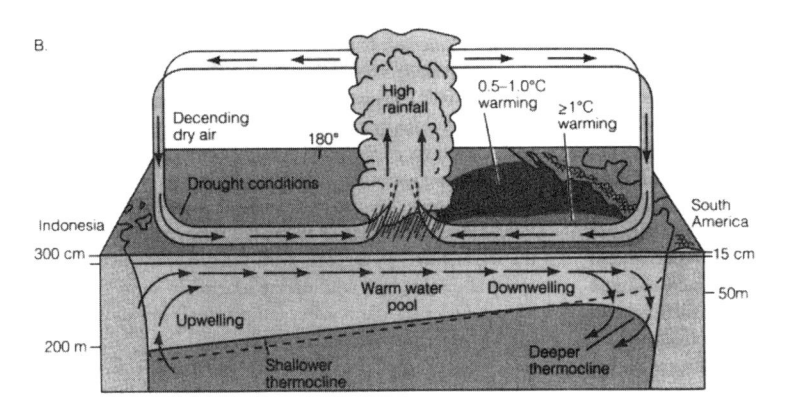

Figure 24.10 The El Niño-Southern Oscillation (ENSO). Source: *Environmental Geology*, Barbara W. Murck, Brian J. Skinner, and Stephen C. Porter, John Wiley, 1996.

The ENSO has had profound effects on weather, wildlife, and society. El Niño years bring drought, forest fires, and heat waves to some regions and storms and floods to others. One of the major El Niño events in the last century occurred in 1982–83; milder events followed in 1986–87 and 1991–95. By analyzing oceanographic and atmospheric data, scientists detected signs of an El Niño early in 1997 and predicted that another event was on its way. The 1997–98 event turned out to be even stronger than the 1982–83 El Niño. In Peru, Chile, and Ecuador, it brought torrential rains, flooding, and failure of the fishing industry. There were also floods in China, East Africa, and southern California. At the other extreme, Indonesia, Australia, southern Africa, and Central America suffered droughts. The weather changes were so severe that Florida, which rarely experiences tornadoes, was hit by a series of killer tornadoes. Estimated losses from the 1997–98 El Niño totaled more than $20 billion, but could have been much higher. The scientists' prediction gave the world advance warning and allowed officials in many countries to build up food reserves and prepare for impending disasters.

MULTIPLE-CHOICE QUESTIONS

In each case, write the number of the word or expression that best answers the question or completes the statement.

1. By which process does moisture leave green plants?
   (1) convection
   (2) condensation
   (3) transpiration
   (4) radiation

2. Bodies of water cool slower than land areas because
   (1) some insolation is converted into potential energy as water evaporates
   (2) bodies of water have a lower density than land areas
   (3) water has a higher specific heat than land
   (4) water is a better reflector of sunlight than land

3. The type of climate that a location has can be determined by comparing the yearly amounts of
   (1) precipitation and potential evapotranspiration
   (2) soil storage and potential evapotranspiration
   (3) precipitation and infiltration
   (4) change in soil storage and stream discharge

Climate

4. Two locations, one in northern Canada and one in the southwestern United States, receive the same amount of precipitation each year. The location in Canada is classified as a humid climate. Why would the location in the United States be classified as an arid climate?
   (1) The yearly distribution of precipitation is different.
   (2) The soil-moisture storage in the southwestern United States is more than that in northern Canada.
   (3) The potential evapotranspiration is greater in the southwestern United States than in northern Canada.
   (4) The vegetation of the southwestern United States is different from that of northern Canada.

5. Omaha, Nebraska, and Eureka, California, are both at 41° North latitude. Omaha is a midcontinent city and Eureka is a coastal city. Why is Eureka warmer in winter and cooler in summer than Omaha?
   (1) The angle of insolation is greater in Eureka than it is in Omaha.
   (2) The duration of insolation in Eureka is longer in winter and shorter in summer than it is in Omaha.
   (3) The prevailing winds in Eureka are from the west and those in Omaha are from the east.
   (4) The ocean water near Eureka is a poorer absorber and reradiator of heat than the land surface surrounding Omaha.

6. Bodies of water have a moderating effect on climate primarily because
   (1) water gains heat more rapidly than land does
   (2) water surfaces are flatter than land surfaces
   (3) water temperatures are always lower than land temperatures
   (4) water temperatures change more slowly than land temperatures do

7. Adjacent water and land masses have the same temperature at sunrise. They are heated by the morning Sun on a clear, calm day. After a few hours, a surface wind develops. Which diagram best represents this wind's direction?

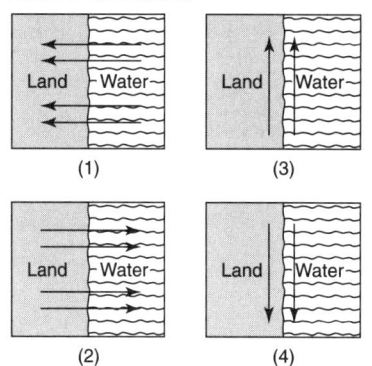

8. The seasonal temperature changes in the climate of New York State are influenced mostly by the
   (1) rotation of Earth on its axis
   (2) changing distance of Earth from the Sun
   (3) changing angle at which the Sun's rays strike Earth's surface
   (4) changing speed at which Earth travels in its orbit around the Sun

9. As a lake's water temperature decreases on a cloudy night, what occurs at the interface between the lake's surface and the air above the lake?
   (1) Energy given up by the lake is lost directly to outer space.
   (2) More energy is gained by the lake than is gained by the air.
   (3) The temperature of the air remains constant.
   (4) Energy is gained by the air from the lake.

10. The map below indicates four locations A, B, C, and D which have the same elevation and latitude. Which location would most likely experience the smallest range of annual temperature?

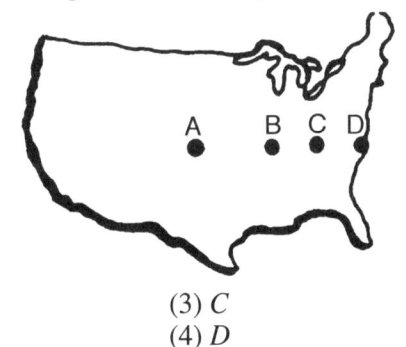

    (1) A   (3) C
    (2) B   (4) D

11. The diagram below shows the positions of the cities of Seattle and Spokane, Washington. Both cities are located at approximately 48° North latitude, and they are separated by the Cascade Mountains.

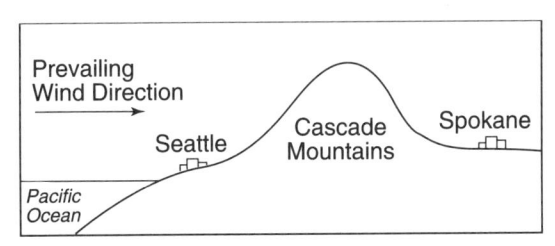

    How does the climate of Seattle compare with the climate of Spokane?
    (1) Seattle hot and dry, Spokane cool and humid
    (2) Seattle hot and humid, Spokane cool and dry
    (3) Seattle cool and humid, Spokane warm and dry
    (4) Seattle cool and dry, Spokane warm and humid

Climate

12. The potential evapotranspiration of an area is dependent upon the amount of evaporation surface and the
    (1) surface runoff   (3) stream discharge
    (2) soil storage   (4) average temperature

13. During which period will potential evapotranspiration usually be the greatest in New York State?
    (1) December 21 to March 21
    (2) March 21 to May 21
    (3) June 21 to August 21
    (4) September 21 to November 21

14. The map below shows the general path of ocean currents in a portion of the Northern Hemisphere. Locations A, B, C, and D are at the shoreline.

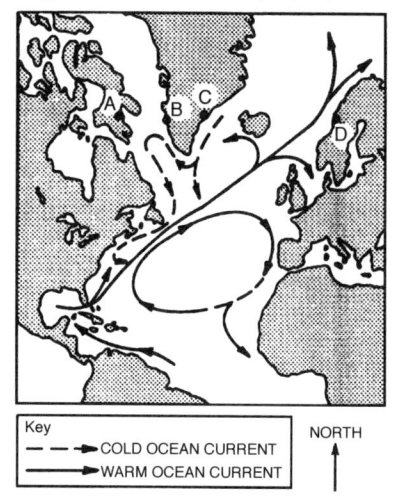

    Key
    - - -▶ COLD OCEAN CURRENT
    ———▶ WARM OCEAN CURRENT
    NORTH ↑

    Which location most likely has the warmest climate?
    (1) A   (3) C
    (2) B   (4) D

Base your answers to questions 15 through 19 on your knowledge of earth science and the diagram below, which represents an imaginary continent on Earth with surface locations *A* through *E*. An ocean surrounds the continent, and two mountain ranges are located as shown.

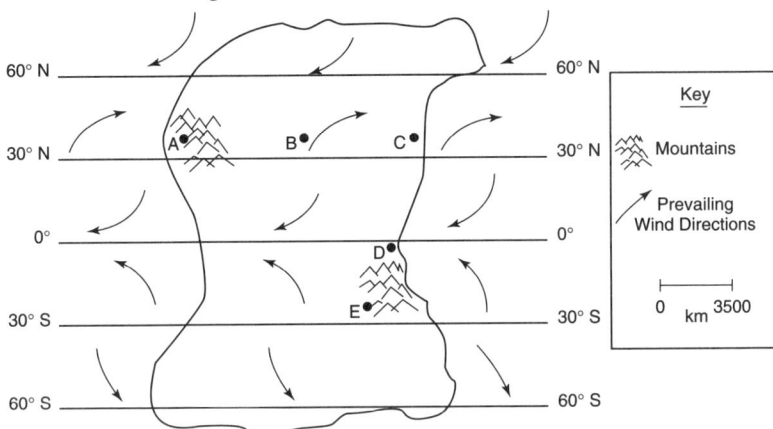

15. At which location would the total yearly potential evapotranspiration be the greatest?
(1) *A* (3) *C*
(2) *B* (4) *D*

16. What type of climate would most likely be found at location *E*?
(1) arid (3) humid
(2) subhumid (4) tropical rain forest

17. Location *A* will probably have a greater
(1) range in yearly temperatures than location *B*
(2) total yearly rainfall than location *B*
(3) altitude of the Sun in the summer than location *B*
(4) variation in the length of daylight hours than location *B*

18. The maximum insolation will occur at location *E* during the month of
(1) March (3) September
(2) June (4) December

Climate

19. Which graph best represents the relationship between temperature and time of year for location *B*?

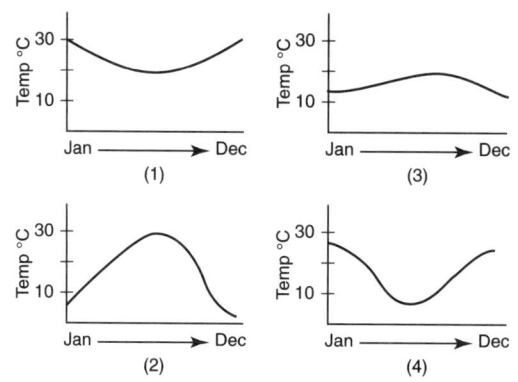

(1) Temp °C vs Jan→Dec (U-shape, high ends ~30, low middle ~15)

(2) Temp °C vs Jan→Dec (peak in middle)

(3) Temp °C vs Jan→Dec (slight bump in middle, low overall)

(4) Temp °C vs Jan→Dec (dip in middle)

20. Which statement best explains why a desert often forms on the leeward side of a mountain range, as shown in the diagram below?

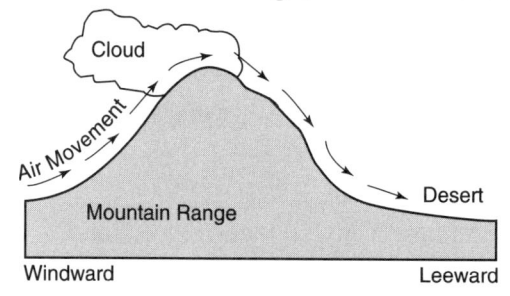

(1) Sinking air compresses and warms.
(2) Sinking air expands and warms.
(3) Rising air compresses and warms.
(4) Rising air expands and warms.

21. The diagram below shows the flow of air over a mountain from point A to point C. Which graph best shows the approximate temperature change of the rising and descending air due to the adiabatic process?

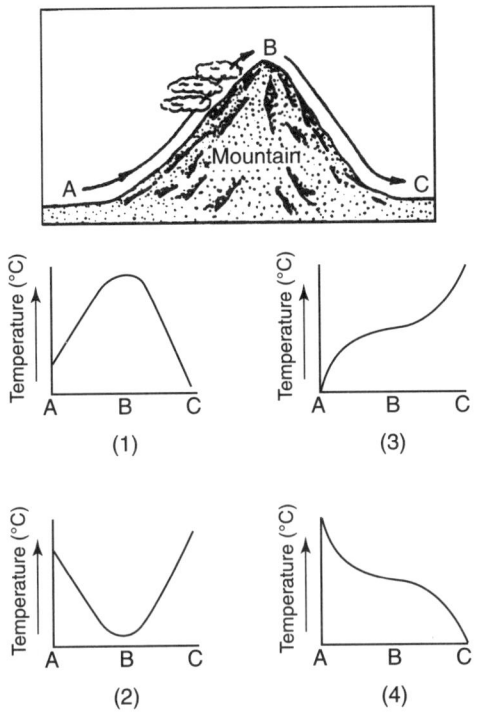

Base your answers to questions 22 through 26 on your knowledge of earth science and the diagram below which shows a mountain. The prevailing wind direction and air temperatures at different elevations on both sides of the mountain are indicated.

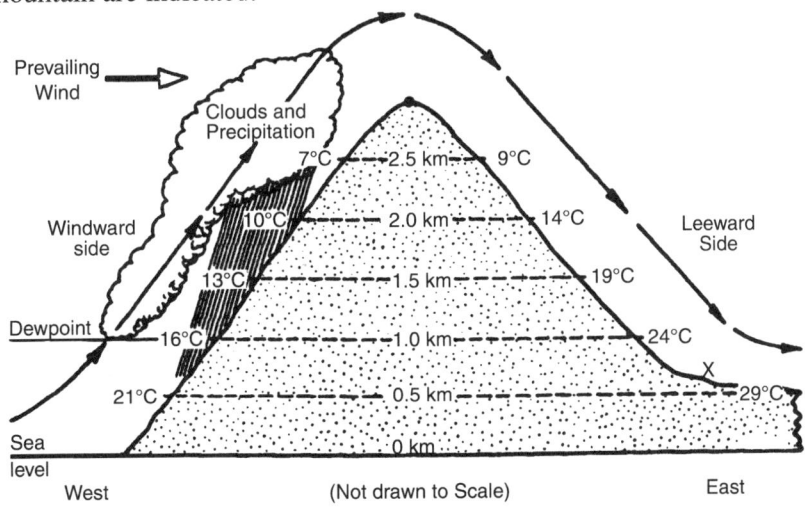

Climate

22. What would be the approximate air temperature at the top of the mountain?
    (1) 12°C
    (2) 10°C
    (3) 0°C
    (4) 4°C

23. On which side of the mountain and at what elevation is the relative humidity probably 100%?
    (1) on the windward side at 0.5 km
    (2) on the windward side at 1.5 km
    (3) on the leeward side at 1.0 km
    (4) on the leeward side at 2.5 km

24. How does the temperature of the air change as the air rises on the windward side of the mountain between sea level and 0.5 kilometer?
    (1) The air is warming because of compression of the air.
    (2) The air is warming because of expansion of the air.
    (3) The air is cooling because of compression of the air.
    (4) The air is cooling because of expansion of the air.

25. Which feature is probably located at the base of the mountain on the leeward side (location X)?
    (1) an arid region
    (2) a jungle
    (3) a glacier
    (4) a large lake

26. The air temperature on the leeward side of the mountain at the 1.5-kilometer level is higher than the temperature at the same elevation on the windward side. What is the probable cause of this difference?
    (1) Heat stored in the ocean keeps the windward side of the mountain warmer.
    (2) The insolation received at sea level is greater on the leeward side of the mountain.
    (3) The air on the windward side of the mountain has a lower adiabatic lapse rate than the air on the leeward side of the mountain.
    (4) Potential energy is lost as rain runs off the windward side of the mountain.

27. The climates of densely populated industrial areas tend to be warmer than similarly located sparsely populated rural areas. From this observation, what can be inferred about the human influence on local climate?
    (1) Local climates are not affected by increases in population density.
    (2) The local climate in densely populated areas can be changed by human activities.
    (3) In densely populated areas, human activities increase the amount of natural pollutants.
    (4) In sparsely populated areas, human activities have stabilized the rate of energy absorption.

28. Some scientists predict that the increase in atmospheric carbon dioxide will cause a worldwide increase in temperature. Which could result from this increase in temperature?
   (1) Continental drift will increase.
   (2) Isotherms will shift toward the Equator.
   (3) Additional landmasses will form.
   (4) Ice caps at Earth's poles will melt.

29. Some scientists believe that high-flying airplanes and the discharge of fluorocarbons from spray cans are affecting the atmosphere. Which characteristic of the atmosphere do these scientists believe is affected?
   (1) composition of the ozone layer of the stratosphere
   (2) wind velocity of the tropopause
   (3) location of continental polar highs
   (4) air movement in the doldrums

Constructed Response Questions

Base your answers to questions 30 and 31 on the data table and profile below. The data table gives the average annual precipitation for locations *A* and *B*. The profile represents a mountain in the western United States. Points *A* and *B* are locations on different sides if the mountain.

DATA TABLE

Location	Average Annual Precipitation (cm)
A	120
B	35

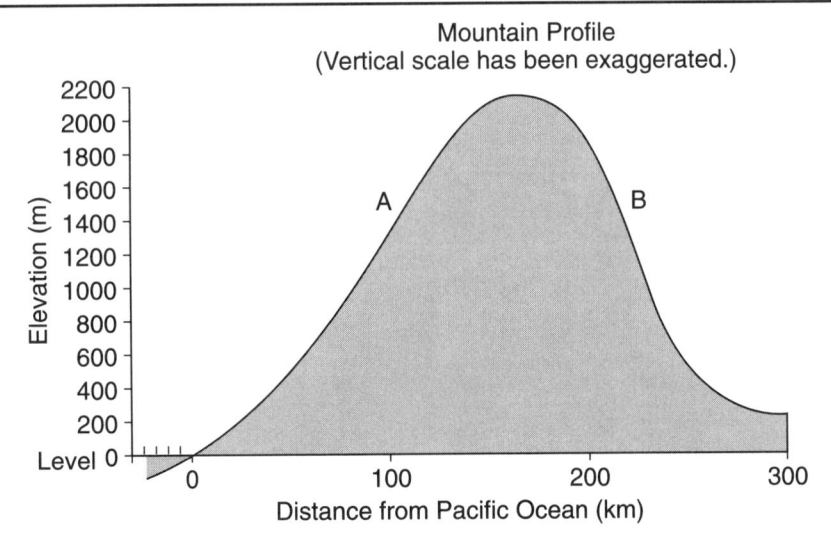

Mountain Profile
(Vertical scale has been exaggerated.)

Climate

30. State the elevation of location A. [1]

31. State one probable reason for the difference in average annual precipitation between location A and location B. [1]

32. (a) What is desertification? [1]
    (b) Describe two ways in which human activities contribute to the process. [2]

33. State two major concerns related to continued greenhouse-effect heating. [2]

34. In one or more complete sentences, explain how monsoons are similar to land and sea breezes. [2]

Base your answers to questions 35 and 36 on the climate graph below. Each graph represents data for a different city in North America. The line graphs connect the average monthly temperatures in degrees Celsius. The bar graphs indicate the average monthly precipitation in millimeters.

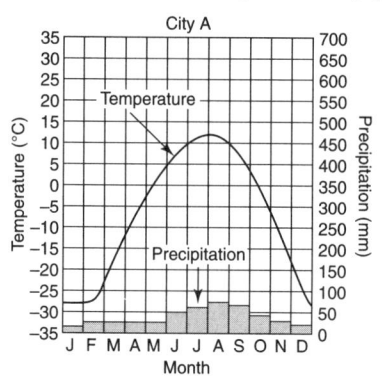

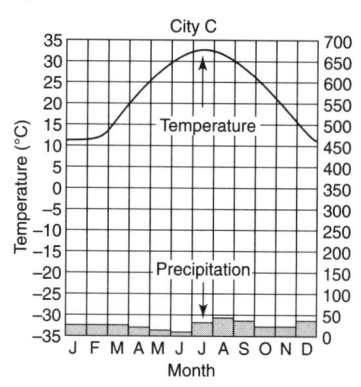

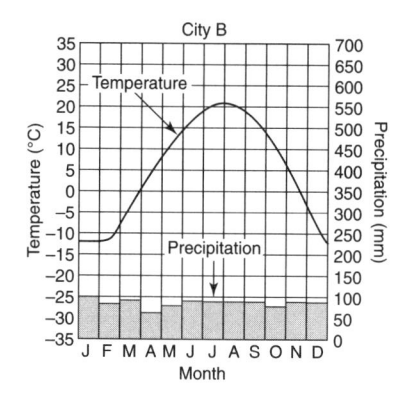

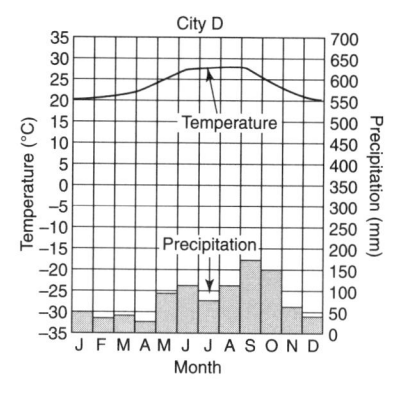

35. State the cities for which winter precipitation is most likely to be snow. [2]

36. List the cities in order of decreasing average yearly precipitation. [2]

37. State the city that is most likely located nearest a large body of water. [1]

EXTENDED CONSTRUCTED RESPONSE QUESTIONS

Base your answers to questions 38 through 41 on the information below and your knowledge of earth science.

The climate of an area is affected by many variables such as elevation, latitude, and distance to a large body of water. The effect of these variables on average surface temperature and temperature range can be represented by graphs on grids that have axes labeled as shown below.

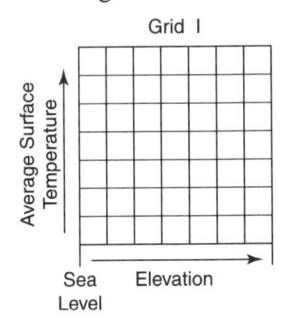

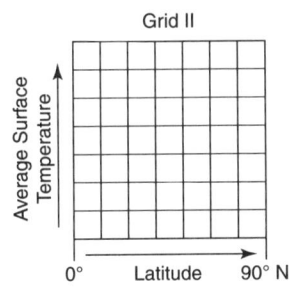

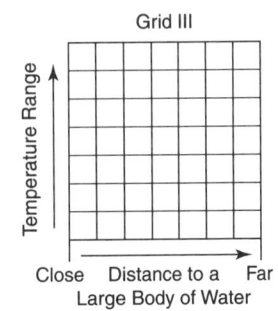

38. On Grid I, draw a line to show the relationship between elevation and average surface temperature. [1]

39. On Grid II, draw a line to show the relationship between latitude and average surface temperature. [1]

40. On Grid III, draw a line to show the relationship between distance to a large body of water and temperature range. [1]

41. The climate at most locations near the Equator is warm and moist.
    (a) Explain why the climate is warm. [1]
    (b) Explain why the climate is moist. [1]

Climate

42. Base your answers to parts (a) through (d) on the chart and map below. The chart represents high and low temperature readings in four United States cities—Savannah, Georgia; Abilene, Texas; Boston, Massachusetts; and Sioux Falls, South Dakota. The map shows the approximate locations of these four cities in the continental United States.

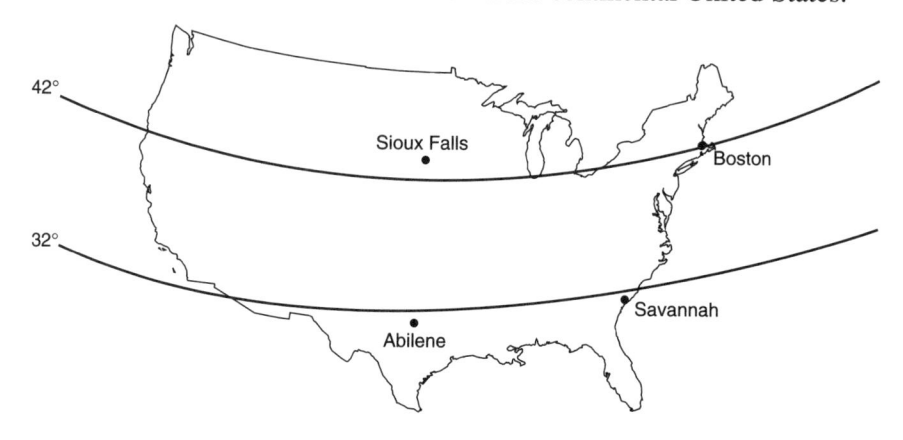

	Boston, MA		Sioux Falls, SD		Savannah, GA		Abilene, TX	
	High	Low	High	Low	High	Low	High	Low
Jan.	35.7	21.6	24.3	3.3	59.7	38.1	54.8	30.8
Feb.	37.5	23.0	29.6	9.7	62.4	41.1	59.7	35.1
Mar.	45.8	31.3	42.3	22.6	70.1	48.3	68.9	43.3
Apr.	55.9	40.2	59.0	34.8	77.5	54.5	77.8	52.9
May	66.6	49.8	70.7	45.9	84.0	62.9	84.4	61.1
June	76.3	59.1	80.5	56.1	88.8	69.2	91.4	68.9
July	81.8	65.1	86.3	62.3	91.1	72.4	95.2	72.7
Aug.	79.8	64.0	83.3	59.4	89.7	72.2	94.5	71.7
Sept.	72.8	56.8	73.1	48.7	85.2	67.8	86.7	65.3
Oct.	62.7	46.9	61.2	36.0	77.5	56.9	77.9	54.8
Nov.	52.2	38.3	43.4	22.6	70.0	48.1	66.3	43.4
Dec.	40.4	26.7	28.0	8.6	62.3	41.0	57.0	33.9
Avg. Yearly Precipitation (in)	41.51		23.86		49.22		24.40	

(a) Complete the table below. [3]

	Boston	Sioux Falls	Savannah	Abilene
Highest summer temperature				
Lowest winter temperature				
Temperature range				

(b) Based on the map above, and the table you just completed, state the relationship between temperature range and distance to a large body of water.

(c) Boston and Sioux Falls are at approximately the same latitude, and therefore receive roughly the same intensity and duration of insolation. In one or more complete sentences, explain why Boston and Sioux Falls have different summer high temperatures and winter low temperatures.

(d) State one reason why Abilene experiences higher temperatures throughout the year than Sioux Falls.

Appendix A

ANSWERS TO REVIEW QUESTIONS

CHAPTER 1

Multiple-Choice Questions

1.	3	5.	4	9.	1	13.	3	17.	2
2.	3	6.	3	10.	4	14.	4	18.	3
3.	4	7.	3	11.	1	15.	4	19.	3
4.	1	8.	4	12.	1	16.	1	20.	1

Constructed Response Questions

21.

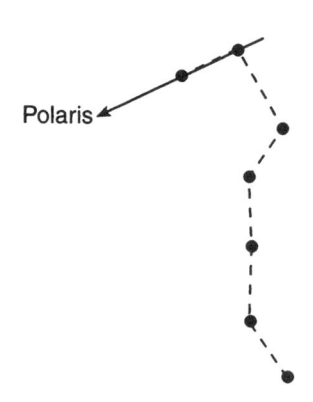

22. 3:00 P.M.
23. Location *D*
24. In 24 hours the Moon would appear slightly to the east of its current position in relation to nearby stars on the celestial sphere.
25. (a) Zenith (b) Celestial meridian (c) Horizon
26. Stars appear to move from east to west at 15° per hour. Therefore, to create an arc of 60° would require 60° ÷ 15°/hr = 4 hr.
27. Real motion is an actual change in position, apparent motion is an "observed" change in position.
28. The stars appear to move because Earth's rotation causes an observer to move in relation to the stars.
29. Stars do not change position in relation to other stars on the celestial sphere; planets do.
    Stars do not appear as disks when viewed with a telescope; planets do.
    Stars twinkle; planets do not.
    Stars are scattered throughout the celestial sphere; all the planets appear near the ecliptic.
30. The North Star is located almost directly over Earth's axis of rotation.

Appendix A: Answers to Review Questions

CHAPTER 2

Multiple-Choice Questions

1.	3	7.	4	12.	2	17.	2	22.	3
2.	4	8.	4	13.	1	18.	1	23.	3
3.	4	9.	1	14.	3	19.	1	24.	4
4.	1	10.	3	15.	1	20.	1	25.	4
5.	2	11.	1	16.	2	21.	1	26.	2
6.	1								

Constructed Response Questions

27. (a) **3** (b) **4** (c) **7** (d) **2** (e) **5**
28. (a) $e = d/l$ (b) $e = 2/8$ (c) $e = 0.25$
    (d) The ellipse is more eccentric than Earth's orbit.
29. (a) **3** (b) **2** (c) **4** (d) **1** (e) **4**
30. Gravitational attraction between the planets and the Sun is the force that keeps the planets in their orbits around the Sun.
31. (a) The movement of the comet through the solar system would not have been possible if solid, crystalline spheres existed; the comet would have shattered the spheres.
    (b) The force of gravitational attraction; Isaac Newton
32. (a) The existence of solid, crystalline celestial spheres
    (b) We do not sense the motion. Falling objects do not follow a curved path and land to the side of the point over which they are released.
    (c) Celestial spheres, or epicycles
33. (a) A planet 0.1 of Earth's mass would exert 0.1 of Earth's gravity if it were the same size as Earth. However, a planet with half the radius of Earth would exert $1/(½)^2 = 1/¼ = 4$ times as much gravitational force (4×0.1), or 0.4 of Earth's gravity. Thus, your weight on that planet would be about 0.4 (a little less than half) of your weight on Earth.
    (b) Mars is about 0.1 of Earth's mass and 0.5 of Earth's diameter. Therefore, an astronaut on Mars would weigh about 0.4 what he or she would weigh on Earth.
34.

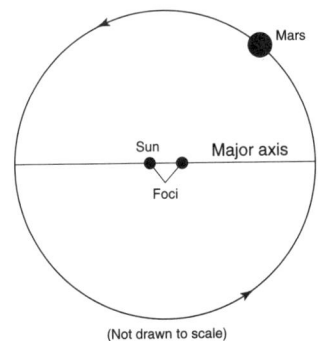

Appendix A: Answers to Review Questions

35. Earth's orbit is less eccentric (more nearly a circle) than Mars's orbit.

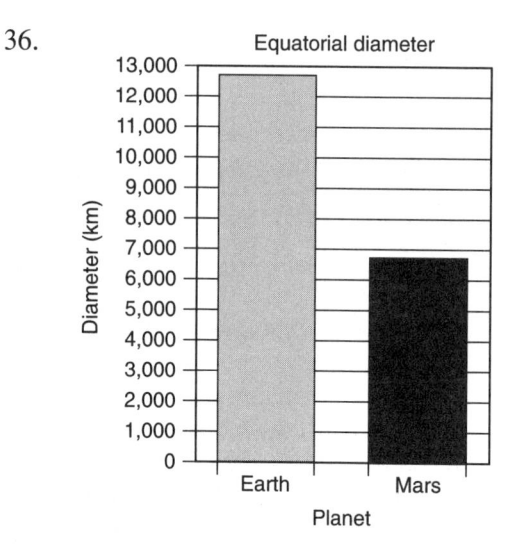

36.

Equatorial diameter

[Bar graph showing Earth with a diameter of approximately 12,700 km and Mars with a diameter of approximately 6,700 km. Y-axis: Diameter (km) from 0 to 13,000. X-axis: Planet]

CHAPTER 3

Multiple-Choice Questions

1.	1	9.	4	17.	2	25.	2	33.	1
2.	2	10.	2	18.	4	26.	4	34.	4
3.	2	11.	1	19.	2	27.	2	35.	1
4.	2	12.	3	20.	2	28.	2	36.	2
5.	3	13.	2	21.	1	29.	2	37.	2
6.	4	14.	1	22.	4	30.	3	38.	4
7.	4	15.	4	23.	4	31.	4	39.	1
8.	1	16.	2	24.	2	32.	3		

Constructed Response Questions

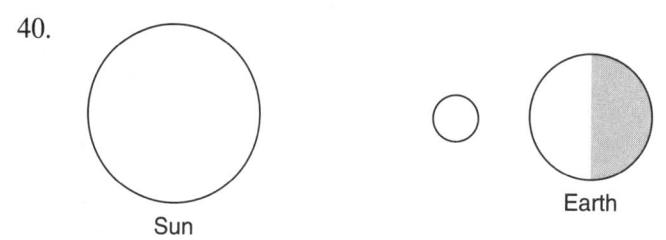

40. [Diagram showing a large circle labeled "Sun", a small circle in the middle, and a medium circle labeled "Earth" with its right half shaded]

(Not drawn to scale)

648

41.

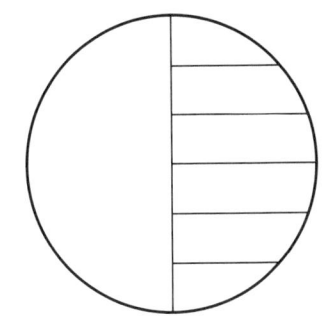

[A circle divided by a vertical line, with the right half containing horizontal lines.]

42. Positions 1 and 5
43. 27⅓ days
44. Refer to Figure 3.16a.
45. The Moon's umbra is never more than 168 miles in diameter; and as Earth rotates, a specific position moves through the umbra at about 1,000 miles per hour. Thus, the maximum time that the Sun can remain eclipsed is about 7.5 minutes; and the period of totality lasts on average, only 2–3 minutes. Earth's umbra, on the other hand, is much larger—about twice the Moon's diameter—and totality may last as long as 1 hour and 40 minutes.
46. Constellations are visible only at night from the portion of Earth facing away from the Sun. As Earth revolves around the Sun, the portion of Earth facing away from the Sun faces different portions of the universe, and thus different constellations are visible.
47. Tides; eclipses; the Coriolis effect; the changing direction of swing of a Foucault pendulum
48. Full Earth

CHAPTER 4

Multiple-Choice Questions

1.	1	10.	1	19.	4	28.	4	37.	2
2.	1	11.	2	20.	4	29.	2	38.	2
3.	1	12.	1	21.	1	30.	4	39.	3
4.	4	13.	2	22.	4	31.	2	40.	4
5.	2	14.	1	23.	1	32.	2	41.	1
6.	2	15.	4	24.	3	33.	2	42.	2
7.	4	16.	3	25.	4	34.	4		
8.	1	17.	4	26.	3	35.	1		
9.	3	18.	2	27.	4	36.	3		

Appendix A: Answers to Review Questions

Constructed Response Questions

43. Northeast
44. 599 meters
45.

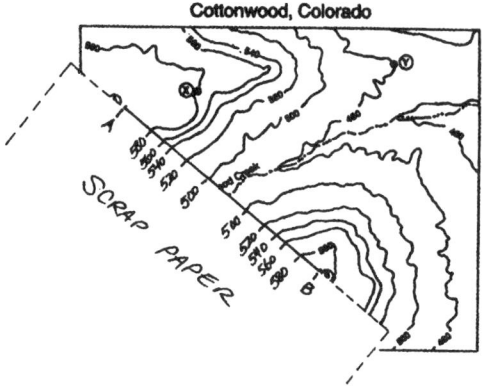

Cottonwood, Colorado

Marking the Edge of the Scrap Paper

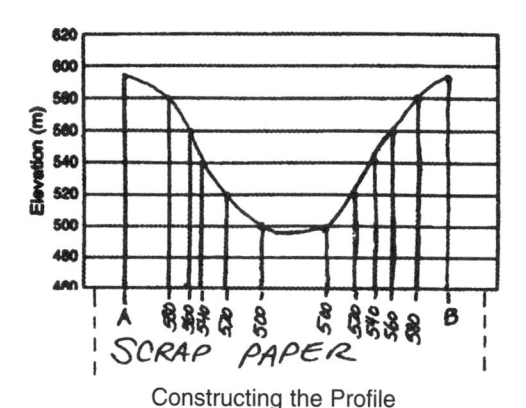

Constructing the Profile

46. (a) Find in the Equations and Proportions section in the *Earth Science Reference Tables* the equation for gradient:

$$\text{gradient} = \frac{\text{change in field value}}{\text{change in distance}}$$

This equation can also be expressed as

$$\text{gradient} = \frac{\Delta \text{ elevation}}{\Delta \text{ distance}}$$

(b) The field value on a topographic map is the elevation, which is shown as isolines called contour lines. Point X lies on the 580-meter contour line, and point Y on the 480-meter contour line. Thus, the change in field value from point X to point Y is 100 meters.

The change in distance from point X to point Y can be found by using the distance scale beneath the map. Mark off the distance between point X and point Y along the edge of a piece of scrap paper. Then place the edge of the scrap paper along the distance scale and measure the distance from X to Y. The distance is 2 kilometers.

Finally, substitute these values in the equation for gradient as follows:

$$\text{gradient} = \frac{580 - 480}{2} = \frac{100}{2} = 50 \frac{\text{meters}}{\text{kilometer}}$$

(c) The gradient is 50 meters/kilometer (50 m/km).

47. Any scientifically correct answer is acceptable; for example, the circular shape of Earth's shadow on the Moon during a lunar eclipse (Earth's shadow could be circular if Earth was shaped like a disk or a cylinder).
48. Latitude is the angular distance north or south of the Equator. In the Northern Hemisphere, latitude can be determined by measuring the altitude of Polaris.
49. Longitude is the angular distance east or west of the Prime Meridian. It can be determined by comparing local time with Greenwich time. Every hour of time difference is equivalent to a 15° difference in longitude. Local time is earlier than Greenwich time to the west of the Prime Meridian and later than Greenwich time to the east of the Prime Meridian.
50. New York City and Rio DeJaneiro are 30° of longitude apart. Since every 15° of longitude is equivalent to a 1-hour time difference, New York City and Rio DeJaneiro have a 2-hour time difference (30/15=2). Since New York City is west of Rio DeJaneiro, its local time is earlier than that of Rio DeJaneiro. Thus, when it is noon in New York City, it is 2:00 P.M. in Rio DeJaneiro.
51. Any scientifically correct answer is acceptable. For example, a topographic map contains contour lines that indicate elevation; road maps do not. Topographic maps show quadrangles bounded by latitude and longitude; road maps typically show political boundaries such as state or county lines.

Appendix A: Answers to Review Questions

Extended Constructed Response Questions

52.

Temperature Field Map (°C)

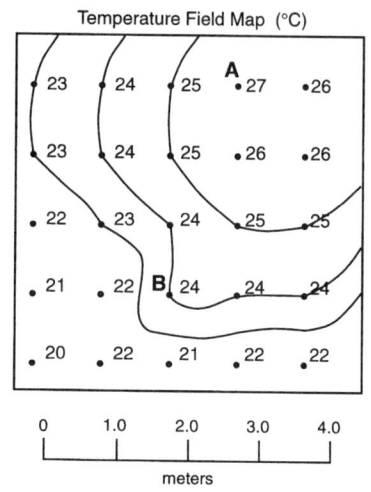

53. (a) $\text{gradient} = \dfrac{\text{change in field value}}{\text{change in distance}}$

Or

$\text{gradient} = \dfrac{\Delta \text{ temperature}}{\Delta \text{ distance}}$

(b) $\text{gradient} = \dfrac{27°C - 24°C}{3m}$ or $\dfrac{3°C}{3m}$

(c) 1°C/meter
Or
1.0°C/m

54. (a)

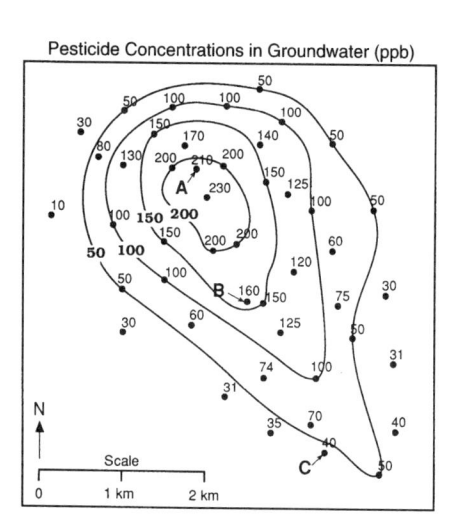

Pesticide Concentrations in Groundwater (ppb)

652

Appendix A: Answers to Review Questions

(b) The highest concentration is near point *A* and the lowest is near point *C*.
Or
The pesticide concentration decreases as you move farther away.

(c) Gradient = $\dfrac{\text{change in field value}}{\text{change in distance}}$

$= \dfrac{160 \text{ ppb} - 40 \text{ ppb}}{2 \text{ km}}$

$= 60 \text{ ppb/km}$

CHAPTER 5

Multiple-Choice Questions

1.	4	8.	2	14.	2	20.	4	26.	2
2.	1	9.	1	15.	1	21.	3	27.	2
3.	1	10.	3	16.	4	22.	3	28.	4
4.	1	11.	4	17.	1	23.	2	29.	3
5.	4	12.	3	18.	1	24.	2	30.	4
6.	3	13.	3	19.	3	25.	4	31.	4
7.	1								

Constructed Response Questions

32. (a) i. **1**   ii. **3**   iii. **2**   iv. **4**
    (b) The mean solar day is longest at position 1 and shortest at position 3.
    (c) Watertown, New York, experiences fewer daylight hours than Miami, Florida, when Earth is at position 1.
33. A sundial records solar time intervals, which vary throughout the year because of changes in Earth's orbital speed. A clock records constant time intervals based on the average length of a solar day. Thus, solar time is sometimes ahead and sometimes behind clock time.

Extended Constructed Response Questions

34. Solar noon always occurs in the middle of the daylight period. Sidereal noon may occur at varying times during Earth's daylight period or even during its darkness period. Thus, the sidereal day is often not "in synch" with our cycle of day and night, making it a poor choice for the basis of civil timekeeping.
35. Simultaneously determine local time and Greenwich Mean Time (GMT). Then calculate the time difference in hours and multiply it by

Appendix A: Answers to Review Questions

15° per hour to determine the difference in longitude. If local time is earlier than GMT, observer's longitude is west of the Prime Meridian; if local time is later than GMT, observer's longitude is east of the Prime Meridian.
36. (a) Refer to Figure 5.6.
    (b) Refer to Figure 5.4.
37. Tokyo is located to the west of the international date line, while New York is located to the east of it. As one travels eastward from Tokyo to New York, the international date line is crossed, and the calendar is set *back* 1 day. This compensates for setting the clock 1 hour *ahead* for each time zone crossed in an easterly direction. Thus, at the precise instant that a fax is sent on December 21 from Tokyo, the date is December 20 in New York.

CHAPTER 6

Multiple-Choice Questions

1.	2	4.	1	7.	3	9.	1	11.	3
2.	4	5.	2	8.	2	10.	4	12.	1
3.	1	6.	2						

Constructed Response Questions

13. Light-gathering power is proportional to the square of the aperture of the telescope. Thus, the light-gathering power of a 200-inch telescope is 40,000 (200^2), and the light-gathering power of a 20-inch telescope is 400 (20^2). Therefore, the light-gathering power of a 200-inch telescope is 1,000 times greater than that of a 20-inch telescope (400,000 ÷ 400 = 1,000).
14. Light-gathering power; resolution; magnification
15. In a refracting telescope a lens is used to create the image that is examined with the objective. The image is formed on the opposite side of the lens from the object. In a reflecting telescope a curved mirror is used to create the image, and the image is on the same side of the mirror as the object. See Figure 6.4.
16. See Figure 6.2d.
17. Telescopes have a much larger aperture than the eye. Therefore, telescopes have much greater light-gathering power, and objects too dim to be seen with the unaided eye are bright enough to be visible through a telescope.
18. Continuous spectra consist of all wavelengths of light dispersed by a prism. In an absorption spectra, wavelengths absorbed as light passes through a gas are missing and appear as black lines. A bright-line spec-

trum consists only of the specific wavelengths of light emitted by a glowing element.
19. The spectral classes of stars are O, B, A, F, G, K, and M. Stars are classified by patterns of visible absorption lines in their spectra.

CHAPTER 7

Multiple-Choice Questions

1.	2	6.	4	11.	4	16.	4	21.	1
2.	4	7.	2	12.	1	17.	2	22.	1
3.	2	8.	4	13.	2	18.	3	23.	1
4.	1	9.	1	14.	4	19.	4	24.	4
5.	2	10.	4	15.	1	20.	2	25.	2

Constructed Response Questions

26. As temperature increases, luminosity increases.
27. Nuclear fusion of hydrogen into helium
28. See Figure 7.10.
29. Fusion of iron absorbs rather than emits energy, and most stars stop fusing and collapse after their cores are converted into iron. Fusion into elements heavier than iron occurs mainly during the violent release of energy of a supernova.
30. Every increase of 1 magnitude is equivalent to an increase in brightness of 2.5. A magnitude-1.0 star is 4 magnitudes brighter than a magnitude-5.0 star, or $2.5 \times 2.5 \times 2.5 \times 2.5 = 39$ times brighter.

Extended Constructed Response Question

31. (a) 1. 15 million years   2. 1 million years   3. 100 million years
    (b) 1. Infrared
        2. When a protostar begins to contract, it is not hot enough to produce light with wavelengths in the visible range. For example, a protostar at a temperature of 3,000°C will emit infrared light that is not visible to the human eye.
    (c) The greater the mass of a newly formed star, the higher its temperature.

Appendix A: Answers to Review Questions

CHAPTER 8

Multiple-Choice Questions

1.	2	8.	1	14.	2	20.	3	26.	2
2.	3	9.	1	15.	2	21.	4	27.	1
3.	4	10.	3	16.	2	22.	4	28.	2
4.	1	11.	4	17.	4	23.	4	29.	4
5.	4	12.	1	18.	1	24.	3	30.	2
6.	4	13.	1	19.	2	25.	2	31.	2
7.	1								

Constructed Response Questions

32. Terrestrial planets: small diameter, short period of revolution
Jovian planets: far from Sun, large mass, low density, many moons, short period of rotation
33. Composition; orbit
34. The disklike shape of the solar system; the division of planets among two types; the compositions of comets, meteorites, and asteroids
35. Stars are self-luminous, planets are not. Stars are much more massive than planets.

Extended Constructed Response Questions

36.

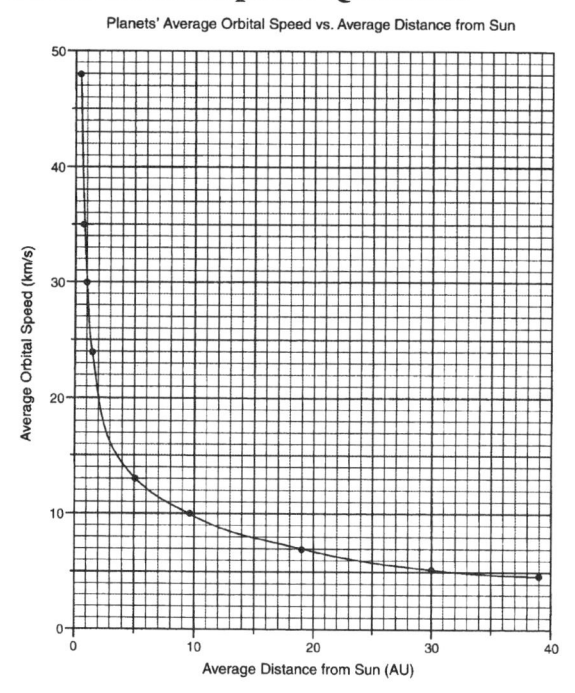

Planets' Average Orbital Speed vs. Average Distance from Sun

[graph: Average Orbital Speed (km/s) vs. Average Distance from Sun (AU)]

37. As distance from the Sun increases, orbital speed decreases.
38. Mars and Jupiter

CHAPTER 9

Multiple-Choice Questions

1.	2	4.	4	7.	4	10.	3	13.	2
2.	3	5.	3	8.	2	11.	4	14.	2
3.	1	6.	1	9.	2	12.	4	15.	1

Constructed Response Questions

16. Background radiation; all galaxies receding
17. (a) Universe contracts, leading to another big bang.
    (b) Universe continues to expand and cool indefinitely.
18. (a) Since the galaxy is 15 billion light-years away, light from the galaxy takes 15 billion years to reach Earth. Thus, the light seen today left the galaxy 15 billion years ago, and we are seeing a 15-billion-year-old image of the galaxy. We will not see the galaxy as it appears now until the light it emits now reaches us in 15 billion years!
    (b) Since our galaxy formed about 15 billion years ago, light from a galaxy 15 billion light-years away allows astronomers to see how that galaxy appeared at about the same time that ours was forming.

Extended Constructed Response Question

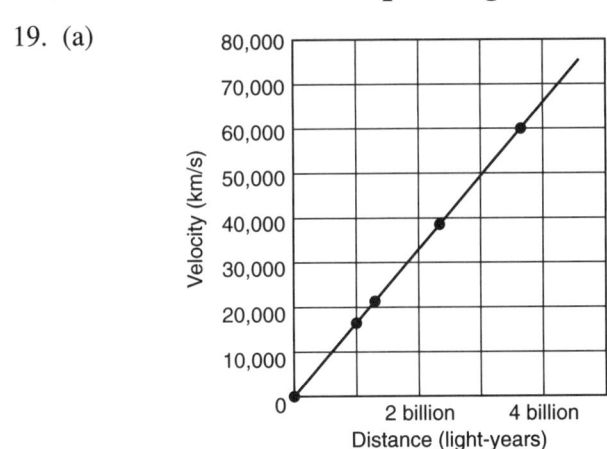

19. (a)

(b) As distance to a galaxy increases, the rate at which the galaxy is receding increases.
(c) No. An observer in any galaxy would observe the same. All galaxies appear to recede from one another because the universe itself is expanding, carrying all galaxies farther apart.

CHAPTER 10

Multiple-Choice Questions

1.	2	5.	2	9.	3	13.	1	17.	2
2.	1	6.	1	10.	4	14.	3	18.	2
3.	1	7.	1	11.	3	15.	1	19.	4
4.	2	8.	1	12.	3	16.	1		

Constructed Response Questions

20. Differentiation occurred while Earth's interior was a molten fluid. In a fluid, two of the forces acting on an object are weight and buoyancy. Weight is a downward force because gravity is attracting the object's mass toward Earth's center. The denser the object, the greater its weight. Buoyancy is an upward force and is equal to the weight of the fluid displaced by the object. If an object is denser than the surrounding fluid, the downward force of weight will be greater than the upward force of buoyancy, and the object will sink downward. If the object is less dense than the surrounding fluid, the downward force of weight will be less than the upward force of buoyancy, and the object will float upward. Thus, denser materials (e.g., iron) sank toward the center of the molten Earth, while less dense materials (e.g., water) floated toward the surface.
21. It is highly unlikely that Earth would have captured such a large object and even less likely that it would go into a circular orbit.
    Or
    Moon rocks are more similar to Earth's mantle than to meteorites like those that would have accreted to form an object the size of the Moon.
22. See Figure 10.2.

Extended Constructed Response Questions

23. (a) 1. Core   2. Mantle   3. Crust
    (b) 1. Core—3,470 km   2. Mantle—2860–2890 km
        3. Crust—10–40 km
    (c) When Earth was mostly molten, denser materials sank toward the center while less dense materials floated toward the surface—a process called differentiation.

Appendix A: Answers to Review Questions

CHAPTER 11

Multiple-Choice Questions

1.	2	6.	4	11.	1	16.	4	21.	3
2.	4	7.	3	12.	1	17.	1	22.	3
3.	3	8.	1	13.	4	18.	3	23.	1
4.	4	9.	3	14.	3	19.	1	24.	2
5.	3	10.	4	15.	2	20.	1	25.	3

Constructed Response Questions

26. Outgassing; cometary impacts
27. As photosynthetic life evolved and proliferated, oxygen was produced in increasing quantities as a by-product of photosynthesis.
28. With water scarce and no oceans, photosynthetic life might not have evolved and Earth might not have developed an oxidizing atmosphere.

Extended Constructed Response Question

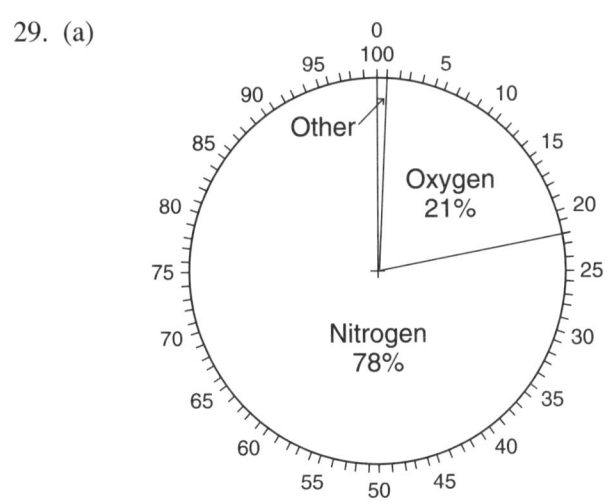

29. (a)

(b) The crust is 94% oxygen by volume, the hydrosphere is 33% oxygen by volume, and the troposphere is 21% oxygen by volume. The oxygen in the troposphere is in the form of molecules of the element oxygen. The oxygen in the crust and hydrosphere is chemically combined with other elements as part of compounds.

(c) Photosynthesis

Appendix A: Answers to Review Questions

CHAPTER 12

Multiple-Choice Questions

1.	1	9.	1	16.	3	23.	3	30.	4
2.	2	10.	2	17.	2	24.	3	31.	2
3.	1	11.	3	18.	4	25.	3	32.	1
4.	4	12.	3	19.	4	26.	1	33.	2
5.	4	13.	3	20.	3	27.	4	34.	3
6.	1	14.	2	21.	2	28.	2	35.	1
7.	3	15.	1	22.	3	29.	1		
8.	3								

Constructed Response Questions

36. (a) Particle shape; packing or sorting
    (b) It would probably not be worthwhile for the resident to drill a well. Clay is impermeable; therefore, after water was pumped out of the well, water would seep back into the well very slowly, if at all.
37. (a) Liquid. The molecules are more closely packed in the liquid than they are in the solid or gas phase.
    (b) The molecules would be spaced even farther apart.
    (c) As the water changed from water to ice, the volume of the water sample would increase.

Extended Constructed Response Questions

38. (a) *A.* Evaporation   *B.* Transpiration   *C.* Condensation
        D. Precipitation   *E.* Runoff   *F.* Infiltration
    (b) Porous, permeable, unsaturated
    (c) Saturated pore spaces, impermeable, steep slope
    (d) The warmer the season, the greater the evaporation.
39. See Figure 12.10.
40. See Figure 12.12.

CHAPTER 13

Multiple-Choice Questions

1.	1	15.	4	29.	1	43.	2	57.	3
2.	4	16.	3	30.	2	44.	2	58.	2
3.	4	17.	1	31.	1	45.	3	59.	2
4.	1	18.	3	32.	4	46.	1	60.	3
5.	4	19.	2	33.	3	47.	4	61.	1
6.	3	20.	3	34.	4	48.	3	62.	2
7.	4	21.	4	35.	2	49.	1	63.	3
8.	3	22.	2	36.	4	50.	3	64.	4
9.	2	23.	4	37.	1	51.	3	65.	1
10.	1	24.	1	38.	2	52.	4	66.	3
11.	4	25.	3	39.	2	53.	4	67.	4
12.	1	26.	3	40.	3	54.	4	68.	1
13.	4	27.	1	41.	4	55.	2	69.	4
14.	2	28.	4	42.	4	56.	1		

Constructed Response Questions

70.

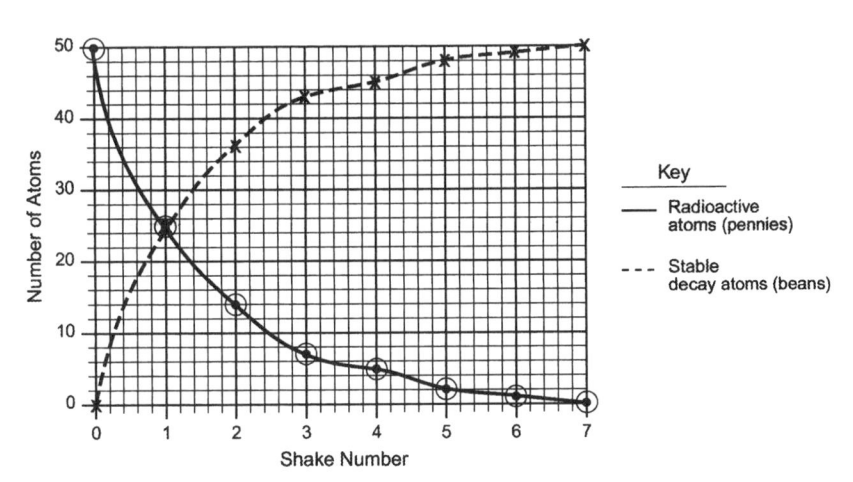

71. 100 years (±20 years)
72. By using similiar fossils
    Or
    Correlate the layers by matching sequences of rock types.

Appendix A: Answers to Review Questions

73.

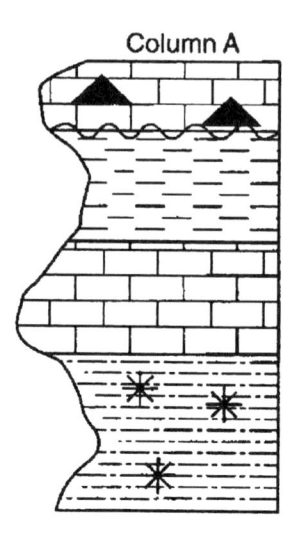

Column A

74. The limestone layer "sticks out" more than other layers.
    Or
    The limestone layer appears less weathered.
75. 540 million years
76. The bold wavy line between X and Y represents a buried erosional surface. After layers *E*, *F*, *G*, and *H* were deposited, they were tilted and uplifted. Then, erosion wore down these layers unevenly, forming the surface marked by line *XY*.
77. Ordovician Period
78. Rock units *E*, *D*, *B*, and *A* have all undergone contact metamorphism and therefore are all older than rock unit *C*. Since rock units *E*, *D*, *B*, and *A* have not been overturned, the principle of superposition applies, and each layer is younger than the one under it and older than the one on top of it. Thus, the order of the rock units from oldest to youngest is *E*, *D*, *B*, *A*, *C*.
79. clay
80. course texture
    Or
    large crystals

Extended Constructed Response Questions

81. Cretaceous Period
82. The basalt is older than 92 million years.
    Or
    Fossils are not normally found in igneous rock.

Appendix A: Answers to Review Questions

CHAPTER 14

Multiple-Choice Questions

1.	2	7.	4	13.	1	19.	1	25.	3
2.	2	8.	1	14.	3	20.	1	26.	1
3.	4	9.	2	15.	3	21.	2	27.	4
4.	1	10.	1	16.	1	22.	4		
5.	1	11.	1	17.	2	23.	1		
6.	3	12.	3	18.	3	24.	4		

Constructed Response Questions

28.

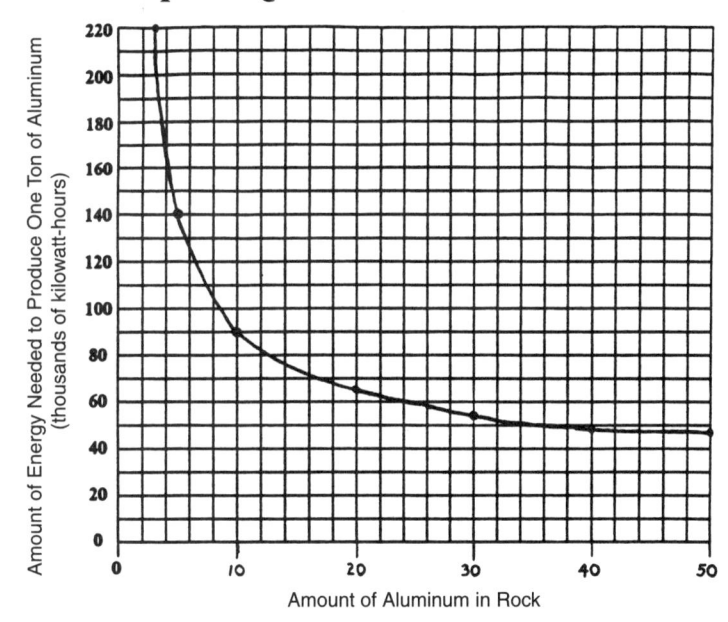

29. 75,000 kilowatt-hours
30. The lower the aluminum content of the ore, the more energy required to extract it.
31. (a) Determine the streak of each sample by rubbing on an unglazed porcelain streak plate. Determine the density of each sample by water displacement. Then refer to the Mineral Identification Chart in the *Earth Science Reference Tables*. The denser mineral sample with a black streak is galena. The less dense mineral sample with a brown streak is hematite.
    (b) Streak and density
32. Many minerals have the same color; few minerals have exactly the same cleavage.
33. Minerals form by natural processes over long periods of time. Once

Appendix A: Answers to Review Questions

used, they may not form again during your lifetime or even during a thousand lifetimes.
34. (a) Copper is used widely as a conductor in electrical circuits.
   (b) The life expectancy of current copper reserves could be extended by recycling, reusing, and exploring for new copper deposits.

Extended Constructed Response Questions

35. Iron is so abundant that known reserves could fill human needs for 205 years. Therefore, iron is inexpensive and recycling is not economical.
36. (a) Manganese
   (b) The United States has no manganese reserves, and manganese is not recycled. Therefore, all manganese is imported from other countries.
37. Manganese is more expensive than zinc because zinc is more readily available. The United States has zinc reserves to last for 11 years and recycles zinc. However, the United States has no manganese reserves and does not recycle manganese. Therefore, the supply of zinc is greater than the supply of manganese.
38. The United States currently imports all chromium, but recycles to decrease the demand. If recycling is banned, the U.S. demand will increase, and so, therefore, will the price of chromium.

CHAPTER 15

Multiple-Choice Questions

1.	1	12.	3	23.	4	34.	2	45.	2
2.	3	13.	4	24.	2	35.	1	46.	4
3.	1	14.	1	25.	1	36.	2	47.	2
4.	1	15.	4	26.	2	37.	3	48.	3
5.	3	16.	3	27.	3	38.	3	49.	2
6.	2	17.	4	28.	3	39.	4	50.	4
7.	4	18.	4	29.	3	40.	2	51.	4
8.	2	19.	1	30.	1	41.	1	52.	1
9.	4	20.	4	31.	3	42.	3	53.	3
10.	2	21.	4	32.	1	43.	1	54.	3
11.	4	22.	4	33.	2	44.	4	55.	2

Constructed Response Questions

56. *A.* Metamorphic   *B.* Extrusive or volcanic   *C.* Conglomerate or breccia or sandstone or siltstone or shale

57. Any smaller-opening screen placed higher would trap particles out of sequence.
Or
Not all particle sizes would be separated.
58. Compaction (or burial) and cementation
59. Sandstone
60. Any scientifically correct answer is acceptable. For example:
Sedimentary rock—melting—magma—solidification—**igneous rock**
Or
Sedimentary rock—heat and pressure—metamorphic rock—melting—magma—solidification—**igneous rock**
61. An igneous rock may melt to form a magma, be weathered and eroded into sediments, or be exposed to heat and pressure to form a metamorphic rock.
62. (a) Metamorphic rock forms when heat or pressure changes existing rock without melting it to form a magma. Once a magma forms, the preexisting rock is destroyed and the rock that forms when the magma solidifies is an igneous rock.
(b) **Magma**—solidification—igneous rock—heat and pressure—**metamorphic rock**

Extended Constructed Response Question

63. (a) An intrusion is insulated by surrounding rock layers so that heat escapes from the magma slowly and slow cooling allows large crystals to form.
(b) Basalt or gabbro
(c) The rocks along the left side of the channel in the diagram are Manhattan schist—metamorphic rock that formed deep underground when pre-existing rock was exposed to great heat and pressure. The rocks along the right side of the channel are sandstones and shales—sedimentary rock that originated as sediments deposited in water at the surface that were later buried and cemented together.
(d) Erosion of overlying layers of rock has exposed the once-buried sill.

Appendix A: Answers to Review Questions

CHAPTER 16

Multiple-Choice Questions

1.	1	11.	4	21.	3	31.	2	41.	3
2.	3	12.	3	22.	4	32.	1	42.	2
3.	4	13.	3	23.	2	33.	1	43.	2
4.	3	14.	3	24.	1	34.	2	44.	4
5.	4	15.	4	25.	2	35.	4	45.	2
6.	4	16.	1	26.	4	36.	2	46.	2
7.	4	17.	4	27.	4	37.	4	47.	3
8.	4	18.	3	28.	1	38.	1	48.	3
9.	2	19.	3	29.	3	39.	4		
10.	2	20.	1	30.	1	40.	2		

Constructed Response Questions

49. Two types of seismic body waves are *P*-waves and *S*-waves. *P*-waves are longitudinal waves that are alternating pulses of compression and expansion of rock in the direction of wave travel. *S*-waves are transverse waves that twist rock back and forth perpendicular to the direction of wave travel.
50. Seismic risk is based on the probability that an earthquake will occur and the potential damage that it will cause. Studies of earthquakes have suggested that they may be periodic. Tens or hundreds of years may elapse between times of seismic activity in a region. Furthermore, regions that were once unpopulated and stood little risk of damage to human life and property may now be densely populated, so that even a minor earthquake would pose a serious risk.
51. Earthquake magnitude is a measure of the energy released by an earthquake. Earthquake intensity is a measure of the impact of an earthquake on human life and property.
52. A scale is a series of numbers. A series is a number sequence in which each term is related to one or more of its predecessors by the same rule. In a linear scale, the rule is additive, that is, every term is obtained by adding a particular number to the previous term. For example, 1; 2 = (1 + 1); 3 = (2 + 1); 4 = (3 + 1), and so on. In a logarithmic scale, the rule is exponential, that is, every term is obtained by increasing the exponent of the previous term by a particular number. For example, 1 = (10^1) = 10; 2 = (10^2) = 100; 3 = (10^3) = 1,000; 4 = (10^4) = 10,000; and so on. As you can see, on a linear scale 4 is 1 greater than 3, but on a logarithmic scale, 4 is 10 times greater than 3, and 100 times greater than 2.

53. The graphs indicate that, when the amount of waste fluid disposed of at the arsenal increases, the earthquake frequency in the Denver area increases.
54. If the hypothesized connection between injection of fluid wastes into wells and earthquake frequency does exist, one would predict an increase in earthquake frequency whenever there is an increase in waste fluid injected. The data in the graphs support this prediction; note that increases in waste fluid injected in late 1961, late 1962, and 1965 were all accompanied by increases in earthquake frequency.
55. Identify several wells near old faults. As a control, eliminate all injection of fluid wastes in some wells; as an experiment, vary the elimination of fluid wastes in others. Then compare the frequencies of earthquakes along all of the associated faults.

Extended Constructed Response Questions

56.

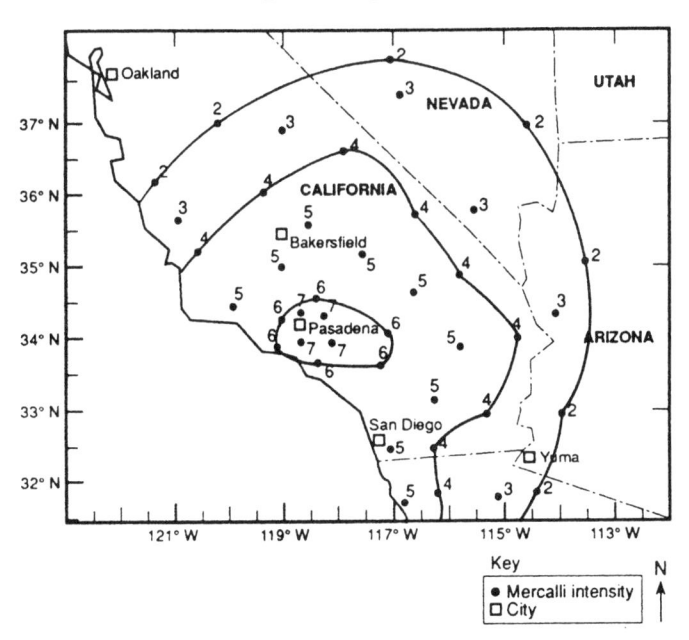

57. Pasadena
58. 35°30'N, 119°W
59. Movement along the San Andreas fault
    Or
    Movement along the boundaries of the North American and Pacific plates

Appendix A: Answers to Review Questions

CHAPTER 17

Multiple-Choice Questions

1.	3	7.	1	13.	3	19.	1	25.	1
2.	1	8.	1	14.	1	20.	4	26.	1
3.	4	9.	4	15.	1	21.	1	27.	1
4.	1	10.	2	16.	1	22.	3	28.	2
5.	1	11.	4	17.	3	23.	4	29.	2
6.	1	12.	2	18.	2	24.	1		

Constructed Response Questions

30. Volcanic eruptions, geothermal activity (geysers, hot springs, etc.), temperature readings taken in deep mines and bore holes
31. Magma is molten rock beneath Earth's surface, lava is molten rock at Earth's surface.
32. (a) Melting point
    Diamond—1,300 km (± 100 km)
    Iron—300 km (± 100 km)
    Quartz—400 km (± 100 km)
    (b) Boiling point
    Diamond—2,800 km (± 100 km)
    Iron—800 km (± 100 km)
    Quartz—500 km (± 100 km)
    (c) The great pressure in the stiffer mantle holds the quartz molecules together and thereby increases the temperatures at which these phase changes occur.
33. Cinder cone—solid particles ejected during a volcanic eruption build up around the vent in a steep, narrow cone.
    Composite cone—large, symmetrical cones consisting of alternate layers of solidified lava and volcanic rock particles are formed by alternating explosive and quiet eruptions.
    Shield cone—a broad, flat cone is formed when lava flows quietly from the vent, spreading out in wide, flat sheets.
    Lava plateau—lava erupts from a fissure and spreads out in wide, thin sheets that blanket the surrounding land.
34. The longer molten rock takes to cool, the more time mineral crystals have to grow in size. Igneous intrusions are insulated by surrounding rock and cool slowly, allowing large mineral crystals to form.
35. The magma associated with explosive eruptions has a composition similar to that of granite and is thick and pasty. The magma associated with quiet eruptions has a composition similar to that of basalt and is thinner and more fluid.

    The material ejected during explosive eruptions typically consists of particles of molten rock that solidify as they "rain" down on the surrounding land. The material ejected during quiet eruptions is fluid rock that spreads out over the surrounding land in sheets and then solidifies.

Extended Constructed Response Questions

36. Layer *I*
37. Layer *B*
38.

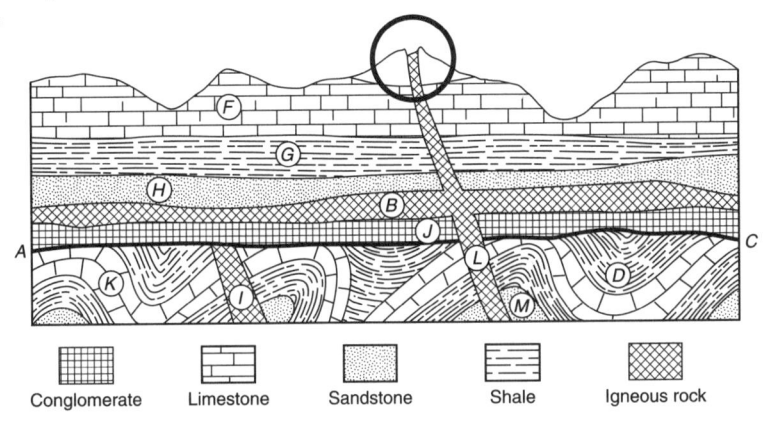

39. The high temperature of the molten rock that intruded to form layer *B* most likely caused the rock in layer *H* to undergo contact metamorphism.
40. Radioactive decay; heat transferred outward from Earth's core

CHAPTER 18

Multiple-Choice Questions

1.	2	10.	4	19.	2	28.	1	37.	1
2.	3	11.	4	20.	1	29.	1	38.	2
3.	4	12.	4	21.	1	30.	1	39.	4
4.	2	13.	2	22.	4	31.	4	40.	2
5.	4	14.	2	23.	2	32.	2	41.	3
6.	4	15.	3	24.	4	33.	4	42.	1
7.	4	16.	3	25.	4	34.	3	43.	4
8.	2	17.	2	26.	1	35.	4	44.	1
9.	4	18.	4	27.	1	36.	1	45.	4

Constructed Response Questions

46. Triassic
47. Correlation of rocks, minerals, and fossils on edges of continents that were once in contact with one another

Appendix A: Answers to Review Questions

48. Northwest
49. As the distance from the Tonga Trench increases, the depth of an earthquake's focus increases.
50. The Indian-Australian Plate and the Pacific Plate
51. The Pacific Plate is moving westward and plunging beneath the Indian-Australian Plate.

Extended Constructed Response Questions

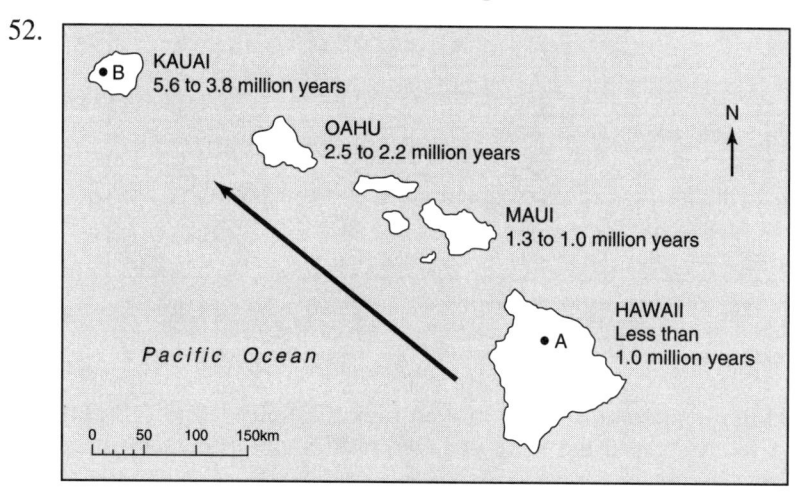

52.

```
KAUAI — 5.6 to 3.8 million years (point B)
OAHU — 2.5 to 2.2 million years
MAUI — 1.3 to 1.0 million years
HAWAII — Less than 1.0 million years (point A)
Pacific Ocean
Scale: 0 50 100 150 km
N
```

53. (a) Rate of change $= \dfrac{\text{change in field value}}{\text{change in time}}$

$$= \dfrac{490 \text{ km}}{4.9 \text{ million years}}$$

$$= \dfrac{100 \text{ km}}{\text{million years}}$$

(b) $\dfrac{100 \text{ million mm}}{\text{million years}} = 100 \text{ mm/year}$

54. Southeast of Hawaii
55. Hawaii, because Diagram II shows that it is currently located over the magma source.
56. Any three of the following: basaltic glass; vesicular basaltic glass; scoria; vesicular basalt; basalt. Quiet eruptions are associated with basaltic lavas and produce extrusive rocks. Thus, one would expect to find extrusive rocks with a basaltic composition on the Hawaiian Islands.

CHAPTER 19

Multiple-Choice Questions

1.	2	8.	4	15.	4	21.	3	27.	2
2.	2	9.	4	16.	2	22.	4	28.	1
3.	1	10.	4	17.	3	23.	3	29.	4
4.	2	11.	3	18.	2	24.	1	30.	2
5.	3	12.	4	19.	4	25.	1	31.	3
6.	4	13.	3	20.	2	26.	2		
7.	4	14.	2						

Constructed Response Questions

32. Rock A is most likely the hardest; its mass decreased by less than 20 grams during the period of shaking, while all of the other rock samples decreased in mass by more than 60 grams.
33. The particles of rocks B and C became rounded by abrasion. With fewer protrusions that could strike one another, the rate at which they weathered decreased.
34. Rock D is a poor choice because it weathers rapidly by abrasion. While initially more expensive, rock A would be a more economical choice because it would last much longer before it would weather away and need to be replaced.
35. Soil horizon A
36. (a) leaching
    (b) The soil in horizon B is less fertile than the soil in horizon A because it contains less organic matter, and its high clay content inhibits the penetration of water and air.
37. Climate; type of bedrock

Appendix A: Answers to Review Questions

Extended Constructed Response Questions

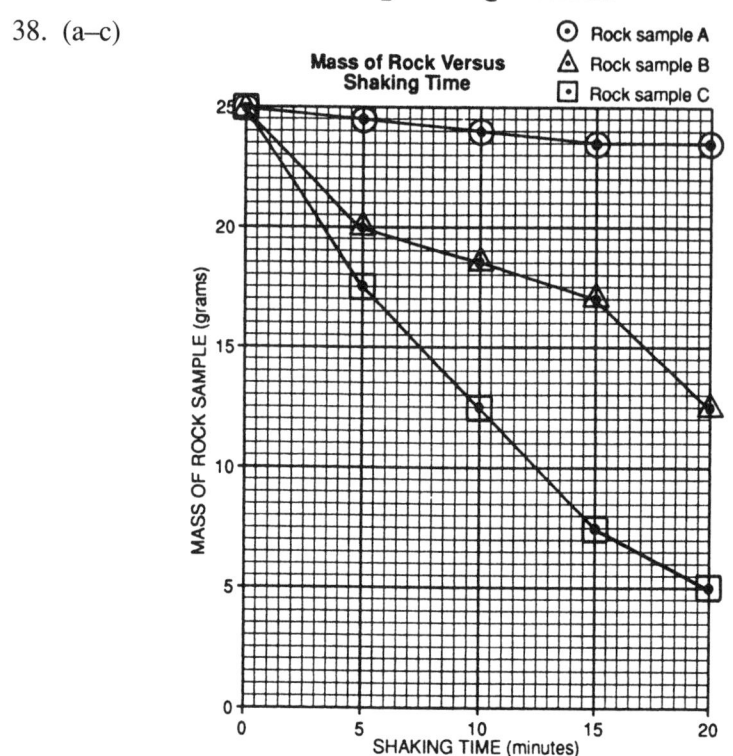

38. (a–c)

Mass of Rock Versus Shaking Time

⊙ Rock sample A
△ Rock sample B
▫ Rock sample C

y-axis: MASS OF ROCK SAMPLE (grams)
x-axis: SHAKING TIME (minutes)

39. Sample A weathered more slowly than samples B and C because sample A is harder and more resistant to erosion.

40. (a) Rate of change in mass = $\dfrac{\text{change in mass}}{\text{change in time}}$

    (b) Rate of change in mass = $\dfrac{25.0 \text{ g} - 5.0 \text{ g}}{20 \text{ min}}$

    (c) Rate of change in mass = 1.0 g/min

41. During the process of abrasion, any part of a rock surface that juts out, such as an edge or corner, is more likely to be struck by another rock particle and be broken off. Thus, after 20 minutes of shaking, the corners and edges of rock material C will be smoother and more rounded than they were before shaking.

CHAPTER 20

Multiple-Choice Questions

1.	3	12.	1	23.	4	34.	1	45.	3
2.	3	13.	4	24.	1	35.	3	46.	4
3.	4	14.	3	25.	3	36.	4	47.	4
4.	3	15.	1	26.	4	37.	4	48.	4
5.	1	16.	2	27.	4	38.	4	49.	3
6.	1	17.	2	28.	4	39.	1	50.	1
7.	2	18.	1	29.	3	40.	1	51.	2
8.	4	19.	2	30.	1	41.	2	52.	2
9.	2	20.	3	31.	4	42.	4	53.	3
10.	4	21.	2	32.	2	43.	4	54.	2
11.	1	22.	4	33.	3	44.	1		

Constructed Response Questions

55. (a) Rate of change $= \dfrac{\text{change in field value}}{\text{change in time}}$

    (b) Rate of change $= \dfrac{300 \text{ m}}{272 \text{ yr}}$

    (c) Rate of erosion = 1.1 m/yr

56. (a) Hypothesis 2
    (b) According to the cross sections, the sequences and compositions of the rock layers at American Falls and Horseshoe Falls are identical. Therefore, the rocks beneath the two falls have the same resistance to erosion. However, according to the map, the Upper Niagara River is much wider at Horseshoe Falls than at American Falls, and the rate of erosion is directly related to the amount of water flowing over the falls. Therefore, hypothesis 2 is better supported by the evidence.
57. An agent of erosion is a medium through which a force (usually gravity) is exerted on sediments, causing them to move. Water, air, and ice are common agents of erosion.
58. Water that infiltrates pore spaces in rock and soil acts as a lubricant, decreasing the amount of friction between material on a slope and the underlying surface. It is friction that holds material in place on a slope against the pull of gravity. If water decreases friction, it may no longer be strong enough to hold the material in place and the material will be pulled downslope by gravity.

Appendix A: Answers to Review Questions

59.

Mass Wasting Process	Slow	Rapid
Creep	✓	
Earthflow	✓	
Rockslide		✓
Mudflow		✓
Slump		✓
Rockfall		✓

60. Any three of the following: Rolling; sliding; bouncing; suspension; solution
61. See Figure 20.11.
62. Wind is able to transport fine particles such as silt and sand, not coarse particles such as pebbles and cobbles. As wind moves across a surface consisting of a mixture of particle sizes, the finer ones are carried away and the coarser ones are left behind. The end result is a surface consisting only of coarse particles—that is, a desert pavement.
63. Valley glaciers are smaller than continental glaciers.
Valley glaciers form between mountain slopes at high elevations; continental glaciers form at high latitudes where temperatures are always cold.
64. Ice forms at a glacier's source and melts at its terminus. If ice is melting at the terminus at the same rate that it is advancing, the glacier can be constantly moving but the position of its terminus will remain unchanged.

Extended Constructed Response Questions

65. The banks on the outside of the meander are being undercut by fast-moving water in the stream. The ground beneath Structure 1 will be undermined, causing it to collapse.
66. As the falls erode back, the supports on either end of the bridge will be undermined. As a result, Structure 2 is in danger of collapse.
67. Any scientifically correct answer is acceptable; for example: The more irregularly shaped the particles, the steeper the angle of repose.
68. If the particles vary in size and shape, there is no way to know which variable causes an observed change in angle of repose.
Or
The amount of each sample that is used is not controlled. The amount of material used might affect the angle of repose.
69. Particle size—sort the particles with a screen sieve before testing.
Particle shape—sort the particles by shape before testing.
Particle density—use only particles that have the same composition.
70. Your experiment should include a hypothesis, and your procedure should involve pouring particles of different sizes into a pile and measuring the angle of repose. You should identify how you will control all variables except particle size and should include several trials for each

particle size. Data should be presented in the form of a table and/or graph, and you should describe data that would support or reject the hypothesis.

CHAPTER 21

Multiple-Choice Questions

1.	3	13.	3	25.	4	37.	3	49.	3
2.	2	14.	1	26.	3	38.	1	50.	2
3.	4	15.	4	27.	2	39.	3	51.	2
4.	1	16.	4	28.	4	40.	2	52.	2
5.	2	17.	4	29.	4	41.	4	53.	1
6.	2	18.	1	30.	2	42.	2	54.	2
7.	3	19.	2	31.	3	43.	4	55.	4
8.	2	20.	3	32.	2	44.	3	56.	1
9.	4	21.	4	33.	1	45.	2	57.	4
10.	2	22.	1	34.	1	46.	4	58.	3
11.	2	23.	1	35.	1	47.	4	59.	1
12.	4	24.	1	36.	3	48.	1		

Constructed Response Questions

60. The bedrock of the Adirondack Mountains consists primarily of metamorphic rocks, such as gneisses and quartzite. Any fossils that may have existed in the bedrock would have been destroyed during metamorphism.
61. Running water (or streams)
62. Devonian (or 362–418 million years old)
63. Trellis Dendritic Radial

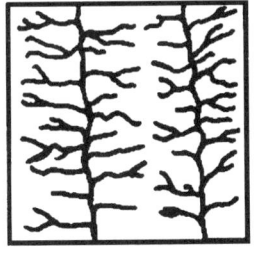

Appendix A: Answers to Review Questions

64. The Adirondack Mountains
65. Vertical sorting occurs when a mixture of sediments settles through a column of water and forms distinct layers, with the roundest, largest, and densest particles on the bottom and the smallest, flattest, and least dense particles at the top.

    Horizontal sorting occurs when a stream gradually slows down over a horizontal distance and there is a gradual reduction in the size of the sediments deposited along the stream bed. The roundest, largest, and densest particles are deposited where stream velocity is highest; and the smallest, flattest, and least dense particles are deposited where stream velocity is lowest.
66. Glacial sediments are transported in, on, and under the ice. When the ice melts, they simply fall to the ground on top of one another. Since the sediments do not "settle" through the ice at different rates, and are not sorted if they fall through air, they are unsorted. Stream sediments, on the other hand, are sorted either as they settle through water, or as the stream decreases in velocity and deposits them.
67. Shape, size and density
68. Your experimental procedure should involve simulating conditions in a stream by agitating a mixture of gravel and water. Your procedure should control all variables except shaking time and should include an experimental control consisting of a mixture of water and gravel that is *not* agitated. The masses of the two gravel samples before and after shaking should be compared. The gravel that undergoes the greater decrease in mass will abrade faster in a stream.
69. The orientation of glacial striations in bedrock; the orientation of the long axis of drumlins; the source region of transported materials

Extended Constructed Response Questions

70. (a) The purpose of a silt screen is to prevent contaminated sediments disturbed by the dredging operation from spreading downstream.
    (b) The PCBs are primarily attached to clay-sized sediments. Clay particles range in diameter from 0.00001 cm to 0.0004 cm. Therefore, a silt screen with openings 0.0006 cm in diameter would not be effective because it would allow contaminated clay particles to escape.
71. The Hudson-Mohawk Lowlands
72. The plume of pollutants would extend even farther downstream if the speed of the river current increased.
73. Sheets of steel locked together would be used when there was a river current.
74. Escaped PCBs could be ingested by organisms and small fish, which, in turn, might be eaten by humans.

CHAPTER 22

Multiple-Choice Questions

1.	3	13.	2	25.	1	37.	3	49.	3
2.	2	14.	1	26.	1	38.	3	50.	1
3.	3	15.	4	27.	3	39.	3	51.	2
4.	4	16.	3	28.	4	40.	2	52.	1
5.	1	17.	4	29.	2	41.	1	53.	4
6.	4	18.	3	30.	2	42.	4	54.	2
7.	3	19.	2	31.	3	43.	1	55.	1
8.	1	20.	4	32.	1	44.	2	56.	3
9.	1	21.	2	33.	3	45.	2	57.	1
10.	1	22.	2	34.	3	46.	3	58.	1
11.	3	23.	3	35.	3	47.	2	59.	3
12.	4	24.	2	36.	2	48.	1		

Constructed Response Questions

60. (a) December
    (b) During January and July, the change in surface temperature is the opposite of the change in duration of insolation. In January, surface temperature decreases while duration of insolation increases; in July, surface temperature increases while duration of insolation decreases. During all other months of the year, surface temperature changes accord with duration of insolation.
    (c) Surface temperatures continue to increase between late June and early August because, even though the duration of insolation is decreasing, it is still greater than the duration of darkness and the angle of insolation is still large. Thus, Earth receives relatively intense radiation for more hours than not each day, and daily energy gain exceeds daily energy loss.
61. When ice crystals form in the atmosphere, latent heat energy is released. This release of energy has a warming effect on the atmosphere.
62. The closer to 90° the angle of insolation, the more intense the radiation striking the ground.
63. Earth is also constantly radiating energy. Overall, Earth is in radiative balance, that is, it loses as much energy as it gains.
64. The angle of insolation is small at the North Pole; therefore, the intensity of the radiation striking the ground is low. Other regions may have shorter durations of insolation, but the high intensity of the radiation they receive more than compensates for any difference in the duration of insolation.

Appendix A: Answers to Review Questions

65. Adding 10 grams of ice at 0°C would be more effective because any energy that the ice absorbed from the drink while it was melting would not result in an increase in temperature. Water, on the other hand, would increase in temperature as it absorbed energy from the drink, thereby decreasing its cooling effect.
66. (a) The energy absorbed during a phase change, as between points B and C or D and E, is used to break molecular bonds, rather than to make molecules move faster. Since molecular speed doesn't increase, temperature doesn't increase.
    (b) D and E
    (c) B and C
    (d) More energy (540 cal/g) is required to evaporate water between points D and E than to melt water (80 cal/g) between points B and C.

Extended Constructed Response Questions

67. (a) CFC-12 has a very long decay time and a large relative greenhouse effect.
    (b) Methane has a relative greenhouse effect of 70, which means that releasing 1 kilogram of methane is equivalent to releasing 70 kilograms of carbon dioxide. Thus, releasing 1 kilogram of methane directly into the atmosphere would contribute more to the greenhouse effect than burning 1 kilogram of methane, thereby releasing 3 kilograms of carbon dioxide.
68. (a) Home B
    (b) The electricity produced by a solar cell is directly related to the amount of energy per unit of surface area it receives. On March 21 at solar noon, sunlight will strike the solar cell on home B more directly than the solar cell on home A. Therefore, the sunlight striking home B's solar cell will be more intense and will produce more electricity.
    (c) If the solar cell on home B was relocated to the north side of the roof, the angle at which sunlight would strike it would be much smaller, the solar cell would receive much less intense radiation, and the output of the solar cell would decrease dramatically.
69. (a) Water has a much higher specific heat than sand, therefore, water must gain more energy than sand in order to heat up.
    (b) Dark colors reflect less solar energy than light colors, and rough surfaces reflect less solar energy than smooth ones. A grassy field reflects less solar energy than a snow-covered field because the grassy field is both darker and rougher.
    (c) According to the chart, an increase in cloud cover would cause more solar energy to be reflected back into space. The effect would be to decrease insolation and slow the rate of global warming.

CHAPTER 23

Multiple-Choice Questions

1.	2	15.	4	29.	1	43.	2	56.	2
2.	2	16.	4	30.	4	44.	1	57.	1
3.	1	17.	3	31.	3	45.	4	58.	4
4.	2	18.	4	32.	2	46.	3	59.	1
5.	4	19.	3	33.	2	47.	3	60.	2
6.	2	20.	3	34.	4	48.	3	61.	3
7.	4	21.	4	35.	3	49.	2	62.	2
8.	1	22.	3	36.	1	50.	2	63.	1
9.	1	23.	1	37.	2	51.	1	64.	3
10.	4	24.	1	38.	1	52.	3	65.	1
11.	4	25.	1	39.	3	53.	3	66.	1
12.	1	26.	4	40.	2	54.	3	67.	1
13.	2	27.	2	41.	3	55.	2	68.	1
14.	1	28.	2	42.	1				

Constructed Response Questions

69. (a) As the front approaches, station E will experience a rapid increase in cloudiness, followed by slower clearing after the front passes.
    (b) Barometric pressure will increase as the front passes station E.
    (c) Temperature will decrease after the front passes station E.
    (d) The denser cool air pushes beneath the warmer air, causing the warm air to rise and cool. When the warm air cools to its dewpoint, condensation occurs and clouds form.

70. (a)

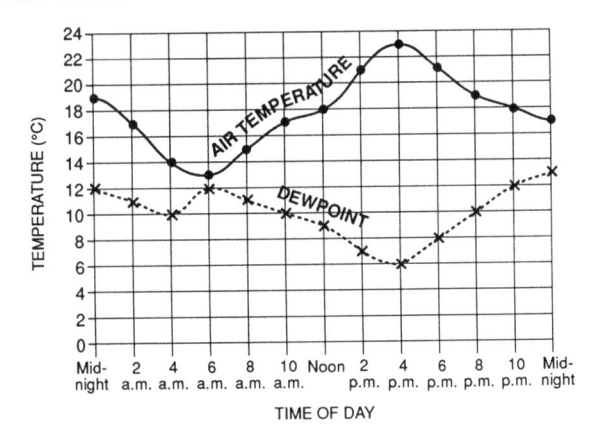

    (b) 4 P.M.

Appendix A: Answers to Review Questions

71. (a) 70°F
(b) Since the isobars are more closely spaced on the July 3 map than on the July 2 map, it can be inferred that the wind speed increased from July 2 to July 3.
(c) When the storm moves over land, the wind speed will decrease.

72. (a)

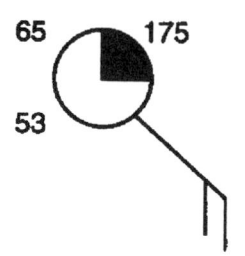

```
        65      175
         \    /
          \  /
           \/
         /\
        /  \
       /    \
      53
           |
           |
          /|
         / |
```

(b) There is partial cloud cover over Oswego.
Or
The skies over Oswego are partly sunny.

Extended Constructed Response Questions

73. (a)

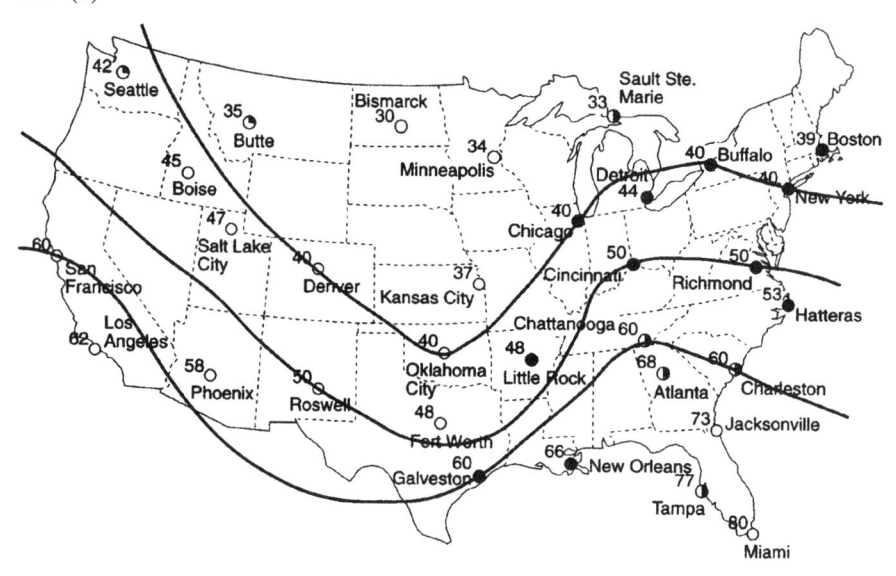

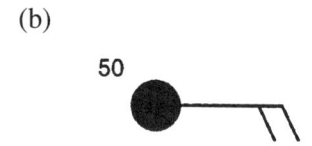

(b)

```
    50
     ●——————
            \
             \
```

(c) Amount of cloud cover

680

Appendix A: Answers to Review Questions

74. (a)

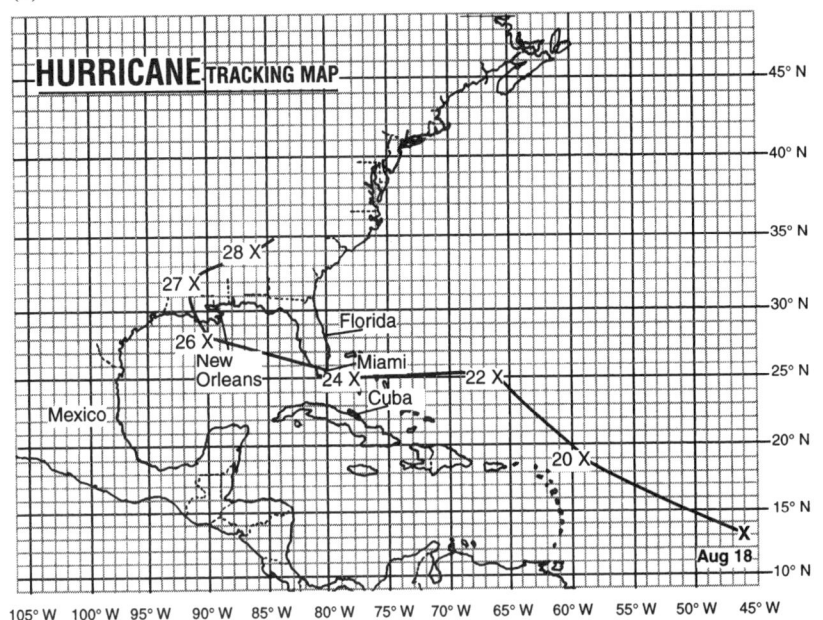

[Hurricane tracking map showing positions labeled Aug 18 through 28 X, tracking from about 13°N 48°W northwestward through the Caribbean near Cuba, across Florida/Miami, to New Orleans and inland.]

(b) The air pressure at Miami was decreasing.
(c) A hurricane will lose its source of energy when it moves over land.
   Or
   Winds decrease because of a lower pressure gradient.
(d) Flooding; high winds; large waves
(e) Examples of acceptable answers:
   Develop and practice a family disaster plan.
   Take a basic first aid, CPR, and/or fire safety course.
   Accumulate an emergency supply of food, water, and medications.
   Obtain an emergency radio.
   Locate gas, water, and electricity shutoffs.
   Plan an evacuation route.
   Place hazardous materials in secure locations.
   Contact local agencies for disaster information.
   Develop plans for farm animals or pets.
   List emergency telephone numbers, and post them prominently.
   Learn the locations of emergency shelters and medical care facilities.

Appendix A: Answers to Review Questions

CHAPTER 24

Multiple-Choice Questions

1.	3	7.	1	13.	3	19.	2	25.	1
2.	3	8.	3	14.	4	20.	1	26.	3
3.	1	9.	4	15.	4	21.	2	27.	2
4.	3	10.	4	16.	1	22.	4	28.	4
5.	4	11.	3	17.	2	23.	2	29.	1
6.	4	12.	4	18.	4	24.	4		

Constructed Response Questions

30. 1,500 meters
31. Location A has a higher average annual precipitation because it is on the windward side of the mountain where rising air cools, causing condensation and precipitation.
Or
At location B, the average annual precipitation is lower than at location A because air is warming by compression; therefore, evaporation is more likely to occur than condensation and/or precipitation.
32. (a) Desertification is the rapid development of deserts caused by the impact of human activities.
(b) One way in which humans contribute to desertification is by overgrazing livestock on dry lands. The resultant loss of protective vegetation leaves fertile topsoil vulnerable to erosion. Another is by clearing land for crops. Intensive crop farming leads to a loss of soil fertility and crop failure. When crops fail, the land is left unplanted for a time and is susceptible to erosion.
33. Rising sea levels; changing global weather and climate patterns
34. Land and sea breezes are due to daily fluctuations in the atmospheric temperature over adjacent land and sea surfaces. Monsoons are also due to fluctuations in the atmospheric temperature over adjacent land and sea surfaces, but these fluctuations are seasonal, not daily.

Extended Constructed Response Questions

35. A and B
36. B, D, A, C
37. D

38.

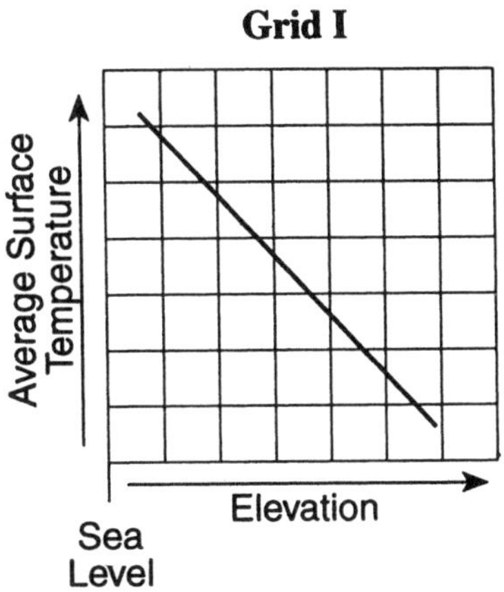
Grid I
Average Surface Temperature
Elevation
Sea Level

39.

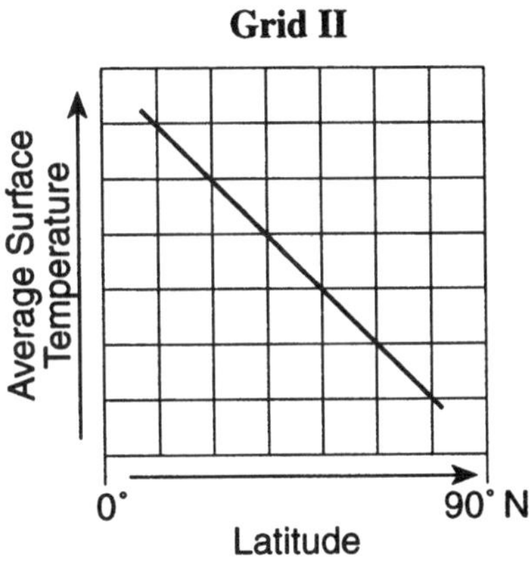
Grid II
Average Surface Temperature
0°
90° N
Latitude

40.

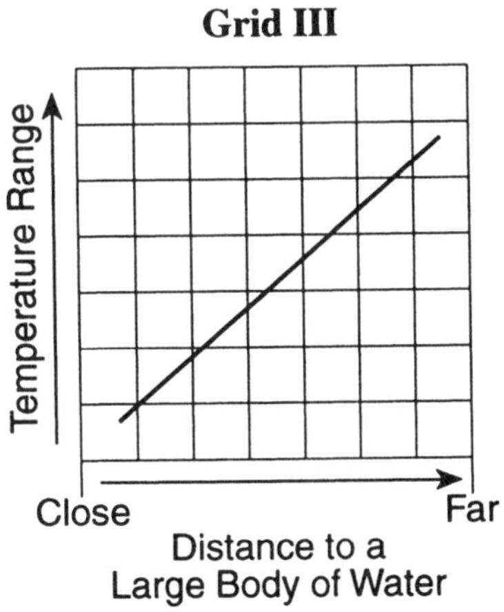
Grid III
Temperature Range
Close
Far
Distance to a Large Body of Water

Appendix A: Answers to Review Questions

41. (a) Locations near the Equator have a warm climate because they receive more direct sunlight.
    (b) Planetary winds converge and rise at the Equator, producing precipitation and a moist climate.

42. (a)

	Boston	Sioux Falls	Savannah	Abilene
Highest summer temperature	81.8	86.3	91.1	95.2
Lowest winter temperature	21.6	24.3	59.7	54.8
Temperature range	60.2	62.0	31.4	40.4

(b) The smaller the distance to a large body of water, the smaller the temperature range.

(c) Water has a higher specific heat than land materials; therefore, when water and land gain the same amount of heat energy, the land will increase in temperature to a greater degree than the water. In the summer, water does not increase in temperature as much as land, and water has a cooling effect on nearby land. Also, in the winter, water does not decrease in temperature as much as land, and water has a warming effect on nearby land. Thus, the presence of a large body of water (the Atlantic Ocean) near Boston has a moderating effect on its temperatures, so that Boston has a smaller yearly temperature range (cooler summers and warmer winters) than Sioux Falls, which is located inland.

(d) Abilene is located at a lower latitude than Sioux Falls and therefore receives more direct sunlight year-round.

Appendix B: EARTH SCIENCE REFERENCE TABLES

PHYSICAL CONSTANTS

Radioactive Decay Data

RADIOACTIVE ISOTOPE	DISINTEGRATION	HALF-LIFE (years)
Carbon-14	$C^{14} \rightarrow N^{14}$	5.7×10^3
Potassium-40	$K^{40} \rightarrow Ar^{40}$ $K^{40} \rightarrow Ca^{40}$	1.3×10^9
Uranium-238	$U^{238} \rightarrow Pb^{206}$	4.5×10^9
Rubidium-87	$Rb^{87} \rightarrow Sr^{87}$	4.9×10^{10}

Specific Heats of Common Materials

MATERIAL	SPECIFIC HEAT (calories/gram·C°)
Water — solid	0.5
Water — liquid	1.0
Water — gas	0.5
Dry air	0.24
Basalt	0.20
Granite	0.19
Iron	0.11
Copper	0.09
Lead	0.03

Properties of Water

Energy gained during melting	80 calories/gram
Energy released during freezing	80 calories/gram
Energy gained during vaporization	540 calories/gram
Energy released during condensation	540 calories/gram
Density at 3.98°C	1.00 gram/milliliter

EQUATIONS

Percent deviation from accepted value:
$$\text{deviation (\%)} = \frac{\text{difference from accepted value}}{\text{accepted value}} \times 100$$

Eccentricity of an ellipse:
$$\text{eccentricity} = \frac{\text{distance between foci}}{\text{length of major axis}}$$

Gradient:
$$\text{gradient} = \frac{\text{change in field value}}{\text{distance}}$$

Rate of change:
$$\text{rate of change} = \frac{\text{change in field value}}{\text{time}}$$

Density of a substance:
$$\text{density} = \frac{\text{mass}}{\text{volume}}$$

2001 EDITION

This edition of the Earth Science Reference Tables should be used in the classroom beginning in the 2000–2001 school year. The first examination for which these tables will be used is the January 2001 Regents Examination in Earth Science.

EURYPTERUS
New York State Fossil

Earth Science Reference Tables — 2001 Edition

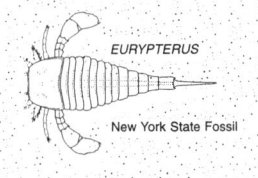

Appendix B: Earth Science Reference Tables

Generalized Landscape Regions of New York State

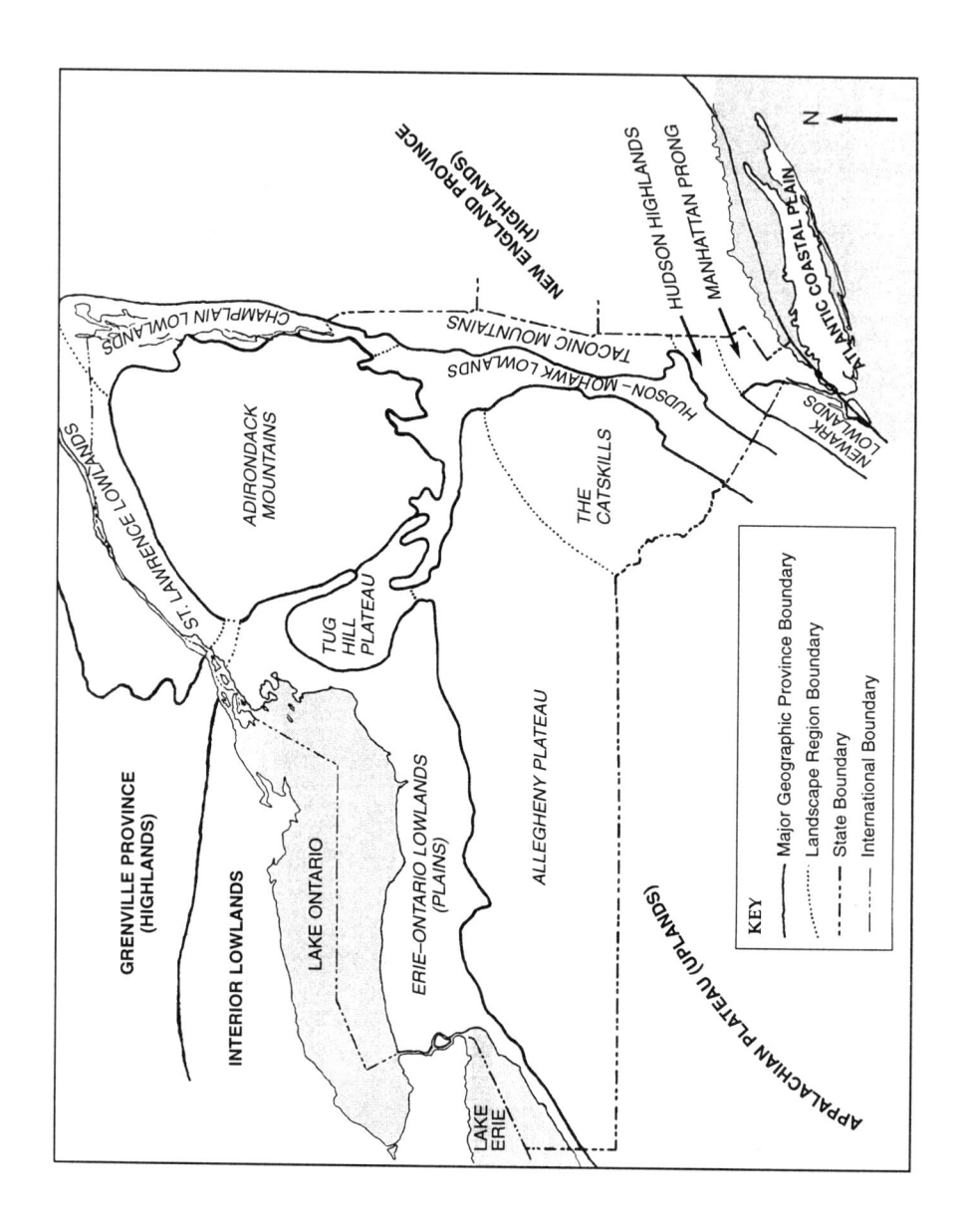

Appendix B: Earth Science Reference Tables

Generalized Bedrock Geology of New York State

modified from
GEOLOGICAL SURVEY
NEW YORK STATE MUSEUM
1989

GEOLOGICAL PERIODS AND ERAS IN NEW YORK

CRETACEOUS, TERTIARY, PLEISTOCENE (Epoch) weakly consolidated to unconsolidated gravels, sands, and clays
LATE TRIASSIC and EARLY JURASSIC conglomerates, red sandstones, red shales, and diabase (in Palisades Sill)
PENNSYLVANIAN and MISSISSIPPIAN conglomerates, sandstones, and shales

- DEVONIAN — limestones, shales, sandstones, and conglomerates
- SILURIAN — Silurian also contains salt, gypsum, and hematite.
- ORDOVICIAN — limestones, shales, sandstones, and dolostones
- CAMBRIAN

CAMBRIAN and EARLY ORDOVICIAN sandstones and dolostones
  Moderately to intensely metamorphosed east of the Hudson River.
CAMBRIAN and ORDOVICIAN (undifferentiated) quartzites, dolostones, marbles, and schists
  Intensely metamorphosed; includes portions of the Taconic Sequence and Cortlandt Complex.
TACONIC SEQUENCE sandstones, shales, and slates
  Slightly to intensely metamorphosed rocks of CAMBRIAN through MIDDLE ORDOVICIAN ages.
MIDDLE PROTEROZOIC gneisses, quartzites, and marbles } Intensely Metamorphosed Rocks
  Lines are generalized structure trends. (regional metamorphism about 1,000 m.y.a.)
MIDDLE PROTEROZOIC anorthositic rocks

Dominantly Sedimentary Origin
Dominantly Metamorphosed Rocks

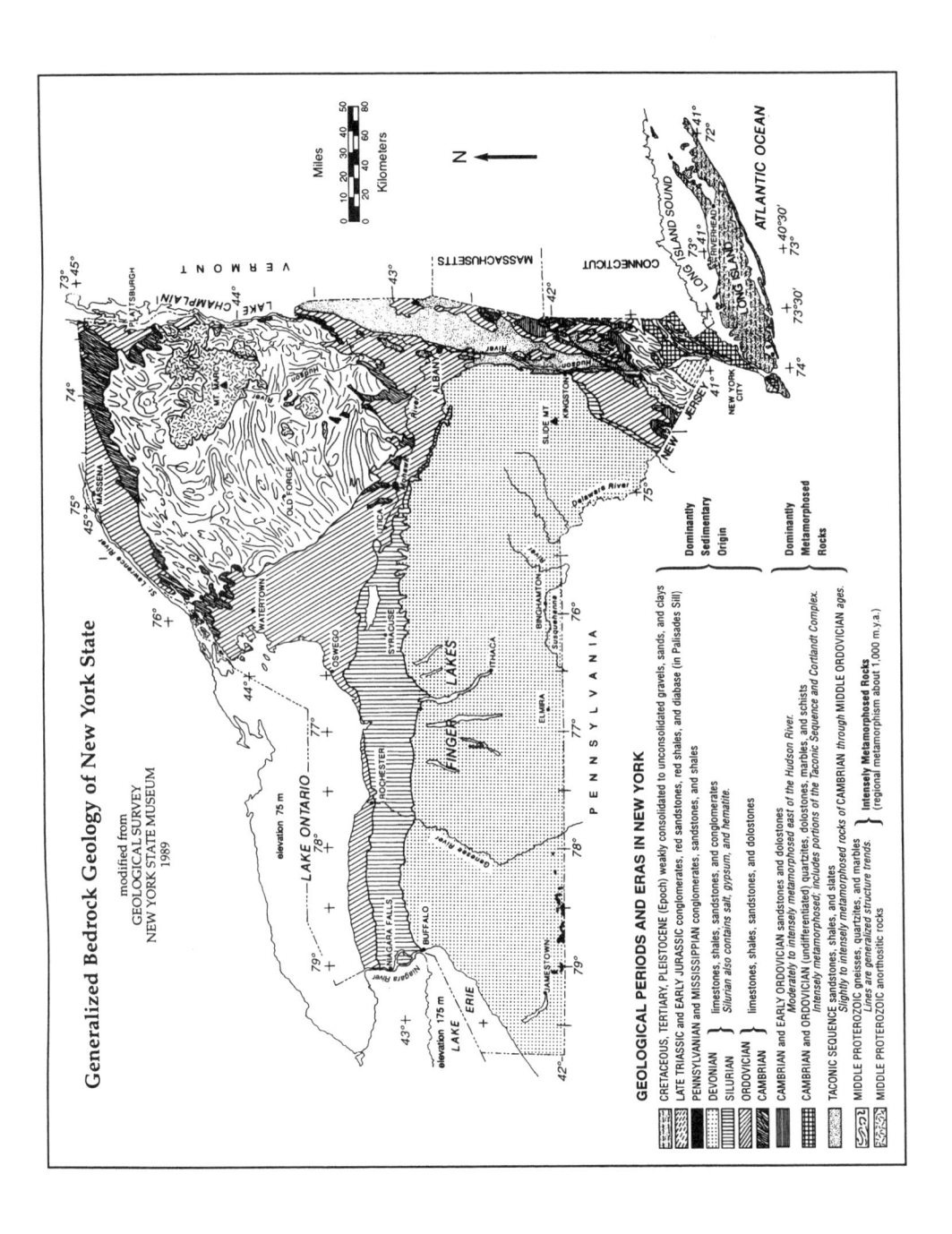

Appendix B: Earth Science Reference Tables

Surface Ocean Currents

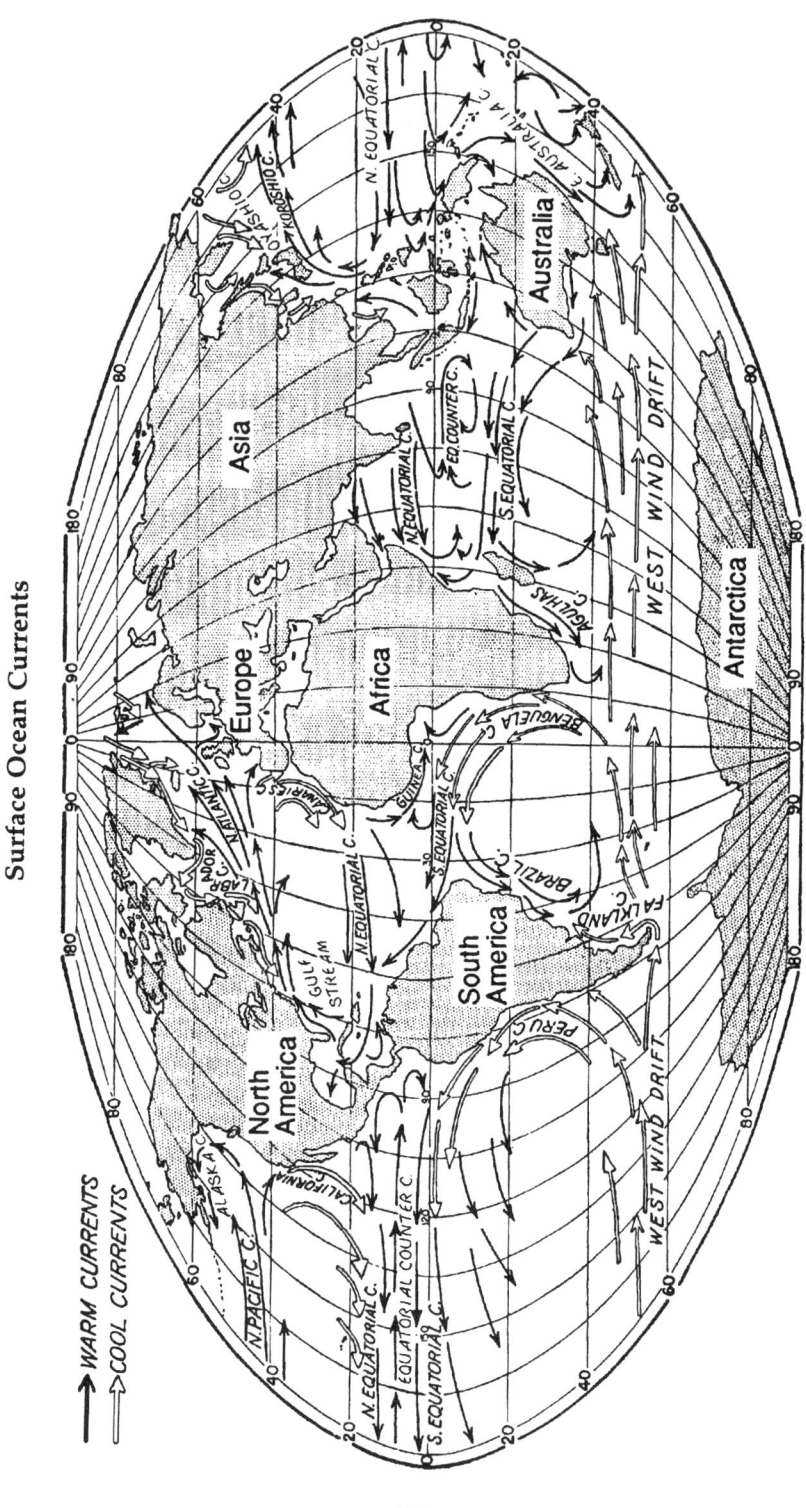

Appendix B: Earth Science Reference Tables

Tectonic Plates

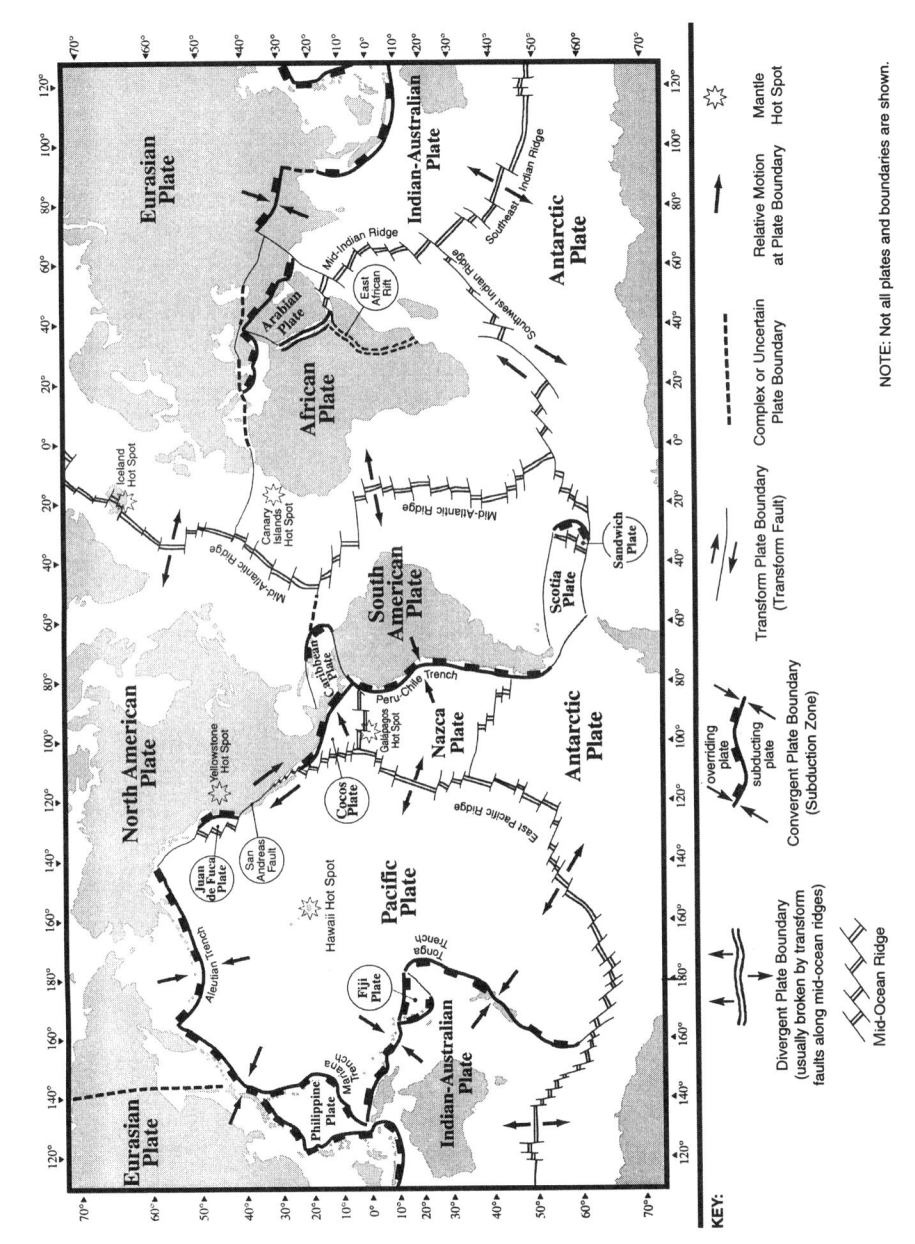

Appendix B: Earth Science Reference Tables

Rock Cycle in Earth's Crust

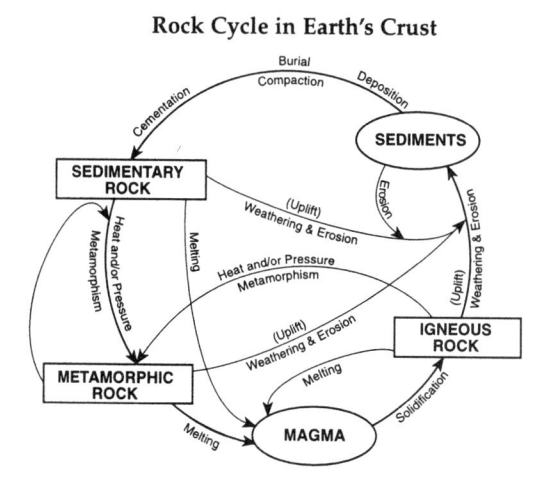

Relationship of Transported Particle Size to Water Velocity

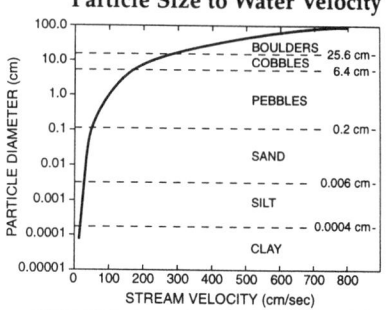

*This generalized graph shows the water velocity needed to maintain, but not start, movement. Variations occur due to differences in particle density and shape.

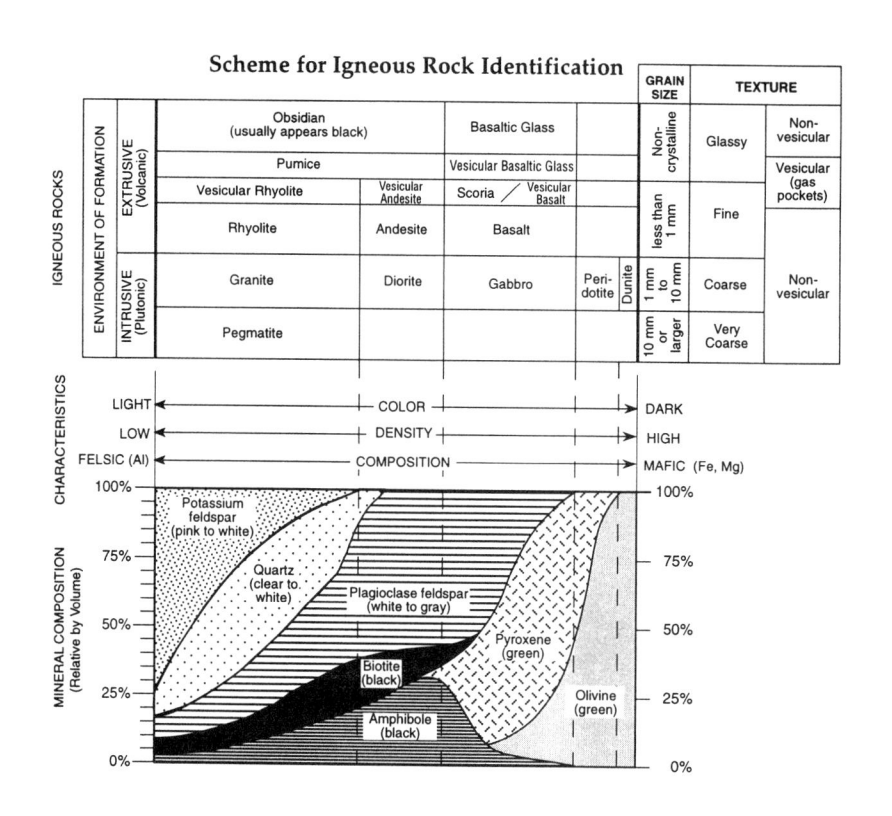

Scheme for Igneous Rock Identification

Appendix B: Earth Science Reference Tables

Scheme for Sedimentary Rock Identification

INORGANIC LAND-DERIVED SEDIMENTARY ROCKS

TEXTURE	GRAIN SIZE	COMPOSITION	COMMENTS	ROCK NAME	MAP SYMBOL
Clastic (fragmental)	Pebbles, cobbles, and/or boulders embedded in sand, silt, and/or clay	Mostly quartz, feldspar, and clay minerals; may contain fragments of other rocks and minerals	Rounded fragments	Conglomerate	
			Angular fragments	Breccia	
	Sand (0.2 to 0.006 cm)		Fine to coarse	Sandstone	
	Silt (0.006 to 0.0004 cm)		Very fine grain	Siltstone	
	Clay (less than 0.0004 cm)		Compact; may split easily	Shale	

CHEMICALLY AND/OR ORGANICALLY FORMED SEDIMENTARY ROCKS

TEXTURE	GRAIN SIZE	COMPOSITION	COMMENTS	ROCK NAME	MAP SYMBOL
Crystalline	Varied	Halite	Crystals from chemical precipitates and evaporites	Rock Salt	
	Varied	Gypsum		Rock Gypsum	
	Varied	Dolomite		Dolostone	
Bioclastic	Microscopic to coarse	Calcite	Cemented shell fragments or precipitates of biologic origin	Limestone	
	Varied	Carbon	From plant remains	Coal	

Scheme for Metamorphic Rock Identification

TEXTURE	GRAIN SIZE	COMPOSITION	TYPE OF METAMORPHISM	COMMENTS	ROCK NAME	MAP SYMBOL
FOLIATED — MINERAL ALIGNMENT	Fine	MICA, QUARTZ, FELDSPAR, AMPHIBOLE, GARNET, PYROXENE	Regional (Heat and pressure increase with depth)	Low-grade metamorphism of shale	Slate	
	Fine to medium			Foliation surfaces shiny from microscopic mica crystals	Phyllite	
				Platy mica crystals visible from metamorphism of clay or feldspars	Schist	
FOLIATED — BANDING	Medium to coarse			High-grade metamorphism; some mica changed to feldspar; segregated by mineral type into bands	Gneiss	
NONFOLIATED	Fine	Variable	Contact (Heat)	Various rocks changed by heat from nearby magma/lava	Hornfels	
	Fine to coarse	Quartz	Regional or Contact	Metamorphism of quartz sandstone	Quartzite	
		Calcite and/or dolomite		Metamorphism of limestone or dolostone	Marble	
	Coarse	Various minerals in particles and matrix		Pebbles may be distorted or stretched	Metaconglomerate	

Appendix B: Earth Science Reference Tables

GEOLOGIC HISTORY OF NEW YORK STATE

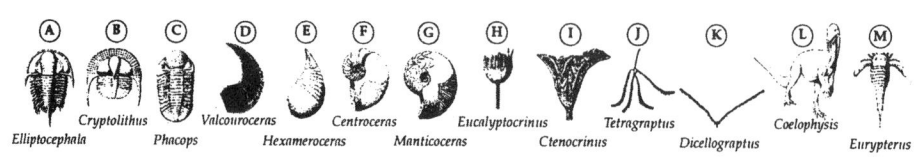

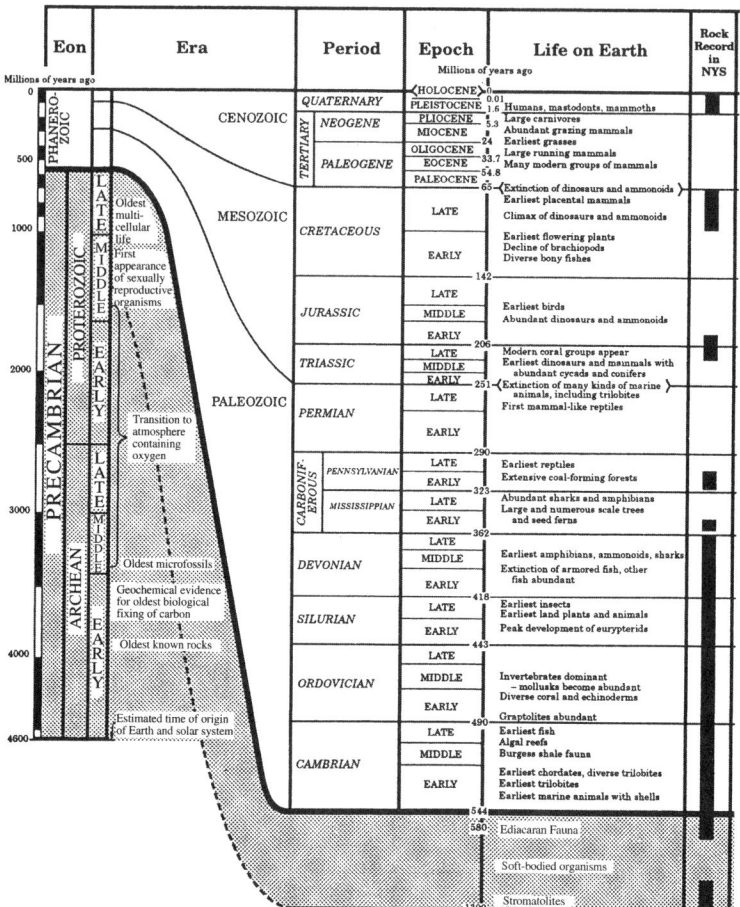

Index Fossils:
- A. *Elliptocephala*
- B. *Cryptolithus* / *Phacops*
- C. *Valcouroceras*
- D. *Centroceras* / *Hexameroceras*
- E. *Eucalyptocrinus* / *Manticoceras*
- F. (Centroceras group)
- G. *Ctenocrinus*
- H. *Tetragraptus*
- I. *Dicellograptus*
- J. *Coelophysis*
- K. (reptile/dinosaur)
- L. (plant)
- M. *Eurypterus*

Eon	Era	Period	Epoch	Life on Earth	Rock Record in NYS
PHANEROZOIC	CENOZOIC	QUATERNARY	HOLOCENE (0.01)		
			PLEISTOCENE (1.6)	Humans, mastodons, mammoths	
		NEOGENE (TERTIARY)	PLIOCENE (5.3)	Large carnivores	
			MIOCENE (24)	Abundant grazing mammals; Earliest grasses	
		PALEOGENE (TERTIARY)	OLIGOCENE (33.7)	Large running mammals	
			EOCENE (54.8)	Many modern groups of mammals	
			PALEOCENE (65)	Extinction of dinosaurs and ammonoids; Earliest placental mammals	
	MESOZOIC	CRETACEOUS	LATE	Climax of dinosaurs and ammonoids	
			EARLY	Earliest flowering plants; Decline of brachiopods; Diverse bony fishes	
		JURASSIC (142)	LATE	Earliest birds	
			MIDDLE	Abundant dinosaurs and ammonoids	
			EARLY (206)		
		TRIASSIC	LATE	Modern coral groups appear	
			MIDDLE	Earliest dinosaurs and mammals with abundant cycads and conifers	
			EARLY (251)	Extinction of many kinds of marine animals, including trilobites	
	PALEOZOIC	PERMIAN	LATE	First mammal-like reptiles	
			EARLY (290)		
		CARBONIFEROUS — PENNSYLVANIAN	LATE	Earliest reptiles	
			EARLY (323)	Extensive coal-forming forests	
		CARBONIFEROUS — MISSISSIPPIAN	LATE	Abundant sharks and amphibians	
			EARLY (362)	Large and numerous scale trees and seed ferns	
		DEVONIAN	LATE	Earliest amphibians, ammonoids, sharks	
			MIDDLE	Extinction of armored fish, other fish abundant	
			EARLY (418)		
		SILURIAN	LATE	Earliest insects; Earliest land plants and animals	
			EARLY (443)	Peak development of eurypterids	
		ORDOVICIAN	LATE		
			MIDDLE	Invertebrates dominant — mollusks become abundant; Diverse coral and echinoderms	
			EARLY (490)	Graptolites abundant	
		CAMBRIAN	LATE	Earliest fish; Algal reefs	
			MIDDLE	Burgess shale fauna	
			EARLY (544)	Earliest chordates, diverse trilobites; Earliest trilobites; Earliest marine animals with shells	
PROTEROZOIC	LATE / MIDDLE / EARLY		(580)	Ediacaran Fauna	
				Soft-bodied organisms	
			(1300)	Stromatolites	
ARCHEAN	LATE / MIDDLE / EARLY			Oldest multicellular life; First appearance of sexually reproductive organisms; Transition to atmosphere containing oxygen; Oldest microfossils; Geochemical evidence for oldest biological fixing of carbon; Oldest known rocks	
				Estimated time of origin of Earth and solar system (4600)	

692

Appendix B: Earth Science Reference Tables

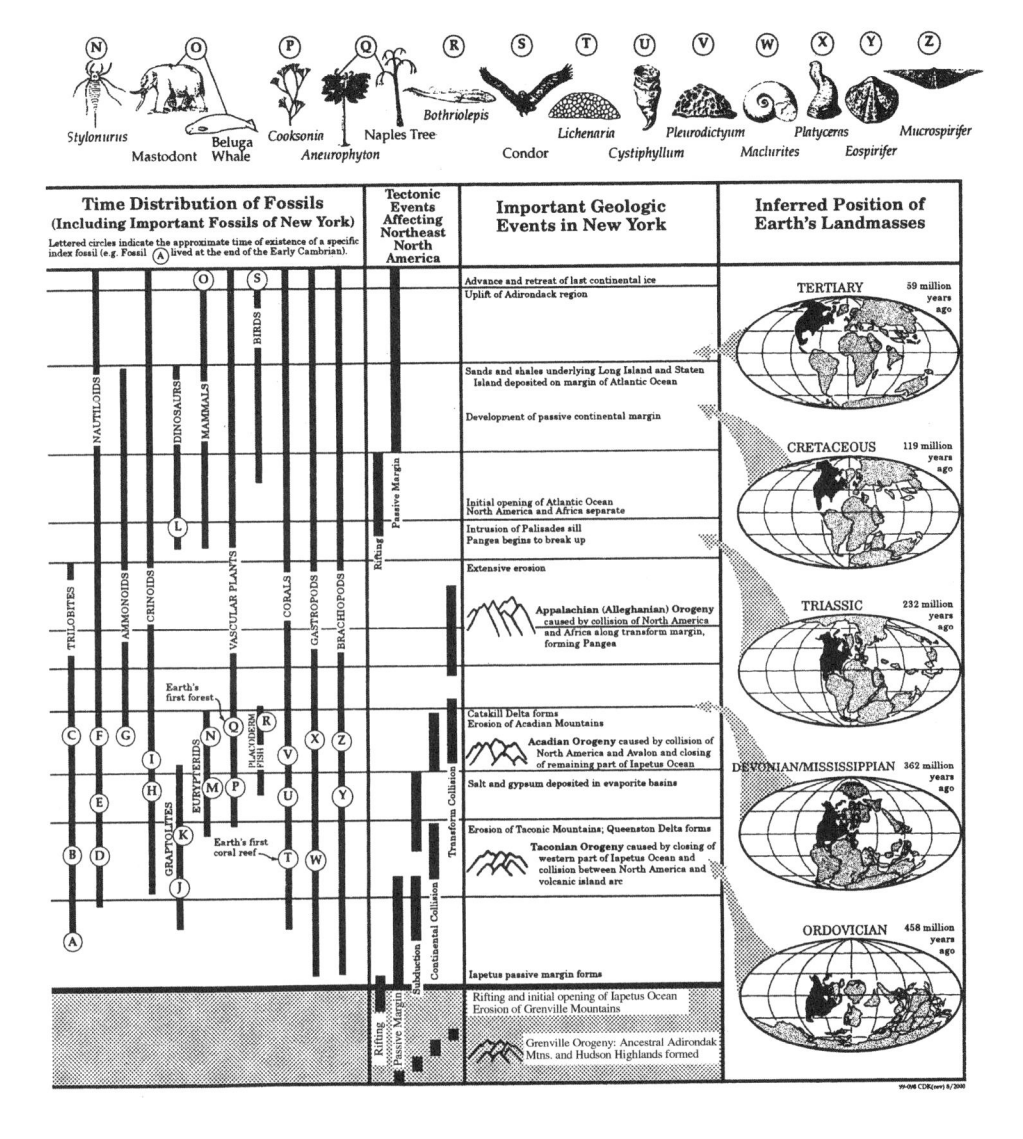

Time Distribution of Fossils
(Including Important Fossils of New York)

Index fossils (labeled N–Z):
- N: Stylonurus
- O: Mastodont
- P: Beluga Whale
- Q: Cooksonia
- R: Aneurophyton (Naples Tree)
- S: Bothriolepis
- T: Lichenaria
- U: Condor
- V: Pleurodictyum
- W: Cystiphyllum
- X: Platyceras
- Y: Maclurites
- Z: Eospirifer / Mucrospirifer

Tectonic Events Affecting Northeast North America

Important Geologic Events in New York

- Advance and retreat of last continental ice
- Uplift of Adirondack region
- Sands and shales underlying Long Island and Staten Island deposited on margin of Atlantic Ocean
- Development of passive continental margin
- Initial opening of Atlantic Ocean; North America and Africa separate
- Intrusion of Palisades sill; Pangea begins to break up
- Extensive erosion
- Appalachian (Alleghanian) Orogeny caused by collision of North America and Africa along transform margin, forming Pangea
- Catskill Delta forms; Erosion of Acadian Mountains
- Acadian Orogeny caused by collision of North America and Avalon and closing of remaining part of Iapetus Ocean
- Salt and gypsum deposited in evaporite basins
- Erosion of Taconic Mountains; Queenston Delta forms
- Taconian Orogeny caused by closing of western part of Iapetus Ocean and collision between North America and volcanic island arc
- Iapetus passive margin forms
- Rifting and initial opening of Iapetus Ocean; Erosion of Grenville Mountains
- Grenville Orogeny: Ancestral Adirondack Mtns. and Hudson Highlands formed

Inferred Position of Earth's Landmasses

- TERTIARY — 59 million years ago
- CRETACEOUS — 119 million years ago
- TRIASSIC — 232 million years ago
- DEVONIAN/MISSISSIPPIAN — 362 million years ago
- ORDOVICIAN — 458 million years ago

693

Appendix B: Earth Science Reference Tables

Inferred Properties of Earth's Interior

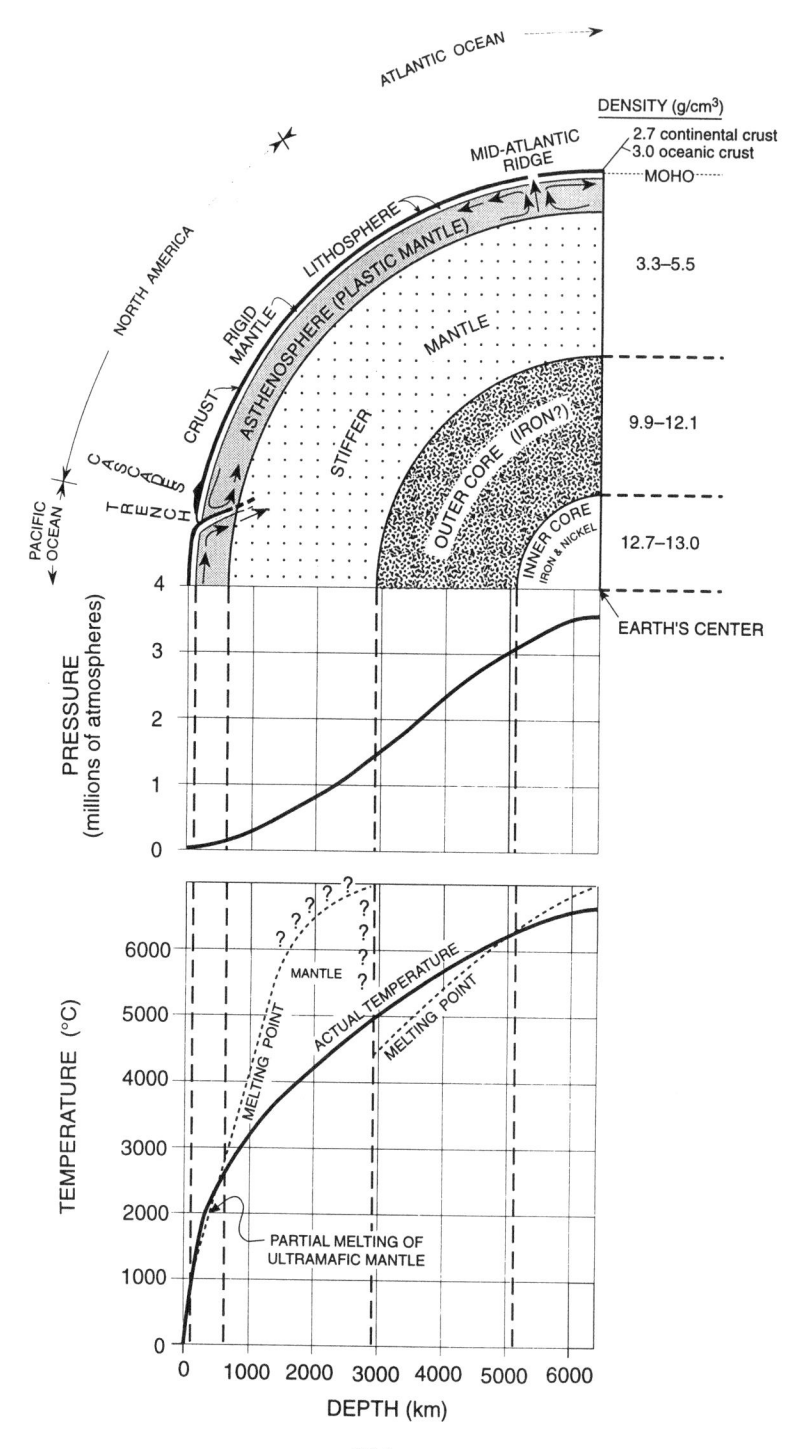

DENSITY (g/cm³)
- 2.7 continental crust
- 3.0 oceanic crust
- MOHO
- 3.3–5.5
- 9.9–12.1
- 12.7–13.0

EARTH'S CENTER

PRESSURE (millions of atmospheres)

TEMPERATURE (°C)

PARTIAL MELTING OF ULTRAMAFIC MANTLE

DEPTH (km)

694

Appendix B: Earth Science Reference Tables

Average Chemical Composition of Earth's Crust, Hydrosphere, and Troposphere

ELEMENT (symbol)	CRUST Percent by Mass	CRUST Percent by Volume	HYDROSPHERE Percent by Volume	TROPOSPHERE Percent by Volume
Oxygen (O)	46.40	94.04	33.0	21.0
Silicon (Si)	28.15	0.88		
Aluminum (Al)	8.23	0.48		
Iron (Fe)	5.63	0.49		
Calcium (Ca)	4.15	1.18		
Sodium (Na)	2.36	1.11		
Magnesium (Mg)	2.33	0.33		
Potassium (K)	2.09	1.42		
Nitrogen (N)				78.0
Hydrogen (H)			66.0	
Other	0.66	0.07	1.0	1.0

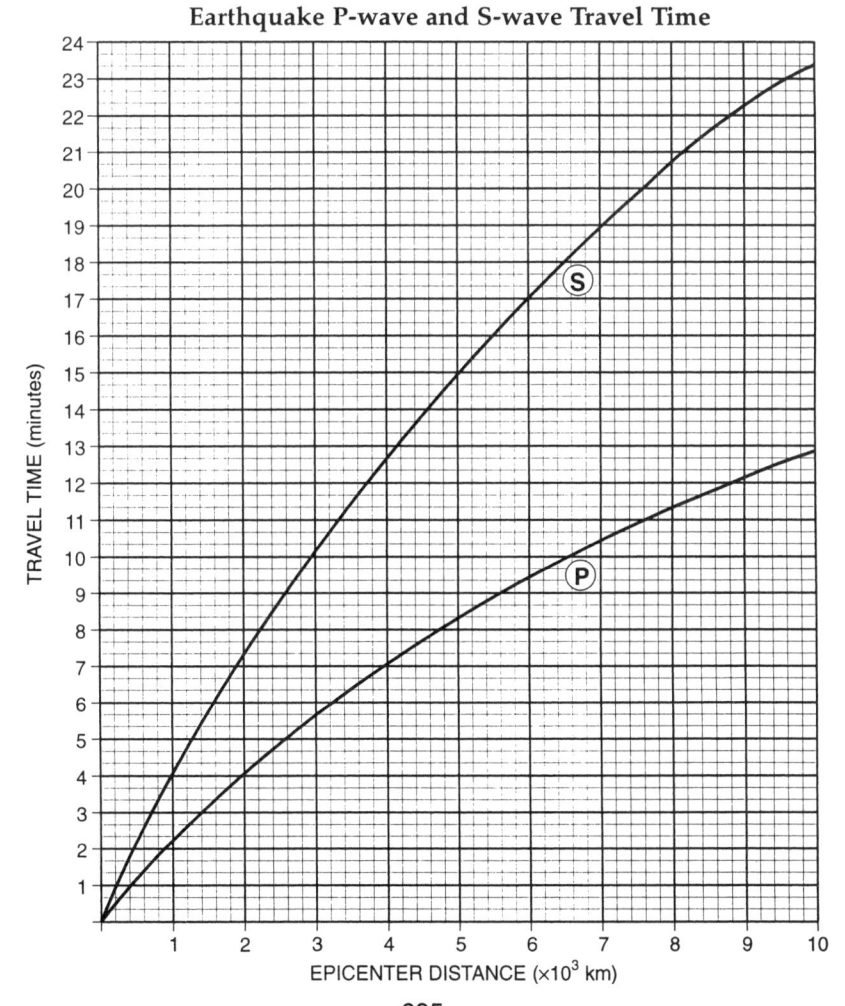

Earthquake P-wave and S-wave Travel Time

Appendix B: Earth Science Reference Tables

Dew point Temperatures (°C)

Dry-Bulb Temperature (°C)	\multicolumn{16}{c}{Difference Between Wet-Bulb and Dry-Bulb Temperatures (C°)}
	0	1	2	3	4	5	6	7	8	9	10	11	12	13	14	15
−20	−20	−33														
−18	−18	−28														
−16	−16	−24														
−14	−14	−21	−36													
−12	−12	−18	−28													
−10	−10	−14	−22													
−8	−8	−12	−18	−29												
−6	−6	−10	−14	−22												
−4	−4	−7	−12	−17	−29											
−2	−2	−5	−8	−13	−20											
0	0	−3	−6	−9	−15	−24										
2	2	−1	−3	−6	−11	−17										
4	4	1	−1	−4	−7	−11	−19									
6	6	4	1	−1	−4	−7	−13	−21								
8	8	6	3	1	−2	−5	−9	−14								
10	10	8	6	4	1	−2	−5	−9	−14	−28						
12	12	10	8	6	4	1	−2	−5	−9	−16						
14	14	12	11	9	6	4	1	−2	−5	−10	−17					
16	16	14	13	11	9	7	4	1	−1	−6	−10	−17				
18	18	16	15	13	11	9	7	4	2	−2	−5	−10	−19			
20	20	19	17	15	14	12	10	7	4	2	−2	−5	−10	−19		
22	22	21	19	17	16	14	12	10	8	5	3	−1	−5	−10	−19	
24	24	23	21	20	18	16	14	12	10	8	6	2	−1	−5	−10	−18
26	26	25	23	22	20	18	17	15	13	11	9	6	3	0	−4	−9
28	28	27	25	24	22	21	19	17	16	14	11	9	7	4	1	−3
30	30	29	27	26	24	23	21	19	18	16	14	12	10	8	5	1

Relative Humidity (%)

Dry-Bulb Temperature (°C)	\multicolumn{16}{c}{Difference Between Wet-Bulb and Dry-Bulb Temperatures (C°)}
	0	1	2	3	4	5	6	7	8	9	10	11	12	13	14	15
−20	100	28														
−18	100	40														
−16	100	48														
−14	100	55	11													
−12	100	61	23													
−10	100	66	33													
−8	100	71	41	13												
−6	100	73	48	20												
−4	100	77	54	32	11											
−2	100	79	58	37	20	1										
0	100	81	63	45	28	11										
2	100	83	67	51	36	20	6									
4	100	85	70	56	42	27	14									
6	100	86	72	59	46	35	22	10								
8	100	87	74	62	51	39	28	17	6							
10	100	88	76	65	54	43	33	24	13	4						
12	100	88	78	67	57	48	38	28	19	10	2					
14	100	89	79	69	60	50	41	33	25	16	8	1				
16	100	90	80	71	62	54	45	37	29	21	14	7	1			
18	100	91	81	72	64	56	48	40	33	26	19	12	6			
20	100	91	82	74	66	58	51	44	36	30	23	17	11	5		
22	100	92	83	75	68	60	53	46	40	33	27	21	15	10	4	
24	100	92	84	76	69	62	55	49	42	36	30	25	20	14	9	4
26	100	92	85	77	70	64	57	51	45	39	34	28	23	18	13	9
28	100	93	86	78	71	65	59	53	47	42	36	31	26	21	17	12
30	100	93	86	79	72	66	61	55	49	44	39	34	29	25	20	16

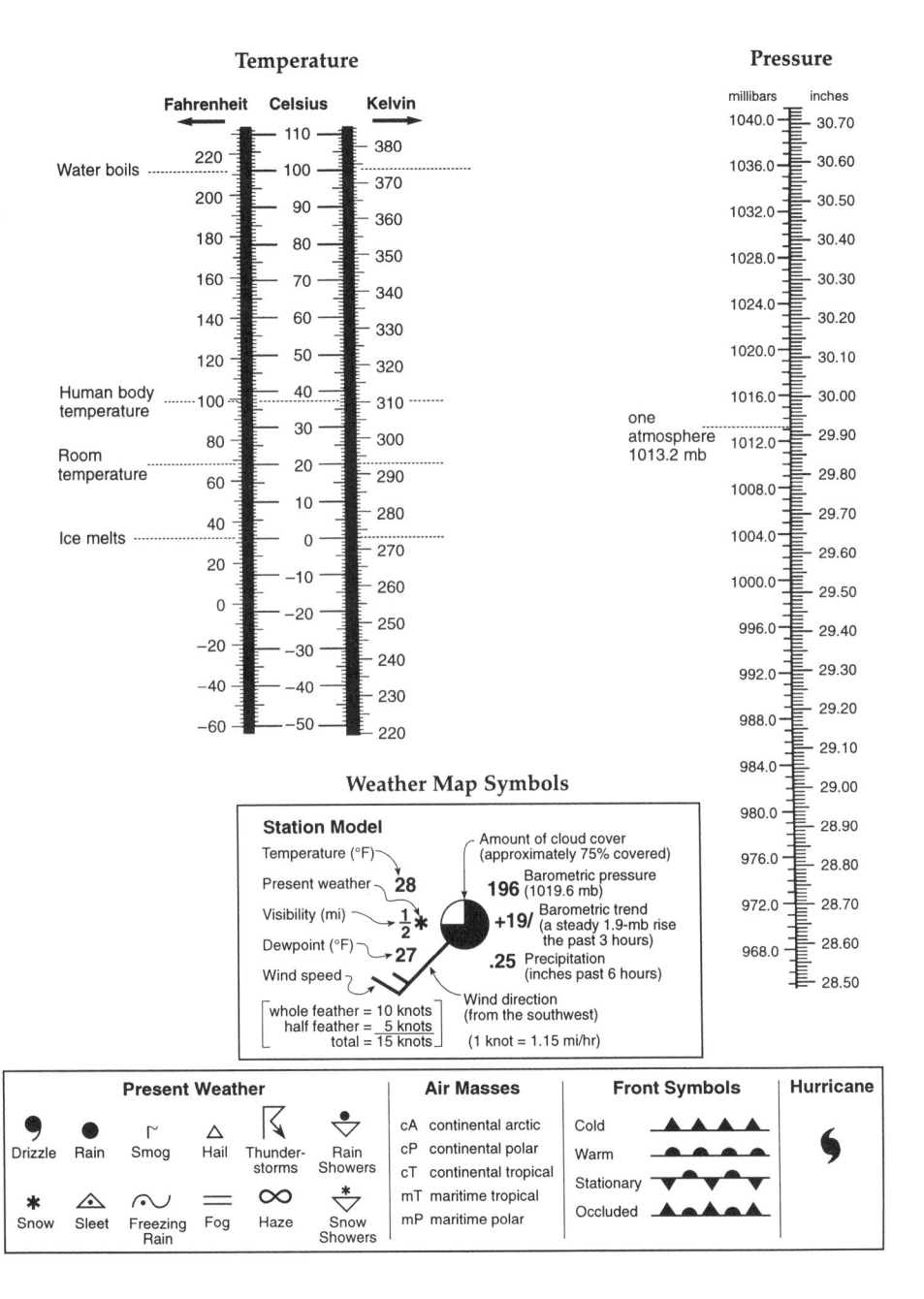

Appendix B: Earth Science Reference Tables

Temperature

Fahrenheit	Celsius	Kelvin
220	110	380
	100 — Water boils	370
200	90	360
180	80	350
160	70	340
140	60	330
120	50	320
100 — Human body temperature	40	310
80	30	300
60 — Room temperature	20	290
40	10	280
	0 — Ice melts	270
20	–10	260
0	–20	250
–20	–30	240
–40	–40	230
–60	–50	220

Pressure

millibars	inches
1040.0	30.70
1036.0	30.60
1032.0	30.50
1028.0	30.40
1024.0	30.30
1020.0	30.20
1016.0	30.10
1012.0 — one atmosphere 1013.2 mb	30.00
1008.0	29.90
1004.0	29.80
1000.0	29.70
996.0	29.60
992.0	29.50
988.0	29.40
984.0	29.30
980.0	29.20
976.0	29.10
972.0	29.00
968.0	28.90
	28.80
	28.70
	28.60
	28.50

Weather Map Symbols

Station Model

- Temperature (°F): 28
- Present weather: (symbol)
- Visibility (mi): 1/2
- Dewpoint (°F): 27
- Wind speed
- Amount of cloud cover (approximately 75% covered)
- Barometric pressure: 196 (1019.6 mb)
- Barometric trend: +19/ (a steady 1.9-mb rise the past 3 hours)
- Precipitation: .25 (inches past 6 hours)
- Wind direction (from the southwest)

whole feather = 10 knots
half feather = 5 knots
total = 15 knots
(1 knot = 1.15 mi/hr)

Present Weather

Symbol	Meaning
Drizzle	Rain	Smog	Hail	Thunderstorms	Rain Showers
Snow	Sleet	Freezing Rain	Fog	Haze	Snow Showers

Air Masses

- cA continental arctic
- cP continental polar
- cT continental tropical
- mT maritime tropical
- mP maritime polar

Front Symbols

- Cold
- Warm
- Stationary
- Occluded

Hurricane

Appendix B: Earth Science Reference Tables

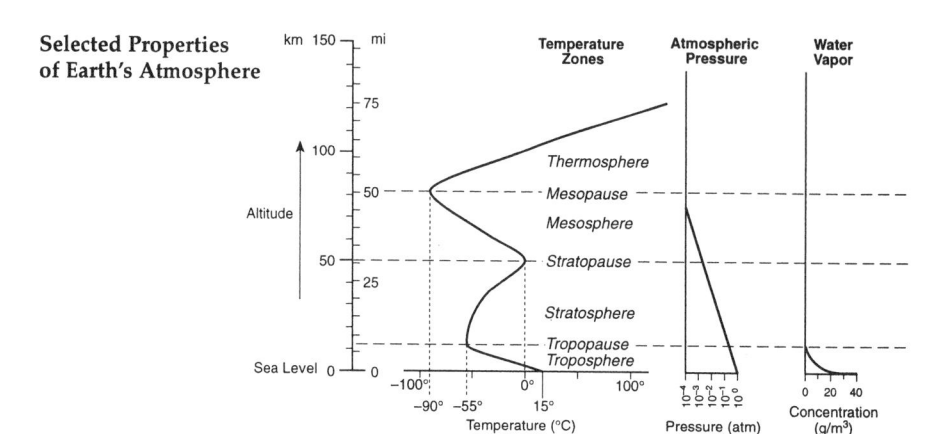

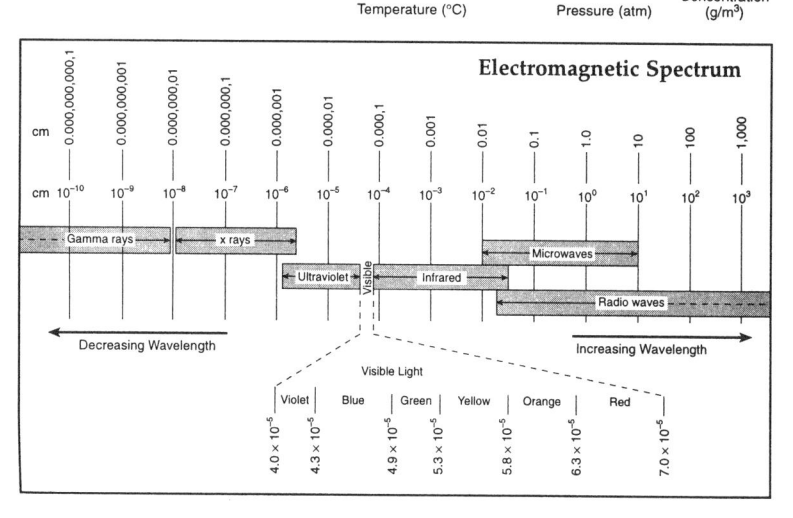

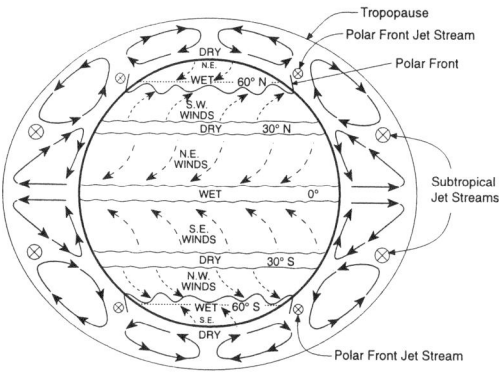

Selected Properties of Earth's Atmosphere

(Graphs showing Altitude vs. Temperature Zones, Atmospheric Pressure, and Water Vapor)

- Thermosphere
- Mesopause
- Mesosphere
- Stratopause
- Stratosphere
- Tropopause
- Troposphere

Temperature (°C): −90°, −55°, 15°
Pressure (atm)
Concentration (g/m³)

Electromagnetic Spectrum

Gamma rays — x rays — Ultraviolet — Visible — Infrared — Microwaves — Radio waves

Decreasing Wavelength ← → Increasing Wavelength

Visible Light:
| Violet | Blue | Green | Yellow | Orange | Red |

4.0×10^{-5}, 4.3×10^{-5}, 4.9×10^{-5}, 5.3×10^{-5}, 5.8×10^{-5}, 6.3×10^{-5}, 7.0×10^{-5}

Planetary Wind and Moisture Belts in the Troposphere

The drawing to the left shows the locations of the belts near the time of an equinox. The locations shift somewhat with the changing latitude of the Sun's vertical ray. In the Northern Hemisphere, the belts shift northward in summer and southward in winter.

Labels on diagram: Tropopause, Polar Front Jet Stream, Polar Front, Subtropical Jet Streams, DRY, WET, N.E. WINDS, S.W. WINDS, S.E. WINDS, N.W. WINDS, 60° N, 30° N, 0°, 30° S, 60° S

698

Appendix B: Earth Science Reference Tables

Luminosity and Temperature of Stars
(Name in italics refers to star shown by a ⊕)

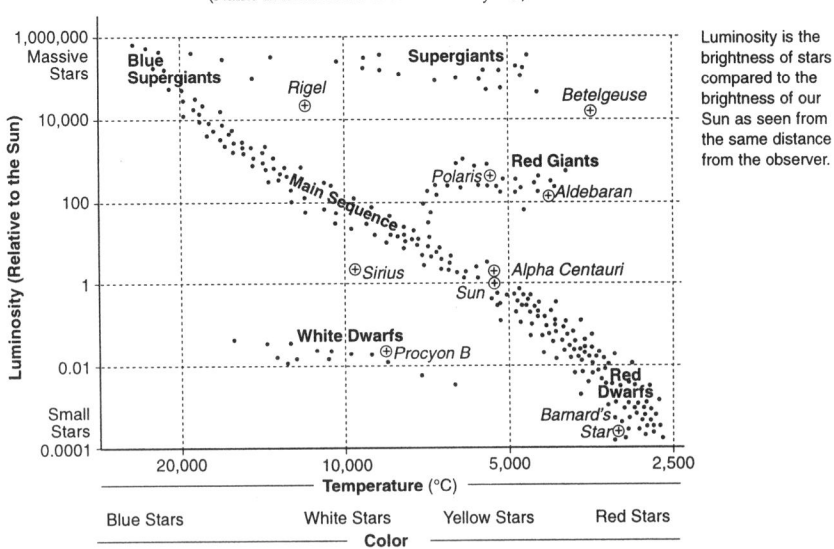

Luminosity is the brightness of stars compared to the brightness of our Sun as seen from the same distance from the observer.

Solar System Data

Object	Mean Distance from Sun (millions of km)	Period of Revolution	Period of Rotation	Eccentricity of Orbit	Equatorial Diameter (km)	Mass (Earth = 1)	Density (g/cm³)	Number of Moons
SUN	—	—	27 days	—	1,392,000	333,000.00	1.4	—
MERCURY	57.9	88 days	59 days	0.206	4,880	0.553	5.4	0
VENUS	108.2	224.7 days	243 days	0.007	12,104	0.815	5.2	0
EARTH	149.6	365.26 days	23 hr 56 min 4 sec	0.017	12,756	1.00	5.5	1
MARS	227.9	687 days	24 hr 37 min 23 sec	0.093	6,787	0.1074	3.9	2
JUPITER	778.3	11.86 years	9 hr 50 min 30 sec	0.048	142,800	317.896	1.3	16
SATURN	1,427	29.46 years	10 hr 14 min	0.056	120,000	95.185	0.7	18
URANUS	2,869	84.0 years	17 hr 14 min	0.047	51,800	14.537	1.2	21
NEPTUNE	4,496	164.8 years	16 hr	0.009	49,500	17.151	1.7	8
PLUTO	5,900	247.7 years	6 days 9 hr	0.250	2,300	0.0025	2.0	1
EARTH'S MOON	149.6 (0.386 from Earth)	27.3 days	27 days 8 hr	0.055	3,476	0.0123	3.3	—

Appendix B: Earth Science Reference Tables

Properties of Common Minerals

LUSTER	HARD-NESS	CLEAVAGE	FRACTURE	COMMON COLORS	DISTINGUISHING CHARACTERISTICS	USE(S)	MINERAL NAME	COMPOSITION*
Metallic Luster	1–2	✔		silver to gray	black streak, greasy feel	pencil lead, lubricants	Graphite	C
Metallic Luster	2.5	✔		metallic silver	very dense (7.6 g/cm³), gray-black streak	ore of lead	Galena	PbS
Metallic Luster	5.5–6.5		✔	black to silver	attracted by magnet, black streak	ore of iron	Magnetite	Fe_3O_4
Metallic Luster	6.5		✔	brassy yellow	green-black streak, cubic crystals	ore of sulfur	Pyrite	FeS_2
Either	1–6.5		✔	metallic silver or earthy red	red-brown streak	ore of iron	Hematite	Fe_2O_3
Nonmetallic Luster	1	✔		white to green	greasy feel	talcum powder, soapstone	Talc	$Mg_3Si_4O_{10}(OH)_2$
Nonmetallic Luster	2		✔	yellow to amber	easily melted, may smell	vulcanize rubber, sulfuric acid	Sulfur	S
Nonmetallic Luster	2	✔		white to pink or gray	easily scratched by fingernail	plaster of paris and drywall	Gypsum (Selenite)	$CaSO_4 \cdot 2H_2O$
Nonmetallic Luster	2–2.5	✔		colorless to yellow	flexible in thin sheets	electrical insulator	Muscovite Mica	$KAl_3Si_3O_{10}(OH)_2$
Nonmetallic Luster	2.5	✔		colorless to white	cubic cleavage, salty taste	food additive, melts ice	Halite	NaCl
Nonmetallic Luster	2.5–3	✔		black to dark brown	flexible in thin sheets	electrical insulator	Biotite Mica	$K(Mg,Fe)_3AlSi_3O_{10}(OH)_2$
Nonmetallic Luster	3	✔		colorless or variable	bubbles with acid	cement, polarizing prisms	Calcite	$CaCO_3$
Nonmetallic Luster	3.5	✔		colorless or variable	bubbles with acid when powdered	source of magnesium	Dolomite	$CaMg(CO_3)_2$
Nonmetallic Luster	4	✔		colorless or variable	cleaves in 4 directions	hydrofluoric acid	Fluorite	CaF_2
Nonmetallic Luster	5–6	✔		black to dark green	cleaves in 2 directions at 90°	mineral collections	Pyroxene (commonly Augite)	$(Ca,Na)(Mg,Fe,Al)(Si,Al)_2O_6$
Nonmetallic Luster	5.5	✔		black to dark green	cleaves at 56° and 124°	mineral collections	Amphiboles (commonly Hornblende)	$CaNa(Mg,Fe)_4(Al,Fe,Ti)_3Si_6O_{22}(O,OH)_2$
Nonmetallic Luster	6	✔		white to pink	cleaves in 2 directions at 90°	ceramics and glass	Potassium Feldspar (Orthoclase)	$KAlSi_3O_8$
Nonmetallic Luster	6	✔		white to gray	cleaves in 2 directions, striations visible	ceramics and glass	Plagioclase Feldspar (Na-Ca Feldspar)	$(Na,Ca)AlSi_3O_8$
Nonmetallic Luster	6.5		✔	green to gray or brown	commonly light green and granular	furnace bricks and jewelry	Olivine	$(Fe,Mg)_2SiO_4$
Nonmetallic Luster	7		✔	colorless or variable	glassy luster, may form hexagonal crystals	glass, jewelry, and electronics	Quartz	SiO_2
Nonmetallic Luster	7		✔	dark red to green	glassy luster, often seen as red grains in NYS metamorphic rocks	jewelry and abrasives	Garnet (commonly Almandine)	$Fe_3Al_2Si_3O_{12}$

*Chemical Symbols:
Al = aluminum
C = carbon
Ca = calcium
Cl = chlorine
F = fluorine
Fe = iron
H = hydrogen
K = potassium
Mg = magnesium
Na = sodium
O = oxygen
Pb = lead
S = sulfur
Si = silicon
Ti = titanium

✔ = dominant form of breakage

Glossary of Earth Science Terms

ablation The melting and evaporation of glacial ice.

ablation debris Material on the surface of a glacier that was suspended in the ice but is now exposed due to ablation.

abrasion The breaking down of rocks by rubbing together.

absolute brightness Any measure of the intrinsic (not dependent on the position or distance of the observer) brightness or luminosity of a celestial object.

absolute humidity The number of grams of water vapor in a cubic meter of air.

absolute magnitude The intrinsic brightness or luminosity of a celestial object expressed as the magnitude that a star would appear to have if it were 10 parsecs away.

absorption spectrum A spectrum in which certain regions are dark or dim because those wavelengths of light were absorbed by atoms between the source and the observer.

abyssal plain Wide, flat areas adjacent to the mid-ocean ridges.

accretion The sticking together of particles to form larger particles.

acid rain Rain that is acidic due to substances that have become dissolved in it.

adiabatic A change that does not involve the addition or removal of heat.

adiabatic lapse rate The rate at which the temperature of air changes due to a decrease in air pressure rather than the addition or removal of heat.

aerobic atmosphere An atmosphere with sufficient oxygen to support aerobic respiration by living things.

agent of erosion A medium set in motion by gravity that can exert a force on sediments causing them to be moved.

air current Air that is moving in a vertical direction.

air mass A large region of air with uniform characteristics at any given level.

air pressure The force exerted on a unit of area by the air.

alluvial fan A flat, fan-shaped mound of sediment deposited where a stream slows down because its gradient suddenly becomes less steep (i.e., where a mountain valley empties onto a plain).

altitude 1. The angle between the horizon and an object seen in the sky with the observer at the vertex. 2. Distance above the earth's surface.

anemometer An instrument that measures wind speed.

angle of repose The greatest angle at which a pile of sediments is stable.

angstrom (Å) A unit of length. There are 10^8 angstroms in a centimeter.

angular unconformity A place where sedimentary layers at one angle are in contact with sedimentary layers at a different angle.

annular eclipse A solar eclipse in which the moon's shadow does not actually reach the surface of the Earth; to an observer on Earth, the sun's disk is obscured by the Moon except for a narrow ring around the perimeter.

anticline An upward, arched fold.

anticyclone A flow of air that is clockwise and outward from a high pressure center.

apparent brightness The brightness of a celestial object as seen by human eyes on Earth.

apparent daily motion The apparent motion of all celestial objects along a circular path at a constant rate of 15° per hour, or one complete circle every day.

apparent magnitude The apparent brightness of a celestial object expressed in the astronomical magnitude scale; the larger the number, the fainter the star.

aquiclude A rock or sediment layer that is impermeable to water.

aquifer A rock or sediment layer in which the pore spaces are saturated with water.

Glossary

arrival time The precise time at which a seismic wave reaches a seismograph and is recorded.

artesian well A well in which water rises to the surface unassisted.

asteroid Solid body having no atmosphere that orbits the sun.

asthenosphere A region of the upper mantle between 100 and 350 kilometers in depth that behaves like a fluid.

astronomical unit The average distance from the Earth to the Sun; 149,598,000 kilometers.

atmosphere The thin shell of gases bound to the Earth by gravity; it consists of gases, water, ice, dust, and other particles.

atmospheric variables The characteristics of the atmosphere that change.

aurora Glowing, often shimmering light forms seen near the poles that are caused by radiation from high altitude air molecules excited by collisions with particles from the Sun.

axis of rotation The central line around which all parts of an object move during rotation.

banding A structure in a metamorphic rock of nearly parallel bands of different textures or minerals; i.e., the nearly parallel bands of light and dark minerals seen in gneiss.

barrier bar A sandbar running parallel to a shoreline that shelters the water behind it from waves.

batholith A large, irregularly shaped pluton covering an area greater than 75 square kilometers.

beach A strip of loose materials along the shore of a body of water formed and washed by waves.

bed A layer of sediment with a particular physical or structural character.

bedrock A general term for the solid rock underlying the soil or loose material covering the Earth's surface.

bench mark A permanent metal marker set into the ground that gives the exact location and precisely measured elevation of that point.

big bang theory The idea that the universe started out with all of its matter in a small volume and then expanded outward in all directions.

body waves Seismic waves that travel through the body of the earth; P-waves and S-waves are body waves.

bright-line spectrum A spectrum consisting of bright, narrow lines of color. Each color corresponds to a specific wavelength of light emitted by an atom when an electron jumps from a specific higher energy level to a specific lower energy level. Since the atoms of each element have a unique structure, each produces a unique sequence of bright lines.

capillary fringe A narrow region just above the water table in which narrow pore spaces are filled with water that is drawn up by capillary action.

capillarity The action by which a fluid, such as water, is drawn upward in narrow spaces as a result of surface tension.

carbonaceous chondrite A stony meteorite containing hydrated claylike minerals and a great variety of organic compounds.

carbonation A chemical reaction in which carbon dioxide combines with a substance.

celestial meridian A meridian on the celestial sphere that passes through the zenith of a given place and terminates at the celestial poles.

celestial objects Objects that can be seen in the sky, but that are beyond the Earth's atmosphere.

celestial sphere An imaginary sphere of infinite radius around Earth's center upon whose "inner surface" all celestial objects appear to be projected. The apparent dome of the visible sky forms half of the celestial sphere.

cementation The binding together of sediment particles by a cementing material.

chaos theory The idea that small, unpre-

Glossary

dictable changes can cause large-scale changes in a system over time.

chemical sediments Particles (usually crystals) that crystallized out of solutions at or near the Earth's surface.

chemical weathering Processes that break down rock by changing its chemical composition.

cinder cone volcano A narrow, steep-sided volcano formed from the accumulation of tephra around the vent.

clastic sediments Rock or mineral fragments formed by the breakdown of rock due to weathering.

cleavage The tendency of a mineral to break along flat surfaces parallel to atomic planes in its crystalline structure.

climate The weather in a region averaged over a long period of time.

closed universe A model of the universe of fixed volume with curved boundaries. If closed, the universe will eventually contract and give rise to another big bang.

cloud base The elevation of the base of a cloud; the elevation at which condensation begins.

cloud ceiling The elevation of the base of a layer of cloud.

cloud cover The fraction of the sky obscured by clouds.

coastal zone The part of a continent that is adjacent to the ocean; it includes features such as the coast, shoreline, beaches, marshes, estuaries, lagoons, and deltas.

comet Bodies composed of meteoroids embedded in ice that revolve around the Sun in highly elliptical orbits.

comet coma The cloud of gas surrounding a comet's nucleus when ice in the nucleus sublimes as the comet approaches the Sun.

comet nucleus The solid center of a comet thought to consist of meteoroids embedded in ice.

comet tail The elongation of the cloud of gas surrounding a comet caused by collisions with particles emitted by the Sun.

compaction The reduction in volume or thickness of a sediment layer due to overlying weight or pressure.

composite cone volcano A large symmetrical volcano consisting of alternating layers of lava flows and tephra.

compound A substance that consists of two or more elements joined together in a definite proportion and have properties different from the elements that compose it.

compression Forces acting toward each other along the same line of action causing rock to be squeezed.

conchoidal fracture A tendency to break along a curved, concave surface that looks somewhat like a mussel shell.

condensation The process by which a gas changes into a liquid.

condensation nuclei Particles suspended in the air that provide a surface upon which condensation can occur.

conic section A family of figures obtained by cutting a circular cone with a plane.

conservation The act of decreasing the consumption of a natural resource, usually by more efficient use.

constellation A group of stars in an area of the sky that form imaginary patterns, supposedly in the form of an animal, person, or thing.

contact metamorphism Changes in rock that result from the extreme heat produced by contact with magma or lava.

continental drift The theory that the continents drift slowly across the Earth's surface, sometimes colliding and breaking into pieces.

continental glacier Immense masses of ice that form at high latitudes where temperatures are always cold and that spread out over the entire land surface rather than being confined to valleys.

continental rise A gently sloping feature between the continental slope and the abyssal plain.

continental shelf The shallow part of the ocean floor adjacent to the coastal zone.

Glossary

continental slope The portion of the ocean floor that slopes steeply from the continental shelf toward the abyssal plain.

continuous spectrum A spectrum that contains light of all colors.

contour interval The difference in elevation between two adjacent contour lines.

convection cell A circular pattern of motion in which warm, moist air rises, expands outward, cools, and sinks back to the surface.

convection current A fluid current set in motion by differences in density within the fluid.

convergent boundary Places where edges of adjacent plates are colliding.

coordinates Two numbers that describe the position of any point on a map in terms of the intersection of two lines.

coordinate system A system for assigning two numbers to every point on a surface.

core The innermost zone of the Earth's interior.

Coriolis effect A deflection of motion from a straight-line path due to the rotation of the Earth.

correlation The process of determining that rock layers in different areas are the same age.

cosmic background radiation A remnant of radiation left over from the original big bang that fills the universe.

cosmology The study of the universe.

crevasse A deep, nearly vertical fissure in glacial ice.

cross-bedding Layers of sediment lying at an angle to the plane in which sediment layers are normally deposited.

crystal A solid whose internal structural pattern is expressed as plane faces that can be seen with the unaided eye.

crystalline Having a definite internal structural pattern.

crystallization The process of forming crystals.

cutoff A new channel formed when a stream cuts through a narrow strip of land between adjacent meanders.

cyclone A flow of air in which motion is counterclockwise and toward a low pressure center.

decarbonation A chemical reaction in which carbon dioxide is given off.

decay product The new element formed when a radioisotope decays.

deferent In Ptolemy's geocentric model of the universe, a circle with the Earth near its center. The center of a planet's epicycle moves around the circumference of a deferent.

deflation The pick up and removal of fine sediments from the Earth's surface by wind.

deflation hollow A shallow depression formed by the removal of sediments by wind.

degassing The process by which water is released from the molecules of minerals when heated to high temperatures.

dehydration A chemical reaction in which water is given off.

delta A flat, fan-shaped mound of sediment deposited where a stream enters a body of standing water and slows down.

density The amount of matter in a given amount of space; usually expressed as the mass per unit of volume of a substance.

desertification The rapid development of deserts caused by the impact of human activities.

desert pavement A continuous layer of sediments too heavy to be moved by wind that remains after finer sediments have been removed by wind, and shields underlying materials from further deflation.

dew Droplets of water that condense on ground surfaces due to radiative cooling of the ground overnight.

dew point The temperature at which the water vapor in air will begin to condense.

differentiation The process by which

Glossary

dike A pluton that cuts across existing rock layers.

disconformity An irregular erosional surface between parallel layers of rock.

divergent boundary Places where edges of adjacent plates are spreading apart.

Doppler effect The shift in wavelength of a spectrum line from its normal position due to relative motion between the source and the observer.

drainage basin The area from which precipitation drains into a stream or system of streams.

drift A general term for all sediments transported and deposited by a glacier.

drizzle Very fine droplets of water falling slowly and close together.

drumlin Low mounds of till having an elongated teardrop shape formed when a glacier rides over a previously deposited pile of sediment.

dry bulb temperature The temperature recorded by a thermometer whose bulb is kept dry.

ductile deformation A permanent change in rock that occurs when stress exceeds the strength of a rock's internal bonds.

dune A mound of sand deposited by wind.

earthflow The slow, downhill movement of a water-saturated layer of soil and vegetation.

earthquake A sudden trembling of the ground.

eccentricity The flatness of an ellipse, expressed as the ratio of the distance between the foci of the ellipse to the length of its major axis.

ecliptic The apparent annual path of the Sun around the sky; the intersection of the plane of Earth's orbit and the celestial sphere.

elastic deformation A temporary change in which rock strains due to stress but returns to its original condition after the stress is removed.

planets develop concentric layers or zones due to density differences.

electromagnetic spectrum A continuum in which electromagnetic waves are arranged in order of wavelength.

electromagnetic wave An oscillation in the electromagnetic field surrounding a particle.

element A substance that cannot be broken down into simpler substances by ordinary chemical means.

ellipse A closed figure obtained by cutting a circular cone with a plane; its appearance is that of a flattened circle.

El Niño A current of warm water flowing southward along the coast of Ecuador, which about every seven to ten years, extends to the coast of Peru, where it interrupts the upwelling of nutrient-rich water, causing plankton and fish to die.

end moraine A pile of till that forms at the melting edge of a glacier.

eon The largest unit of geologic time.

epicenter The point on the Earth's surface directly above the focus of an earthquake.

epicycle In Ptolemy's geocentric model of the universe, the circular orbit of a planet. The center of an epicycle travels around the circumference of another circle called the deferent.

epoch The unit of geologic time into which periods are divided.

equator A circle on which all points are midway between the Earth's poles.

equatorial diameter Diameter of the Earth measured from a point on the Equator through the center of the Earth to the Equator (12,756 kilometers).

equinox Either of the two times during a year when the Sun crosses the celestial Equator and when the length of day and night are approximately equal.

era The unit of geologic time into which eons are divided.

erosion Any process that transports sediments.

erratic An isolated boulder deposited by a glacier.

escape velocity The smallest velocity a body must attain in order to escape

Glossary

from the gravitational attraction of another body.

evaporite Sediments, rocks, or minerals formed when water containing dissolved minerals evaporates.

evapotranspiration The combined processes of evaporation and transpiration, the two mechanisms by which water is returned to the atmosphere.

evolution The idea that current life-forms have developed from earlier, different life-forms.

exfoliation The scaling off, or peeling, of successive layers from the surface of rocks.

extrusive igneous rock A rock formed by the rapid cooling of lava at the surface of the Earth.

faulting A sudden movement of rock along planes of weakness in the Earth's crust called faults.

felsic Rocks rich in the minerals feldspar and silica (quartz).

field A region of space in which a measurable quantity is present at every point.

field map Represents any quantity that varies in a region of space.

floodplain A broad, flat strip of land adjacent to a stream consisting of sediments deposited when the stream is in flood.

focus The point where rock first breaks or moves in an earthquake.

focus of an ellipse One of two points along the major axis of an ellipse with the characteristic that the total distance from one focus to any point on the ellipse and back to the other focus is always the same.

fog A cloud whose base is at ground level.

fold A bend or warp in layered rock.

fossil Any remains, trace, or imprint of a living thing that has been preserved in the Earth's crust.

fossil fuel A fuel derived from the remains of once-living things; e.g., coal, petroleum, and natural gas.

Foucault pendulum A freely swinging pendulum whose path appears to change direction relative to the Earth's surface in a predictable manner due to the Earth's rotation.

fracture The tendency not to break along any particular direction.

fragmental sedimentary rock A rock composed of rock fragments cemented or compacted together in a solid mass.

frame of reference A specific point in time or space that serves as the basis of comparison for describing change; in a coordinate system, the coordinate axes, in terms of which position is specified.

frequency The number of waves that pass a given point in a unit of time.

front The boundary between two adjacent air masses.

frost Ice crystals that form on ground surfaces by sublimation due to radiative cooling of the ground overnight.

frost action The breaking down of rock by the force exerted by water expanding as it freezes into ice.

galaxy A system consisting of hundreds of billions of stars.

gas giant Planets that have thick atmospheres surrounding a small, rocky core.

geocentric model A model of the universe in which the Earth is at the center and all other objects revolve around the Earth.

geologic column The correlation of rocks worldwide into a single sequence showing their relative ages.

geologic time scale The division of the Earth's past into a sequence of time units.

geothermal gradient The rate at which temperatures within Earth rise with depth.

glacial advance The forward and downslope movement of a glacier that results when accumulation exceeds ablation.

glacial basin Depressions formed when fast-moving glaciers remove large quantities of rock material plucked from the bedrock surface.

glacial polish A bedrock surface smoothed by abrasion as a glacier flows over it.

glacial retreat A decrease in the length of a glacier that results from ablation exceeding accumulation.

glacial striations Long, narrow grooves inscribed in bedrock by rock fragments embedded in a glacier as the glacier moves over the bedrock.

glacial till Unsorted sediment deposited by a glacier.

glacier A solid mass of ice formed by the compaction of accumulated snow that flows downslope and outward under the influence of gravity.

glaze Rain that forms a layer of ice as soon as it comes into contact with below-freezing surfaces.

global wind belt A region of prevailing winds caused by a difference in air pressure between adjacent regions on a global scale.

graded bedding A layer of sediments in which the particle size changes gradually from large on the bottom to small on the top.

gradient The rate at which a field changes; usually measured in terms of change in field value per unit of distance.

gravitational differentiation The process by which the material from which the solar system formed became separated into layers according to density, because atoms of lighter elements do not exert as strong a gravitational attraction as atoms of heavier elements.

gravity The force of attraction that exists between all matter.

great circle A circle that bisects the Earth.

greenhouse effect The process by which solar radiation that is re-radiated at a longer wavelength is absorbed by carbon dioxide, water vapor, and other gases in the atmosphere rather than escaping into space.

ground moraine A thin, widespread layer of till deposited beneath a glacier.

groundmass A mass of fine mineral crystals surrounding the large, isolated phenocrysts in a porphyry.

gully A long, narrow trough eroded in soil by running water.

hail Balls of ice with an internal structure of concentric layers of ice and snow formed when ice pellets fall through turbulent air and are alternately carried through layers of air that are above and below freezing.

half-life The time required for half of a mass of radioactive atoms to decay.

hanging valley A valley of a tributary glacier whose floor is above, or hanging over, the floor of the main glacier that it feeds.

hardness A mineral's ability to resist being scratched.

heliocentric model A model of the universe in which the Sun is at the center of our solar system and the planets revolve around the Sun.

Hertzsprung-Russell Diagram A plot of the luminosity versus spectral type for a group of stars; also called the H-R Diagram.

horizon The visible line of demarcation between land or sea and sky.

horizontal sorting The gradual change in the size of sediments deposited horizontally along the bed of a stream as the water slows down.

humidity The amount of water vapor in the air.

humus The dark, highly decomposed organic matter in soil.

hurricane Large, cyclonic storms in which the winds exceed 119 kilometers per hour.

hydration A chemical reaction in which water combines with a substance.

hydrogen bonding A weak bond between adjacent water molecules due to the attraction between the slightly positive end of one water molecule and the

Glossary

slightly negative end of another water molecule.

hydrolysis A reaction in which water separates into hydrogen ions and hydroxide ions that replace ions in a mineral.

hydrosphere All of the water that rests on the lithosphere; it includes all oceans, lakes, streams, groundwater, and ice.

ice age A period of time during which conditions exist that cause ice sheets more than 1 million square kilometers in area and thousands of meters thick to develop on nonpolar continents.

iceberg A mass of floating ice that detached from a glacier into a body of water.

igneous intrusion A mass of magma that penetrates an opening in an existing rock and solidifies.

igneous rock Rock formed by the cooling and crystallization of molten minerals (magma or lava).

immature soil A soil in which boundaries between forming horizons are indistinct.

impact crater A circular depression formed on a surface by the impact of a projectile, such as a meteorite or a comet.

in solution Sediments that are transported as a consequence of being dissolved in the water flowing in a stream.

index fossil A fossil of an organism that had distinctive body features and was abundant over a wide geographical area, but only existed for a short period of time.

infiltrate The process by which water seeps into the ground and through interconnected pore spaces.

inorganic Non-living and not derived from a living thing.

insolation Solar radiation that reaches the surface of the Earth; short for incoming solar radiation.

interglacial Time periods during an ice age when glaciers retreat or even disappear.

international date line The 180° meridian; it marks the transition from one date to another.

intrusive igneous rock A rock formed by the slow cooling of magma within the earth.

ion An atom or molecule in which the electrical charges are unbalanced resulting in a net positive or negative electrical charge.

isobar A line connecting points of equal air pressure.

isoline Lines connecting points of equal field value.

isostasy The idea that the Earth's crust floats on hot fluid rock in the mantle.

isosurface Surfaces on which every point has the same field value.

isotherm A line connecting points of equal temperature on a field map.

isotope Varieties of the same element that differ slightly in mass.

joint A fracture in a rock along which movement has not occurred.

kettle A depression formed when a block of ice buried in till melts.

key bed Well-defined, easily identifiable rock layers that have distinctive characteristics or fossil content that allow(s) them to be used in correlation.

Kuiper Belt A thick belt of debris left over from the formation of the outer planets, extending from Neptune's orbit (29 AU) to the outer edge of Pluto's orbit (50 AU). Considered the probable source of short-period comets, over 60 objects with diameters over 100 kilometers have been detected orbiting the Sun within this region. Pluto and its moon Charon are considered Kuiper Belt objects.

laccolith A pluton formed when magma intrudes between rock layers and pushes upward and solidifies creating

Glossary

a rock structure that has a flat bottom and an arched top.

lagoon The sheltered body of water between a barrier bar and the mainland.

land breeze A local wind blowing from land toward an adjacent body of water because nighttime cooling increases the air pressure over the land.

landform Physical features of Earth's surface that have a characteristic shape and are produced by natural processes.

landscape A region in which the landforms are related by their structure, the processes that formed them, and the way in which they developed.

landscape region A grouping of landscapes with similar relief, stream patterns and soil associations.

latent heat of fusion The amount of heat energy given off or absorbed in causing a change between the liquid and solid states that does not result in a change in temperature.

latent heat of vaporization The amount of heat energy given off or absorbed in causing a change between the liquid and gaseous states that does not result in a change in temperature.

lateral moraine A mound of till deposited along the sides of a glacier.

latitude The angular distance north or south of the equator.

Laurentide ice sheet The glacier that advanced several times across the northern United States during the Pleistocene epoch.

lava Magma (molten rock) that has emerged onto the Earth's surface.

lava plateau A flat sheet of igneous rock formed when a very fluid lava erupts from a long fissure instead of from a central vent.

leeward The side of a feature facing away from the wind.

lens A transparent object of regular form that changes the direction of travel of light waves that pass through it.

levee A low, thick, ridgelike deposit along the banks of a stream consisting of coarse sediments deposited when a stream overflows its banks.

light year The distance light travels in one year; roughly 9.5 trillion kilometers, or 6 trillion miles.

lithification The process of converting sediments into coherent, solid rock.

lithosphere Earth's solid outer layer of soil and solid brittle rock extending from the surface to a depth of about 100 kilometers; it includes the crust and the upper mantle.

loess A fine, buff-colored sediment deposited by wind.

longitude The angular distance east or west of the prime meridian.

longitudinal waves Waves in which the medium oscillates parallel to the direction of wave motion.

longshore current An ocean current flowing parallel to a shore caused by the approach of waves at an angle to the coast.

luminosity The total amount of energy radiated by a star in 1 second.

lunar eclipse An event in which the Moon passes into the Earth's shadow during the full moon phase.

luster The way light reflects from the surface of a mineral.

mafic Rocks rich in minerals containing magnesium and Fe (iron).

magma Molten rock beneath the Earth's surface.

magma chamber A pocket of molten magma in the crust or upper mantle.

magnetosphere The region surrounding Earth that is influenced by its magnetic field

magnitude scale The astronomical system for measuring the brightness of a star; the higher the magnitude, the fainter the star.

main sequence A diagonal line across the H-R diagram on which the majority of stars lie; the main phase of the life of a star.

mantle The zone of the Earth's interior

Glossary

between the crust and the core.
map A model of the Earth's surface.
map scale The ratio between distance on a map and actual distance on the ground; usually expressed as a ratio.
mass The amount of matter in an object.
mass wasting The downhill movement of sediments under the direct influence of gravity.
mature soil A soil in which distinct horizons have formed.
meander A curving loop in a stream channel.
meridian Semicircles on which all points have the same longitude.
mesopause The boundary between the mesosphere and the thermosphere.
mesosphere The layer of the atmosphere ranging from 30 to 50 kilometers; temperatures in this layer decrease with altitude.
metamorphic rock Rocks that form as a result of physical and chemical changes in existing rocks.
metamorphism The processes that change the physical and chemical properties of existing rocks.
meteor The streak of light seen in the sky as a meteoroid is vaporized by frictional heating as it falls through the atmosphere.
meteorite The portion of a meteoroid that survives to strike the ground.
meteoroid A chunk of matter moving through space.
mid-ocean ridge A chain of undersea mountains running through the center of an ocean.
Milky Way galaxy The galaxy in which our solar system is located.
mineral A naturally occurring, inorganic, crystalline solid with a composition that can be expressed by a chemical symbol or a formula.
mineral composition The minerals present in a rock.
mixture Consisting of one or more substances that retain their characteristic properties and can be separated by relatively simple means.

modified Mercalli scale A system that measures the strength of an earthquake based upon perception of motion by human observers and damage to structures built by humans.
Moho The boundary between the dense rock of the mantle and the less dense rock of the crust. It was named for the seismologist Andrija Mohorovicic, who recognized its existence after analyzing seismic wave behavior inside the Earth.
Mohorovicic discontinuity The boundary between the Earth's crust and mantle. It marks the interface between rocks within the Earth of two different average densities.
monocline A fold in which only one side of the fold is inclined.
moraine A deposit of unsorted sediments deposited directly from glacial ice.
mountain Any part of Earth's crust that projects at least 300 m (1,000 ft) above the surrounding land, has a limited summit area (as opposed to a plateau), steep sides, and considerable bare-rock surface.
mudflow The rapid downhill flow of a mixture of rock fragments, soil, and water.

natural selection The process by which organisms that are better suited to their environment survive and reproduce while poorly suited organisms do not.
neap tide A tidal cycle in which there is very little difference between the water level at high and low tides; caused by the canceling out effect of the Moon's gravity acting at right angles to the Sun's.
nebula A cloud of interstellar matter.
neutrino A fundamental particle produced in certain nuclear reactions that has no charge and no mass, but which does have momentum.
neutron star An incredibly dense (10^{16}-10^{18} kg/m^3) star whose interior is composed mostly of neutrons
nonconformity A place where sedimen-

Glossary

tary rock lies directly on another type of rock, such as igneous or metamorphic rock.

nuclear fusion A nuclear reaction in which the nuclei of atoms combine; during fusion, some of the mass of the combining nuclei is converted into energy.

numerical forecasting Weather forecasting based upon mathematical models of the physical behavior of the atmosphere.

oasis A place where deflation has lowered the surface in a desert to the water table allowing plants to take root and grow.

oblate spheroid A slightly flattened sphere; the Earth's shape.

ocean-floor spreading The idea that the ocean floors were spreading sideways away from the mid-ocean ridges.

Oort Cloud A vast shell of icy objects (i.e., comets) surrounding the solar system between 50,000 and 100,000 AU; thought to be the source of long-period comets.

open universe A model of the universe with infinite volume and no boundaries. If open, the universe will continue expanding forever.

orbit The path along which a satellite revolves around a primary.

orbital velocity The smallest velocity a body must attain in order to enter a stable orbit around another body.

ordinary well A hole dug in the ground so that it penetrates the water table in which water accumulates and can be pumped to the surface.

organic sediments Particles produced by the life activities of plants or animals.

origin time The time at which an earthquake occurred.

original horizontality, law of The observation that sediments deposited in water form flat level layers.

orographic effect The net increase in temperature of a parcel of air that is forced over a mountain range due to adiabatic cooling on the windward side followed by adiabatic heating on the leeward side.

outgassing The release of gases and water vapor from molten rocks, leading to the formation of Earth's atmosphere and oceans.

outwash plain A flat plain next to the leading edge of a glacier formed by the deposition of sediments from meltwater streams.

oxbow lake A crescent-shaped body of water formed when a cutoff isolates the water in a meander from the main channel of the stream.

oxidation A chemical reaction in which a substance combines with oxygen.

oxidizing atmosphere An atmosphere that contains enough oxygen to cause elements and compounds to oxidize, but not enough to support aerobic life.

ozone A gas whose molecules consist of three atoms of oxygen joined together.

P-waves Longitudinal seismic waves produced by an earthquake that travel the fastest and are therefore the first to be recorded by a seismograph.

Pacific ring of fire A narrow zone of active volcanoes surrounding the Pacific Ocean.

paleomagnetism A record of the orientation of the Earth's magnetic field preserved in the crystalline structure of igneous rock containing magnetic minerals.

parallel Circles on which all points have the same latitude north or south of the Equator.

parsec The distance at which a star's parallax would be 1 second of arc. 1 parsec (pc) = 206,265 AU = 3.26 ly.

parting The tendency to break along flat surfaces that follow structural weaknesses.

path of totality The path of the Moon's umbra over the surface of the Earth during a solar eclipse.

pedalfer Soils rich in aluminum and iron compounds.

pedocal Soils rich in calcium compounds.

penumbra The portion of a shadow in which only part of the light has been blocked, so the light is dimmed but not totally absent.

period The unit of geologic time into which eras are divided.

period of revolution The length of time it takes for a satellite to complete one revolution around its primary.

permeability The rate at which water can infiltrate a material.

phases of matter The form in which matter may exist; solid, liquid, and gas.

phases of the Moon The changing shape of the illuminated portion of the Moon that can be seen by an observer on Earth.

phenocryst Isolated large crystals in a rock.

photosynthesis The process by which cells convert water and carbon dioxide into carbohydrates and oxygen using light as an energy source.

physical weathering Processes that break down rock without changing their chemical composition.

physiographic province A large-scale region in which all parts are similar in geologic structure and climate, and thus has had a unified history of landform development.

phytoplankton Plankton that carry out photosynthesis.

plain A flat area at low elevation.

planet A body that is at least partly solid, orbits the Sun, and emits mainly reflected sunlight.

planetesimal One of the small bodies from which planets formed.

plant action The breaking down of rock by the force exerted by the growth of plant roots and stems or by chemicals produced as a by-product of plant growth.

plateau A large, flat region elevated more than 150–300 m (500–1000 ft) above the surrounding land or above sea level.

plate tectonics The theory that the Earth's crust consists of large, rigid slabs of rock, or plates, resting on a fluidlike layer of denser rock and that these plates slide around causing their edges to collide, separate, and slide past each other.

Pleistocene The time period, which began 1.6 million years ago, during which climates grew colder worldwide.

plucking The tearing loose of rock frozen to the base of a glacier as the glacier moves along.

pluton A rock structure formed by the solidification of an igneous intrusion.

polar diameter Diameter of the Earth measured from a pole through the center of the Earth to the opposite pole (12,714 kilometers).

polar front A boundary between polar and tropical air that forms in the prevailing westerly global wind belts.

Polaris A star whose position is almost directly over the Earth's north geographic pole.

pollutant The particular materials or forms of energy that are harmful to humans.

pollution The concentration of any material or energy form that is harmful to humans.

porosity The percentage of empty space in a material.

porphyry A rock consisting of large crystals called phenocrysts embedded in a mass of fine mineral grains.

potential evapotranspiration The maximum amount of water that could be returned to the atmosphere by evapotranspiration under a given set of conditions.

precession The very slow change in the orientation of a rotating body's axis of rotation.

precipitate A solid that crystallizes out of a solution and settles to the bottom of the solution.

precipitation Condensed moisture that falls to the ground.

Glossary

pressure unloading The release of stress on rock as the stress of the weight of overlying rock is removed as the latter is worn away, or the removal of the stress as the weight of the ice in a glacier is removed as the glacier melts away.

primary The body a satellite orbits.

prime meridian The zero meridian, or starting point, for measuring the angular distance east or west of any other meridian.

profile What a cross section of the land between two points would look like viewed from the side.

protostar A collapsing cloud of dust and gas destined to become a star.

radiation curve A plot of intensity versus wavelength of the radiation emitted by a source, e.g. a star. The resulting graph illustrates the rate at which a source is emitting energy at different wavelengths.

radiative balance The condition in which the Earth's average level of heat energy remains constant over a long period of time.

radioactive decay The process by which unstable isotopes break apart spontaneously giving off energy, subatomic particles, or both.

radioisotope Isotopes that are unstable and undergo radioactive decay.

recharge Water that seeps into the ground and replaces soil moisture that was lost during a period of drought.

red giant A large red star that has consumed most of its hydrogen and developed a helium core; heat from contraction of the helium core triggered fusion in the outer shell of hydrogen, causing the star to expand to hundreds of times its initial size, and its outer surface to cool.

reducing atmosphere An atmosphere with so little oxygen that elements and compounds do not oxidize.

refraction The bending of light as it passes from one medium to another.

regional metamorphism Changes in rock over an extensive area that occur due to the pressure and high temperatures associated with either deep burial or movements of the Earth's crust.

relative age The age of an object or event relative to another object or event.

relative humidity The ratio of water vapor in the air to the maximum amount of water vapor the air can hold at that temperature.

relief The differences in the height of landforms in an area.

renewable resource A material that can be renewed, or restored to its original form after being used, by means of a natural process; i.e., pure water that is polluted by human use is purified by the natural processes of the water cycle.

replacement reaction A chemical reaction in which one element replaces another element in a molecule.

residual soil A soil formed by the weathering of bedrock in place.

respiration The oxidation of food within living cells by which the chemical energy of food molecules is released in a series of metabolic steps involving the consumption of oxygen and the liberation of carbon dioxide and water.

retrograde motion An apparent motion of a planet in a direction opposite to its normal motion.

revolution The motion of one body around another body.

Richter scale A system that measures the strength of an earthquake by the amount of motion of the Earth's crust that occurs during the earthquake.

rill Tiny grooves eroded in soil by a trickling stream of water flowing downhill.

rock The naturally formed, solid material that makes up the Earth's crust.

rock cycle A sequence of events that shows how rocks are formed, changed, destroyed, and reformed from the same parent material.

rock-forming minerals The one hundred or so common minerals that make up more than 95 percent of the rock in the

Glossary

Earth's crust.

rockfall Mass wasting process in which rock fragments broken loose by weathering fall from cliffs or bounce by leaps down steep slopes.

rockslide Mass wasting process in which rock fragments slide downhill.

rotation Motion in which every part of an object is moving in a circular path around a central line called the axis of rotation.

runoff Runoff is precipitation that does not evaporate or sink into the ground, but runs downhill along the Earth's surface.

S-waves Transverse seismic waves produced by an earthquake that travel slower than *P*-waves and are therefore the second to be recorded by a seismograph.

salinity The total amount of dissolved matter in seawater expressed in parts per thousand.

saltation A type of motion in which sediments transported in a stream move along the stream bed in a series of bounces, hops, or leaps.

sandbar A long, narrow pile of sand deposited in open water.

satellite A solid body that orbits another body; i.e., the Moon is a satellite of the Earth.

sea breeze A local wind blowing from a body of water toward land that develops because daytime heating lowers the air pressure over the land.

sea cliff A steep cliff formed by the undercutting of steep slopes by waves and subsequent collapse of the overhanging material.

sea stack An isolated pillar of resistant rock that was detached from a headland by wave erosion.

sea terrace A flat platform of sediments deposited by wave action on the ocean floor adjacent to a steep coastline.

sea waves Wind-generated sea waves that are directly affected by the wind.

sediment Solid particles that have been transported and then deposited by agents of erosion.

sedimentary rock Rock formed by the lithification of sediments.

seismic wave A vibration of the Earth's crust produced by faulting.

seismogram A line recorded on paper by a seismograph that represents the motion during an earthquake.

seismograph A device that detects the motion of the Earth's crust during an earthquake.

seismology The study of earthquakes and their effects.

shadow zone A region in which no seismic waves from an earthquake are received by seismograph stations because the waves have been refracted or blocked as they traveled through the Earth.

shear Forces acting along different lines of action causing rock to twist or tear.

shield volcano A wide, gently sloping volcano formed from successive layers of solidified lava flows.

sidereal day The time from when a star crosses an observer's meridian until the same star next crosses the observer's meridian.

sidereal month The time it takes the Moon to complete one 360° orbit around Earth, measured relative to a distant star; 27.32166 mean solar days, or 27 days, 7 hours, forty-three minutes and 11 seconds.

silicates Minerals composed of compounds of silicon and oxygen together with other elements; the most abundant mineral group on Earth.

silicon tetrahedron An ion of silicon joined with four ions of oxygen in the shape of a tiny tetrahedron (pyramid-like shape).

sill A pluton that forms between, and parallel to, existing rock layers.

sleet Clear pellets of ice that form when raindrops freeze as they fall through layers of air at below-freezing temperatures.

slump Mass wasting process in which a mass of bedrock or soil slides down-

Glossary

ward along a curved plane of weakness.

snow Hexagonal crystals of ice or needlelike ice crystals formed by the sublimation of water vapor in the atmosphere.

soil The accumulation of loose, weathered material that covers much of the land surface of the Earth.

soil association A group of two or more soils occurring together in a characteristic pattern in a given geographical area.

soil creep The invisibly slow, downhill movement of soil.

soil horizon Recognizable layers in soil with characteristic properties.

soil moisture storage Water that is retained in the soil after water seeps into the ground.

soil moisture zone A moist layer of soil in which water clinging to soil particles supplies plant roots with water.

soil profile A cross section of soil from surface to bedrock.

solar day The time it takes Earth to rotate from one solar noon until the next solar noon.

solar eclipse An event in which the Moon passes directly between the Earth and the Sun, casting a shadow on the Earth and partly or completely blocking an observer's view of the Sun.

solar noon The moment when the Sun is at its highest point in the sky for the day.

solar wind The flow of gas from the Sun out through the solar system; near Earth the solar wind travels at speeds of 600–1000 km/s.

solstice The two times of the year when the Sun's altitude at noon reaches a maximum or a minimum. The summer solstice in the Northern Hemisphere occurs about June 21, when the Sun is in the zenith at the Tropic of Cancer; the winter solstice occurs about December 21, when the Sun is over the Tropic of Capricorn. The summer solstice is the longest day of the year and the winter solstice is the shortest.

source region The geographical region in which an air mass originated.

specific gravity The density of a mineral compared to the density of water.

specific heat capacity The number of degrees the temperature of one gram of a material will increase if one calorie of heat is added to it.

spectrograph An instrument for making a photographic record of a spectrum.

spectrum The band of colors formed when a beam of white light is passed through a prism; each wavelength of light in the beam is refracted at a slightly different angle, causing the light to be arrayed in order of its constituent wavelengths.

spit A sandbar attached at one end to the mainland.

spring A place where groundwater seeps out of the ground.

spring tide A tidal cycle in which there are very high and very low tides; caused by the additive effect of the Sun and Moon's gravity acting together along the same line of action.

station model A shorthand symbolic representation of the weather conditions at a particular location.

statistical forecasting Weather forecasting based upon the statistical analysis of historical weather data.

stellar parallax The change in the position of a near object relative to distant objects as an observer's position changes.

stock An irregularly shaped pluton covering an area less than 75 square kilometers.

strain The change in size or shape of a rock due to stress.

stratopause The boundary between the stratosphere and the mesosphere.

stratosphere The layer of the atmosphere ranging from about 6 to 30 kilometers above the Earth's surface; it contains ozone that absorbs ultraviolet light; temperatures in this layer rise

Glossary

with altitude.

streak The color of a powdered mineral; created by rubbing a mineral against an unglazed piece of porcelain.

stream A body of water that moves under the influence of gravity to progressively lower levels in a narrow, but clearly defined channel.

stream banks The sides of the channel containing the water of a stream.

stream bed The bottom of the channel containing the water of a stream.

stream channel Clearly defined path along which the water in a stream flows.

stream gradient The angle between the stream bed and the horizontal; the slope of the stream.

stream load The sediments transported by the water flowing in a stream.

stream mouth The site at which a stream flows into a standing body of water.

stream pattern The pattern formed by the system of streams in an area as they flow down slopes and join with other streams, e.g. dendritic, trellis, radial, rectangular, annular.

stream source The site at which the water in a stream originates.

stress Forces acting on a rock.

subduction The sliding of a denser ocean plate beneath a less dense continental plate resulting in the melting of the ocean plate as it plunges into the hot mantle.

sublimation The process by which a solid changes directly into a gas without passing through a liquid stage.

supercluster A cluster of galaxy clusters.

supergiant star An exceptionally luminous star that is 10 to 1000 times the Sun's diameter.

supernova A rare stellar explosion in which the entire outer envelope of a star is blown away in a violent outburst, leaving behind a dense core.

superposition The idea that the bottom layer of a sedimentary series is the oldest, unless it was overturned or had older rock thrust over it, because the bottom layer was deposited first.

surf Waves that are breaking along a shoreline.

suspended load The sediments carried in suspension in the water flowing in a stream.

swells Long wavelength sea waves that settle into a uniform pattern once they leave the area of turbulent winds that formed them.

syncline A downward, valleylike fold.

synodic month The time between successive new moon phases; 29.530588 mean solar days, or 29 days, 12 hours, 44 minutes and three seconds.

synoptic forecasting Weather forecasting based upon analysis of a sequence of synoptic weather maps and charts.

synoptic weather map A map that summarizes weather data from many locations using station models.

system A group of things or processes that interact to perform some function.

talus A mound of rock fragments that accumulates at the base of a rock cliff that is weathering.

technology The application of knowledge to do things.

telescope An instrument for collecting electromagnetic radiation and producing magnified images of distant objects.

temperature A specific degree of hotness or coldness as indicated on or referred to a standard scale.

tension Forces acting away from each other along the same line of action causing rock to stretch.

tephra A collective name for particles formed when globs of lava ejected during an eruption cool and solidify as they fall to the ground.

terminal moraine A mound of till deposited along the leading edge of a glacier.

terrigenous Sediments derived from the land or a continent.

texture The size, shape, and arrangement of mineral crystals or grains in a rock.

Glossary

thermocline A layer of seawater in which temperature drops sharply with depth.

thermosphere The uppermost layer of the atmosphere ranging from about 50 to 75 kilometers in altitude where it grades into space; this layer contains the ionosphere, which consists of charged particles or ions that deflect harmful radiation and reflect radio waves.

thunderstorm A vigorous, turbulent convection cell in the atmosphere resulting in a transient, violent storm of thunder and lightning, often accompanied by rain and sometimes hail.

time zone A region in which the same time is used throughout, instead of the local time at each place within it.

topographic map A field map in which the field value is elevation above sea level; it shows the three-dimensional shape of the Earth's surface in two dimensions.

tornado A rotating column of air whirling at speeds of up to 800 kilometers per hour, usually accompanied by a funnel-shaped downward extension of a cumulonimbus cloud.

transform boundary Places where edges of plates are sliding laterally past each other.

transpiration The process by which plants release water into the atmosphere through their leaves.

transported soil A soil that formed from material that was deposited in a region after being transported from another place.

transverse waves Waves in which the medium oscillates perpendicular to the direction of wave motion.

travel time The time it takes for a seismic wave to travel from the epicenter to a seismograph.

trench A deep crevice in the ocean floor produced by downward warping of the ocean floor where plates collide.

tropopause The boundary between the troposphere and the stratosphere.

troposphere The layer of the atmosphere ranging from the Earth's surface to about 6 kilometers in altitude; it contains most of the air in the atmosphere; temperatures in this layer decrease with altitude.

tsunami Waves caused by undersea movements of the Earth's crust such as slumping, earthquakes, or volcanic eruptions.

umbra The portion of a shadow in which all light has been blocked.

unconformity A break or gap in the sequence of a series of rock layers.

uniformitarianism The idea that the Earth's features have formed gradually by natural processes that are still occurring, not by instantaneous creation or by catastrophic events.

usage Moisture lost from the soil during a period of drought.

valley glacier A glacier that forms between mountain slopes at high elevations.

velocity The speed and direction in which something is moving.

vent The central opening of a volcano out of which lava erupts.

ventifact A rock particle that has developed a flat side facing the prevailing winds due to abrasion of its exposed surface.

vertical sorting The separation and deposition of sediments in successive layers caused by differences in the settling rate of the sediment particles; particles that settle fastest form the bottom-most layer, whereas those that settle most slowly form the top layer.

vesicular A rock texture characterized by cavities formed by gas bubbles escaping from a lava as it cools and solidifies.

visibility The horizontal distance through which the eye can distinguish objects.

volcanic glass A rock formed by cooling that was so rapid that no crystalline structure was able to form; such rocks

Glossary

volcanism have a glassy appearance.
volcanism The processes by which magma rises into the crust and is extruded onto the Earth's surface.
volcano The opening in the Earth's crust through which lava emerges and the moundlike structure formed by the accumulation of lava.

walking the outcrop Physically following layers of rock from one place to another in order to correlate them.
water cycle The process by which the Earth's water cycles in and out of the atmosphere.
watershed The area drained by a stream and its tributaries.
water table The top of the subsurface zone in which the pore spaces in a soil are saturated with water.
wave cyclone Warm and cold fronts that move in a counterclockwise direction around a low pressure center.
wavelength The distance between two successive wave crests or troughs.
weather The present condition of the atmosphere at any location.
weathering The breakdown of rocks into smaller particles by natural processes.

weight The gravitational force exerted on an object by the Earth.
Wentworth Scale A scheme for classifying and naming sediment particles according to their size.
wet-bulb temperature The temperature recorded by a thermometer whose bulb is kept wet; it is usually lower than dry-bulb temperature due to evaporative cooling.
white dwarf A star, below the main sequence in the H-R diagram, which is small and dense; a dying star that has collapsed to about Earth's size and is slowly cooling off.
wind Air that is moving horizontally.
windward The side of a feature facing the wind.

zenith A point directly overhead to an observer.
zone of aeration The subsurface zone in which the pore spaces in a soil are filled mainly with air.
zone of saturation The subsurface zone in which the pore spaces in a soil are saturated with water.

Examination June 2003
Physical Setting/Earth Science

PART A

Answer all questions in this part.

Directions (1–35): For *each* statement or question, write in the spaces provided the *number* of the word or expression that, of those given, best completes the statement or answers the question. Some questions may require the use of the *Earth Science Reference Tables*.

1 The planetary winds in Earth's Northern Hemisphere generally curve to the right due to Earth's

　(1) orbit around the Sun
　(2) spin on its axis
　(3) magnetic field
　(4) force of gravity　　　　　　　　　　　　1____

2 The redshift of light from distant galaxies provides evidence that the universe is

　(1) shrinking, only
　(2) expanding, only
　(3) shrinking and expanding in a cyclic pattern
　(4) remaining the same size　　　　　　　　2____

3 Which of these characteristics identify an Earth surface that is likely to be the best absorber of insolation?
   (1) light colored and smooth
   (2) light colored and rough
   (3) dark colored and smooth
   (4) dark colored and rough

4 Which phase change requires water to gain 540 calories per gram?
   (1) solid ice melting
   (2) liquid water freezing
   (3) liquid water vaporizing
   (4) water vapor condensing

5 The diagram below shows the positions of the Moon and the Sun at sunset during an evening in New York State. Points A, B, C, and D represent positions along the western horizon.

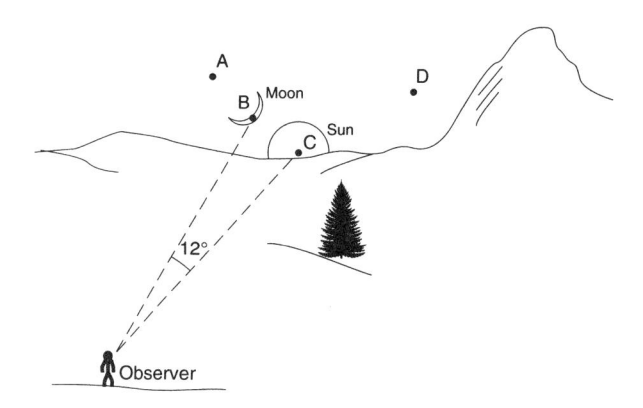

At sunset on the following evening, the Moon will be located at position
   (1) A
   (2) B
   (3) C
   (4) D

6 Which diagram best illustrates how air rising over a mountain produces precipitation?

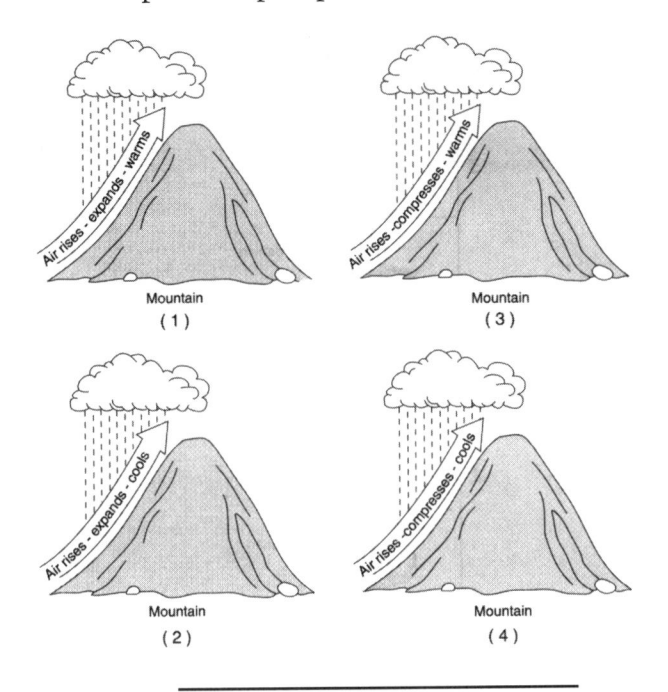

(1) Air rises - expands - warms — Mountain
(3) Air rises - compresses - warms — Mountain
(2) Air rises - expands - cools — Mountain
(4) Air rises - compresses - cools — Mountain

6_____

7 A student used a sling psychrometer to measure the humidity of the air. If the relative humidity was 65% and the dry-bulb temperature was 10°C, what was the wet-bulb temperature?

(1) 5°C
(2) 7°C
(3) 3°C
(4) 10°C

7_____

8 A gradual increase in atmospheric carbon dioxide would warm Earth's atmosphere because carbon dioxide is a

(1) poor reflector of ultraviolet radiation
(2) good reflector of ultraviolet radiation
(3) poor absorber of infrared radiation
(4) good absorber of infrared radiation   8_____

9 Why are the beaches that are located on the southern shore of Long Island often considerably cooler than nearby inland locations on hot summer afternoons?

(1) A land breeze develops due to the lower specific heat of water and the higher specific heat of land.
(2) A sea breeze develops due to the higher specific heat of water and the lower specific heat of land.
(3) The beaches are closer to the Equator than the inland locations are.
(4) The beaches are farther from the Equator than the inland locations are.   9_____

Base your answers to questions 10 and 11 on the chart below, which shows the geologic ages of some well-known fossils.

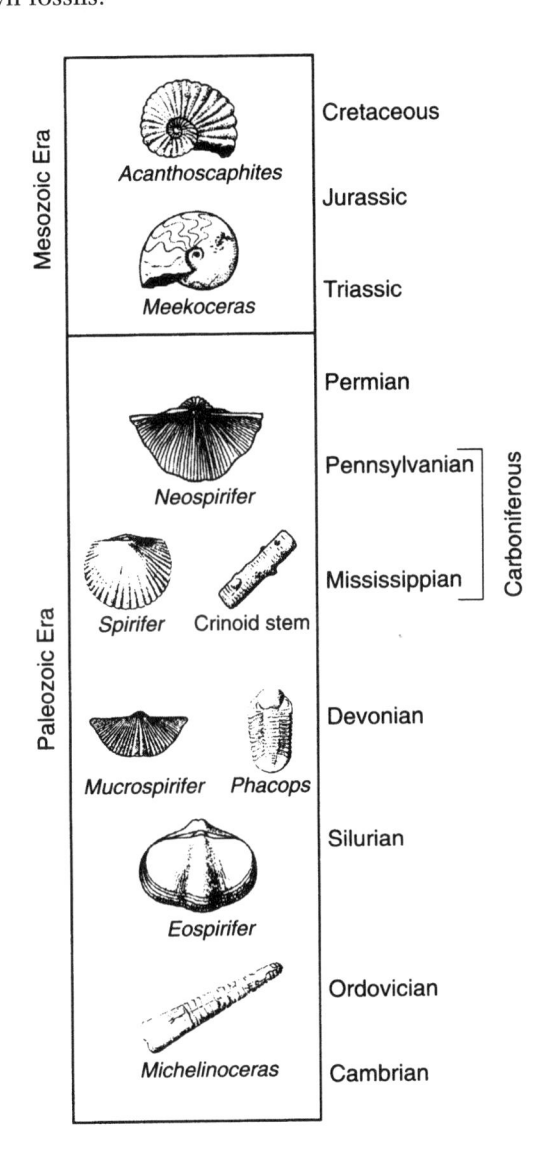

Era	Fossil	Period
Mesozoic Era	Acanthoscaphites	Cretaceous
		Jurassic
	Meekoceras	Triassic
Paleozoic Era		Permian
	Neospirifer	Pennsylvanian ⎤
		⎥ Carboniferous
	Spirifer Crinoid stem	Mississippian ⎦
	Mucrospirifer Phacops	Devonian
	Eospirifer	Silurian
		Ordovician
	Michelinoceras	Cambrian

10 The *Spirifer*, Crinoid stem, and *Neospirifer* fossils might be found in some of the surface bedrock of which New York State landscape region?

(1) the Allegheny Plateau southeast of Jamestown
(2) the Catskills near Slide Mountain
(3) the Adirondack Mountains near Mt. Marcy
(4) the Erie-Ontario Lowlands northeast of Niagara Falls   10____

11 Which New York State fossil is found in rocks of the same period of geologic history as *Meekoceras*?

(1) Condor   (3) *Eurypterus*
(2) Placoderm fish   (4) *Coelophysis*   11____

12 The flowchart below shows part of Earth's water cycle. The question marks indicate a part of the flowchart that has been deliberately left blank.

Precipitation → Runoff → Ocean → ??? → Water vapor

Which process should be shown in place of the question marks to best complete the flowchart?

(1) condensation   (3) evaporation
(2) deposition   (4) infiltration   12____

13 Which weather-station model shows an air pressure of 993.4 millibars?

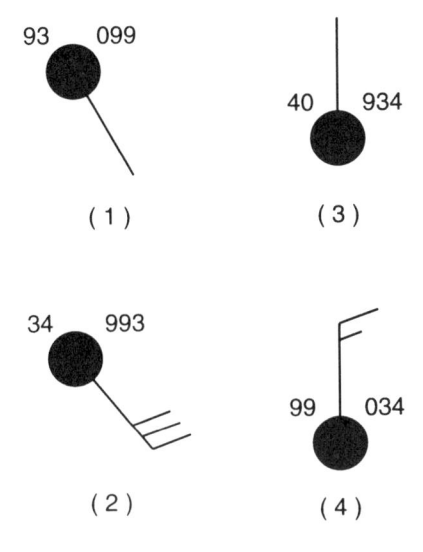

(1) 93 / 099

(3) 40 / 934

(2) 34 / 993

(4) 99 / 034

13____

14 An Earth science student observed the following weather conditions in Albany, New York, for 2 days: The first day was warm and humid with southerly winds. The second day, the temperature was 15 degrees cooler, the relative humidity had decreased, and wind direction was northwest. Which type of air mass most likely had moved into the area on the second day?

(1) continental tropical
(2) continental polar
(3) maritime tropical
(4) maritime polar

14____

15 A sample of wood found in an ancient tomb contains 25% of its original carbon-14. The age of this wood sample is approximately

(1) 2,800 years
(2) 5,700 years
(3) 11,400 years
(4) 17,100 years

15____

16 Which set of conditions would produce the most runoff of precipitation?

(1) gentle slope and permeable surface
(2) gentle slope and impermeable surface
(3) steep slope and permeable surface
(4) steep slope and impermeable surface

17 Which map view best shows the movement of surface air around a low-pressure system in the Northern Hemisphere?

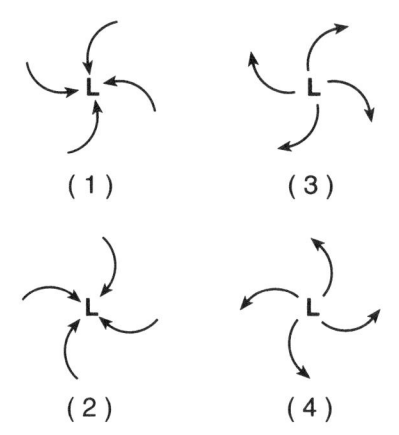

18 The surface bedrock of a region of eastern New York State is shale. Which statement best explains why the soil that covers the shale in this region contains abundant garnet and gneiss pebbles?

(1) Volcanic lava flowed over the shale bedrock.
(2) A meteor impact scattered garnet and gneiss pebbles over the area.
(3) The soil consists of rock materials transported to this region by agents of erosion.
(4) The soil formed from the chemical and physical weathering of shale.

Base your answers to questions 19 and 20 on the satellite image below, which shows cloud patterns associated with weather fronts over the United States on a certain day. The states of Nebraska (NE) and New York (NY) have been labeled.

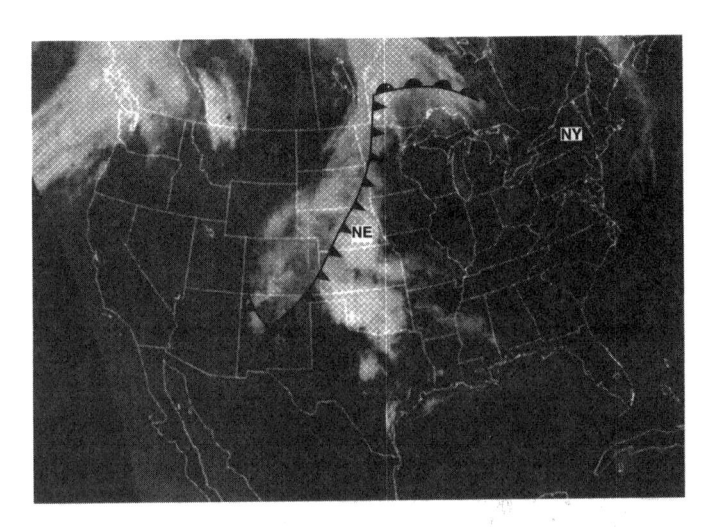

19 At the time this satellite image was taken, what were the weather conditions in New York State?

(1) clear skies with no precipitation
(2) mostly cloudy in the northern part of the State and clear in the southern part
(3) cloudy with heavy precipitation
(4) very cloudy with no precipitation   19_____

20 Which type of front was producing the weather in Nebraska when this image was taken?

(1) cold front   (3) stationary front
(3) warm front   (4) occluded front   20_____

Examination June 2003

Base your answers to questions 21 and 22 on the graph below, which shows the changes in relative humidity and air temperature during a spring day in Washington, D.C.

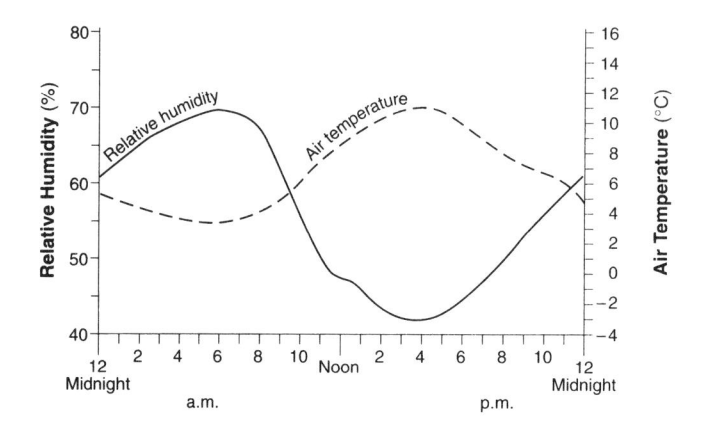

21 Which statement best describes the relationship between relative humidity and air temperature as shown by the graph?

(1) Relative humidity decreases as air temperature decreases.
(2) Relative humidity decreases as air temperature increases.
(3) Relative humidity increases as air temperature increases.
(4) Relative humidity remains the same as air temperature decreases.   21_____

22 What were the relative humidity and air temperature at noon on this day?

(1) 47% and 32°F   (3) 47% and 48°F
(2) 65% and 32°F   (4) 65% and 48°F   22_____

23 Landscapes will undergo the most chemical weathering if the climate is

(1) cool and dry
(3) warm and dry
(3) cool and wet
(4) warm and wet

23 _____

24 A huge undersea earthquake off the Alaskan coastline could produce a

(1) tsunami
(3) hurricane
(2) cyclone
(4) thunderstorm

24 _____

25 Which rock is foliated, shows mineral alignment but not banding, and contains medium-sized grains of quartz and pyroxene?

(1) phyllite
(3) gneiss
(2) schist
(4) quartzite

25 _____

26 The cross section below shows a V-shaped valley and the bedrock beneath the valley.

Which agent of erosion is responsible for cutting most V-shaped valleys into bedrock?

(1) surface winds
(3) glacial ice
(2) running water
(4) ocean waves

26 _____

27 The geologic cross section below shows a hillslope and the rock layers that underlie it.

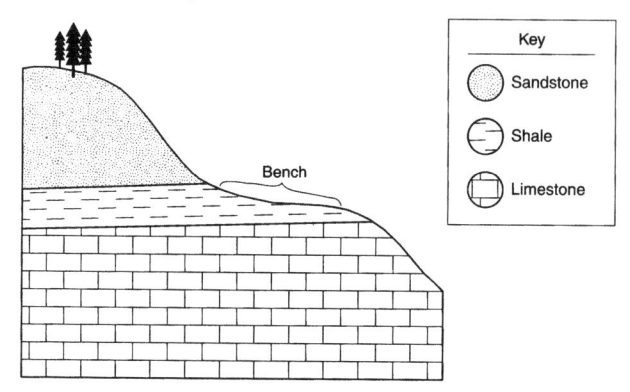

Which difference between the sandstone, shale, and limestone layers caused the formation of the relatively gently sloped section labeled "bench"?

(1) rock age
(2) fossil content
(3) resistance to weathering
(4) amount of uranium-238

27____

28 Which graph best represents the range of particle sizes that can be carried by a glacier?

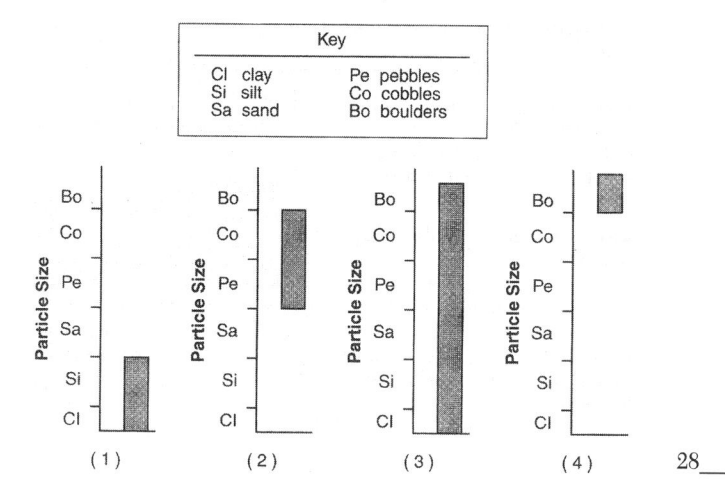

28____

Base your answers to questions 29 and 30 on the diagram below, which shows three minerals with three different physical tests, A, B, and C, being performed on them.

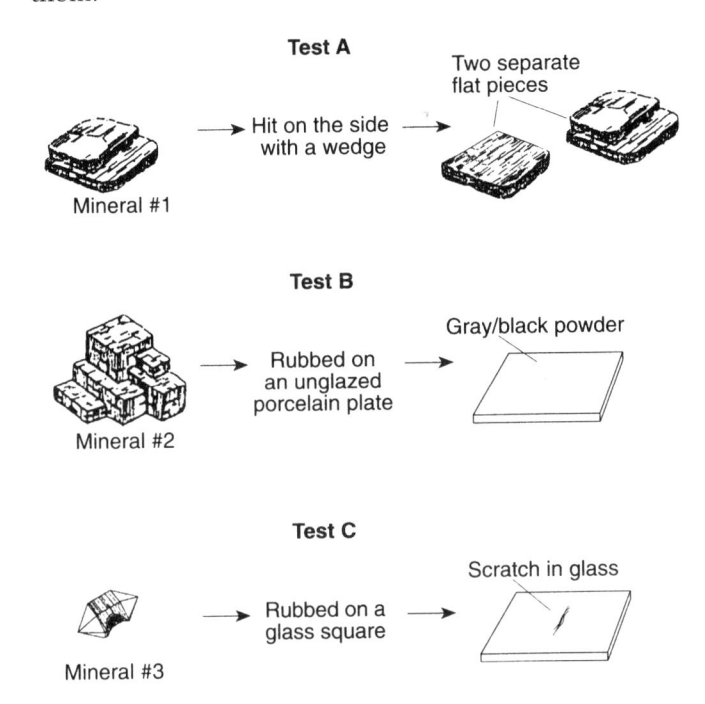

Test A

Mineral #1 → Hit on the side with a wedge → Two separate flat pieces

Test B

Mineral #2 → Rubbed on an unglazed porcelain plate → Gray/black powder

Test C

Mineral #3 → Rubbed on a glass square → Scratch in glass

29 Which sequence correctly matches each test, A, B, and C, with the mineral property tested?

(1) A—cleavage; B—streak; C—hardness
(2) A—cleavage; B—hardness; C—streak
(3) A—streak; B—cleavage; C—hardness
(4) A—streak; B—hardness; C—cleavage

29 ____

30 The results of all three physical tests shown are most useful for determining the

(1) rate of weathering of the minerals
(2) identity of the minerals
(3) environment where the minerals formed
(4) geologic period when the minerals formed   30_____

31 An air temperature of 95°C most often exists in which layer of the atmosphere?

(1) troposphere   (3) mesosphere
(2) stratosphere   (4) thermosphere   31_____

32 During the intrusion of the Palisades Sill, contact metamorphism changed sandstone and shale into

(1) diorite   (3) limestone
(2) marble   (4) hornfels   32_____

33 Which process most likely formed a layer of the sedimentary rock, gypsum?

(1) precipitation from seawater
(2) solidification of magma
(3) folding of clay-sized particles
(4) melting of sand-sized particles   33_____

34 The diagram below shows a stream flowing past points X and Y. If the velocity of the stream at point X is 100 centimeters per second, which statement best describes the sediments being transported past these points?

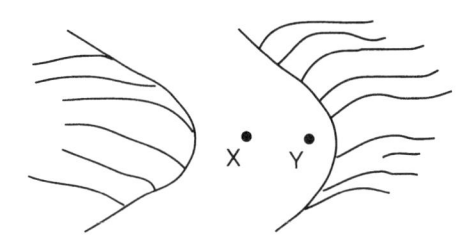

(1) At points X and Y, only clay is being transported.
(2) At points X and Y, only sand, silt, and clay are being transported.
(3) Some pebbles being transported at point Y are bigger than those being transported at point X.
(4) Some pebbles and cobbles are being transported at points X and Y, but not sand, silt, or clay.

34____

35 Specific mass extinction of living organisms and global climatic changes in geologic history are inferred by most scientists to have been caused by

(1) the impact of asteroids or large meteors on Earth's surface
(2) the gravitational pull of the Sun on Earth's surface
(3) large energy surges from the surface of the Sun
(4) earthquakes occurring along crustal plate boundaries

35____

Examination June 2003

PART B–1

Answer all questions in this part.

Directions (36–50): For *each* statement or question, write in the spaces provided the *number* of the word or expression that, of those given, best completes the statement or answers the question. Some questions may require the use of the *Earth Science Reference Tables*.

Base your answers to questions 36 through 38 on the data table below, which gives information collected at seismic stations A, B, C, and D for the same earthquake. Some of the data has been deliberately omitted.

Seismic Station	P-Wave Arrival Time	S-Wave Arrival Time	Difference in Arrival Times	Distance to Epicenter
A	08:48:20	No S-waves arrived		
B	08:42:00		00:04:40	
C	08:39:20		00:02:40	
D	08:45:40			6,200 km

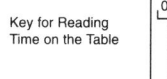

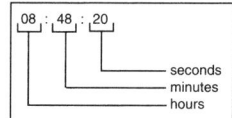

Key for Reading Time on the Table: 08 : 48 : 20 — seconds, minutes, hours

36 What is the most probable reason for the absence of S-waves at station *A*?

(1) S-waves cannot travel through liquids.
(2) S-waves were not generated at the epicenter.
(3) Station *A* was located on solid bedrock.
(4) Station *A* was located too close to the epicenter.   36_____

37 What is the approximate distance from station *C* to the earthquake epicenter?

(1) 3,200 km   (3) 1,600 km
(2) 2,400 km   (4) 1,000 km   37_____

734

38 How long did it take the P-wave to travel from the epicenter of the earthquake to seismic station D?
(1) 00:46:30 (3) 00:17:20
(2) 00:39:20 (4) 00:09:40   38____

Base your answers to questions 39 and 40 on the map below, which shows the latitude and longitude of five observers, A, B, C, D, and E, on Earth.

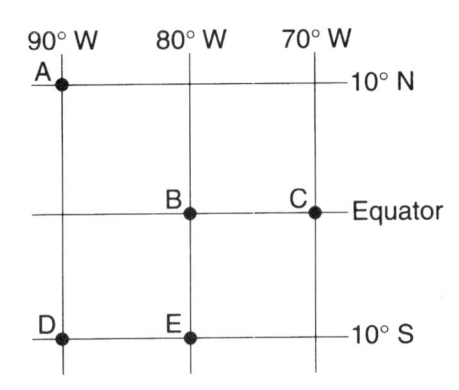

39 What is the altitude of *Polaris* (the North Star) above the northern horizon for observer A?
(1) 0°   (3) 80°
(2) 10° (4) 90°   39____

40 Which two observers would be experiencing the same apparent solar time?
(1) A and C   (3) B and E
(2) B and C   (4) D and E   40____

735

Examination June 2003

Base your answers to questions 41 though 43 on the diagram below, which shows a model of the apparent path and position of the Sun in relation to an observer at four different locations, A, B, C, and D, on Earth's surface on the dates indicated. The zenith (z) and the actual position of the Sun in the model at the time of the observation are shown. [The zenith is the point directly over the observer.]

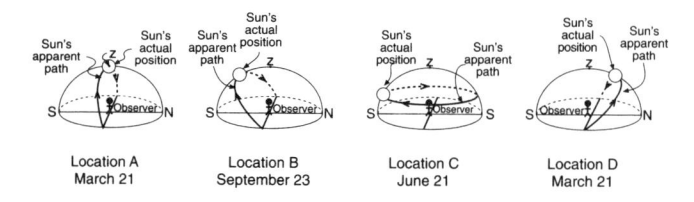

Location A — March 21
Location B — September 23
Location C — June 21
Location D — March 21

41 According to the Sun's actual position shown in the diagrams, the most intense insolation is being received by the observer at location

(1) A
(2) B
(3) C
(4) D

41_____

42 Where on Earth's surface is the observer at location C located?

(1) at the Equator
(2) at the South Pole
(3) at the North Pole
(4) in Oswego, New York

42_____

43 From sunrise to sunset at location B, the length of the observer's shadow will

(1) increase, only
(2) decrease, only
(3) increase, then decrease
(4) decrease, then increase

43_____

Base your answers to questions 44 through 46 on the map below, which shows the location of mid-ocean ridges and the age of some oceanic bedrock near these ridges. Letters A through D are locations on the surface of the ocean floor.

Age of Rocks on the Sea Bottom Relative to Ridges

——— Mid-ocean ridges
• • • • • • • • 10 million years old
— — — — 40 million years old
•—•—•—• 60 million years old

44 What is the most probable age, in millions of years, of the bedrock at location *B*?

(1) 5 (3) 48
(2) 12 (4) 62   44____

45 Rising convection currents in the asthenosphere would most likely be under location

(1) *A* (3) *C*
(2) *B* (4) *D*   45____

46 The age of oceanic bedrock on either side of a mid-ocean ridge is supporting evidence that at the ridges, tectonic plates are

(1) diverging   (3) locked in place
(2) converging   (4) being subducted   46____

Base your answers to questions 47 and 48 on the geologic cross section below. The large cone-shaped mountain on Earth's surface is a volcano. Letters A, B, and C represent certain rocks.

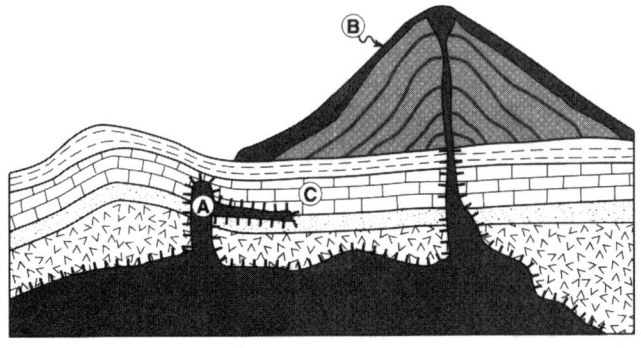

Key
- Igneous rock A and B
- Gabbro
- Limestone
- Contact metamorphism
- Sandstone
- Shale
- Ash layers

47 Which statement correctly describes the relative ages of rocks A and C and gives the best supporting evidence from the cross section?

(1) A is younger than C, because A is a lower sedimentary rock layer.
(2) A is younger than C, because the intrusion of A metamorphosed part of rock layer C.
(3) A is older than C, because A has older index fossils.
(4) A is older than C, because the intrusion of A cuts across rock layer C.

47_____

48 Rock B is most likely which type of igneous rock?

(1) granite
(2) peridotite
(3) pegmatite
(4) basalt

48_____

Base your answers to questions 49 and 50 on the diagram below, which shows sunlight entering a room through the same window at three different times on the same winter day.

Sunrise   Noon   Sunset

49 The apparent change in the Sun's position shown in the diagram is best explained by

(1) the Sun rotating at a rate of 15° per hour
(2) Earth rotating at a rate of 15° per hour
(3) the Sun's axis tilted at an angle of $23\frac{1}{2}°$
(4) Earth's axis tilted at an angle of $23\frac{1}{2}°$ 49____

50 This room is located in a building in New York State. On which side of the building is the window located?

(1) north   (3) east
(2) south   (4) west   50____

Examination June 2003

PART B-2

Answer all questions in this part.

Directions (51–60): Record your answers in the spaces provided. Some questions may require the use of the *Earth Science Reference Tables*.

Base your answers to questions 51 through 54 on the topographic map below. Points A, B, Y, and Z are reference points on the topographic map. The symbol △ 533 represents the highest elevation on Aurora Hill.

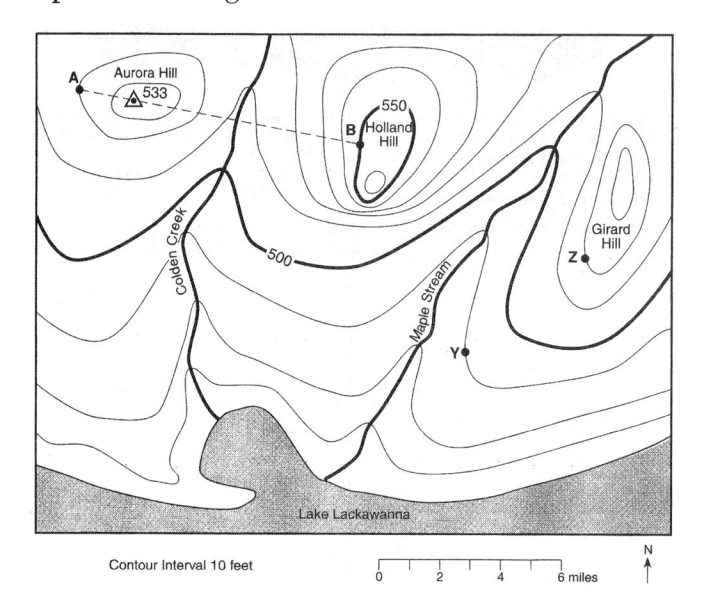

51 State the general compass direction in which Maple Stream is flowing. [1]

_____

52 Calculate the gradient between points Y and Z on the map, and label the answer with the correct units. [2]

Gradient = _____

53 Describe the evidence shown on the map that indicates that the southern side of Holland Hill has the steepest slope. [1]

_____

_____

54 On the grid provided below, construct a topographic profile from point A to point B by following the directions below.

  a Plot the elevation along line AB by marking with an **X** *each* point where a contour line is crossed by line AB. Points A and B have been plotted for you. [2]

  b Complete the profile by correctly connecting the plotted points with a smooth, curved line. [1]

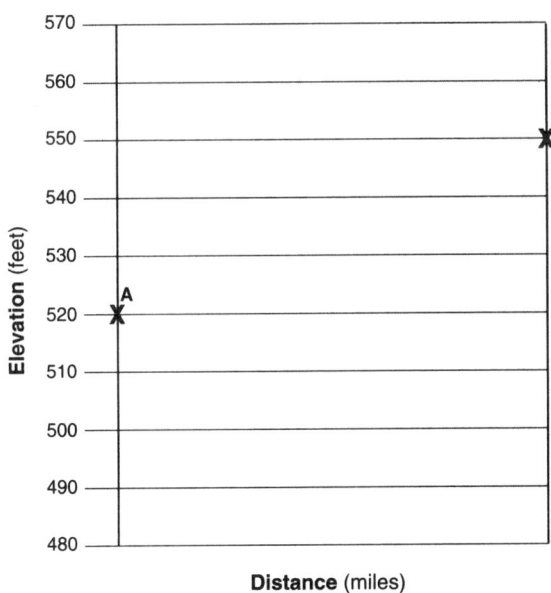

Distance (miles)

55 The cross section below illustrates the normal pattern of sediments deposited where a stream enters a lake. Letter X represents a particular type of sediment.

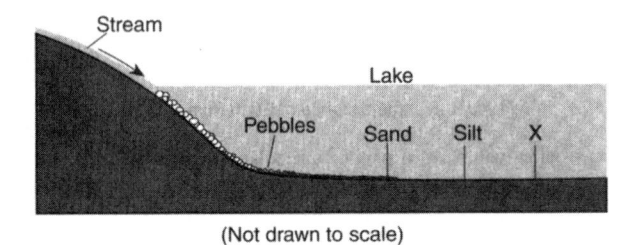

a Briefly explain why deposition of sediment usually occurs where a stream enters a lake. [1]

_____
_____

b Name the type of sediment most likely represented by letter X. [1]

_____

Base your answers to questions 56 and 57 on the temperature field map provided below. The map shows air temperatures, in degrees Fahrenheit, recorded at the same time at weather stations across North America. The air temperature at location A has been deliberately left blank.

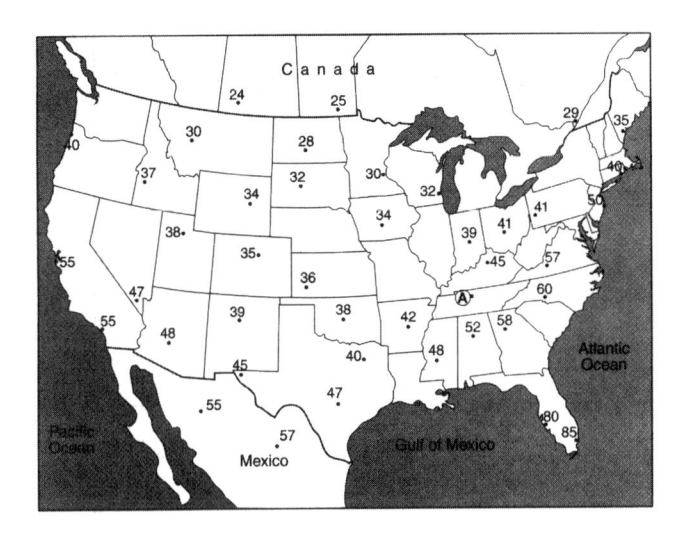

56 On the map provided above, use smooth, curved solid lines to draw the 30°F, 40°F, and 50°F isotherms. [2]

57 What is the most probable air temperature at location A? [1]

_____ °F

Base your answers to questions 58 through 60 on the information, data table, and diagram below and on your knowledge of Earth science.

Astronomers have discovered strong evidence for the existence of three large extrasolar (outside our solar system) planets that orbit *Upsilon Andromedae*, a star located 44 light years from Earth. The three planets are called planet B, planet C, and planet D. Some of the information gathered about these three new planets is shown in the table below. The period of revolution for planet C has been deliberately left blank.

Characteristics of Planets B, C, and D Orbiting Star *Upsilon Andromedae*

Planet	Mass	Distance from Upsilon Andromedae	Period of Revolution
B	$\frac{3}{4}$ of the mass of Jupiter	0.06 AU	4.6 Earth days
C	2 times the mass of Jupiter	0.83 AU	
D	4 times the mass of Jupiter	2.50 AU	3.5 to 4.0 Earth years

[1 AU = average distance of Earth from the Sun]

The diagram below compares a part of our solar system to the *Upsilon Andromedae* planetary system. Planet distances from their respective star and the relative size of each planet are drawn to scale. [The scale for planet distances is not the same scale used for planet size.]

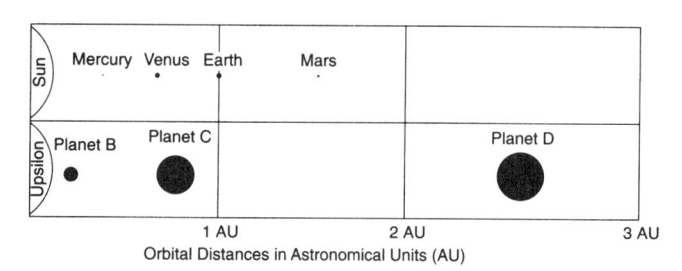

58 Planet *D*'s diameter is 10 times greater than Earth's diameter. What planet in our solar system has a diameter closest in size to the diameter of planet *D*? [1]

_____

59 As planet *B* travels in its orbit, describe the change in orbital velocity of planet *B* as the distance between *Upsilon Andromedae* and planet *B decreases*. [1]

_____
_____

60 If our solar system had a planet located at the same distance from the Sun as planet *C* is from *Upsilon Andromedae*, what would be its approximate period of revolution? [1]

_____ days

PART C

Answer all questions in this part.

Directions (61–75): Record your answers in the spaces provided below. Some questions may require the use of the *Earth Science Reference Tables*.

Base your answers to questions 61 and 62 on the information below and on your knowledge of Earth science.

Howe Caverns

Many scientists believe that the formation of the rocks in which Howe Caverns is now found began millions of years ago. At that time, an ocean covered the eastern region of New York State. Hundreds of feet of calcium carbonate ($CaCO_3$) sediments were deposited in layers along the edge of this ocean. These layers eventually formed the sedimentary rock limestone, which makes up the walls of today's Howe Caverns.

Much later, tectonic forces raised this region of New York State above sea level exposing the rock to weathering and erosion. These tectonic forces cracked the thick limestone, creating pathways for groundwater to infiltrate and gradually increase the size of the cracks. Eventually some of the larger cracks provided pathways for the underground stream, which carved the winding passages of Howe Caverns seen today.

61 State *two* processes that caused these sediments to become limestone. [2]

Process 1: _____

Process 2: _____

62 Identify one method that could be used to determine that the walls of Howe Caverns are made of limestone. [1]

_____
_____

Base your answers to questions 63 through 66 on the passage and map below and on your knowledge of Earth science. The passage provides some information about the sediments under Portland, Oregon, and the map shows where Portland is located.

Bad seismic combination under Portland: Earthquake faults and jiggly sediment

Using a technique called seismic profiling, researchers have found evidence of ancient earthquake faults under Portland, Oregon. The faults may still be active, a USGS [United States Geological Survey] seismologist will announce tomorrow.

The research also turned up a 250-foot deep layer of silt and mud, deep under the city, which may have been caused by a catastrophic ice dam break some 15,000 years ago.

The two findings could together mean bad news, as soft sediment is known to amplify ground shaking during strong earthquakes. In the 1989 San Francisco earthquake, much of the damage to buildings was caused by liquefaction, a shaking and sinking of sandy, water-saturated soil along waterways. . . .

—Robert Roy Britt
excerpted from
"Bad sesimic combination under Portland: Earthquake faults and jiggly sediment"
explorezone.com 05/03/99

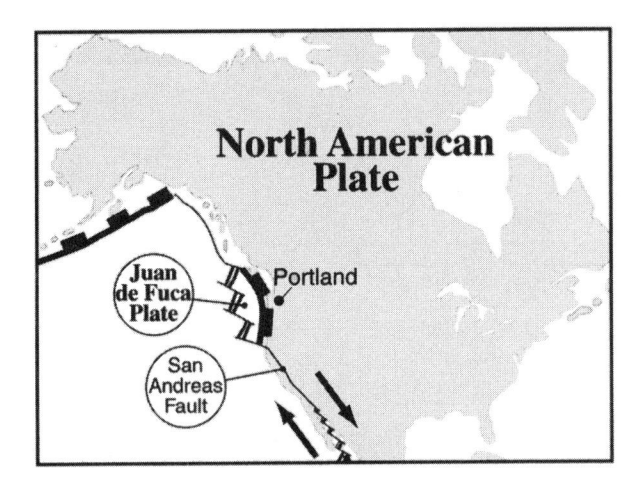

63 Explain why Portland is likely to experience a major earthquake. [1]

_____
_____

64 Why is the presence of a layer of silt and mud deep under the city a danger to Portland? [1]

_____
_____

65 Describe one precaution that can be taken to prevent or reduce property damage in preparation for a future earthquake in Portland. [1]

_____
_____

66 What type of tectonic plate boundary is shown at the San Andreas Fault? [1]

_____

Base your answers to questions 67 and 68 on the diagram of the ellipse below.

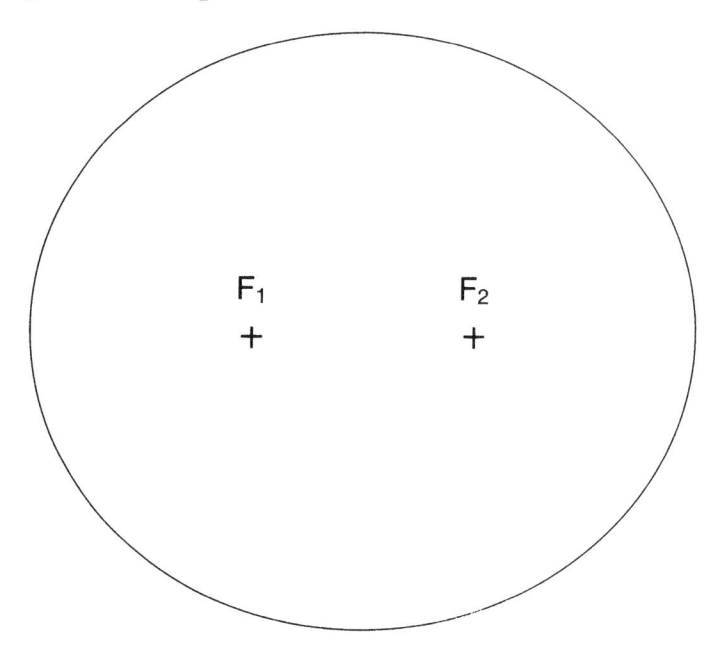

67 Calculate the eccentricity of the ellipse to the *nearest thousandth*. [1]

Eccentricity = _____

68 State how the eccentricity of the given ellipse compares to the eccentricity of the orbit of Mars. [1]

_____
_____
_____

Base your answers to questions 69 through 72 on your knowledge of Earth science and on the table below, which lists the seven brightest stars, numbered 1 through 7, in the constellation Orion. This constellation can be seen in the winter sky by an observer in New York State. The table shows the celestial coordinates for the seven numbered stars of Orion.

Location of the Seven Brightest Stars in Orion		
Star Number	Celestial Longitude (measured in hours)	Celestial Latitude (measured in degrees)
1	5.9	+7.4
2	5.4	+6.3
3	5.2	−8.2
4	5.8	−9.7
5	5.7	−1.9
6	5.6	−1.2
7	5.5	−0.3

69 On the grid provided below, graph the data shown in the table by following the steps below.

a Mark with an **X**, the position of *each* of the seven stars. Write the number of the plotted star beside each **X**. The first star has been plotted for you. [2]

b Show the apparent shape of Orion by connecting the Xs in the following order:

$$5 - 1 - 2 - 7 - 3 - 4 - 5 - 6 - 7 \quad [1]$$

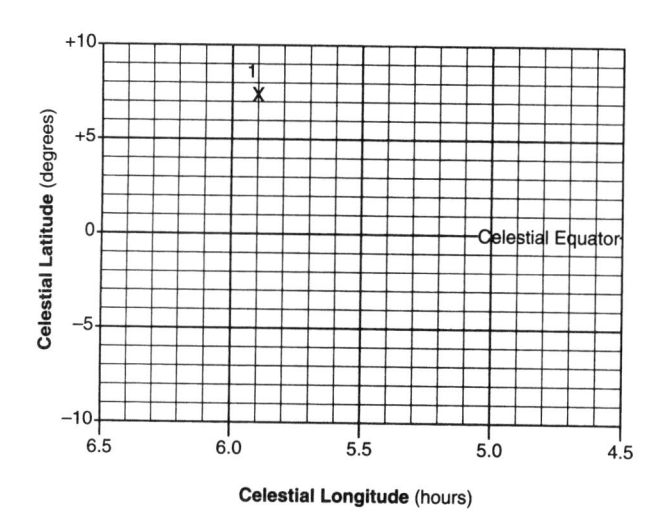

70 Star 1 plotted on the grid is the star *Betelgeuse*. Star 3 plotted on the grid is the star *Rigel*. How do the temperature and luminosity of *Betelgeuse* compare to the temperature and luminosity of *Rigel*? [1]

71 The seven stars of the constellation Orion that were plotted are located within our galaxy. Name the galaxy in which the plotted stars of Orion are located. [1]

_____

72 State one reason why an observer in New York State can never observe the constellation Orion at midnight during July but can observe the constellation Orion at midnight during January. [1]

_____

_____

Base your answers to questions 73 through 75 on your knowledge of Earth science and on the data table below, which shows the industrial uses of wollastonite, a mineral mined in the eastern Adirondack Mountains of New York State.

Industrial Uses of Wollastonite in the United States

Industrial Uses of Wollastonite	Percent of Total Use
Plastics	37
Ceramics	28
Metallurgy	10
Paint	10
Asbestos substitute	9
Miscellaneous	6

Examination June 2003

73 On the pie graph provided below, complete the graph to show the percent of *each* industrial use of wollastonite. Label *each* section of the pie graph with its industrial use. The percent for Miscellaneous and for Asbestos substitute has been drawn and labeled for you. [2]

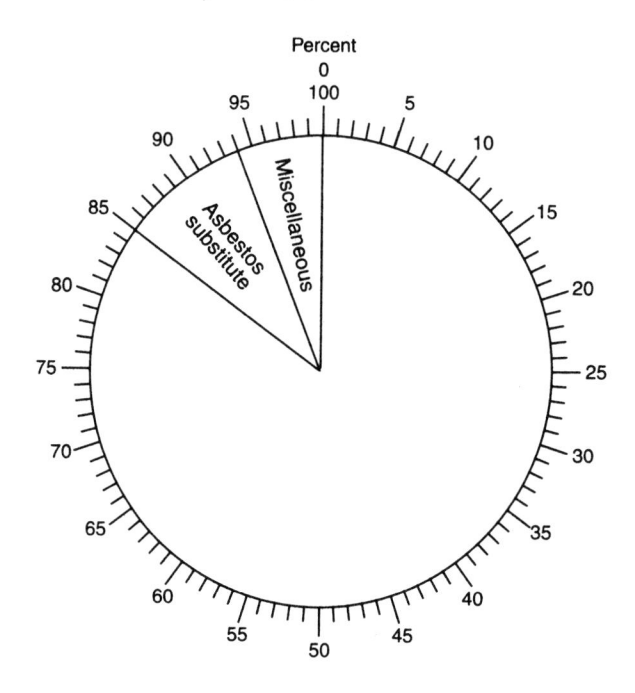

74 Wollastonite forms during the intense metamorphism of a sandy limestone. The expression below shows part of the process that results in the formation of wollastonite.

$$\underset{\text{Mineral 1}}{CaCO_3} + \underset{\text{Mineral 2}}{SiO_2} \xrightarrow{\text{Metamorphism}} \underset{\text{Wollastonite}}{CaSiO_2} + \underset{\text{Carbon dioxide}}{CO_2}$$

a Name the *two* minerals involved in the formation of wollastonite. [1]

_____ and _____

b What *two* conditions normally cause intense metamorphism? [1]

_____ and _____

75 Identify the geologic age of the New York State Adirondack Mountain bedrock in which wollastonite deposits are found. [1]

_____

Answers June 2003
Physical Setting/Earth Science

Answer Key

PART A

1. 2	8. 4	15. 3	22. 3	29. 1
2. 2	9. 2	16. 4	23. 4	30. 2
3. 4	10. 1	17. 1	24. 1	31. 4
4. 3	11. 4	18. 3	25. 2	32. 4
5. 1	12. 3	19. 1	26. 2	33. 1
6. 2	13. 3	20. 1	27. 3	34. 3
7. 2	14. 2	21. 2	28. 3	35. 1

PART B–1

36. 1	39. 2	42. 3	45. 3	48. 4
37. 3	40. 3	43. 4	46. 1	49. 2
38. 4	41. 1	44. 3	47. 2	50. 2

Index

A

Abrasion, 422, 466
Absolute magnitude, 140
Actual age, 248–249
Adiabatic changes, 576
Age, 248–249
   determining, 258–260
Air:
   composition, 210
   humidity, 569–574
   masses, 588–589
   movements, 574–575
   pressure, 567–569
      and humidity, 577
      and temperature, 575–577
      and winds, 579–582
   temperature, 566–567
      and humidity, 578–579
Alfonsine Tables, 22
Alluvial fans, 494–495
Amino acids, 245
Ancient environments, 264
Andesite, 315
Angle of:
   insolation, 535–536
   repose, 445
Angular unconformities, 250
Animal action, 422–423
Annual:
   cycles, 535–536
   traverse of the constellations, 45
Annular:
   eclipse, 52
   stream pattern, 508
Anticline, 392
Anticyclones, 591–592
Apparent:
   magnitude, 139
   motion, 5–6
Aquicludes, 233
Archean eon, 254
Aretes, 469
Aristotle, 18–20
Artesians, 233–234
Asterism, minerals, 291
Asteroids, 169
Asthenosphere, 399–400, 402
Astronomy:
   early, 1–17
      celestial:
         coordinate system, 8–9
         meridian, 7
         motion, 4–6
         objects, 3–4
         sphere, 6–10
      constellations, 5
      Dubhe, 8
      horizon, 7
      Jupiter, 10
      Mars, 10
      Merak, 8
      Mercury, 10
      Moon, 10
      planets, 10
      Polaris, 8
      Saturn, 10
      sky:
         motion in, 2
         observing, 1–6
      stars, 10
         pointer, 8
      Sun, 10
         path, 9–10
      Venus, 10
      zenith, 7
      zodiac, 9
   modern:
      development of, 18–40
         Alfonsine Tables, 22
         Aristotle, 18–20
         astronomical unit, 28
         Brahe, Tycho, 24–25
         Copernicus, 22–24
         Galilei, Galileo, 29
         gravity, 29–31
         heliocentric model, 18–40
         Kepler, Johannes, 24–28
            law of equal areas, 27
            law of planetary motion, 25–28
         Moon, 29
         Newton, Isaac, 29–31
         orbital motion, 30–31
         parallax, 19–20, 23
         planetary motion:
            Kepler's law, 25–28
            velocity, 27
         Ptolemy, 18, 21–22
         retrograde motion, 21, 24
         revolution, period of, 27–28
         spheres:
            crystalline, 19–20
            within spheres, 19
         universal gravitation, 29–30
   tools, 121–134
      lenses, 123–124

Index

Mount Palomar Observatory, 124
observation instruments, 130–131
spectrograph, 129
spectrophotometer, 129
spectroscopes, 125–129
   brightline spectrum, 127
   continuous spectrum, 126
   dispersion, 126
   prism, 125
   Roy G. Biv, 125
   spectral line, 127
telescopes, 124–125
Atmosphere, 530–564
  aerobic, 209
  circulation, 542–544
  development, 208–210
  entering, 540–547
  circulation, 542–544
  Coriolis effect, 544–547
  conduction, 540
  convection, 540
  latent heat, 541–542
  radiation, 540–541
  formation, 207–208
  insolation, 535–537
    angle of insolation, 535–536
    annual cycles, 535–536
    color, 537–538
    daily cycles, 535–536
    Earth's surface, 537–540
    duration, 536–537
    latent heat of water, 539–540
    latitude, 535, 537
    season, 536
    specific heat capacity, 538–539
    texture, 537–538
  layers, 211–212
  lost, 206
  origin, 206–210
  oxidizing, 208–209
  reducing, 208
  solar radiation, 531–535
    electromagnetic spectrum, 532
    electromagnetic waves, 531
    global radiation budget, 534
    radiative balance, 534–535
  structure, 210–212
  transparency, 575
  variables, 566
Autocatalysis, 246
Axis, 73
  of rotation, 42

B

Banding, 324
Barite, 296
Barometer, 568
Barrier bars, 497
Basalt, 312, 315
Basaltic glass, 314
Batholiths, 380
Bayer, Johann, 136
Beaches, 495–496
Bench marks, 389
Berms, 495
Big Bang Theory, 186–188
Biogenic sediments, 319
Black hole, 151
Borates, 296
Boundaries:
  convergent, 400
  divergent, 400
  transform, 400–401
Brahe, Tycho, 24–25
Breezes, 581–582
Brightline spectrum, 127
Brightness, stellar, 138–140

C

Calcite, 296
Callisto, 167
Capillarity, 225
  action, 233
  fringe, 232
Carbohydrates, 245
Carbon-14, 260
Carbonates, 296
Carbonation, 426
Cartesian system, 73
Catalyst, 246
Catastrophic theory, 159
Celestial:
  coordinate system, 8–9
  meridian, 7
  motion, 4–5
  objects, 3–4
  sphere, 6–10
Celestite, 296
Celsius, 567
Cementation, 316
Cenozoic era, 254
Chaos theory, 595
Characteristics, 136–144
Charon, 161
Chatoyancy, minerals, 291
Chemical:
  formulas, 286
  sediments, 319
  symbol, 286
  tests, minerals, 291
  weathering, 424–427
Chromates, 296
Cinder cone volcano, 379
Cinnabar, 296

Index

Circles, great, 75–77
Cirques, 465, 469
Clastic sediments, 318–319
Cleavage, minerals, 289
Climate, 427–428, 618–645
   definition, 618
   effects, 627–628
   factors, 619–626
   and landscapes, 509–510
   past, continental drift theory, 396–397
   types, 626–627
Climatic trends, 628–633
Clouds, 625–626
   ceiling, 575
   cover, 575
   formation, 582–584
Clusters, 183–184
Coastal processes, 456–458
Cold fronts, 589–590
Color, 537–538
   minerals, 288
   stars, 140–144
Cometary impacts, 207–208
Comets, 171
Compaction, 316
Composite cone, 379
Compounds, 286
Condensation, 229, 569–570, 582
Conduction, 376, 540
Constellations, 5
   annual traverse, 45
Continental:
   drift theory, 393–397
      contemporary organisms, 395
      crust, 394
      fossils, 395
      geological structures, 395
      past climates, 396–397
   glaciers, 464–466
Contiuous spectrum, 126
Contour lines, 80
Convection, 376, 540
   currents, 452
Coordinate system, 73
   celestial, 8–9
Copernicus, 22–24
Core, 356–357
Coriolis effect, 43–44, 544–547, 582
Correlation, rock layers, 250
Corundum, 296
Cosmic background radiation, 188
Cosmology, 182–194
   Big Bang Theory, 186–188
   clusters, 183–184
   cosmic background radiation, 188
   Doppler effect, 185–186
   light-year, 184

Olber's Paradox, 183
superclusters, 184
universe:
   closed, 189
   expanding, 185–186
   future, 188
   open, 188
   structure, 183–185
Crust, 356
   dynamic, 389–392
Crustal movements:
   causes, 391
   effects, 391–392
   evidence, 389–390
Crystalline:
   form, 287
   spheres, 19–20
   structure, 227
   texture, 323
Currents, 459
   convection, 453
   equatorial, 453
   longshore, 457–458
   ocean, 452–454
   wind-driven, 453–454
Cyclones, 591–592
Cyclonic storm tracks, 625–626

D

Daily cycles, 535–536
Day, 98–106
   sidereal, 101–103
   solar, 101–103
Deferents, 21
Deltas, 494–495
Dendritic stream pattern, 507
Density, 324
Deposition, 488–529
   glaciers, 499–505
      deposits, 500–501
      drumlins, 503
      erratics, 503–504
      Finger Lakes, 504
      glacial drift, 502
      glacial lakes, 504–505
      glacial transport, 499–500
      kettle lake, 504
      landforms, 502–505
      moraines, 502
      outwash plains, 504
   gravity, 491–493
      earthflow, 492
      mass movements, effects of, 492–493
      mudflow, 492
      rapid mass movements, deposition, 491–492
      rockfalls, 491

Index

slow mass movements, deposition, 492
slump, 492
soil creep, 492
talus, 491
moving water, 493–497
alluvial fans, 494–495
barrier bars, 497
beaches, 495–496
berms, 495
deltas, 494–495
floodplains, 493–494
lagoons, 497
levees, 493–494
oceans, deposition, 495–497
oxbow lakes, 493
sandbar, 496
streams, deposition, 493
tombolos, 496
particles, characteristics, 490
sediment, bedding, 491
sorting, 489
speed of medium, 489
wind, 497–499
dunes, 498–499
loess, 498
Deposits, 501
Desertification, 629
Dessication, 316
Dew, 584
point, 571, 574
Diabase, 312
Dikes, 379
Diorite, 315
Disconformities, 250
Dispersion, 126
Displaced structures, 389
DNA, 245–247
Domolmite, 296
Doppler effect, 185–186
Double refraction, minerals, 291
Drainage basin, 506
Draper, Henry, 129
Drizzle, 584
Drumlins, 503
Ductile deformation, 392
Dunes, 498–499
Dunite, 315

E

Earth, 163–165
atmosphere, 206–218
origin, 206–210
aerobic, 209
air, composition, 210
cometary impacts, 207–208
development, 208–210
formation, 207–208
layers, 211–212
lost, 206
origin, 206–210
outgassing, 207
oxidizing, 208–209
reducing, 208
structure, 210–212
coordinate system, 66–97
core, 196
crust, 196
continental drift theory, 394
crustal movements:
causes, 391
compression, 391
shear, 391
stress, 391
tension, 391
effects, 391–392
deformation, 391–392
elastic limit, 391
faults, 392–393
folds, 392
fracture, 391–392
joints, 392
evidence, 389–390
dynamic, 389–392
lithosphere, 389
hydrosphere, 219–242
interior structure, 355–359
core, 356–357
crust, 356
mantle, 356–357
magnet, 197–198
mantle, 196
motions:
effects, 46–54
heliocentric, 41–65
constellations, annual traverse, 45
Coriolis effect, 43–44
Earth-Moon-Sun System, 50–54
eclipses, 51–53
Foucault Pendulum, 42–43
Moon, phases, 50–51
parallelism, Earth's axis of rotation, 44
Polaris, 45
precession, 45
revolution, 44–45
rotation, 42–44
parallelism, 44
Sun's path, 46–49
tides, 53–54
Vega, 45
origins, 195–205
birth, 195–196
differentiation, 196–197

Index

layers, 196–197
shape, 67–71
  eclipses, 68
  evidence, 68–71
  gravity, measurements, 70–71
  photographs from space, 69
  Polaris, altitude, 69–70
  roundness, 67–68
  smoothness, 68
size, determining, 71–72
surface, 537–540
  mapping, 72–84
Earthflow, 447, 492
Earthquakes, 341–372, 389
causes, 342
definition, 341–342
effects, 352–354
elastic rebound theory, 342
epicenter, 342
  determining distance, 348–349
  determining location, 349–350
measuring, 351–352
  Mercalli scales, 351, 353
  Richter magnitude scale, 351
P-waves, 347
seismic waves, 342, 345–346
seismogram, 346–351
  analyzing, 347–351
seismograph, 345–346
S-waves, 347
waves, 342–347
  behavior, 343–344
  frequency, 344
  length, 344
  longitudinal waves, 345
  motion, 342
  periodic waves, 344
  refraction, 344
  transverse waves, 345
  types, 345
  velocity, 344
zones, 354–355
Eccentricity, 25
Eclipses, 51–53
annular, 52
evidence of Earth's shape, 68
lunar, 52–53
solar, 51–52
El Niño, 628, 630–633
Elastic:
deformation, 392
rebound theory, 342
Electromagnetic:
spectrum, 532
waves, 531
Elements, 286
Elevation, 625

field maps, 80
Ellipse, 25–26
Energy, entering atmosphere, 540–547
circulation, 542–544
Coriolis effect, 544–547
conduction, 540
convection, 540
latent heat, 541–542
radiation, 540–541
Environments, ancient, 264
Eons, 254
Epicenter, 342
determining distance, 348–349
determining location, 349–350
Epicycles, 21–22
Epochs, 255
Equator, 74
Equatorial currents, 453
Equinoxes, 110
Eras, 254
Eratoshenes' method, 71–72
Erosion, 443–487
agents of, 444–472
  glaciers, 461–469, 472
    abrasion, 466
    aretes, 469
    cirque, 465, 469
    continental glaciers, 464–466
    glacial erosion, 466–467
    glacial grooves, 467–468
    glacial polish, 467
    glacial striations, 467
    hanging valleys, 468–469
    horns, 469
    ice movement, 462–464
    plucking, 466–467
    rock drumlins, 468
    types, 464–466
    U-shaped valleys, 468
    valley glaciers, 465
  gravity, 444
    angle of repose, 445
    earthflow, 447
    mass wasting, 445–448, 469–470
      rapid, 446–447
      slow, 447–448
    mudflow, 447
    rockfalls, 446
    slump, 446–447
    soil creep, 447–448
  water in motion, 448–459
    coastal processes, 456–458
    convection currents, 452
    equatorial currents, 453
    longshore currents, 457–458
    oceans, 452–459
      currents, 452–454

Index

raindrops, 448
runoff, 448
shoreline erosion, 458–459
streams, 449, 470–471
   erosion, 450
   life cylce, 450–452
   transport, 449–450
swells, 456
tsunamis, 456
waves, 454–456, 471
   crest, 455
   height, 455
   length, 455
   refraction, 456–457
   trough, 455
   wind-generated, 454–456, 458–459
   wind-driven currents, 453–454
wind, 459–461, 471–472
   currents, 459
   cut notch, 458
   sea:
      cave, 459
      cliff, 459
      terrace, 459
  evidence of, 444
Erratics, 503–504
Europa, 167
Evaporation, 226, 229, 569–570
Evaporative cooling, 226–227
Evapotranspiration, 624
Evolution, 261
Exfoliation, 422
Exposure, 428–429
Extrusive rocks, 311–312
Eyepiece, 124

F

Fahrenheit, 567
Faults, 249
Feldspar, 313
Field maps, 78–84, 585–586
  contour lines, 80
  direction, 82
  elevation field maps, 80
  gradient, 79
  isolines, 78
  isosurfaces, 79
  profiles, 83
  scale, 81–82
  symbols, 80–81
  topographic, 80
    conventions, 83–84
Finger Lakes, 504
Flame color, minerals, 291
Floodplains, 493–494
Focal length, 123

Focal point, 123
Focus, 25
Fog, 584
Folded rock layers, 389
Foliation, 324
Fossils, 390
  continental drift theory, 395
  fuels, conservation, 297–298
  index, 252
Foucault Pendulum, 42–43
Fracture, minerals, 289–290
Frame of reference, 73
Freezing, 227
Frequency, 344
Frigid zone, 620
Frontal cyclones, 591–592
Fronts, 588–589
Frost, 584
  action, 421–422
Fusion, 144–145

G

Gabbro, 311–312, 315
Galaxies, 130–131, 158
Galena, 296
Galilei, Galileo, 29
Ganymede, 167
Geocentric model, 1–17
Geocentric universe:
  Aristotle, 19–20
  Ptolemy, 21–22
Geological:
  column, 253
  record, 247–264
  structures, continental drift theory, 395
  time scale, 253–255
Geothermal gradient, 373–374
Glacial:
  deposits, types, 500–501
  drift, 502
  erosion, 466–467
  grooves, 467–468
  lakes, 504–505
  polish, 467
  striations, 467
  transport, 499–500
Glaciers, 461–469, 472
  deposition, 499–505
    landforms, 502–505
  types, 464–466
Global:
  climatic trends, 628–633
  radiation budget, 534
  warming, 629–630
  wind belts, 581–582
Glucose, 245
Gradient, 79, 587

Index

Granite, 311, 315
Gravitational differentiation, 170
Gravity, 29–31, 146–147
   agent of erosion, 444
   measurements, 70–71
Greenwich Mean Time, 105
Groundmass, 312
Groundwater, 229, 231–234
   aquicludes, 233
   artesians, 233–234
   capillary action, 233
   capillary fringe, 232
   permeability, 232
   porosity, 231
   soil moisture zone, 232
   water table, 232–233
   wells, 233
   zone of saturation, 232
Gypsum, 296

H

Hail, 584
Half-life, 258–260
Halides, 296
Halite, 320
Halley's Comet, 162
Hanging valleys, 468–469
Hardness, minerals, 288–289
Heat:
   capacity, 226
   internal, Earth's:
      transfer, 375–376
         conduction, 376
         convection, 376
         radiation, 376
      sources, 374–375
         original heat, 374–375
         radioactive decay, 375
         tidal friction, 375
Heliocentric:
   Earth motions, 41–65
   model, 18–40
Hematite, 296
Hertzsprung-Russell Diagram, 142–144
Hipparchus, 139
Horizon, 7
Horns, 469
Hot spots, 401–402
Hours, 99–100
Human perception, 121–124
   limits, 122
Humidity, 569–574
   air, 569–574
   relative, 570, 573
Hurricanes, 597
Hutton, James, 247
Huygens, Christian, 167

Hydration, 425–426
Hydrogen bonds, 224
Hydrolysis, 425–426
Hydrosphere:
   groundwater, 229, 231–234
   origin, 219–242
      groundwater, 229
      outgassing, 221
      runoff, 229
      seawater, 222–224
         composition, 222–224
         dissolved gases, 224
         inorganic matter, 222–223
         organic matter, 223
         particulates, 224
      water, 220–222
         liquid, 225–226
         solid, 227–228
         states, 225–228
         structure, 224–225
         vapor, 226–227
Hydroxides, 296

I

Ice movement, 462–464
   glaciers, 462–464
Igneous rock, 249, 309
   charcteristics, 310
   formation, 309–310
   identification, 313–315
   mineral composition, 313
   porphyrys, 312–313
   texture, 311–313
Impact events, 171–173
Index fossils, 252
Immature soil, 431
Inorganic matter, 285–286
Insolation, 535–537
   angle of, 535–536
   annual cycles, 535–536
   color, 537–538
   daily cycles, 535–536
   duration, 536–537
   Earth's surface, 537–540
   latent heat of water, 539–540
   latitude, 535, 537
   season, 536
   specific heat capacity, 538–539
   texture, 537–538
Internal heat transfer, 375–376
International Date Line, 105–106
Interplanetary bodies, 169–174
Intrusions, 379
Intrusive rocks, 311
Io, 167
Isobars, 586
Isolines, 78, 585

Index

Isosurfaces, 79
Isotherms, 586
Isotopes, 258–260

J
Jovian planets, 160, 163, 166–169
Jupiter, 10, 161, 163–164, 166–167

K
Kelvin, 567
Kepler, Johannes, 24–28
    law of equal areas, 27
    law of planetary motion, 25–28
Kettle lake, 504
Kettles, 504
Key beds, 252
Kirchhoff, Gustav, 127
Kuiper belt, 161–162

L
Laccoliths, 380
Lagoons, 497
Land breezes, 581–582
Land distribution, 620–621
Landforms, 505
Landscape development, 505–512
    annular stream pattern, 508
    climate and landscapes, 509–510
    dendritic stream pattern, 507
    drainage basin, 506
    landforms, 505
    landscape regions, 510–512
        New York State, 511–512
    mountains, 505–506
    plains, 506
    plateaus, 506
    radial stream pattern, 507
    rectangular stream pattern, 508
    relief, 505
    soil associations, 508
    stream drainage system, 506
    stream patterns, 506–508
    topography, 505
    trellis stream pattern, 508
    tributary, 506
Latent heat, 541–542
    of fusion, 228
    of vaporization, 226
    of water, 539–540
Latitude, 535, 537, 620
    determining, 77–78
    -longitude coordinate system, 74–77
    Sun's path changes, 48–49
Lava, 309, 378
    plateau, 379
Laws:
    equal areas, 27
    original horizontality, 248
    planetary motion, 25–28
Layers, distortion, 324
Lenses, 123–124
Levees, 493–494
Lichens, 426
Life:
    amino acids, 245
    carbohydrates, 245
    definition, 244
    DNA, 245–247
    evolution, 261
    geological record, 247–264
    glucose, 245
    history, geological record, 247–264
    lipids, 245–246
    metabolism, 244
    natural selection, 261–264
    nucelotides, 245
    nucleic acids, 245
    organic molecules, 244–246
    origin, 243–283
        theories, 246–247
    proteins, 245
    RNA, 245–247
    starches, 245
    water, role, 244–246
Light, 123–124
    refracted, 123
    visible, 122
    -year, 184
Limbs, 392
Lipids, 245–246
Lithification, 316–317
Lithosphere, 389, 399, 402
Local time, 103–106
Loess, 498
Long-range weather forecasting, 594–595
Longitude, 74–77
    determining, 77–78
    meridians, of, 75
Longitudinal waves, 345
Longshore currents, 457–458
Luminescence, minerals, 291
Luminosity, 138, 141, 143
Lunar eclipse, 52–53
Luster, minerals, 288

M
Magma, 309
    origin, 377
Magnetism, minerals, 291
Magnetite, 296
Magnetosphere, 197–198
Magnitudes, 139
    absolute, 140
    apparent, 139

Index

scale, 139
Main sequence, 142–143
Mantle, 356
Map:
    direction, 82
    profiles, 83
    scale, 81–82
    symbols, 80–81
Mapping, 66–94
    axis, 73
    Cartesian system, 73
    coordinate systems, 73
    equator, 74
    field maps, 78–84
        contour lines, 80
        direction, 82
        elevation field maps, 80
        gradient, 79
        isolines, 78
        isosurfaces, 79
        profiles, 83
        scale, 81–82
        symbols, 80–81
        topographic maps, 80
            conventions, 83–84
    frame of reference, 73
    great circles, 75
    latitude:
        determining, 77–78
        -longitude coordinate system, 74–77
    longitude, 74–77
        determining, 77–78
        meridians, of, 75
    meridians, 75–77
        of longitude, 75
        prime, 76
    origin, 73
    parallels, 74
    polar coordinate system, 73
    prime meridian, 76
Marcasite, 296
Mariner spacecrafts, 165
Mars, 10, 163–166
Mass, 147–151
    movements, effects of, 492–493
    wasting, 445–448, 469–470
        rapid, 446–447
        slow, 447–448
Mature soil, 431
Merak, 8
Mercalli scales, 351, 353
Mercury, 10, 163–165
Meridians, 75–77
    celestial, 7
    of longitude, 75
    prime, 76
Mesozoic era, 254

Metabolism, 244
Metamorphic rocks, 309, 321–326
    characteristics, 323–325
    classification, 325
    formation, 321–322
        chemical activity, 322
        heat, 321
        pressure, 322
    identification, 325–326
    texture, 323–324
Metamorphism, 322–323
    types, 323
Meteor, 169–171
Meteorite, 169–171
Meteoroids, 169–171
Mid-ocean ridges, 397
Milky Way, 130–131, 158
Miller, Stanley, 246
Minerals, 284–307
    asterism, 291
    barite, 296
    borates, 296
    building blocks of rocks, 292–293
    calcite, 296
    carbonates, 296
    celestite, 296
    changes, 324–325
    chatoyancy, minerals, 291
    chemical:
        formulas, 286
        symbol, 286
        tests, 291
    chromates, 296
    cinnabar, 296
    cleavage, 289
    color, 288
    composition, 429
    compounds, 286
    conservation, 297
    corundum, 296
    crystalline form, 287
    definition, 285–287
    domolmite, 296
    double refraction, minerals, 291
    elements, 286
    flame color, minerals, 291
    fracture, minerals, 289–290
    galena, 296
    groups, 295–297
        minor, 296–297
    gypsum, 296
    halides, 296
    hardness, 288–289
    hematite, 296
    hydroxides, 296
    identifying, 287–293
    inorganic matter, 285–286

Index

luminescence, 291
luster, 288
magnetism, 291
magnetite, 296
marcasite, 296
nitrates, 296
oxides, 296
oxygen, 296
parting, 289
phosphates, 296
piezoelectricity, 291
play of colors, 291
polyatomic ions, 296
pyrite, 296
pyroelectricity, 291
resource distribution, global implications, 298–299
silicates, 293–295
special properties, 291
specific gravity, 290–291
sphalerite, 296
stibnite, 296
streak, 288
sulfates, 296
sulfides, 296
Mohorovicic, Andrija, 356
Monoclines, 392
Monsoons, 622–623
Month, 106–108
   sidereal, 107–108
   synodic, 107–108
Moon, 10, 29, 165
   eclipses, 51–53
   origins, 199–201
      Apollo missions, 199
      giant impact, 199–201
   phases, 50–51
Moraines, 502
Motion:
   apparent, 5–6
   celestial, 4–5
   Newton's laws, 29–31
   orbital, 30–31
   planetary, Kepler's law, 25–28
   real, 5–6
   retrograde, 21, 24
   sky, 2
Mount Palomar Observatory, 124
Mountains, 505–506
   ranges, 625
Mt. Etna, 380
Mt. Vesuvius, 380
Mudflow, 447, 492

N

Natural Selection, 261–264
Neap tides, 54
Nebula, 146–147
Nebulae, 131
Nebular theory, 159–160
Neptune, 161, 163–164, 168
Neutrinos, 150
Neutrons star, 150
Newton, Isaac, 29–31
Nitrates, 296
Nonconformities, 250
Nucleotides, 245
Nuclear fusion, 144–145
Nucleic acids, 245
Numerical weather forecasting, 594

O

Objective, 124
Observations, instruments, 130–131
Obsidian, 312, 314
Occluded fronts, 591
Oceans, 452–459
   currents, 452–454, 624
   deposition, 495–497
   floor spreading theory, 397–399
Olber's Paradox, 183
Oort cloud, 162
Optical axis, 123
Orbital motion, 30–31
Orbits, 25–27
Organic:
   molecules, 244–246
   sediments, 319
Original:
   heat, 374–375
   horizontality, 248
Orion, 163
Orographic effect, 625
Outcrop, walking, 250–251
Outgassing, 207, 221
Outwash plains, 504
Overturned, 392
Oxbow lakes, 493
Oxidation, 424–425
Oxides, 296
Oxygen, 296

P

Paleomagnetism, 397–398
Paleozoic era, 254
Parallax, 19–20, 23, 136–137
   parsec, 137–138
Parallelism, Earth's axis of rotation, 44
Parallels, 74
Parent:
   material, 431
   rock, 322
Parsec, 137–138

Index

Particle:
    characteristics, 490
    size, 428
Parting, minerals, 289
Path of totality, 51
Pedalfers, 432
Pegmatite, 311, 315
Penumbra, 51
Perception, human, 121–124
    limits, 122
Peridotite, 315
Periodic waves, 344
Periods, 254–255
    of revolution, 27–28
Permeability, 232
Phanerozoic eon, 254
Phenocrysts, 312
Phosphates, 296
Physical weathering, 420–424
Piezoelectricity, minerals, 291
Plains, 506
Planetary motion:
    Kepler's law, 25–28
    velocity, 27
Planetesimals, 160
Planets, 10, 163–164
    Earth, 163–165
    jovian planets, 163, 166–169
    Jupiter, 161, 163–164, 166–167
    Mars, 163–166
    Mercury, 163–165
    Neptune, 161, 163–164, 168
    Pluto, 161, 163–164, 168–169
    Saturn, 161, 163–164, 167–168
    terrestrial planets, 163–166
    Uranus, 161, 163–164, 168
    Venus, 163–165
Plant action, 422–423, 426
Plate:
    boundaries, interactions, 400–401
    tectonics, 387–419
        theory, 399–404
            asthenosphere, 399–400, 402
            boundaries, 400–401
            hot spots, 401–402
            key ideas, 402–404
            lithosphere, 399, 402
            rigidity, 399
Plateaus, 506
Play of colors, minerals, 291
Plucking, 466–467
Pluto, 161, 163–164, 168–169
Plutonic rock, 311
Plutons, 379
Polar:
    coordinate system, 73
    easterlies, 582
    front, 592
Polaris, 8, 45
    altitude, 69–70
Polyatomic ions, 296
Porosity, 231
Precambrian eon, 254
Precession, 45
Precipitation, 229, 584
Predictions, 21–22
Pressure:
    air, 567–569
    belts, 621–622
    unloading, 424
Prevailing winds, 581
Prime meridian, 76
Principle of superposition, 248
Prism, 125
Proteins, 245
Proterozoic eon, 254
Protostar, 146–147
Ptolemy, 18, 21–22
Pumice, 312, 315
P-waves, 347
Pyrite, 296
Pyroelectricity, minerals, 291

Q
Quartz, 313

R
Radial stream pattern, 507
Radiation, 376, 540–541
    curve, 140
Radiative balance, 534–535
Radioactive decay, 375
Radioisotopes, 260
Rain, 584
Raindrops, 448
Rapid mass movements, deposition, 491–492
Real motion, 5–6
Rectangular stream pattern, 508
Recumbent folds, 392
Red dwarfs, 148
Reflecting telescope, 124
Refracted light, 123
Refracting telescope, 124
Refraction, 344
Relative:
    age, 248–249
    humidity, 570, 573
Relief, 505, 625
Residual soil, 431
Resource distribution, global implications, 298–299
Retrograde motion, 21, 24
Revolution, 44–45

Index

period of, 27–28
Rhyolite, 315
Richter magnitude scale, 351
RNA, 245–247
Rock:
  age, 258–260
  cycle, 326–327
  drumlins, 468
  folded layers, 389
  igneous, 249
  layers, correlation, 250
  rigidity, 399
  tilted layers, 389
  young, 397
Rockfalls, 446, 491
Rocks, 308–340
  banding, 324
  crystalline texture, 323
  definition, 309
  density, 324
  extrusive rocks, 311–312
  foliation, 324
  igneous rock, 309
    characteristics, 310
    formation, 309–310
    identification, 313–315
    mineral composition, 313
    porphyrys, 312–313
    texture, 311–313
  intrusive rocks, 311
  lava, 309
  layers, distortion, 324
  magma, 309
  metamorphic rocks, 309, 321–326
    characteristics, 323–325
    classification, 325
    formation, 321–322
      chemical activity, 322
      heat, 321
      pressure, 322
    identification, 325–326
    texture, 323–324
  metamorphism, 322–323
  types, 323
  mineral:
    building blocks, 292–293
    composition, changes, 324–325
  parent rock, 322
  physical characteristics, matching, 251–253
  sedimentary, 309, 316–321
    cementation, 316
    characteristics, 317–318
    chemical sediments, 319
    compaction, 316
    dessication, 316
    identification, 320–321
    lithification, 316–317
    organic sediments, 319
    particles, types, 318–320
  types, 309–326
  volcanic glasses, 311
  Wentworth scale, 319
Rotation, 42–44
  axis of, 42
    parallelism, 44
  Coriolis effect, 43–44
  evidence of, 42–43
  Foucault Pendulum, 42–43
Roy G. Biv, 125
Runoff, 229, 448

S

Sandbar, 496
Satellites, 163
Saturn, 10, 161, 163–164, 167–168
Scale, map, 81–82
Sea:
  breezes, 581–582
  cave, 459
  cliff, 459
  terrace, 459
Season, 536
  Sun's path, 47–48, 108–109
Seawater, 222–224
  composition, 222–224
  dissolved gases, 224
  inorganic matter, 222–223
  organic matter, 223
  origin, 222
  particulates, 224
Sediment, 430
  bedding, 491
  shallow water, 390
Sedimentary rock, 309, 316–321
  characteristics, 317–318
  identification, 320–321
  layers, high elevation, 389
  particles, types, 318–320
Seismic waves, 342, 345–346
Seismogram, 346–351
  analyzing, 347–351
Seismograph, 345–346
Shield volcano, 379
Shoreline erosion, 458–459
Short-lived giants, 149–151
Sidereal:
  day, 101–103
  month, 107–108
Silicates, 293–295
Sills, 379
Sky:
  motion in, 2
  observing, 1–6

Index

Sleet, 584
Slow mass movements, deposition, 492
Slump, 446–447, 492
Snow, 584
Soil, 430–432
  associations, 508
  creep, 447–448, 492
  formation, 420–442
    immature soil, 431
    mature soil, 431
    parent material, 431
    pedalfers, 432
    residual soil, 431
    sediments, 430
    soil profile, 431
    transported soil, 430
  moisture zone, 232
Solar:
  day, 101–103
    mean, 103
  eclipse, 51–52
  noon, 101
  radiation, 531–535
    electromagnetic spectrum, 532
    electromagnetic waves, 531
    global radiation budget, 534
    radiative balance, 534–535
  system, 157–181
    asteroids, 169
    comets, 171
    evolution, 159–163
      catastrophic theory, 159
      nebular theory, 159–160
    exploration, 174
    galaxy, 158
    gravitational differentiation, 170
    impact events, 171–173
    interplanetary bodies, 169–174
    jovian planets, 160
    kuiper belt, 161–162
    meteor, 169–171
    meteorite, 169–171
    meteoroids, 169–171
    Milky Way, 158
    moon, 165
    origin, 159–163
    planetesimals, 160
    planets, 163–164
      Earth, 163–165
      jovian planets, 163, 166–169
      Jupiter, 161, 163–164, 166–167
      Mars, 163–166
      Mercury, 163–165
      Neptune, 161, 163–164, 168
      Pluto, 161, 163–164, 168–169
      Saturn, 161, 163–164, 167–168
      terrestrial planets, 163–166

      Uranus, 161, 163–164, 168
      Venus, 163–165
    structure, 163–174
Solstices, 109–110
Solution, 425–426
Sorting, 489
Southern Oscillation, 630–633
Space, photographs from, 69
Special properties, minerals, 291
Specific gravity, minerals, 290–291
Specific heat capacity, 538–539
Spectral line, 127
Spectrograph, 129
Spectrophotometer, 129
Spectroscopes, 125–129
  brightline spectrum, 127
  continuous spectrum, 126
  dispersion, 126
  prism, 125
  Roy G. Biv, 125
  spectral line, 127
Speed of medium, 489
Sphalerite, 296
Spheres:
  crystalline, 19–20
  within spheres, 19
Spring tides, 54
Starches, 245
Stars, 10, 130, 135–156
  black hole, 151
  brightness, 138–140
  characteristics, 136–144
  color, 140–144
  distance to, 136–138
  evolution, 144–151
  gravity, 146–147
  Hertzsprung-Russell Diagram, 142–144
  luminosity, 138, 141, 143
  magnitudes, 139
  main sequence, 142–143
  mass, 147–151
  names, 136
  nebula, 146–147
  neutrinos, 150
  nuclear fusion, 144–145
  origin, 144–151
  parallax, 136–137
  pointer, 8
  protostar, 146–147
  radiation curve, 140
  red dwarfs, 148
  short-lived giants, 149–151
  Sun-class stars, 148–149
  supernova, 150
  temperature, 140–144
Stationary fronts, 590
Statistical weather forecasting, 593–594

Index

Stellar brightness, 138–140
Stibnite, 296
Streak, minerals, 288
Streams, 449, 470–471
   deposition, 493
   drainage system, 506
   erosion, 450
   life cycle, 450–452
   patterns, 506–508
   transport, 449–450
Structures, displaced, 389
Sublimation, 582
Sulfates, 296
Sulfides, 296
Sun, 10, 163–164
   -class stars, 148–149
   eclipses, 51–52
   path, 9–10
      changing, 47–49, 111–112
      seasons, 108–109
Sunrise, 100
Sunset, 100
Superclusters, 184
Supernova, 150
Superposition, 248
Surface tension, 224–225
S-waves, 347
Swells, 456
Symbols, map, 80–81
Syncline, 392
Synodic months, 107–108
Synoptic weather:
   forecasting, 593
   maps, 585

T

Talus, 491
Telescopes, 124–125
Temperate zone, 620
Temperature, 619
   air, 566–567
   changes, 423–424
   star, 140–144
Tephra, 378
Terrestrial planets, 163–166
Texture, 537–538
Thermometer, 566
Thunderstorms, 595–596
Tidal friction, 375
Tides, 53–54
   neap, 54
   spring, 54
Tilted rock layers, 389
Time, 98–120, 430
   day, 98–106
   eons, 254
   epochs, 255
   equinoxes, 110
   eras, 254
   geological time scale, 253–255
   Greenwich Mean Time, 105
   hours, 99–100
   International Date Line, 105–106
   local, 103–106
   month, 106–108
   periods, 254–255
   seasons, Sun's path, 108–109
   sidereal:
      day, 101–103
      month, 107–108
   solar:
      day, 101–103
         mean, 103
      noon, 101
   solstices, 109–110
   Sun's path:
      changing, 111–112
      seasons, 108–109
      sunrise, 100
      sunset, 100
   synodic months, 107–108
   year, 108–113
      measuring, 112–113
   zones, 104–105
Titan, 167
Tombolos, 496
Topographic maps, 80
   conventions, 83–84
Topography, 505
Totality, path of, 51
Tornadoes, 596
Torrid zone, 620
Trade winds, 581, 621
Transpiration, 230, 624
Transported soil, 430
Transverse waves, 345
Trellis stream pattern, 508
Trenches, 398–399
Tributary, 506
Tsunamis, 456

U

Umbra, 51
Unconformities, 250
Uniformitarianism, 247
Universal gravitation, 29–30
Universe:
   closed, 189
   expanding, 185–186
   future, 188
   open, 188
   structure, 183–185
   theories of origin, 182–194
Uranus, 161, 163–164, 168

Urey, Harold, 246
U-shaped valleys, 468

V
Valley glaciers, 465
Valleys:
    hanging, 468–469
    U-shaped, 468
Van Allen belts, 198
Vega, 45
Vegetative cover, 624–625
Velocity, planetary motion, 27
Venus, 10, 163–165
Vesicular rocks, 315
Vessicles, 246
Visibility, 575
Volcanic:
    eruptions, 389
    glasses, 311
    rock, 311–312
Volcanoes, 373–387
    batholiths, 380
    cinder cone, 379
    composite cone, 379
    dikes, 379
    formation, 378
    geothermal gradient, 373–374
    heat, Earth's, internal:
        sources, 374–375
        transfer, 375–376
    intrusions, 379
    intrusive activities, 379–380
    laccoliths, 380
    lava, 378
    plateau, 379
    magma, origin, 377
    Mt. Etna, 380
    Mt. Vesuvius, 380
    plutons, 379
    shield volcano, 379
    sills, 379
    structures, 378–379
    tephra, 378
von Fraunhofer, Joseph, 126

W
Warm fronts, 590
Wasting, mass, 445–448, 469–470
Water:
    capillarity, 225
    condensation, 229
    crystalline structure, 227
    cycle, 229–230
    distribution, 620–621
    evaporation, 226, 229
    evaporative cooling, 226–227
    freezing, 227
    heat capacity, 226
    hydrogen bonds, 224
    latent heat of:
        fusion, 228
        vaporization, 226
    liquid, 225–226
    motion, erosion, 448–459
    origin, 220–222
    precipitation, 229
    role in life, 244–246
    solid, 227–228
    states, 225–228
    structure, 224–225
    surface tension, 224–225
    table, 232–233
    transpiration, 230
    universal solvent, 228–229
    vapor, 226–227
Waves, 454–456, 471
    behavior, 343–344
    crest, 455
    cut notch, 458
    cyclones, 591–592
    height, 455
    motion, 342
    refraction, 456–457
    trough, 455
    types, 345
    velocity, 344
    wind-generated, 454–456, 458–459
Wavelength, 344, 455
Weather, 565–616
    adiabatic changes, 576
    air:
        humidity, 569–574
        masses, 588–589
        movements, 574–575
        pressure, 567–569
            and humidity, 577
            and temperature, 575–577
            and winds, 579–582
        temperature, 566–567
            and humidity, 578–579
    anticyclones, 591–592
    atmospheric:
        transparency, 575
        variables, 566
    barometer, 568
    breezes, 581–582
    Celsius, 567
    chaos theory, 595
    cloud:
        ceiling, 575
        cover, 575
        formation, 582–584
    cold fronts, 589–590
    condensation, 569–570, 582

Index

Coriolis effect, 582
cyclones, 591–592
dew, 584
　point, 571, 574
drizzle, 584
evaporation, 569–570
Fahrenheit, 567
field maps, 585–586
fog, 584
forecasts, 588–595
frontal cyclones, 591–592
fronts, 588–589
frost, 584
global wind belts, 581–582
gradient, 587
hail, 584
hazardous, emergency preparedness, 598–599
humidity, 569–574
hurricanes, 597
isobars, 586
isolines, 585
isotherms, 586
kelvin, 567
land breezes, 581–582
long-range weather forecasting, 594–595
maps, 585–587
numerical weather forecasting, 594
occluded fronts, 591
polar:
　easterlies, 582
　front, 592
precipitation, 584
prevailing winds, 581
rain, 584
relative humidity, 570, 573
sea breezes, 581–582
sleet, 584
snow, 584
stationary fronts, 590
statistical weather forecasting, 593–594
sublimation, 582
synoptic:
　forecasting, 593
　maps, 585
thermometer, 566
thunderstorms, 595–596
tornadoes, 596
trade winds, 581
visibility, 575
warm fronts, 590
warnings, 598–599
watches, 598–599
wave cyclones, 591–592
Weathering, 420–442
　chemical, 424–427
　　carbonation, 426
　　hydration, 425–426
　　hydrolysis, 425–426
　　lichens, 426
　　oxidation, 424–425
　　solution, 425–426
　factors affecting, 427–430
　　climate, 427–428
　　exposure, 428–429
　　mineral composition, 429
　　particle size, 428
　　time, 430
　physical, 420–424
　　abrasion, 422
　　animal action, 422–423
　　exfoliation, 422
　　frost action, 421–422
　　plant action, 422–423, 426
　　pressure unloading, 424
　　temperature, changes, 423–424
　products of, 430–432
　　soils, 430–432
Wegener, Alfred, 251, 393–394
Wentworth scale, 319
Wind:
　belts, 621–622
　deposition, 497–499
　-driven currents, 453–454
　erosion, 459–461, 471–472

Y
Year, 108–113
　measuring, 112–113
Young rock, 397

Z
Zenith, 7
Zodiac, 9
Zone of:
　saturation, 232
　soil moisture, 232

It's finally here—
online Regents exams from the experts!

Welcome to Barronsregents.com

The ultimate Regents test-prep site for students and online resource for teachers.

With ***www.barronsregents.com*** *you can now take Regents exams online!*

- Take complete practice tests by date, or choose questions by topic
- All questions answered with detailed explanations
- Instant test results let you know where you need the most practice

Online Regents exams are available in the following subjects:

- Biology–The Living Environment
- Chemistry–The Physical Setting
- Earth Science–The Physical Setting
- Global History and Geography
- Seq. Math I, II, & III
- Math A & Math B
- Physics–The Physical Setting
- U.S. History & Government

Getting started is a point and a click away!

Only $11.95 for a subscription...includes all of the subjects listed above!

Teacher and classroom rates also available.

For more subscription information and to try our demo
Visit our site at ***www.barronsregents.com***

BARRON'S

7/03 (#105)

NOTES